Analytical Fluid Dynamics

SECOND EDITION

Analytical Fluid Dynamics

SECOND EDITION

George Emanuel

Professor
Department of Mechanical and Aerospace Engineering
University of Texas, Arlington
Arlington, Texas

CRC Press
Boca Raton London New York Washington, D.C.

Library of Congress Cataloging-in-Publication Data

Emanuel, George.
Analytical fluid dynamics / George Emanuel. 2nd ed.
p. cm.
Includes bibliographical references and index.
ISBN 0-8493-9114-8 (alk. paper)
1. Fluid dynamics. I. Title.
QA911 .E432000
532′.05—dc21

99-089453
CIP

Direct all inquiries to CRC Press LLC, 2000 N.W. Corporate Blvd., Boca Raton, Florida 33431.

International Standard Book Number 0-8493-9114-8
Library of Congress Card Number 99-089453
Printed in the United States of America 1 2 3 4 5 6 7 8 9 0
Printed on acid-free paper

Dedication

Dedicated with love to my wife and companion, Lita

Preface

The objectives of this edition remain the same as the first. The analysis and formulation are provided for a variety of selected topics in inviscid and viscous fluid dynamics, it is hoped with physical insight. In part, this means formulating the appropriate equations and then transforming them into a suitable form for the specific flow under scrutiny. The approach is applied to viscous boundary layers, shock waves, Prandtl–Meyer flow, etc. Sometimes a solution is obtained; other times a final answer requires numerical computation. Of crucial interest, however, is the analytical process itself and the coinciding physical interpretation.

A more in-depth coverage of topics is favored compared to a broad one that bypasses crucial or difficult details. At the graduate level, I believe an intensive approach is preferable. The book tries to avoid too much repetition of undergraduate course material. Of course, some repetition is both useful and unavoidable. When it occurs, however, the level and manner of treatment are different, often markedly so, from those at the undergraduate level. I have attempted whenever possible to point out the assumptions and limitations of the topic under discussion. Conversely, an attempt is made to discuss why a particular topic is worthy of study. For instance, a solution may be useful as a first (or initial value) estimate for CFD calculations. The rate of convergence is usually accelerated by having a reasonable initial flow field. Analytical solutions, such as those provided by the substitution principle, can be used to verify Euler or Navier–Stokes codes. An analytical approach often yields suitable first estimates for parameters of interest. In this regard, some of the homework problems are designed to give the student practice in obtaining "back-of-the-envelope" solutions. My personal motivation, however, still remains the beauty and elegance of analytical fluid dynamics (AFD).

As mentioned in the preface to the first edition, much of the material in that edition was unique. This is even truer for this edition, where all of the added material is unique to this text. The chapters covering a calorically imperfect gas flow, sweep, shock wave interference with an expansion, unsteady one-dimensional flow, and the force and moment analysis are new. In addition, the thermodynamic chapter is largely new as are Appendices B and C. The chapters that remain from the first edition have been revised to improve the clarity of the presentation.

When appropriate, topics where future research is warranted are pointed out. Fluid dynamics, including the AFD specialty, is very much alive and growing. Consequently, not everything in this text is complete or polished. A variety of major topics are not discussed. These topics include turbulent flow, CFD, experimental methods, etc., that are major subjects in themselves.

I owe a debt of gratitude to the many friends and colleagues who have contributed to this undertaking, especially past and present students. It is indeed a pleasure to acknowledge their comments and assistance. I particularly thank Dr. Jose Rodriguez, Professor Frank K. Lu, and Professor Milton Van Dyke for his comments on Chapter 23. I am especially in debt to Susan Houck for her superb typing and preparation of the manuscript.

Contents

Part I: Basic Concepts

Part II: Advanced Gas Dynamics

Chapter 9 Calorically Imperfect Flows

Chapter 10 Sweep

Chapter 11 Interaction of an Expansion Wave with a Shock Wave

Chapter 12 Unsteady, One-Dimensional Flow

Part III: Viscous/Inviscid Fluid Dynamics

Chapter 13 Coordinate Systems and Related Topics

Part V: Laminar Boundary-Layer Theory for Steady Two-Dimensional or Axisymmetric Flow

Chapter 18 Incompressible Flow over a Flat Plate

Chapter 19 Large Reynolds Number Flow

Chapter 20 Incompressible Boundary-Layer Theory

Chapter 21 Compressible Boundary-Layer Theory

Chapter 22 Supersonic Boundary-Layer Examples

Chapter 23 Second-Order Boundary-Layer Theory

Appendices

Part I

Basic Concepts

Outline of Part I

We embark on an in-depth study of fluid dynamics by discussing a variety of topics in a more general manner than encountered at the undergraduate level. Some of the topics are familiar to you, e.g., the Euler and Navier–Stokes equations, and the first and second laws of thermodynamics. One purpose of this text is to prepare you for intensive courses in computational fluid dynamics, turbulence, high-speed flow, rarefied gas dynamics, and so on. A second objective is to help you understand the fluid dynamic journal literature. Last, but not least, I hope to convey some of the intellectual fascination that abounds in our subject.

In this text, we often are not concerned with solutions to specific flow problems, although such solutions are often used to illustrate the theory. Specific flows also regularly appear in the homework problems and represent an essential element of this text. Nevertheless, we are primarily concerned with general features of inviscid and viscous fluid flows.

This is especially true for Part I, which provides many of the basic concepts. The first chapter is primarily concerned with establishing the Eulerian formulation, the constitutive relations, and several integral relations that are needed in later chapters. The conservation equations for mass, momentum, and energy are derived in the second chapter, while a general formulation for thermodynamics is provided in Chapter 3. The final chapter of Part I discusses general properties of a fluid flow that are not based on the conservation equations or the second law of thermodynamics. Such properties are referred to as kinematic and they include Kelvin's equations and the Helmholtz vortex theorems.

While some of the topics in Part I date from the very origin of fluid mechanics, much of the contents have a more recent origin. Indeed, since fluid dynamics is still evolving, some of the material is the result of recent research. Even topics of some antiquity, such as the second law, will appear new to you.

One reason well-known topics may appear different is our systematic use of vector and tensor analysis. Some background in these topics is presumed. A summary of the pertinent vector and tensor equations is provided in Appendix A. The trend toward an ever greater reliance on these and other analytical tools has been evident for some time. This trend stems from the need for a more flexible mathematical language for the increasing complexity of fluid dynamics. Once you become familiar with these topics, their utilization for our subject becomes indispensable.

Many scientists, mathematicians, and engineers have contributed to fluid dynamics over its long history. The amount of material that could be covered far exceeds my grasp of it or what can be covered in this text. (The following remarks are not limited to Part I, but hold for the overall text.) Self-imposed limitations are therefore essential. The first of these is that the fluid, gas or liquid, is easily deformable. We, therefore, deal with that branch of continuum mechanics that does not include solids. As a rule we shall assume the fluid is

(i) isotropic in its properties; fluids with polymers, rheological fluids, etc. are excluded;
(ii) not ionized, chemically reacting, diffusionally mixing, or a multi-phase fluid; and
(iii) not close to its critical point.

[In Chapter 3, when discussing thermodynamics, we are more general and do not always assume items (ii) and (iii).]

Another major restriction is that the fluid behaves as a continuous medium. This implies that the mean-free path of the molecules in a gas, or the mean distance between molecules in a liquid, is many orders of magnitude smaller than the smallest characteristic length of physical interest. Under a wide variety of conditions of practical importance, this assumption is fully warranted.

Our final assumptions are that relativistic effects and quantum mechanics can be safely ignored. This would not be the case, for instance, with liquid helium, which is a quantum fluid, or in jets emanating from astrophysical bodies.

All of the above assumptions, at one time or another, would require reconsideration. For instance, when a meteor is entering the atmosphere the surrounding air is chemically reacting and ionized during part of its downward trajectory. Similarly, an orbiting satellite, at a relatively low altitude, experiences the drag of a free-molecular flow. Nevertheless, the vast majority of applications that fluid dynamicists deal with still adhere to the foregoing assumptions.

The above exclusions are usually treated in more advanced courses, like those dealing with the dynamics of real gases or rarified flows. This is certainly true for turbulence; hence, we will not be concerned with turbulent flows. Our discussion, however, will not be restricted to incompressible fluid dynamics, since compressible flows, including those with shock waves, are of fundamental importance. We shall also often focus on vorticity for both incompressible and compressible flows.

1 Background Discussion

1.1 PRELIMINARY REMARKS

As always in engineering, we need to reduce the subject to quantifiable terms. This means that solvable equations need to be established. The relevant equations can be subdivided into three categories. In the first, we have the mechanical equations that express conservation of mass and the momentum equation, which is based on Newton's second law. In the second category we have the first and second laws of thermodynamics. The first law expresses conservation of energy, while the second law is a constraint on any physically realizable process.

The foregoing laws are of great power and generality. (Nevertheless, they do not always hold, e.g., when nuclear reactions occur as in fission or fusion. In this circumstance, conservation of mass holds in a modified form.) The final group of relations is not nearly as general. They are referred to as constitutive equations. For example, Fourier's heat conduction equation and the perfect gas thermal state equation are in this category. The relation between stress and the rate of deformation is similarly a constitutive relation. These relations are not universal but provide the properties for a specific class of substances and hold for a specific class of physical processes. At any rate, they are essential; without them the more general laws do not constitute a closed mathematical system. Closure of the system thus requires a proper number of consistent constitutive relations. Taken as a whole, the complete set of equations is referred to as the governing equations. By way of contrast, the three equations dealing with mass, momentum, and energy are referred to as conservation equations.

This chapter is devoted to a discussion of the Euler and Lagrange formulations in fluid dynamics. We then consider the stress tensor and the relation between this tensor and the rate-of-deformation tensor. We conclude by discussing a Newtonian fluid, Fourier's equation, the constitutive relations, and certain useful integral relations.

1.2 EULER AND LAGRANGE FORMULATIONS

Eulerian Formulation

There are two ways to formulate the equations of fluid dynamics: the Eulerian and Lagrangian approaches. In the Eulerian formulation, which we discuss first, the position vector $\vec{r}$ and time t are the independent variables. Thus, any scalar, such as the pressure, can be written as

$$p = p(\vec{r}, t) \tag{1.1}$$

while a vector, e.g., the fluid velocity $\vec{w}$, becomes

$$\vec{w} = \vec{w}(\vec{r}, t) \tag{1.2}$$

The Eulerian approach provides a field representation, in terms $\vec{r}$ and t, for any variable of interest. For example, a differential change in the pressure is provided by

$$dp = \frac{\partial p}{\partial \vec{r}} \cdot d\vec{r} + \frac{\partial p}{\partial t} dt \tag{1.3}$$

where the first term on the right side is the directional derivative of p in the direction $d\vec{r}$. Suppose we introduce Cartesian coordinates x_i and their corresponding orthonormal basis $\hat{1}_i$. Then $\vec{r}$ and $d\vec{r}$ are given by

$$\vec{r} = x_i \hat{1}_i \tag{1.4}$$

$$d\vec{r} = dx_i \hat{1}_i \tag{1.5}$$

where the repeated index summation convention is used. We also adhere to the convenient convention of not writing Cartesian coordinates as x^i, which would be the proper contravariant tensor notation. With Equation (1.5), we can write dp as

$$dp = \frac{\partial p}{\partial x_i} dx_i + \frac{\partial p}{\partial t} dt$$

or as

$$\frac{dp}{dt} = \frac{\partial p}{\partial t} + \frac{dx_i}{dt}\frac{\partial p}{\partial x_i} \tag{1.6}$$

The velocity is given by

$$\vec{w} = \frac{d\vec{r}}{dt} = \frac{dx_i}{dt}\hat{1} = w_i \hat{1}_i \tag{1.7}$$

where the w_i are the Cartesian velocity components, while the gradient of the pressure is provided by the del operator

$$\nabla p = \frac{\partial p}{\partial x_i}\hat{1}_i \tag{1.8}$$

Hence, Equation (1.6) reduces to

$$\frac{dp}{dt} = \frac{\partial p}{\partial t} + \vec{w} \cdot \nabla p \tag{1.9}$$

We shall utilize a notation, first introduced by George Stokes, to define the operator

$$\frac{D}{Dt} = \frac{\partial}{\partial t} + \vec{w} \cdot \nabla \tag{1.10}$$

which is called the substantial or material derivative. This definition is independent of any specific coordinate system. With tensor analysis, the del operator can be defined for any general curvilinear coordinate system; it is not restricted to Cartesian coordinates as in Equation (1.8). The substantial derivative also can be applied to vector quantities. For instance, when applied to the position vector, we have

$$\frac{D\vec{r}}{Dt} = \frac{\partial \vec{r}}{\partial t} + \vec{w} \cdot \nabla \vec{r}$$

where

$$\frac{\partial \vec{r}}{\partial t} = 0 \tag{1.11}$$

since $\vec{r}$ and t are independent variables. The gradient of $\vec{r}$ is

$$\nabla \vec{r} = \hat{1}_j \frac{\partial \vec{r}}{\partial x_j} = \hat{1}_j \hat{1}_i \frac{\partial x_i}{\partial x_j} = \hat{1}_j \hat{1}_i \delta_{ij} = \hat{1}_i \hat{1}_i = \overleftrightarrow{I} \tag{1.12}$$

where δ_{ij} is the Kronecker delta. Thus, $\nabla \vec{r}$ is a dyadic; in fact, it is the unit dyadic $\overleftrightarrow{I}$. We thereby obtain (see Appendix A)

$$\vec{w} \cdot \nabla \vec{r} = \vec{w} \cdot \overleftrightarrow{I} = \vec{w}$$

and $D\vec{r}/Dt$ becomes

$$\frac{D\vec{r}}{Dt} = \vec{w} \tag{1.13}$$

As a second example, let us determine the acceleration $\vec{a}$, which is given by

$$\vec{a} = \frac{D\vec{w}}{Dt} = \frac{\partial \vec{w}}{\partial t} + \vec{w} \cdot \nabla \vec{w} \tag{1.14}$$

The dot product on the right side can be interpreted as $\vec{w} \cdot (\nabla \vec{w})$, which involves the dyadic $\nabla \vec{w}$, or as $(\vec{w} \cdot \nabla)\vec{w}$, which does not involve a dyadic. With tensor analysis one can show that both interpretations yield the same result; the second one is usually preferred because of its greater simplicity. In Cartesian coordinates, e.g., we have

$$(\vec{w} \cdot \nabla)\vec{w} = \left(w_i \hat{1}_i \cdot \hat{1}_k \frac{\partial}{\partial x_k}\right) w_j \hat{1}_j = \left(w_i \frac{\partial}{\partial x_i}\right) \vec{w}_j \hat{1}_j = w_i \frac{\partial w_j}{\partial x_i} \hat{1}_j \tag{1.15}$$

An alternate expression for $\vec{a}$, of considerable utility, is based on the vector identity (see Appendix A, Table 5)

$$\nabla(\vec{A} \cdot \vec{B}) = \vec{A} \cdot \nabla \vec{B} + \vec{B} \cdot \nabla \vec{A} + \vec{A} \times (\nabla \times \vec{B}) + \vec{B} \times (\nabla \times \vec{A}) \tag{1.16}$$

where $\vec{A}$ and $\vec{B}$ are arbitrary vectors. We set $\vec{B} = \vec{A}$, to obtain

$$\nabla(\vec{A} \cdot \vec{A}) = 2\vec{A} \cdot \nabla \vec{A} + 2\vec{A} \times (\nabla \times \vec{A})$$

or

$$\vec{A} \cdot \nabla \vec{A} = \nabla\left(\frac{1}{2}A^2\right) - \vec{A} \times (\nabla \times \vec{A})$$

where

$$A^2 = \vec{A} \cdot \vec{A}$$

We now utilize

$$\vec{A} = \vec{w}$$

and

$$\nabla \times \vec{w} = \vec{\omega} \tag{1.17}$$

where $\vec{\omega}$ is the vorticity, to obtain

$$\vec{w} \cdot \nabla \vec{w} = \nabla\left(\frac{1}{2}w^2\right) + \vec{\omega} \times \vec{w}$$

The acceleration is therefore given by

$$\vec{a} = \frac{\partial \vec{w}}{\partial t} + \nabla\left(\frac{1}{2}w^2\right) + \vec{\omega} \times \vec{w} \tag{1.18}$$

in any coordinate system.

The substantial derivative has an important physical interpretation. It provides the time rate of change of any fluid quantity, scalar or vector, following a fluid particle. This viewpoint is apparent in Equation (1.13), where the time rate of change of the position of a fluid particle equals its velocity. Thus, the pressure of a given fluid particle changes in accordance with Equation (1.9). The substantial derivative consists of two terms. The first of these, $\partial(\ )/\partial t$, provides the changes at a fixed position due to any unsteadiness in the flow. For a steady flow, this term is zero. The second term, $\vec{w} \cdot \nabla$, is referred to as the convective derivative. It represents the changes that occur with position at a fixed time. This term is generally nonzero in a steady or unsteady flow.

Lagrangian Formulation

As mentioned, the Eulerian formulation provides a field description of a flow. The Lagrange formulation provides a particle description. Suppose a fluid particle has the location $\vec{r} = \vec{r}_o$ at $t = t_o$. In the Lagrangian approach the independent variables are $\vec{r}_o$ and t. Thus, the position of the fluid particle at time t is given by

$$\vec{r} = \vec{r}(\vec{r}_o, t) \tag{1.19}$$

where $\vec{r}_o$ is the particle's position at time t_o

$$\vec{r}_o = \vec{r}(\vec{r}_o, t_o)$$

and $\vec{r}_o$ is a fixed label on the particle as it moves. In this formulation, the velocity and acceleration are

$$\vec{w} = \frac{\partial \vec{r}}{\partial t}, \qquad \vec{a} = \frac{\partial^2 \vec{r}}{\partial t^2}$$

where $\vec{r}_o$ is kept fixed in both derivatives.

The two formulations can be related by assuming we know $\vec{w}(\vec{r},t)$ in the Eulerian description. We then integrate Equation (1.13) subject to the initial condition

$$\vec{r} = \vec{r}_o \quad \text{at} \quad t = t_o \tag{1.20}$$

The solution is then the Lagrangian description, Equation (1.19).

The Lagrangian approach is widely used in mechanics; e.g., consider a marble rolling down an inclined plane under the influence of gravity. The problem is solved by first establishing a differential equation for the motion of the marble. The solution of this equation provides the position of the marble as a function of time and its initial position.

There are several reasons for not utilizing the Lagrangian description. First, we generally are not interested in the actual location of a fluid particle, whereas, as engineers, we are interested in the pressure and velocity, since these provide the pressure and shear stress forces on a body. Second, obtaining $\vec{r}(\vec{r}_o,t)$ represents a greater effort than is required for obtaining p and $\vec{w}$. Finally, the Lagrangian approach is cumbersome for a viscous flow. We, therefore, follow a well-established tradition and hereafter focus on the Eulerian description.

Before leaving this topic, recall that the substantial derivative follows a fluid particle. While the concept is Lagrangian, the derivative itself is Eulerian, since $\vec{r}$ and t, not $\vec{r}_o$ and t, are the independent variables.

Pathlines and Streamlines

The trajectory of a fluid particle is called a pathline or particle path. This is found by integrating Equation (1.13) subject to the initial condition, Equation (1.20). We shall not discuss a different type of curve called a streakline. (This is a particle path that originates at a fixed position.) More important than either pathlines or streaklines are the streamlines. Streamlines are curves, which at a given instant are tangent to the velocity field. In an unsteady flow, pathlines, streaklines, and streamlines are all different. In a steady flow, however, they all coincide.

Let $d\vec{r}$ be tangent to the velocity and therefore tangent to a streamline. Then $d\vec{r}$ satisfies

$$d\vec{r} \times \vec{w} = 0 \tag{1.21a}$$

or with Cartesian coordinates

$$\begin{bmatrix} \hat{1}_1 & \hat{1}_2 & \hat{1}_3 \\ dx_1 & dx_2 & dx_3 \\ w_1 & w_2 & w_3 \end{bmatrix} = 0$$

On expanding this relation, we obtain

$$(w_3 dx_2 - w_2 dx_3)\hat{1}_1 - (w_3 dx_1 - w_1 dx_3)\hat{1}_2 + (w_2 dx_1 - w_1 dx_2)\hat{1}_3 = 0$$

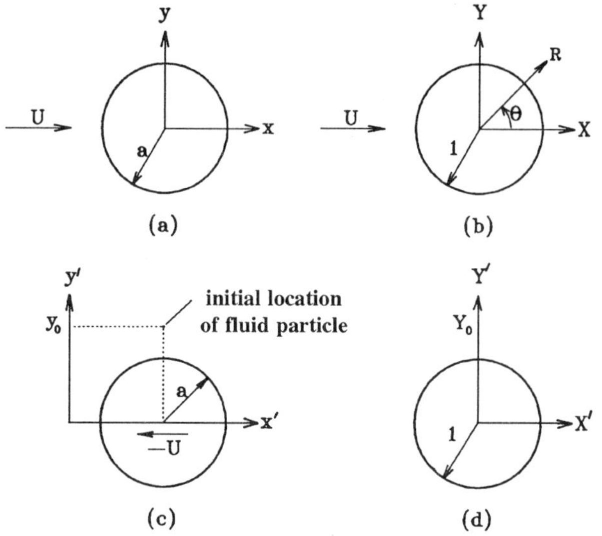

FIGURE 1.1 Coordinate systems associated with flow about a circular cylinder; (a) and (b) are for steady flow; (c) and (d) are for unsteady flow.

or, in scalar form,

$$\frac{dx_1}{w_1} = \frac{dx_2}{w_2} = \frac{dx_3}{w_3} \tag{1.21b}$$

The solution of these two ordinary differential equations provides the streamline curves, subject to a given initial condition. Recall that the streamlines are tangent to the velocity field at a given instant of time. Thus, if the w_i are time-dependent, the t variable is treated as a fixed parameter during the integration of these equations.

Illustrative Example

As an example, we first determine the streamline equation for steady, two-dimensional cross flow about a circular cylinder of radius a, as sketched in Figure 1.1(a). (Later, the unsteady flow pathlines are found.) In addition, we assume a uniform freestream, with speed U and an incompressible, inviscid flow without circulation. Hence, the cylinder is not subjected to either a lift or drag force. From elementary aerodynamic theory, we obtain the x and y velocity components as

$$\frac{u}{U} = 1 + \frac{Y^2 - X^2}{(X^2 + Y^2)^2}, \quad \frac{v}{U} = -\frac{2XY}{(X^2 + Y^2)^2} \tag{1.22}$$

where $X = (x/a)$ and $Y = (y/a)$. Since the flow is two dimensional, we need to integrate only one of the equations in (1.21b), written as

$$\frac{dx}{u} = \frac{dy}{v}$$

to obtain the equation for the streamlines. The equations in (1.22) are substituted into this differential equation, with the result

$$\frac{dX}{dY} = -\frac{Y^2 - X^2 + (X^2 + Y^2)^2}{2XY}$$

To separate variables, cylindrical coordinates, shown in Figure 1.1(b), are introduced as

$$X = R\cos\theta, \qquad Y = R\sin\theta$$

to obtain

$$\frac{R^2 + 1}{(R-1)(R+1)R}dR = -\cot\theta\, d\theta$$

The method of partial fractions is now used for the left side, with the result

$$\int_{Y_o}^{R}\left(\frac{1}{2}\frac{R^2}{R-1} + \frac{1}{2}\frac{1}{R-1} + \frac{1}{2}\frac{R^2}{R+1} + \frac{1}{2}\frac{1}{R+1} - R - \frac{1}{R}\right)dR = -\int_{\pi/2}^{\theta}\cot\theta\, d\theta$$

where a point Y_o on the Y axis is used for the lower limit and, at this point, $\theta = \pi/2$. As a result of the integration, we obtain

$$\frac{Y_o}{Y_o^2 - 1}\frac{R^2 - 1}{R} = \frac{1}{\sin\theta}$$

By returning to X,Y coordinates, the streamline equation simplifies to

$$X^2 + Y^2 = \frac{Y}{Y - Y_\infty} \tag{1.23a}$$

where Y_∞ is the streamline ordinate at $X \to \pm\infty$. Figure 1.2(a) shows a typical streamline pattern. The two special Y values are related by

$$Y_\infty = Y_o - \frac{1}{Y_o} \tag{1.23b}$$

where $X_o = 0$ and $Y_o \geq 1$ for any streamline outside the cylinder. (There is a related streamline pattern inside the cylinder.)

The solution, Equation (1.23a), can also be obtained, with negligible effort, from the stream function (defined in Chapter 5) equation

$$\psi = Uy\left(1 - \frac{a^2}{x^2 + y^2}\right)$$

where $Y_\infty = \psi(\pm\infty, Y_\infty)/(aU)$, and from the fact that a stream function is constant along streamlines in a steady flow. Only in special cases, however, is a stream function available, whereas our purpose is to illustrate how Equations (1.21b) are generally utilized.

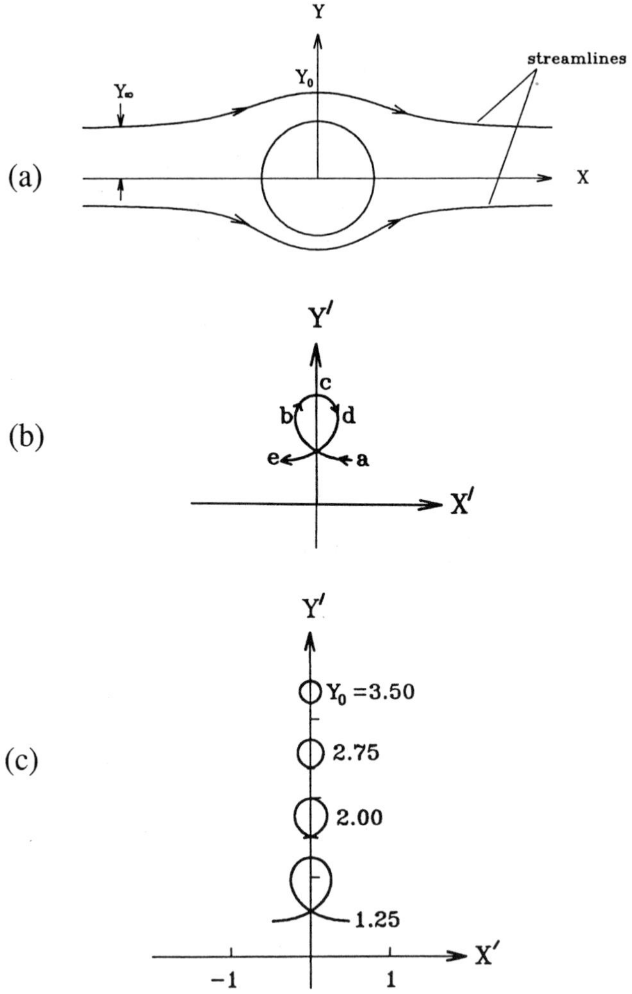

FIGURE 1.2 Streamlines (a) and pathlines, (b) and (c), are for flow about a circular cylinder.

The determination of the pathlines in an unsteady flow is more difficult. Moreover, the physical interpretation of a pathline solution is far from trivial. As indicated in Figure 1.1(c), the same problem is considered, but now the cylinder is moving to the left, with a speed $-U$, into a fluid that is quiescent far from the cylinder. A prime is used to denote unsteady variables, and our goal is to determine the trajectory of a fluid particle. It is analytically convenient to fix the initial condition for the particle directly over the center of the cylinder with $t' = 0$ and $y' = y_o$, as shown in Figure 1.1(c). Consequently, a full trajectory requires the particle's position for both positive and negative time. The "initial condition" phrase therefore does not refer to the particle's position when $t' \rightarrow -\infty$.

This flow is essentially the same as the steady flow case; only our viewpoint is different. In a steady flow, we move with the cylinder, whereas in the unsteady case we have a fixed (laboratory) coordinate system. It is convenient to again introduce nondimensional variables

$$X' = \frac{x'}{a}, \quad Y' = \frac{y'}{a}, \quad T' = \frac{U}{a}t'$$

and use a Galilean transformation

$$x' = x - Ut, \quad y' = y, \quad t' = t, \quad u' = u - U, \quad v' = v$$

to convert the steady flow velocity field into the unsteady one. Equations (1.22) thus become

$$\frac{u'}{U} = \frac{Y'^2 - (X' - T')^2}{[(X' + T')^2 + Y'^2]^2}, \quad \frac{v'}{U} = -\frac{2(X' + T')Y'}{[(X' + T')^2 + Y'^2]^2}$$

The center of the cylinder is at $x = y = 0$, or

$$X' + T' = 0, \quad Y' = 0$$

Hence, the initial condition for a fluid particle is

$$X' = 0, \quad Y' = Y_o \quad \text{when} \quad T' = 0$$

with $Y_o \geq 1$. The X',Y' coordinate system is therefore shifted to the left or right until the position of the particle of interest is located at $X' = 0$ when $T' = 0$. When T' is sufficiently negative, the particle is upstream of the center of the cylinder, which is at a positive X' value. Remember that when the particle is above the cylinder's center, $T' = X' = 0$. Similarly, when T' is sufficiently positive, X' is negative. This behavior is illustrated in Figure 1.2(b), where point a is the location of a particle when $T' \to -\infty$, while point e is the location when $T' \to +\infty$. In this figure, the center of the cylinder moves from $X' \to \infty$, $T' \to -\infty$ to $X' \to -\infty$, $T' \to \infty$, whereas the lateral motion of a particle is finite. The one exception is a particle with $Y_\infty = 0$; this particle wets the cylinder.

At its initial location, when $T' = 0$, the velocity components of the particle are

$$\left(\frac{u'}{U}\right)_o = \frac{1}{Y_o^2}, \quad \left(\frac{v'}{U}\right)_o = 0$$

Thus, the particle, at this time, is moving in the positive X' direction, as indicated by point c in Figure 1.2(b). For a particle far upstream of the cylinder, we have

$$X' > 0, \quad T' << 0, \quad Y' \cong Y_\infty, \quad \frac{u'}{U} < 0, \quad \frac{v'}{U} > 0$$

When the particle is far downstream of the cylinder, we have

$$X' < 0, \quad T' >> 0, \quad Y' \cong Y_\infty, \quad \frac{u'}{U} < 0, \quad \frac{v'}{U} < 0$$

and the cylinder is to the left of the particle. Far from the cylinder, in either X' direction, the particle moves in the negative X' direction. The sign change in u', which occurs when the particle is near the cylinder, is discussed shortly. Note that Y_∞ and Y_o are still related by Equation (1.23b).

We are now ready to utilize Equation (1.13), written as

$$\frac{dx'}{dt'} = u', \quad \frac{dy'}{dt'} = v'$$

for the particle paths. In contrast to the streamline situation, we have one additional differential equation to solve. In terms of nondimensional variables these equations become

$$\frac{dX'}{dT'} = \frac{Y'^2 - (X' + T')^2}{[(X' + T')^2 + Y'^2]^2}, \quad \frac{dY'}{dT'} = -\frac{2(X' + T')Y'}{[(X' + T')^2 + Y'^2]^2} \tag{1.24}$$

After Equation (1.23a) is transformed, it also represents a particle path. In other words,

$$(X' + T')^2 + Y'^2 = \frac{Y'}{Y' - Y_\infty} \tag{1.25}$$

is a first integral of Equations (1.24). This can be demonstrated by differentiating this equation with respect to T' and eliminating dX'/dT' and dY'/dT' to obtain an identity. We next utilize this equation to eliminate $X' + T'$ from the dY'/dT' equation, with the result

$$\frac{dT'}{dT'} = \pm\frac{2(1 + Y_\infty Y' - Y'^2)^{1/2}(Y' - Y_\infty)^{3/2}}{Y'^{1/2}}$$

where a $\pm$ sign is introduced when the square root of $(X' + T')^2$ is taken. The plus sign holds when $T' < 0$, while the minus sign holds when $T' > 0$.

The above differential equation is integrated from the initial condition, $Y' = Y_o$ when $T' = 0$, to obtain

$$T' = \pm\frac{1}{2}\int_{Y'}^{Y_o}\left(\frac{Y'}{1 + Y_\infty Y' - Y'^2}\right)^{1/2}\frac{dY'}{(Y' - Y_\infty)^{3/2}}$$

Since

$$1 + Y_\infty Y' - Y'^2 = (Y_o - Y')\left(Y' + \frac{1}{Y_o}\right)$$

the integral can be written in a standard form as

$$T' = \pm\frac{1}{2}\int_{Y'}^{Y_o}\left[\frac{Y'}{(Y_o - Y')(Y' - Y_\infty)^3\left(Y' + \frac{1}{Y_o}\right)}\right]^{1/2} dY'$$

This quadrature can be evaluated in terms of elliptic integrals of the first, F, and second, E, kinds, defined as (*Handbook of Mathematical Functions*, 1972)

$$F(\phi\backslash\alpha) = \int_0^{\phi}\frac{d\theta}{(1 - \sin^2\alpha\sin^2\theta)^{1/2}}$$

$$E(\phi\backslash\alpha) = \int_0^{\phi}(1 - \sin^2\alpha\sin^2\theta)^{1/2}d\theta$$

where θ is a dummy integration variable. With the aid of a table of elliptic integrals (Gradshteyn and Ryzhik, 1980; No. 47, p. 272), one can show that the final form for T' is then

$$T' = \pm Y_o \left\{ F(\phi \backslash \alpha) - E(\phi \backslash \alpha) + \left[\frac{Y_o Y'(Y_o - Y')}{(Y' - Y_\infty)(1 + Y_o Y')} \right]^{1/2} \right\}$$

where ϕ and α are given by

$$\phi(Y') = \sin^{-1}\left[Y_o^{3/2} \left(\frac{Y_o - Y'}{1 + Y_o Y'} \right)^{1/2} \right]$$

$$\alpha = \sin^{-1}\frac{1}{Y_o^2}$$

This relation, in conjunction with Equation (1.25), represents the pathlines in an implicit form. In other words, given Y_o (or Y_∞) and Y', these two equations determine X' and T'.

Figure 1.2 shows, to scale, the expected streamline pattern in (a) and the pathline pattern in (b) and (c), where all patterns are symmetric about the Y or Y' axis. The arrows on the streamlines and pathlines indicate increasing time or the direction of the velocity.

Along $a - b - c$ in Figure 1.2(b), T' is negative, and the center of the cylinder is at the origin when the fluid particle is at point c, where T' is zero. At point a, T' equals $-\infty$, while at point e, T' is $+\infty$. (The value of X'_a is the subject of Problem 1.7.) For any other point on $a - b - c$, the center of the cylinder is to the right of the fluid particle. In this regard, it is useful to note that a particle is upstream of the cylinder's center when $X' + T' < 0$ and downstream otherwise. This result stems from the Galilean transformation, $X = X' + T'$. At points b and d, u' is zero, while at point c, v' is zero. One exception to part of this discussion is a particle with $Y_\infty = 0$ and $X' > 0$, which ultimately wets the cylinder's surface. Otherwise, all other fluid particles have similar trajectories, including the loop.

Along $c - d - e$, $T' \geq 0$ and the particle is downstream of the cylinder's center. Consequently, along $a - b$ the particle is being pushed by the cylinder and $u' \leq 0$, while along $d - e$ the particle is being pulled by the cylinder, and again $u' \leq 0$. When the particle is close to the cylinder along $b - c - d$, there is a transition region between the pushing and pulling where $u' \leq 0$. In this region, v' changes sign. As evident in Figure 1.2(c), the size of the loop depends on Y_∞ (or Y_o). Particles with a small Y_∞ value, which initially are close to the X' axis, have a relatively large loop. This is caused by the cylinder imparting a large velocity to the particle as it is shoved aside.

A particle experiences a horizontal drift (Darwin, 1953) as a result of the cylinder's motion, given by

$$\Delta = X'_a - X'_e = 2X'_a$$

As shown in Problem 1.7, Δ becomes infinite when $Y_\infty \to 0$ and goes to zero as $Y_\infty \to \infty$. This displacement, or drift, also occurs in the steady flow case, since the particles that pass close to the cylinder are retarded more than those that pass at a distance. As shown in Problem 5.22 in Chapter 5, along a given pathline the change in kinetic energy balances the work done in moving a fluid particle. Because viscosity is not present, the work done on adjacent pathlines or streamlines is not related. Consequently, the change in displacement Δ with Y_∞ does not involve any dissipative work.

As you might imagine, the streamlines and pathlines for flow about a sphere are similar to that of a cylinder. Both types of patterns are also considered in Problems 5.23 and 5.24, where a Galilean transformation is again convenient for the unsteady spherical case.

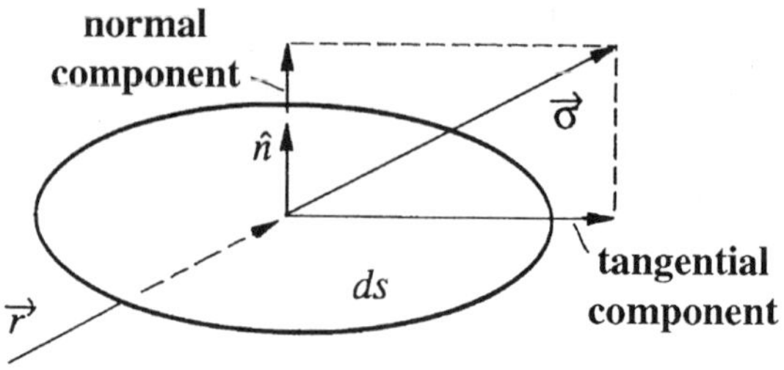

FIGURE 1.3 Schematic of a surface force.

1.3 THE STRESS TENSOR

We now turn our attention to the two types of forces that can act on an arbitrary infinitesimal fluid element or particle. One of these is the body force, e.g., the force due to gravity or an electromagnetic field. By definition, a body force is one that acts throughout a volume, as is the case with gravity, where

$$\vec{F}_b = \vec{g} \tag{1.26}$$

and $\vec{g}$ is the acceleration due to gravity. This force per unit volume is $\rho\vec{g}$, where ρ is the density. Hence, a per unit volume body force is proportional to the density. There are other apparent or effective body forces in a coordinate system that is rotating or accelerating, like the centripetal and Coriolis forces that also are proportional to the density. (These forces are discussed in Section 2.5.) The electromagnetic force depends on the net charges, not on the bulk density; however, it is properly treated as a body force, since the charges are usually distributed throughout the fluid medium. We will not be concerned with this type of force.

By definition, a surface force is one that is proportional to the amount of surface area it acts upon. The surface of interest need not be a real surface, such as the surface of a droplet, but a conceptual one, such as that surrounding an infinitesimal fluid particle. The simplest example of a surface force is the one due to hydrostatic pressure. There are also surface forces that act at real surfaces, such as an interfacial force at a phase boundary. We will not deal with this type of force.

An analytical description of a surface force is not nearly as simple as Equation (1.26). For this description, we utilize a differential surface area ds, whose spatial orientation is provided by a unit normal vector $\hat{n}$, as indicated in Figure 1.3. The surface force per unit area, $\vec{\sigma}$, that acts on ds is generally not in the plane of the surface. As indicated in the figure, $\vec{\sigma}$ will have a component along $\hat{n}$ and a tangential component in the plane of the surface. Since $\vec{\sigma}$ is per unit area, the actual force on ds is $\vec{\sigma}ds$. We call $\vec{\sigma}$ the stress vector; the component along $\hat{n}$ results in the normal stress, while the component in the plane of the surface results in the shear stress.

In general, $\vec{\sigma}$ is a function of both position and surface orientation; i.e.,

$$\vec{\sigma} = \vec{\sigma}(\vec{r}, \hat{n}) \tag{1.27}$$

The stress vector can be related to a second-order tensor that depends on $\vec{r}$ but not on $\hat{n}$. To show this, we need Newton's third law, which states that for every action (force) there is an equal but opposite reaction. Hence, we have

$$\vec{\sigma}(\vec{r}, \hat{n}) = -\vec{\sigma}(\vec{r}, -\hat{n}) \tag{1.28}$$

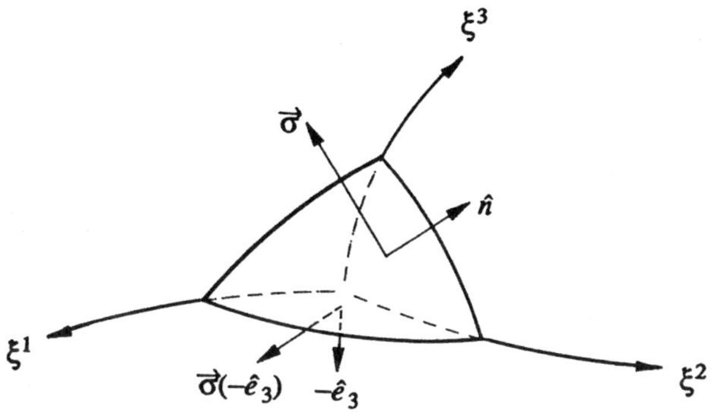

FIGURE 1.4 Effect of a stress vector on an infinitesimal tetrahedron.

For the subsequent discussion, it will be convenient to introduce orthogonal curvilinear coordinates ξ^i and the corresponding orthonormal basis $\hat{e}_i$. Consider an infinitesimal tetrahedron as shown in Figure 1.4. The outward unit normal vectors to the surfaces coplanar with the ξ^i coordinates are $-\hat{e}_i$. Let $\vec{\sigma}_i$ be the outward facing stress vector on these surfaces; i.e.,

$$\vec{\sigma}_i = -\vec{\sigma}(\vec{r}, -\hat{e}_i) \tag{1.29}$$

Note that $\vec{\sigma}_i$ is a vector, not a component of $\vec{\sigma}(\vec{r},\hat{n})$. Shortly, we will relate these two vectors. By virtue of Equation (1.28), we have

$$\vec{\sigma}(\vec{r}, -\hat{e}_i) = -\vec{\sigma}(\vec{r}, \hat{e}_i) \tag{1.30}$$

For the tetrahedron, let Δs be the slant face surface area, Δs_i the surface area normal to ξ^i, and Δv the volume of the tretrahedron. This volume is given by

$$\Delta v = (1/3)\Delta h \Delta s$$

where Δh is the normal distance from the origin to the slant face. With the aid of vector analysis, the various surface areas can be related by

$$\Delta s_i \hat{e}_i = (\Delta s)\hat{n} \tag{1.31}$$

Since the basis is orthonormal, we have

$$\hat{e}_i \cdot \hat{e}_j = \delta_{ij} \tag{1.32}$$

As a consequence, when we multiply Equation (1.31) with $\hat{e}_j$, we obtain

$$\Delta s_i \delta_{ij} = (\Delta s)\hat{e}_j \cdot \hat{n}$$

or

$$\Delta s_i = (\hat{n} \cdot \hat{e}_i)\Delta s \tag{1.33}$$

Newton's second law for the mass, $\rho\Delta v$, within the tetrahedron can be written as

$$(\rho\Delta v)\vec{a} = \vec{\sigma}(\vec{r}, \hat{n})\Delta s - \vec{\sigma}(\vec{r}, \hat{e}_i)\Delta s_i + (\rho\Delta v)\vec{F}_b$$

where $\vec{a}$ is the mass's acceleration, ρ is the density, and the right side represents the four surface forces and the body force that act on the tetrahedron. We now replace Δs_i with Equation (1.33), Δv with $\Delta h\Delta s/3$, and $\vec{\sigma}(\vec{r},\hat{e}_i)$ with Equations (1.29) and (1.30), to obtain

$$(1/3)\Delta h(\Delta s)\rho\vec{a} = \vec{\sigma}\Delta s - \vec{\sigma}_i(\hat{n} \cdot \hat{e}_i)\Delta s + (1/3)\Delta h(\Delta s)\rho\vec{F}_b$$

We assume $\vec{a} - \vec{F}_b$ remains finite as the tetrahedron shrinks to a point. In this limit, $\Delta h \to 0$, and we obtain

$$\vec{\sigma} = (\hat{n} \cdot \hat{e}_i)\vec{\sigma}_i \quad \textbf{(1.34)}$$

where the right side contains three terms since i is summed over.

As previously indicated, the stress depends on the force vector $\vec{\sigma}$ and the vector $\hat{n}$ that prescribes the orientation of the surface area on which $\vec{\sigma}$ acts. For a given coordinate system, this dependence can be reduced to two sets of vectors, $\vec{\sigma}_i$ and $\hat{e}_i$. The stress is therefore a second-order tensor, which can be written in dyadic form as

$$\overleftrightarrow{\sigma} = \hat{e}_i\vec{\sigma}_i \quad \textbf{(1.35)}$$

where a dyadic is just the juxtaposition of two vectors. As a consequence, Equation (1.34) becomes

$$\vec{\sigma}(\vec{r}, \hat{n}) = \hat{n} \cdot \overleftrightarrow{\sigma}(\vec{r}) \quad \textbf{(1.36)}$$

The second-order stress tensor $\overleftrightarrow{\sigma}$ is thus related to the force vector $\vec{\sigma}$ and helps provide the explicit dependence of $\vec{\sigma}$ on $\hat{n}$. In other words, $\overleftrightarrow{\sigma}$ is independent of the orientation of the surface. We now define the component form of $\vec{\sigma}_i$ and $\overleftrightarrow{\sigma}$ as

$$\vec{\sigma}_i = \sigma_i^j\hat{e}_j \quad \textbf{(1.37a)}$$

$$\overleftrightarrow{\sigma} = \sigma^{ij}\hat{e}_i\hat{e}_j \quad \textbf{(1.37b)}$$

In a Cartesian coordinate system, σ^{ij} is written as σ_{ij}. Also note that the right side of Equation (1.37b) consists of nine terms in contrast to Equation (1.35), which contains only three. By comparing these two equations, we obtain

$$\sigma^{ij} = \sigma_i^j \quad \textbf{(1.38)}$$

while Equation (1.37a) yields the contravariant result

$$\vec{\sigma}_i = \sigma^{ij}\hat{e}_j \quad \textbf{(1.39)}$$

These last equations express the fact that σ^{ij} represents the stress on an area perpendicular to the ξ^i coordinate and in the jth direction.

The stress vector $\vec{\sigma}$ at $\vec{r}$ is determined by the nine components of $\overleftrightarrow{\sigma}$ and the normal $\hat{n}$ to the surface ds. Not all of the components are independent of each other. We have already utilized two conditions, namely, the action equals reaction principle and Newton's second law. The components of $\overleftrightarrow{\sigma}$, however, are subject to a third condition that requires the resultant moment of these forces, about any point, to vanish. This condition will be examined in Section 2.6, where it results in $\overleftrightarrow{\sigma}$ being a symmetric tensor,

$$\sigma^{ij} = \sigma^{ji} \qquad \text{or} \qquad \sigma_{ij} = \sigma_{ji} \tag{1.40}$$

in which case $\overleftrightarrow{\sigma}$ has only six independent components. In this circumstance, Equation (1.36) can be written as

$$\vec{\sigma} = \hat{n} \cdot \overleftrightarrow{\sigma} = \overleftrightarrow{\sigma} \cdot \hat{n} \tag{1.41}$$

If $\overleftrightarrow{\sigma}$ is not symmetric, then $\hat{n} \cdot \overleftrightarrow{\sigma} \neq \overleftrightarrow{\sigma} \cdot \hat{n}$.

1.4 RELATION BETWEEN STRESS AND DEFORMATION-RATE TENSORS

Let us assume a constant velocity field and ignore gravity. In this circumstance $\overleftrightarrow{\sigma}$, and therefore $\vec{\sigma}$, has no dependence on $\vec{r}$. Furthermore, the fluid possesses no shearing motion and no shear stresses. In a Cartesian coordinate system, σ_{ij} can be written as

$$\sigma_{ij} = (\text{constant})\,\delta_{ij} \tag{1.42}$$

(As mentioned earlier, we use the covariant component form for vectors and tensors when the coordinates are Cartesian.) Equation (1.42) guarantees no shear stress; i.e., a nonzero shear stress requires $\sigma_{ij} \neq 0$ for some $i \neq j$.

Our frequent use of a Cartesian coordinate system requires a word of explanation. One can show, using the Gram–Schmidt procedure of vector analysis, that any vector basis can be replaced by an orthonormal basis. This new basis, in turn, can be replaced with a Cartesian one. These replacements are performed when convenient and result in no loss of generality, since the laws of physics are independent of the choice of a coordinate system. It will prove convenient for us to use Cartesian coordinates for some of the subsequent derivations. As noted, there is no loss in generality in doing this.

Equation (1.42) means that the normal stress is independent of the orientation of the surface ds as given by $\hat{n}$. This is the case for the stress due to the hydrostatic pressure p, which varies with $\vec{r}$ but not with $\hat{n}$. We therefore write this equation as

$$\sigma_{ij} = -p\,\delta_{ij} \tag{1.43}$$

where, by convention, a compressive stress is negative, hence producing the minus sign. A fluid motion with a nonzero velocity gradient will possess normal stresses that are not equal to the negative of the hydrostatic pressure.

We now subtract the hydrostatic pressure term from $\overleftrightarrow{\sigma}$ to obtain the *viscous* stress tensor

$$\tau_{ij} = \sigma_{ij} - (-p\,\delta_{ij}) = \sigma_{ij} + p\,\delta_{ij}$$

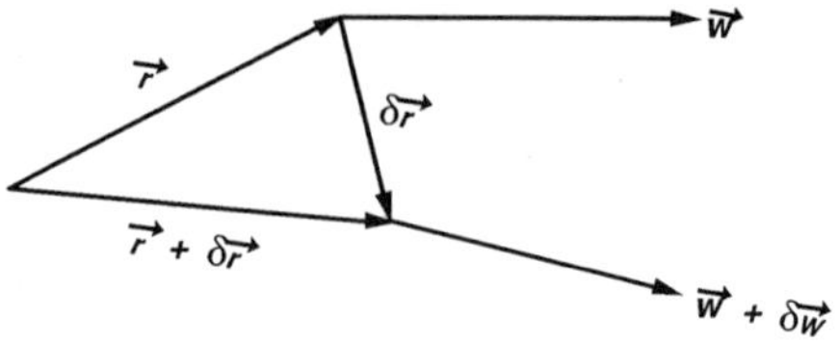

FIGURE 1.5 Strain rate schematic.

or more generally

$$\overleftrightarrow{\tau} = \overleftrightarrow{\sigma} + p\overleftrightarrow{I} \tag{1.44}$$

The viscous stress tensor $\overleftrightarrow{\tau}$ is nonzero only if the fluid possesses a nonzero relative motion. It is $\overleftrightarrow{\tau}$ that we relate to the rate of deformation. (In a solid, $\overleftrightarrow{\tau}$ depends on the deformation itself and not the rate of deformation. This trivial-sounding difference represents the demarcation between solid and fluid mechanics.) We further require that the dependence of $\overleftrightarrow{\tau}$ on the rate of deformation be independent of the choice of the coordinate system.

To help fix ideas, we observe that the motion of a fluid can be decomposed into four types of motion: uniform translation, solid-body rotation, extensional strain or dilatation, and a shear strain. The first two types of motion produce no relative motion; hence, $\overleftrightarrow{\tau}$ should depend only on the dilatation and shearing motions. Consequently, $\overleftrightarrow{\tau}$ cannot depend on $\vec{w}$ or its components, i.e., on the translational motion; however, $\overleftrightarrow{\tau}$ can depend on derivatives of the velocity components.

We now consider the relative motion of two adjacent fluid particles that are separated by a small distance $\delta\vec{r}$, as shown in Figure 1.5. At some instant, the particles have velocities $\vec{w}$ and $\vec{w} + \delta\vec{w}$, where $\delta\vec{w}$ becomes $d\vec{w}$ as $\delta\vec{r} \to d\vec{r}$. We evaluate $d\vec{w}$ by writing the Taylor series

$$\vec{w}(\vec{r} + d\vec{r}) = \vec{w}(\vec{r}) + d\vec{r} \cdot (\nabla\vec{w}) + \cdots$$

to obtain

$$d\vec{w} = \vec{w}(\vec{r} + d\vec{r}) = \vec{w}(\vec{r}) + d\vec{r} \cdot (\nabla\vec{w}) \tag{1.45}$$

The rightmost term is just the directional derivative of $\vec{w}$ in the $d\vec{r}$ direction, and $\nabla\vec{w}$ is the velocity gradient tensor.

The evaluation of $d\vec{w}$ requires decomposing $d\vec{r} \cdot \nabla\vec{w}$ in accordance with the above discussion. It is evident that this quantity does not depend on any uniform translational motion, since $\vec{w}$ appears only in the gradient. However, we must still subtract any solid-body rotation from $d\vec{w}$. To accomplish this, we observe that any second-order tensor can be written as the sum of symmetric and anti-symmetric tensors. Hence, we write

$$\nabla\vec{w} = \overleftrightarrow{\varepsilon} + \overleftrightarrow{\omega} \tag{1.46}$$

The symmetric tensor $\overleftrightarrow{\varepsilon}$, called the rate-of-deformation tensor, is given by

$$\overleftrightarrow{\varepsilon} = (1/2)[\nabla\vec{w} + (\nabla\vec{w})^t] \tag{1.47a}$$

where $(\ )^t$ denotes the transposition operation. For example, we write $\nabla\vec{w}$ in Cartesian coordinates as

$$\nabla\vec{w} = \frac{\partial w_i}{\partial x_j}\hat{\imath}_j\hat{\imath}_i$$

Then

$$(\nabla\vec{w})^t = \frac{\partial w_j}{\partial x_i}\hat{\imath}_j\hat{\imath}_i$$

and $\overleftrightarrow{\varepsilon}$ becomes

$$\overleftrightarrow{\varepsilon} = \frac{1}{2}\left(\frac{\partial w_i}{\partial x_j} + \frac{\partial w_j}{\partial x_i}\right)\hat{\imath}_j\hat{\imath}_i \quad \textbf{(1.47b)}$$

The antisymmetric part of $\nabla\vec{w}$ is the rotation tensor

$$\overleftrightarrow{\omega} = \frac{1}{2}[\nabla\vec{w} - (\nabla\vec{w})^t] \quad \textbf{(1.48a)}$$

which becomes in Cartesian coordinates

$$\overleftrightarrow{\omega} = \frac{1}{2}\left(\frac{\partial w_i}{\partial x_j} - \frac{\partial w_j}{\partial x_i}\right)\hat{\imath}_j\hat{\imath}_i \quad \textbf{(1.48b)}$$

Observe that $\overleftrightarrow{\varepsilon}$ and $\overleftrightarrow{\omega}$ sum to $\nabla\vec{w}$. With Equation (1.46), we see that $d\vec{w}$ is

$$d\vec{w} = d\vec{r}\cdot\overleftrightarrow{\varepsilon} + d\vec{r}\cdot\overleftrightarrow{\omega} \quad \textbf{(1.49)}$$

We need to interpret the $d\vec{r}\cdot\overleftrightarrow{\omega}$ term. For this, consider a body whose sole motion is that of solid-body rotation with a constant angular velocity $\vec{\omega}_{\text{rot}}$. From mechanics, the velocity $\vec{w}$ at point $\vec{r}$ of the body is provided by the cross product

$$\vec{w} = \vec{\omega}_{\text{rot}} \times \vec{r} \quad \textbf{(1.50)}$$

We take the curl of both sides to obtain

$$\nabla\times\vec{w} = \nabla\times(\vec{\omega}_{\text{rot}}\times\vec{r}) = 2\vec{\omega}_{\text{rot}}$$

where the rightmost equality stems from the use of standard vector identities in Table 5 of Appendix A. (Although $\nabla\times\vec{w}$ is the vorticity, this observation is irrelevant to the present discussion.) This relation is now multiplied by $d\vec{r}$, with the result

$$\vec{\omega}_{\text{rot}}\times d\vec{r} = \frac{1}{2}(\nabla\times\vec{w})\times d\vec{r}$$

With the aid of Equations (1.5) and (1.7), Cartesian coordinates are used to evaluate the right side as

$$\vec{\omega}_{\text{rot}} \times d\vec{r} = \frac{1}{2}\left(\frac{\partial w_i}{\partial x_j} - \frac{\partial w_j}{\partial x_i}\right) dx_j \, \hat{\imath}_i$$

However, with the use of Equation (1.48b) we observe that

$$d\vec{r} \cdot \overleftrightarrow{\omega} = \overleftrightarrow{\omega}_{\text{rot}} \times d\vec{r} \tag{1.51}$$

The $d\vec{r} \cdot \overleftrightarrow{\omega}$ term in Equation (1.49) is thus associated with a solid-body rotation and does not contribute to the viscous stress tensor. This means that $\overleftrightarrow{\tau}$ can only depend on the rate-of-deformation tensor $\overleftrightarrow{\varepsilon}$. This tensor, however, is symmetric with six independent components. These components can be further subdivided into those producing a shearing motion and those responsible for a dilatation or extensional strain. This later strain is given by the trace of $\overleftrightarrow{\varepsilon}$,

$$\varepsilon_{ii} = \frac{\partial w_i}{\partial x_i} = \nabla \cdot \vec{w} \tag{1.52}$$

where i is summed over. The three independent off-diagonal components produce only a shearing motion.

1.5 CONSTITUTIVE RELATIONS

As indicated at the end of the last section, we will relate $\overleftrightarrow{\tau}$ to $\overleftrightarrow{\varepsilon}$. There are constitutive equations whose coefficients are transport properties. By means of these equations, we express the unique characteristics of a gaseous or liquid substance. For our purposes, it suffices to view these equations as empirical, i.e., based on experiment although they can be justified for a gas by kinetic theory.

In this section, we derive equations for the viscous stress tensor $\overleftrightarrow{\tau}$ and the heat flux vector $\vec{q}$. For $\overleftrightarrow{\tau}$, which is discussed first, we utilize a Newtonian fluid assumption, while Fourier's equation will be used for the heat flux. We return to $\overleftrightarrow{\sigma}$ only later in the discussion. As will become apparent, both the Newtonian approximation and Fourier's equation are based on the same assumptions, namely, isotropy and a linear relation. For an introductory discussion of non-Newtonian fluid mechanics, see DeKee and Wissbrun (1998).

NEWTONIAN FLUID

Again following Stokes, we postulate a linear relation between $\overleftrightarrow{\tau}$ and $\overleftrightarrow{\varepsilon}$. We further assume an isotropic fluid in which a coordinate rotation or interchange of the axis leaves the stress, rate-of-deformation relation unaltered. A fluid that adheres to both assumptions is called Newtonian. Gases, except under extreme conditions such as a shock wave, and most common liquids very accurately satisfy the Newtonian approximation. Liquids containing long-chain polymers do not satisfy this approximation as accurately.

Each of the above tensors has nine scalar components. The linear assumption means that each $\overleftrightarrow{\tau}$ component is proportional to the nine components of $\overleftrightarrow{\varepsilon}$; hence, there are 81 scalar coefficients that relate the two tensors. These coefficients are the components of a fourth-order tensor, since $3^4 = 81$. With a subscript notation, the linear relation is

$$\tau_{ij} = c_{ijmn}\varepsilon_{mn} \tag{1.53}$$

where c_{ijmn} is called the fourth-order viscosity coefficient tensor. The most general form for an isotropic fourth-order tensor is

$$c_{ijmn} = A\delta_{ij}\delta_{mn} + B\delta_{im}\delta_{jn} + C\delta_{in}\delta_{jm}$$

where A, B, and C are the only coefficients remaining out of the original 81. Consequently, we have

$$\begin{aligned}\tau_{ij} &= A\delta_{ij}\delta_{mn}\varepsilon_{mn} + B\delta_{im}\delta_{jn}\varepsilon_{mn} + C\delta_{in}\delta_{jm}\varepsilon_{mn} \\ &= A\delta_{ij}\varepsilon_{mm} + B\varepsilon_{ij} + C\varepsilon_{ji}\end{aligned}$$

Since $\overleftrightarrow{\varepsilon}$ is symmetric, this further simplifies to

$$\tau_{ij} = A\delta_{ij}\varepsilon_{mm} + (B + C)\varepsilon_{ij}$$

We now utilize Equation (1.52) and introduce the notation μ and λ for the first and second viscosity coefficients, respectively:

$$\mu = \frac{1}{2}(B + C), \qquad \lambda = A$$

Normally, μ is simply referred to as the viscosity or shear viscosity. The final form for the viscous stress tensor is

$$\tau_{ij} = 2\mu\varepsilon_{ij} + \lambda\delta_{ij}\nabla\cdot\vec{w} \tag{1.54a}$$

In tensor notation, we have

$$\overleftrightarrow{\tau} = 2\mu\overleftrightarrow{\varepsilon} + \lambda(\nabla\cdot\vec{w})\overleftrightarrow{I} \tag{1.54b}$$

where $\overleftrightarrow{I}$ is the unit dyadic.

By incorporating the pressure stresses, we obtain the familiar result for the stress tensor components

$$\sigma_{ij} = -p\delta_{ij} + \mu\left(\frac{\partial w_i}{\partial x_j} + \frac{\partial w_j}{\partial x_i}\right) + \lambda\delta_{ij}\frac{\partial w_k}{\partial x_k} \tag{1.55a}$$

while in tensor notation, this becomes

$$\overleftrightarrow{\sigma} = (-p + \lambda\nabla\cdot\vec{w})\overleftrightarrow{I} + 2\mu\overleftrightarrow{\varepsilon} \tag{1.55b}$$

We emphasize that these equations are restricted to a Newtonian fluid.

When initially discussing isotropy, we stated that $\overleftrightarrow{\tau}$ (or $\overleftrightarrow{\sigma}$) should be invariant with respect to an interchange of axis. This is accomplished for $\overleftrightarrow{\tau}$ by interchanging the i and j subscripts in the viscous terms in Equation (1.55a). Observe that the interchange does not alter these terms.

In Section 2.2, we show that for an incompressible flow, conservation of mass becomes

$$\nabla \cdot \vec{w} = 0 \tag{1.56}$$

In this circumstance, the term in Equations (1.54) and (1.55) containing λ is zero; therefore, the second viscosity coefficient plays no role in an incompressible flow.

To appreciate the role of λ for a compressible flow, we take the trace of σ_{ij} and divide by three. An average of these normal stresses is thus obtained:

$$\frac{1}{3}\sigma_{ii} = -p + \left(\lambda + \frac{2}{3}\mu\right) = -p + \mu_b \nabla \cdot \vec{w}$$

where

$$\mu_b = \lambda + \frac{2}{3}\mu \tag{1.57}$$

is the bulk viscosity. We replace λ with μ_b in Equation (1.54b), with the result

$$\overleftrightarrow{\tau} = 2\mu\left(\overleftrightarrow{\varepsilon} - \frac{1}{3}\nabla \cdot \vec{w}\overleftrightarrow{I}\right) + \mu_b \nabla \cdot \vec{w}\overleftrightarrow{I} = 2\mu\mathring{\overleftrightarrow{\varepsilon}} + \mu_b \nabla \cdot \vec{w}\overleftrightarrow{I}$$

where the $\mathring{\overleftrightarrow{\varepsilon}}$ tensor has a zero trace. Although this tensor has a zero trace, its diagonal elements are generally nonzero with only their sum being zero. This is called the rate-of-shear tensor, since it provides the viscous stresses associated only with a shearing motion. Consequently, the bulk viscosity term provides the viscous stresses due to a dilatational motion. The shear and dilatational stresses are caused by the attractive and repulsive forces between molecules and the collisional relaxation of the rotational and vibrational energy modes of polyatomic molecules, respectively.

Ultrasonic absorption measurements show μ_b to be zero for a low-density monatomic gas, in accordance with kinetic theory. For certain polyatomic molecules, such as CO_2, μ_b can be much larger than μ. In any case, the second law of thermodynamics requires $\mu_b \geq 0$, as will be shown in Chapter 3. Stokes originally hypothesized that

$$\mu_b = 0, \qquad \lambda = -\frac{2}{3}\mu \tag{1.58}$$

for all gases. This hypothesis is exact for a monatomic gas (except at a very high density) and can be used for an incompressible flow when the value of λ is irrelevant. The approximation, however, is often used for compressible flows of polyatomic gases. For instance, ultrasonic measurements yield $\mu_b \cong (2\mu/3)$ for air at 293 K and 1 atm. This is a very small value for μ_b and frequently can be neglected. Nevertheless, we will not invoke this hypothesis anywhere in the subsequent analysis. A more comprehensive physical discussion of bulk viscosity can be found in Section 13.4.

Fourier's Equation

As discussed in Section 1.1, we neglect the transport of energy by molecular diffusion, chemical reactions, or radiative heat transfer. For heat conduction, a linear relation between the heat flux $\vec{q}$ and the temperature gradient is assumed as

$$\vec{q} = \overleftrightarrow{\kappa} \cdot \nabla T$$

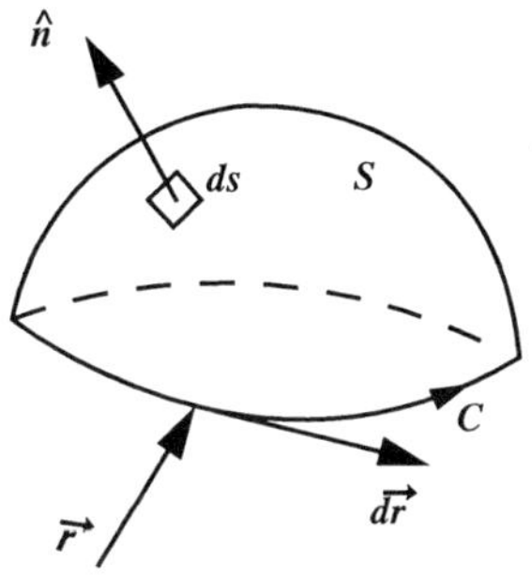

FIGURE 1.6 Cap bordered by a simple closed curve.

where T is the temperature and $\overleftrightarrow{\kappa}$ is a second-order thermal conductivity tensor. Note that $\vec{q}$ is not necessarily oriented in the same direction as ∇T. (This difference in orientation occurs in crystals.) Although referred to as the heat flux, $\vec{q}$ actually provides the rate of heat transfer having units of energy per unit area per unit time. (The thermal conductivity tensor has units of energy per unit length per degree Kelvin per unit time.) In Cartesian component form, the heat flux becomes

$$q_i - \kappa_{ij} \frac{\partial T}{\partial x_j}$$

For an isotropic fluid, the κ_{ij} tensor is given by

$$\kappa_{ij} = -\kappa \delta_{ij}$$

where κ is the conventional coefficient of thermal conductivity. (For notational consistency, transport coefficients are denoted with Greek symbols; hence, we use κ for the thermal conductivity instead of the more common k symbol.) We thus obtain the standard form for Fourier's equation

$$\vec{q} = -\kappa \nabla T \tag{1.59}$$

where the minus sign ensures that heat flows from hot to cold when $\kappa > 0$.

By way of summary, we observe that we have obtained the constitutive relations, Equations (1.55b) and (1.59), for the stress tensor and heat flux vector. These equations contain only three scalar coefficients, μ, λ, and κ. This is a remarkable simplification since the total number of scalar coefficients in the two equations has been reduced from 90. These three coefficients are viewed as known functions of the temperature and density. This dependence is often empirically established or it may come from the kinetic theory of gases.

1.6 INTEGRAL RELATIONS

A number of integral equations will be needed in the subsequent analysis. The first few of these are standard vector relations; we state them without proof. The first one is the Stokes theorem:

$$\oint_C \vec{A} \cdot d\vec{r} = \int_S \hat{n} \cdot (\nabla \times \vec{A})\, ds \tag{1.60}$$

where $\vec{A}(\vec{r})$ is an arbitrary vector function. The surface S is an open surface, or cap, bounded by a simple closed curve C as shown in Figure 1.6. The vector $d\vec{r}$ is tangent to C, while $\hat{n}$ is a unit vector normal to S. The Stokes theorem is useful for converting a line integral into a surface integral, or vice versa.

We write the Gauss divergence theorem in a generalized form

$$\int_V \nabla \begin{bmatrix} \phi \\ \cdot\vec{A} \\ \times\vec{A} \end{bmatrix} dv = \oint_S \hat{n} \begin{bmatrix} \phi \\ \cdot\vec{A} \\ \times\vec{A} \end{bmatrix} ds \tag{1.61}$$

where the volume V is fully enclosed by S, $\hat{n}$ is an outward unit normal vector to S, $\phi(\vec{r})$ is an arbitrary scalar function, and $\vec{A}(\vec{r})$ is again an arbitrary vector function. This result is actually three separate equations combined into a single convenient form. Of these three equations, we shall make explicit use of the first two. Note that the surface integral is a double integral while the volume integral is a triple integral. We shall also need a dyadic version of the middle equation, given by

$$\int_V \nabla \cdot \overleftrightarrow{\Phi} dv = \oint_S \hat{n} \cdot \overleftrightarrow{\Phi} ds \tag{1.62}$$

where $\overleftrightarrow{\Phi}$ is an arbitrary dyadic in three-dimensional space.

The next standard relation is the integral definition of the divergence operation, given by

$$\nabla \cdot \vec{A} = \lim_{\delta v \to 0} \frac{1}{\delta v} \oint_{\delta s} \hat{n} \cdot \vec{A} ds \tag{1.63}$$

where δv is a small volume bounded by δs. This relation is easily derived from Equation (1.61). There are various extensions or generalizations to Equations (1.60) and (1.63) (see Appendix A, Tables 4 and 6) that are not considered, since they will not be needed.

Leibniz's Rule

Suppose the integrand and one or both integration limits of a one-dimensional integral depend on a parameter t. If the integral is differentiated with respect to t, Leibniz's rule provides

$$\frac{d}{dt} \int_{x_i(t)}^{x_2(t)} \psi(x,t) dx = \int_{x_1(t)}^{x_2(t)} \frac{\partial \psi}{\partial t} dx + \psi(x_2(t),t) \frac{dx_2}{dt} - \psi(x_1(t),t) \frac{dx_1}{dt}$$

In this mathematical identity t is not necessarily time, but this identification provides us with a suitable physical interpretation. Thus, dx_i/dt represents the speed with which the end points move, while the $\psi(x_i,t)dx_i/dt$ terms represent the flux of ψ across the end points.

We will need a three-dimensional version of Leibniz's rule.* For this, we introduce a volume V of finite magnitude with a surface S that encloses V. Let $\vec{w}_s(\vec{r},t)$ be the velocity with which S moves where $\vec{r}$ denotes a point on S, and let $\hat{n}$ be the outward unit normal vector to S. For the desired generalization, we need to evaluate the amount of ψ that crosses S due to its motion. The amount that crosses a differential area ds of S, per unit time, is

$$\psi \vec{w}_s \cdot \hat{n} ds$$

* I am indebted to Professor M. L. Rasmussen for suggesting the Reynolds' transport theorem derivation, which starts with Leibniz's three-dimensional rule.

When $\vec{w}_s \cdot \hat{n} > 0$, the flux $\psi \vec{w}_s$ of ψ leaves V; thus, the net flux of ψ that crosses S is

$$\oint_S \psi \vec{w}_s \cdot \hat{n} ds$$

Consequently, the rule in three dimensions is

$$\frac{d}{dt}\int_{V(t)} \psi dv = \int_{V(t)} \frac{\partial \psi}{\partial t} dv + \oint_{S(t)} \psi \vec{w}_s \cdot \hat{n} ds \tag{1.64}$$

where the volume integral on the right represents the change in ψ that occurs within V, while the surface integral accounts for the transport of ψ across the moving surface S. Note that if $\vec{w}_s \cdot \hat{n}$ is everywhere positive on S, then V and S increase with time. In this circumstance, the surface integral is positive when $\psi > 0$ and contributes toward an increasing value for the volume integral on the left side. Equation (1.64) does not stem from fluid dynamics; it is purely mathematical. In this regard, observe that $\vec{w}_s$ is not necessarily related to a fluid velocity and ψ has not been identified with any fluid property.

Reynolds' Transport Theorem

We now assume $S(t)$ moves with the fluid velocity

$$\vec{w}_s = \vec{w} \tag{1.65}$$

As a consequence, a fluid particle initially within V will remain within V, and a particle outside of V remains outside. Thus, $V(t)$ contains a fixed amount of mass, referred to as a material volume, and is equivalent to a closed thermodynamic system. In this circumstance, it is appropriate to replace the time derivative on the left side of Equation (1.64) with the substantial derivative. With this change, Equation (1.64) becomes

$$\frac{D}{Dt}\int_{V(t)} \psi dv = \int_{V(t)} \frac{\partial \psi}{\partial t} dv + \int_{S(t)} \psi \vec{w} \cdot \hat{n} ds \tag{1.66}$$

which is the first version of the transport theorem. The quantity $\psi(\vec{r},t)$ can be a scalar, vector, or higher order tensor, and the terms on the right side have the same physical interpretation as those on the right side of Equation (1.64).

If, instead of Equation (1.65), we set $\vec{w}_s = 0$, the volume and bounding surface are usually referred to as a fixed control volume (CV) and control surface (CS). Since mass may cross the control surface, this is an open system. Equation (1.64) now reduces to

$$\frac{d}{dt}\int_{CV} \psi dv = \int_{CV} \frac{\partial \psi}{\partial t} dv \tag{1.67}$$

where ψ should be continuous within the control volume. The need for this proviso becomes evident by setting $\psi = \rho$ for a control volume containing two flows that are separated by a moving shock wave.

A relation between an open and closed system is obtained by equating V and S with CV and CS, respectively, at a given instant of time. By subtracting Equation (1.67) from (1.66), we obtain

$$\frac{D}{Dt}\int_V \psi dv = \frac{d}{dt}\int_{CV} \psi dv + \oint_{CS} \psi \vec{w} \cdot \hat{n} ds \tag{1.68}$$

where the left side refers to a moving material volume and the right side refers to a fixed control volume, which is an open system.

When ψ is a scalar (or a higher-order tensor; see Problem 1.6), the surface integral in Equation (1.66) can be written as

$$\oint_S \psi \vec{w} \cdot \hat{n} ds = \oint_S (\hat{n} \cdot \vec{w}) \psi ds = \oint_S \hat{n} \cdot (\vec{w}\psi) ds$$

where $(\vec{w}\psi)$ is a dyadic if ψ is a vector. By means of the divergence theorem, Equation (1.61), we obtain

$$\oint_S \psi \vec{w} \cdot \hat{n} ds = \int_V \nabla \cdot (\vec{w}\psi) dv$$

and Equation (1.66) becomes

$$\frac{D}{Dt}\int_V \psi dv = \int_V \left[\frac{\partial \psi}{\partial t} + \nabla \cdot (\vec{w}\psi)\right] dv \tag{1.69a}$$

The divergence term can be expanded as

$$\nabla \cdot (\vec{w}\psi) = \vec{w} \cdot \nabla \psi + \psi \nabla \cdot \vec{w}$$

to yield

$$\begin{aligned}\frac{D}{Dt}\int_V \psi dv &= \left[\int_V \frac{\partial \psi}{\partial t} + \vec{w} \cdot \nabla \psi + \psi \nabla \cdot \vec{w}\right] dv \\ &= \int_V \left(\frac{D\psi}{Dt} + \psi \nabla \cdot \vec{w}\right) dv \end{aligned} \tag{1.69b}$$

Equations (1.66) and (1.69) are alternate forms of the transport theorem, where Equation (1.69b) is utilized in Section 2.3.

As an illustration, set $\psi = 1$ and replace V with a small volume δv, to obtain

$$\frac{D}{Dt}\int_{\delta v} dv = \int_{\delta v} \nabla \cdot \vec{w} dv$$

from Equation (1.69b). The integral on the left side is just δv. We thus have

$$\frac{D\delta v}{Dt} = \int_{\delta v} \nabla \cdot \vec{w} dv = \oint_{\delta s} \hat{n} \cdot \vec{w} ds$$

where the divergence theorem is again utilized. Equation (1.63) is now used to finally obtain

$$\frac{1}{\delta v}\frac{D\delta v}{Dt} = \lim_{\delta v \to 0}\frac{1}{\delta v}\oint_{\delta s}\hat{n}\cdot\vec{w}ds$$

or

$$\frac{1}{\delta v}\frac{D(dv)}{Dt} = \nabla\cdot\vec{w} \tag{1.70}$$

where the left side is referred to as the dilatation. It represents the rate of volumetric strain of a fluid particle.

Equation (1.70) enables us to provide still another interpretation to Equation (1.69b). For this, we proceed as follows:

$$\frac{D}{Dt}\int_V \psi dv = \int_V \frac{D}{Dt}(\psi dv) = \int_V\left(\frac{D\psi}{Dt}dv + \psi\frac{Ddv}{Dt}\right)$$
$$= \int_V\left(\frac{D\psi}{Dt}dv + \psi\nabla\cdot\vec{w}dv\right) = \int_V\left(\frac{D\psi}{Dt} + \psi\nabla\cdot\vec{w}\right)dv$$

We again obtain Equation (1.69b); however, observe that the $\nabla\cdot\vec{w}$ term stems from the change with time of the differential volume of a fluid particle. This change increases or decreases the ψ content of V even if ψ is constant.

REFERENCES

Abramowitz, M. and Stegun, I.A., eds., *Handbook of Mathematical Functions*, NBS Applied Mathematical Series 55, 1972.
Darwin, C., "Note on Hydrodynamics," *Proc. Camb. Philos. Soc.* 49, 342 (1953).
DeKee, D. and Wissbrun, "Polymer Rheology," *Physics Today*, 24–29 (June 1998).
Gradshteyn, I.S. and Ryzhik, I.M., *Tables of Integrals, Series, and Products*, Academic Press, New York, 1980.

PROBLEMS

1.1 Consider an unsteady velocity field given by

$$\vec{w} = a_1x_3\hat{\imath}_1 + (x_1 - a_2t)\hat{\imath}_2 + a_3x_1\hat{\imath}_3$$

where the a_i are constants.

(a) Derive the vorticity $\vec{\omega}$.

(b) Determine the pathlines, where the initial condition

$$x_i = x_{i,o}, \qquad t = 0$$

is utilized.

(c) Determine the streamlines using the same initial condition.

1.2 Use

$$\vec{w} = b(t)x\,\hat{1}_x - b(t)y\,\hat{1}_y + 0\,\hat{1}_z$$

to determine the pathlines and streamlines. Construct the velocity $\vec{w}$ and acceleration $\vec{a}$ from the pathline equations.

1.3 Use spherical coordinates

$$\xi^1 = r, \quad \xi^2 = \theta, \quad \xi^3 = \theta$$

with

$$\xi^1 = (x^2 + y^2 + z^2)^{1/2}$$

$$\xi^2 = \cos^{-1}\left[\frac{z}{(x^2 + y^2 + z^2)^{1/2}}\right]$$

$$\xi^3 = \tan^{-1}\left(\frac{y}{x}\right)$$

to develop equations for

$$\vec{e}_i, \quad g_{ij}, \quad |g|, \quad g^{ij}, \quad \hat{e}_i, \quad \vec{e}^{\,j}, \quad \overleftrightarrow{I}, \quad \Gamma^k_{ij}$$

where g_{ij} is the fundamental metric and Γ^k_{ij} is the Christoffel symbol (see Appendix A).

1.4 Continue with Problem 1.3 for spherical coordinates and write the equations for

$$\nabla\psi, \quad \frac{D\psi}{Dt}, \quad \nabla\vec{w}, \quad \frac{D\vec{w}}{dt}, \quad \overleftrightarrow{\varepsilon}, \quad \nabla\cdot w, \quad \overleftrightarrow{\omega}, \quad \overleftrightarrow{\sigma}$$

where ψ is a scalar. Write the velocity as

$$\vec{w} = v_r\hat{e}_r + v_\theta\hat{e}_\phi + v_\phi\hat{e}_\phi = w^i\vec{e}_i$$

Use the v_i components and the $\hat{e}_i$ as the basis for ∇ and the subsequent computations.

1.5 Suppose the velocity in Problem 1.4 is given by

$$\vec{w} = f(r, t)\hat{e}_r$$

in spherical coordinates. Determine $D\vec{w}/Dt$, $\overleftrightarrow{\varepsilon}$, $\overleftrightarrow{\omega}$, and $\overleftrightarrow{\omega}$.

1.6 Start with Equation (1.66) where ψ is not restricted to being a scalar function. Set

$$\psi = \vec{A}$$

and derive the vector counterparts to Equations (1.69a) and (1.69b).

1.7 Consider the unsteady flow about a circular cylinder as discussed in Section 1.2.

(a) Determine an equation for X_a'.

(b) Show that the displacement Δ satisfies

$$\Delta \to \infty, \qquad Y_\infty = 0$$
$$\Delta = 0, \qquad Y_\infty \to \infty$$

(c) Determine asymptotic formulas for u' and v' as $T' \to -\infty$.

1.8 Show that for a Newtonian fluid the surface force can be written as

$$\vec{\sigma} = \mu\hat{n}\cdot[\nabla\vec{w} + (\nabla\vec{w})'] + [\lambda(\nabla\cdot\vec{w}) - p]\hat{n}$$

2 The Conservation Equations

2.1 PRELIMINARY REMARKS

The equations that govern the motion of a fluid consist of conservation equations, auxiliary relations, and initial and boundary conditions. The first group is preeminent; it includes conservation of mass, relations for linear and angular momentum, an energy equation, and the second law of thermodynamics. These equations are most simply obtained in integral form, from which their differential version can be deduced. Auxiliary equations are principally the constitutive relations that provide equations for the thermodynamic and transport properties. Equations (1.54b) and (1.59) are in this category. They also include equations for properties of interest, such as the vorticity, rate of entropy production, and the skin friction and heat transfer at a wall. Indeed, these last two items are principal engineering concerns. Any fluid flow, of course, is incomplete unless conditions fixing the configuration, such as upstream boundary conditions, wall conditions, etc., are prescribed.

This chapter is devoted to a derivation of equations that govern the transport of mass, momentum, and energy. Because of their importance, alternative forms are obtained for each of these equations. We also discuss a number of related topics including a second transport theorem, the role of an inertial frame for the equations, conservation of angular momentum, and viscous dissipation.

All of the conservation equations are derived from a single viewpoint. A conservation principle is first invoked that involves the time rate of change of some flow property that occurs in a material volume of fluid of fixed mass. We thus make use of Reynolds' transport theorem, Equation (1.69b). The quantity ψ will take on, in turn, five values:

$$\rho, \quad \rho\vec{w}, \quad \rho\vec{r}\times\vec{w}, \quad \rho e, \quad \rho w^2/2$$

where e is specific internal energy and ρ is the density. Associated with these values are the fluxes of mass, linear momentum, angular momentum, internal energy, and kinetic energy, respectively.

A material volume, rather than a fixed control volume, is utilized because the resulting derivation is mathematically simpler and physically clearer. This will be evident for all the conservation principles. Furthermore, a material volume coincides with a closed system that is used for the first law of thermodynamics. In each case, the results of the analysis will be in the form of a vectorial equation and thus independent of any specific coordinate system. However, for purposes of clarity, we often write individual terms or equations in Cartesian form.

2.2 MASS EQUATION

The principle of mass conservation states that the mass of fluid in V moves with the fluid velocity; i.e., V is a material volume. We set $\psi = \rho$ and write the principle as

$$\frac{D}{Dt}\int_V \rho dv = 0 \tag{2.1}$$

Equation (1.69b) now yields

$$\int_V \left(\frac{D\rho}{Dt} + \rho\nabla\cdot\vec{w}\right)dv = 0$$

or, since V is arbitrary,

$$\frac{D\rho}{Dt} + \rho\nabla \cdot \vec{w} = 0 \tag{2.2a}$$

An alternate form

$$\frac{\partial\rho}{\partial t} + \nabla \cdot (\rho\vec{w}) = 0 \tag{2.2b}$$

is obtained by expanding the $D\rho/Dt$ term. Either of these relations is referred to as the continuity equation. Although they are scalar equations, when written in vectorial form they are independent of any specific coordinate system.

The incompressible condition that a fluid particle has a constant density is

$$\frac{D\rho}{Dt} = 0 \tag{2.3}$$

In combination with Equation (2.2a) this yields Equation (1.56), which is an alternate condition for an incompressible fluid.

2.3 TRANSPORT THEOREM

We derive a general integral equation that is based on the Reynolds transport theorem and the continuity equation. In view of Problem 1.6, let ψ be a scalar or a vector. Replace ψ with $\rho\psi$ in Equation (1.69b) to obtain

$$\begin{aligned}\frac{D}{Dt}\int_V \rho\psi dv &= \int_V \left[\frac{D(\rho\psi)}{Dt} + \rho\psi\nabla \cdot \vec{w}\right] dv \\ &= \int_V \left(\rho\frac{D\psi}{Dt} + \psi\frac{D\rho}{Dt} + \rho\psi\nabla \cdot \vec{w}\right) dv \\ &= \int_V \left[\rho\frac{D\psi}{Dt} + \psi\left(\frac{D\rho}{Dt} + \rho\nabla \cdot \vec{w}\right)\right] dv\end{aligned}$$

By continuity, a second transport theorem is obtained as

$$\frac{D}{Dt}\int_V \rho\psi dv = \int_V \rho\frac{D\psi}{Dt} dv \tag{2.4}$$

that will be of considerable utility.

In later discussion, we will find that the various conservation principles involve only three types of integral terms:

$$\frac{D}{Dt}\int_V (\ )dv, \quad \int_V (\ )dv, \quad \oint_S (\ )\hat{n}ds$$

The first of these is converted to a volume integral, without the external substantial derivative, by means of Equation (2.4). The surface integral is also converted to a volume integral by means of the divergence theorem. By combining the various volume integrals, a single integral of the form

$$\int_V (\quad)dv = 0$$

is obtained. By virtue of the arbitrariness of V, the integrand is zero. This integrand, when set equal to zero, is a partial differential equation that represents a conservation principle.

2.4 LINEAR MOMENTUM EQUATION

Newton's second law of motion states that the time rate of change of linear momentum, $\rho\vec{w}$, of a material volume equals the applied forces. These are the body force $\vec{F}_b$, given by Equation (1.26) for gravity, and the surface stress vector, $\vec{\sigma}$, given by Equation (1.36). In the derivation, we do not assume a symmetric $\overleftrightarrow{\sigma}$; thus, $\vec{\sigma}$ equals $\hat{n}\cdot\overleftrightarrow{\sigma}$ but does not necessarily equal $\overleftrightarrow{\sigma}\cdot\hat{n}$. Newton's second law can therefore be written as

$$\frac{D}{Dt}\int_V \rho\vec{w}dv = \int_V \rho\vec{F}_b dv + \oint_S \vec{\sigma}ds$$

which simplifies to

$$\int_V \rho\frac{D\vec{w}}{Dt}dv = \int_V \rho\vec{F}_b dv + \oint_S \hat{n}\cdot\overleftrightarrow{\sigma}ds$$
$$= \int_V \rho\vec{F}_b dv + \int_V \nabla\cdot\overleftrightarrow{\sigma}dv$$

The differential form then is

$$\rho\frac{D\vec{w}}{Dt} = \nabla\cdot\overleftrightarrow{\sigma} + \rho\vec{F}_b \tag{2.5}$$

where the left side represents mass times acceleration, while the right side provides the vector sum of the applied surface and body forces, per unit volume.

To simplify the divergence term, we use Equation (1.44), which yields

$$\nabla\cdot\overleftrightarrow{\sigma} = \nabla\cdot\overleftrightarrow{\tau} - \nabla\cdot(p\overleftrightarrow{I}) = \nabla\cdot\overleftrightarrow{\tau} - \nabla p\cdot\overleftrightarrow{I} - p\nabla\cdot\overleftrightarrow{I}$$
$$= \nabla\cdot\overleftrightarrow{\tau} - \nabla p \tag{2.6}$$

since

$$\vec{A}\cdot\overleftrightarrow{I} = \vec{A}, \qquad \nabla\cdot\overleftrightarrow{I} = 0 \tag{2.7}$$

for any vector $\vec{A}$. Equation (2.5) now becomes

$$\rho \frac{D\vec{w}}{Dt} = -\nabla p + \nabla \cdot \overleftrightarrow{\tau} + \rho \vec{F}_b \tag{2.8}$$

A further simplification of the divergence term occurs if a Newtonian fluid, Equation (1.54b), is assumed. We thereby obtain

$$\begin{aligned}\nabla \cdot \overleftrightarrow{\tau} &= (\nabla\lambda \cdot \overleftrightarrow{I})(\nabla \cdot \vec{w}) + \lambda[\nabla(\nabla \cdot \vec{w})] \cdot \overleftrightarrow{I} + \lambda(\nabla \cdot \vec{w})\nabla \cdot \overleftrightarrow{I} + 2(\nabla\mu) \cdot \overleftrightarrow{\varepsilon} + 2\mu \cdot \overleftrightarrow{\varepsilon} \\ &= 2\mu\nabla \cdot \overleftrightarrow{\varepsilon} + 2\nabla\mu \cdot \overleftrightarrow{\varepsilon} + \lambda\nabla(\nabla \cdot \vec{w}) + (\nabla\lambda)(\nabla \cdot \vec{w})\end{aligned} \tag{2.9a}$$

where the last result utilizes Equations (2.7). In a Cartesian coordinate system, this equation becomes

$$\nabla \cdot \overleftrightarrow{\tau} = \left[\mu\left(\frac{\partial^2 w_i}{\partial x_i \partial x_j} + \frac{\partial^2 w_j}{\partial x_i^2}\right) + \frac{\partial \mu}{\partial x_i}\left(\frac{\partial w_i}{\partial x_j} + \frac{\partial w_j}{\partial x_i}\right) + \lambda \frac{\partial^2 w_i}{\partial x_i \partial x_j} + \frac{\partial \lambda}{\partial x_j}\frac{\partial w_i}{\partial x_i}\right]\hat{1}_j \tag{2.9b}$$

Any additional simplification of $\nabla \cdot \overleftrightarrow{\tau}$ would require further assumptions or approximations. The most common of these would be an incompressible flow [Equation (1.56)], Stokes' hypothesis [Equation (1.58)], or the assumption of constant values for μ and λ.

2.5 INERTIAL FRAME

In the derivation considered thus far, an inertial frame or coordinate system has been tacitly assumed. In particular, Newton's second law requires that the velocity and acceleration of a fluid particle be measured in a stationary, or inertial, frame; thus, the origin of a curvilinear coordinate system ξ^i should be at the center of mass of the universe and should rotate with the average angular momentum of the universe. Since neither this mass nor the angular momentum is known, a simpler definition for an inertial frame could be based on several very distant galaxies. However, a practical alternative to either approach is to define an inertial system as one in which Newton's second law, in the form

$$\vec{F} = m\vec{a} \tag{2.10}$$

holds. In this circumstance, unbarred variables are used to denote the system.

We consider a second coordinate system $\bar{\xi}^j$, located at a distance $\vec{R}(t)$ from the origin of the inertial system, which rotates with a solid-body motion whose angular velocity is $\vec{\omega}_{\text{rot}}(t)$. Barred variables denote this system, as shown in Figure 2.1.

We are not especially concerned with a transformation of coordinates between ξ^i and $\bar{\xi}^j$, as this transformation is not essential for the discussion; however, Sections 6.4 and 6.6 will utilize this type of transformation. These later sections provide a more general approach than that considered here. Time in the two systems is assumed to be the same, as are most scalar quantities like pressure and density. These scalar quantities are thus invariant under any transformation that takes you from one system to the other. (Two exceptions to scalar invariance are w^2 and the stagnation enthalpy, which is defined in Section 2.9.)

What is needed are relations that account for the relative translational and rotational motions of the two systems. These effects are dealt with by assuming $\vec{R}$ and $\vec{\omega}_{\text{rot}}$ are known functions of time. Once derived, these relations will provide a transformation of vector quantities between the two systems.

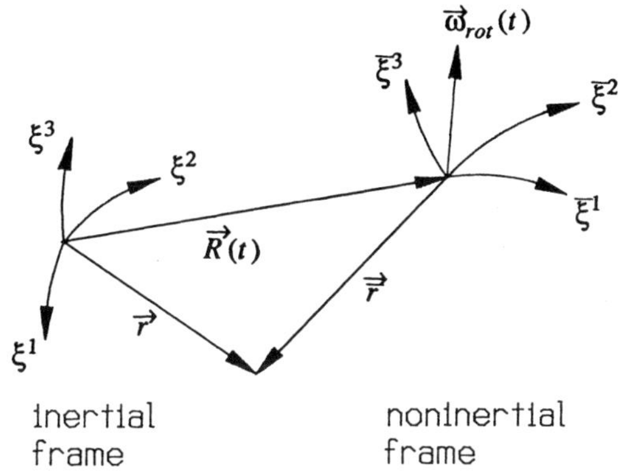

FIGURE 2.1 Schematic relation between inertial and noninertial frames.

The basic scalar equations of continuity and energy, derived later, are also invariant under this transformation. We will prove this assertion for continuity shortly. However, the linear momentum equation is a vector relation and therefore is not invariant. It is for this reason that a noninertial system is discussed in terms of Newton's second law.

Transformation of the Momentum Equation

The position vectors in the two systems are related by

$$\vec{r} = \vec{\bar{r}} + \vec{R} \tag{2.11}$$

as shown in Figure 2.1. The velocity $\vec{w}$ in the inertial system equals the sum of $\vec{\bar{w}}$ and velocities due to the translational and rotational motions of the noninertial system relative to the inertial one; i.e.,

$$\vec{w} = \vec{\bar{w}} + \frac{d\vec{R}}{dt} + \vec{w}_{\text{rot}} \times \vec{\bar{r}} \tag{2.12}$$

Similarly, the acceleration is given by

$$\vec{a} = \vec{\bar{a}} + \frac{d^2\vec{R}}{dt^2} + 2\vec{\omega}_{\text{rot}} \times \vec{\bar{w}} + \vec{\omega}_{\text{rot}} \times \left(\vec{\omega}_{\text{rot}} \times \vec{\bar{r}}\right) + \frac{d\vec{\omega}_{\text{rot}}}{dt} \times \vec{\bar{r}} \tag{2.13}$$

where the four rightmost terms provide the acceleration of the noninertial system due to its translational and rotational motion relative to the inertial system. [See Problem 2.13 for the derivation of Equations (2.12) and (2.13).] Thus, in a noninertial frame the momentum equation has the form

$$\rho\left[\left(\overline{\frac{D\vec{\bar{w}}}{Dt}}\right) + \frac{d^2\vec{R}}{dt^2} + 2\vec{\omega}_{\text{rot}} \times \vec{\bar{w}} + \vec{\omega}_{\text{rot}} \times \left(\vec{\omega}_{\text{rot}} \times \vec{\bar{r}}\right) + \frac{d\vec{\omega}_{\text{rot}}}{dt} \times \vec{\bar{r}}\right] = -\nabla p + \nabla \cdot \overleftrightarrow{\tau} + \rho\vec{F}_b$$

where the acceleration can be written as

$$\overline{\frac{D\vec{\bar{w}}}{Dt}} = \frac{\partial\vec{\bar{w}}}{\partial t} + \nabla\frac{\bar{w}^2}{2} + \vec{\bar{\omega}} \times \vec{\bar{w}}$$

Observe that the del operator is associated with the inertial system. If, for example, both systems use Cartesian coordinates, we then have

$$\nabla\phi = \bar{\nabla}\phi$$
$$\nabla\cdot\vec{A} = \bar{\nabla}\cdot\vec{A}$$
$$\nabla\times\vec{A} = \bar{\nabla}\times\vec{A}$$

since only spatial derivatives are involved. Furthermore, the applied forces are the same in the two systems, with the result

$$-\nabla p + \nabla\cdot\overleftrightarrow{\tau} + \rho\vec{F}_b = -\bar{\nabla}p + \bar{\nabla}\cdot\overleftrightarrow{\bar{\tau}} + \rho\vec{\bar{F}}_b$$

Hence, aside from the addition of the four acceleration terms, the rest of the momentum equation has the same form as Equation (2.8).

While $\nabla\times\vec{A} = \bar{\nabla}\times\vec{A}$ and $\nabla\times\vec{A} = \bar{\nabla}\times\vec{\bar{A}}$ if $\vec{A}$ is a force (see Problem 2.16), the later relation generally does not hold. This occurs when $\vec{A}$ is the vorticity; see Problem 2.4. Thus, $\vec{\omega}$ is not invariant between the inertial and noninertial systems.

Discussion

If an experiment or analysis is being performed in a noninertial frame, then the momentum equation should be written with the additional acceleration terms. The surface of the earth is such a frame, due to its rotation about its axis as well as its translational motion relative to our galaxy. However, with a few exceptions, the contributions of $\vec{R}$ and $\vec{\omega}_{\text{rot}}$ are totally negligible, and Equation (2.8) is utilized with excellent results.

Several exceptions do occur where noninertial effects are important. These include the motion of the atmosphere, the motion of ocean currents, and the analysis of rotating machinery when rotating at a large angular speed. These situations are appreciably simplified by assuming

$$\frac{d\vec{R}}{dt} = 0, \qquad \vec{\omega}_{\text{rot}} = \text{constant}$$

and only rotation about an axis with a constant angular velocity is being considered. With this, Equations (2.11) to (2.13) simplify to

$$\vec{r} = \vec{\bar{r}}$$
$$\vec{w} = \vec{\bar{w}} + \vec{\omega}_{\text{rot}}\times\vec{r}$$
$$\vec{a} = \vec{\bar{a}} + 2\vec{\omega}_{\text{rot}}\times\vec{\bar{w}} + \vec{\omega}_{\text{rot}}\times(\vec{\omega}_{\text{rot}}\times\vec{r})$$

where $2\vec{\omega}_{\text{rot}}\times\vec{\bar{w}}$ is the Coriolis acceleration and $\vec{\omega}_{\text{rot}}\times(\vec{\omega}_{\text{rot}}\times\vec{r})$ is the centripetal acceleration.

If the barred system is fixed to a vehicle (car, aircraft, missile, etc.) that is linearly accelerating or decelerating, then the $d^2\vec{R}/dt^2$ term is required.

Transformation of the Continuity Equation

Let us now discuss the invariance of the continuity equation. We first note that the time rate of change of a scalar quantity, following a fluid particle, is the same in the two systems. We therefore have

$$\frac{D\rho}{Dt} = \overline{\frac{D\rho}{Dt}}$$

With the aid of Equation (2.12), the divergence of the velocity is

$$\nabla \cdot \vec{w} = \overline{\nabla} \cdot \vec{w} = \overline{\nabla} \cdot \left(\vec{\overline{w}} + \frac{d\vec{R}}{dt} + \vec{\omega}_{\text{rot}} \times \vec{\overline{r}} \right) = \overline{\nabla} \cdot \vec{\overline{w}} + \overline{\nabla} \cdot \frac{d\vec{R}}{dt} + \overline{\nabla} \cdot \left(\vec{\omega}_{\text{rot}} \times \vec{\overline{r}} \right)$$

Since $\vec{R}$ is a function only of t, we have

$$\overline{\nabla} \cdot \frac{d\vec{R}}{dt} = 0$$

We use the vector identity

$$\nabla \cdot (\vec{A} \times \vec{B}) = \vec{B} \cdot (\nabla \times \vec{A}) - \vec{A} \cdot (\nabla \times \vec{B})$$

to show that

$$\overline{\nabla} \cdot \left(\vec{\omega}_{\text{rot}} \times \vec{\overline{r}} \right) = \vec{\overline{r}} \cdot (\overline{\nabla} \times \vec{\omega}_{\text{rot}}) - \vec{\omega}_{\text{rot}} \cdot \left(\overline{\nabla} \times \vec{\overline{r}} \right)$$

However, we have

$$\overline{\nabla} \times \vec{\omega}_{\text{rot}} = 0, \qquad \overline{\nabla} \times \vec{\overline{r}} = 0$$

so that

$$\nabla \cdot \vec{w} = \overline{\nabla} \cdot \vec{\overline{w}}$$

Consequently, Equation (2.2a) for continuity transforms into

$$\overline{\frac{D\rho}{Dt}} + \rho \overline{\nabla} \cdot \vec{\overline{w}} = 0$$

and continuity is invariant under the transformation, as mentioned earlier.

2.6 ANGULAR MOMENTUM EQUATION

Consider a particle of mass m at an arbitrary position $\vec{r}$. As a result of an applied force $\vec{F}$, the particle experiences a moment, $m\vec{r} \times \vec{F}$, which is a vector called the torque, that is perpendicular to $\vec{r}$ and $\vec{F}$. For the angular momentum of a fluid particle, we similarly encounter the moments $\vec{r} \times \vec{\sigma}$ and

$\vec{r} \times \vec{F}_b$ as a result of the applied surface and body forces. The principle of angular momentum states that the time rate of change of the angular momentum of the material in volume V equals the moments of the applied forces. In an inertial frame, where Equation (2.10) holds, this principle becomes

$$\frac{D}{Dt}\int_V \rho\vec{r} \times \vec{w} dv = \int_S \vec{r} \times \vec{\sigma} ds + \int_V \rho\vec{r} \times \vec{F}_b dv \tag{2.14}$$

where $\rho\vec{r} \times \vec{w}$ is the angular momentum of a fluid particle relative to an arbitrary origin.

We utilize Equation (2.4) with $\psi = \vec{r} \times \vec{w}$ to obtain, for the left side of Equation (2.14)

$$\frac{D}{Dt}\int_V \rho\vec{r} \times \vec{w} dv = \int_V \rho\frac{D(\vec{r} \times \vec{w})}{Dt} dv$$

Equation (1.13) is used with the result

$$\frac{D(\vec{r} \times \vec{w})}{Dt} = \frac{D\vec{r}}{Dt} \times \vec{w} + \vec{r} \times \frac{D\vec{w}}{Dt} = \vec{w} \times \vec{w} + \vec{r} \times \frac{D\vec{w}}{Dt} = \vec{r} \times \frac{D\vec{w}}{Dt}$$

The left side of Equation (2.14) thus becomes

$$\frac{D}{Dt}\int_V \rho\vec{r} \times \vec{w} dv = \int_V \rho\vec{r} \times \frac{D\vec{w}}{Dt} dv \tag{2.15}$$

For the surface integral, Equation (1.36) is used, to obtain

$$\vec{r} \times \vec{\sigma} = -\vec{\sigma} \times \vec{r} = -(\hat{n} \cdot \overleftrightarrow{\sigma}) \times \vec{r} = -\hat{n} \cdot (\overleftrightarrow{\sigma} \times \vec{r})$$

where the last step stems from the associativity law for a multiplication involving vectors and dyadics (see Table 6 in Appendix A). We now use the divergence theorem, Equation (1.62), with $\overleftrightarrow{\Phi} = \overleftrightarrow{\sigma} \times \vec{r}$, to obtain

$$\int_S \vec{r} \times \vec{\sigma} ds = -\int_S \hat{n} \cdot (\overleftrightarrow{\sigma} \times \vec{r}) ds = -\int_V \nabla \cdot (\overleftrightarrow{\sigma} \times \vec{r}) dv$$

With this relation and Equation (2.15), Equation (2.14) becomes

$$\int_V \rho\vec{r} \times \frac{D\vec{w}}{Dt} dv = -\int_V \nabla \cdot (\overleftrightarrow{\sigma} \times \vec{r}) dv + \int_V \rho\vec{r} \times \vec{F}_b dv$$

or

$$\rho\vec{r} \times \frac{D\vec{w}}{Dt} = -\nabla \cdot (\overleftrightarrow{\sigma} \times \vec{r}) + \rho\vec{r} \times \vec{F}_b \tag{2.16}$$

This relation can be compared with the linear momentum equation by multiplying Equation (2.5) with $\vec{r}$ and subtracting the result from Equation (2.16), to yield

$$\nabla \cdot (\overleftrightarrow{\sigma} \times \vec{r}) + \vec{r} \times (\nabla \cdot \overleftrightarrow{\sigma}) = 0 \tag{2.17}$$

As shown in Problem 2.3, this relation is an identity providing $\overleftrightarrow{\sigma}$ is symmetric. This symmetry condition is

$$\overleftrightarrow{\sigma} = \overleftrightarrow{\sigma}^t \tag{2.18a}$$

or, in indicial notation,

$$\sigma_{ij} = \sigma_{ji} \tag{2.18b}$$

If $\overleftrightarrow{\sigma}$ is symmetric, then by Equation (1.44) $\overleftrightarrow{\tau}$ is symmetric. We thus conclude that conservation of angular momentum does not yield an independent equation other than the symmetry of $\overleftrightarrow{\sigma}$. We, therefore, have derived the result originally anticipated by Equation (1.40).

There are several circumstances when Equations (2.18a,b) are not forthcoming (Rae, 1976). In certain non-Newtonian liquids the molecules may have a coupling moment. In this case, Equation (2.14) is incomplete and, consequently, $\overleftrightarrow{\sigma}$ is not symmetric.

2.7 ENERGY EQUATION

TOTAL ENERGY EQUATION

Several equivalent forms for the energy equation are derived. The first is called the total energy equation and, conceptually, it is the simplest of the group. The total energy per unit volume consists of its internal energy, ρe, and its kinetic energy, $\rho w^2/2$. The time rate of change of the total energy of a material volume V then is

$$\frac{D}{Dt}\int_V \rho\left(e + \frac{1}{2}w^2\right)dv = \int_V \rho \frac{D}{Dt}\left(e + \frac{1}{2}w^2\right)dv \tag{2.19}$$

In addition, the fluid in V experiences heat transfer with the surroundings, and the rate of work done on V by surface and body forces needs to be included. As we know from mechanics, the work done on a particle is $\vec{F} \cdot \hat{t}d\ell$, where $\hat{t}$ is a unit vector tangent to the particle path and ℓ is arc length along the path. The rate of doing work, or the power, is then $\vec{F} \cdot \vec{w}$, where $\vec{w}$ equals $\hat{t}(d\ell/dt)$. These terms, which will appear on the right side of the energy equation, are

$$-\oint_S \vec{q} \cdot \hat{n}ds, \qquad \oint_S \vec{w} \cdot \vec{\sigma}ds, \qquad \oint_V \rho\vec{w} \cdot \vec{F}_b dv$$

The first integral provides the rate of heat transfer to V from the surroundings, while the other integrals provide the rate of work done on the material volume.

Our governing energy principle states that the time rate of change of the total energy of a material volume is due to the rate of heat transfer to V and the rate of work done on the material volume.

We therefore have

$$\int_V \rho \frac{D}{Dt}\left(e + \frac{1}{2}w^2\right) dv = -\oint_S \vec{q} \cdot \hat{n} ds + \oint_S \vec{w} \cdot \vec{\sigma} ds + \oint_V \rho \vec{w} \cdot F_b dv$$

where Equation (2.19) is utilized on the left side. The surface integrals on the right side are transformed with the aid of the divergence theorem as follows:

$$\oint_S \vec{q} \cdot \hat{n} ds = \oint_V \nabla \cdot \vec{q} dv$$

$$\vec{w} \cdot \vec{\sigma} = \vec{\sigma} \cdot \vec{w} = (\hat{n} \cdot \overleftrightarrow{\sigma}) \cdot \vec{w} = \hat{n} \cdot (\overleftrightarrow{\sigma} \cdot \vec{w})$$

$$\oint_S \vec{w} \cdot \vec{\sigma} ds = \oint_S \hat{n} \cdot (\overleftrightarrow{\sigma} \cdot \vec{w}) ds = \oint_V \nabla \cdot (\overleftrightarrow{\sigma} \cdot \vec{w}) dv$$

The above equations combine to yield

$$\int_V \rho \frac{D}{Dt}\left(e + \frac{1}{2}w^2\right) dv = -\int_V \nabla \cdot \vec{q} dv + \int_V \nabla \cdot (\overleftrightarrow{\sigma} \cdot \vec{w}) dv + \int_V \rho \vec{w} \cdot \vec{F}_b dv$$

or, in differential form,

$$\rho \frac{D}{Dt}\left(e + \frac{1}{2}w^2\right) = \nabla \cdot (\overleftrightarrow{\sigma} \cdot \vec{w} - \vec{q}) + \rho \vec{w} \cdot \vec{F}_b \quad \textbf{(2.20)}$$

for the total energy equation. With Cartesian coordinates the divergence term can be written as

$$\nabla \cdot (\overleftrightarrow{\sigma} \cdot \vec{w} - \vec{q}) = \frac{\partial}{\partial x_i}(\sigma_{ij} w_{ij} - q_i)$$

Equation (2.20) relates the changes in total energy to the rates of work and heat transfer. Observe that gravitational potential energy is effectively represented by the $\vec{F}_b$ term. As a consequence, the total energy does not contain the potential energy due to gravity; some authors include the gravitational potential energy as part of the total energy. In this circumstance, $\vec{F}_b$ would not be associated with gravity.

Kinetic Energy Equation

Let us now multiply Equation (2.5) by $\vec{w}$ and use Equation (1.18) for the acceleration, to obtain

$$\rho \vec{w} \cdot \left[\frac{\partial \vec{w}}{\partial t} + \nabla \frac{w^2}{2} - \vec{w} \times (\nabla \times \vec{w})\right] = \vec{w} \cdot (\nabla \cdot \overleftrightarrow{\sigma}) + \rho \vec{w} \cdot \vec{F}_b$$

$$\rho\left[\vec{w} \cdot \frac{\partial \vec{w}}{\partial t} + \vec{w} \cdot \nabla \frac{w^2}{2} - \vec{w} \cdot \vec{w} \times (\nabla \times \vec{w})\right] = \vec{w} \cdot (\nabla \cdot \overleftrightarrow{\sigma}) + \rho \vec{w} \cdot \vec{F}_b$$

We utilize the relations

$$\vec{w} \cdot \frac{\partial \vec{w}}{\partial t} = \frac{\partial}{\partial t}\left(\frac{w^2}{2}\right)$$

$$\frac{\partial}{\partial t}\frac{w^2}{2} + \vec{w} \cdot \nabla \frac{w^2}{2} = \frac{D}{Dt}\frac{w^2}{2}$$

$$\vec{w} \cdot \vec{w} \times (\nabla \times \vec{w}) = 0$$

to arrive at

$$\rho \frac{D}{Dt}\frac{w^2}{2} = \vec{w} \cdot (\nabla \cdot \overleftrightarrow{\sigma}) + \rho \vec{w} \cdot \vec{F}_b \tag{2.21}$$

This result is called the kinetic energy equation and is arrived at independently of Equation (2.20). The right side represents the rates of doing work on a particle of fluid by the surface and body forces. The left side represents the time rate of change of kinetic energy, per unit volume, experienced by a fluid particle. Note that heat transfer and internal energy are not involved.

Internal Energy Equation

We next subtract Equation (2.21) from Equation (2.20), with the result

$$\rho \frac{De}{Dt} = -\nabla \cdot \vec{q} + \nabla \cdot (\overleftrightarrow{\sigma} \cdot \vec{w}) - \vec{w} \cdot (\nabla \cdot \overleftrightarrow{\sigma})$$

From tensor analysis, we have the decomposition (see Problem 2.5)

$$\nabla \cdot (\overleftrightarrow{\sigma} \cdot \vec{w}) = \vec{w} \cdot (\nabla \cdot \overleftrightarrow{\sigma}) + \overleftrightarrow{\sigma} : (\nabla \vec{w})^t \tag{2.22}$$

where the double dot product is defined in Table 6 of Appendix A. Observe that $\vec{w} \cdot (\nabla \cdot \overleftrightarrow{\sigma})$ represents the rate of work done by the resultant of the surface forces and accounts for the rate of translational work. The double dot product term represents the rate of work associated with both the normal (hydrostatic) and viscous stresses. Hence, the left side of Equation (2.22) provides the net rate at which work is done on a fluid particle. The internal energy equation thus becomes

$$\rho \frac{De}{Dt} = -\nabla \cdot \vec{q} + \vec{\sigma} : (\nabla \vec{w})^t \tag{2.23}$$

and only the rate of work associated with $\overleftrightarrow{\sigma} : (\nabla \vec{w})^t$ remains. In Cartesian coordinates, this term is evaluated by first writing the velocity gradient and its transposition as

$$\nabla \vec{w} = \frac{\partial w_m}{\partial x_n} \hat{\imath}_n \hat{\imath}_m \tag{2.24a}$$

$$\nabla \vec{w}^t = \frac{\partial w_m}{\partial x_n} \hat{\imath}_m \hat{\imath}_n \tag{2.24b}$$

We therefore obtain the relatively simple result

$$\overleftrightarrow{\sigma}:(\nabla\vec{w})^t = \sigma_{ij}\,\hat{\imath}_i\hat{\imath}_j : \frac{\partial w_m}{\partial x_n}\hat{\imath}_m\hat{\imath}_n = \sigma_{ij}\frac{\partial w_m}{\partial x_n}(\hat{\imath}_j\cdot\hat{\imath}_m)(\hat{\imath}_i\cdot\hat{\imath}_n)$$

$$= \sigma_{ij}\frac{\partial w_m}{\partial x_n}\delta_{jm}\delta_{in} = \sigma_{ij}\frac{\partial w_j}{\partial x_i} \tag{2.25}$$

Observe that Equation (2.23) has no body force term. From this equation and Equation (2.21), we see that the body force rate of work term can alter only the kinetic energy of a fluid particle, not its internal energy.

It is useful to evaluate the double dot product term in a different manner than that given by Equation (2.25). We introduce Equation (1.44), to obtain

$$\overleftrightarrow{\sigma}:(\nabla\vec{w})^t = (\overleftrightarrow{\tau}-p\overleftrightarrow{I}):(\nabla\vec{w})^t = \overleftrightarrow{\tau}:(\nabla\vec{w})^t - p\overleftrightarrow{I}:(\nabla\vec{w})^t$$

With Equation (2.24b), the double dot product in the rightmost term becomes

$$\overleftrightarrow{I}:(\nabla\vec{w})^t = \hat{\imath}_i\hat{\imath}_i : \frac{\partial w_m}{\partial x_n}\hat{\imath}_m\hat{\imath}_n = \frac{\partial w_m}{\partial x_n}(\hat{\imath}_i\cdot\hat{\imath}_m)(\hat{\imath}_i\cdot\hat{\imath}_n) = \frac{\partial w_m}{\partial x_n}\delta_{im}\delta_{in} = \frac{\partial w_i}{\partial x_i}$$

In this regard, we note that the dilatation can be written as

$$\frac{\partial w_i}{\partial x_i} = \nabla\cdot\vec{w} = -\frac{1}{\rho}\frac{D\rho}{Dt} = \varepsilon_{ii} \tag{2.26}$$

where Equation (2.2a) is used. We therefore have

$$\overleftrightarrow{\sigma}:(\nabla\vec{w})^t = \overleftrightarrow{\tau}:(\nabla\vec{w})^t + \frac{p}{\rho}\frac{D\rho}{Dt} \tag{2.27a}$$

$$= \overleftrightarrow{\tau}:(\nabla\vec{w})^t - \frac{p}{v}\frac{Dv}{Dt} \tag{2.27b}$$

where the specific volume v is given by

$$v = \frac{1}{\rho} \tag{2.28}$$

The double dot product term on the right-hand side of Equation (2.27a) is called the viscous dissipation, while the other term represents pdv work. It is evaluated in the next section, after which we complete the discussion of the energy equation.

2.8 VISCOUS DISSIPATION

The double dot product (viscous dissipation) term on the right side of Equations (2.27a,b) is denoted as Φ. Since $\overleftrightarrow{\varepsilon}$ is symmetric while $\overleftrightarrow{\omega}$ is antisymmetric, we have from Equation (1.46)

$$(\nabla\vec{w})^t = \overleftrightarrow{\varepsilon}^t + \overleftrightarrow{\omega}^t = \overleftrightarrow{\varepsilon}^t - \overleftrightarrow{\omega} \tag{2.29}$$

As we have shown, $\overleftrightarrow{\tau}$ is a symmetric tensor that yields for Φ

$$\Phi = \overleftrightarrow{\tau} : (\nabla\vec{w})' = \overleftrightarrow{\tau} : (\overleftrightarrow{\varepsilon} - \overleftrightarrow{\omega}) = \overleftrightarrow{\tau} : \overleftrightarrow{\varepsilon} \tag{2.30}$$

since the double dot product of a symmetric dyadic and an antisymmetric dyadic is zero (see Problem 2.7). The form $\overleftrightarrow{\tau} : \overleftrightarrow{\varepsilon}$ for Φ is a convenient and commonly encountered one, although it does require $\overleftrightarrow{\tau}$ to be symmetric.

Further simplification utilizes the Newtonian assumption for $\overleftrightarrow{\tau}$. We now have

$$\Phi = \overleftrightarrow{\tau} : \overleftrightarrow{\varepsilon} = 2\mu\overleftrightarrow{\varepsilon} : \overleftrightarrow{\varepsilon} + \lambda(\nabla \cdot \vec{w})\overleftrightarrow{I} : \overleftrightarrow{\varepsilon}$$

Without loss of generality, we can write $\overleftrightarrow{\varepsilon}$ in terms of an orthonormal basis as

$$\overleftrightarrow{\varepsilon} = \varepsilon_{ij}\hat{e}_i\hat{e}_j \tag{2.31}$$

We then have

$$\overleftrightarrow{I} : \overleftrightarrow{\varepsilon} = \varepsilon_{ii} = \nabla \cdot \vec{w}$$

$$\overleftrightarrow{\varepsilon} : \overleftrightarrow{\varepsilon} = \varepsilon_{ij}\varepsilon_{km}(\hat{e}_j \cdot \hat{e}_k)(\hat{e}_i \cdot \hat{e}_m) = \varepsilon_{ij}\varepsilon_{km}\delta_{jk}\delta_{im}$$

$$= \varepsilon_{ij}\varepsilon_{ij} = \varepsilon_{ii}\varepsilon_{ii} + 2(\varepsilon_{12}^2 + \varepsilon_{23}^2 + \varepsilon_{31}^2)$$

where the 2 stems from symmetry. We thus have the result

$$\Phi = 2\mu\varepsilon_{ii}\varepsilon_{ii} + 4\mu(\varepsilon_{12}^2 + \varepsilon_{23}^2 + \varepsilon_{31}^2) + \lambda(\varepsilon_{ii})^2 \tag{2.32a}$$

where

$$\varepsilon_{ii}\varepsilon_{ii} = \sum_{i=1}^{3}\varepsilon_{ii}^2, \qquad (\varepsilon_{ii})^2 = \left(\sum_{i=1}^{3}\varepsilon_{ii}\right)^2$$

With Cartesian coordinates, the rate-of-deformation tensor is given by Equation (1.47b). With this equation, we obtain the familiar result

$$\Phi = \mu\left[2\sum_{i=1}^{3}\left(\frac{\partial w_i}{\partial x_i}\right)^2 + \left(\frac{\partial w_1}{\partial x_2} + \frac{\partial w_2}{\partial x_1}\right)^2 + \left(\frac{\partial w_2}{\partial x_3} + \frac{\partial w_3}{\partial x_2}\right)^2 + \left(\frac{\partial w_3}{\partial x_1} + \frac{\partial w_1}{\partial x_3}\right)^2\right] + \lambda\left(\sum_{i=1}^{3}\frac{\partial w_i}{\partial x_i}\right)^2 \tag{2.32b}$$

for the viscous dissipation. Observe that the λ term is zero if the flow is incompressible, and that Φ is proportional to a series of terms of the form $(\partial w_i/\partial x_j)(\partial w_m/\partial x_n)$. Thus, Φ is significant when the gradient of the flow speed is substantial. The two most prominent examples are the flow interior to a shock wave and in a high-speed boundary layer. In the case of a normal shock wave with speed w_1 in the x_1 direction, we readily obtain

$$\Phi = (2\mu + \lambda)\left(\frac{\partial w_1}{\partial x_1}\right)^2 = \left(\frac{4}{3}\mu + \mu_b\right)\left(\frac{\partial w_1}{\partial x_1}\right)^2$$

Consequently, there is dissipation due to the bulk viscosity, which is additive with the shear viscosity dissipation. For a two-dimensional boundary layer with $w_3 = 0$, $\partial(\ )/\partial x_3 = 0$, and ignoring the small gradients in the x_1 flow direction, we have

$$\Phi \cong \mu\left[2\left(\frac{\partial w_2}{\partial x_2}\right)^2 + \left(\frac{\partial w_1}{\partial x_1}\right)^2\right] + \lambda\left(\frac{\partial w_2}{\partial x_2}\right)^2$$

$$= \mu\left(\frac{\partial w_1}{\partial x_2}\right)^2 + \left(\frac{4}{3}\mu + \mu_b\right)\left(\frac{\partial w_2}{\partial x_2}\right)^2$$

In general, the first term on the right dominates. This term represents the shearing motion of the fluid in a viscous boundary layer.

2.9 ALTERNATE FORMS FOR THE ENERGY EQUATION

With the aid of Equations (2.27b) and (2.30), Equation (2.23) is written for the internal energy as

$$\rho\frac{De}{Dt} = -\nabla\cdot\vec{q} + \Phi - \frac{p}{v}\frac{Dv}{Dt}$$

or as

$$\frac{De}{Dt} = -p\frac{Dv}{Dt} + v\Phi - v\nabla\cdot\vec{q} \tag{2.33}$$

It is conceptually useful to compare this relation to the first law of thermodynamics

$$de = \delta w + \delta q \tag{2.34}$$

when the thermodynamic system is a fluid particle. A delta is used to indicate an inexact differential; for example, δw is the differential work done on the system and δq is the differential heat transfer into the system. Observe that $\vec{q}$ and q have opposite signs and different units; $\vec{q}$ is per unit area per unit time, while q is per unit mass. If the process under consideration is reversible, we have the familiar result

$$\delta w_{\text{rev}} = -pdv, \qquad \delta q_{\text{rev}} = Tds \tag{2.35a,b}$$

where Tds represents the reversible heat transfer that crosses the system boundary, and p and s are the thermodynamic pressure and specific entropy, respectively, for a system in equilibrium.

With a fluid particle as the system, we write Equation (2.34) as a rate equation

$$\frac{de}{dt} = \frac{\delta w}{dt} + \frac{\delta q}{dt} \tag{2.36}$$

Equation (2.33) is similarly written as

$$\frac{de}{dt} = -p\frac{dv}{dt} + v\Phi - v\nabla\cdot\vec{q}$$

where dt represents the differential time change following a fluid particle. By comparing these two relations, we obtain

$$\delta w = -pdv + v\Phi dt \tag{2.37a}$$

We presume the pressures in Equations (2.35a) and (2.37a) are the same. In other words, the hydrostatic and thermodynamic pressures are, henceforth, equal to each other. Equation (2.37a) means that the work δw can be decomposed into reversible and irreversible terms, where the latter term is associated with the viscous stresses.

The foregoing comparison also yields

$$\delta q = -v(\nabla \cdot \vec{q})dt \tag{2.37b}$$

which provides the connection between q and $\vec{q}$. We now write

$$\dot{q} = \frac{\delta q}{dt}$$

and let δv, δs, and $\delta m(= \rho\delta v)$ represent the differential volume, surface area, and mass of a fluid particle, respectively. Equation (2.37b) is multiplied by ρdv and integrated over δv. With the aid of the divergence theorem, we have

$$\dot{q}\int_{\delta v} \rho dv = \int_{\delta v} \rho\dot{q}dv = -\oint_{\delta s} \hat{n}\cdot\vec{q}\,ds$$

or, as an alternative to Equation (2.37b),

$$\dot{q} = -\frac{1}{\delta m}\oint_{\delta s} \hat{n}\cdot\vec{q}\,ds$$

Thus, $-\dot{q}\delta m$ equals the net flux of $\vec{q}$ across the surface of the fluid particle. Section 3.7 will discuss the possibility of decomposing δq into reversible and irreversible terms, as was done for the work.

Enthalpy Form of the Energy Equation

Another alternate form of the energy equation replaces e with the enthalpy, h, by means of

$$e = h - \frac{p}{\rho} \tag{2.38a}$$

$$\frac{De}{Dt} = \frac{Dh}{Dt} - \frac{1}{\rho}\frac{Dp}{Dt} + \frac{p}{\rho^2}\frac{D\rho}{Dt} \tag{2.38b}$$

Consequently, Equations (2.23) and (2.27a) become

$$\rho\frac{Dh}{Dt} = \frac{Dp}{Dt} + \Phi - \nabla\cdot\vec{q} \tag{2.39}$$

for the enthalpy. This is the initial form for the energy equation most often used in later chapters.

We assume a Newtonian fluid for our final version of the energy equation. The first divergence term in Equation (2.20) is written as

$$\begin{aligned}\nabla \cdot (\overleftrightarrow{\sigma} \cdot \vec{w}) &= \nabla \cdot (\overleftrightarrow{\tau} \cdot \vec{w} - p\overleftrightarrow{I} \cdot \vec{w}) = \nabla \cdot (\overleftrightarrow{\tau} \cdot \vec{w} - p\vec{w}) \\ &= \nabla \cdot (\overleftrightarrow{\tau} \cdot \vec{w}) - \vec{w} \cdot \nabla p - p\nabla \cdot \vec{w} \\ &= \nabla \cdot (\overleftrightarrow{\tau} \cdot \vec{w}) - \vec{w} \cdot \nabla p + \frac{p}{\rho}\frac{D\rho}{Dt}\end{aligned} \tag{2.40}$$

The stagnation, or total, enthalpy is introduced as

$$h_o = h + \frac{1}{2} w^2 \tag{2.41}$$

and for the substantial derivative in Equation (2.20), we have

$$e + \frac{1}{2} w^2 = h - \frac{p}{\rho} + \frac{1}{2} w^2 = h_o - \frac{p}{\rho}$$

$$\frac{D}{Dt}\left(e + \frac{1}{2} w^2\right) = \frac{D}{Dt}\left(h_o - \frac{p}{\rho}\right) = \frac{Dh_o}{Dt} - \frac{1}{\rho}\frac{Dp}{Dt} + \frac{p}{\rho^2}\frac{D\rho}{Dt} \tag{2.42}$$

With the aid of Equations (2.40) and (2.42), Equation (2.20) becomes

$$\rho \frac{Dh_o}{Dt} = \frac{\partial p}{\partial t} + \nabla \cdot (\overleftrightarrow{\tau} \cdot \vec{w} - \vec{q}) + \rho\vec{w} \cdot \vec{F}_b \tag{2.43}$$

for the stagnation enthalpy. This version of the energy equation is especially convenient in gas dynamics where $\overleftrightarrow{\tau}$, $\vec{q}$, and $\vec{F}_b$ are assumed to be zero.

REFERENCE

Rae, W.J., "Flows with Significant Orientational Effects," *AIAA J.* 14, 11, (1976).

PROBLEMS

2.1 Evaluate $\nabla \times (\rho\vec{F}_b)$ when $\vec{w}$ = constant.

2.2 Use Cartesian coordinates to show that

$$\hat{n} \cdot (\overleftrightarrow{\sigma} \times \vec{r}) \neq (\overleftrightarrow{\sigma} \times \vec{r}) \cdot \hat{n}$$

even if $\overleftrightarrow{\sigma}$ is symmetric. The term on the left side occurs in the derivation of Equation (2.16).

2.3 Derive Equation (2.17) when $\overleftrightarrow{\sigma}$ is symmetric.

2.4 Show that the vorticity is given by

$$\nabla \times \vec{w} = \bar{\nabla} \times \vec{w} + 2\vec{\omega}_{\text{rot}}$$

where an overbar is associated with a noninertial system. Consequently, the vorticity is not invariant when transforming from an inertial to a noninertial system, or vice versa.

2.5 Prove Equation (2.22).

2.6 Use Cartesian coordinates to evaluate

$$\nabla \cdot (\rho \vec{w}\vec{w})$$

Show that it equals

$$\rho(\vec{w} \cdot \nabla)\vec{w} + \vec{w}\nabla \cdot (\rho \vec{w})$$

but does not equal

$$\rho \vec{w}(\nabla \cdot \vec{w}) + \vec{w}\nabla \cdot (\rho \vec{w})$$

2.7 **(a)** Evaluate

$$\overleftrightarrow{I} : \overleftrightarrow{\Phi}$$

for

$$\overleftrightarrow{\Phi} = \phi_{jk} \hat{1}_j \hat{1}_k$$

(b) Let $\overleftrightarrow{\Phi}_s$ and $\overleftrightarrow{\psi}_a$ be symmetric and antisymmetric dyadics, respectively. Evaluate

$$\overleftrightarrow{\Phi}_s : \overleftrightarrow{\psi}_a$$

(c) Evaluate

$$\overleftrightarrow{\Phi}_s : \overleftrightarrow{\Phi}_s$$

These relations are used in the viscous dissipation discussion in Section 2.8.

2.8 In spherical coordinates the divergence of the dyadic

$$\overleftrightarrow{\psi} = \psi^{is}\vec{e}_i\vec{e}_s$$

can be written as

$$\nabla \cdot \overleftrightarrow{\psi} = \psi_r\hat{e}_r + \psi_\theta\hat{e}_\theta + \psi_\phi\hat{e}_\phi$$

where

$$\psi_r = \frac{\partial \psi^{rr}}{\partial r} + \frac{\partial \psi^{\theta r}}{\partial \theta} + \frac{\partial \psi^{\phi r}}{\partial \phi} + \frac{2}{r}\psi^{rr} + \frac{\psi^{\theta r}}{\tan\theta} - r\psi^{\theta\theta} - r\sin^2\theta\psi^{\phi\phi}$$

$$\psi_\theta = r\left[\frac{\partial \psi^{r\theta}}{\partial r} + \frac{\partial \psi^{\theta\theta}}{\partial \theta} + \frac{\partial \psi^{\phi\theta}}{\partial \phi} + \frac{3}{r}\psi^{r\theta} + \frac{\psi^{\theta\theta}}{\tan\theta} + \frac{1}{r}\,\psi^{\theta r} - \sin\theta\,\cos\theta\,\psi^{\phi\phi}\right]$$

$$\psi_\phi = r\sin\theta\left[\frac{\partial \psi^{r\phi}}{\partial r} + \frac{\partial \psi^{\theta\phi}}{\partial \theta} + \frac{\partial \psi^{\phi\phi}}{\partial \phi} + \frac{3}{r}\,\psi^{r\phi} + \frac{2}{\tan\theta}\,\psi^{\theta\phi} + \frac{1}{r}\,\psi^{\phi r} + \frac{\psi^{\phi\theta}}{\tan\theta}\right]$$

Note that ψ^{rr}, $\psi^{\theta r}$, ..., are the dyadic components when the dyads are $\vec{e}_r\vec{e}_r$, $\vec{e}_\theta\vec{e}_r$,....

(a) Evaluate $\nabla \cdot \overleftrightarrow{I}$ in spherical coordinates.

(b) Determine $\overleftrightarrow{\tau}$ in terms of the $\hat{e}_r$, $\hat{e}_\theta$, and $\hat{e}_\phi$ basis.

(c) Write the velocity as

$$\vec{w} = v_r\hat{e}_r + v_\theta\hat{e}_\theta + v_\phi\hat{e}_\phi$$

Assume a Newtonian fluid and write

$$\nabla \cdot \overleftrightarrow{\tau} = \alpha_r\hat{e}_r + \alpha_\theta\hat{e}_\theta + \alpha_\phi\hat{e}_\phi$$

Determine the α_i in terms of $\nabla \cdot \vec{w}$, μ, λ, v_i, and various first- and second-order partial derivatives with respect to r, θ, and ϕ. To simplify the analysis use

$$\nabla \cdot (\psi\overleftrightarrow{\Phi}) = \psi\nabla \cdot \overleftrightarrow{\Phi} + (\nabla\psi) \cdot \overleftrightarrow{\Phi}$$

where ψ and $\overleftrightarrow{\Phi}$ are an arbitrary scalar and dyadic, respectively.

2.9 Utilize the results of Problems 1.4 and 2.8 and derive the five scalar conservation equations, without a body force, in spherical coordinates. Assume a Newtonian fluid and Fourier's equation, but do not assume that μ, λ, or κ are constants. Simplify the notation by utilizing previously derived expressions for $\nabla \cdot \vec{w}$, $D(\ )/Dt$, $\nabla \cdot \overleftrightarrow{\tau}$, and Φ.

2.10 Consider stratified Couette flow, where properties can only vary with y. The flow is steady and two dimensional, the lower wall is fixed, and the upper wall moves in its own plane at a constant speed U_∞. The gas need not be perfect; μ, λ, and κ are arbitrary functions of T; ignore body forces; and the pressure is a constant.

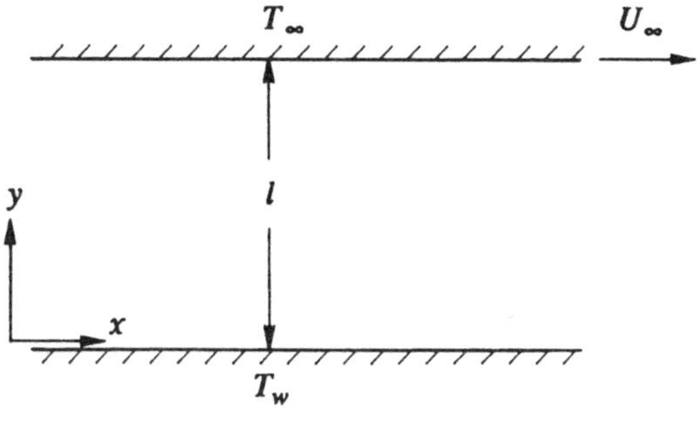

(a) Use the conservation equations to establish two first-order ODEs; one of the ODEs is for dw/dy and the other for dT/dy. Indicate what quantities in the ODEs are constants.

(b) Obtain a second-order ODE for T that does not involve w.

(c) Show that the vorticity and viscous dissipation can be written as

$$\vec{\omega} = -\frac{\tau_w}{\mu}\hat{1}_z, \quad \Phi = \frac{\tau_w^2}{\mu}$$

where τ_w is the shear stress at the lower wall, given by $(\mu dw/dy)_w$.

2.11 Assume a Newtonian fluid and no body forces.

(a) Start with Equations (2.8), (2.9a), and (4.19) and use various vector identities to derive an equation for $D\vec{\omega}/dt$.

(b) Use the results of part (a) to evaluate $D\vec{\omega}/dt$ when ρ and μ are constants. Be sure to eliminate $\overleftrightarrow{\varepsilon}$ in favor of $\vec{w}$.

2.12 Consider a body that is immersed in a steady, viscous flow of infinite extent. Let S_b be the surface of the body and S_{CV} be a control volume surface that encloses S_b.

(a) Neglect body forces and derive the equation

$$\vec{F} = -\oint_{S_b} \vec{\sigma} ds = \oint_{S_{CV}} (\vec{\sigma} - \rho\vec{w}\vec{w} \cdot \hat{n}) ds$$

for the force acting on the body, where $\hat{n}$ is the outward unit normal vector to S_{CV}.

(b) Similarly, show that the moment $\vec{L}$ acting on the body is

$$\vec{L} = -\oint_{S_b} \vec{r} \times \vec{\sigma} ds = \oint_{S_{CV}} \vec{r} \times (\vec{\sigma} - \rho\vec{w}\vec{w} \cdot \hat{n}) ds$$

2.13 Let $\vec{e}_i$ and $\vec{\bar{e}}_i$ be the basis vectors for the fixed and rigidly rotating coordinate systems of Section 2.5, respectively. We then have

$$\frac{d\vec{e}_i}{dt} = 0, \qquad \frac{d\vec{\bar{e}}_i}{dt} = \vec{\omega}_{\text{rot}} \times \vec{\bar{e}}_i$$

and

$$\vec{w} = \frac{dx^i}{dt}\vec{e}_i, \qquad \vec{\bar{w}} = \frac{d\bar{x}^i}{dt}\vec{\bar{e}}_i$$

$$\vec{a} = \frac{d^2x^i}{dt^2}\vec{e}_i, \qquad \vec{\bar{a}} = \frac{d^2\bar{x}^i}{dt^2}\vec{\bar{e}}_i$$

where

$$\vec{r} = x^i\vec{e}_i, \qquad \vec{\bar{r}} = \bar{x}^i\vec{\bar{e}}_i$$

(a) Start with Equation (2.11) and derive Equation (2.12).

(b) Derive Equation (2.13).

2.14 Consider a viscous, irrotational (i.e., $\nabla \times \vec{w} = 0$) flow without body forces. Assume Fourier's equation, a Newtonian fluid, and constant values for ρ, μ, κ, and the specific heat at constant volume c_v.

(a) Set $\vec{w} = \nabla\phi$, and derive equations for ϕ, $\nabla \cdot \overleftrightarrow{\varepsilon}$, p, and Φ.

(b) Start with the internal energy equation and derive a PDE for T. Simplify your results whenever possible.

(c) What is unusual about the answers for parts (a) and (b)?

2.15 Consider an unsteady, one-dimensional motion with x, t as the independent variables and p, ρ, h, and u (flow speed) as the dependent variables.

(a) Write the appropriate governing equations ignoring body forces, then simplify these equations by assuming a Newtonian fluid, Fourier's equation, and $\mu' \equiv 2\mu + \lambda$.

(b) Assume a thermally and calorically perfect gas; use a zero subscript to indicate a constant dimensional quantity in a quiescent gas and a unity subscript for a small, non-dimensional, perturbation variable. Linearize the governing equations by introducing

$$z = \frac{x}{\ell}, \quad \tau = \frac{a_o t}{\ell} = \left(\gamma \frac{p}{\rho}\right)^{1/2} \frac{t}{\ell}, \quad u = a_o u_1, \quad p = p_o(1 + p_1), \quad \rho = \rho_o(1 + \rho_1)$$

where a is the speed of sound and ℓ is a characteristic length. To standardize the notation, utilize the ratio of specific heats $\gamma = (c_p/c_v)$ and

$$Re = \text{Reynolds number} = \frac{\rho a \ell}{\mu'}$$

$$Pe = \text{Peclet number} = \frac{\rho a \ell}{\mu'} \frac{\mu' c_p}{\kappa}$$

where c_p is the specific heat at constant pressure. Evaluate Re_o/ℓ and Pe_o/ℓ for air at 300 K.

(c) Use the results of part (b) to derive a single, fifth-order PDE for u_1. Then determine the u_1 equations when only heat conduction or viscosity occur.

2.16 Use Cartesian coordinates to show that the viscous force

$$\nabla \cdot \overset{\leftrightarrow}{\tau} = \overline{\nabla} \cdot \overset{\leftrightarrow}{\overline{\tau}}$$

is the same in the inertial and noninertial systems.

3 Classical Thermodynamics

3.1 PRELIMINARY REMARKS

The classical modifier means, e.g., that statistical mechanics, quantum mechanics, and critical point theory are not considered. In fact, in the early part of this chapter the theory is discussed as it existed, more or less, at the start of the 20th century. More advanced concepts, however, are introduced that are useful in later chapters. Since this is not a thermodynamic text, the presentation is selective and occasionally sketchy.

Initially, the emphasis is on a simple, closed system. A closed system, of course, has a constant mass and does not involve molecular diffusion. We also ignore gravitational effects and interfacial tension. Despite these limitations, the elegant theory is nevertheless adequate for most fluid dynamic applications. If the flow is incompressible, then thermodynamics is of marginal utility. It is of central importance, however, when compressibility effects cannot be ignored.

The next three sections discuss the combined first and second laws, thermodynamic potential functions, and an open system. This material represents conventional classical thermodynamics. Section 3.5, however, provides a new, elegant approach for the coupling between thermodynamics and fluid dynamics. This procedure is particularly beneficial for flows with complicated equations of state. Concepts from the earlier sections are applied to a compressible liquid or solid in Section 3.6. The chapter concludes with a discussion of the second law that focuses on the physics of entropy production.

3.2 COMBINED FIRST AND SECOND LAWS

Introductory Discussion

We start with the basic equation

$$de = Tds - pdv \tag{3.1}$$

which is often called the combined first and second laws. In this relation, e, T, S, p, and v are the specific internal energy, absolute temperature, specific entropy, pressure, and specific volume, respectively. The equation stems from the assumption of a reversible process consisting of heat transfer

$$(\delta q)_{rev} = Tds \tag{3.2a}$$

and boundary work

$$(\delta w)_{rev} = -pdv \tag{3.2b}$$

where q and w are per unit mass and are positive when they increase the internal energy of the system. As before, it is convenient to denote with a δ the differential of a nonstate variable. Since all parameters in Equation (3.1) are intensive state variables, the equation is actually free of the reversible process assumption. It thus applies to every state point in a fluid flow, provided each infinitesimal region is viewed as a simple closed system in equilibrium.

State Properties

Any intensive state variable can be written as a function of any other two intensive state variables. Hence, we can write

$$e = e(s, v) \tag{3.3}$$

and

$$de = \frac{\partial e}{\partial s_v} ds + \frac{\partial e}{\partial v_s} dv \tag{3.4}$$

where the subscript notation denotes a fixed variable in the partial derivative. By comparing Equations (3.1) and (3.4), we have

$$T = \frac{\partial e}{\partial s_v}, \qquad p = -\frac{\partial e}{\partial v_s} \tag{3.5a,b}$$

Thus, if we start with a relation of the form of Equation (3.3), the temperature and pressure are *defined* by the above partial derivatives. Relation (3.3) is referred to as a potential function for the thermodynamic system.

Potential functions play an important role in thermodynamics. All thermodynamic information about the system is available in Equation (3.3) and is obtained by performing various partial derivatives, such as Equations (3.5). This aspect will be amply illustrated later in this chapter. The choice of variables is important; e.g.,

$$e = e(T, v) \tag{3.6}$$

is a legitimate thermodynamic equation, but it does not encompass the full amount of information available in Equation (3.3) and is not a potential function. Nevertheless, it is quite useful; e.g., it results in

$$de = \frac{\partial e}{\partial T_v} dT + \frac{\partial e}{\partial v_T} dv = c_v dT + \frac{\partial e}{\partial v_T} dv \tag{3.7}$$

where c_v is the specific heat at constant volume. The other partial derivative is provided by the reciprocity relation (Emanuel, 1987)

$$\frac{\partial e}{\partial v_T} = T\frac{\partial p}{\partial T_v} - p \tag{3.8}$$

which is a special form of the more general Maxwell equations (Emanuel, 1987). It is especially useful when the thermal equation of state (eos) has the form

$$p = p(T, v) \tag{3.9}$$

as is usually the case. In this circumstance, reciprocity can be integrated with respect to v:

$$e(T, v) = f(T) + \int \left(T \frac{\partial p}{\partial T}_v - p \right) dv \quad \textbf{(3.10)}$$

where f is a function of integration, and T is held fixed in the integrand of the indefinite integral. When an eos in the form of Equation (3.9) is known, the reciprocity equation represents a constraint on the functional form for $e(T,v)$. As we see, this form includes a function of just the temperature that is associated with the caloric eos, and a term, or set of terms, that is fully defined by the thermal eos. If the above equation is differentiated with respect to T, we have

$$c_v = \frac{e}{\partial T}_v = c_v^{\circ}(T) + \frac{\partial}{\partial T}_v \int \left(T \frac{\partial p}{\partial T}_v - p \right) dv \quad \textbf{(3.11)}$$

where $c_v^{\circ} = df/dT$, or

$$f = \int c_v^{\circ}(T) dT + \text{constant} \quad \textbf{(3.12)}$$

and c_v°, or f, can be viewed as the caloric eos. The c_v° parameter provides the translational, rotational, vibrational, and electronic contributions to the specific heat for the atom or molecule under consideration, while the rightmost term in Equation (3.11) provides the thermal eos contribution. (A second c_v° interpretation is provided shortly.) If $T(\partial p/\partial T)_v - p$ and its partial derivative with respect to T are continuous functions of v, as is the case for a single-phase fluid, then the order of differentiation and integration in Equation (3.11) can be interchanged. We thereby obtain

$$c_v = c_v^{\circ} + T \int \frac{\partial^2 p}{\partial T^2} dv \quad \textbf{(3.13)}$$

where v is held fixed when performing the partial derivatives but T is held fixed when performing the integration.

Perfect Gas

The theory of compressible flows centers around the eos of a thermally perfect gas

$$p = \frac{RT}{v} \quad \textbf{(3.14)}$$

where R is the gas constant. Reciprocity yields

$$\frac{\partial e}{\partial v_T} = 0 \quad \textbf{(3.15a)}$$

or

$$c_v = c_v^{\circ}(T) \quad \textbf{(3.15b)}$$

(An ideal gas is defined as having a temperature dependent c_v°.) There is no thermal state equation contribution to c_v when Equation (3.14) holds (or with the van der Waals thermal eos). The density $\rho = (1/v)$ is introduced, such that

$$p = \rho RT \tag{3.16}$$

In view of the last two equations, c_v° is also referred to as the thermally perfect, low-density gas, specific heat contribution to c_v. If c_v° is a constant, the gas is said to be calorically perfect. When a gas is both thermally and calorically perfect, for brevity it is simply referred to as perfect.

We introduce the enthalpy

$$h = e + pv \tag{3.17}$$

and assume a perfect gas. In this case, we have the well-known relations

$$c_v = \frac{R}{\gamma - 1}, \quad c_p = \frac{\gamma R}{\gamma - 1}, \quad c_p - c_v = R$$
$$e = c_v T, \quad h = c_p T, \quad \frac{s}{R} = \ln(vT^{1/(\gamma-1)}) + \text{constant} \tag{3.18}$$

where c_p is the constant-pressure specific heat, and $\gamma(= c_p/c_v)$ is the ratio of specific heats, which also is a constant.

3.3 POTENTIAL FUNCTIONS

We mentioned that e is a potential function when written in terms of s and v. There are many other potential functions, each of which contains the same thermodynamic information concerning the system as does any other potential function. Potential functions are thus equivalent to each other; they differ only in terms of analytical convenience. One of the basic postulates of thermodynamics is that the entropy is a monotonically increasing function of the internal energy (Emanuel, 1987). In this case, Equation (3.3) is theoretically invertible as

$$s = s(e, v) \tag{3.19}$$

where s is a potential when written in terms of e and v. By differentiation,

$$ds = \frac{\partial s}{\partial e_v} de + \frac{\partial s}{\partial v_e} dv \tag{3.20}$$

which combines with Equation (3.1) to yield

$$\left(1 - T\frac{\partial s}{\partial e_v}\right) de = \left(T\frac{\partial s}{\partial v_e} - p\right) dv \tag{3.21}$$

Since de and dv are both arbitrary, we obtain

$$\frac{1}{T} = \frac{\partial s}{\partial e_v}, \quad \frac{p}{T} = \frac{\partial s}{\partial v_e} \tag{3.22}$$

which are the entropy counterpart to Equations (3.5); i.e., these relations define p and T in the entropy representation. Suppose both s and e are considered to be functions of T and v. We can then write

$$\frac{\partial s}{\partial e_v} = \left(\frac{\frac{\partial s}{\partial T_v} dT + \frac{\partial s}{\partial v_T} dv}{\frac{\partial e}{\partial T_v} dT + \frac{\partial e}{\partial v_T} dv} \right)_v = \frac{\frac{\partial s}{\partial T_v}}{\frac{\partial e}{\partial T_v}} = \frac{1}{c_v} \frac{\partial s}{\partial T_v} \tag{3.23}$$

or

$$c_v = T \frac{\partial s}{\partial T_v} \tag{3.24}$$

As was done with the reciprocity equation, this relation can be integrated with respect to T:

$$s(T, v) = g(v) + \int \frac{c_v(T, v)}{T} dT \tag{3.25}$$

where g is a function of integration and v is held fixed in the integrand.

Equation (3.1) is now solved for p

$$p = T \frac{ds}{dv} - \frac{de}{dv}$$

and T is held fixed in the two derivatives

$$p = T \frac{\partial s}{\partial v_T} - \frac{\partial e}{\partial v_T} \tag{3.26}$$

With the aid of reciprocity, we have

$$\frac{\partial s}{\partial v_T} = \frac{\partial p}{\partial T_v} \tag{3.27}$$

When applied to Equation (3.25), this yields a first-order, ordinary differential equation

$$\frac{dg}{dv} = \frac{\partial p}{\partial T_v} - \int \frac{\partial c_v}{\partial v_T} \frac{dT}{T} \tag{3.28}$$

for g. The first term on the right is determined by the thermal eos, while the integral term is provided by Equation (3.13). Thus, the combination of Equations (3.25) and (3.28) provides $s(T,v)$.

Additional potential functions are generated by Legendre transformations. For instance, suppose we want a potential function of the form $\phi = \phi(s,p)$. The Legendre transformation that does this is written as (Emanuel, 1987)

$$\phi(s, p) = e(s, v) + pv \tag{3.29}$$

where v and e can be viewed as functions of s and p, and ϕ is recognized as the enthalpy. Thus, two other potentials are

$$h = h(s, p) = e + pv \tag{3.30}$$

and

$$f = f(T, v) = e - Ts \tag{3.31}$$

where f is the Helmholtz potential or free energy. [This f is not related to the one in Equation (3.12).] For the enthalpy, we write

$$dh = \frac{\partial h}{\partial s_p} ds + \frac{\partial h}{\partial p_s} dp = de + vdp + pdv = Tds + vdp \tag{3.32}$$

where Equation (3.1) is again used. Hence, we have

$$T = \frac{\partial h}{\partial s_p}, \qquad v = \frac{\partial h}{\partial p_s} \tag{3.33}$$

In a similar manner, we obtain

$$s = -\frac{\partial f}{\partial T_v}, \qquad p = -\frac{\partial f}{\partial v_T} \tag{3.34a,b}$$

First derivative equations of a potential function, such as Equations (3.5), (3.22), (3.33), and (3.34), are called state equations.

3.4 OPEN SYSTEM

In thermodynamics there are two distinctly different open system concepts. In the first, the system is open because there is a velocity flux of mass crossing its boundary. This type of open system is conventional in fluid dynamics. In the second case, a fluid element is envisioned in which mass may cross the boundary of the element because of diffusion. There may also be a phase change or a change in chemical composition taking place inside the element. In this chapter, we are concerned with the second type of open system.

For an open system, Equation (3.1) generalizes to

$$de = Tds - pdv + \sum \bar{\mu}_i dn_i \tag{3.35}$$

where $\bar{\mu}_i$ is the chemical potential of species i (or phase i) per kmole of species i, n_i is the number of kmoles of species i (or phase i) per kgram of mass of the mixture, and the summation is over all species (or phases) present in the system. The n_i variable is called the mole-mass ratio of species i or phase i. Hence, $\bar{\mu}_i$ has units of energy per kmole of species i or phase i. The summation term provides the change in the specific internal energy associated with diffusion, chemical changes, and phase changes when s and v are constants. This equation was first introduced by J.W. Gibbs, an American scientist of the 19th century. It is of fundamental importance in chemistry and chemical engineering.

There are two limiting cases for this equation. In the first, dn_i is zero and the composition is fixed. In the second,

$$\sum \bar{\mu}_i dn_i = 0 \tag{3.36}$$

and the changing composition is in chemical equilibrium. Both cases reduce Equation (3.35) to Equation (3.1), which tacitly assumed a fixed composition. For a fluid system in motion, the first case is referred to as a frozen flow. The second case is referred to as shifting equilibrium or local thermodynamic equilibrium. In this situation, the rate at which compositional changes occur, relative to the fluid motion, is rapid enough to maintain the fluid locally in chemical equilibrium. For a reacting system, the chemical composition is established by the above relation.

The internal energy potential function counterpart of Equation (3.3) is

$$e = e(s, v, n_1, n_2, \ldots) \tag{3.37}$$

This relation is differentiated as

$$de = \frac{\partial e}{\partial s}_{v,\, n_i} ds + \frac{\partial e}{\partial v}_{s,\, n_i} dv + \frac{\partial e}{\partial n_1}_{s,\, v,\, n_j} dn_i + \cdots \tag{3.38}$$

where the n_j subscript indicates that all the n_j are held fixed, except for the n_i variable in the partial derivative. The state equations become

$$T = \frac{\partial e}{\partial s}_{v,\, n_1}, \qquad p = -\frac{\partial e}{\partial v}_{s,\, n_1}, \qquad \bar{\mu}_i = \frac{\partial e}{\partial n_i}_{s,\, v,\, n_j} \tag{3.39}$$

When some or all of the dn_i are not zero, Equation (3.35) has an integrated form

$$e = Ts - pv + \sum \bar{\mu}_i n_i \tag{3.40}$$

that is quite useful.

General Ideal Gas Mixture

Molar quantities, such as $\bar{\mu}_i$ and the molar specific heat at constant pressure for species i, $\bar{c}_{pi}$, are written with an overbar. It is simplest to start with a per mole formula for the entropy of species i (Emanuel, 1987)

$$\bar{s}_i(T, p_i) = \bar{s}_{io} - \bar{R} \ln\left(x_i \frac{p}{p_o}\right) + \int_{T_o}^{T} \bar{c}_{pi}(T') \frac{dT'}{T'} \tag{3.41}$$

where an o subscript denotes a reference state value, $\overline{R}$ is the universal gas constant, x_i is the mole fraction of species i, $p_i(= x_i p)$ is the partial pressure of species i, and a prime denotes a dummy integration variable. To convert to a per unit mass basis, the relations

$$R = \frac{\overline{R}}{W}, \quad \sum n_1 = \frac{1}{W}, \quad p = \frac{\overline{R}T}{Wv}, \quad x_i = \frac{n_i}{\sum n_i} \tag{3.42}$$

are useful, where W is the mixture molecular weight. The specific entropy for the mixture, in terms of T, v, n_1, n_2, …, is given by

$$s = \sum n_i \bar{s}_i = \sum n_i \bar{s}_{io} - R\ln\left(\frac{\overline{R}T}{p_o v}\right) - \overline{R}\sum n_i \ln n_i + \sum n_i \int_{T_o}^{T} \bar{c}_{pi}\frac{dT'}{T'} \tag{3.43a}$$

where the reference pressure p_o is usually taken as 1 bar or 1 atm. Similarly, for the internal energy, we write

$$e(T, v, n_1, n_2, \ldots) = \sum n_i \bar{e}_{io} + \sum n_i \int_{T_o}^{T} \bar{c}_{vi}(T')dT' \tag{3.43b}$$

where

$$\bar{c}_{vi} = \bar{c}_{vi}^{o}(T), \qquad \bar{c}_{pi} = \bar{c}_{vi} + \overline{R}$$

Equations (3.43) constitute an e or s potential function for the system, where the temperature is viewed as a parameter that theoretically can be eliminated.

The Helmholtz potential function can be written as

$$\begin{aligned} f(T, v, n_1, n_2, \ldots) = e - Ts &= \sum n_i(\bar{e}_{io} - T\bar{s}_{io}) - RT\ln v + \overline{R}T\sum n_i \ln n_i \\ &+ \sum n_i \int_{T_o}^{T}\left(1 - \frac{T}{T'}\right)\bar{c}_{vi}(T')dT' \end{aligned} \tag{3.44}$$

Equation (3.34b) yields

$$p = -\frac{\partial f}{\partial v_{T, n_i}} = \frac{RT}{v} \tag{3.45}$$

which is the thermally perfect (ideal) gas eos. The chemical potential $\bar{\mu}_i$ is given by the state equation

$$\bar{\mu}_i = \frac{\partial f}{\partial n_{iT, v, n_j}} \tag{3.46}$$

and results in the ideal gas relation

$$\bar{\mu}_i = \bar{e}_{io} - T\bar{s}_{io} + \overline{R}T\left(1 + \ln\frac{n_i}{v}\right) + \int_{T_o}^{T}\left(1 - \frac{T}{T'}\right)\bar{c}_{vi}dT' \tag{3.47}$$

Harmonic Oscillator Model

An equation of state with the form of Equation (3.9) and an equation for $c_v^\circ(T)$ are sufficient to fully establish the thermodynamics of a simple system. This includes determining a coexistence curve, the vapor pressure, etc. Hence, these two relations are equivalent to a potential function, such as the Helmholtz potential function. This assertion is evident by examining Equation (3.10), with the c_v° integral replacing f, and Equations (3.25) and (3.28). This result, it should be noted, does not extend to other forms of the eos, such as $p = p(s, v)$.

Although not specific to an open system, it is convenient at this time to consider a calorically imperfect gas model for c_v°. It is often used with an ideal gas. For a monatomic gas or a thermally perfect gas, e.g., air at room temperature, c_v° is a constant. In this circumstance, we can use

$$c_v^\circ = \frac{1}{2}(5 + \delta) R \tag{3.48}$$

where

$$\begin{aligned} \delta &= -2, \quad \text{monatomic molecule} \\ &= 0, \quad \text{linear polyatomic} \\ &= 1, \quad \text{nonlinear polyatomic} \end{aligned} \tag{3.49}$$

Relation (3.48) assumes the translational and rotational modes of the atom or molecule are fully excited but the vibrational (if there is any) and electronic modes are not excited. For many room-temperature gases (He, Ar, H_2, O_2, N_2, air, CO, NO), these assumptions are warranted.

For larger molecules at room temperature and the foregoing diatomics at higher temperatures, the vibrational mode, or modes, starts to become active, and c_v° now has the form

$$c_v^\circ = \frac{1}{2}(5 + \delta)R + c_{vv}(T) \tag{3.50}$$

where c_{vv} represents the vibrational mode(s) contribution. This result then combines with the thermal eos contribution to yield the specific heat of Equation (3.11). A common, generally accurate, approximation for c_{vv} is called the harmonic oscillator model. In this model, each vibrational mode is approximated with equally spaced, quantized vibrational levels. The model typically loses accuracy only at elevated temperatures when the level spacing decreases and molecular dissociation becomes significant. This calorically imperfect gas model can be used for a pure diatomic gas, such as N_2, in conjunction with a thermally perfect gas eos. As such, it greatly extends the range of validity of this eos. This approach is the basis of Chapter 9, which examines gas dynamic flows of a calorically imperfect gas. It can also be used with diatomic, or larger molecules, that are not thermally perfect, as occurs in the vicinity of the coexistence curve. The model has one other important attribute. The c_v integrals that appear in the equation for the internal energy, Equation (3.12), and in the entropy equation, Equation (3.25), can be analytically evaluated.

The equilibrium internal energy, e_v, of a single vibrational mode is represented by the harmonic oscillator model as

$$e_v = \frac{RT_v}{e^{T_v/T} - 1} \tag{3.51}$$

where T_v is the characteristic vibrational temperature of the mode. If the vibrational levels are widely spaced apart, then T_v has a large value, e.g., $T_v = 3352$ K for N_2. At low and high temperatures,

e_v has the limits

$$\begin{aligned} e_v &= 0, \qquad T/T_v \to 0 \\ &= RT, \qquad T/T_v \to \infty \end{aligned} \tag{3.52}$$

From Equation (3.51), the specific heat at constant volume of the mode is

$$c_{vv} = \frac{de_v}{dT} = R\left(\frac{\theta_v}{\sinh\theta_v}\right)^2 \tag{3.53}$$

where

$$\theta_v = \frac{T_v}{2T} \tag{3.54}$$

Suppose a molecule has N atoms, with $N \geq 3$. The molecule will have more than one vibrational mode, with a number of distinct characteristic vibrational temperatures. Because of possible symmetry of the molecular structure, some modes may be repeated; i.e., they have a degeneracy g_v. Thus, c_{vv} is given by

$$c_{vv} = R\sum_v g_v\left(\frac{\theta_v}{\sinh\theta_v}\right)^2 \tag{3.55}$$

where the summation is over those modes with distinct characteristic vibrational temperatures. (See Emanuel, 1987, p. 37, for a list of T_v and g_v values for an assortment of molecules.) Since the total number of vibrational modes is $3N - 5 - \delta$, we have

$$\sum_v g_v = 3N - 5 - \delta \tag{3.56}$$

Thus, c_{vv} is bounded by

$$0 < c_{vv} < (3N - 5 - \delta)R \tag{3.57}$$

in accord with the T/T_v limits in Equations (3.52). Finally, c_v° is written as

$$c_v^\circ = \frac{1}{2}(5 + \delta)R + R\sum_v g_v\left(\frac{\theta_v}{\sinh\theta_v}\right)^2 \tag{3.58}$$

We can obtain several properties of c_v° that will be useful in some of the subsequent chapters. The derivative is given by

$$\frac{dc_v^\circ}{dT} = 4R\sum_v \frac{g_v}{T_v}\left(\frac{\theta_v}{\sinh\theta_v}\right)^3 (\theta_v\cosh\theta_v - \sinh\theta_v) \tag{3.59}$$

The integral that appears in the internal energy and enthalpy equations is

$$\frac{1}{R}\int_1^{T/T_o} c_v^\circ d\left(\frac{T'}{T_o}\right) = \frac{1}{2}(5+\delta)\left(\frac{T}{T_o}-1\right) + \sum_v g_v\theta_{vo}(\coth\theta_v - \coth\theta_{vo}) \tag{3.60}$$

where T_o is a reference temperature and

$$\theta_{vo} = \frac{T_v}{2T_o} \tag{3.61}$$

The integral that appears in the entropy equation is

$$\frac{1}{R}\int_1^{T/T_o} c_v^\circ \frac{dT'}{T'} = \frac{1}{2}(5+\delta)\ln\frac{T}{T_o} + \sum_v g_v(\theta_v\coth\theta_v - \coth\theta_{vo}) + \ln\left[\prod_v\left(\frac{\sinh\theta_{vo}}{\sinh\theta_v}\right)^{g_v}\right] \tag{3.62}$$

3.5 COUPLING TO FLUID DYNAMICS

Introductory Discussion

This section deals with the coupling between thermodynamics and fluid dynamics, especially when the thermodynamic model is relatively complex. Thus, for a perfect gas or a constant density flow, this section is unnecessary. A distinction must be made here between a flow that is incompressible because the density is a constant and one where the maximum value of the Mach number is small compared to unity. If the latter case arises in a gas flow with complicated thermodynamics, then this section remains relevant.

The Euler equations directly involve only p, ρ, and h. Thus, a thermal eos of the form

$$h = h(p, \rho) \tag{3.63}$$

yields a mathematically closed system. Since the transport properties in the Navier–Stokes equations are usually temperature dependent, these equations may also require

$$T = T(p, \rho) \tag{3.64}$$

for a mathematically closed system. These two relations represent the same thermodynamic surface; hence, care must be exercised that they are thermodynamically compatible with each other. Of course, other thermodynamic properties are almost always of interest. These include the specific heats, the speed of sound a, the entropy, etc. As compared to a perfect gas, the governing equations are far more complicated even with a van der Waals fluid and a calorically perfect equation of state. In many real gas flows, however, an adequate representation of the physics requires imperfect caloric and thermal state equations of considerable complexity. In this circumstance, an analytical solution is out of the question, while a computational solution may involve enormous difficulty. One area where this type of difficulty arises is in dense gas flows (Argrow, 1996). This type of flow occurs in a region on the vapor side of the coexistence curve near the critical point and involves large molecules with very large specific heats. A second area is in modeling the detonation wave that propagates through a condensed phase explosive (Mader, 1979; Fickett and Davis, 1979) such as TNT. On the downstream side of the normal shock that propagates through a solid or liquid explosive, the density is comparable to the solid or liquid density ahead of the shock. As a minimum, it is therefore essential to include the covolume in the thermal eos when modeling the gas-phase shock.

To computationally deal with these types of flows, Swesty (1996) introduces a bi-quintic interpolation scheme, in conjunction with the use of Maxwell's equations to construct thermodynamically self-consistent eos tables. In this approach, some tabular data must be available at the start. Alternatively, Merkle et al. (1998) utilize

$$h = h(p, T), \qquad \rho = \rho(p, T)$$

instead of Equations (3.63) and (3.64). These functions must be known, compatible with each other, and differentiable.

A new approach (at the time of writing) is introduced that is a generalization of the one by Merkle et al. It starts with the use of a Helmholtz potential function, whose derivatives provide the thermodynamic properties, including state equations. This approach has been standard practice for some time in the computerized modeling of thermodynamic properties (Emanuel, 1987, Section 8.5). In this case, the Helmholtz potential equation, which may contain as many as 100 constants, accurately represents the vapor and liquid properties of a real fluid. This is especially important when the state of the fluid is in the vicinity of the coexistence curve.

The use of a Helmholtz potential means that the governing fluid dynamic equations should be formulated in terms of v, or ρ, and T. Other parameters, such as p, h, s, …, are obtained by differentiation without recourse to a cumbersome iterative or interpolation evaluation process. Because of its consistency and completeness, Maxwell's equations and the Clapeyron–Clausius equation, for the vapor pressure, are unnecessary. The approach possesses considerable generality; e.g., files containing f and its derivatives need to be established only once for a given fluid. A liquid/vapor two-phase flow is readily handled. Moreover, the use of temperature as a variable is convenient for representing transport properties.

For pedagogical reasons, an inverse approach is utilized in which the potential is found starting with several well-known thermal eos. We also demonstrate how the potential is applied to a liquid/vapor mixture. Finally, we illustrate how thermodynamic and CFD models can be efficiently coupled so that much of the thermodynamic computation is post-processed.

Helmholtz Potential

As will become apparent, the Helmholtz potential, Equation (3.31), is particularly advantageous for our task. As previously noted, the first derivatives of f provide

$$p = -f_v, \qquad s = -f_T \tag{3.65a,b}$$

where f_v denotes $\partial f/\partial v$ with T held fixed, and f_T is $\partial f/\partial T$ with v held fixed. The variables to be held fixed in a partial derivative are indicated only when required by clarity. Aside from the above equations, other parameters of interest are

$$e = \text{specific internal energy} = f - Tf_T \tag{3.66a}$$

$$h = \text{specific enthalpy} = f - Tf_T - vf_v \tag{3.66b}$$

$$c_v = \text{constant volume specific heat} = -Tf_{TT} \tag{3.66c}$$

$$c_p = \text{constant pressure specific heat} = \frac{T}{f_{vv}}(f_{Tv}^2 - f_{TT}f_{vv}) \tag{3.66d}$$

$$\kappa_T = \text{isothermal compressibility} = -\frac{1}{v}\frac{\partial v}{\partial p_T} = \frac{1}{v f_{vv}} \tag{3.66e}$$

$$\kappa_s = \text{adiabatic compressibility} = -\frac{1}{v}\frac{\partial v}{\partial p_s} = -\frac{1}{v}\frac{f_{TT}}{f_{Tv}^2 - f_{TT} f_{vv}} \tag{3.66f}$$

As discussed in Menikoff and Plohr (1989), thermodynamic stability requires that

$$\frac{1}{c_v} \geq \frac{1}{c_p} \geq 0 \tag{3.67a}$$

$$\frac{1}{\kappa_s} \geq \frac{1}{\kappa_T} \geq 0 \tag{3.67b}$$

One can show, however, that these conditions reduce to

$$0 \geq f_{TT}, \qquad f_{vv} \geq 0 \tag{3.68a,b}$$

when using a Helmholtz potential. This remarkable simplification is, in fact, typical for this potential. For example, the spinodal curve (Emanuel, 1987) is given by

$$f_{vv} = 0 \tag{3.69}$$

The speed of sound a is given by

$$a^2 = \frac{\partial p}{\partial \rho_s} = \frac{v}{\kappa_s} \tag{3.70}$$

where κ_s is provided by Equation (3.66f). The fundamental derivative of gas dynamics Γ (Emanuel, 1996) is defined as

$$\Gamma = 1 + \frac{\rho}{2a^2}\frac{\partial a^2}{\partial \rho_s} \tag{3.71a}$$

This parameter is important in dense gas flows (Argrow, 1996). One can show that this relation reduces to

$$\Gamma = -\frac{v^3}{2a^2 f_{TT}^3}\left(-f_{Tv}^3 f_{TTT} + 3 f_{TT} f_{Tv}^2 f_{TTv} - 3 f_{TT}^2 f_{Tv} f_{Tvv} + f_{TT}^3 f_{vvv}\right) \tag{3.71b}$$

which is an unusually simple result for this parameter. This is the only parameter to be discussed that requires third derivatives of f.

The foregoing relations demonstrate that first, second, and third derivatives, i.e.,

$$f_T, f_v$$
$$f_{TT}, f_{Tv}, f_{vv}$$
$$f_{TTT}, f_{TTv}, f_{Tvv}, f_{vvv}$$

are required. The higher-order derivatives are best performed with symbolic manipulation software. Once the equations for the derivatives are established for a given f, they can be stored for subsequent and repeated use. Indeed, a wide variety of such files can be established. These would include files for different Helmholtz potential functions, including equations for f for a wide variety of fluids, as well as files based on Equations (3.65) and (3.66) and other equations still to be developed.

The coexistence curve is given by

$$p(T, v_g) = p(T, v_f) \tag{3.72a}$$

$$\mu_g(T, v_g) = \mu_f(T, v_f) \tag{3.72b}$$

where g and f subscripts denote the saturated vapor and liquid states, respectively. The chemical potential for a pure substance, per unit mass, can be written as

$$\mu = f - v f_v \tag{3.73}$$

With the aid of this relation, Equations (3.72) simplify to

$$f_v(T, v_g) = f_v(T, v_f) \tag{3.74a}$$

$$(f - v f_v)_g = (f - v f_v)_f \tag{3.74b}$$

The variables are T, v_g, and v_f; hence, these equations provide $v_g = v_g(T)$ and $v_f = v_f(T)$, which can be numerically evaluated and stored. To determine if a state point is on or inside the coexistence curve, we first require that its temperature satisfy $T \le T_c$, where a c subscript denotes a critical point value. Secondly, we require that $v_f \le v \le v_g$, where v_f and v_g stem from Equations (3.74) and are evaluated at T. By occasionally monitoring the state of the fluid, it is easy to establish whether it is a liquid, vapor, or a two-phase mixture.

Equations (3.74) are exact in the vicinity of a critical point. At a critical point, however, a fluid's representation is nonanalytic and none of the subsequent analytic equations of state is valid. This difficulty is restricted to the immediate vicinity of the critical point (Emanuel, 1996).

Ideal Gas

We begin with an ideal gas

$$p = \frac{RT}{v}, \qquad c_v = c_v(T) \tag{3.75a,b}$$

where $c_v(T)$ is a known function. Equation (3.65a) yields

$$f = g(T) - RT \ln \frac{v}{v_o} \tag{3.76}$$

where g is a function of integration. We next have

$$c_v = -T f_{TT} = -T g'' \tag{3.77}$$

where a prime denotes differentiation with respect to T. Repeated integrations provide

$$g(T) = -\int_{T_o}^{T} dT'' \int_{T_o}^{T''} \frac{c_v(T')}{T'} dT' \tag{3.78a}$$

The double integral version of Dirichlet's more general formula

$$\int_a^x dx_2 \int_a^{x_2} f(x_1)dx_1 = \int_a^x (x-t)f(t)dt \tag{3.79}$$

is utilized, to obtain

$$g = \int_{T_o}^{T} \left(1 - \frac{T}{T'}\right) c_v(T')dT' \tag{3.78b}$$

Consequently, the Helmholtz function for an ideal gas is

$$f = -RT \ln \frac{v}{v_o} + \int_{T_o}^{T} \left(1 - \frac{T}{T'}\right) c_v(T')dT' \tag{3.80}$$

In conjunction with Equation (3.71b), this yields

$$\Gamma = \frac{1}{2\frac{c_v}{R}\left(1 + \frac{c_v}{R}\right)} \left[\left(1 + \frac{c_v}{R}\right)\left(1 + 2\frac{c_v}{R}\right) - T \frac{c_v'}{c_v}\right] \tag{3.81}$$

If c_v is independent of the temperature, then Γ reduces to $(\gamma + 1)/2$.

van der Waals Fluid

If, for a given eos, e and s are known as functions of T and v, then

$$f = e - Ts \tag{3.82}$$

directly yields the Helmholtz potential. For a van der Waals eos, this process yields

$$f = e_c + RT_c\left(-\frac{s_c}{R} T_r + \frac{9}{8} \frac{v_r - 1}{v_r} - T_r \ln \frac{3v_r - 1}{2} + g_r\right) \tag{3.83}$$

where

$$g_r = \int_1^{T_r} \left(1 - \frac{T_r}{T_r'}\right) c_v^{\circ}(T_r')dT_r' \tag{3.84}$$

The r subscript denotes a reduced variable. Note that, at the critical point,

$$g_r(1) = 0, \qquad f_c = e_c - T_c s_c \tag{3.85a,b}$$

Without loss of generality, c_v can be written as $c_v^\circ(T) + \hat{c}_v(v, T)$, where $\hat{c}_v$ is determined by the thermal eos, and c_v° is an arbitrary function of the temperature. Thus, c_v° appears in f; its functional form is not determined by f. Over a limited temperature range, c_v° can be modeled as a polynomial. Alternatively, a representation whose accuracy is consistent with a Helmholtz potential is provided by the harmonic oscillator approximation, Equation (3.58). In terms of this approximation, g_r is provided by Equations (3.60) and (3.62) as

$$g_r = R\left(\frac{1}{R}\int_1^{T_r} c_v^\circ dT_r' - \frac{T_r}{R}\int_1^{T_r} c_v^\circ \frac{dT_r'}{T_r'}\right) \tag{3.86}$$

with $T_o = T_c$.

Clausius-II Fluid

For the Clausius-II eos

$$p_r = \frac{4T_r}{A} - \frac{3}{T_r B^2} \tag{3.87a}$$

$$A = 1 + 4Z_c(v_r - 1) \tag{3.87b}$$

$$B = 1 + \frac{8}{3} Z_c(v_r - 1) \tag{3.87c}$$

we obtain (Emanuel, 1987)

$$f = e_c + RT_c\left[\frac{9}{4} - \left(\frac{9}{8} + \frac{s_c}{R}\right)T_r - \frac{9}{8T_r B} - T_r \ln A + g_r(T_r)\right] \tag{3.88}$$

Here, Z_c is the critical value of the compressibility factor:

$$Z_c = \left(\frac{pv}{RT}\right)_c \tag{3.89}$$

Martin–Hou Fluid

As a final example, we mention the Martin–Hou eos (Martin et al., 1959)

$$p = \frac{RT}{v-b} + \sum_{i=2}^{5} \frac{\hat{f}_i(T)}{(v-b)^i} \tag{3.90}$$

where b is the covolume,

$$\hat{f}_i = A_i + B_i T + C_i e^{-kT_r} \tag{3.91}$$

and the A_i, B_i, C_i, and k are constants. The same procedure used to obtain the Helmholtz potential for an ideal gas results in

$$f = RT \ln\left(\frac{v-b}{v+b}\right) + \sum_{i=2}^{5} \frac{\hat{f}_i}{(i-1)(v-b)^{i-1}} + \int_{T_o}^{T} \left(1 - \frac{T}{T'}\right) c_v^\circ(T')dT' \tag{3.92}$$

Mixture Region

For equilibrium mixture states, a new state variable, the quality x, is introduced as

$$x = \frac{v - v_f}{v_g - v_f} \tag{3.93a}$$

where the conventional definition is

$$v = x v_g + (1-x) v_f \tag{3.93b}$$

We thus have

$$\frac{\partial x}{\partial T_v} = -\frac{x v_g' + (1-x) v_f'}{v_g - v_f} \tag{3.94a}$$

$$\frac{\partial x}{\partial v_T} = \frac{1}{v_g - v_f} \tag{3.94b}$$

$$\frac{\partial v}{\partial T_x} = x v_g' + (1-x) v_f' \tag{3.95a}$$

$$\frac{\partial v}{\partial x_T} = v_g - v_f \tag{3.95b}$$

Let χ represent v, e, h, or s. Then χ is linear with respect to x; i.e.,

$$\chi = x\chi_g + (1-x)\chi_f \tag{3.96}$$

where χ_g and χ_f are functions only of the temperature. In view of Equation (3.82), χ also represents the Helmholtz potential. Equations (3.95) can be replaced with

$$\frac{\partial \chi}{\partial T_x} = x\chi_g' + (1-x)\chi_f' \tag{3.97a}$$

$$\frac{\partial \chi}{\partial x_T} = \chi_g - \chi_f \tag{3.97b}$$

where T and x, instead of T and v, are viewed as the independent variables in the mixture region.

The various v and T derivatives of χ are evaluated with the aid of Jacobian theory (Appendix B). For instance, write

$$\chi_v = \frac{\partial \chi}{\partial v_T} = \frac{\dfrac{\partial(\chi, T)}{\partial(x, T)}}{\dfrac{\partial(v, T)}{\partial(x, T)}} = \frac{\dfrac{\partial \chi}{\partial x_T}}{\dfrac{\partial v}{\partial x_T}} = \frac{\chi_g - \chi_f}{v_g - v_f} \tag{3.98}$$

Hence, χ_v is a function only of T. Further v derivatives of χ_v are zero, while the T derivative of χ_v is straightforward. Thus, χ_{vv}, χ_{vvv}, and χ_{Tvv} are zero. The quantity χ_T becomes

$$\chi_T = \frac{\partial \chi}{\partial T_v} = \frac{\dfrac{\partial(\chi, v)}{\partial(x, T)}}{\dfrac{\partial(T, v)}{\partial(x, T)}} = x\chi_g' + (1-x)\chi_f' - \frac{\chi_g - \chi_f}{v_g - v_f}[xv_g' + (1-x)v_f'] \tag{3.99}$$

where the right-hand side only depends on T and x. The derivative

$$\chi_{Tv} = \frac{\partial \chi}{\partial v_T}\left[\chi_g' - \chi_f' - \frac{\chi_g - \chi_f}{v_g - v_f}(v_g' - v_f')\right] = \frac{\chi_g' - \chi_f'}{v_g - v_f} - \frac{\chi_g - \chi_f}{(v_g - v_f)^2}(v_g' - v_f') \tag{3.100}$$

can be more easily obtained directly from Equation (3.98). As with χ_v, the right side only depends on T. Finally, the χ_{TT} derivative becomes

$$\begin{aligned}
\chi_{TT} &= x\chi_g'' + (1-x)\chi_f'' - \chi_{vT}[xv_g' + (1-x)v_f'] - \chi_v[xv_g'' + (1-x)v_f''] \\
&\quad - \frac{xv_g' + (1-x)v_f'}{v_g - v_f}[\chi_g' - \chi_f' - \chi_v(v_g' - v_f')] \\
&= x(\chi_g'' + v_g''\chi_v) + (1-x)(\chi_f'' + v_f''\chi_v) - 2\,\frac{xv_g'' + (1-x)v_f'}{v_g - v_f}[\chi_g' - \chi_f' - \chi_v(v_g' - v_f')]
\end{aligned} \tag{3.101}$$

Again, the coefficient of x and of $1 - x$ only depends on T. The only derivatives not explicitly evaluated, so far, are χ_{TTT} and χ_{TTv}, which can be done with symbolic manipulation software.

The derivatives $f_T, f_v, f_{TT}, \ldots$, in the single-phase theory are replaced with their x,T counterparts. For instance, s_{TT} and f_{TT} are respectively replaced with the right-hand side of Equation (3.101) with χ equaling s and then f. This is done for equilibrium mixture states. Actually, single-phase metastable states can occur between the coexistence and spinodal curves. These subcooled or supersaturated states are accessed by extending, without alteration, the foregoing single-phase formulation into these regions. The switchover from a metastable state to an equilibrium one is discontinuous; e.g., a condensation shock may occur.

Equations (3.74) should be checked whenever a discontinuity, such as a contact surface or shock wave, is encountered. Otherwise, these equations need frequent monitoring only when approaching a coexistence curve.

As a simple example, consider c_v in the mixture region

$$c_{v,mx} = \frac{\partial e}{\partial T_v} = \frac{\partial}{\partial T_v}[e_f + x(e_g - e_f)] = xc_{vg} + (1-x)c_{vf} - \frac{e_g - e_f}{v_g - v_f}[xv_g' + (1-x)v_f'] \quad \textbf{(3.102)}$$

where Equation (3.99) is used to differentiate e. The jump discontinuity in the equilibrium value of c_v on the liquid side ($x = 0$) is

$$c_{v,mx,f} - c_{vf} = \frac{e_g - e_f}{v_g - v_f} v_f' \quad \textbf{(3.103a)}$$

while on the vapor side ($x = 1$), we have

$$c_{v,mx,g} - c_{vg} = \frac{e_g - e_f}{v_g - v_f} v_g' \quad \textbf{(3.103b)}$$

Since $v_f' > 0$ and $v_g' < 0$, the jump in c_v is negative (positive) when $x = 0$ ($x = 1$). Consequently, $(\partial c_{v,mx}/\partial x)_T$ can have a substantial positive value.

Many parameters experience a jump discontinuity on the coexistence curve in their equilibrium value. Thus,

$$c_v,\ c_p,\ a,\ \kappa_T,\ \kappa_s,\ \Gamma \quad \textbf{(3.104)}$$

have a discontinuity. Incidentally, a^2 and Γ simplify to

$$a_{mx}^2 = -\frac{v^2 f_{Tv}^2}{f_{TT}} \quad \textbf{(3.105a)}$$

$$\Gamma_{mx} = \frac{v}{2f_{TT}^2}(3f_{TT}f_{TTv} - f_{Tv}f_{TTT}) \quad \textbf{(3.105b)}$$

where the various derivatives are still to be replaced with their x,T counterparts. In stating that Γ experiences a finite jump discontinuity on the coexistence curve, we differ with Menikoff and Plohr (1989), where it is stated that Γ has a δ-function singularity. One exception to this discussion is the critical point. For instance, for all real fluids, c_v, c_p, and Γ become positively infinite at the critical point (Emanuel, 1996).

Coupling to CFD

For conciseness, the discussion is limited to the one-dimensional, unsteady Euler equations. This limitation, however, is only for convenience. We start with the conservation equations in a dimensional, conservative form:

$$\frac{\partial \rho}{\partial t} + \frac{\partial(\rho w)}{\partial x} = 0 \quad \textbf{(3.106a)}$$

$$\frac{\partial(\rho w)}{\partial t} + \frac{\partial}{\partial x}(p + \rho w^2) = 0 \tag{3.106b}$$

$$\frac{\partial(-p + \rho h_0)}{\partial t} + \frac{\partial}{\partial x}(\rho w h_0) = 0 \tag{3.106c}$$

where w is the flow speed, and

$$h_0 = h + \frac{1}{2}w^2 \tag{3.107}$$

The only thermodynamic parameters that appear are ρ, p, and h. These depend on T and v by means of

$$\rho = \frac{1}{v}, \qquad p = -f_v, \qquad h = f - Tf_T - vf_v \tag{3.108a,b,c}$$

Consequently, the second- and higher-order derivatives previously discussed need not be evaluated during a CFD calculation. The independent variables are x and t, while v, T, and w are the dependent variables. In turn, v and T determine ρ, f, f_v, and f_T, where f, f_v, and f_T are provided by *explicit* algebraic equations. Thus, f, f_v, and f_T can be efficiently evaluated innumerable times during a computation. Step size control for numerical stability may require the speed of sound, which requires second derivatives of f. This evaluation, however, need not be done at every time step and at every grid point.

Equations (3.106) can be written in vector form as

$$\frac{\partial p}{\partial t} + \frac{\partial Q}{\partial x} = 0 \tag{3.109}$$

where

$$P = \begin{bmatrix} \dfrac{1}{v} \\ \dfrac{w}{v} \\ \dfrac{1}{v}\left(f - Tf_T + \dfrac{1}{2}w^2\right) \end{bmatrix} \tag{3.110a}$$

$$Q = \begin{bmatrix} w/v \\ -f_v + w^2/v \\ \dfrac{w}{v}\left(f - Tf_T - vf_v + \dfrac{1}{2}w^2\right) \end{bmatrix} \tag{3.110b}$$

Decoding the third element of P is the only iterative step. Other parameters, such as p, s, and Γ, can be post-processed after the CFD computation is completed. In the mixture region, v is replaced with the quality as a dependent variable. The change from single phase to a mixture, whether at the coexistence curve or interior to it, may require damping if numerical instability is to be avoided.

3.6 COMPRESSIBLE LIQUID OR SOLID

A constant-density approximation is often used for a liquid or solid. This usually suffices in engineering practice. Situations arise, however, where the liquid or solid is relatively compressible and a constant density approximation is inadequate. For instance, liquid H_2 in the space shuttle main engine feed system enters a turbopump at 2.41×10^6 Pa and 26 K and leaves at 4.74×10^7 Pa and 56 K. During the compression process the liquid hydrogen has a density change of more than 40% (Kolcio and Helmicki, 1996). These authors demonstrate that a liquid eos, based on the approximations

$$\kappa_T = \text{constant} \tag{3.111a}$$

$$\beta = \text{thermal expansion coefficient}$$

$$= \frac{1}{v}\frac{\partial v}{\partial T_p} = \text{constant} \tag{3.111b}$$

is appropriate. Our discussion, however, is based on an earlier analysis by Kestin (1979). Other pertinent references are Flory et al. (1964), Macdonald (1969), and Boushehri and Mason (1993).

We start with

$$v = v(p, T) \tag{3.112}$$

$$dv = \frac{\partial v}{\partial p_T} dp + \frac{\partial v}{\partial T_p} dT$$

$$= -v\kappa_T dp + v\beta\, dT$$

or

$$d \ln v = -\kappa_T dp + \beta dT \tag{3.113}$$

Since $\ln v$ is a state property, we have

$$-\frac{\partial \kappa_T}{\partial T_p} = \frac{\partial \beta}{\partial p_T} \tag{3.114}$$

which is satisfied if κ_T and β are constants. In this circumstance, integration of Equation (3.113) yields

$$\ln \frac{v}{v_o} = \beta(T - T_o) - \kappa_T(p - p_o)$$

or

$$p = p_o + \frac{\beta}{\kappa_T}(T - T_o) - \frac{1}{\kappa_T}\ln\frac{v}{v_o} \tag{3.115}$$

as a thermal eos with the form of Equation (3.9).

Reciprocity now results in

$$\frac{\partial e}{\partial v_T} = T\frac{\partial p}{\partial T_v} - p = \frac{\beta}{\kappa_T}T_o - p_o + \frac{1}{\kappa_T}\ln\frac{v}{v_o} \tag{3.116}$$

After integration, we have

$$e = e_o\frac{v}{v_o} + \frac{v}{\kappa_T}\ln\frac{v}{v_o} + \int_{T_o}^{T} c_v^\circ dT' \tag{3.117a}$$

where

$$e_o = -\left(p_o - \frac{\beta}{\kappa_T}T_o + \frac{1}{\kappa_T}\right)v_o \tag{3.117b}$$

The constant volume specific heat is given by

$$c_v = \frac{\partial e}{\partial T_v} = c_v^\circ \tag{3.118}$$

The constant-pressure specific heat is provided by the general thermodynamic relation for a pure substance (Emanuel, 1987):

$$c_p = c_v + vT\frac{\beta^2}{\kappa_T} \tag{3.119}$$

Equation (3.25) becomes

$$s = g(v) + \int_{T_o}^{T}\frac{c_v}{T'}dT' \tag{3.120}$$

Equation (3.27) is used, to obtain

$$\frac{dg}{dv} = \frac{\beta}{\kappa_T}$$

which integrates to

$$g = \frac{\beta}{\kappa_T}v + \text{constant}$$

We thus have the rather simple result for the entropy

$$s = s_o + \frac{\beta}{\kappa_T}(v - v_o) + \int_{T_o}^{T}\frac{c_v}{T'}dT' \tag{3.121}$$

A self-consistent thermodynamic model is obtained in which β, κ_T, and $c_v(T)$ are empirically established. With e and s known, the enthalpy and Helmholtz potential are easily found. Application of Equation (3.72a) for the coexistence curve yields

$$v_g = v_f \tag{3.122}$$

which means the approximation, Equations (3.111), is incapable of producing a proper coexistence curve. This defect is important, e.g., if cavitation is of interest. Problem 3.3 presents an alternate approach that may not have this coexistence curve difficulty.

3.7 SECOND LAW

INTRODUCTORY DISCUSSION

In contrast to Chapter 2, it is convenient and useful to provide a fairly general second law formulation that also considers radiative heat transfer, chemical reactions, and molecular diffusion. We still assume a continuum fluid and Fourier's equation for conductive heat transfer, but a Newtonian fluid is not assumed until later. A thermally perfect gas is utilized only for illustrative purposes. Our discussion is partly based on Argrow et al. (1987).

For over a century, the second law has been recognized as a fundamental law of nature; nevertheless, it is not extensively used in fluid dynamics. Unlike the conservation equations, it is not required for obtaining a flow field solution. Typical illustrations of its limited usage are in ruling out expansion shock waves for a perfect gas, establishing criteria for transport coefficients, and in one-dimensional flows known as Rayleigh and Fanno flows, which are discussed in Section 13.5. Toward the end of this section, however, we will discuss the relevance of the second law to CFD, where its significance can be overlooked. As indicated, the second law is bypassed because CFD algorithms do not utilize it when obtaining a numerical solution, even though the solution may be in violation of it.

Because of the introduction of processes not considered in Chapter 2, a cursory reexamination of the energy equation is provided. After this, a general form of the second law is derived and discussed. The standard assumptions invoked in the first two chapters are reintroduced and the second law is limited to a viscous, heat-conducting flow. This section concludes with a derivation of bounds for the viscosity and thermal conductivity coefficients.

GENERAL FORM OF THE ENERGY EQUATION

Equation (2.39), written as

$$\rho \frac{D\hat{h}}{Dt} = \frac{D\hat{p}}{Dt} - \nabla \cdot \vec{q} + \Phi \tag{3.123}$$

is a suitable starting point, with the enthalpy $\hat{h}$ and pressure $\hat{p}$ now including radiative contributions. These are additive; i.e.,

$$\hat{h} = h + e_R, \qquad \hat{p} = p + p_R$$

where h and p have their usual definitions, e_R is the radiative energy density, and p_R is the radiative stress tensor. (Vincenti and Kruger, 1965, provide a more detailed discussion of radiation and chemistry than can be given here.) Except in special situations, such as in the interior of stars, these radiative contributions are quite negligible in comparison with h and p. The reason for this is that e_R and p_R are given by equations that involve a very small $1/c$ multiplicative factor, where c is the speed of light. We therefore neglect e_R and p_R and Equations (2.39) and (3.123) are then identical.

On the other hand, the radiative heat flux, $\vec{q}_R$, does not involve the $1/c$ factor and often is not negligible. It is given by

$$\vec{q}_R(\vec{r}, t) = \int_0^\infty d\nu \int_0^{4\pi} I_\nu \hat{\ell}\, d\Omega \tag{3.124}$$

where ν is the frequency of the radiation, I_ν is the radiative specific intensity at frequency ν in the differential solid angle $d\Omega$, and the unit vector $\hat{\ell}$ specifies the direction of propagation for I_ν. The specific intensity I_ν is provided by a radiative transport equation. In general, it is a function of $\vec{r}$, t, and ν, and at a given point of the flow field, I_ν depends on which way $\hat{\ell}$ is pointing.

Let y_α be the mass fraction of chemical species α in a fluid containing a total of N distinct species, and h_α be the specific enthalpy of species α. For the specific enthalpy of the mixture, we have

$$h = \sum_{\alpha=1}^{N} y_\alpha h_\alpha \tag{3.125a}$$

where

$$\sum_{\alpha=1}^{N} y_\alpha = 1 \tag{3.125b}$$

If the mixture is not dilute or is a mixture of thermally perfect gases, then h_α needs to be replaced with the partial molar enthalpy of species α. However, the dilute assumption is sufficient for the gaseous flows typically encountered in engineering.

Two processes, molecular diffusion and chemical reactions, govern $y_\alpha(\vec{r}, t)$. For diffusion, we need the mass diffusion flux vector for species α, $\vec{j}_\alpha$, which is given by

$$\vec{j}_\alpha = \rho y_\alpha(\vec{w}_\alpha - \vec{w}) \tag{3.126}$$

where $\vec{w}_\alpha(\vec{r},t)$ is the average velocity of species α and ρy_α is the density of species α. Observe that

$$\sum_\alpha \vec{j}_\alpha = 0 \tag{3.127}$$

is required in order for $\vec{w}$ to be the mass-averaged velocity of the mixture. Thus, $\vec{j}_\alpha$ provides the diffusional flux of species α relative to the average mixture velocity. For notational convenience, we hereafter write the species summation without the 1 and N limits; these limits are understood. (See Bird et al., 1960, for further details on the constitutive equations for mass diffusion.)

For the chemical reactions, we introduce the time rate of change of the mass of species α, $\dot{\omega}_\alpha$, due to reactions. Thus, $\dot{\omega}_\alpha$ is a sum over all reactive processes that alter y_α and requires a detailed knowledge of the actual chemical kinetics that are present in the mixture. The two processes can be combined into a rate equation

$$\rho \frac{Dy_\alpha}{Dt} = \rho \dot{\omega}_\alpha - \nabla \cdot \vec{j}_\alpha, \qquad \alpha = 1,\ldots,N \tag{3.128}$$

for y_α. We view the $\rho \dot{\omega}_\alpha$ term as source or sink terms for species α. The rest of the equation is analogous to continuity; e.g., the $\nabla \cdot \vec{j}_\alpha$ term arises from applying the divergence theorem. This term provides the time rate of change of the mass of species α in a fluid particle due to diffusion of species α into or out of the particle. Both $\dot{\omega}_\alpha$ and $\nabla \cdot \vec{j}_\alpha$ may be positive or negative according to whether or not these processes are increasing or decreasing the mass of α in the fluid particle.

The heat flux vector in Equation (2.39) now consists of a conductive contribution given by Fourier's equation, a radiative contribution, and a mass transfer contribution, given by

$$\sum_\alpha h_\alpha \vec{j}_\alpha$$

We thus have

$$\vec{q} = -\kappa \nabla T + \vec{q}_R + \sum_\alpha h_\alpha \vec{j}_\alpha \tag{3.129}$$

Shortly, we shall need another flux vector $\vec{q}^*$, defined by

$$\vec{q} = \vec{q}^* + \sum_\alpha \mu_\alpha \vec{j}_\alpha \tag{3.130a}$$

or

$$\vec{q}^* = -\kappa \nabla T + \sum_\alpha (h_\alpha - \mu_\alpha) \vec{j}_\alpha + \vec{q}_R \tag{3.130b}$$

where μ_α is the chemical potential of species α, per unit mass of species α. For a mixture of thermally perfect gases, the chemical potential of species α is (Emanuel, 1987)

$$\mu_\alpha = h_\alpha - Ts_\alpha \tag{3.131}$$

where s_α is the specific entropy of species α. Thus, the mass transfer term in Equation (3.130b) becomes

$$\sum_\alpha (h_\alpha - \mu_\alpha) \vec{j}_\alpha = T \sum_\alpha s_\alpha \vec{j}_\alpha \tag{3.132}$$

and provides in $\vec{q}^*$ the diffusive flux of entropy in contrast to the enthalpy flux contained in $\vec{q}$. [We note that Equation (3.131), which is for a thermally perfect gas, is not essential in the subsequent analysis.]

The reformulation of the energy equation is completed by noting that

$$\Phi = \overleftrightarrow{\tau} : (\nabla \vec{w})^t \tag{3.133}$$

in which case, $\overleftrightarrow{\tau}$ need not be a symmetric tensor. The energy equation is still provided by Equation (3.123) but with the carets deleted. In this formulation, the mass fractions y_α are determined by Equation (3.128), $\vec{j}_\alpha$ is determined by Equation (3.126), $\vec{q}$ is given by Equation (3.129), and $\vec{q}_R$ is determined by Equation (3.124). The dependent variables are p, ρ, T, $\vec{w}$, I_ν, and y_α, while $\vec{r}$, t, and $\hat{\ell}$ are the independent variables. Additional relations, of course, are also required for $\vec{w}_\alpha$, h_α, $\dot{\omega}_\alpha$, μ_α, etc.

General Form of the Second Law and Entropy Production

The first part of the second law postulates a thermodynamic state variable, the entropy, defined by Equation (2.35b) as

$$ds = \frac{(\delta q)_{\text{rev}}}{T} \tag{3.134}$$

This definition was originally for a simple, closed system, where $(\delta q)_{\text{rev}}$ is the reversible heat transfer that crosses the system boundary and T is the absolute temperature of the surrounding medium at the boundary. From this definition and the first law of thermodynamics, the relation for a closed system

$$Tds = dh - \frac{dp}{\rho}$$

is obtained. This equation can be extended to an open system by writing it as

$$Tds = dh - \frac{dp}{\rho} - \sum_\alpha \mu_\alpha dy_\alpha \tag{3.135}$$

If h and p are held constant, then $-\mu_\alpha/T$ provides the entropy change associated with a compositional change dy_α in species α. Thus, the chemical potential enables us to consider mass transfer across the boundary as well as compositional changes within the system. This extension is essential if chemical reactions, phase changes, or diffusion are present. Moreover, the heat transfer in Equation (3.134) is now not restricted to conduction but may encompass any reversible heat transfer process including mass diffusion, chemical reactions, and the transport of radiative energy.

For the thermodynamic system, we utilize an infinitesimal fluid particle that moves with velocity $\vec{w}$ but may have diffusional fluxes at its boundary. In this circumstance, we can replace the thermodynamic derivatives in Equation (3.135) with the substantial derivative, to obtain

$$T\frac{Ds}{Dt} = \frac{Dh}{Dt} - \frac{1}{\rho}\frac{Dp}{Dt} - \sum_\alpha \mu_\alpha \frac{Dy_\alpha}{Dt} \tag{3.136a}$$

We now use Equation (3.123), without the carets, to eliminate Dh/Dt and Equation (3.128) to eliminate Dy_α/Dt, with the result

$$T\frac{Ds}{Dt} = \frac{1}{\rho}\left(\frac{Dp}{Dt} - \nabla\cdot\vec{q} + \Phi\right) - \frac{1}{\rho}\frac{Dp}{Dt} - \sum_\alpha \mu_\alpha\left(\dot{\omega}_\alpha - \frac{1}{\rho}\nabla\cdot\vec{j}_\alpha\right)$$

or

$$\rho\frac{Ds}{Dt} = \frac{1}{T}\left(\Phi - \rho\sum_\alpha \mu_\alpha\dot{\omega}_\alpha - \nabla\cdot\vec{q} + \sum_\alpha \mu_\alpha \nabla\cdot\vec{j}_\alpha\right) \tag{3.136b}$$

This relation provides the rate of change of entropy of a fluid particle, where part of this change is due to the transport of entropy across the system boundary, as given by $(\delta q)_{\text{rev}}/T$. In order to focus on entropy transport, in contrast to energy transport, we replace $\vec{q}$ with $\vec{q}^{\,*}$. This alteration is conveniently accomplished by writing the two rightmost terms in the above equation as

$$
\begin{aligned}
-\nabla\cdot\vec{q}+\sum_\alpha \mu_\alpha\nabla\cdot\vec{j}_\alpha &= -\nabla\cdot\left(\vec{q}^{\,*}+\sum_\alpha \mu_\alpha\vec{j}_\alpha\right)+\sum_\alpha \mu_\alpha\nabla\cdot\vec{j}_\alpha\\
&= -\nabla\cdot\vec{q}^{\,*}-\sum_\alpha \vec{j}_\alpha\cdot\nabla\mu_a-\sum_\alpha \mu_\alpha\nabla\cdot\vec{j}_\alpha+\sum_\alpha \mu_\alpha\nabla\cdot\vec{j}_\alpha\\
&= -\nabla\cdot\vec{q}^{\,*}-\sum_\alpha \vec{j}_\alpha\cdot\nabla\mu_\alpha
\end{aligned}
\tag{3.137}
$$

to obtain

$$
\rho\frac{Ds}{Dt} = \frac{1}{T}\left[\Phi-\sum_\alpha(\rho\mu_\alpha\dot{\omega}_\alpha+\vec{j}_\alpha\cdot\nabla\mu_\alpha)\right]-\frac{1}{T}\nabla\cdot\vec{q}^{\,*} \tag{3.136c}
$$

In Equation (3.134), the temperature is that of the surface of the system. In $\nabla\cdot\vec{q}^{\,*}/T$, the temperature is that of the interior of the fluid particle; hence, this term is not the counterpart of $(\delta q)_{\text{rev}}/T$. We observe that the quantity we seek is provided by the surface integral

$$
\oint_{\delta s}\hat{n}\cdot\left(\frac{\vec{q}^{\,*}}{T}\right)ds
$$

where δs is the surface area of a fluid particle. By means of Equation (1.63), we have

$$
\nabla\cdot\left(\frac{\vec{q}^{\,*}}{T}\right) = \lim_{\delta v\to 0}\frac{1}{\delta v}\oint_{\delta s}\hat{n}\cdot\left(\frac{\vec{q}^{\,*}}{T}\right)ds \tag{3.138}
$$

Thus, $\nabla\cdot(\vec{q}^{\,*}/T)$ corresponds to the entropy transport into or out of the particle. From the definition of $\vec{q}^{\,*}$, Equation (3.130b), we observe that heat conduction, mass transfer, and radiative heat transfer contribute to the entropy transport, but that viscous effects and chemical reaction do not. As observed in the previous section, the mass transfer contribution is associated with an entropy flux

$$
\frac{1}{T}\sum_\alpha(h_\alpha-\mu_\alpha)\vec{j}_\alpha = \sum_\alpha s_\alpha\vec{j}_\alpha \tag{3.139}
$$

when μ_α is replaced by Equation (3.131).

We introduce $\nabla\cdot(\vec{q}^{\,*}/T)$ into Equation (3.136c) by means of the identity

$$
\nabla\cdot\left(\frac{\vec{q}^{\,*}}{T}\right) = \frac{1}{T}\nabla\cdot\vec{q}^{\,*}+\vec{q}^{\,*}\cdot\nabla\frac{1}{T} = \frac{1}{T}\nabla\cdot\vec{q}^{\,*}-\frac{1}{T^2}\vec{q}^{\,*}\cdot\nabla T
$$

to obtain

$$
-\frac{1}{T}\nabla\cdot\vec{q}^{\,*} = -\nabla\frac{\vec{q}^{\,*}}{T}-\frac{1}{T^2}\vec{q}^{\,*}\cdot\nabla T \tag{3.140}
$$

With these relations, Equation (3.136c) becomes

$$\rho\frac{Ds}{Dt} = \rho\dot{s}_{\text{irr}} - \nabla\cdot\frac{\vec{q}^{\,*}}{T} \tag{3.141}$$

where the rate of entropy production, per unit volume of the mixture, is given by

$$\rho\dot{s}_{\text{irr}} = \frac{1}{T}\left[\Phi - \frac{1}{T}\vec{q}^{\,*}\cdot\nabla T - \sum_{\alpha}(\rho\mu_{\alpha}\dot{\omega}_{\alpha} + \vec{j}_{\alpha}\cdot\nabla\mu_{\alpha})\right] \tag{3.142}$$

Our presentation of the first part of the second law, which started with Equation (3.134), has culminated in the above equations. The second part of the law states that

$$\dot{s}_{\text{irr}} = 0 \tag{3.143a}$$

for a reversible process and that

$$\dot{s}_{\text{irr}} > 0 \tag{3.143b}$$

for an irreversible process. Although $\dot{s}_{\text{irr}}$ is nonnegative, Ds/Dt can be negative, since the divergence term can have either sign.

The rate of entropy production, $\dot{s}_{\text{irr}}$, represents the irreversible processes that occur within a fluid particle. Hence, the entropy changes are due to internal processes and entropy transport into or out of the particle. With the assistance of the methods discussed in Section 2.3, Equation (3.141) can be written as

$$\frac{D}{Dt}\int_{V}\rho s dv = \int_{V}\rho\dot{s}_{\text{irr}}dv - \oint_{S}\hat{n}\cdot\left(\frac{\vec{q}^{\,*}}{T}\right)ds \tag{3.144}$$

By way of illustration, consider a mixture of thermally perfect gases that are diffusionally mixing. With this as the only process, we have

$$\vec{q}^{\,*} = \sum_{\alpha}(h_{\alpha} - \mu_{\alpha})\vec{j}_{\alpha} = T\sum_{\alpha}s_{\alpha}\vec{j}_{\alpha}$$

or

$$\frac{\vec{q}^{\,*}}{T} = \sum_{\alpha}s_{\alpha}\vec{j}_{\alpha} \tag{3.145}$$

where Equation (3.131) is utilized. Consequently, an $\hat{n}\cdot\vec{j}_{\alpha}$ appears in the rightmost term in Equation (3.144). Equation (3.142) now reduces to

$$\begin{aligned}\rho\dot{s}_{\text{irr}} &= \frac{1}{T}\left(-\frac{1}{T}\vec{q}^{\,*}\cdot\nabla T - \sum_{\alpha}\vec{j}_{\alpha}\nabla\cdot\mu_{\alpha}\right) = -\frac{1}{T}\left[\sum_{\alpha}s_{\alpha}j_{\alpha}\cdot\nabla T + \sum_{\alpha}\vec{j}_{\alpha}\cdot\nabla(h_{\alpha} - Ts_{\alpha})\right]\\ &= \sum_{\alpha}\vec{j}_{\alpha}\cdot\left(\nabla s_{\alpha} - \frac{1}{T}\nabla h_{\alpha}\right)\end{aligned} \tag{3.146}$$

Thus, a dot product involving $\vec{j}_{\alpha}$ appears in each of the terms on the right-hand side.

Viscous effects and chemical reactions enter through Φ and $\dot{\omega}_\alpha$, respectively, and are internal processes that appear only in $\dot{s}_{irr}$. Evidently, the contribution of viscous dissipation to $\dot{s}_{irr}$ is zero when both μ and μ_b are zero or when the velocity gradient is zero. The contribution from reactions is zero when the flow is chemically frozen, i.e., all $\dot{\omega}_\alpha$ are zero, or when the reactions are in equilibrium, in which case the sum $\Sigma\mu_\alpha\dot{\omega}_\alpha$ is zero. Conductive heat transfer, radiative heat transfer, and mass diffusion, through $\vec{q}^*$, appear in both $\dot{s}_{irr}$ and in $\nabla \cdot (\vec{q}^*/T)$. The analysis has not ruled out body forces or restricted the acceleration, $D\vec{w}/Dt$, in any manner. [Remember that $\vec{F}_b$ does not appear in Equation (2.39).] A purely accelerative inviscid flow or the work due to a body force does not result in any (irreversible) production of entropy.

Because q is *not* a state variable, we write $(\delta q_{rev})/dt$ as $\dot{q}_{rev}$ rather than as Dq_{rev}/Dt, which would be wrong, when following a fluid particle. Thus, Equation (3.134) can be written as

$$T\frac{Ds}{Dt} = \dot{q}_{rev} \tag{3.147}$$

and $T(Ds/Dt)$ represents the amount of heat gained, per unit time and per unit mass, by a fluid particle undergoing a reversible process. By comparing this result with Equation (3.141), we have

$$\dot{q}_{rev} = -\frac{T}{\rho}\nabla \cdot \frac{\vec{q}^*}{T} \tag{3.148}$$

for a reversible process, where $\dot{s}_{irr} = 0$. This relation is not equivalent to Equation (2.37b), which holds for an irreversible or reversible process of a closed, simple system.

For a general process, $T(Ds/Dt)$ still represents the heat transfer to a fluid particle. We see from Equation (3.141) that the heat gained is due to $\dot{s}_{irr}$ and $\vec{q}^*$. It is tempting to partition the heat transfer between reversible and irreversible contributions, as was done with the work [see Equation (2.37a)]. Such a decomposition, however, is not justified, as can be seen from the following argument. Consider a process consisting solely of irreversible conductive heat transfer. In this case, we have

$$\vec{q}^* = \vec{q} = -\kappa\nabla T \tag{3.149}$$

which appears in both the $\dot{s}_{irr}$ term and the divergence term on the right side of Equation (3.141). Appearing, as it does, in both terms, the heat transfer $\vec{q}$ cannot be split into reversible and irreversible components.

Suppose a flow field has been analytically or numerically determined. Then $\dot{s}_{irr}(\vec{r},t)$ is given by Equation (3.142), which is an algebraic equation. At all times and at all points of the flow field, Equations (3.143) must hold; i.e., $\dot{s}_{irr}$ cannot be negative. With a known flow field solution, we can determine $\dot{s}_{irr}(\vec{r},t)$, and with the aid of Equation (3.141), we can determine $s(\vec{r},t)$. Thus, $\dot{s}_{irr}$ and s can be evaluated while a solution is being (numerically) obtained or evaluated afterward.

As previously indicated, our formulation is particularly suitable for ensuring that CFD codes provide a solution that does not violate the second law. There are a variety of natural processes, such as in a chemically reacting boundary layer, that will produce entropy. There are also a variety of sources for numerically produced entropy, including artificial viscosity or damping terms, errors due to roundoff and truncation, numerical instabilities, and the presence of discontinuities like a shock wave. The entropy production from some of these numerical sources, such as in the vicinity of a discontinuity, can result in both negative and positive entropy production (Cox and Argrow, 1993).

It has become common practice to compute a steady-flow solution by using the nonsteady equations of motion. During the lengthy nonsteady computation, the numerically produced entropy may gradually accumulate. The validity of the computation is then uncertain unless the numerically produced entropy, at termination, is still small in comparison to that produced by natural processes. Apparently, it is possible for the numerically produced entropy to become sufficiently negative to cause a violation of the second law (Powell et al., 1987).

Second Law for a Viscous, Heat Conducting Flow

We return to our basic assumptions of Chapter 1 and suppose only viscous stresses and conductive heat transfer are present. In this circumstance, Equations (3.141) and (3.142) become

$$\rho \frac{Ds}{Dt} = \rho \dot{s}_{\text{irr}} - \nabla \cdot \frac{\vec{q}}{T} \tag{3.150a}$$

$$\rho \dot{s}_{\text{irr}} = \frac{1}{T}\left(\Phi - \frac{1}{T}\vec{q} \cdot \nabla T\right) \tag{3.151a}$$

where $\vec{q}$ is now the conductive heat flux. If we further assume Fourier's equation and a Newtonian fluid, we have

$$\rho \frac{Ds}{Dt} = \rho \dot{s}_{\text{irr}} + \nabla \cdot \left(\frac{\kappa}{T}\nabla T\right) \tag{3.150b}$$

$$\rho \dot{s}_{\text{irr}} = \frac{1}{T}\left(\Phi + \frac{\kappa}{T}(\nabla T)^2\right) \tag{3.151b}$$

where Φ is provided by Equation (2.32a) or (2.32b).

The second law requires $\dot{s}_{\text{irr}} \geq 0$ for any realizable process. By detailed balancing, this must hold individually for the heat transfer and the viscous work. We, therefore, require

$$\kappa \geq 0, \qquad \Phi \geq 0 \tag{3.152}$$

For heat conduction, the second law is satisfied providing the coefficient of thermal conductivity is nonnegative.

From Equation (2.32b), it would appear that the $\Phi \geq 0$ condition is satisfied if μ and λ are nonnegative. While μ must be nonnegative, λ may be negative as is evident from the Stokes hypothesis, which presumes $\lambda = -(2\mu/3)$.

To determine the minimum allowed value for λ, we consider a purely dilatational motion, in which case

$$\frac{\partial w_i}{\partial x_j} = \delta_{ij} a \tag{3.153}$$

where a is a constant. In such a flow, all shearing stress terms are zero. For this flow, Equation (2.32b) readily yields

$$\Phi = \mu[2(3a^2)] + \lambda(3a)^2 = 9\left(\lambda + \frac{2}{3}\mu\right)a^2 = 9\mu_b a^2 \tag{3.154}$$

where the bulk viscosity, μ_b, is defined by Equation (1.57). However, any constraint on μ, λ, and κ must be independent of the assumed flow model, since these parameters are material properties. A necessary condition, therefore, for $\Phi \geq 0$ is that $\mu_b \geq 0$.

Observe from Equation (2.32b) that the shearing velocity derivatives occur only in the μ part of Φ and then always in squared terms. Hence, the μ term is a minimum when these shearing derivatives are zero and Φ is similarly minimized. This leaves only the dilatation terms, which are dealt with in the above paragraph. Thus, $\mu \geq 0$ and $\mu_b \geq 0$ are the necessary and sufficient conditions for $\Phi \geq 0$. In turn, we see that the second law simply requires

$$\kappa \geq 0, \qquad \mu \geq 0, \qquad \mu_b \geq 0 \tag{3.155}$$

for a flow where Fourier's equation and a Newtonian fluid are utilized. Once these relations are satisfied, the corresponding conservation equations of Chapter 2 cannot yield an analytical solution that violates the second law. Of course, a numerical solution is a different matter, since numerical processes, such as artificial damping, are involved that are not present in the conservation equations.

The above discussion does not conflict with the earlier remarks about applying the second law to an expansion shock or to Fanno and Rayleigh flows. The first of these applications is to the algebraic jump conditions across a shock, while the others are averaged, one-dimensional flows. In none of these instances are the conservation equations of Chapter 2 being directly utilized.

Like Φ, $\dot{s}_{\text{irr}}$ is proportional to the square or product of gradient terms. It is, therefore, significant in the flow interior to a shock wave or in a high-speed boundary layer. However, $\dot{s}_{\text{irr}}$ is of second order for a slightly perturbed uniform or quiescent flow; see Problem 3.9. As mentioned, this is a consequence of $\dot{s}_{\text{irr}}$ being proportional to the square or product of gradient terms. Thus, when these terms are negligible, as is often the case in fluid dynamics, we have

$$\dot{s}_{\text{irr}} = 0, \qquad \frac{Ds}{Dt} = 0 \tag{3.156}$$

The entropy of a fluid particle is now a constant and the flow is referred to as isentropic. These two relations are basic for the analysis in Part II.

REFERENCES

Argrow, B.M., "Computational Analysis of Dense Gas Shock Tube Flow," *Shock Waves* 6, 241 (1996).

Argrow, B.M., Emanuel, G., and Rasmussen, M.L., "Entropy Production in Nonsteady General Coordinates," *AIAA J.* 25, 1629 (1987).

Bird, R.B., Stewart, W.E., and Lightfoot, E.N., *Transport Properties*, John Wiley, New York, 1960, Section 18.4.

Boushehri, A. and Mason, E.A., "Equation of State for Compressed Liquids and Their Mixtures from the Cohesive Energy Density," *Int. J. Thermophys.* 14, 685 (1993).

Cox, R.A. and Argrow, B.M., "Entropy Production in Finite-Difference Schemes," *AIAA J.* 31, 210 (1993).

Emanuel, G., *Advanced Classical Thermodynamics*, AIAA Education Series, Washington, D.C., 1987.

Emanuel, G., "Analysis of a Critical Point with Applications to Fluid Dynamics," The University of Oklahoma, The School of Aerospace and Mechanical Engineering, AME Report 96–1 (1996).

Fickett, W. and Davis, W.C., *Detonation*, University of California Press, Berkeley, 1979.

Flory, P.J., Orwoll, R.A., and Vrij, A., "Statistical Thermodynamics of Chain Molecule Liquids. I. An Equation of State for Normal Paraffin Hydrocarbons," *J. Am. Chem. Soc.* 86, 3507 (1964).

Kestin, J., *A Course in Thermodynamics*, Vol. II, Revised Printing, Hemisphere Pub. Co., 1979, pp. 262–268.

Kolcio, K. and Helmicki, A.J., "Development of Equations of State for Compressible Liquids," *J. Prop.* 12, 213 (1996).

Macdonald, J.R., "Review of Some Experimental and Analytical Equations of State," *Rev. Modern Phys.* 41, 316 (1969).

Mader, C.L., *Numerical Modeling of Detonations*, University of California Press, Berkeley, 1979.
Martin, J.J., Kapoor, R.M., and Nevers, N.D., "An Improved Equation of State for Gases," *A.I.Ch.E. J.* 5, 159 (1959).
Menikoff, R. and Plohr, B.J., "The Riemann Problem for Fluid Flow of Real Materials," *Rev. Modern Phys.* 61, 75 (1989).
Merkle, C.L., Sullivan, J.Y., Buelow, P.E.O., and Venkateswaran, S., "Computation of Flows with Arbitrary Equations of State," *AIAA J.* 36, 515 (1998).
Powell, K.G., Murman, E.M., Perez, E.S., and Baron, J.R., "Total Pressure Loss in Vortical Solutions of the Conical Euler Equations," *AIAA J.* 25, 360 (1987).
Swesty, F.D., "Thermodynamically Consistent Interpolation for Equation of State Variables," *J. Comp. Phys.* 127, 118 (1996).
Vincenti, W.G. and Kruger, C.H., Jr., *Introduction to Physical Gas Dynamics*, John Wiley, New York, 1965.

PROBLEMS

3.1 Start with Equation (3.40) and obtain the Gibbs–Duhem equation

$$sdT - vdp + \sum_i n_i d\bar{\mu}_i = 0$$

Obtain the integrated form, comparable to Equation (3.40), for the enthalpy and its corresponding Gibbs–Duhem equation.

3.2 Write P and Q, Equations (3.110), explicitly in terms of T_r, v_r, w, and constants for a van der Waals fluid. Do not assume a constant value for c_v°. Simplify your results.

3.3 Assume a liquid or solid satisfies the following relations:

$$\beta_s = \frac{1}{v}\frac{\partial v}{\partial s_p} = \text{constant}, \qquad \kappa_s = \text{constant}$$

(a) Obtain a state equation of the form $p = p(v,s)$.
(b) Use Equation (3.1) to obtain equations of the form $e = e(s,v)$ and $T = T(s,v)$. These equations will involve an arbitrary function $g(s)$.
(c) Obtain $c_v(s,v)$ and $p = f(T,v)$.
(d) Derive the equations for the coexistence curve.

3.4 Consider a perfect gas between two infinite parallel walls. All quantities depend only on x, and only heat conduction and radiative heat transfer occur. With the additional gray-gas approximation, we can write for the radiation

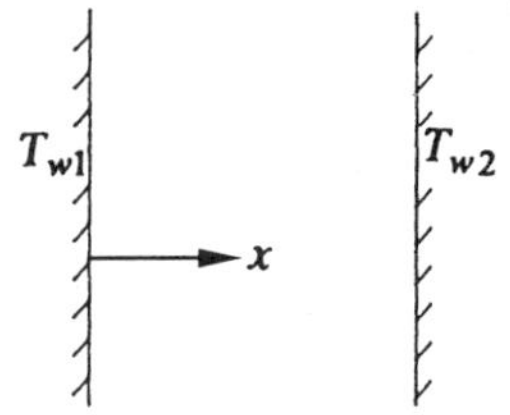

$$\vec{q}_R = q_R[x, T(x)]\hat{1}_x$$

$$q_R = 2\sigma\left[T_{w1}E_3(\eta) + \int_0^\eta T^4 E_2(\eta - \tilde{\eta})d\tilde{\eta} - \int_\eta^\infty T^4 E_2(\tilde{\eta} - \eta)d\tilde{\eta}\right]$$

where the E_i are the exponential integral functions.
(a) Derive an algebraic equation for $\dot{s}_{\text{irr}}$ except for the two integrals that appear in q_R.
(b) Assume a continuous temperature and $T_{w2} > T_{w1}$. Determine the sign of $\dot{s}_{\text{irr}}$.
(c) Derive an algebraic equation for ds/dx and show that s varies linearly with x only if $T\kappa$ equals a constant.

3.5 Continue with Problem 2.9 and further assume $v_\theta = v_\phi = 0$, and that the flow, including μ, λ, and κ, depends only on r and t.

(a) Evaluate

$$\frac{D\psi}{Dt}, \quad \nabla \cdot \vec{w}, \quad \alpha_i, \quad \Phi, \quad \dot{s}_{\text{irr}}$$

where ψ is an arbitrary scalar, the α_i are defined in Problem 2.8, and $\dot{s}_{\text{irr}}$ includes only heat conduction and viscous processes.

(b) Neglect body forces and write the conservation equations in scalar form.

(c) Assume the flow is steady, ρ, μ, κ, and c_v are constants, and e equals $c_v T$. Determine a solution of the equations in part (b) for v_r and p, and establish a differential equation for $T(r)$.

3.6 As shown in the sketch, we have a stratified Couette flow of a perfect gas, where

$$c_p = \frac{\gamma R}{\gamma - 1}, \quad \frac{\mu}{\mu_\infty} = \frac{T}{T_\infty}, \quad \frac{\kappa}{\kappa_\infty} = \frac{T}{T_\infty}$$

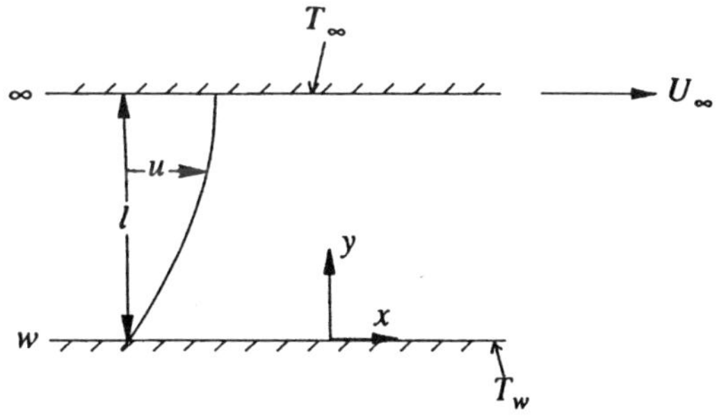

This is a steady flow with a constant pressure whose velocity is given by

$$\vec{w} = u(y)\hat{1}_x$$

The solution for $u(y)$ and $T(y)$ can be written as

$$Y = \frac{y}{\ell} = \frac{2}{1 + b + \frac{a}{3}} V\left[b + \frac{1}{2}(1 - b + a)V - \frac{a}{3}V^2\right]$$

$$\theta = \frac{T - T_w}{T_\infty - T} = \frac{1}{1 - b} V(1 - b + a - aV)$$

where

$$V = \frac{\mu}{U_\infty}, \quad Pr = \frac{c_p \mu_\infty}{\kappa_\infty}, \quad M_\infty = \frac{U_\infty}{(\gamma R T_\infty)^{1/2}}, \quad a = \frac{\gamma - 1}{2} Pr M_\infty^2, \quad b = \frac{T_w}{T_\infty}$$

Show that $\dot{s}_{\text{irr}}$ is nonnegative and discuss the possibility that the second law may place restrictions on the values for a and b.

3.7 Consider a steady, two-dimensional, parallel viscous flow, where the velocity is in the x-direction and all flow properties vary only with y, and assume Fourier's equation and a compressible Newtonian fluid. Further assume the flow is thermodynamically reversible. Use the second law to determine the solution for $\vec{w}$ and T.

3.8 Consider a steady, viscous flow without body forces. The fluid is between two concentric infinitely long rotating cylinders, as shown in the sketch. Use a cylindrical coordinate system and assume a Newtonian fluid and Fourier's equation.

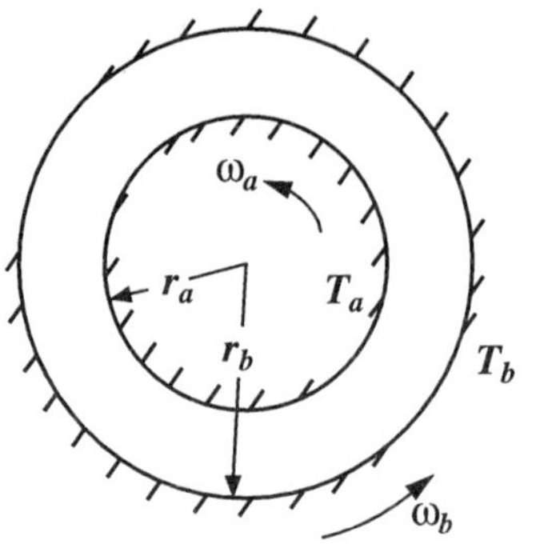

(a) Derive the appropriate governing differential equations assuming a variable density and that the flow depends only on r.

(b) Assume constant values for μ, λ, and κ, and a perfect gas. Determine algebraic equations for $w(r)$ and $T(r)$, where there is no velocity slip or temperature jump at either wall. Determine a quadrature solution for $p(r)$.

(c) Determine an algebraic equation for $\dot{s}_{irr}$.

3.9 Determine an equation for $\dot{s}_{irr}$ to second order using the first-order perturbation results of Problem 2.15.

3.10 Consider the throat region of a supersonic nozzle whose cross-sectional area is

$$\frac{A}{A^*} = 1 + \frac{x^2}{\ell^2}$$

where A^* is the throat area, x is distance along the symmetry axis, and ℓ is a reference length. Assume steady, inviscid, one-dimensional flow of a perfect gas. Also assume all transport properties are constants. Utilize standard isentropic relations to evaluate $\dot{s}_{irr}$ at the throat in terms of the transport properties, γ, R, ℓ, and the stagnation conditions T_o and p_o. The standard isentropic equations are

$$M = \frac{w}{a}$$

$$a = (\gamma R T)^{1/2}$$

$$X = 1 + \frac{\gamma - 1}{2} M^2$$

$$\frac{T}{T_o} = X^{-1}$$

$$\frac{p}{p_o} = X^{-\gamma/(\gamma-1)}$$

$$\frac{A}{A^*} = \left(\frac{2}{\gamma+1}\right)^{\frac{\gamma+1}{2(\gamma-1)}} \frac{1}{M} X^{\frac{\gamma+1}{2(\gamma-1)}}$$

3.11 Two solid bars of cross-sectional area A are brought into thermal contact at time $t = 0$, as shown in the sketch, where x_I and x_{II} are positive lengths. Let c be the specific heat, and assume the following properties or parameters for both bars are constant and are known:

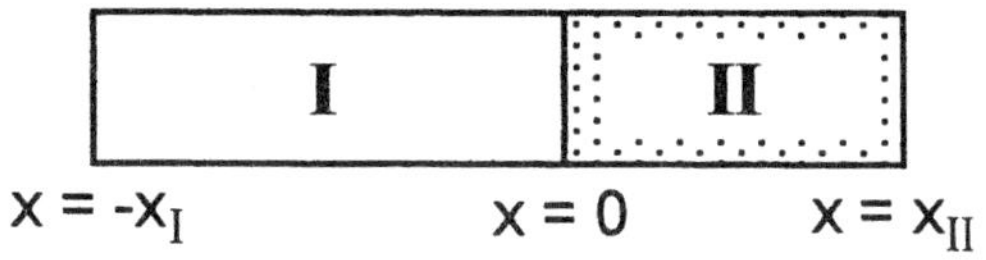

$$\rho,\ c,\ \kappa,\ A,\ T_i,\ x_I,\ X_{II}$$

where $A_I = A_{II}$, and $\rho_I \neq \rho_{II}$, ..., and i and f denote initial and final conditions. The external surface of the bars is well insulated. Assume $T_{Ii} > T_{IIi}$ and that only one-dimensional heat conduction occurs.

(a) Determine the final equilibrium temperature T_f in terms of the above constants. To standardize notation, utilize

$$\alpha = \text{thermal diffusivity} = \frac{\kappa}{\rho c}, \qquad \beta = \frac{(\rho c x)_I}{(\rho c x)_{II}}$$

where $\alpha_I \neq \alpha_{II}$.

(b) Establish an energy equation for $T(x,t)$ and the appropriate boundary and initial conditions. Note that at $x = 0$ we have contact resistance, where only the heat flux is continuous.

(c) For each bar, determine the specific entropy change $s(x,t)–s_i$ in terms of the temperature. Determine the total entropy change ΔS undergone by the two bars.

(d) Obtain a solution for $T(x,t)$ for both bars at small time, or, equivalently, set $x_I, x_{II} \to \infty$. In this circumstance, a similarity solution is possible, which is much simpler than a separation of variable solution, which would be required for finite x_I, x_{II} values. A new boundary condition at $x = 0$, $T_I = T_{II}$, needs to be added when $t > 0$.

(e) Use the solution of part (d) to obtain $\dot{s}_{\text{irr}}$ for bar I.

3.12 Develop a thermodynamic model for a Martin–Hou fluid. In other words, establish equations for p, s, e, h, c_v, c_p, a^2, and Γ that are functions of only T, v, and $c_v^\circ(T)$. Do not assume a harmonic oscillator model. Simplify your answers as much as possible.

4 Kinematics

4.1 PRELIMINARY REMARKS

We examine a number of general results whose validity does not depend on the conservation equations or the second law of thermodynamics. Such results are termed kinematic. Their generality is such that they hold for incompressible as well as compressible flows, steady or unsteady flows, viscous and inviscid flows, and flows with body forces. On the surface, some of these theorems appear to be limited to incompressible flows; however, this restriction is not accurate. They did, however, originate for incompressible flows in the middle of the 19th century, primarily by Kelvin and Helmholtz.

One important application of these theorems is in wing theory, where the development of lift is closely associated with shed vortices. Aside from this, these theorems are not widely utilized. As is the case with the second law, they are not needed for obtaining a flow field solution. Nevertheless, they are of fundamental importance, as is the second law.

Before embarking on a discussion of these theorems, it is convenient to introduce common definitions and to prepare some of the mathematical apparatus that will be needed. This is done in the next section. The remaining two sections are devoted to the Kelvin and Helmholtz theorems.

4.2 DEFINITIONS

FIELD LINES

Let C be a curve in space and let $d\vec{r}$ be tangent to C. Consider an arbitrary vector field $\vec{F}(\vec{r},t)$ that is single valued. The curve C is then a field line of $\vec{F}$ if $d\vec{r}$ is parallel to $\vec{F}$. This condition can be expressed in two equivalent ways:

$$d\vec{r} \times \vec{F} = 0, \quad d\vec{r} = dg\vec{F} \tag{4.1a,b}$$

where g is a scalar function. In a steady flow with $\vec{F} = \vec{w}$, the streamlines are field lines. (If the flow is unsteady, the pathlines are the $\vec{F} = \vec{w}$ field lines.) With Cartesian coordinates, Equation (4.1a) yields Equation (1.21b) for the streamlines. These same equations are obtained from Equation (4.1b) with $\vec{F} = \vec{w}$ by writing $|d\vec{r}|/|\vec{w}|$ for the differential dg.

Let C' be a second curve in space that is not tangent to any field line. The surface consisting of the field lines that pass through C' is called a vector surface. If C' is a simple closed curve, the vector surface is a vector tube. A filament is a vector tube in which C' has shrunk to a point.

As shown in Figure 4.1, a tube can have two different types of simple curves on its surface. The curve C' is referred to as reducible, since it can be shrunk to a point without leaving the surface. The curve C is called irreducible, because such shrinking cannot be done.

SOLENOIDAL VECTORS

A divergence-free vector field $\vec{F}$ is referred to as solenoidal; i.e.,

$$\nabla \cdot \vec{F} = 0 \tag{4.2}$$

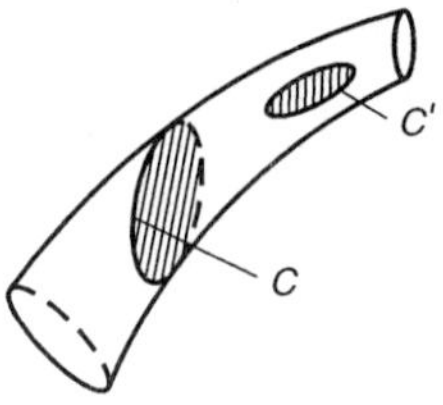

FIGURE 4.1 Vector tube with reducible (C') and irreducible (C) curves on its surface.

By means of the divergence theorem, we have

$$\int_V \nabla \cdot \vec{F} dv = \oint_S \hat{n} \cdot \vec{F} ds = 0$$

for a solenoidal field, providing S is a closed surface. We apply this result to a slice of a vector tube. As shown in Figure 4.2, the tube is enclosed by three surfaces

$$S = S_1 + S_2 + S_3$$

where $\hat{n}$ is an outward pointing unit normal vector. Let $\vec{F}$ be solenoidal; hence,

$$\oint_S \hat{n} \cdot \vec{F} ds = \int_{S_1} \hat{n} \cdot \vec{F} ds + \int_{S_2} \hat{n} \cdot \vec{F} ds + \int_{S_3} \hat{n} \cdot \vec{F} ds = 0$$

Since $\hat{n} \cdot \vec{F} = 0$ on S_3, this simplifies to

$$\int_{S_2} \hat{n} \cdot \vec{F} ds = \int_{S_1} (-\hat{n}) \cdot \vec{F} ds \tag{4.3}$$

where $-\hat{n}$ on S_1 points into the tube.

For an open surface S, such as the shaded areas in Figure 4.1, and a solenoidal vector field, $\vec{F}$, the integral

$$\int_S \hat{n} \cdot \vec{F} ds \tag{4.4}$$

is zero when S is bounded by a reducible curve. When S is bounded by an irreducible curve, the integral is not zero. In this later case, the integral provides the strength of a vector tube or a filament, if C shrinks to a point. Furthermore, Equation (4.3) provides the important result that the strength is constant along the tube or filament at a given instant of time. This constant strength is independent of the inclination angle that curve C in Figure 4.1 has relative to the tube's axis.

Assume $\vec{F}$ is solenoidal and is tangent to a vector tube. We presume the surface in integral (4.4) is approximately perpendicular to the tube's axis; i.e., S is bounded by an irreducible curve, such as C_1 in Figure 4.2. Further, suppose the cross-sectional area of S decreases along the axis of the tube. In this circumstance, for the integral to remain constant, the magnitude of $\vec{F}$ must increase as the area of S decreases. One consequence is that the winds are most intense in a tornado where its diameter is narrowest.

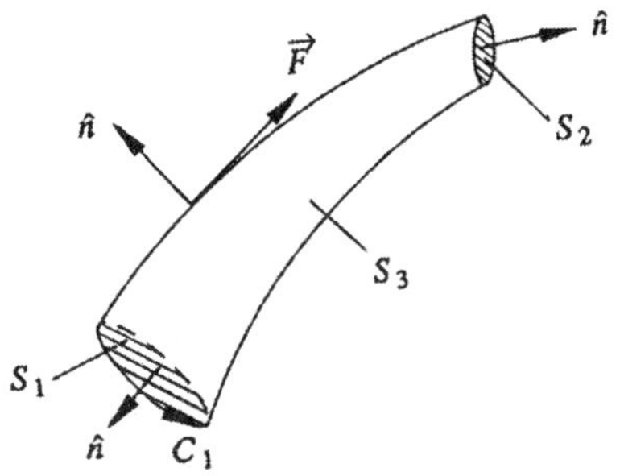

FIGURE 4.2 Nomenclature for a sliced vector tube.

In view of Equation (4.2), any solenoidal field can be written in terms of a vector potential $\vec{A}$ as

$$\vec{F} = \nabla \times \vec{A} \tag{4.5}$$

since

$$\nabla \cdot (\nabla \times \vec{A}) = 0$$

is a vector identity. Vector potentials are not often utilized in fluid dynamics because one vector, with three scalar components, is merely replaced by another vector with an equal number of scalar components.

The simplest example of a solenoidal field is $\vec{F} = \vec{\omega}$, since the vorticity is defined as $\nabla \times \vec{w}$. Another example occurs in incompressible flow, where $\vec{F} = w$ is solenoidal, since continuity yields Equation (4.2). In this case, a vector tube is called a streamtube, a filament is a streamline, and the strength of a streamtube is

$$\int_S \hat{n} \cdot \vec{w} ds = \text{constant}$$

which is the streamtube's volumetric flow rate. In a compressible flow, $\vec{F} = \rho\vec{w}$ is solenoidal if the flow is steady. In this circumstance, the strength of a streamtube is

$$\int_S \hat{n} \cdot (\rho\vec{w}) ds = \dot{m}$$

where $\dot{m}$ is the constant mass flow rate through the tube.

Irrotational Vectors

A second type of vector field is referred to as irrotational if it satisfies

$$\nabla \times \vec{F} = 0 \tag{4.6}$$

Since

$$\nabla \times (\nabla \phi) = 0$$

is a vector identity, an irrotational field can be written in terms of a potential function ϕ as

$$\vec{F} = \nabla\phi \tag{4.7}$$

In contrast to a vector potential, the replacement of $\vec{F}$ with ϕ is highly advantageous, since three scalar functions are now replaced by one in a three-dimensional flow.

The simplest example of an irrotational vector field occurs when the vorticity $\vec{\omega}$ is zero. From Equation (1.17), we have the condition for an irrotational velocity field

$$\vec{\omega} = \nabla\times\vec{w} = 0 \tag{4.8}$$

and $\vec{w}$ can be replaced by a scalar potential that is called a velocity potential.

Helmholtz's Theorem

Under appropriate conditions, such as being single valued, any vector function $\vec{F}$ can be resolved into solenoidal and irrotational components. In this circumstance, we have

$$\vec{F} = \vec{F}_s + \vec{F}_i$$

where

$$\nabla\cdot\vec{F}_s = 0, \quad \nabla\times\vec{F}_i = 0$$

For the solenoidal vector, we use

$$\vec{F}_s = \nabla\times\vec{A}$$

while the irrotational vector is replaced with

$$\vec{F}_i = \nabla\phi$$

Hence, $\vec{F}$ can be written as

$$\vec{F} = \nabla\times\vec{A} + \nabla\phi \tag{4.9}$$

which is known as Helmholtz's theorem. Notice that the three scalar components of $\vec{F}$ are determined by ϕ and the three scalar components of $\vec{A}$. As a consequence, one additional scalar condition can be imposed for a unique decomposition of $\vec{F}$. Without loss of generality, we require $\vec{A}$ to be solenoidal:

$$\nabla\cdot\vec{A} = 0$$

An equation for ϕ is obtained by taking the divergence of Equation (4.9), with the result

$$\nabla\cdot\vec{F} = \nabla\cdot(\nabla\phi) = \nabla^2\phi \tag{4.10}$$

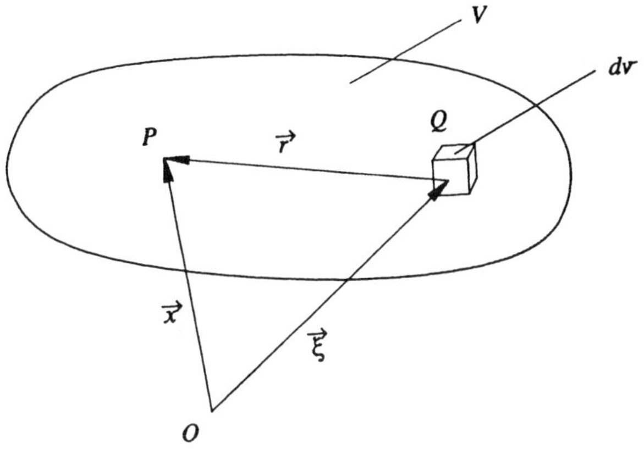

FIGURE 4.3 Region of integration V for the vector and scalar potential solutions.

which is a Poisson equation for ϕ. (We assume $\nabla \cdot \vec{F}$ is a known algebraic function.) As shown in Figure 4.3, we consider a volume V containing points P and Q whose location with respect to an arbitrary origin is given by $\vec{x}$ and $\vec{\xi}$, respectively. The solution of Equation (4.10) can be written as

$$\phi(\vec{x}) = -\frac{1}{4\pi}\int_V \frac{\nabla \cdot \vec{F}}{r}\,dv \tag{4.11}$$

where r is the distance between points P and Q, while the coordinates of $\nabla \cdot \vec{F}$ and dv are provided by $\vec{\xi}$, and V is a region where $\nabla \cdot \vec{F}$ is not zero.

We arrive at an equation for the vector potential $\vec{A}$ by taking the curl of Equation (4.9), with the result

$$\nabla \times \vec{F} = \nabla \times (\nabla \times \vec{A}) = (\nabla \cdot \vec{A}) - \nabla^2 \vec{A} = -\nabla^2 \vec{A}$$

since $\vec{A}$ is solenoidal. This is also a Poisson equation, which has the solution

$$\vec{A}(\vec{x}) = \frac{1}{4\pi}\int_V \frac{\nabla \times \vec{F}}{r}\,dv$$

Helmholtz's theorem is especially useful when $\vec{F}$ is the velocity. The irrotational part $\vec{w}_i$ of the velocity yields

$$\nabla \times \vec{w}_i = 0, \qquad \vec{w}_i = \nabla\phi$$

where ϕ is a velocity potential. The solenoidal part results in

$$\nabla \cdot \vec{w}_s = 0 \qquad \vec{w}_s = \nabla \times \vec{A}$$

and the vorticity is given by

$$\vec{\omega} = \nabla \times \vec{w} = \nabla \times (\vec{w}_i + \vec{w}_s) = \nabla \times \vec{w}_s$$

This decomposition is computationally effective for analyzing inviscid transonic flow over airfoils and wings (Dang and Chen, 1989). In this application, the flow is irrotational everywhere except

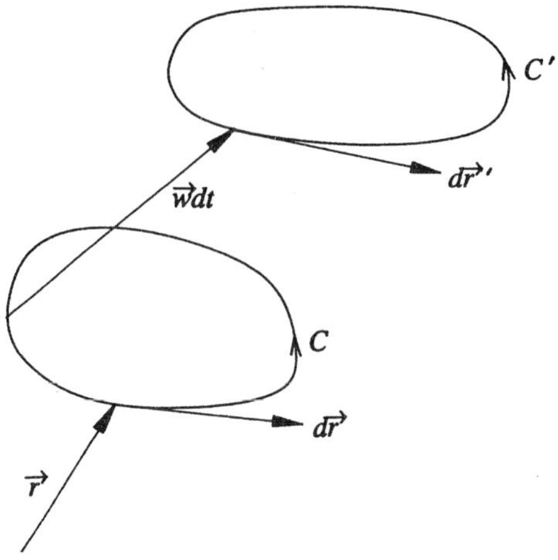

FIGURE 4.4 Circulation schematic.

in the region downstream of the curved shock wave that first appears above the upper surface of the airfoil or wing. Only in this region are $\vec{w}_i$ and $\vec{w}_s$ both nonzero.

4.3 KELVIN'S EQUATION AND VORTICITY

The circulation Γ is defined as the integral along a closed curve C:

$$\Gamma = \oint_C \vec{w} \cdot d\vec{r} \tag{4.12}$$

where $d\vec{r}$ is tangent to C and points in the counterclockwise direction (see Figure 4.4). (In aerodynamics the circulation is defined as $-\Gamma$ in order that $\Gamma > 0$ should correspond to a positive lift.)

We now evaluate the change in Γ following the fluid particles initially located on curve C at time t, as shown in Figure 4.4. In the analysis, C is fixed in space and time, while C' consists of the same fluid, at time $t + dt$, initially located on C. Since we are following fluid particles, we can use the substantial derivative to obtain

$$\frac{D\Gamma}{Dt} = \oint_C \frac{D\vec{w}}{Dt} \cdot d\vec{r} + \oint_C \vec{w} \cdot \frac{D}{Dt}(d\vec{r})$$

where the two operations

$$\frac{D}{Dt}\oint_C = \oint_C \frac{D}{Dt}$$

commute, since curve C is fixed in space and time. For the rightmost integral, we have

$$\frac{D}{Dt}(d\vec{r}) = d\left(\frac{Dr}{Dt}\right) = d\vec{w} \tag{4.13}$$

$$\vec{w} \cdot D\frac{(d\vec{r})}{Dt} = \vec{w} \cdot d\vec{w} = d\left(\frac{w^2}{2}\right)$$

and, consequently,

$$\oint_C \vec{w} \cdot \frac{D}{Dt}(d\vec{r}) = \oint_C d\left(\frac{w^2}{2}\right) = 0$$

We thus obtain Kelvin's equation:

$$\frac{D\Gamma}{Dt} = \oint_C \frac{D\vec{w}}{Dt} \cdot dr \tag{4.14}$$

The integral on the right is evaluated in Section 4.4 and still differently in Section 5.4. We note that $D\vec{w}/Dt$ is the acceleration in this later evaluation.

Stokes' theorem, Equation (1.60), relates the circulation to the vorticity. The theorem yields

$$\Gamma = \oint_C \vec{w} \cdot d\vec{r} = \int_S \hat{n} \cdot (\nabla \times \vec{w}) ds = \int_S \hat{n} \cdot \vec{\omega} ds \tag{4.15}$$

where S is the open surface that caps C. This relation can be differentiated, obtaining

$$d\Gamma = \hat{n} \cdot \vec{\omega} ds \tag{4.16}$$

Thus, $d\Gamma$ equals the component of $\vec{\omega}$ that is normal to ds as curve C shrinks to a point. This relation represents the connection between the circulation and the vorticity. To clarify the connection, consider a region of flow in which $\vec{\omega} = 0$. Then $d\Gamma = 0$ and the circulation is a constant, which need not be zero, in the region. On the other hand, if the vorticity is nonzero, then the magnitude of $d\Gamma$, at each point of the region, depends on the orientation of $\hat{n}$. For instance, if $\hat{n}$ is perpendicular to $\vec{\omega}$, then $d\Gamma = 0$ at this point. Thus, $d\Gamma/ds$ can have any value from $-\omega$ to $+\omega$ at the point in question.

Shear Layer

It is convenient, at this time, to discuss a shear layer, i.e., a parallel, two-dimensional flow with a shearing motion, as sketched in Figure 4.5. We first evaluate the circulation about the $ABCD$ path, per unit depth of flow. Along AD and BC, we have

$$\vec{w} \cdot d\vec{r} = 0$$

while $\vec{\omega}$ is constant on AB and CD. We thus obtain

$$\Gamma = \oint_{ABCD} \vec{w} \cdot d\vec{r} = w_1\overline{AB} - w_2\overline{DC} = (w_1 - w_2)\overline{AB} \tag{4.17}$$

where the $\overline{AB}$ and $\overline{DC}$ distances are equal.

We can use Equation (4.15) and

$$\hat{n} = \hat{1}_z, \quad S = \overline{AB}\,\Delta y, \quad \vec{\omega} = \omega \hat{1}_z$$

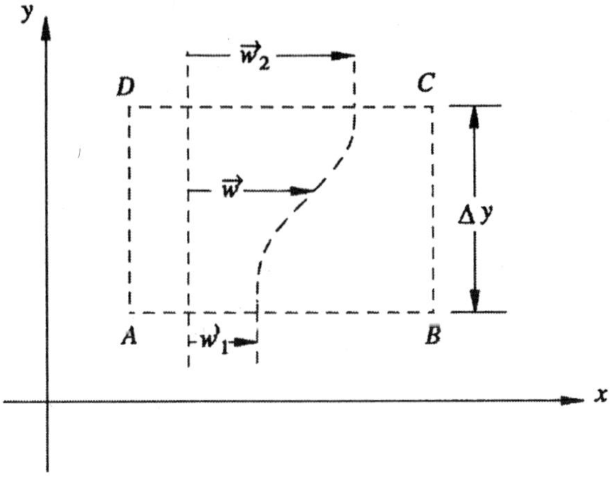

FIGURE 4.5 Schematic of a shear layer.

to also obtain for Γ

$$\Gamma = \overline{AB}(\Delta y)\omega$$

This relation is equated with Equation (4.17), to obtain

$$\omega = \frac{w_1 - w_2}{\Delta y}$$

or

$$\vec{\omega} = \frac{w_1 - w_2}{\Delta y}\hat{1}_z$$

A better way of writing this result is to use a unit vector $\hat{n}(=\hat{1}_y)$ that is normal to the shear layer

$$\vec{\omega} = \frac{1}{\Delta y}\hat{n} \times (\vec{w}_2 - \vec{w}_1) \tag{4.18}$$

in a right-handed system. We now let Δy shrink to dy, thereby obtaining a vortex sheet or layer, as illustrated in Figure 4.6. Let ds be the surface area and dv be the volume, $dyds$, of an infinitesimal piece of the sheet. A surface vorticity $\vec{\omega}_s$ is defined by

$$\vec{\omega}_s ds = \vec{\omega} dv$$

With the aid of Equation (4.18), this becomes

$$\vec{\omega}_s = \vec{\omega} dy = \hat{n} \times (\vec{w}_2 - \vec{w}_2)$$

Vortex layers are common, especially in supersonic flows. The strength and orientation of such layers is provided by $\vec{\omega}_s$, whose physical interpretation is clear.

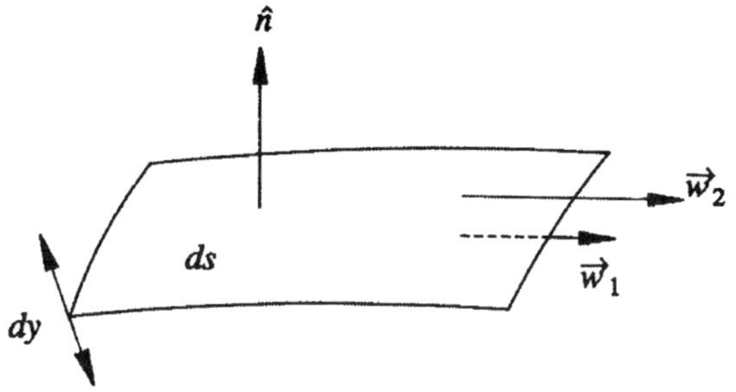

FIGURE 4.6 Vortex sheet schematic.

4.4 HELMHOLTZ VORTEX THEOREMS

THE FIRST THEOREM

We focus on the case where the vector in Equations (4.1) is the vorticity

$$\vec{F} = \vec{\omega} = \nabla \times \vec{w}$$

Thus, $\vec{\omega}$ is solenoidal, and for the irreducible curve C_1 in Figure 4.2,

$$\Gamma = \int_{C_1} \vec{w} \cdot d\vec{r} = \int_{S_1} \hat{n} \cdot \vec{\omega} ds \neq 0$$

while for a reducible curve on the S_3 surface, $\Gamma = 0$. Hence, the circulation Γ is the strength of the vortex and is constant along the tube at a given instant of time. This is Helmholtz's first vortex theorem.

Frequently, vorticity is found to be concentrated in narrow tubes, or filaments, that are referred to as vortex filaments or vortices. As we have just shown, the strength of a vortex filament or tube is constant along its length at a given instant of time. It is worth noting that this is not a result of any conservation law.

THE SUBSTANTIAL DERIVATIVE OF THE VORTICITY

Before embarking on a derivation of Helmholtz's second theorem, we obtain an important result that is needed in the derivation. This is a kinematic formula for the substantial derivative of the vorticity. The derivation begins by evaluating the curl of the acceleration:

$$\begin{aligned} \nabla \times \vec{a} &= \nabla \times \frac{D\vec{w}}{Dt} = \nabla \times \left(\frac{\partial \vec{w}}{\partial t} + \vec{\omega} \times \vec{w} + \nabla \frac{w^2}{2} \right) \\ &= \frac{\partial}{\partial t}(\nabla \times \vec{w}) + \nabla \times (\vec{\omega} \times \vec{w}) = \frac{\partial \vec{\omega}}{\partial t} + \nabla \times (\vec{\omega} \times \vec{w}) \end{aligned}$$

where Equation (1.18) is utilized along with the identity

$$\nabla \times \nabla \left(\frac{1}{2} w^2 \right) = 0$$

The vector identity (see Appendix A, Table 5)

$$\nabla \times (\vec{A} \times \vec{B}) = \vec{A}(\nabla \cdot \vec{B}) - \vec{B}(\nabla \cdot \vec{A}) + \vec{B} \cdot (\nabla \vec{A}) - \vec{A} \cdot (\nabla \vec{B})$$

is used with

$$\vec{A} = \vec{\omega}, \quad \vec{B} = \vec{w}$$

to obtain

$$\nabla \times (\vec{\omega} \times \vec{w}) = \vec{\omega}(\nabla \cdot \vec{w}) - \vec{w}(\nabla \cdot \vec{\omega}) + \vec{w} \cdot (\nabla \vec{\omega}) - \vec{\omega} \cdot (\nabla \vec{w})$$

Another vector identity is

$$\nabla \cdot (\nabla \times \vec{A}) = 0$$

thus yielding $\nabla \cdot \vec{\omega} = 0$ and

$$\nabla \times (\vec{\omega} \times \vec{w}) = \vec{\omega}(\nabla \cdot \vec{w}) + \vec{w} \cdot (\nabla \vec{\omega}) - \vec{\omega} \cdot (\nabla \vec{w})$$

With this relation, we obtain

$$\nabla \times \vec{a} = \frac{\partial \vec{\omega}}{\partial t} + \vec{w} \cdot (\nabla \vec{\omega}) + (\nabla \cdot \vec{w})\vec{\omega} - \vec{\omega} \cdot (\nabla \vec{w}) = \frac{D\vec{\omega}}{Dt} + (\nabla \cdot \vec{w})\vec{\omega} - \vec{\omega} \cdot (\nabla \vec{w})$$

or

$$\frac{D\vec{\omega}}{Dt} = \nabla \times \vec{a} + \vec{\omega} \cdot (\nabla \vec{w}) - (\nabla \cdot \vec{w})\vec{\omega} \tag{4.19}$$

for the substantial derivative of $\vec{\omega}$. In the second vortex theorem, we actually need

$$\vec{\omega} \times \frac{D\vec{\omega}}{Dt} = \vec{\omega} \times (\nabla \times \vec{a}) + \vec{\omega} \times [\vec{\omega} \cdot (\nabla \vec{w})] = \vec{\omega} \times (\nabla \times \vec{a}) - [\vec{\omega} \cdot (\nabla \vec{w})] \times \vec{\omega} \tag{4.20}$$

where the rightmost term, which resembles a triple scalar product, is not zero because $\nabla \vec{\omega}$ is a dyadic and not a vector (see Problem 4.2). We could, of course, replace $\vec{a}$ with the momentum equation. The result, however, would no longer be kinematic.

The Second and Third Theorems

In general, the fluid particles that constitute a vortex filament at one instant of time will not constitute the filament at a later instant of time. The second theorem of Helmholtz provides the condition for the fluid particles to remain fixed with the filament. When this occurs, the filament is a material line. For a vortex filament to be a field line, $\vec{\omega}$ must be tangent to the filament and Equation (4.1a) can be used with $\vec{F} = \vec{\omega}$. For the field line to also be a material line, we must have

$$\frac{D}{Dt}(d\vec{r} \times \vec{\omega}) = 0$$

or

$$\frac{D}{Dt}(d\vec{r} \times \vec{\omega}) = \frac{D(d\vec{r})}{Dt} \times \vec{\omega} + d\vec{r} \times \frac{D\vec{\omega}}{Dt} = 0$$

Equations (1.45) and (4.13) combine to yield

$$\frac{D(d\vec{r})}{Dt} = d\vec{w} = d\vec{r} \cdot (\nabla \vec{w})$$

where the rightmost term is the directional derivative of $\vec{w}$ in the $d\vec{r}$ direction. Equation (4.1b) is also used, with $\vec{F} = \vec{\omega}$, for the condition that $\vec{\omega}$ is tangent to a filament. We thus obtain, with the aid of Equation (4.20),

$$[d\vec{r} \cdot (\nabla \vec{w})] \times \vec{\omega} + d\vec{r} \times \frac{D\vec{\omega}}{Dt} = 0$$

$$dg\left\{[\vec{\omega} \cdot (\nabla \vec{w})] \times \vec{\omega} + \vec{\omega} \times \frac{D\vec{\omega}}{Dt}\right\} = 0$$

$$[\vec{\omega} \cdot (\nabla \vec{w})] \times \vec{\omega} + \vec{\omega} \times (\nabla \times \vec{a}) - [\vec{\omega} \cdot (\nabla \vec{w})] \times \vec{\omega} = 0$$

or

$$\vec{\omega} \times (\nabla \times \vec{a}) = 0 \tag{4.21}$$

When this condition holds, the vortex filaments are material lines.

We need to interpret this condition in a physically meaningful way. Suppose the acceleration $\vec{a}$ is an irrotational vector field so that

$$\vec{a} = \frac{D\vec{w}}{Dt} = \nabla \phi \tag{4.22}$$

We then have

$$\nabla \times \vec{a} = \nabla \times (\nabla \phi) = 0 \tag{4.23}$$

and Equation (4.21) is satisfied. However, Equation (4.14) yields

$$\frac{D\Gamma}{Dt} = \oint_C \frac{D\vec{w}}{Dt} \cdot d\vec{r} = \oint_C \vec{a} \cdot d\vec{r} = \oint_C \nabla \phi \cdot d\vec{r} = \oint_C d\phi = 0 \tag{4.24}$$

Hence, an irrotational acceleration implies both Equation (4.21) and $D\Gamma/Dt = 0$. We thus have Helmholtz's second theorem, which states that the vortex filaments are material lines in a circulation-preserving flow, i.e., a flow in which the circulation along a material line is constant with time.

Equation (4.21) is also satisfied if the velocity field is irrotational:

$$\vec{\omega} = 0$$

and again vortex filaments are material lines. Suppose the irrotationality condition holds throughout the flow except along distinct vortex filaments or surfaces that are constructed of filaments, such as a vortex tube. For purposes of simplicity, consider an otherwise irrotational flow that contains a single vortex tube, like the one shown in Figure 4.1. The circulation Γ is zero for a reducible curve, such as C', and finite for an irreducible curve, such as C. These curves, however, need not lie on the vortex tube; thus, any curve that does not enclose the tube will have a zero circulation, while any curve that encloses it once will have the same constant value Γ as curve C. In view of this, we have $D\Gamma/Dt$ as zero, both for the reducible and irreducible curves. We, therefore, again obtain the second vortex theorem.

Helmholtz's third theorem states that in a flow in which the vortex filaments are material lines, the strength of all vortex tubes does not change with time. This result differs from the first theorem, which holds at a given instant of time.

Another consequence of these theorems is that any vortex filament either closes on itself or terminates at a boundary. This result holds for the filaments in an otherwise irrotational flow and is useful for understanding the vortex filaments shed from the wings of a moving aircraft. The gradual decay of the filaments is due to viscous stresses, since vortices are associated with a shearing motion. Thus, the vortices shed from the wing tips of an aircraft during takeoff or landing decay with time. A landing strip is safe for reuse only after this decay is largely completed.

REFERENCE

Dang, T.Q. and Chen, L.-T., "Euler Correction Method for Two- and Three-Dimensional Transonic Flows," *AIAA J.* 27, 1377 (1989).

PROBLEMS

4.1 Consider a steady, parallel inviscid flow of a perfect gas in which the pressure is constant. Between $y = -\delta$ and $y = \delta$ the velocity profile has a linear variation, as shown in the sketch. Similarly, the stagnation enthalpy h_o has a linear variation between h_{o1} and h_{o2} in this region. Determine the variation with y, in this region, for the entropy s and vorticity ω.

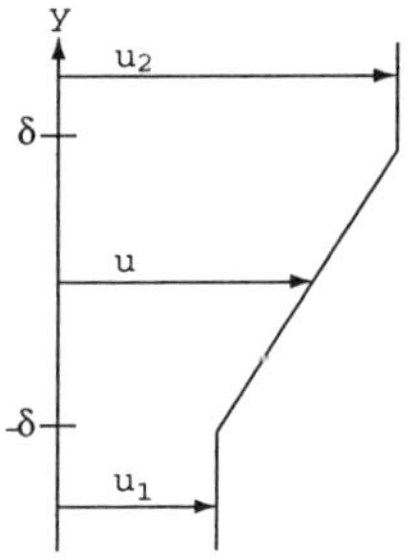

4.2 In the derivation of the second Helmholtz vortex theorem, the term

$$\vec{\omega} \times [\vec{\omega} \cdot (\nabla \vec{\omega})]$$

appears in Equation (4.20). Evaluate this term using curvilinear coordinates. The term has the appearance of a triple scalar product with two elements that are the same; nevertheless, show that, in general, it is not zero.

4.3 Utilize the results of Problem 2.11(b) to obtain a simple equation for $\nabla \times \vec{a}$ in terms of the vorticity. In addition to the assumptions required for Problem 2.11(b), what additional assumptions yield $\nabla \times \vec{a} = 0$?

4.4 Utilize Helmholtz's theorem, Equation (4.9), to decompose

$$\vec{F} = [A(x_1) + B(x_2)]\hat{\imath}_1 + [B(x_2) + C(x_3)]\hat{\imath}_2 + [C(x_3) + A(x_1)]\hat{\imath}_3$$

into a solenoidal vector $\vec{F}_s$ and an irrotational vector $\vec{F}_i$, where $\hat{\imath}_i$ is a Cartesian basis. Do not impose the uniqueness condition associated with $\vec{F}_s$.

4.5 **(a)** Assume that the body force $\vec{F}_b$ is given by a scalar potential and the fluid is Newtonian to obtain

$$\frac{D\vec{\omega}}{Dt} = \frac{1}{\rho^2}\nabla\rho\times\nabla p-\frac{1}{\rho^2}\nabla\rho\times\{\vec{F}^s+\nabla[\lambda(\nabla\cdot\vec{w})]\}+\frac{1}{\rho}\nabla\times\vec{F}^s+\vec{\omega}\cdot(\nabla\vec{w})-(\nabla\cdot\vec{w})\vec{\omega}$$

where $\vec{F}^s$ is the surface force on a fluid particle associated with the shear viscosity; i.e.,

$$\vec{F}^s = 2\nabla\cdot(\mu\overleftrightarrow{\varepsilon})$$

(b) Show that

$$\rho\frac{D\vec{\omega}}{Dt} = \nabla\times\vec{F}^s$$

holds for a two-dimensional, incompressible flow.

(c) With a constant kinematic viscosity $\nu(=\mu/\rho)$ and $\vec{w} = u\hat{\imath}_x+v\hat{\imath}_y$, write the steady-flow scalar version of the part (b) result, where only ν, u, v, and derivatives of ω appear.

4.6 Utilize Helmholtz's theorem to decompose

$$\vec{F} = A(x_1)B(x_2)C(x_3)\hat{n}$$

where

$$\hat{n} = \alpha_i\hat{\imath}_i, \quad \alpha_i = \text{constant}$$

into a solenoidal vector $\vec{F}_s$ and an irrotational vector $\vec{F}_i$. Do not impose the uniqueness condition associated with $\vec{F}_s$, and simplify your answers.

4.7 Consider a steady, incompressible, two-dimensional, homogeneous flow with no body force. A homogeneous flow is one where the velocity gradient has constant components; i.e.,

$$\begin{bmatrix}\frac{\partial w_1}{\partial x_1} & \frac{\partial w_1}{\partial x_2}\\ \frac{\partial w_2}{\partial x_1} & \frac{\partial w_2}{\partial x_2}\end{bmatrix} = \begin{bmatrix} c & c_1\\ c_2 & -c\end{bmatrix}$$

where the x_i are Cartesian coordinates and c, c_1, and c_2 are constants.

(a) Determine w_1 and w_2 in terms of x_1 and x_2.

(b) Determine the scalar vorticity, ω.

(c) Determine p in terms of the x_i.

(d) Determine the algebraic equation for an arbitrary streamline.

4.8 Consider a steady, or unsteady, incompressible flow of a Newtonian fluid with constant transport coefficients and a negligible body force.

(a) Determine a vector equation for $\vec{w}$ such that the momentum equation has no viscous force term.
(b) Introduce Cartesian coordinates and simplify your part (a) answer.
(c) What are the equations that determine $\vec{w}$ and p/ρ?
(d) Introduce a vector potential $\vec{A}$ for $\vec{w}$. Determine the governing equation for $\vec{A}$. Also determine the equation for $\vec{\omega}$ and the equation for p/ρ, all in terms of $\vec{A}$.
(e) Evaluate the helicity density

$$H_d = \vec{w} \cdot \vec{\omega}$$

in terms of the Cartesian components A_i of $\vec{A}$. Simplify your answers whenever possible.

4.9 From Problem 4.8, part (d), we know that a vector potential for $\vec{w}$ must satisfy

$$(\nabla \cdot \nabla)\nabla \times \vec{A} = 0$$

(a) Show that one possible form for $\vec{A}$ is

$$\vec{A} = \vec{A}_o + \overleftrightarrow{B} \cdot \vec{r} + \left(\overleftrightarrow{\overleftrightarrow{C}} \cdot \vec{r}\right) \cdot \vec{r}$$

where $\vec{A}_o$, $\overleftrightarrow{B}$, and $\overleftrightarrow{\overleftrightarrow{C}}$ are constants. The Cartesian components of the triadic satisfy

$$C_{ijk} = C_{ikj}$$

(b) Use $\vec{A}$ to determine $\vec{\omega}$ and the helicity H_d.
(c) Determine p/ρ for a steady flow.

Part II

Advanced Gas Dynamics

Outline of Part II

Because of the practical and historical importance of the Euler equations, Part II investigates a number of their general properties. Some of these, like the Bernoulli equations and irrotational flow, are familiar to you. Thus, Chapter 5 is concerned with easily derivable consequences of the Euler equations, e.g., Crocco's equation. Shock waves are treated in a general manner in the next chapter, and hodograph transformation theory is discussed in Chapter 7. Chapter 8 discusses a transformation, called the substitution principle, that is quite different from the hodograph transformation. Basic gas dynamic theory is extended to a calorically imperfect gas in Chapter 9. The caloric equation of state is based on the harmonic oscillator model, and isentropic flow, oblique shock waves, etc., are analyzed. Sweep in a supersonic flow is discussed in Chapter 10 within the context of an oblique shock and a Prandtl–Meyer expansion. The next chapter considers the interaction process when a Prandtl–Meyer expansion interferes with an upstream oblique shock. In this circumstance, the shock possesses a discontinuity in its curvature where the expansion first interacts with the shock. The last chapter in this part of the book discusses unsteady, one-dimensional flow, as typically occurs in a shock tube. Special features in this chapter are an interior ballistic presentation and the Riemann function method.

Although the emphasis is on general properties of the Euler equations, a number of realistic flow fields are discussed either in the text or in various homework problems. These include spiral flow, Prandtl–Meyer flow, and the flow in a convergent/divergent nozzle. Quite often rotational flow fields are of primary interest. Part II can be viewed as containing advanced topics in gas dynamics.

5 Euler Equations

5.1 PRELIMINARY REMARKS

The Euler equations are readily obtained from the governing equations of Chapter 2 by setting

$$\mu = \lambda = \kappa = 0 \tag{5.1}$$

The resulting equations represent an inviscid, adiabatic flow. Usually, the flow is simply referred to as inviscid because κ and μ are not independent parameters but are related by the Prandtl number

$$Pr = \frac{\mu c_p}{\kappa} \tag{5.2}$$

where Pr is finite and positive. The assumption of $\mu = 0$ therefore implies $\kappa = 0$ and vice versa. It is useful to note that an adiabatic flow is generally not isothermal, and vice versa.

The vast bulk of a large Reynolds number flow field satisfies the Euler equations. Those regions where these equations do not hold have large gradients where viscous stresses and heat conduction are significant. These regions include wall boundary layers, free shear layers, and the flow internal to a shock wave. A solution of the Euler equations is usually essential before a viscous boundary-layer solution, for example, can be obtained.

The next section discusses the Euler equations and their associated boundary and initial conditions. Sections 5.3 and 5.4 then consider the Bernoulli equations and vorticity, respectively. The Bernoulli equations relate the pressure to the velocity and therefore are of primary importance in evaluating the pressure force on a surface or body. The last three sections assume a steady flow and consider the special cases associated with two-dimensional or axisymmetric flow and when natural coordinates are utilized. A treatment of a swirling flow is developed as part of the analysis of axisymmetric flow.

The treatment in this chapter can be viewed as fundamental, inasmuch as it is devoid of solutions to specific problems. However, Problem 5.4 treats parallel flow, Problems 5.8 and 5.21 deal with spiral flow, Problems 5.12 and 5.16 provide the forces on a contoured duct, while Problems 5.22, 5.23, and 5.24 are concerned with the incompressible flow about a cylinder or sphere. A variety of problems (5.10, 5.11, 5.13, 5.14, and 5.15) are concerned with vorticity or $\vec{\omega}/\rho$. These problems include a derivation of Beltrami's equation and various forms for the curl of the acceleration, $\nabla \times \vec{a}$, which is utilized in the second vortex theorem.

5.2 EQUATIONS — INITIAL AND BOUNDARY CONDITIONS

Conservation of mass, Equation (2.2b), is unaltered by Equations (5.1). With Equation (1.18) and $\vec{\tau} = 0$, the momentum equation reduces to

$$\frac{\partial \vec{w}}{\partial t} + \nabla \frac{w^2}{2} + \vec{\omega} \times \vec{w} = -\frac{1}{\rho} \nabla p + \vec{F}_b \tag{5.3}$$

With $\vec{q} = 0$, the energy equation, Equation (2.43), simplifies to

$$\frac{Dh_o}{Dt} = \frac{1}{\rho}\frac{\partial p}{\partial t} + \vec{w} \cdot \vec{F}_b \tag{5.4}$$

The foregoing relations have ρ, p, h, and $\vec{w}$ as dependent variables, since $\vec{F}_b$ is presumed known. The conservation equations are therefore insufficient in number for determining these variables. To obtain a closed system of equations, they are supplemented with a relation, like an incompressible assumption, or by several thermodynamic state equations. In this case, we typically use a thermally and calorically perfect gas given by

$$p = \rho RT, \qquad h = \frac{\gamma R}{\gamma - 1}T \tag{5.5a,b}$$

where γ is the constant ratio of specific heats and R is the specific gas constant. In SI units, R is evaluated in terms of the universal gas constant $\bar{R}$ and the molecular weight W as

$$R = \frac{\bar{R}}{W} = \frac{8314(\text{J/kmol-K})}{W(\text{kg/kmol})} \tag{5.6}$$

The unsteady form of the Euler equations is subject to a number of conditions, the first of which is a set of initial conditions. At some instant of time, enough details of the flow field must be known so that a solution can be obtained a short time later. Our approach, although limited to the Euler equations, is more general in concept. The equations are written with the unsteady derivatives, one per equation, on the left sides. The initial conditions then constitute whatever information is required for evaluating the right sides of the equations.

We thus write the Euler equations in the form

$$\frac{\partial \rho}{\partial t} = -\nabla \cdot (\rho \vec{w}) \tag{5.7a}$$

$$\frac{\partial \vec{w}}{\partial t} = -\nabla \frac{w^2}{2} - (\nabla \times \vec{w}) \times \vec{w} - \frac{1}{\rho}\nabla p + \vec{F}_b \tag{5.7b}$$

$$\frac{\partial h_o}{\partial t} - \frac{1}{\rho}\frac{\partial p}{\partial t} = -\vec{w} \cdot \nabla h_o + \vec{w} \cdot \vec{F}_b \tag{5.7c}$$

The energy equation requires further revision by introducing the definition of h_o and thermodynamic state relations. For example, with the assumption of a perfect gas, we obtain

$$h_o = \frac{\gamma}{\gamma - 1}\frac{p}{\rho} + \frac{1}{2}w^2$$

$$\frac{\partial h_o}{\partial t} = \frac{\gamma}{\gamma - 1}\frac{1}{\rho}\frac{\partial p}{\partial t} - \frac{\gamma}{\gamma - 1}\frac{p}{\rho^2}\frac{\partial \rho}{\partial t} + \vec{w} \cdot \frac{\partial \vec{w}}{\partial t}$$

$$= \frac{\gamma}{\gamma - 1}\frac{1}{\rho}\frac{\partial p}{\partial t} + \frac{\gamma}{\gamma - 1}\frac{p}{\rho^2}\nabla \cdot (\rho \vec{w}) - \vec{w} \cdot \nabla \frac{w^2}{2} - \frac{1}{\rho}\vec{w} \cdot \nabla p + \vec{w} \cdot \vec{F}_b$$

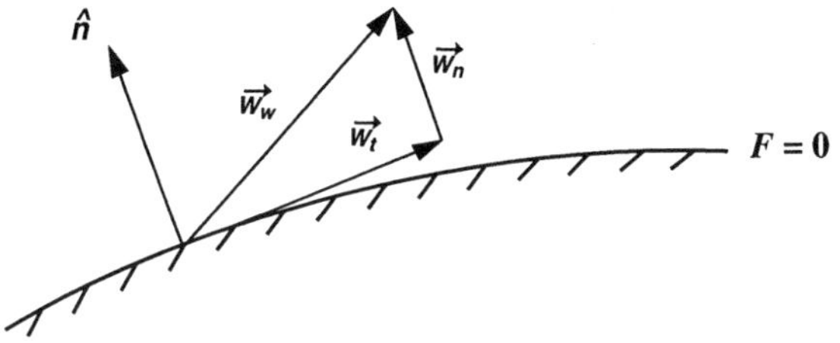

FIGURE 5.1 Schematic of a wall that is moving into a fluid.

where Equations (5.7a) and (5.7b) are used. With this result, Equation (5.7c) becomes

$$\frac{\partial p}{\partial t} = -\gamma p \nabla \cdot \vec{w} - \vec{w} \cdot \nabla p \tag{5.8}$$

(A more direct derivation is obtained by assuming a homentropic flow.) Once ρ, p, and $\vec{w}$ are known as functions of position, at some instant of time, the right sides of Equations (5.7a), (5.7b), and (5.8) provide values for ρ, p, and $\vec{w}$ at a slightly later time. To start an unsteady solution, we need ρ, p, and $\vec{w}$ as a function of position at a given time.

If a steady flow solution is sought, the initial condition is replaced by an upstream condition, which is usually a uniform flow.

Boundary Conditions

In addition to an initial condition, the solution must satisfy a velocity tangency condition at all impermeable walls. Let

$$F = F(\vec{r}, t) = 0 \tag{5.9}$$

be the surface of one such wall, which may be moving. Then the velocity $\vec{w}_w$of a fluid particle that is adjacent to the wall consists of a component $\vec{w}_n$ that is normal to the wall and a component $\vec{w}_t$ that is tangential to the wall. It is useful to introduce a nondimensional unit vector

$$\hat{n} = \frac{\nabla F}{|\nabla F|} \tag{5.10}$$

that is normal to the wall, as shown in Figure 5.1. We then have

$$\vec{w}_n = w_n \hat{n}, \qquad \vec{w}_t \cdot \hat{n} = 0 \tag{5.11}$$

where

$$\vec{w}_w = \vec{w}_n + \vec{w}_t$$

The tangential component $\vec{w}_t$ is arbitrary, while the $\vec{w}_n$ component must match the component of the velocity of the wall that is perpendicular to the wall. This velocity component, $\vec{w}_n$, is obtained from

$$\frac{DF}{Dt} = \frac{\partial F}{\partial t} + \vec{w}_w \cdot \nabla F = 0$$

or

$$\frac{\partial F}{\partial t} + \vec{w}_n \cdot \nabla F = 0 \tag{5.12}$$

since $\vec{w}_w \cdot \nabla F$ is zero. If the wall is motionless, we have

$$\vec{w}_t \cdot \nabla F = 0$$

or, better yet,

$$\vec{w}_w \cdot \hat{n} = 0 \tag{5.13}$$

and the fluid velocity is tangential to the wall.

When $\partial F/\partial t$ is not zero, an explicit equation is obtained for w_n by using Equations (5.10) through (5.12), with the result

$$\frac{\partial F}{\partial t} + |\nabla F| w_n \hat{n} \cdot \hat{n} = \frac{\partial F}{\partial t} + |\nabla F| w_n = 0$$

or

$$w_n = -\frac{1}{|\nabla F|}\frac{\partial F}{\partial t}$$

Hence, the velocity of the fluid that is adjacent to the wall is

$$\vec{w}_w = \vec{w}_t - \frac{\hat{n}}{|\nabla F|}\frac{\partial F}{\partial t} = \vec{w}_t - \frac{\nabla F}{|\nabla F|^2}\frac{\partial F}{\partial t} \tag{5.14}$$

Thus, if the wall is moving into the fluid with speed $-(\partial F/\partial t)/|\nabla F|$, then $\vec{w}_w$ is oriented into the fluid as indicated in Figure 5.1.

5.3 BERNOULLI'S EQUATIONS

There are two important integrals of the governing equations that typically go under the name of a Bernoulli equation. Both require an inviscid, adiabatic flow, i.e., the Euler equations, and both require that the body force be a conservative force field. In this case, $\vec{F}_b$ is provided by the gradient of a scalar function

$$\vec{F}_b = -\nabla\Omega \tag{5.15}$$

Gravity is this type of force field, with Ω given by

$$\Omega = -g_o\frac{r_o^2}{r}$$

where r is measured from the center of the Earth and g_o is the magnitude of $\vec{g}$ when $r = r_o$. The quantity Ω is the potential energy due to gravitational attraction, per unit mass.

Steady-Flow Bernoulli Equation

The first version of Bernoulli's equation further assumes a steady flow and utilizes the energy equation, which is

$$\vec{w} \cdot \nabla h_o = -\vec{w} \cdot \nabla \Omega$$

or

$$\vec{w} \cdot \nabla\left(h + \frac{1}{2}w^2 + \Omega\right) = 0$$

In a steady flow, this relation is the condition that $h_o + \Omega$ be constant along a streamline, since

$$\frac{D}{Dt}(h_o + \Omega) = \frac{\partial}{\partial t}(h_o + \Omega) + \vec{w} \cdot \nabla(h_o + \Omega) = \vec{w} \cdot \nabla(h_o + \Omega) = 0$$

We therefore obtain

$$h + \frac{1}{2}w^2 + \Omega = C \tag{5.16}$$

where C is constant along a streamline but may vary from streamline to streamline. The constant is generally evaluated in the upstream flow; if this flow is uniform and Ω is a constant, then C is the same constant for all streamlines that originate in the upstream flow. Equation (5.16) can be viewed as an energy equation that provides h once w and Ω are known as functions of position.

Irrotational Flow Bernoulli Equation

Most often, Bernoulli's equation refers to the form that is based on the momentum equation. The flow need not be steady; instead, it is assumed to be irrotational. The momentum equation then becomes

$$\frac{\partial \vec{w}}{\partial t} + \nabla \frac{w^2}{2} = -\frac{1}{\rho}\nabla p - \nabla \Omega \tag{5.17a}$$

In view of irrotationality, a velocity potential function $\phi(\vec{r}, t)$ can be introduced as

$$\vec{w} = \nabla \phi \tag{5.18}$$

Since the gradient and time derivative operations commute, we have

$$\frac{\partial \vec{w}}{\partial t} = \frac{\partial}{\partial t}\nabla \phi = \nabla \frac{\partial \phi}{\partial t}$$

and Equation (5.17a) can be written as

$$\nabla\left[\frac{\partial \phi}{\partial t} + \frac{1}{2}(\nabla \phi)^2 + \Omega\right] = -\frac{1}{\rho}\nabla p \tag{5.17b}$$

If the flow is incompressible, we then have

$$\nabla\left[\frac{\partial\phi}{\partial t}+\frac{1}{2}(\nabla\phi)^2+\frac{p}{\rho}+\Omega\right] = 0$$

which immediately integrates to

$$\frac{\partial\phi}{\partial t}+\frac{1}{2}(\nabla\phi)^2+\frac{p}{\rho}+\Omega = C(t) \tag{5.19}$$

By setting $\phi = \tilde{\phi}+\int C dt$, the function of integration $C(t)$ can be absorbed into the $\partial\phi/\partial t$ term. This relation is the incompressible, irrotational form of Bernoulli's equation. (With the same assumptions, Problem 2.14 shows that it holds for a viscous flow when the viscosity μ is a constant. Keep in mind, however, that viscous flows are generally rotational.) The equation provides the pressure once ϕ and Ω are known. If the incompressible flow is also steady, we have

$$p+\frac{1}{2}\rho w^2+\rho\Omega = p_\infty+\frac{1}{2}\rho w_\infty^2+\rho\Omega_\infty = \text{constant} \tag{5.20}$$

where the infinity subscript denotes a freestream condition, and where the Ω terms account for the potential energy associated with gravity. This is the most common form encountered for Bernoulli's equation.

We would like to obtain a relation similar to Equation (5.19) that holds for a compressible flow. In this circumstance, we require the pressure to be a thermodynamic function of ρ and s; i.e.,

$$p = p(\rho,s) \tag{5.21}$$

A barotropic flow is assumed in which either ρ is constant or p depends only on ρ:

$$p = p(\rho) \tag{5.22}$$

The constant density assumption would then yield Equation (5.19) and we, henceforth, exclude this case from the discussion. In view of Equations (5.21) and (5.22), the barotropic assumption is equivalent to a constant entropy or

$$\nabla s = 0 \tag{5.23}$$

in which case the flow is referred to as homentropic. Flows that satisfy Equations (5.22) and (5.23) are quite common. For example, for a perfect gas, Equation (5.22) can be written as

$$\frac{p}{p_0} = \left(\frac{\rho}{\rho_0}\right)^\gamma \tag{5.24}$$

where the zero subscript denotes a stagnation or reference condition. We are familiar with those flows where this relation is utilized.

With the barotropic assumption, the thermodynamic relation for the enthalpy

$$dh = Tds + \frac{dp}{\rho} \tag{5.25a}$$

simplifies to

$$dh = \frac{dp}{\rho} \tag{5.25b}$$

Upon integration, we have

$$h = \int \frac{dp}{\rho} + \text{constant} \tag{5.26}$$

Recall that the directional derivative of a scalar, ψ, is given by

$$d\psi = d\vec{r} \cdot \nabla \psi$$

where $d\vec{r}$ provides the direction, which is arbitrary. With the aid of this relation, two different equations that involve the dp derivative are written as

$$dp = d\vec{r} \cdot \nabla p$$

$$\frac{dp}{\rho} = d\int \frac{dp}{\rho} = d\vec{r} \cdot \nabla \int \frac{dp}{\rho}$$

Eliminate dp, to obtain

$$\frac{1}{\rho}(d\vec{r} \cdot \nabla p) = d\vec{r} \cdot \nabla \int \frac{dp}{\rho}$$

Since the $d\vec{r}$ vector is arbitrary, we can cancel the $d\vec{r}$ factor with the result

$$\frac{\nabla p}{\rho} = \nabla \int \frac{dp}{\rho} = \nabla h \tag{5.27}$$

Thus, for a barotropic flow, $\nabla p/\rho$ equals ∇h.

The above discussion shows that the derivatives in Equation (5.25b) can be replaced by the gradient operator. A key step in the proof is that $d\vec{r}$ is an arbitrary vector, thereby allowing the cancellation of $d\vec{r}$. The replacement of the thermodynamic derivative with ∇ is not limited to Equation (5.25b) but holds in general, e.g., to Equation (5.25a).

With Equation (5.27), Equation (5.17b) integrates to

$$\frac{\partial \phi}{\partial t} + \frac{1}{2}(\nabla \phi)^2 + h + \Omega = C(t) \tag{5.28}$$

which is the sought-after compressible counterpart to Equation (5.19). The only new assumption required for this result is that the flow is barotropic. As in Equation (5.19), $C(t)$ can be absorbed into the $\partial\phi/\partial t$ term.

Discussion

The two versions of Bernoulli's equations are quite distinct. The energy version holds only along streamlines in a steady flow. However, when $\Omega = 0$, as is the case in a gaseous flow, h_o is usually a constant throughout the flow field, not just along individual streamlines.

On the other hand, the momentum version holds throughout an unsteady, irrotational flow field. This version would appear to be more general than the first. This is an exaggeration, however, as can be seen by examining a steady flow downstream of a curved shock. The energy version holds on both sides of the shock, while the momentum version does not apply downstream of the shock. In this region, as we will show, the flow is not irrotational, barotropic, or homentropic. Application of the momentum version is therefore of no use in a flow where curved shock waves are prevalent.

As shown by Problem 5.13(d), Equation (5.27) can be used to develop the following version of Helmholtz's second vortex theorem. In an inviscid, barotropic flow with a conservative body force, vortex lines are material lines. Moreover, if the fluid is initially irrotational, it will remain irrotational. Of course, if the flow encounters a curved shock, it will not be barotropic or irrotational downstream of the shock.

5.4 VORTICITY

Vorticity has been a favorite topic of fluid dynamicists for quite a long time (Vazsonyi, 1945). It is important in the analysis of several naturally occurring flows, such as dust devils, tornadoes, and hurricanes. It is also of interest in the study of transitional and turbulent flows. When a boundary layer separates, it may roll up into a vortex or, if it reattaches, a vortical recirculation region forms (Délery, 1992).

Chapter 4 introduced the topic and provides in Problem 4.5 a fundamental equation for the rate of change of the vorticity. The first term on the right side of this equation is zero if the flow is barotropic. This term, therefore, alters the vorticity anywhere downstream of a curved shock wave. There are two terms associated with the shear viscosity that combine as

$$\nabla \times \left(\frac{\vec{F}^s}{\rho} \right)$$

This is just the curl of the shear force per unit mass. The term associated with the second viscosity coefficient can be written as

$$-\frac{\nabla \rho}{\rho^2} \times \nabla[\lambda(\nabla \cdot \vec{w})] = \frac{\lambda}{\rho^3} \nabla \rho \times \left(\nabla \frac{D\rho}{Dt} \right)$$

if λ is a constant. This result can be further expanded by using the general result

$$\nabla \frac{D\chi}{Dt} = \frac{D\nabla\chi}{Dt} + \nabla\chi \cdot \nabla\vec{w} + \nabla\chi \times \vec{\omega}$$

where χ is any scalar. In an incompressible flow, the foregoing terms in $D\vec{\omega}/Dt$ are zero, except for the shear force term.

If the flow is inviscid and barotropic, then only the two terms

$$\vec{\omega} \cdot (\nabla \vec{w}) - (\nabla \cdot \vec{w})\vec{\omega}$$

remain. In this circumstance, an initially irrotational flow will remain irrotational. As mentioned, this result does not apply downstream of a curved shock, where the flow is not barotropic. If the flow is two dimensional or axisymmetric (without swirl, discussed shortly), then $\vec{\omega}$ is perpendicular to the plane of the flow. In this situation, $\vec{\omega} \cdot (\nabla \vec{w})$ is zero if the flow is two-dimensional, but nonzero if the flow is axisymmetric and, again, is oriented normal to the plane of the flow. The term, $(\nabla \cdot \vec{w})\vec{\omega}$, is zero only if the flow is irrotational or incompressible.

Crocco's Equation

In contrast to Equation (5.20), which dates from the 18th century, Crocco's equation dates from the 1930s. Furthermore, this equation is quite general; it does not require an irrotational or barotropic flow. In its original form, a steady, inviscid flow is assumed, assumptions which here are not invoked. We start with the thermodynamic relation

$$\nabla h = T \nabla s + \frac{1}{\rho} \nabla p \tag{5.29a}$$

or

$$-\frac{1}{\rho} \nabla p = T \nabla s - \nabla h = T \nabla s - \nabla h_o + \nabla \frac{w^2}{2} \tag{5.29b}$$

The pressure gradient is eliminated from Equation (2.8), with the result

$$\frac{\partial \vec{w}}{dt} + \vec{\omega} \times \vec{w} = T \nabla s - \nabla h_o + \frac{1}{\rho} \nabla \cdot (\overleftrightarrow{\tau} + \vec{F}_b) \tag{5.30a}$$

which is Crocco's equation in vector form. This relation is most often used for an inviscid, supersonic flow with no body force; i.e.,

$$\frac{\partial \vec{w}}{dt} + \vec{\omega} \times \vec{w} = T \nabla s - \nabla h_o \tag{5.30b}$$

Problem 5.25 addresses the changes required in a noninertial frame, while Wu and Hayes (1958) extend it to a chemically reacting mixture.

A streamline version of Equation (5.30b) can be obtained by multiplying by $\vec{w}$ to yield

$$\vec{w} \cdot \frac{\partial \vec{w}}{dt} + \vec{w} \cdot (\vec{\omega} \times \vec{w}) = T \vec{w} \cdot \nabla s - \vec{w} \cdot \nabla h_o$$

With the aid of

$$\vec{w} \cdot \frac{\partial \vec{w}}{dt} = \frac{\partial}{\partial t} \frac{w^2}{2}, \qquad \vec{w} \cdot (\vec{\omega} \times \vec{w}) = 0$$

we have the streamline (pathline in an unsteady flow) form

$$\frac{\partial}{\partial t}\frac{w^2}{2} = T\vec{w} \cdot \nabla s - \vec{w} \cdot \nabla h_o \quad \textbf{(5.31)}$$

of the equation. Although the vorticity does not appear in this relation, it is useful, as will be evident in the next section.

Discussion

Equation (5.30b) is important for understanding the role of vorticity in an inviscid flow field. The most common type of inviscid flow field encountered is a steady one where the upstream flow at infinity is uniform. Equation (5.16) then yields

$$h + \frac{1}{2}w^2 = h_o \quad \textbf{(5.32)}$$

where h_o is a constant throughout the flow field, since the upstream flow is uniform. In this case, we refer to the flow field as homenergetic. (In fluid mechanics, the prefixes *iso-* and *homo-* mean constant along streamlines and constant throughout a region of flow, respectively.)

In a homenergetic flow

$$\nabla h_o = 0 \quad \textbf{(5.33)}$$

and for a steady flow, Crocco's equations become

$$\vec{\omega} \times \vec{w} = T\nabla s, \qquad \vec{w} \cdot \nabla s = 0 \quad \textbf{(5.34a,b)}$$

Thus, the entropy is constant along streamlines but may vary from streamline to streamline. As noted at the end of Chapter 3, this type of flow is called isentropic. When the upstream flow is uniform, the entropy is constant in some region that includes the far upstream flow and in this region the flow is then homentropic. Equation (5.34a) further implies that if the flow field is rotational (i.e., $\vec{\omega} \neq 0$), then it cannot be homentropic.

Beltrami Flow

A steady, inviscid flow with a conservative body force that satisfies

$$\vec{\omega} \times \vec{w} = 0$$

is called a Beltrami flow (Morino, 1986). We ignore the trivial case where $\vec{w} = 0$ and note that a possible solution is $\vec{\omega} = 0$ and the flow is irrotational. An alternative solution is

$$\vec{\omega} = g(\vec{r})\vec{w}$$

where g is an arbitrary scalar function. Let us examine this equation for the possibility that $\vec{\omega} \neq 0$. Kelvin's equation and the momentum equation yield

$$\frac{D\Gamma}{Dt} = \oint_C \frac{D\vec{w}}{Dt} \cdot d\vec{r} = -\oint_C \left(\frac{1}{\rho}\nabla p + \nabla\Omega\right) \cdot d\vec{r} = -\oint_C \frac{dp}{\rho}$$

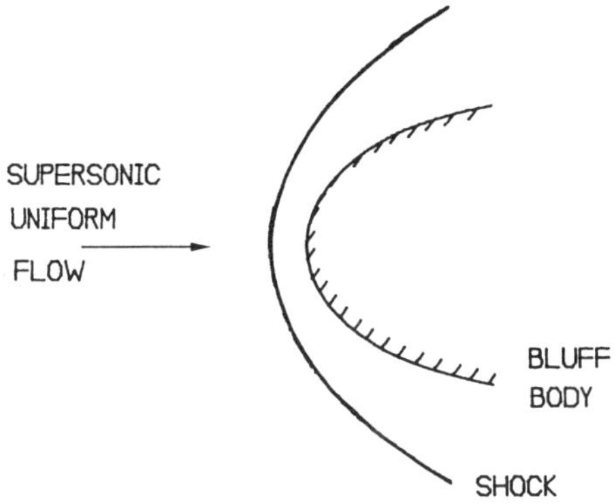

FIGURE 5.2 Curved shock wave in a uniform supersonic flow.

We now assume a barotropic flow so that the rightmost integral is zero. Since the orientation of $\hat{n}$ in Equation (4.16) is arbitrary, we see that $\vec{\omega} = 0$; thus, a barotropic Beltrami flow is irrotational. Moreover, the barotropic assumption implies that $\nabla s = 0$ and the flow is homentropic. By Crocco's vector equation, with no body force, we see that the flow is also homenergetic.

Thus, a steady, inviscid, barotropic flow is quite simple. Some of our conclusions, however, no longer hold downstream of a curved shock wave, as in Figure 5.2. There is a jump in entropy across a shock, and the magnitude of the jump depends on the slope of the shock wave relative to the upstream velocity. Hence, with a constant value for the entropy upstream of the shock, there is a variation in the value of the entropy on its downstream side. The downstream flow is therefore isentropic but not homentropic.

For the flow under discussion, Equation (5.33) holds on both sides of a curved shock, and Crocco's vector and scalar equations, on the downstream side, reduce to Equations (5.34). In this region, neither ∇s nor $\vec{\omega}$ is zero and the flow is rotational. Thus, a curved shock wave generates vorticity. Furthermore, the flow downstream of a curved shock further alters the vorticity, since ω is not a constant along streamlines. (See the Beltrami equation problems at the end of the chapter.)

Helicity

Another rotational quantity of interest is the helicity density (Levy et al., 1990)

$$H_d = \vec{w} \cdot \vec{\omega}$$

or the helicity

$$H = \int \vec{w} \cdot \vec{\omega} dv$$

which is a volume integral. In a two-dimensional or axisymetric inviscid flow, H_d is zero. This would not be the case in a transitional or turbulent flow, where the helicity concept is of primary interest (Hunt and Hussain, 1991).

5.5 STEADY FLOW

In this section, we assume a steady, inviscid flow with no body forces. The governing equations can be written as

$$\nabla \cdot (\rho \vec{w}) = 0 \tag{5.35a}$$

$$\nabla \frac{w^2}{2} + (\nabla \times \vec{w}) \times \vec{w} = -\frac{1}{\rho} \nabla p \tag{5.35b}$$

$$\vec{w} \cdot \nabla h_o = 0 \tag{5.35c}$$

These represent five scalar equations whose dependent variables are p, ρ, and w_i. Crocco's two equations reduce to

$$\vec{\omega} \times \vec{w} = T\nabla s - \nabla h_o, \qquad \vec{w} \cdot \nabla s = 0 \tag{5.36a,b}$$

and therefore the flow is isentropic.

A word of caution is necessary. Although s may be constant along streamlines, the flow need not be homentropic (or barotropic). When s varies from streamline to streamline, p is not just a function of ρ and Equation (5.21) should be utilized. Similarly, h_o is constant along streamlines but may vary from streamline to streamline, in which case the flow is not homenergetic.

We call h_o the stagnation enthalpy, since it equals h whenever the flow stagnates. Other stagnation quantities can be defined, provided an isentropic restriction is also imposed. Imagine a fluid particle being brought to rest isentropically; i.e.,

$$\frac{Ds}{Dt} = 0 \tag{5.37}$$

(This process does not require a steady flow.) If the flow involves dissipative processes, such as viscous stresses, heat conduction, or chemical reactions, then the isentropic process embodied in Equation (5.37) is hypothetical. In this circumstance, envision a fluid particle that is removed from its dissipative surroundings before undergoing an isentropic deceleration.

In view of Equation (5.36b), the steady Euler equations are consistent with the requirement of an isentropic process. Consequently, the distinction between a real and hypothetical isentropic process disappears and Equation (5.37) is redundant with Equation (5.36b).

We now show that ρ_o, p_o, and h_o are constant along streamlines in a steady, inviscid (or Euler) flow. Observe that Equation (5.35a) can be written as

$$\rho \nabla \cdot \vec{w} + \vec{w} \cdot \nabla \rho = 0$$

Hence, continuity becomes

$$\frac{D\rho}{Dt} = -\rho \nabla \cdot \vec{w} = \vec{w} \cdot \nabla \rho$$

The stagnation density can be defined by

$$\rho_o = \lim_{\vec{w} \to 0} \rho$$

where the entropy is held constant while taking this limit. With $\vec{w} \to 0$, we obtain

$$\frac{D\rho_o}{Dt} = \lim_{\vec{w}\to 0} \vec{w} \cdot \nabla \rho = 0 \tag{5.38}$$

and ρ_o is constant along streamlines in a steady, inviscid flow. The steady flow caveat is essential as will be shown shortly; $D\rho_o/Dt$ is generally not zero in an unsteady, inviscid flow.

If we multiply Equation (5.35b) by $\vec{w}$, we obtain

$$\vec{w} \cdot \nabla \frac{w^2}{2} = -\frac{1}{\rho} \vec{w} \cdot \nabla p = -\frac{1}{\rho}\frac{Dp}{Dt}$$

or

$$\frac{Dp}{Dt} = -\rho \vec{w} \cdot \nabla \frac{w^2}{2}$$

Hence, $\vec{w} \to 0$ also yields

$$\frac{Dp_o}{Dt} = 0 \tag{5.39}$$

while Equation (5.35c) directly shows that $Dh_o/Dt = 0$ also holds.

In a steady Euler flow, there are several differences between h_o and the other stagnation quantities. Equation (5.32) provides an explicit equation for h_o. Explicit relations, for instance, for ρ_o and p_o require a thermodynamic model, such as thermal and caloric state equations. A second distinction is that h_o is constant across shock waves, whereas ρ_o and p_o are not.

It is worth reemphasizing that the above discussion requires a steady flow. In an unsteady, inviscid flow, the energy equation, Equation (2.43), can be written as

$$\frac{Dh_o}{Dt} = \frac{1}{\rho}\frac{\partial p}{\partial t} \tag{5.40}$$

and h_o is not constant following a fluid particle. A similar result holds for ρ_o and p_o. We also observe that in a viscous (or heat-conducting), steady flow, stagnation conditions again are not constant along streamlines.

If a steady, inviscid flow is also homentropic, as is often the case, then all stagnation conditions are constant in this region of the flow. In other words, h_o, ρ_o, and p_o are constant throughout the region and not just along streamlines.

Thus far, we have avoided the use of a specific gas model. To obtain further results, assume a perfect gas. In this case, it can be shown that

$$\frac{h_o}{h} = \frac{T_o}{T} = X \tag{5.41a}$$

$$\frac{\rho_o}{\rho} = X^{1/(\gamma-1)} \tag{5.41b}$$

$$\frac{p_o}{p} = X^{\gamma/(\gamma-1)} \tag{5.41c}$$

where γ is the (constant) ratio of specific heats, and X, the Mach number M, and speed of sound a are given by

$$X = 1 + \frac{\gamma - 1}{2} M^2 \tag{5.42}$$

$$M = \frac{w}{a} \tag{5.43}$$

$$a^2 = \frac{\gamma p}{\rho} \tag{5.44}$$

Equations (5.41) are some of the well-known isentropic equations of gas dynamics. In an isentropic flow, the various stagnation quantities are constant along streamlines. They will vary from streamline to streamline unless the flow is homentropic (or barotropic).

5.6 TWO-DIMENSIONAL OR AXISYMMETRIC FLOW

It is useful to introduce the assumption that the flow field is either two-dimensional or axisymmetric. This simplification is often warranted; most exact solutions or well-known flows fall into this category. A principal objective will be to derive a form of the governing equations that simultaneously holds for both types of flow fields.

Two-Dimensional Flow

We begin by utilizing a Cartesian coordinate system in which the velocity is written as Equation (1.7). The usual operations involving the del operator apply. For instance, the vorticity and the vorticity term in the acceleration are

$$\begin{aligned}\vec{\omega} = \nabla \times \vec{w} &= \begin{vmatrix} \hat{\imath}_1 & \hat{\imath}_2 & \hat{\imath}_3 \\ \dfrac{\partial}{\partial x_1} & \dfrac{\partial}{\partial x_2} & \dfrac{\partial}{\partial x_3} \\ w_1 & w_2 & w_3 \end{vmatrix} \\ &= \left(\frac{\partial w_3}{\partial x_2} - \frac{\partial w_2}{\partial x_3}\right)\hat{\imath}_1 + \left(\frac{\partial w_1}{\partial x_3} - \frac{\partial w_3}{\partial x_1}\right)\hat{\imath}_2 + \left(\frac{\partial w_2}{\partial x_1} - \frac{\partial w_1}{\partial x_2}\right)\hat{\imath}_3\end{aligned} \tag{5.45}$$

$$\begin{aligned}\vec{\omega} \times \vec{w} = {} & \left[w_3\left(\frac{\partial w_1}{\partial x_3} - \frac{\partial w_3}{\partial x_1}\right) - w_2\left(\frac{\partial w_2}{\partial x_1} - \frac{\partial w_1}{\partial x_2}\right)\right]\hat{\imath}_1 \\ & + \left[w_1\left(\frac{\partial w_2}{\partial x_1} - \frac{\partial w_1}{\partial x_2}\right) - w_3\left(\frac{\partial w_3}{\partial x_2} - \frac{\partial w_2}{\partial x_3}\right)\right]\hat{\imath}_2 \\ & + \left[w_2\left(\frac{\partial w_3}{\partial x_2} - \frac{\partial w_2}{\partial x_3}\right) - w_1\left(\frac{\partial w_1}{\partial x_3} - \frac{\partial w_3}{\partial x_1}\right)\right]\hat{\imath}_3\end{aligned} \tag{5.46}$$

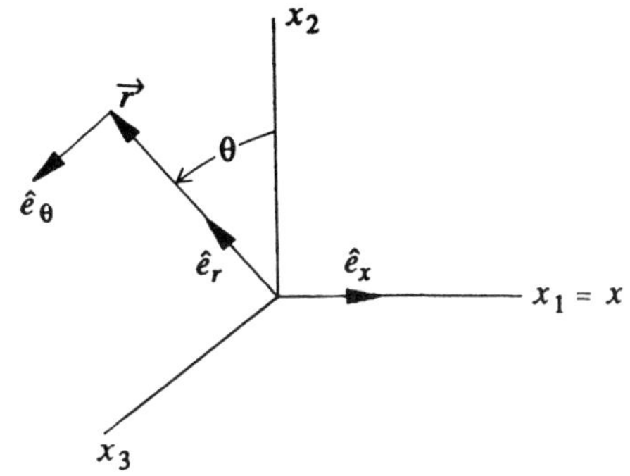

FIGURE 5.3 Cartesian and cylindrical coordinate system in which $\hat{e}_\theta$ and $\hat{e}_r$ are in the x_2,x_3 plane.

For a two-dimensional flow, we set

$$\frac{\partial}{\partial x_3} = 0 \tag{5.47}$$

and occasionally replace x_1, x_2 with x,y and w_1, w_2 with u,v for notational simplicity. Equation (5.47) does not mean w_3 is zero; it could be a constant. When w_3 is a nonzero constant, the flow field, as a whole, has a uniform motion in the x_3 direction. In either case, Equations (5.45) and (5.46) simplify to

$$\vec{\omega} = \left(\frac{\partial w_2}{\partial x_1} - \frac{\partial w_1}{\partial x_2}\right)\hat{1}_3 \tag{5.48}$$

$$\vec{\omega} \times \vec{w} = w_2\left(\frac{\partial w_1}{\partial x_2} - \frac{\partial w_2}{\partial x_1}\right)\hat{1}_1 + w_1\left(\frac{\partial w_2}{\partial x_1} - \frac{\partial w_1}{\partial x_2}\right)\hat{1}_2 \tag{5.49}$$

Axisymmetric Flow

Figure 5.3 shows the cylindrical coordinate system to be used for axisymmetric flows. In this system, $\hat{e}_x$, $\hat{e}_r$, and $\hat{e}_\theta$ form an orthonormal basis, and θ, $\hat{e}_\theta$, and $\hat{e}_r$ are in the x_2,x_3 plane. The scale factors, h_x, h_r, and h_θ and a few of the common del operations are

$$h_x = 1, \qquad h_r = 1, \qquad h_\theta = r \tag{5.50}$$

$$\nabla = \hat{e}_x\frac{\partial}{\partial x} + \hat{e}_r\frac{\partial}{\partial r} + \frac{\hat{e}_\theta}{r}\frac{\partial}{\partial \theta} \tag{5.51}$$

$$\vec{A} = A_i\hat{e}_i = A_x\hat{e}_x + A_r\hat{e}_r + A_\theta\hat{e}_\theta \tag{5.52}$$

$$\nabla \cdot \vec{A} = \frac{\partial A_x}{\partial x} + \frac{1}{r}\frac{\partial}{\partial r}(rA_r) + \frac{1}{r}\frac{\partial A_\theta}{\partial \theta} \tag{5.53}$$

$$\nabla \times \vec{A} = \frac{1}{h_x h_r h_\theta}\begin{vmatrix} h_x\hat{e}_x & h_r\hat{e}_r & h_\theta\hat{e}_\theta \\ \frac{\partial}{\partial x} & \frac{\partial}{\partial r} & \frac{\partial}{\partial \theta} \\ h_x A_x & h_r A_r & h_\theta A_\theta \end{vmatrix} = \frac{1}{r}\begin{vmatrix} \hat{e}_x & \hat{e}_r & \hat{e}_\theta \\ \frac{\partial}{\partial x} & \frac{\partial}{\partial r} & \frac{\partial}{\partial \theta} \\ A_x & A_r & rA_\theta \end{vmatrix}$$

$$= \frac{1}{r}\left(\frac{\partial}{\partial r}(rA_\theta) - \frac{\partial A_r}{\partial \theta}\right)\hat{e}_x + \left(\frac{1}{r}\frac{\partial A_x}{\partial \theta} - \frac{\partial A_\theta}{\partial x}\right)\hat{e}_r + \left(\frac{\partial A_r}{\partial x} - \frac{\partial A_x}{\partial r}\right)\hat{e}_\theta \tag{5.54}$$

Let x be the symmetry axis of the flow field, so that

$$\frac{\partial}{\partial \theta} = 0 \tag{5.55}$$

We write the velocity as

$$\vec{w} = w_i\hat{e}_i = w_x\hat{e}_x + w_r\hat{e}_r + w_\theta\hat{e}_\theta \tag{5.56}$$

and note that w_θ need not be zero for an axisymmetric flow. In fact, this type of swirling axisymmetric flow is of some importance. Swirl occurs in tubular flow in some heat exchangers, tornadoes have swirl, and nozzle flows with swirl have been of interest (Chang and Merkle, 1989). We therefore include the possibility of swirl and write the axisymmetric counterparts of Equations (5.48) and (5.49) as

$$\vec{\omega} = \frac{1}{r}\frac{\partial(rw_\theta)}{\partial r}\hat{e}_x - \frac{\partial w_\theta}{\partial x}\hat{e}_r + \left(\frac{\partial w_r}{\partial x} - \frac{\partial w_x}{\partial r}\right)\hat{e}_\theta \tag{5.57}$$

$$\begin{aligned}\vec{\omega}\times\vec{w} &= \begin{vmatrix} \hat{e}_x & \hat{e}_r & \hat{e}_\theta \\ \omega_x & \omega_r & \omega_\theta \\ w_x & w_r & w_\theta \end{vmatrix} = -\left[w_\theta\frac{\partial w_\theta}{\partial x} + w_r\left(\frac{\partial w_r}{\partial x} - \frac{\partial w_x}{\partial r}\right)\right]\hat{e}_x \\ &+ \left[w_x\left(\frac{\partial w_r}{\partial x} - \frac{\partial w_x}{\partial r}\right) - \frac{w_\theta}{r}\frac{\partial(rw_\theta)}{\partial r}\right]\hat{e}_r \\ &+ \left[\frac{w_r}{r}\frac{\partial(rw_\theta)}{\partial r} + w_x\frac{\partial w_\theta}{\partial x}\right]\hat{e}_\theta \end{aligned} \tag{5.58}$$

A steady, homenergetic flow of a perfect gas is now assumed. The governing axisymmetric equations, in scalar form, are

$$\frac{\partial(\rho w_x)}{\partial x} + \frac{1}{r}\frac{\partial}{\partial r}(r\rho w_r) = 0$$

$$\frac{\partial}{\partial x}\left(\frac{w^2}{2}\right) - w_\theta\frac{\partial w_\theta}{\partial x} - w_r\left(\frac{\partial w_r}{\partial x} - \frac{\partial w_x}{\partial r}\right) + \frac{1}{\rho}\frac{\partial p}{\partial x} = 0$$

$$\frac{\partial}{\partial r}\left(\frac{w^2}{2}\right) + w_x\left(\frac{\partial w_r}{\partial x} - \frac{\partial w_x}{\partial r}\right) - \frac{w_\theta}{r}\frac{\partial}{\partial r}(rw_\theta) + \frac{1}{\rho}\frac{\partial p}{\partial r} = 0$$

$$\frac{w_r}{r}\frac{\partial(rw_\theta)}{\partial r} + w_x\frac{\partial w_\theta}{\partial x} = 0$$

$$\frac{\gamma}{\gamma - 1}\frac{p}{\rho} + \frac{1}{2}w^2 = h_o$$

where h_o is a constant and

$$w^2 = w_x^2 + w_r^2 + w_\theta^2 \tag{5.59}$$

These equations can be written more simply as

$$\frac{\partial(r\rho w_x)}{\partial x} + \frac{\partial}{\partial r}(r\rho w_r) = 0 \tag{5.60a}$$

$$w_x\frac{\partial w_x}{\partial x} + w_r\frac{\partial w_x}{\partial r} + \frac{1}{\rho}\frac{\partial p}{\partial x} = 0 \tag{5.60b}$$

$$w_x\frac{\partial w_r}{\partial x} + w_r\frac{\partial w_r}{\partial r} - \frac{w_\theta^2}{r} + \frac{1}{\rho}\frac{\partial p}{\partial r} = 0 \tag{5.60c}$$

$$w_x\frac{\partial(rw_\theta)}{\partial x} + w_r\frac{\partial(rw_\theta)}{\partial r} = 0 \tag{5.60d}$$

$$\frac{\gamma}{\gamma - 1}\frac{r}{\rho} + \frac{1}{2}w^2 = h_o \tag{5.60e}$$

For steady, axisymmetric flow, the substantial derivative of a scalar or a vector is

$$\frac{D}{Dt} = \vec{w} \cdot \nabla = w_x\frac{\partial}{\partial x} + w_r\frac{\partial}{\partial r} + \frac{w_\theta}{r}\frac{\partial}{\partial \theta}$$

There are two exceptions to Equation (5.55) given by (see Appendix A, Table 4)

$$\frac{\partial \hat{e}_r}{\partial \theta} = \hat{e}_\theta, \qquad \frac{\partial \hat{e}_\theta}{\partial \theta} = -\hat{e}_r \tag{5.61}$$

(The other seven basis derivatives are zero.) Thus, the substantial derivative of any scalar ϕ is

$$\frac{D\phi}{Dt} = w_x\frac{\partial \phi}{\partial x} + w_r\frac{\partial \phi}{\partial r} \tag{5.62}$$

while the acceleration is

$$\frac{D\vec{w}}{Dt} = \vec{w}\cdot\nabla\vec{w} = \left(w_x\frac{\partial w_x}{\partial x} + w_r\frac{\partial w_x}{\partial r}\right)\hat{e}_x$$
$$+\left(w_x\frac{\partial w_r}{\partial x} + w_r\frac{\partial w_r}{\partial r} - \frac{w_\theta^2}{r}\right)\hat{e}_r + \left(w_x\frac{\partial w_\theta}{\partial x} + w_r\frac{\partial w_\theta}{\partial r} - \frac{w_r w_\theta}{r}\right)\hat{e}_\theta$$

where the w_θ^2/r and $w_r w_\theta/r$ terms stem from Equations (5.61). The acceleration can therefore be written as

$$\frac{D\vec{w}}{Dt} = \frac{Dw_x}{Dt}\hat{e}_x + \left(\frac{Dw_r}{Dt} - \frac{w_\theta^2}{r}\right)\hat{e}_r + \left(\frac{Dw_\theta}{Dt} - \frac{w_r w_\theta}{r}\right)\hat{e}_\theta$$
$$= \frac{Dw_x}{Dt}\hat{e}_x + \left(\frac{Dw_r}{Dt} - \frac{w_\theta^2}{r}\right)\hat{e}_r + \frac{1}{r}\frac{D(rw_\theta)}{Dt}\hat{e}_\theta \qquad \textbf{(5.63)}$$

By comparing with Equations (5.60b to d), we have

$$\frac{Dw_x}{Dt} + \frac{1}{\rho}\frac{\partial p}{\partial x} = 0 \qquad \textbf{(5.64a)}$$

$$\frac{Dw_r}{Dt} - \frac{w_\theta^2}{r} + \frac{1}{\rho}\frac{\partial p}{\partial r} = 0 \qquad \textbf{(5.64b)}$$

$$\frac{D(rw_\theta)}{Dt} = 0 \qquad \textbf{(5.64c)}$$

for the momentum equations. Notice that in an inviscid flow with swirl, rw_θ is constant for a fluid particle.

Combined Two-Dimensional and Axisymmetric Flow

We are now in a position to combine the two-dimensional and axisymmetric governing equations, in scalar form, by defining

$$\sigma = \begin{cases} 0, & \text{two-dimensional flow} \\ 1, & \text{axisymmetric flow} \end{cases} \qquad \textbf{(5.65)}$$

$$\begin{array}{lll} x \to x_1, & y, r \to x_2, & z, \theta \to x_3 \\ h_1 = 1, & h_2 = 1, & h_3 = x_2^\sigma \\ \hat{\imath}_1, \hat{e}_x \to \hat{\imath}_1, & \hat{\imath}_2, \hat{e}_r \to \hat{\imath}_2, & \hat{\imath}_3, \hat{e}_\theta \to \hat{\imath}_3 \\ u, w_x \to w_1, & v, w_r \to w_2, & w_z, w_\theta \to w_3 \end{array} \qquad \textbf{(5.66)}$$

where z, $\hat{\imath}_3$, and w_z are in the plane that is perpendicular to a two-dimensional flow. Observe that the orthonormal basis $\hat{\imath}_j$ is not Cartesian when the flow is axisymmetric. With Equations (5.65) and (5.66), we finally obtain

$$\frac{\partial(x_2^\sigma \rho w_1)}{\partial x_1} + \frac{\partial(x_2^\sigma \rho w_2)}{\partial x_2} = 0 \qquad \textbf{(5.67a)}$$

$$\frac{Dw_1}{Dt} + \frac{1}{\rho}\frac{\partial p}{\partial x_1} = 0 \qquad \textbf{(5.67b)}$$

$$\frac{Dw_2}{Dt} - \frac{\sigma w_3^2}{x_2} + \frac{1}{\rho}\frac{\partial p}{\partial x_2} = 0 \qquad \textbf{(5.67c)}$$

$$\frac{D}{Dt}(x_2^\sigma w^3) = 0 \qquad \textbf{(5.67d)}$$

$$\frac{\gamma}{\gamma-1}\frac{p}{\rho} + \frac{1}{2}w^2 = h_o \qquad \textbf{(5.67e)}$$

where

$$\frac{D}{Dt} = w_1\frac{\partial}{\partial x_1} + w_2\frac{\partial}{\partial x_2}, \qquad w^2 = w_i w_i \qquad \textbf{(5.68)}$$

In this notation, the vorticity is

$$\vec{\omega} = \frac{\sigma}{x_2}\frac{\partial(x_2^\sigma w_3)}{\partial x_2}\hat{\imath}_1 - \sigma\frac{\partial w_3}{\partial x_1}\hat{\imath}_2 + \left(\frac{\partial w_2}{\partial x_1} - \frac{\partial w_1}{\partial x_2}\right)\hat{\imath}_3 \qquad \textbf{(5.69)}$$

These equations allow for w_z = constant and for swirl. A two-dimensional flow has sweep when w_z is nonzero, and the projection of the streamlines in the x_1,x_3 plane is curved. Several flows of interest with w_z equal to a nonzero constant occur in supersonic gas dynamics (see Chapter 10). When the flow is axisymmetric with swirl, the streamlines are roughly helical. In either case, the Mach number, based on the velocity that includes the w_z or w_θ component, exceeds the Mach number for the symmetry plane flow field. Thus, the flow in the symmetry plane may be subsonic, while the three-dimensional flow may be supersonic. It is worth mentioning that Problem 5.8 deals with spiral flow, which is a flow with swirl. Problem 5.21 provides the helical counterpart to the spiral flow of Problem 5.8.

Stream Function

The above equations are utilized to examine several common simplifications or assumptions. For instance, a stream function ψ can be introduced by setting

$$\frac{\partial \psi}{\partial x_1} = -\rho w_2 x_2^\sigma, \qquad \frac{\partial \psi}{\partial x_2} = \rho w_1 x_2^\sigma \qquad \textbf{(5.70)}$$

which satisfies continuity for a steady flow. These relations are used to replace w_1 and w_2, and $D(\,)/Dt$ becomes

$$\frac{D}{Dt} = \frac{1}{\rho x_2^\sigma}\left(\frac{\partial\psi}{\partial x_2}\frac{\partial}{\partial x_1} - \frac{\partial\psi}{\partial x_1}\frac{\partial}{\partial x_2}\right) \tag{5.71}$$

The physical meaning of the stream function can be found by first determining the equation for a streamline, Equation (1.21a). For this, we need

$$d\vec{r} = h_i dq^i \hat{e}_i = h_i dx_i \hat{i}_i \tag{5.72}$$

so that

$$d\vec{r} \times \vec{w} = \begin{vmatrix} \hat{i}_i & \hat{i}_2 & \hat{i}_3 \\ dx_1 & dx_2 & x_2^\sigma dx_3 \\ w_1 & w_2 & w_3 \end{vmatrix} = 0$$

In scalar form, we have

$$\frac{dx_1}{w_1} = \frac{dx_2}{w_2} = x_2^\sigma \frac{dx_3}{w_3} \tag{5.73}$$

which is in accordance with Equation (1.21b). If $w_3 = 0$, then $dx_3 = 0$ and the dx_3 term is deleted. However, we also have

$$\begin{aligned} \psi &= \psi(x_1, x_2) \\ d\psi &= \frac{\partial\psi}{\partial x_1}dx_1 + \frac{\partial\psi}{\partial x_2}dx_2 = \rho x_2^\sigma(-w_2 dx_1 + w_1 dx_2) \\ &= -\rho x_2^\sigma w_1 w_2\left(\frac{dx_1}{w_1} - \frac{dx_2}{w_2}\right) \end{aligned}$$

Thus, $d\psi = 0$ along streamlines, and we have the conventional interpretation that ψ is constant on streamlines or stream surfaces.

In a three-dimensional steady flow, two stream functions, ψ_1 and ψ_2, are necessary. One can then show that the relation

$$\rho\vec{w} = (\nabla\psi_1)\times(\nabla\psi_2)$$

satisfies continuity. This approach, however, is seldom used. There is a second type of stream function called a Crocco stream function (Emanuel, 1986) and it also is seldom used.

VELOCITY POTENTIAL FUNCTION AND THE GAS DYNAMIC EQUATION

As we know, an irrotational flow is a commonly encountered situation. From Equation (5.69), we obtain

$$\frac{\partial w_1}{\partial x_2} = \frac{\partial w_2}{\partial x_1} \tag{5.74}$$

when the flow is two-dimensional or axisymmetric, and

$$\frac{\partial (x_2^{\sigma} w_3)}{\partial x_2} = 0, \qquad \frac{\partial w_3}{\partial x_1} = 0$$

when the swirling flow is axisymmetric. These latter relations yield for the swirl component

$$w_3 = \frac{c_1}{x_2^{\sigma}} \tag{5.75}$$

which also holds when the flow is two dimensional, and where c_i, hereafter, denotes integration constants. This result is in accord with Equation (5.67d).

In accord with Section 4.2, a velocity potential function can be introduced as

$$\vec{w} = \nabla \phi \tag{5.76a}$$

or in component form,

$$w_1 = \frac{\partial \phi}{\partial x_1}, \qquad w_2 = \frac{\partial \phi}{\partial x_2}, \qquad w_3 = \frac{1}{x_2^{\sigma}} \frac{\partial \phi}{\partial x_3} \tag{5.76b}$$

With swirl or sweep, $\partial\phi/\partial x_3$ is not zero but equals c_1. By integrating the w_3 equation, we obtain

$$\phi = c_1 x_3 + c_2 + \bar{\phi}(x_1, x_2)$$

Since

$$w_1 = \frac{\partial \phi}{\partial x_1}, \qquad w_2 = \frac{\partial \phi}{\partial x_2} \tag{5.77}$$

Equation (5.74), for irrotationality, is satisfied.

The governing equations can be replaced with a single partial differential equation (PDE), where ϕ is the dependent variable. We start with continuity and write

$$x_2^{\sigma} \frac{\partial}{\partial x_1}(\rho \phi_{x_1}) + \frac{\partial}{\partial x_2}(x_2^{\sigma} \rho \phi_{x_2}) = 0$$

where the convenient shorthand notation

$$\phi_{x_i} = \frac{\partial \phi}{\partial x_i}$$

is used. The derivatives are expanded to yield

$$x_2^\sigma(\rho\phi_{x_1x_1} + \phi_{x_1}\rho_{x_1}) + \sigma\rho\phi_{x_2} + x_2^\sigma(\rho\phi_{x_2x_2} + \phi_{x_2}\rho_{x_2}) = 0$$

or

$$\phi_{x_1x_1} + \phi_{x_2x_2} + \frac{\rho_{x_1}}{\rho}\phi_{x_1} + \frac{\rho_{x_2}}{\rho}\phi_{x_2} + \frac{\sigma\phi_{x_2}}{x_2} = 0 \tag{5.78}$$

The x_1 and x_2 momentum equations have the form

$$\phi_{x_1}\phi_{x_1x_2} + \phi_{x_2}\phi_{x_1x_2} + \frac{1}{\rho}p_{x_1} = 0 \tag{5.79a}$$

$$\phi_{x_1}\phi_{x_1x_2} + \phi_{x_2}\phi_{x_2x_2} - \frac{\sigma c_1^2}{x_2^3} + \frac{1}{\rho}p_{x_2} = 0 \tag{5.79b}$$

In order to replace the density-containing factors in Equation (5.78), we further assume homenergetic flow of a perfect gas. Thus, Equation (5.44) provides the speed of sound and Equation (5.67e) becomes

$$a^2 = a_o^2 - \frac{\gamma-1}{2}\left(\phi_{x_1}^2 + \phi_{x_2}^2 + \frac{c_1^2}{x_2^{2\sigma}}\right) \tag{5.80}$$

where

$$a_o^2 = (\gamma-1)h_o = \gamma\frac{p_o}{\rho_o}$$

Since the flow is steady, irrotational, and homenergetic, Crocco's equation yields Equation (5.23) and the flow is also homentropic or barotropic. Consequently, Equation (5.22) yields

$$p_{x_i} = \frac{dp}{d\rho}\rho_{x_i} = a^2\rho_{x_i}, \qquad i = 1, 2 \tag{5.81}$$

since the speed of sound is defined by

$$a^2 = \left(\frac{\partial p}{\partial \rho}\right)_s \tag{5.82}$$

Equations (5.79) and (5.81) are solved for ρ_{x_1}/ρ, with the result

$$\frac{\rho_{x_1}}{\rho} = -\frac{1}{a^2}(\phi_{x_1}\phi_{x_1x_1} + \phi_{x_2}\phi_{x_1x_2})$$

$$\frac{\rho_{x_2}}{\rho} = -\frac{1}{a^2}\left(\phi_{x_1}\phi_{x_1x_2} + \phi_{x_2}\phi_{x_2x_2} - \frac{\sigma c_1^2}{x_2^3}\right)$$

These relations are substituted into Equation (5.78), to yield

$$(a^2 - \phi_{x_1}^2)\phi_{x_1x_1} - 2\phi_{x_1}\phi_{x_2}\phi_{x_1x_2} + (a^2 - \phi_{x_2}^2)\phi_{x_2x_2} + \left(a^2 + \frac{c_1^2}{x_2^2}\right)\frac{\sigma}{x_2}\phi_{x_2} = 0 \qquad \textbf{(5.83)}$$

where Equation (5.80) provides a^2. This relation is referred to as the gas dynamic equation; it is a second-order, quasilinear PDE, even when swirl or sweep (with $c_1 \neq 0$) is present. (A quasilinear PDE is one where the equation is linear in its highest-order derivatives, which here are second order.) When an equation is quasilinear and hyperbolic, the method-of-characteristics can be used to obtain a solution. Once ϕ is known, w_1 and w_2 are provided by Equation (5.77), while a^2 is obtained from Equation (5.80). Since the flow is barotropic, the pressure is given by

$$p = \left(\frac{\rho_o a^2}{\gamma p_o^{1/\gamma}}\right)^{\gamma/(\gamma-1)}$$

Because the flow is steady and irrotational, a Bernoulli equation can also be used. As shown by Problem 5.17, an analogous gas dynamic equation can be obtained for a rotational flow by introducing the stream function defined by Equations (5.70). Part (d) of this problem demonstrates that the rotational gas dynamic equation is also quasilinear. Problem 5.27 involves the derivation of an unsteady, three-dimensional gas dynamic equation (Pai, 1959).

5.7 NATURAL COORDINATES

Natural coordinates are occasionally used for a two-dimensional flow. Although we will also consider their axisymmetric form, without swirl, they are rarely used in this situation because of a $1/x_2$ factor that appears in the continuity equation. In either case, the resulting equations, called the intrinsic equations of motion (Serrin, 1959), have the simplest form possible for the steady Euler equations. Their physical interpretation, in this form, is relatively straightforward.

As shown in Figure 5.4, the coordinates consist of ξ_1, which is along the streamlines, and ξ_2, which is orthogonal to ξ_1 and in the plane of the flow. A third coordinate, ξ_3, is orthogonal to both ξ_1 and ξ_2; it equals x_3 when the flow is two-dimensional and is the aximuthal angle when the flow is axisymmetric. (The ξ_i coordinates are not natural coordinates, which are defined in the next paragraph.) The figure also shows the orthonormal basis $\hat{e}_i$ and the angle θ of the velocity relative to the x_1-axis. This angle will prove quite useful in the analysis.

Since the ξ_i are orthogonal, the h_i scale factors can be introduced as (Emanuel, 1986)

$$\frac{1}{h_2}\frac{\partial h_1}{\partial \xi_2} = -\frac{\partial \theta}{\partial \xi_1}, \qquad \frac{1}{h_1}\frac{\partial h_2}{\partial \xi_1} = \frac{\partial \theta}{\partial \xi_2}, \qquad h_3 = x_2^\sigma \qquad \textbf{(5.84)}$$

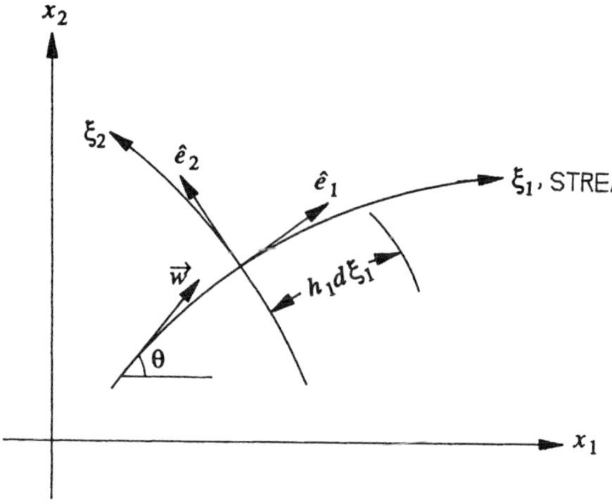

FIGURE 5.4 Natural coordinates.

Let s and n be natural coordinates, where s is arc length along a streamline and n is the arc length normal to a streamline and in the plane of the flow. Consequently, we have (see Figure 5.4 for ∂s)

$$\frac{\partial}{\partial s} = \frac{1}{h_1}\frac{\partial}{\partial \xi_1}, \qquad \frac{\partial}{\partial n} = \frac{1}{h_2}\frac{\partial}{\partial \xi_2} \tag{5.85}$$

For ξ_1 and ξ_2 to provide a legitimate coordinate system, they must satisfy a compatibility condition

$$\frac{\partial^2 \theta}{\partial \xi_1 \partial \xi_2} = \frac{\partial^2 \theta}{\partial \xi_2 \partial \xi_1}$$

In conjunction with Equations (5.84), this yields the Gauss equation

$$\frac{\partial}{\partial \xi_1}\left(\frac{1}{h_1}\frac{\partial h_2}{\partial \xi_1}\right) + \frac{\partial}{\partial \xi_2}\left(\frac{1}{h_2}\frac{\partial h_1}{\partial \xi_2}\right) = 0$$

for the scale factors.

The transformation between Cartesian and the ξ_i coordinates can be written as

$$x_i = x_i(\xi_1, \xi_2), \quad i = 1, 2 \tag{5.86}$$

There is no need to consider ξ_3, since

$$\xi_3 = x_3, \qquad \sigma = 0$$
$$\xi_3 = \phi, \qquad \sigma = 1$$

where ϕ is the aximuthal angle. Once the transformation is known, the scale factors are given by (Emanuel, 1986)

$$h_j = \left[\left(\frac{\partial x_1}{\partial \xi_i}\right)^2 + \left(\frac{\partial x_2}{\partial \xi_i}\right)^2\right]^{1/2}, \quad j = 1, 2$$

The $\hat{e}_1$ and $\hat{e}_2$ vectors can be evaluated without an explicit knowledge of Equations (5.86). This is fortunate, since these equations are generally not known *a priori*; their determination requires a simultaneous solution of the flow field and several PDEs for the x_i. (This is still another reason natural coordinates are rarely encountered.) For the $\hat{e}_i$ evaluation, we utilize

$$\hat{e}_i \cdot \hat{e}_j = \delta_{ij}, \quad i, j = 1, 2$$
$$\vec{w} = w\hat{e}_1 = u\hat{i}_1 + v\hat{i}_2 = w\cos\theta\hat{i}_1 + w\sin\theta\hat{i}_2 \tag{5.87}$$

where δ_{ij} is the Kronecker delta. From these relations, we easily obtain

$$\hat{e}_1 = \cos\theta\hat{i}_1 + \sin\theta\hat{i}_2$$
$$\hat{e}_2 = -\sin\theta\hat{i}_1 + \cos\theta\hat{i}_2 \tag{5.88a}$$

or their inverse

$$\hat{i}_1 = \cos\theta\hat{e}_1 - \sin\theta\hat{e}_2$$
$$\hat{i}_2 = \sin\theta\hat{e}_1 + \cos\theta\hat{e}_2 \tag{5.88b}$$

For the del operator, we have (see Appendix A)

$$\nabla = \frac{\hat{e}_i}{h_i}\frac{\partial}{\partial\xi_i}$$

With $\partial(\ )/\partial\xi_3 = 0$ (except for $\hat{e}_1$ and $\hat{e}_2$) and Equation (5.85), we obtain

$$\nabla = \frac{\hat{e}_1}{h_1}\frac{\partial}{\partial\xi_1} + \frac{\hat{e}_2}{h_2}\frac{\partial}{\partial\xi_2} + \frac{\hat{e}_3}{h_3}\frac{\partial}{\partial\xi_3} = \hat{e}_1\frac{\partial}{\partial s} + \hat{e}_2\frac{\partial}{\partial n} + \frac{\hat{e}_3}{x_2^\sigma}\frac{\partial}{\partial\xi_3} \tag{5.89}$$

We will also need the $\partial\hat{e}_j/\partial\xi_i$ derivatives, obtained from Equation (5.88a) as

$$\frac{\partial\hat{e}_1}{\partial\xi_i} = -\sin\theta\frac{\partial\theta}{\partial\xi_i}\hat{i}_1 + \cos\theta\frac{\partial\theta}{\partial\xi_i}\hat{i}_2 = \frac{\partial\theta}{\partial\xi_i}\hat{e}_2, \quad i = 1, 2 \tag{5.90a}$$

$$\frac{\partial\hat{e}_2}{\partial\xi_i} = -\cos\theta\frac{\partial\theta}{\partial\xi_i}\hat{i}_1 - \sin\theta\frac{\partial\theta}{\partial\xi_i}\hat{i}_2 = -\frac{\partial\theta}{\partial\xi_i}\hat{e}_1, \quad i = 1, 2 \tag{5.90b}$$

For the derivation of the continuity equation, given by $\nabla \cdot (\rho w\hat{e}_1) = 0$, we also need (see Appendix A, Table 4)

$$\frac{\partial\hat{e}_1}{\partial\xi_3} = \frac{1}{h_1}\frac{\partial h_3}{\partial\xi_1}\hat{e}_3 = \frac{\sigma}{h_1}\frac{\partial x_2}{\partial\xi_1}\hat{e}_3 = \sigma\sin\theta\hat{e}_3 \tag{5.90c}$$

where (Emanuel, 1986)

$$\frac{\partial x_2}{\partial \xi_1} = h_1 \sin\theta$$

The foregoing relations result in the following form for the Euler equations:

$$\frac{\partial(\rho w)}{\partial s} + \rho w \frac{\partial \theta}{\partial n} + \frac{\sigma \rho w \sin\theta}{x_2} = 0 \tag{5.91a}$$

$$w\frac{\partial w}{\partial s} + \frac{1}{\rho}\frac{\partial p}{\partial s} = 0 \tag{5.91b}$$

$$w^2\frac{\partial \theta}{\partial s} + \frac{1}{\rho}\frac{\partial p}{\partial n} = 0 \tag{5.91c}$$

$$h + \frac{1}{2}w^2 = h_o(n) \tag{5.91d}$$

where $h_o(n)$ can be evaluated in the upstream flow. Observe the $1/x_2$ factor in Equation (5.91a) that was mentioned at the start of this section. When the flow is axisymmetric ($\sigma = 1$), x_2 is an unknown function of s and n. Note the simple form for the momentum equations. Since the radius of curvature of a streamline is $-(\partial\theta/\partial s)^{-1}$, the transverse pressure gradient is equal to ρw^2 divided by this radius. Thus, the transverse pressure gradient balances the angular momentum. When the streamlines are curved as indicated in Figure 5.4, $\partial\theta/\partial s < 0$ and $\partial p/\partial n$ is positive.

Also of interest is the vorticity, which is evaluated as

$$\vec{\omega} = \nabla \times \vec{w} = \frac{1}{h_1 h_2 h_3}\begin{vmatrix} h_1\hat{e}_1 & h_2\hat{e}_2 & h_3\hat{e}_3 \\ \frac{\partial}{\partial \xi_1} & \frac{\partial}{\partial \xi_2} & \frac{\partial}{\partial \xi_3} \\ h_1 w_1 & 0 & 0 \end{vmatrix}$$

$$= \frac{1}{h_1 h_2 h_3}\left[h_2\hat{e}_2\frac{\partial}{\partial \xi_3}(h_1 w) - h_3\hat{e}_3\frac{\partial}{\partial \xi_2}(h_1 w)\right] \tag{5.92}$$

However, we have

$$\frac{\partial}{\partial \xi_3}(h_1 w) = 0$$

and

$$\frac{\partial h_1}{\partial \xi_2} = -h_2\frac{\partial \theta}{\partial \xi_1} = -h_1 h_2\frac{\partial \theta}{\partial s}$$

with the result

$$\vec{\omega} = \left(-\frac{\partial w}{\partial n} + w\frac{\partial \theta}{\partial s}\right)\hat{e}_3 \tag{5.93a}$$

or

$$\vec{\omega} = -\left(\frac{\partial w}{\partial n} + \frac{1}{\rho w}\frac{\partial p}{\partial n}\right)\hat{e}_3 \tag{5.93b}$$

These relations can also be obtained from Equation (5.69) with $w_3 = 0$.

THREE-DIMENSIONAL FLOW

Section 20 of Serrin (1959) utilizes the Serret–Frenet equations in order to extend natural coordinates to a three-dimensional flow. These equations have the form (Struik, 1950)

$$\begin{aligned} \frac{\partial \hat{t}}{\partial s} &= \kappa \hat{n} \\ \frac{\partial \hat{n}}{\partial s} &= -\kappa \hat{t} + \tau \hat{b} \\ \frac{\partial \hat{b}}{\partial s} &= -\tau \hat{n} \end{aligned} \tag{5.94}$$

where $\hat{t}$, $\hat{n}$, and $\hat{b}$ represent an orthonormal basis. The vector $\hat{t}$ is tangent to the streamlines, $\hat{n}$ is normal to the streamlines and lies in the osculating plane, and $\hat{b}$ is the binormal vector, which equals $\hat{t} \times \hat{n}$. (Each point of each streamline has its own osculating plane. At a given point, this is the plane defined by a circle that is tangent to the streamline at the point in question. The vectors $\hat{t}$ and $\hat{n}$ are in the osculating plane with $\hat{n}$ pointing toward the center of the tangent circle.) The parameters κ and τ are the curvature and torsion, respectively, of a particular streamline and depend on the arc length s along the streamline. One elegant result of formulating the three-dimensional Euler equations in this manner is that Equations (5.91b) and (5.91c) still hold, while the momentum equation in the binormal direction reduces to a zero pressure gradient in this direction (Serrin, 1959).

For an orthogonal system, ξ_j, the derivatives of the basis vectors are given by (see Appendix A, Table 4)

$$\frac{\partial \hat{e}_i}{\partial \xi_j} = \sum_{k \neq i}\left(\frac{\delta_{jk}}{h_i}\frac{\partial h_j}{\partial \xi_i} - \frac{\delta_{ij}}{h_k}\frac{\partial h_i}{\partial \xi_k}\right)\hat{e}_k$$

where we take ξ_1 to be along a streamline. For this coordinate, we have

$$\frac{\partial \hat{e}_1}{\partial \xi_1} = -\frac{1}{h_2}\frac{\partial h_1}{\partial \xi_2}\hat{e}_2 - \frac{1}{h_3}\frac{\partial h_1}{\partial \xi_3}\hat{e}_3, \quad \frac{\partial \hat{e}_2}{\partial \xi_1} = \frac{1}{h_2}\frac{\partial h_1}{\partial \xi_2}\hat{e}_1, \quad \frac{\partial \hat{e}_3}{\partial \xi_1} = \frac{1}{h_3}\frac{\partial h_1}{\partial \xi_3}\hat{e}_1 \tag{5.95}$$

where the left sides are converted to s derivatives by multiplying by $1/h_1$. Since both bases are orthonormal, we can set

$$\begin{aligned}\hat{t} &= \hat{e}_1\\ \hat{n} &= -\cos\alpha\hat{e}_2 + \sin\alpha\hat{e}_3\\ \hat{b} &= -\sin\alpha\hat{e}_2 + \cos\alpha\hat{e}_3\end{aligned} \tag{5.96}$$

where the angle α is a function of the ξ_j coordinates. Thus, $\hat{n}$ and $\hat{b}$ are, respectively, rotated about a streamline by α starting from $-\hat{e}_2$ and $-\hat{e}_3$. By comparing Equations (5.95) and (5.96) with Equations (5.94), we obtain

$$\begin{aligned}\tan\alpha &= -\frac{h_2}{h_3}\frac{\dfrac{\partial h_1}{\partial \xi_3}}{\dfrac{\partial h_1}{\partial \xi_2}}\\ \kappa &= -\frac{1}{h_1 h_3}\frac{\partial h_1}{\partial \xi_3}\csc\alpha\\ \tau &= -\frac{1}{h_1}\frac{\partial \alpha}{\partial \xi_1} = -\frac{\partial \alpha}{\partial s}\end{aligned} \tag{5.97}$$

For a two-dimensional or axisymmetric flow,

$$\alpha = 0, \qquad \tau = 0, \qquad \hat{n} = -\hat{e}_2, \qquad \hat{b} = -\hat{e}_3$$

and κ is indeterminate, since the derivative, $\partial h_1/\partial \xi_3$, is zero. For a three-dimensional flow, Equations (5.97) provide the link between the h_i scale factors and the curvature κ and torsion τ of a streamline. (See Emanuel, 1993, for the natural coordinate formulation for an unsteady, viscous flow.)

REFERENCES

Chang, C. and Merkle, C., "Viscous Swirling Nozzle Flow," AIAA 89-0280 (1989).

Délery, J.M., "Physics of Vortical Flows," *J. Aircraft* 29, 856 (1992).

Emanuel, G., *Gasdynamics: Theory and Applications*, AIAA Education Series, Washington, D.C., 1986.

Emanuel, G., "Unsteady Natural Coordinates for a Viscous Compressible Flow," *Phys. Fluids* A5, 294 (1993).

Greitzer, E.M., "A New Exact Solution for Compressible Swirling Flow," *Aero. Quart.* 26, 297 (1975).

Hunt, J.C.R. and Hussain, F., "A Note on Velocity, Vorticity and Helicity of Inviscid Fluid Elements," *J. Fluid Mech.* 229, 569 (1991).

Levy, Y., Degani, D., and Seginer, A., "Graphical Visualization of Vortical Flows by Means of Helicity," *AIAA J.* 28, 1347 (1990).

Morino, L., "Material Contravariant Components: Vorticity Transport and Vortex Theorems," *AIAA J.* 24, 526 (1986).

Pai, S.-I., *Introduction to the Theory of Compressible Flow*, D. Van Nostrand Co., New York, 1959, p.76.

Serrin, J., "Mathematical Principles of Classical Fluid Mechanics," in *Encyclopedia of Physics,* edited by S. Flügge, Vol. VIII/1, Springer-Verlag, Berlin, 1959.

Struik, D.J., *Differential Geometry*, Addison-Wesley Press, Cambridge, MA, 1950.

Vazsonyi, A., "On Rotational Flows," *Quart. Appl. Math.* 3, 29 (1945).

Wu, C.S. and Hayes, W.D., "Crocco's Vorticity Law in a Non-Uniform Material," *Quart. Appl. Math.* 16, 81 (1958).

PROBLEMS

5.1 Consider unsteady, one-dimensional flow with x as the spatial coordinate.

(a) Use the Euler equations, without a body force, and write the governing equations in scalar form using h instead of h_o.

(b) With $h = h(p,\rho)$, eliminate the enthalpy and derive three equations whose left sides consist only of ρ_t, u_t, and p_t.

(c) Introduce the Mach number $M(= u/a)$ and speed of sound, which can be defined as

$$a^2 = \frac{\rho h_\rho}{1 - \rho h_p}$$

Use $a = a(p,\rho)$ to eliminate the a_t and a_x derivatives and derive equations where only ρ_t, p_t, and M_t appear on the left sides.

(d) Simplify your answer to part (c) by assuming a perfect gas. Write your results in terms of ratios, such as M_t/M, M_x/M, p_t/p, ...

5.2 Derive Equation (5.26), Bernoulli's energy equation, starting with Equation (2.21). List the assumptions required for the derivation.

5.3 **(a)** Consider an unsteady, spherically symmetric flow without a body force. In this flow

$$\vec{w} = w(r, t)\hat{e}_r$$

Write the Euler equations in scalar form using r and t as the independent variables and w, p, ρ, and h_o as the dependent variables.

(b) Is the flow irrotational?

(c) What can you conclude about $\partial w/\partial t$ if the flow is homentropic? For part (a) you may want to use results from Problem 2.9.

5.4 Start with the steady Euler equations without a body force and derive the most general solution possible for a parallel flow, i.e., a flow where the velocity is given by

$$\vec{w} = w_1(x_1, x_2, x_3)\hat{1}_1$$

Your solution should be algebraic and involve several functions of integration whose particular arguments (i.e., x_1, x_2, or x_3) should be specified.

5.5 For a steady flow without a body force, the Euler momentum equation along a streamline can be written as Equation (5.91b). Assume a perfect gas; introduce the Mach number and the isentropic relations, such as Equation (5.41). Derive an equation for the stagnation pressure p_o.

5.6 Assume a steady, two-dimensional or axisymmetric (without swirl), homenergetic flow of a perfect gas. Derive the following two PDEs:

$$\frac{1}{2}\frac{\partial}{\partial x}[u^2 - (\gamma - 1)v^2] + \gamma v \frac{\partial u}{\partial y} + \gamma \frac{p}{\rho^2}\frac{\partial p}{\partial x} = 0$$

$$\frac{1}{2}\frac{\partial}{\partial y}[-(\gamma - 1)u^2 + v^2] + \gamma v \frac{\partial v}{\partial x} + \gamma \frac{p}{\rho^2}\frac{\partial p}{\partial y} = 0$$

where u and v are the usual Cartesian velocity components.

5.7 Transform the equations of Problem 5.6 such that a stream function ψ and ρ are the dependent variables.

5.8 Cylindrical coordinates, as shown in the sketch, are used for a spiral flow. This is a steady, homenergetic solution of the Euler equations in which there is no body force or flow dependence on x_3 or θ.

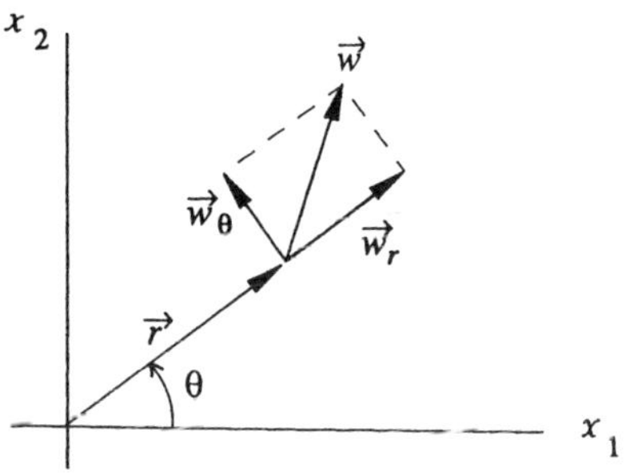

(a) Assume a perfect gas, and derive the spiral flow solution for w_θ, w_r, p, and ρ. Do not assume a homentropic flow.

(b) Introduce the mass flow rate $\dot{m}$, circulation Γ, nondimensional parameter k, and η, M_c, and X variables

$$k = \frac{\dot{m}}{\rho_o \Gamma}, \qquad M_c^2 = \frac{w_r^2 + w_\theta^2}{a^2}, \qquad \eta = \frac{2\pi a_o r}{\Gamma}, \qquad X = 1 + \frac{\gamma - 1}{2} M_c^2$$

Write the solution of part (a) for p/p_o, ρ/ρ_o, w_r/a_o, w_θ/a_o, and η in terms of γ, k, M_c, and X.

(c) Is the flow irrotational?

(d) Determine the equations for a streamline in terms of quadratures of the form

$$\theta = \theta^* + \int_1^{M_c^2} f(M_c;\gamma, k)dM_c^2, \qquad \eta = \eta^* + \int_1^{M_c^2} g(M_c;\gamma, k)dM_c^2$$

where θ^* and η^* are the θ,η values when $M_c = 1$.

5.9 Consider a compressible, steady, two-dimensional inviscid flow. With the velocity written as

$$\vec{w} = u\hat{\imath}_1 + v\hat{\imath}_2$$

the potential function ϕ and stream function ψ are defined by

$$\phi_x = u, \qquad \phi_y = v$$

$$\psi_x = -\frac{\rho}{\rho_o}v, \qquad \psi_y = -\frac{\rho}{\rho_o}u$$

where a zero subscript represents a constant stagnation quantity. Start by writing

$$\phi = \phi(x, y), \qquad \psi = \psi(x, y)$$

and derive equations for dx and dy in terms of ψ_θ and ψ_w, where θ is the angle $\vec{w}$ has relative to the x-axis. Your final answer for dx should have the form

$$dx = f\left(\theta, w, \frac{\rho}{\rho_o}, \psi_\theta, \psi_w\right)dw + g\left(\theta, w, \frac{\rho}{\rho_o}, \psi_\theta, \psi_w\right)d\theta$$

with a similar equation for dy.

5.10 **(a)** With $\vec{F}_b$ given by a potential, derive inviscid equations for $D\vec{\omega}/Dt$ and for $D(\vec{\omega}/\rho)/Dt$. Your result for $D\vec{\omega}/Dt$ is a special case of Problem 4.5.

(b) Simplify the $D\vec{\omega}/Dt$ result when the flow is homentropic.

(c) Use the results of part (a), without the homentropic assumption of part (b), to determine $\partial\omega/\partial s$ for a steady, inviscid flow, where s is the streamline arc length. The only derivatives in your answer should appear as the Jacobian $\partial(\omega,\rho)/\partial(s,n)$ (see Appendix B). What is the implication of this result for a two-dimensional flow?

5.11 Continue with Problem 5.10 using natural coordinates for a steady flow without a body force.

(a) Use S for entropy and write Crocco's vector equation in scalar form. Solve one of these equations for ω.

(b) Compare your result for ω with Equation (5.93b) and resolve any differences.

5.12 A liquid is in steady flow through a horizontal duct with a bend, as shown in the sketch. Determine the F_x and F_y components of the force experienced by the duct in terms of p_1, ρw_1^2, A_1/A_1, $A_1 \sin\alpha_1$, $A_2 \sin\alpha_2$, $A_1 \cos\alpha_1$, and $A_2 \cos\alpha_2$.

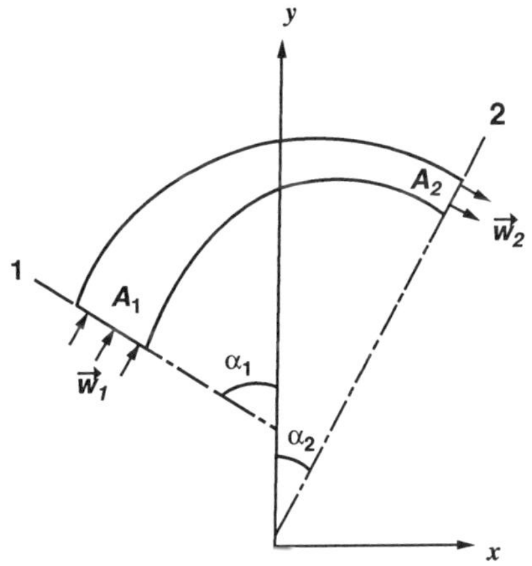

5.13 **(a)** Evaluate

$$\nabla \times \vec{a} = \nabla \times \frac{D\vec{w}}{dt}$$

for a barotropic flow with a conservative body force.

(b) Utilize continuity and Equation (4.19) to derive Beltrami's equation

$$\frac{D}{Dt}\left(\frac{\vec{\omega}}{\rho}\right) = \frac{1}{\rho}\vec{\omega} \cdot (\nabla\vec{\omega}) + \frac{1}{\rho}\nabla \times \vec{a}$$

(c) Verify that

$$\frac{\vec{\omega}}{\rho} = \frac{\vec{\omega}_o}{\rho} \cdot \nabla_{\vec{r}_o} \vec{r}$$

is an exact solution of Beltrami's equation for a barotropic flow. Here the $(\,)_o$ subscript denotes the flow at $t = 0$ and the $\vec{r}_o$ subscript means the derivative is with respect to Lagrangian coordinates; see Equation (1.19).

(d) Use the part (c) equation to show that a vortex filament is a material line in a barotropic flow.

5.14 Recall that vortex filaments are material lines when $\nabla \times \vec{a} = 0$. Beltrami's equation in Problem 5.13(b) provides the dynamics of $\vec{\omega}$ when Helmholtz's second theorem does not apply. In this circumstance there are several different ways to evaluate $\nabla \times \vec{a}$.

(a) Use the Euler momentum equation to show that

$$\nabla \times a = \frac{1}{\rho^2}(\nabla\rho) \times (\nabla p)$$

and

$$\nabla \times \vec{a} = (\nabla T) \times (\nabla s)$$

(b) Assume steady, homenergetic flow and show that

$$\nabla \times \vec{a} = T\left[\vec{w}\left(\vec{\omega} \cdot \nabla \frac{1}{T}\right) - \vec{\omega}\frac{D}{Dt}\frac{1}{T}\right]$$

(c) Use the part (b) result to show that Beltrami's equation can be written as

$$\frac{D}{Dt}\left(\frac{\vec{\omega}}{\rho T}\right) = \frac{\vec{\omega}}{\rho} \cdot \nabla \frac{\vec{w}}{T}$$

(d) Assume two-dimensional or axisymmetric flow, without swirl, and obtain the scalar version of the part (c) equation.

5.15 Assume steady flow with no body force.

(a) With the aid of Problem 5.14(a), show that

$$\frac{\vec{\omega}}{\rho} \cdot \nabla s$$

is constant along a streamline. Note that $\vec{\omega} \cdot \nabla s$ is trivially zero when the flow is two-dimensional or axisymmetric and without swirl.

(b) For an axisymmetric flow with swirl, evaluate $(\vec{\omega}/\rho) \cdot \nabla s$ under the assumptions of part (a).

5.16 Start with the initial statement of Newton's second law in Section 2.4 and assume a steady, inviscid flow without a body force.

(a) Derive the momentum theorem

$$\oint_{CS} (p\hat{n} + \rho \vec{w}\vec{w} \cdot \hat{n})ds = 0$$

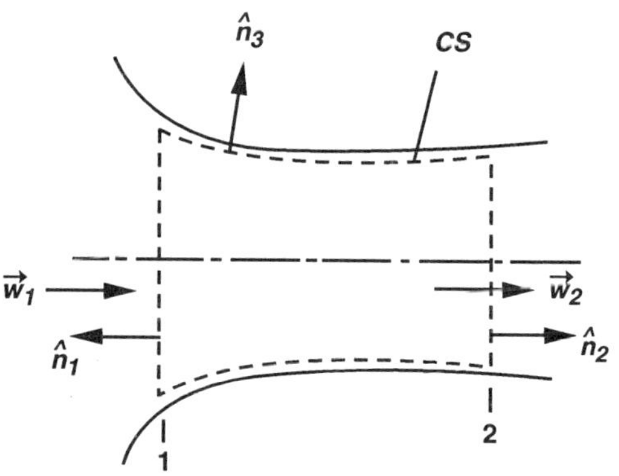

where the oval symbol reminds us that CS is an enclosing control surface whose outward unit normal vector is $\hat{n}$, and where the CS does not contain any internal body (see adjoining sketch).

(b) Apply the result of part (a) to a one-dimensional flow in a duct as illustrated in the sketch. By definition, the thrust $\Im$ is the force exerted on the fluid in the control volume. Show that $\Im = F_2 - F_1$ where the impulse function is $F = pA + \rho w^2 A$.

5.17 Consider steady, homenergetic, two-dimensional or axisymmetric flow, without swirl, of a perfect gas. Do not assume homentropic (or irrotational) flow.

(a) Show that the vorticity can be written as

$$\omega = \rho T x_2^{\sigma} \frac{ds}{d\psi}$$

where s is the entropy and ψ is a stream function.

(b) Derive the following equations:

$$w_1 s_{x_2} - w_2 s_{x_1} = \rho w^2 x_2^\sigma \frac{ds}{d\psi}$$

$$w_1 p_{x_2} - w_2 p_{x_1} = \frac{1}{x_2^\sigma}(w_1^2 \psi_{x_1x_1} + 2w_1 w_2 \psi_{x_1x_1} + w_2^2 \psi_{x_2x_2})$$

$$w_1 \rho_{x_2} - w_2 \rho_{x_1} = \frac{1}{a^2}(w_1 p_{x_2} - w_2 p_{x_1}) - \frac{\gamma - 1}{\gamma R} \rho^2 w^2 x_2^\sigma \frac{ds}{d\psi}$$

(c) Utilize the above results to obtain the rotational form of the gas dynamic equation

$$\left(1 - \frac{w_1^2}{a^2}\right)\psi_{x_1x_1} - 2\frac{w_1 w_2}{a^2}\psi_{x_1x_2} + \left(1 - \frac{w_2^2}{a^2}\right)\psi_{x_2x_2} - \frac{\sigma}{x_2}\psi_{x_2} = -x_2^{2\sigma}[1 + (\gamma - 1)M^2]\rho^2 T \frac{ds}{d\psi}$$

where M is the Mach number. Of course, this relation also holds for an irrotational flow by setting $ds/d\psi$ equal to zero.

(d) Eliminate $\rho^2 T$ and w_i/a from the above gas dynamic equation to obtain

$$[\psi_{x_1}^2 + (1 - M^2)\psi_{x_2}^2]\psi_{x_1x_1} + 2M^2 \psi_{x_1}\psi_{x_2}\psi_{x_1x_2} + [(1 - M^2)\psi_{x_1}^2 + \psi_{x_2}^2]\psi_{x_1x_2}$$
$$- \frac{\sigma}{x_2}(\psi_{x_1}^2 + \psi_{x_2}^2)\psi_{x_2} = -\frac{1}{\gamma R}(\psi_{x_1}^2 + \psi_{x_2}^2)^2 \frac{ds}{d\psi}$$

where M^2 is given by

$$\frac{M^2}{\left(1 - \frac{\gamma - 1}{2}M^2\right)^{(\gamma+1)/(\gamma-1)}} = \frac{\psi_{x_1}^2 + \psi_{x_2}^2}{\gamma \rho_o p_o x_2^{2\sigma}} \exp[2(s - s_o)/R]$$

With the assumption that $s = s(\psi)$ is a known function, e.g., as established by upstream conditions, these two equations show that the rotational gas dynamic equation is also quasilinear.

5.18 The thermodynamics of a van der Waals gas is based on the thermal equation of state

$$p = \frac{\rho R T}{1 - \beta\rho} - \alpha\rho^2$$

where α, β, and $R(=\bar{R}/W)$ are constants. By introducing reduced variables

$$p = p_c p_r, \qquad \rho = \rho_c \rho_r, \qquad T = T_c T_r$$

where a c subscript denotes a critical point value

$$p_c = \frac{\alpha}{27\beta^2}, \qquad \rho_c = \frac{1}{3\beta}, \qquad T_c = \frac{8\alpha}{27\beta R}$$

we obtain

$$p_r = \frac{8\rho_r T_r}{3-\rho_r} - 3\rho_r^2$$

For this gas, the specific heat c_v is an arbitrary function only of the temperature; for purposes of simplicity, assume c_v to be a constant. As a consequence, the reduced enthalpy, entropy, and speed of sound are

$$h_r = \frac{h-h_r}{RT_c} = \frac{c_v}{R}(T_r - 1) + \frac{3T_r}{3-\rho_r} + \frac{9}{4}(1-\rho_r) - \frac{3}{2}$$

$$s_r = \frac{s-s_c}{R} = \ell n\left(\frac{3-\rho_r}{2\rho_r}\right) + \frac{c_v}{R}\ell n T_r$$

$$a_r^2 = \frac{a^2}{a_c^2} = \frac{a^2}{\frac{2}{3}\frac{R}{c_v}\frac{\alpha}{\beta}} = \left(1+\frac{c_v}{R}\right)\frac{4T_r}{(3-\rho_r)^2} - \frac{c_v}{R}\rho_r$$

Consider steady, homenergetic, inviscid flow of a van der Waals gas with a constant value for c_v. Assume ρ_r, T_r, and $M(= w/a)$ are known at some point in the flow. Explain how you would determine the stagnation quantities ρ_{ro}, T_{ro}, p_{ro}, and h_{ro} at this location.

5.19 Consider a steady, homenergetic axisymmetric flow, with swirl, of a perfect gas. Do not assume homentropic (or irrotational) flow.

(a) Show that the swirl velocity component w_3 satisfies

$$\frac{Dw_3}{Dt} = -\frac{w_2 w_3}{x_2}$$

(b) Show that Crocco's equation is compatible with Equations (5.67b to e).

5.20 Consider a steady, inviscid, homenergetic, rotational flow with no body force.

(a) Show that the momentum equation can be written as

$$\vec{\omega} \times \vec{w} = T\nabla s$$

(b) Define a vector $\vec{A}$ as

$$\vec{A} = \vec{\omega} + \rho\vec{w}$$

and show that $\vec{A}$ can be written as

$$\vec{A} = (\nabla\chi)\times(\nabla s)$$

where χ is a scalar function.

(c) Show that χ is given by

$$\vec{w}\cdot\nabla\chi = T$$

5.21 This problem is a generalization of Problem 5.8 (Greitzer, 1975). Continue to assume a steady, homenergetic flow of a perfect gas, replace x_3 with x, and assume the velocity component $w_x(r, \theta, x)$ is not zero, while p, ρ, w_θ, and w_r are still only functions of r.

(a) Determine the solution for w_θ, w_r, w_x, ρ, and p utilizing the nondimensional parameters and variables of Problem 5.8 along with the actual Mach number $M(=w/a)$ and write ρ_{oc}, p_{oc}, ... only when M_c is involved.

(b) Is the flow irrotational?

(c) As in Problem 5.8, determine the equations for a streamline.

5.22 Consider unsteady, incompressible flow about a circular cylinder as discussed in Section 1.2.

(a) Use the solution given for u' and v' to determine $D(wh'^2/U^2)/DT'$. Sketch this substantial derivative vs. $X' + T'$.

(b) Determine the velocity potential function $\phi(X', Y', T')$.

(c) Use ϕ and the appropriate form of Bernoulli's equation to determine the pressure.

(d) Start with Equation (2.21) and derive the nondimensional form of the energy equation for this unsteady flow. Provide a physical interpretation for your result. Show that the results of parts (a) and (c) satisfy the energy equation.

5.23 Use the same assumptions and type of notation introduced in the illustrative example of Section 1.2, only now the flow is around a sphere instead of a cylinder. We utilize the axisymmetric coordinates of Figure 5.3, rather than spherical coordinates, because the Galilean transformation, which is needed for the unsteady case, is simpler in terms of these coordinates. With these coordinates, the velocity components for incompressible, steady flow about a sphere of radius a are

$$\frac{w_x}{U} = 1 + \frac{a^3}{2}\frac{r^2 - 2x^2}{(r^2 + x^2)^{5/2}}$$

$$\frac{w_r}{U} = -\frac{3}{2}\frac{a^3 xr}{(r^2 + x^2)^{5/2}}$$

$$\frac{w_\theta}{U} = 0$$

(a) Use the axisymmetric equivalent of Equation (1.21b) to determine the equation for a streamline that passes through the point

$$R = R_\infty, \qquad X = \pm\infty$$

where $R = (r/a)$.

(b) Determine the equations for the unsteady flow pathlines. Hence, show that T' is given by the quadrature

$$T' = \pm\frac{2}{3}\int_{R'/R_\infty}^{R_o/R_\infty} \frac{z^3 dz}{(z^3 - z)^{4/3}[1 - R_\infty^2(z^3 - z)^{2/3}]^{1/2}}$$

where z is a dummy variable and $R' = R_o$ when $X' + T' = 0$.

(c) Is the velocity field for the steady flow irrotational?

(d) Show that the vorticity is invariant under an arbitrary Galilean transformation.

5.24 Continue with Problem 5.23 and numerically determine the streamlines and pathlines for flow about a sphere. Generate tables and plots for $R_\infty = 0.8732, 1.871, 2.683, 3.459$.

5.25 Use the material in Section 2.5 to derive Crocco's equation for a noninertial system.

5.26 Consider steady, inviscid, incompressible flow about an arbitrary body whose shape is given by

$$F = F(\vec{r}) = 0$$

Far upstream of the body, the velocity $\vec{w}_\infty$ is uniform. Use a Cartesian coordinate system x_i with x_1 aligned with $\vec{w}_\infty$. Since the flow is irrotational, introduce a velocity potential $\phi(\vec{r})$.

(a) Determine the governing equation for ϕ and appropriate boundary conditions.

(b) Determine the differential equations, in terms of ϕ, for the streamlines.

(c) Introduce a Galilean transformation such that the upstream flow is quiescent and determine $\phi'(\vec{r}\,', t')$ in terms of ϕ. Use this result to derive the differential equations for the unsteady flow particle paths.

5.27 Consider inviscid, irrotational, barotropic flow without a body force. Derive an unsteady, three-dimensional gas dynamic equation using Cartesian coordinates.

5.28 Assume unsteady, incompressible, but possibly stratified, inviscid flow with a body force. Introduce a vector potential $\vec{A}$ for the velocity and a condition that makes $\vec{A}$ unique. Derive vector equations for continuity, momentum, and energy in terms of $\vec{A}$. Use the stagnation enthalpy form for the energy equation, and simplify your answers as much as possible.

6 Shock Wave Dynamics

6.1 PRELIMINARY REMARKS

The most distinctive feature of a supersonic flow is shock waves. They were discovered theoretically by Rankine (1870) and Hugoniot (1877) over a century ago. Ernst Mach was the first to demonstrate their existence by publishing in the 1880s schlieren photographs of a bullet in supersonic flight. (See Van Dyke, 1982, for an enlargement of one of these photographs.) This remarkable picture shows the bow shock, recompression shock, and turbulent wake. Nevertheless, shock wave theory developed slowly until the end of World War II. At the time of the war, only the basic fundamentals were known; this material is usually covered in an undergraduate compressible flow course. After the war, the pace of discovery quickened, spurred on by interest in supersonic flight, nuclear explosions, and the reentry physics of long-range missiles.

A wide range of shock wave topics is still under investigation. These topics range from the internal structure of a shock to shock wave reflection, refraction, diffraction, and interference (see Chapter 19 of Emanuel, 1986). Shock waves occur in both steady and unsteady flows. They may be associated with a wide variety of physical gas phenomena, e.g., chemical reactions that change the shock into a detonation wave. Interaction phenomena are also important, especially shock wave boundary-layer interaction. Flows with shock waves are frequently discussed in later chapters.

These shock-related phenomena have one feature in common: the associated flow fields are usually difficult to analyze. This is apparent even in CFD, where predicting a complex system of interacting shock waves remains a research topic (Moretti, 1987).

Simple planar or conical shocks are relatively easily handled, both analytically and computationally. In complex supersonic flow fields, however, multiple shocks typically occur constituting a shock system. There has been some success in analyzing such a flow, most notably with shock wave boundary-layer interaction.

The discussion in this chapter presumes familiarity with the basic concepts of a shock wave, including the equations for an oblique shock wave in a steady flow of a perfect gas. We also assume in this and subsequent chapters some familiarity with the theory of characteristics (see Section 12.4). Instead of examining the details of a specific flow, such as shock wave reflection, we concentrate on topics of a more general nature. We therefore begin, in the next section, by formulating the jump conditions for a shock in an unsteady, three-dimensional flow without assuming a perfect gas. The rest of the chapter is largely devoted to evaluating the derivatives of various quantities, such as pressure and velocity, just downstream of the shock. Readers not interested in these derivatives may skip Sections 6.3 to 6.6. These sections, however, discuss topics whose significance goes beyond the derivatives. Section 6.4 provides a general shock-fitted coordinate transformation that may be useful when the upstream flow is nonuniform and unsteady. Second, the vorticity just downstream of a shock is evaluated in a general manner at the end of the chapter.

Derivatives both normal and tangential to the shock are obtained. Derivatives in any arbitrary direction can then be found. This facet is illustrated in Section 6.3 by obtaining the derivatives in the streamline and Mach line directions. In general, this type of systematic theory should prove useful in future analytical studies, in checking the accuracy of computer codes, and in the development of new CFD algorithms for solving flow fields with shock waves.

Because of its complexity, the basis of our discussion was first published in 1988 by Emanuel and Liu. In the past, theoretical shock wave studies were typically done in a piecemeal fashion. A particular topic would be focused on, such as nonequilibrium vibrational excitation of a diatomic molecule downstream of a shock. Our development of a theory of some generality is timely, especially in view of the continuing interest in hypersonic fluid dynamics. Most applications of the theory will probably be associated with complex flow fields that are best analyzed with CFD, especially shock-fitting routines. Nevertheless, shock waves will play an important and ubiquitous role in many subsequent chapters, especially Chapter 11.

The analysis, however, is not without its restrictions. A large Reynolds number flow is the most important of these. It is essential in order to treat the shock as a surface of zero thickness across which the variables of interest can change discontinuously. (At a relatively low Reynolds number, the shock thickness is no longer negligible compared to a body dimension.) We assume the shock, over some region of its surface, to be smooth and differentiable. A triple point, or line, where shocks intersect is thereby excluded from consideration. (The analysis of a triple point is the subject of Problem 6.7.) We also assume the gas is in thermodynamic equilibrium on both sides of the shock. As a consequence, we do not consider detonation waves, ionizing shock waves, or radiative transfer across the wave. We presume the upstream flow field and the shape of the shock are known quantities.

The foregoing assumptions are viewed as basic. Within the context of these assumptions, two versions of the derivative theory are developed. The first of these is the simpler; it utilizes the basic assumptions in conjunction with a second set. The latter group presumes steady, two-dimensional or axisymmetric flow of a perfect gas in which the upstream flow is uniform. Most simple flow fields that contain a shock satisfy these more restrictive assumptions. This first version is given in Section 6.3 and has the important advantage of providing explicit results for all derivatives. This version is further developed at the end of Section 6.4.

Only the basic assumptions are invoked in the second version; hence, the gas need not be perfect; the flow field, including the shock, may be unsteady and the upstream flow may be nonuniform. This level of complexity is encountered in the flow field downstream of the shuttle orbiter's bow shock during atmospheric reentry. Additional shock waves are present downstream of the curved bow shock where these shocks have a nonuniform upstream flow. A second example is the phenomenon, called buzz, which is caused by an oscillating shock wave system at the inlet of a supersonic diffuser. Finally, we note that the flow in the test section of a supersonic, cryogenic wind tunnel is generally not a thermally and calorically perfect gas, since the thermodynamic state in the freestream is close to the coexistence curve of the gas.

An essential feature of this second version is the introduction in Section 6.4 of an orthogonal coordinate system that is contoured to the shock and moves with it. Section 6.5 then provides the tangential derivatives, while Section 6.6 provides the normal derivatives.

6.2 JUMP CONDITIONS

Basis Vector System and Shock Velocity

Only the basic assumptions are utilized in this section. A fixed Cartesian coordinate system x_i and its corresponding basis $\hat{1}_i$ are introduced. The shock wave surface, which may be in motion, is represented by

$$F = F(x_i, t) = 0 \tag{6.1}$$

Conditions just upstream and just downstream of the surface are denoted with subscripts 1 and 2, respectively. In a more conventional treatment, the upstream flow is uniform and steady and the "just" upstream qualification is unnecessary. In our analysis, the $\vec{w}_j$ velocities are written as

$$\vec{w}_j = w_{j,i}(x_k, t)\hat{1}_i, \quad j = 1, 2 \tag{6.2}$$

where the x_k and t satisfy Equation (6.1). The arbitrary sign of F is chosen so that

$$\vec{w}_i \cdot \nabla F \geq 0 \tag{6.3}$$

for some region of the shock's surface. For this region, the flow will primarily be in the direction of ∇F.

The analysis is sometimes a local one, in that it focuses on an arbitrary point of the shock. For this point, we shall ultimately relate the downstream conditions to those upstream of the shock. As previously stated, we assume Equation (6.1) is available. Of course, from a CFD point of view, the shock's location is generally not known but must be found. For our purposes, however, the assumption is warranted and the resulting jump and derivative relations hold, whether or not the shock's location is actually known.

A unit vector $\hat{n}$ is defined as

$$\hat{n} = \hat{n}(x_i,t) = \frac{\nabla F}{|\nabla F|} \tag{6.4}$$

that is normal to the shock and, in view of Equation (6.3), is oriented in the downstream direction. We also introduce a vector component $\vec{w}_s$ of the shock wave's velocity that is normal to the shock wave's surface. As will become evident, a tangential shock wave velocity component does not enter into the analysis. In accordance with Equation (5.12), we obtain $\vec{w}_s$ from

$$\frac{\partial F}{\partial t} + \vec{w}_s \cdot \nabla F = 0$$

With $\vec{w}_s$ proportional to $\hat{n}$, we have

$$\vec{w}_s = w_s \hat{n} = -\frac{1}{|\nabla F|}\frac{\partial F}{\partial t}\hat{n} \tag{6.5}$$

From the viewpoint of an observer moving with the shock, only the velocity of the gas $\vec{w}_j^*$ relative to the shock is significant. These velocities are defined by

$$\vec{w}_j^* = \vec{w}_j - \vec{w}_s = \vec{w}_j - w_s\hat{n}, \quad j = 1, 2 \tag{6.6}$$

When $w_s < 0$, the shock is moving into the upstream flow, and there is an increase in the component of $\vec{w}_1^*$ that is normal to the shock.

The vectors $\hat{n}$ and $\vec{w}_1^*$ define a unique plane. Aside from several exceptions, each point of the shock contains such a plane. One exception is a triple point, where shocks intersect. Another exception is when the shock is normal to $\vec{w}_1^*$. Aside from these exceptions, conservation of momentum shows that $\vec{w}_2^*$ lies in this plane. Equations (6.6) then state that the $\vec{w}_j$ also lie in the plane.

It is convenient to define a unit vector $\hat{t}$ that is tangent to the shock, as shown in Figure 6.1. We thus have an orthonormal basis, $\hat{n}$, $\hat{t}$, and $\hat{b}$, where the binormal $\hat{b}$ is perpendicular to $\hat{n}$ and $\hat{t}$. This basis moves with the shock, where $\hat{t}$ is in the special $\hat{n}, \vec{w}_1^*$ plane.

An important reason for our particular $\hat{n}$ and $\hat{t}$ definitions is to be able to introduce the θ and β angles. These are conventionally used with a steady, planar, oblique shock wave. As shown in Figure 6.1, θ is the acute angle between $\vec{w}_2^*$ and $\vec{w}_1^*$ and β is the acute angle between the shock

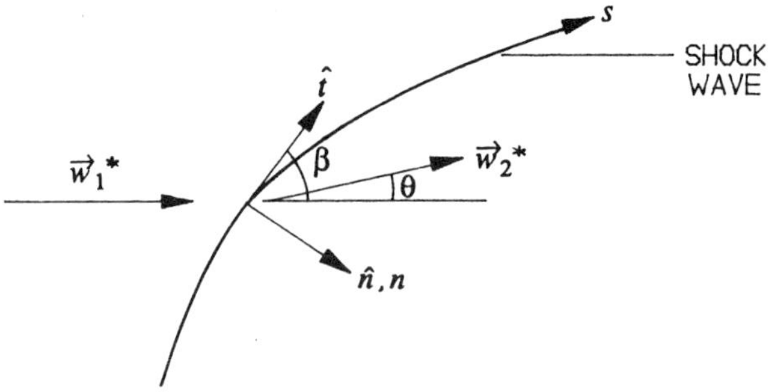

FIGURE 6.1 Section through a shock that contains both $\vec{w}_1^*$ and $\hat{n}$. The $\hat{b}$ vector is perpendicular to the plane of the page.

and $\vec{w}_1^*$. These angles will play a major role in the subsequent analysis. They are defined only in the $\hat{n},\vec{w}_1^*$ plane and their values change from point to point on the shock's surface. When the shock is normal to $\vec{w}_1^*$, θ and β are then 0 and 90°, respectively.

Although F and $\vec{w}_1^*$ are presumed known in terms of a Cartesian coordinate system, the $\hat{n}, \hat{t}$, and $\hat{b}$ system is far more convenient for the subsequent analysis. For instance, from Figure 6.1, we obtain

$$\vec{w}_1^* = w_1^*(\sin\beta\hat{n} + \cos\beta\hat{t}) \tag{6.7a}$$

$$\vec{w}_2^* = w_2^*[\sin(\beta-\theta)\hat{n} + \cos(\beta-\theta)\hat{t}] \tag{6.7b}$$

while Equations (6.6) yield

$$\vec{w}_1 = w_1^*\sin(\beta + w_s)\hat{n} + w_1^*\cos\beta\,\hat{t} \tag{6.8a}$$

$$\vec{w}_2 = [w_2^*\sin(\beta-\theta) + w_s]\hat{n} + w_2^*\cos(\beta-\theta)\hat{t} \tag{6.8b}$$

Alternatively, an explicit form for $\vec{w}_1^*$ is

$$\vec{w}_1^* = \vec{w}_{1,i}^* = \hat{\imath}_i = \vec{w}_1 - w_s\hat{n} = w_{1,i}\,\hat{\imath}_i + \frac{\partial F}{\partial t}\frac{\nabla F}{|\nabla F|^2} \tag{6.9}$$

where the quantities on the right side are known functions of x_i and t. We therefore view $\vec{w}_1^*$ as a known velocity.

The binormal is given by

$$\hat{b} = \frac{\vec{w}_1^* \times \hat{n}}{w_1^*\cos\beta} \tag{6.10a}$$

where the denominator converts $\vec{w}_1^* \times \hat{n}$ into a unit vector, and $\hat{t}$ is given by

$$\hat{t} = \hat{n} \times \hat{b} \tag{6.10b}$$

The angle β is defined by

$$\sin\beta = \frac{\vec{w}_1^* \cdot \nabla F}{w_1^* |\nabla F|} \tag{6.11}$$

A relation for θ is discussed later in this section.

We have evaluated $\hat{b}$ and $\hat{t}$ in terms of $\vec{w}_1^*$. For the later discussion, it is useful to also evaluate these vectors in terms of $\vec{w}_1$. For $\hat{b}$, we write

$$\hat{b} = \frac{\vec{w}_1^* \times \hat{n}}{w_1^* \cos\beta} = \frac{(\vec{w}_1 - w_s\hat{n}) \times \hat{n}}{w_1^* \cos\beta} = \frac{\vec{w}_1^* \times \hat{n}}{w_1^* \cos\beta} = \frac{1}{I} \vec{w}_1 \times \nabla F \tag{6.12}$$

where

$$I = w_1^* |\nabla F| \cos\beta = [w_1^{*2} |\nabla F|^2 - (\vec{w}_1^* \cdot \nabla F)^2]^{1/2}$$

in view of Equation (6.11). With the use of the notation

$$F_t = \frac{\partial F}{\partial t}, \quad F_{x_i} = \frac{\partial F}{\partial x_i}$$

and Equation (6.9), we obtain

$$w_{1,i}^* = w_{1,i} + \frac{F_t F_{x_i}}{|\nabla F|^2}$$

$$\vec{w}_1^* \cdot \nabla F = \vec{w}_1 \cdot \nabla F + F_t$$

Hence, I becomes

$$I = \left[\left(w_1^2 + 2\frac{F_t}{|\nabla F|^2} \vec{w}_1 \cdot \nabla F + \frac{F_t^2}{|\nabla F|^2}\right) |\nabla F|^2 - (\vec{w}_1 \cdot \nabla F)^2 - 2F_t \vec{w}_1 \cdot \nabla F - F_t^2\right]^{1/2}$$

$$I = [w_1^2 |\nabla F|^2 - (\vec{w}_1 \cdot \nabla F)^2]^{1/2} \tag{6.13}$$

Thus, the equation for $\hat{b}$ is unaltered if $\vec{w}_1^*/w_1^*$ is replaced with $\vec{w}_1/w_1$. Finally, we note that $\hat{t}$ can be written as

$$\hat{t} = \hat{n} \times \hat{b} = \frac{\nabla F}{|\nabla F|} \times \frac{\vec{w}_1 \times \nabla F}{I} = \frac{1}{I|\nabla F|} [|\nabla F|^2 \vec{w}_1 - (\vec{w}_1 \cdot \nabla F) \nabla F] \tag{6.14}$$

Consequently, the $\hat{n}, \hat{t}, \hat{b}$ basis can be defined using either $\vec{w}_1^*$ or $\vec{w}_1$.

Conservation Equations

The same conservation principles that led to the governing equations are applied to a differential volume element that contains a piece of the shock. Application of these principles then yields the jump conditions, whereby flow conditions on the two sides of the shock are related. For these equations, we will need the substantial derivative

$$\left(\frac{DF}{Dt}\right)_j = \frac{\partial F}{\partial t} + \vec{w}_j \cdot \nabla F, \quad j = 1, 2 \tag{6.15}$$

which has a different value on each side of the shock.

Conservation of the flux of mass across the shock is given by

$$\left(\rho \frac{DF}{Dt}\right)_1 = \left(\rho \frac{DF}{Dt}\right)_2 \tag{6.16}$$

This relation can be understood by writing the left side as

$$\left(\rho \frac{DF}{Dt}\right)_1 = \rho_1 \left(\frac{\partial F}{\partial t} + \vec{w} \cdot \nabla F\right) = \rho_1 |\nabla F| \left(\frac{1}{|\nabla F|} \frac{\partial F}{\partial t} + \hat{n} \cdot \vec{w}_1\right)$$

with a similar result for the right side. Conservation of mass flux, Equation (6.16), now becomes

$$\frac{\rho_1}{|\nabla F|} \frac{\partial F}{\partial t} + \rho_1 \hat{n} \cdot \vec{w}_1 = \frac{\rho_2}{|\nabla F|} \frac{\partial F}{\partial t} + \rho_2 \hat{n} \cdot \vec{w}_2$$

The two $\hat{n}$ terms represent the mass flux across the shock, as if it were steady, while the two $\partial F/\partial t$ terms provide the contribution from a moving shock. Recall that $|\nabla F|^{-1}(\partial F/\partial t)$ also appeared in Equation (5.14), where it represents the normal component of the velocity of a moving wall.

In a similar manner, we write the flux equations that represent conservation of the normal component of momentum across the shock, conservation of the tangential momentum component, and conservation of energy. We thereby obtain

$$\left[p|\nabla F|^2 + \rho \left(\frac{DF}{Dt}\right)^2\right]_1 = \left[p|\nabla F|^2 + \rho \left(\frac{DF}{Dt}\right)^2\right]_2 \tag{6.17}$$

$$\left(\rho \vec{w} \cdot \hat{t} \frac{DF}{Dt}\right)_1 = \left(\rho \vec{w} \cdot \hat{t} \frac{DF}{Dt}\right)_2 \tag{6.18}$$

$$\left[h|\nabla F|^2 + \frac{1}{2}\left(\frac{DF}{Dt}\right)^2\right]_1 = \left[h|\nabla F|^2 + \frac{1}{2}\left(\frac{DF}{Dt}\right)^2\right]_2 \tag{6.19}$$

Equations (6.16) to (6.19) are the jump conditions in a general form. (In addition, the second law requires the entropy conditions, $s_2 \geq s_1$. We do not list it, since it is not directly utilized in the subsequent analysis.) This form, however, is not a convenient one. We would like explicit equations for p_2, ρ_2, h_2, and w_2^*, which are viewed as the unknowns.

With the aid of Equations (6.4) to (6.6), the substantial derivatives that appear in the jump conditions now become

$$\left(\frac{DF}{Dt}\right)_j = \frac{\partial F}{\partial t} + (\vec{w}_j^* + w_s\hat{n}) \cdot \nabla F = \frac{\partial F}{\partial t} + \vec{w}_j^* \cdot \nabla F - \frac{1}{|\nabla F|}\frac{\partial F}{\partial t}\frac{\nabla F \cdot \nabla F}{|\nabla F|}$$

$$= \overline{w}_j^* \cdot \nabla F, \qquad j = 1, 2$$

Hence, the conditions simplify to

$$(\rho\vec{w}^* \cdot \hat{n})_1 = (\rho\vec{w}^* \cdot \hat{n})_2 \tag{6.20a}$$

$$[p + \rho(\vec{w}^* \cdot \hat{n})^2]_1 = [p + \rho(\vec{w}^* \cdot \hat{n})^2]_2 \tag{6.20b}$$

$$(\vec{w}^* \cdot \hat{t})_1 = (\vec{w}^* \cdot \hat{t})_2 \tag{6.20c}$$

$$\left[h + \frac{1}{2}(\vec{w}^* \cdot \hat{n})^2\right]_1 = \left[h + \frac{1}{2}(\vec{w}^* \cdot \hat{n})^2\right]_2 \tag{6.20d}$$

These are the jump conditions in a frame fixed to the shock. We could have started with these relations in preference to Equations (6.16) to (6.19). Their algebraic form is a result of the shock containing no mass. When a shock is of finite thickness and has mass, the analysis requires differential equations, in part, because viscous stresses and heat conduction must be included.

Explicit Solution

In order to evaluate the dot products that appear in Equations (6.20a to d), we use Equations (6.7a,b), with the result

$$\vec{w}_1^* \cdot \hat{t} = w_1^* \cos\beta \tag{6.21a}$$

$$\vec{w}_1^* \cdot \hat{n} = w_1^* \sin\beta \tag{6.21b}$$

$$\vec{w}_2^* \cdot \hat{t} = w_2^* \cos(\beta - \theta) \tag{6.21c}$$

$$\vec{w}_2^* \cdot \hat{n} = w_2^* \sin(\beta - \theta) \tag{6.21d}$$

From Equation (6.20c) for the velocity tangency condition, we have

$$w_2^* = w_1^* \frac{\cos\beta}{\cos(\beta - \theta)} \tag{6.22}$$

This relation cannot be used for a normal shock, since the ratio of cosines is indeterminant. (The normal shock formulation is discussed in a subsequent illustrative example.)

Conservation of mass flux, Equation (6.20a), yields

$$\rho_1 w_1^* \sin\beta = \rho_1 w_2^* \sin(\beta-\theta)$$

In combination with Equation (6.22), this becomes

$$\rho_2 = \rho_1 \frac{\tan\beta}{\tan(\beta-\theta)} \tag{6.23}$$

From Equation (6.20b), we have

$$\begin{aligned} p_2 &= p_1 + (\rho w^{*2})_1 \sin^2\beta - (\rho w^{*2})_2 \sin^2(\beta-\theta) \\ &= p_1 + (\rho w^{*2})_1 \left[\sin^2\beta - \frac{\sin\beta\cos\beta\sin(\beta-\theta)}{\cos(\beta-\theta)}\right] \\ &= p_1 + (\rho w^{*2})_1 \frac{\sin\beta\sin\theta}{\cos(\beta-\theta)} \end{aligned} \tag{6.24}$$

where Equations (6.22) and (6.23) are used. From conservation of energy, we obtain

$$\begin{aligned} h_2 &= h_1 + \frac{1}{2} w_1^{*2}\beta - \frac{1}{2} w_2^{*2}\sin^2(\beta-\theta) \\ &= h_1 + \frac{1}{2} w_1^{*2}\left[\sin^2\beta - \frac{\cos^2\beta\sin^2(\beta-\theta)}{\cos^2(\beta-\theta)}\right] \\ &= h_1 + \frac{1}{2} w_1^{*2} \frac{\sin(2\beta-\theta)\sin\theta}{\cos^2(\beta-\theta)} \end{aligned} \tag{6.25}$$

Only the basic assumptions have been utilized in deriving Equations (6.22) to (6.25). The downstream variables w_2^*, ρ_2, p_2, and h_2 are explicitly provided by these equations. The equations hold for unsteady, three-dimensional shocks and do not assume a perfect gas. Note the absence of the ratio of specific heats. The variables are not normalized, since their upstream counterparts are functions of position on the shock surface and of time. The downstream velocity w_2^* is then provided by Equation (6.8b). The variables on the right sides consist of β, θ, ρ_1, p_1, h_1, and w_1^*. Except for θ, these quantities are presumed known, where β is given by Equation (6.11).

To evaluate θ, a thermodynamic state equation involving ρ, p, and h needs to be introduced. Since the enthalpy is present in only one jump condition, the most convenient form for this relation is

$$h = h(p, \rho) \tag{6.26}$$

Equation (6.25) thus becomes

$$h(p_2, \rho_2) = h(p_1, \rho_1) + \frac{1}{2} w_1^{*2} \frac{\sin\theta\sin(2\beta-\theta)}{\cos^2(\beta-\theta)} \tag{6.27}$$

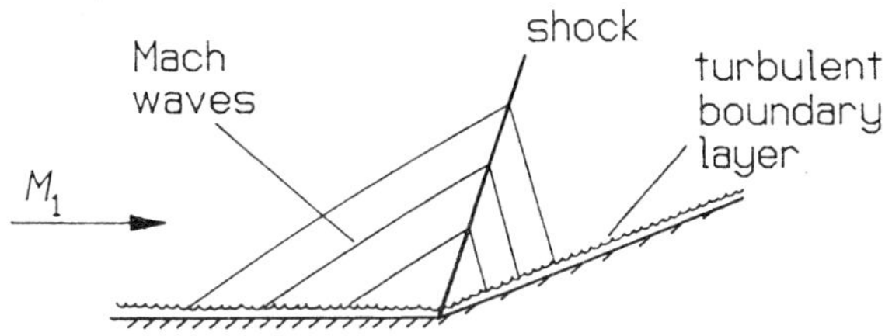

FIGURE 6.2 Oblique shock caused by a sharp wall turn. The Mach waves emanate from a turbulent boundary layer.

Equations (6.23) and (6.24) can now be used to eliminate ρ_2 and p_2 from $h(p_2,\rho_2)$. The result would be an implicit equation for θ. For alternate approaches to treating real gas shock wave phenomena, see Vincenti and Kruger (1965), Zel'dovich and Raizer (1966), and Zucrow and Hoffman (1976).

For a steady shock wave in a perfect gas, Problem 6.1 shows that Equation (6.27) reduces to the oblique shock equation

$$\tan\theta = \cot\beta \frac{M_1^2 \sin^2\beta - 1}{1 + \{[(\gamma+1)/2] - \sin^2\beta\}M_1^2} \quad \textbf{(6.28)}$$

where the upstream Mach number is

$$M_1 = \frac{w_1^*}{(\gamma p_1/\rho_2)^{1/2}}$$

In this case, the equation for θ is explicit. An explicit equation for β, referred to as the inversion formula, is provided by Appendix C. Problem 6.4 develops the van der Waals counterpart to Equation (6.28).

Illustrative Example

A shock wave may be unsteady for any of several reasons. The most obvious cause would be an unsteady upstream flow. Another weak source of unsteadiness would stem from a turbulent boundary layer, as sketched in Figure 6.2. The noise generated by the unsteady outer edge of the boundary layer travels along Mach waves until these waves impinge on the shock. As indicated in the figure, the Mach waves from both the upstream and downstream walls travel toward the shock. A third source would be an oscillating wall downstream of the shock. For instance, imagine that the downstream wall in Figure 6.2 is vibrating. The shock wave will then vibrate in response to the wall's motion.

If the shock is sufficiently intense, the flow downstream of it is subsonic. In this circumstance, disturbances can propagate in an upstream direction, thereby causing the shock to become unsteady. This mechanism is involved with the buzz phenomenon mentioned in the previous section. For instance, consider an axisymmetric, supersonic inlet with a single centrally located cone. During buzz, which typically occurs with a frequency of about 10 to 20 Hz, there is a single, detached, normal shock when the shock system is in its most upstream position. When in its downstream position, it is a multiple system of oblique shock waves, where the upstream-most shock is conical and is attached to the apex of the cone.

To illustrate the theory, we examine the sinusoidal oscillation

$$F = x - b\sin(2\pi\kappa t) = 0$$

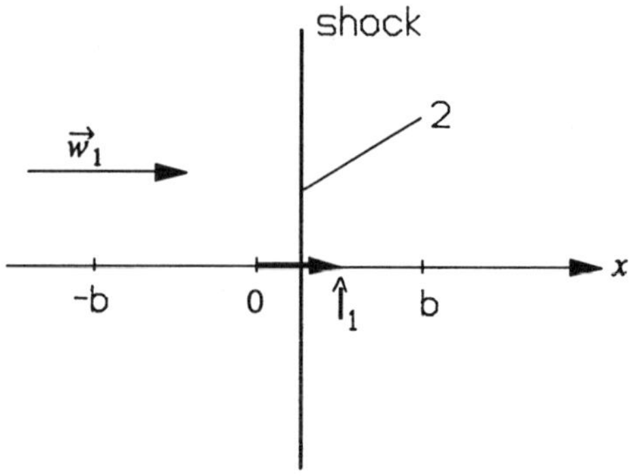

FIGURE 6.3 Schematic of an unsteady normal shock.

of a normal shock, where the amplitude b and frequency κ are constants. The upstream velocity (see Figure 6.3)

$$\vec{w} = w_1 \hat{1}_1 \tag{6.29}$$

is taken as steady and uniform.

For this flow, we readily establish that $\hat{n} = \hat{1}_1$,

$$\vec{w}_1 \cdot \nabla F = w_1 > 0$$

in accordance with Equation (6.3), and $\hat{t}$ and $\hat{b}$ are unnecessary. The shock velocity is

$$\vec{w}_s = w_s \hat{1}_1 = -\frac{\hat{n}}{|\nabla F|}\frac{\partial F}{\partial t} = 2\pi b\kappa \cos(2\pi\kappa t)\hat{1}_1$$

When the shock is moving to the right, $w_s > 0$ and M_2 exceeds the value

$$\overline{M}_2 = \left(\frac{1 + (\gamma - 1)M_1^2/2}{\gamma M_1^2 - (\gamma - 1)/2}\right)^{1/2}, \quad w_s = 0 \tag{6.30}$$

it would have if the shock were motionless. Thus, when $w_s > 0$, the shock is weaker than if it were stationary. Although $M_2 > \overline{M}_2$, M_2 cannot exceed unity, since the motion is due to downstream disturbances that propagate upstream in the subsonic flow.

When the shock is moving to the left, $w_s < 0$, it is stronger than its stationary value, and $M_2 < \overline{M}_2$. Furthermore, M_2 may be negative if w_s is sufficiently negative. (The M_1 and M_2 Mach numbers are with respect to a laboratory frame, not a shock-fixed frame.) As sketched in Figure 6.4, part of the time we can view the flow as being driven leftward by a fictitious piston, where the $\vec{w}_2$ velocity is negative. Whenever $\vec{w}_2 = 0$, the flow is similar to that downstream of the reflected shock wave that occurs in shock tube flow.

For a normal shock $\theta = 0°$, $\beta = 90°$, and the velocity tangency condition cannot be used. Moreover, Equations (6.23) to (6.25) are indeterminate. This difficulty is avoided by using the initial equations for ρ_2, p_2, and h_2. For example, ρ_2 is given by

$$\rho_2 = \rho_1 \frac{w_1 - w_s}{w_2 - w_s} \tag{6.31a}$$

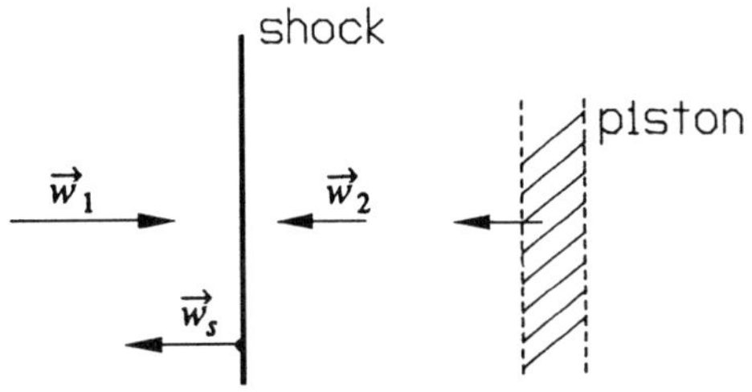

FIGURE 6.4 Piston analogy for the shock in Figure 6.3.

Similarly, we have for p_2 and h_2

$$p_2 = p_1 + \rho_1(w_1 - w_s)(w_1 - w_2) \tag{6.31b}$$

$$h_2 = h_1 + \frac{1}{2}w_1^{*2} - \frac{1}{2}w_2^{*2} = h_1 + \frac{1}{2}(w_1^* - w_2^*)(w_1^* + w_2^*)$$

$$= h_1 + \frac{1}{2}(w_1 - w_2)(w_1 + w_2 - 2w_s) \tag{6.31c}$$

To proceed with the analysis, a perfect gas is assumed and the enthalpy equation becomes

$$\frac{p_2}{\rho_2} = \frac{p_1}{\rho_1} + \frac{\gamma - 1}{2\gamma}(w_1 - w_2)(w_1 + w_2 - 2w_s) \tag{6.32}$$

From Equations (6.31a) and (6.31b), we have

$$\frac{p_2}{\rho_2} = \frac{p_1}{\rho_1}\frac{w_2 - w_s}{w_1 - w_s} + (w_1 - w_2)(w_2 - w_s)$$

After eliminating p_2/ρ_2, we obtain for w_2

$$w_2 = \frac{\gamma - 1}{\gamma + 1}w_1 + \frac{2\gamma}{\gamma + 1}\frac{p_1/\rho_1}{w_1 - w_s} + \frac{2}{\gamma + 1}w_s \tag{6.33}$$

Thus, w_2 equals a steady term, a w_s term that is proportional to $(2\pi\kappa t)$, and a term with this cosine in a denominator.

As usual with a perfect gas, it is convenient to introduce the Mach numbers

$$M_1 = \frac{w_1}{a_1} = \frac{w_1}{(\gamma p_1/\rho_1)^{1/2}} \tag{6.34a}$$

$$M_2 = \frac{w_2}{a_2} = \frac{w_2}{(\gamma p_2/\rho_2)^{1/2}} \tag{6.34b}$$

$$M_s = \frac{w_s}{a_1} = 2\pi\left(\frac{\rho_1}{\gamma p_1}\right)^{1/2} b\kappa\cos(2\pi\kappa t) \tag{6.34c}$$

where M_s, the shock wave Mach number, is negative whenever w_s is negative. In order to obtain explicit results, a relation is needed between the upstream and downstream sound speeds. We multiply Equation (6.32) with γ to obtain

$$a_2^2 = a_1^2 + \frac{\gamma-1}{2}(w_1 - w_2)(w_1 + w_2 - 2w_s)$$

With the aid of Equation (6.33), w_2 is eliminated with the result

$$\frac{a_2^2}{a_1^2} = 1 + \frac{2(\gamma-1)}{(\gamma+1)^2}\frac{[(M_1 - M_s)^2 - 1][\gamma(M_1 - M_s)^2 + 1]}{(M_1 - M_s)^2}$$

or

$$\frac{a_2}{a_1} = \frac{2}{(\gamma+1)}\frac{\left[\gamma(M_1 - M_s)^2 - \frac{\gamma-1}{2}\right]^{1/2}\left[1 + \frac{\gamma-1}{2}(M_1 - M_s)^2\right]^{1/2}}{M_1 - M_s} \tag{6.35}$$

These are the usual jump condition formulas for a_2/a_1 with M_1 replaced by $M_1 - M_s$.

Equation (6.33) is now written as

$$a_2 M_2 = \frac{\gamma-1}{\gamma+1}a_1 M_1 + \frac{2}{\gamma+1}\frac{a_1}{M_1 - M_s} + \frac{2}{\gamma+1}a_1 M_s$$

or

$$M_2 = \frac{2}{\gamma+1}\frac{1 + (M_1 - M_s)\left(\frac{\gamma-1}{2}M_1 + M_s\right)}{M_1 - M_s}\frac{a_1}{a_2}$$

Equation (6.35) is utilized to eliminate a_1/a_2, with the result

$$M_2 = \frac{1 + (M_1 - M_s)\left(\frac{\gamma-1}{2}M_1 + M_s\right)}{\left[1 + \frac{\gamma-1}{2}(M_1 + M_s)^2\right]^{1/2}\left[\gamma(M_1 - M_s)^2 - \frac{\gamma-1}{2}\right]^{1/2}} \tag{6.36}$$

which reduces to Equation (6.30) when $M_s = 0$. This relation provides the time dependence of M_2 through M_s, which is given by Equation (6.34c). While the shock speed and M_s are simple sinusoids, the variation of M_2, p_2, …, is not as simple.

From the denominator of Equation (6.36), a real finite solution for M_2 requires

$$(M_1 - M_s)^2 > \frac{\gamma-1}{2\gamma}$$

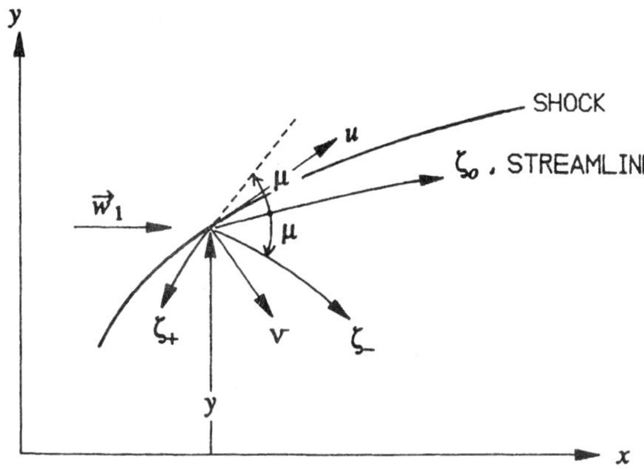

FIGURE 6.5 Streamline (ζ_o), left-running (ζ_+) and right-running (ζ_-) characteristic directions on the downstream side of a shock.

Since $M_1 - M_s$ can be written as

$$M_1 - M_s = M_1\left[1 - 2\pi\left(\frac{b\kappa}{w_1}\right)\cos(2\pi\kappa t)\right]$$

the left side of the inequality is a minimum when the cosine is unity. As a consequence, we have

$$\left(1 - 2\pi\frac{b\kappa}{w_1}\right)^2 > \frac{\gamma - 1}{2\gamma}\frac{1}{M_1^2}$$

The right side is always well below unity and small values for $b\kappa/w_1$ readily satisfy the inequality. Nevertheless, there is a range of values of $2\pi(b\kappa/w_1)$, centered about unity, for which a real solution is not obtained, and the postulated sinusoidal flow does not occur.

6.3 STEADY, TWO-DIMENSIONAL OR AXISYMMETRIC FLOW

The analysis of Section 5.6 is extended by presuming a steady, two-dimensional or axisymmetric flow of a perfect gas that contains a shock wave. In addition, we also assume no swirl and a uniform upstream flow. To simplify the notation, x,y coordinates are introduced; see Figure 6.5, where x is along the symmetry axis and y is the radial coordinate in an axisymmetric flow. We first derive the jump conditions, next the tangential derivatives on the downstream side of the shock, and, finally, the normal derivatives. The utility of the theory is then illustrated by providing the derivatives in the streamline and Mach line directions, and in establishing whether a reflected wave is compressive or expansive. The theory is further developed at the end of Section 6.4.

Jump Conditions

For this flow, the shape of the shock can be written as

$$F = f(x) - y = 0 \tag{6.37}$$

and the freestream velocity is

$$\vec{w}_1 = w_1\hat{i}_1 \tag{6.38}$$

where w_1 is a constant and the $\hat{\imath}_i$ bases are defined by Equations (5.66). The gradient of F is

$$\nabla F = f'\hat{\imath}_1 - \hat{\imath}_2, \quad f' = \frac{df}{dx} \tag{6.39}$$

and Equation (6.3) is satisfied. For $\hat{n}$, $\hat{t}$, and $\hat{b}$, we readily obtain

$$\hat{n} = \frac{f'\hat{\imath}_1 - \hat{\imath}_2}{(1+f'^2)^{1/2}} \tag{6.40a}$$

$$\hat{t} = \frac{\hat{\imath}_1 + f'\hat{\imath}_2}{(1+f'^2)^{1/2}} \tag{6.40b}$$

$$\hat{b} = -\hat{\imath}_3 \tag{6.40c}$$

In addition, we have

$$\beta = \tan^{-1} f', \quad w_s = 0 \tag{6.41}$$

Thus, Equations (6.22) to (6.25) become

$$w_2 = w_1 \frac{\cos\beta}{\cos(\beta-\theta)} \tag{6.42a}$$

$$\rho_2 = \rho_1 \frac{\tan\beta}{\tan(\beta-\theta)} \tag{6.42b}$$

$$p_2 = p_1 + (\rho w^2)_1 \frac{\sin\beta\sin\theta}{\cos(\beta-\theta)} \tag{6.42c}$$

$$h_2 = h_1 + \frac{1}{2} w_1^2 \frac{\sin\theta\sin(2\beta-\theta)}{\cos^2(\beta-\theta)} \tag{6.42d}$$

for the jump conditions.

We next introduce the perfect gas assumption and the Mach numbers provided by Equations (6.34a,b). To further simplify the notation, we also define

$$\begin{aligned} m &= M_1^2 \\ w &= (M_1 \sin\beta)^2 \\ X &= 1 + \frac{\gamma-1}{2} w \\ Y &= \gamma w - \frac{\gamma-1}{2} \end{aligned} \tag{6.43}$$

(The above w should not be confused with w_1 or w_2.)

For the later analysis, it is convenient to use velocity components u and v that are tangential and normal to the shock, respectively, as shown in Figure 6.5. These components are related to w_1 and w_2 by means of

$$u_1 = u_2 = w_1 \cos\beta = w_2 \cos(\beta - \theta) \qquad \textbf{(6.44a)}$$

$$v_1 = w_1 \sin\beta, \qquad v_2 = w_2 \sin(\beta - \theta) \qquad \textbf{(6.44b)}$$

Equations (6.42a to d), in combination with perfect gas thermodynamic state equations, then yield the jump conditions in terms of u and v. These equations are summarized in Part 1 of Appendix D, which shows several Mach number functions, since these appear in later equations. For the same reason, we also list $\sin(\beta - \theta)$ and $\cos(\beta - \theta)$. The equation for $\tan\theta$ easily reduces to Equation (6.28).

The equations in Appendix D1 are arrived at by replacing the normal component of the Mach numbers, M_{1n} and M_{2n}, with

$$M_{1n} = M_1 \sin\beta, \qquad M_{2n} = M_2 \sin(\beta - \theta) \qquad \textbf{(6.45)}$$

in the standard equations for a normal shock. The results appear different because $\sin(\beta - \theta)$ has been systematically eliminated. For instance, the usual equation for M_2 can be written as

$$M_2^2 = \frac{1}{\sin^2(\beta - \theta)} \frac{1 + \frac{\gamma - 1}{2} M_1^2 \sin^2\beta}{\gamma M_1^2 \sin^2\beta - \frac{\gamma - 1}{2}} = \frac{1}{\sin^2(\beta - \theta)} \frac{X}{Y}$$

With the $\sin(\beta - \theta)$ relation in the appendix, this yields the listed M_2^2 equation.

Tangential Derivatives

For the tangential and normal derivatives, we introduce the arc lengths s and n, which are, respectively, along and normal to the shock; see Figure 6.1. As in boundary-layer theory, the normal coordinate n is well defined only at the shock. This limitation poses no difficulties, since we are concerned only with derivatives just downstream of the shock.

The tangential derivatives are obtained by differentiating the equations in Appendix D1 with respect to s. The resulting derivatives are proportional to β', where

$$\beta' = \frac{d\beta}{ds} = \frac{f''}{(1 + f'^2)^{3/2}} \qquad \textbf{(6.46)}$$

which stems from Equation (6.41). It is worth noting that the curvature of the shock is $-\beta'$. It is also useful for the latter discussion to include the derivative of the Mach angle μ, defined by

$$\mu = \sin^{-1}\frac{1}{M} \qquad \textbf{(6.47)}$$

To illustrate how Appendix D2 is obtained, we derive the derivative of the stagnation pressure, $(\partial p_o/\partial s)_2$, starting with $p_{o,2}$ in Appendix D1 as follows:

$$\left(\frac{\partial p_0}{\partial s}\right)_2 = \frac{2}{\gamma(\gamma+1)}\frac{(\rho w^2)_1}{m}\left\{\left(1+\frac{\gamma-1}{2}M_2^2\right)^{\gamma/(\gamma-1)}\frac{\partial Y}{\partial s}\right.$$

$$\left.+\frac{\gamma}{\gamma-1}Y\left(1+\frac{\gamma-1}{2}M_2^2\right)^{[\gamma/(\gamma-1)]-1}\frac{\gamma-1}{2}\left(\frac{\partial M^2}{\partial s}\right)_2\right\}$$

$$-\frac{2}{\gamma(\gamma+1)}\frac{(\rho w^2)_1}{m}\left(1+\frac{\gamma-1}{2}M_2^2\right)^{\gamma/(\gamma-1)}\left[2\gamma m\beta'\sin\beta\cos\beta\right.$$

$$\left.-\frac{\gamma}{2}\frac{\gamma}{1+\frac{\gamma-1}{2}M_2^2}\frac{(\gamma+1)^2}{2}\left(1+\frac{\gamma-1}{2}m\right)(1+\gamma w^2)\frac{\beta' m\sin\beta\cos\beta}{X^2Y^2}\right]$$

$$=\frac{2}{\gamma+1}(\rho w^2)_1\left(1+\frac{\gamma-1}{2}M_2^2\right)^{\gamma/(\gamma-1)}\beta'\sin\beta\cos\beta\left(2-\frac{1+\gamma w^2}{wX}\right)$$

$$=-\frac{2}{\gamma+1}(\rho w^2)_1\left(1+\frac{\gamma-1}{2}M_2^2\right)^{\gamma/(\gamma-1)}\frac{\beta'(w-1)^2}{mX\tan\beta}$$

Observe that $(\partial M^2/\partial s)_2$ is used in the above derivation. Although σ does not appear in Appendices D1 and D2, all results hold for an axisymmetric shock. Except for $(\partial u/\partial s)_2$, the listed derivatives are proportional to $\cos\beta$, which means they are zero when the shock is perpendicular to the freestream velocity.

Normal Derivatives

The Euler equations are needed for the normal derivatives. In particular, we need these equations in a scalar form and with orthogonal coordinates, where one coordinate is along the shock and the other is normal to it. Emanuel (1986, Section 13.3) derives these equations in this form, and Table 6.1 provides the change to our notation. The minus sign that appears with ∂n and v stems

TABLE 6.1
Transformation to the Current Notation

Emanuel (1986)	Present Notation
$h_1\partial\xi_1$	∂s
$h_2\partial\xi_2$	$-\partial n$
x_2	y
v_1	u
v_2	$-v$
κ_1	$-\beta'$
κ_2	0
θ	β

from the downstream orientation of the n coordinate, while the curvature κ_2 of this coordinate is zero, since the n coordinate is straight at the shock. Thus, the version of the Euler equations, given in the next paragraph, applies only to the flow field just downstream of the shock. For this reason, the θ of Emanuel (1986), which is the angle between ξ_1 and x_1, now becomes β.

For notational simplicity, we suppress the subscript 2 that should appear on all variables and derivatives and write the Euler equations as

$$\frac{\partial(\rho u)}{\partial s} + \frac{\partial(\rho v)}{\partial n} + \beta'\rho v - \frac{\sigma\rho}{y}(u \sin\beta - v\cos\beta) = 0$$

$$u\frac{\partial u}{\partial s} + v\frac{\partial u}{\partial n} + \beta' uv + \frac{1}{\rho}\frac{\partial p}{\partial s} = 0$$

$$u\frac{\partial v}{\partial s} + v\frac{\partial v}{\partial n} - \beta' u^2 + \frac{1}{\rho}\frac{\partial p}{\partial s} = 0$$

$$\left(u\frac{\partial}{\partial s} + v\frac{\partial}{\partial n}\right)\left[\frac{\gamma}{\gamma-1}\frac{p}{\rho} + \frac{1}{2}(u^2+v^2)\right] = 0 \qquad \textbf{(6.48)}$$

Remember that u and v are the velocity components shown in Figure 6.5. The value of the u, v, p, and ρ variables and their s derivatives are known from Appendices D1 and D2. For instance, for the $\partial(\rho u)/\partial s$ term in continuity, we use

$$\frac{\partial(\rho u)}{\partial s} = \rho\frac{\partial u}{\partial s} + u\frac{\partial \rho}{\partial s} = \left(\frac{\gamma+1}{2}\rho_1\frac{w}{X}\right)(-w_1\beta'\sin\beta) + (w_1\cos\beta)\left[(\gamma+1)\rho_1\frac{\beta' m\sin\beta\cos\beta}{X^2}\right]$$

$$= (\gamma+1)(\rho w)_1\frac{\beta'\sin\beta}{X^2}\left(-\frac{1}{2}wX + m\cos^2\beta\right)$$

$$= (\gamma+1)(\rho w)_1\frac{\beta'\sin\beta}{X^2}\left(m - \frac{3}{2}w - \frac{\gamma-1}{4}w^2\right)$$

Consequently, Equations (6.48) are four linear, inhomogeneous, algebraic equations for $\partial u/\partial n$, $\partial v/\partial n$, $\partial p/\partial n$, and $\partial\rho/\partial n$. The solution of these equations, which was obtained with the assistance of the MACSYMA code (Rand, 1984), is given in Appendix D3. The g_i, which appear in these equations, are functions only of γ and w; they are listed in Appendix D4 with numerical values given in Table 6.2 when $\gamma = 1.4$.

Examination of Appendix D3 shows that the normal derivatives, with the exception of those for u and p_o, are proportional to two terms, one of which is linear in β' while the other is linear in $\sigma\cos\beta/y$. This latter term stems from the axisymmetric ($\sigma = 1$) term in continuity, where y is the radial distance from the symmetry axis to the shock wave. For a detached axisymmetric shock, the σ terms are indeterminate on the symmetry axis, since both β and y are zero. In this circumstance, the ratio $\cos\beta/y$ can be shown to equal the curvature of the shock wave on the symmetry axis (see Problem 6.2).

In the $w = 1$ limit, the shock becomes a Mach wave. In this limit, some of the terms in Appendix D3 are indeterminate, since β' and $w - 1$ go to zero. A more interesting limit is the hypersonic one. If the shock wave is normal or nearly normal to the upstream flow, the rightmost terms in Appendix D4 dominate, and the X, Y, and $w - 1$ factors simplify in an obvious manner. Another hypersonic limit is for slender bodies, when w is of order unity, and the g_i, X, Y, and $w - 1$ factors do not simplify. Nevertheless, the equations do simplify because of the presence of m, which approaches infinity.

TABLE 6.2
The g_i vs. w When $\gamma = 1.4$

w	g_1	g_2	g_3	g_4	g_5	g_6	g_7	g_8
1.0	−0.2000E + 01	0.3333E + 01	0.5760E + 01	−0.5760E + 01	0.9600E + 01	−0.9600E + 01	0.2765E + 01	−0.2765E + 01
1.2	−0.1867E + 01	0.4806E + 01	0.6269E + 01	−0.7830E + 01	0.1246E + 02	−0.1557E + 02	0.3705E + 02	−0.4628E + 02
1.4	−0.1733E + 01	0.6556E + 01	0.6835E + 01	−0.1022E + 02	0.1562E + 02	−0.2336E + 02	0.4760E + 02	−0.7126E + 02
1.6	−0.1600E + 01	0.8583E + 01	0.7459E + 01	−0.1298E + 02	0.1906E + 02	−0.3318E + 02	0.5913E + 02	−0.1031E + 03
1.8	−0.1467E + 01	0.1089E + 02	0.8141E + 01	−0.1613E + 02	0.2278E + 02	−0.4522E + 02	0.7142E + 02	−0.1423E + 03
2.0	−0.1333E + 01	0.1347E + 02	0.8880E + 01	−0.1972E + 02	0.2680E + 02	−0.5967E + 02	0.8429E + 02	−0.1890E + 03
4.0	0.0000E + 00	0.5458E + 02	0.1944E + 02	−0.8748E + 02	0.8280E + 02	−0.3798E + 03	0.2022E + 03	−0.1019E + 04
6.0	0.1333E + 01	0.1235E + 03	0.3576E + 02	−0.2388E + 03	0.1676E + 03	−0.1156E + 04	0.1649E + 03	−0.1783E + 04
8.0	0.2667E + 01	0.2201E + 03	0.5784E + 02	−0.5100E + 03	0.2812E + 03	−0.2583E + 04	−0.2210E + 03	−0.3141E + 03
10.0	0.4000E + 01	0.3446E + 03	0.8568E + 02	−0.9378E + 03	0.4263E + 03	−0.4857E + 04	−0.1149E + 04	0.7020E + 04
15.0	0.7333E + 01	0.7772E + 03	0.1805E + 03	−0.2931E + 04	0.9056E + 03	−0.1553E + 05	−0.7112E + 04	0.8804E + 05
20.0	0.1067E + 02	0.1383E + 04	0.3113E + 03	−0.6701E + 04	0.1568E + 04	−0.3575E + 05	−0.2070E + 05	0.3611E + 06
30.0	0.1733E + 02	0.3117E + 04	0.6809E + 03	−0.2185E + 05	0.3432E + 04	−0.1171E + 06	−0.8283E + 05	0.2247E + 07
40.0	0.2400E + 02	0.5545E + 04	0.1194E + 04	−0.5094E + 05	0.6016E + 04	−0.2733E + 06	−0.2117E + 06	0.7765E + 07
50.0	0.3067E + 02	0.8667E + 04	0.1852E + 04	−0.9853E + 05	0.9320E + 04	−0.5287E + 06	−0.4316E + 06	0.1993E + 08

Derivatives along Streamlines and Mach Lines

To illustrate this theory, we obtain the differential operators along streamlines and Mach lines, which are sketched in Figure 6.5. (These operators may prove useful for irrotational or rotational method-of-characteristic codes.) Streamlines are denoted as ζ_o and have an angle θ relative to the x-axis, while the left-running (ζ_+) Mach lines have an angle $\mu + \theta$ and the right-running (ζ_-) lines have a positive angle $\mu - \theta$. The variables ζ_o and $\zeta_\pm$ represent arc lengths in their respective directions. Small disturbances propagate along the left-running characteristics in the $-\zeta_+$ direction, i.e., toward the shock wave. These characteristics then reflect from the shock as ζ_- and ζ_o characteristics and signals propagate along them in these directions.

In view of the above definitions, we have

$$\left(\frac{\partial s}{\partial \zeta_o}\right)_2 = \cos(\beta - \theta), \quad \left(\frac{\partial n}{\partial \zeta_o}\right)_2 = \sin(\beta - \theta)$$

where $\beta - \theta$ is the acute angle between a streamline and the shock wave. Hence, the streamline derivative is

$$\left(\frac{\partial}{\partial \zeta_o}\right)_2 = \left(\frac{\partial s}{\partial \zeta_o}\frac{\partial}{\partial s} + \frac{\partial n}{\partial \zeta_o}\frac{\partial}{\partial n}\right)_2 = \cos(\beta - \theta)\left(\frac{\partial}{\partial s}\right)_2 + \sin(\beta - \theta)\left(\frac{\partial}{\partial n}\right)_2 \tag{6.49}$$

where the sine and cosine coefficients are provided in Appendix D1.

Downstream of a shock, the stagnation pressure should be constant along a streamline. As a check on the theory, we evaluate

$$\begin{aligned}\left(\frac{\partial p_o}{\partial \zeta_o}\right)_2 &= \cos(\beta - \theta)\left(\frac{\partial p_o}{\partial s}\right)_2 + \sin(\beta - \theta)\left(\frac{\partial p_o}{\partial n}\right)_2 \\ &= (\rho w^2)_1\left[1 + \left(\frac{\gamma+1}{2}\frac{m \sin\beta\cos\beta}{X}\right)^2\right]^{-1/2} \frac{\beta'(w-1)^2}{X^2}\left(1 + \frac{\gamma-1}{2}M_2^2\right)^{\gamma/(\gamma-1)} \\ &\quad \times\left[-\frac{\gamma+1}{2} m \sin\beta\cos\beta\left(\frac{2}{\gamma+1}\frac{1}{m\tan\beta}\right) + \cos^2\beta\right] = 0\end{aligned}$$

and obtain zero, as expected.

For the Mach line directions, we will need the sines and cosines of $\mu + \beta - \theta$ and $\mu - \beta + \theta$. These are the angles that the ζ_- and ζ_+ characteristics have with respect to the shock. As an example, one of the sines is evaluated:

$$\begin{aligned}\sin(\mu + \beta - \theta) &= \sin(\beta - \theta)\cos\mu + \cos(\beta - \theta)\sin\mu \\ &= \left[1 + \left(\frac{\gamma+1}{2}\frac{m\sin\beta\cos\beta}{X}\right)^2\right]^{-1/2}\left[\frac{(M_2^2 - 1)^{1/2}}{M_2} + \frac{\gamma+1}{2}\frac{m\sin\beta\cos\beta}{M_2 X}\right] \\ &= \frac{(M_2^2-1)^{1/2} + \frac{\gamma+1}{2}\frac{m\sin\beta\cos\beta}{X}}{M_2\left[1 + \left(\frac{\gamma+1}{2}\frac{m\sin\beta\cos\beta}{X}\right)^2\right]^{1/2}} \\ &= \left(\frac{Y}{X}\right)^{1/2}\frac{(M_2^2-1)^{1/2} + \frac{\gamma+1}{2}\frac{m\sin\beta\cos\beta}{X}}{1 + \left(\frac{\gamma+1}{2}\frac{m\sin\beta\cos\beta}{X}\right)^2}\end{aligned}$$

where it is not convenient to eliminate $(M_2^2-1)^{1/2}$. Note that M_2 must equal or exceed unity for a real valued result. In a similar manner, we obtain

$$\cos(\mu+\beta-\theta) = \left(\frac{Y}{X}\right)^{1/2} \frac{\dfrac{\gamma+1}{2}\dfrac{m\sin\beta\cos\beta}{X}(M_2^2-1)^{1/2}-1}{1+\left(\dfrac{\gamma+1}{2}\dfrac{m\sin\beta\cos\beta}{X}\right)^2}$$

$$\sin(\mu-\beta+\theta) = \left(\frac{Y}{X}\right)^{1/2} \frac{\dfrac{\gamma+1}{2}\dfrac{m\sin\beta\cos\beta}{X}-(M_2^2-1)^{1/2}}{1+\left(\dfrac{\gamma+1}{2}\dfrac{m\sin\beta\cos\beta}{X}\right)^2}$$

$$\cos(\mu-\beta+\theta) = \left(\frac{Y}{X}\right)^{1/2} \frac{\dfrac{\gamma+1}{2}\dfrac{m\sin\beta\cos\beta}{X}(M_2^2-1)^{1/2}+1}{1+\left(\dfrac{\gamma+1}{2}\dfrac{m\sin\beta\cos\beta}{X}\right)^2}$$

For the right-running characteristic direction, we now have

$$\frac{\partial s}{\partial \zeta_-} = \cos(\mu+\beta-\theta), \quad \frac{\partial n}{\partial \zeta_-} = \sin(\mu+\beta-\theta)$$

and

$$\begin{aligned}\frac{\partial}{\partial \zeta_-} &= \frac{\partial s}{\partial \zeta_-}\frac{\partial}{\partial s}+\frac{\partial n}{\partial \zeta_-}\frac{\partial}{\partial n}\\ &= \left(\frac{Y}{X}\right)^{1/2}\left[1+\left(\frac{\gamma+1}{2}\frac{m\sin\beta\cos\beta}{X}\right)^2\right]^{-1}\left\{\left[\frac{\gamma+1}{2}\frac{m\sin\beta\cos\beta}{X}(M_2^2-1)^{1/2}+1\right]\frac{\partial}{\partial s}\right.\\ &\quad \left.+\left[(M_2^2-1)^{1/2}-\frac{\gamma+1}{2}\frac{m\sin\beta\cos\beta}{X}\right]\frac{\partial}{\partial s}\right\}\end{aligned} \tag{6.50a}$$

For the left-running characteristic direction, we similarly obtain

$$\begin{aligned}\frac{\partial}{\partial \zeta_+} &= \frac{\partial s}{\partial \zeta_+}\frac{\partial}{\partial s}+\frac{\partial n}{\partial \zeta_+}\frac{\partial}{\partial n}\\ &= -\cos(\mu-\beta+\theta)\frac{\partial}{\partial s}+\sin(\mu-\beta+\theta)\frac{\partial}{\partial n}\\ &= -\left(\frac{Y}{X}\right)^{1/2}\left[1+\left(\frac{\gamma+1}{2}\frac{m\sin\beta\cos\beta}{X}\right)^2\right]^{-1}\left\{\left[\frac{\gamma+1}{2}\frac{m\sin\beta\cos\beta}{X}(M_2^2-1)^{1/2}+1\right]\frac{\partial}{\partial s}\right.\\ &\quad \left.+\left[(M_2^2-1)^{1/2}-\frac{\gamma+1}{2}\frac{m\sin\beta\cos\beta}{X}\right]\frac{\partial}{\partial n}\right\}\end{aligned} \tag{6.50b}$$

While Equations (6.49) and (6.50) are not particularly simple, they nevertheless provide explicit relations for the derivatives of interest.

Wave Reflection from a Shock Wave

We mentioned earlier that left-running Mach lines, or characteristics, in part, reflect from the downstream side of a shock as a wave consisting of right-running Mach lines. This is the case in the upper half-plane as sketched in Figure 6.5; for the lower half-plane, the two families are reversed. The reflected wave is an expansion wave if the Mach lines diverge from each other when they are traced in the general flow direction. If the lines converge, the wave is compressive. Moreover, converging Mach lines that start to overlap must be replaced with a developing shock wave. Thus, an internal shock can form in a supersonic flow containing converging Mach lines of the same family. In this situation, flow conditions just upstream of the internal shock are nonuniform. This process results in the downstream shock system that appears in the jet emanating from an underexpanded nozzle (Emanuel, 1986, Section 19.4).

It is useful to know whether or not the reflected wave is compressive or expansive. This wave may impinge on the body and alter the boundary layer. For instance, if the wave is compressive, it may not be strong enough to induce boundary-layer separation, but it can hasten the transition process.

The family of right-running ζ_- characteristics is referred to as a C_- wave. The slope of these characteristics, just downstream of the shock, is $\mu - \theta$ relative to the x-axis (see Figure 6.5). As we travel along the shock, in the downstream direction, the wave is compressive (expansive) if the positive angle, $\mu - \theta$, increases (decreases). (When $\mu - \theta$ increases, the ζ_- characteristics tend to converge.) Thus, the C_+ wave, which consists of the ζ_+ characteristics, reflects from the shock wave as a compression if

$$\frac{d(\mu-\theta)}{ds} > 0 \qquad \text{or} \qquad \frac{d(\mu-\theta)}{d\beta} < 0$$

where the second form is analytically more convenient. Remember that $d\beta/ds$ is negative if the C_+ wave is expansive.

It is hopelessly complicated to attempt to derive an equation for $d(\mu - \theta)/d\beta$ without the assistance of the theory in this section. With this theory, the derivation is simple; we start with Equation (6.47) and obtain

$$\frac{d\mu}{dM} = -\frac{1}{M(M^2-1)^{1/2}}$$

Consequently, we can write

$$\frac{d(\mu-\theta)}{d\beta} = \frac{\dfrac{d\mu}{ds} - \dfrac{d\theta}{ds}}{\dfrac{d\beta}{ds}} = -\frac{1}{\beta'}\left[\frac{1}{M(M^2-1)^{1/2}}\frac{dM}{ds} + \frac{d\theta}{ds}\right]$$

With the assistance of Appendix D, we directly obtain

$$\frac{d(\mu-\theta)}{d\beta} = \frac{(\gamma+1)^{3/2}\left(1+\dfrac{\gamma-1}{2}m\right)(1+\gamma w^2)m\sin\beta\cos\beta}{2X^{5/2}Y^{1/2}\left[1+\left(\dfrac{\gamma+1}{2}\dfrac{m\sin\beta\cos\beta}{X}\right)^2\right][(\gamma+1)mw+2+(\gamma-3)w-2\gamma w^2]^{1/2}}$$
$$-\frac{\dfrac{\gamma+1}{2}m(1+w)+1-2w-\gamma w^2}{(\gamma+1)\left(1+\dfrac{\gamma+1}{4}m\right)w+1-2w-\gamma w^2} \qquad \textbf{(6.51)}$$

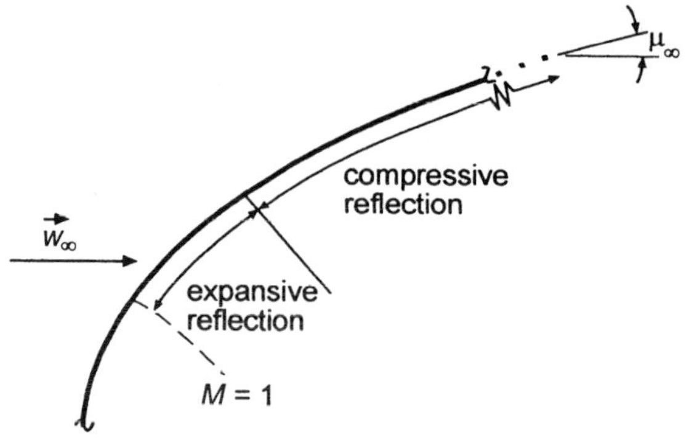

FIGURE 6.6 Expansive and compressive regions downstream of a shock when $\gamma = 1.4$ and $M_\infty > 1.59$.

This is an exact result that does not depend on the nature of the incoming C_+ wave, which may be expansive or compressive. Moreover, it is independent of whether or not the flow is two-dimensional or axisymmetric. It is also independent of the local shock wave curvature, $-\beta'$, since the β' factors cancel. Although complicated, the right side depends only on γ, M_1, and β; hence, the influence of the incident wave is limited to its effect on the wave angle β.

For the detached shock pictured in Figure 5.2, the flow between the shock and body in the nose region is subsonic. This region is bordered by a curved sonic line that intersects the shock where its slope is β^*. The above discussion, of course, only holds when β is less than β^*. A relation for β^* is obtained by setting $M_2^2 - 1 = 0$ in Appendix D1; i.e.,

$$(\gamma + 1)mw^* + 2 + (\gamma - 3)w^* - 2\gamma(w^*)^2 = 0$$

This result also means that $d(\mu - \beta)/d\beta$ is positively infinite at the sonic point. The above relation becomes

$$\gamma(M_1 \sin\beta^*)^4 - \frac{1}{2}[\gamma - 3 + (\gamma + 1)M_1^2](M_1 \sin\beta^*)^2 - 1 = 0$$

which yields

$$\sin\beta^* = \left(\frac{1}{4\gamma M_1^2}\left\{\gamma - 3 + (\gamma + 1)M_1^2 + [17\gamma^2 - 6\gamma + 9 + 2(\gamma - 3)(\gamma + 1)M_1^2 + (\gamma + 1)^2 M_1^4]^{1/2}\right\}\right)^{1/2} \tag{6.52}$$

With $\gamma = 1.4$ and $1.5 \le M_1 \le \infty$, β^* is in the 61° to 67.79° range, where the last value occurs when M_1 is infinite.

As is generally the case, let the incoming wave be an expansion, thereby weakening the shock. In a blunt body flow, the C_+ wave originates on the sonic line. In any case, detailed calculations with $\gamma = 1.4$ show that the reflected wave is expansive, for all β values, when $M_1 < 1.59$. For larger M_1 values, there is a range of β values

$$\mu_1 \le \beta \le \begin{cases} 39°, & M_1 = 1.59 \\ 38°, & 2 \le M_1 \le 4 \\ 36°, & M_1 = 6 \\ 35°, & M_1 = 8 \end{cases} \tag{6.53}$$

for which the reflected wave is compressive. The compressive β region starts at $M_1 = 1.59$, where $\mu_1(1.59) = 39°$. For larger β values, the reflected wave is expansive. Consequently, for a freestream Mach number in excess of 1.59, both types of reflection processes are typically present, as sketched in Figure 6.6. Note that the compressive reflections occur downstream, where the shock is relatively weak.

Hypersonic small disturbance theory is now briefly discussed. In this theory, we have the limit

$$M_\infty \to \infty, \qquad K_\beta = M_\infty \sin\beta = O(1)$$

With

$$X = 1 + \frac{\gamma - 1}{2} K_\beta^2, \qquad Y = \gamma K_\beta^2 - \frac{\gamma - 1}{2}$$

Equation (6.51) yields, to leading order,

$$\frac{d(\mu - \theta)}{d\beta} = \frac{\gamma - 1}{\gamma + 1}\frac{1 + \gamma K_\beta^4}{K_\beta^2 X^{1/2} Y^{1/2}} - \frac{2}{\gamma + 1}\frac{1 + K_\beta^2}{K_\beta^2} \tag{6.54a}$$

For instance, when K_β is unity, this yields

$$\frac{d(\mu - \theta)}{d\beta} = -\frac{2(3 - \gamma)}{\gamma + 1} \tag{6.54b}$$

and the reflected wave is compressive.

6.4 COORDINATE TRANSFORMATION

ξ_j COORDINATES

We return to the theory in Section 6.2 which utilizes only the basic assumptions. A key step in the subsequent analysis is the introduction of a coordinate system ξ_j that is fixed to the shock wave. This system is orthogonal and is a function of both the x_i and t; i.e.,

$$\xi_j = \xi_j(x_i, t), \quad j = 1, 2, 3 \tag{6.55a}$$

$$\tau = t \tag{6.55b}$$

In Section 6.6, the distinction between the two time variables t and τ becomes important. For instance, ξ_j and τ are both independent variables so that

$$\frac{\partial \xi_j}{\partial \tau} = 0$$

On the other hand, $\partial\xi_j/\partial t$ is generally not zero.

We let ξ_1 and ξ_2 conform to the shock's surface; hence, ξ_3 is normal to the shock. More precisely, ξ_1 and ξ_2 are tangent to $\hat{t}$ and $\hat{b}$, respectively, and increase in the direction for which these vectors are positive. Consequently, the coordinate system is aligned with the possibly nonuniform flow just upstream of the shock. For ξ_3, we have a simple choice

$$\xi_3 = F(x_i,t) \tag{6.56}$$

where F is given by Equation (6.1) and $\xi_3 = 0$ at the location of the shock. Although the ξ_j would appear to be well defined only at the shock's surface, we will find a transformation, Equation (6.55a), that is not so restricted.

For an arbitrary point in space, we have

$$d\vec{r} = \frac{\partial \vec{r}}{\partial \xi_j} d\xi_j = \vec{e}_j d\xi_j = \hat{\imath}_i dx_i \tag{6.57}$$

The $\vec{e}_j$ basis, for a right-handed system, is then given by

$$\vec{e}_1 = h_1\hat{t}, \qquad \vec{e}_2 = h_2\hat{b}, \qquad \vec{e}_3 = h_3\hat{n} \tag{6.58a}$$

while the unit vectors are

$$\hat{e}_1 = \hat{t}, \qquad \hat{e}_2 = \hat{b}, \qquad \hat{e}_3 = \hat{n} \tag{6.58b}$$

The h_j scale factors, which may be time dependent, will need to be determined. Conventional tensor notation would write ξ_j as ξ^j, since $\vec{e}_j$ is the basis for the ξ^j. In the interest of notational simplicity, we have not done this.

The scalar product is evaluated as

$$a_{ij} = \hat{\imath}_i \cdot \hat{e}_j$$

with the assistance of Equations (6.14), (6.12), and (6.4), to obtain

$$a_{i1} = \hat{\imath}_i \cdot \hat{t} = \frac{|\nabla F|^2 w_{1,i} - (\vec{w}_1 \cdot \nabla F)F_{x_i}}{|\nabla F| I} \tag{6.59a}$$

$$a_{i2} = \hat{\imath}_i \cdot \hat{b} = \frac{1}{I}\varepsilon_{ijk} w_{1,j} F_{x_k} \tag{6.59b}$$

$$a_{i3} = \hat{\imath}_i \cdot \hat{n} = \frac{F_{x_i}}{|\nabla F|} \tag{6.59c}$$

where ε_{ijk} is the alternating symbol defined in Table 2 of Appendix A. We hereafter view the a_{ij} as known functions of the x_k and t. The rightmost of Equations (6.57) is multiplied with $\vec{e}_j$, with the result

$$d\xi_j = \frac{a_{ij}}{h_j} dx_i \qquad \text{(no } j \text{ sum)}$$

This is compared with

$$d\xi_j = \frac{\partial \xi_j}{\partial x_i} dx_i$$

at a fixed instant of time, to obtain

$$\frac{\partial \xi_j}{\partial x_i} = \frac{a_{ij}}{h_j}, \qquad \text{(no } j \text{ sum)} \tag{6.60}$$

which is a central result of this section.

Equation (6.56) yields

$$\frac{\partial \xi_3}{\partial x_i} = F_{x_i}$$

On the other hand, Equation (6.60) for $j = 3$ and Equation (6.59c) provide

$$\frac{\partial \xi_3}{\partial x_i} = \frac{a_{i3}}{h_3} = \frac{F_{x_i}}{h_3|\nabla F|}$$

By comparing the two $\partial\xi_3/\partial x_i$ relations, we have

$$h_3 = \frac{1}{|\nabla F|} \tag{6.61}$$

Thus, only $j = 1$ and 2 need further consideration. Integration of Equation (6.60) for these two j values then yields the desired transformation equations. This integration, however, first requires finding h_1 and h_2.

Scale Factors

The scale factors are not arbitrary. They are established by the requirement that Equation (6.60) be integrable, which is assured if the compatibility condition (Stoker, 1969)

$$\frac{\partial^2 \xi_j}{\partial x_k \partial x_m} = \frac{\partial^2 \xi_j}{\partial x_m \partial x_k}, \qquad j = 1, 2, 3 \quad m \neq k \tag{6.62}$$

is satisfied. Gauss' equation in Section 5.7 stems from this condition. Since ξ_3 is given by Equation (6.56), it satisfies the compatibility condition and only $j = 1,2$ need to be considered. For each j, this equation represents three equations. In combination with Equation (6.60), these become

$$a_{mj}\frac{\partial q_j}{\partial x_k} - a_{kj}\frac{\partial q_j}{\partial x_m} = \frac{\partial a_{mj}}{\partial x_k} - \frac{\partial a_{kj}}{\partial x_m}$$

where $q_j = \ell n h_j$. When written out, we have

$$a_{2j}\frac{\partial q_j}{\partial x_1} \quad a_{1j}\frac{\partial q_i}{\partial x_2} = \frac{\partial a_{2j}}{\partial x_1} - \frac{\partial a_{1j}}{\partial x_2} \tag{6.63a}$$

$$a_{3j}\frac{\partial q_j}{\partial x_1} - a_{1j}\frac{\partial q_j}{\partial x_3} = \frac{\partial a_{3j}}{\partial x_1} - \frac{\partial a_{1j}}{\partial x_3} \tag{6.63b}$$

$$a_{3j}\frac{\partial q_j}{\partial x_2} - a_{2j}\frac{\partial q_j}{\partial x_3} = \frac{\partial a_{3j}}{\partial x_2} - \frac{\partial a_{2j}}{\partial x_3} \tag{6.63c}$$

for $j = 1$ and 2. With $\partial q_j/\partial x_i$ as unknowns, the determinant of the left side is zero. Hence, elimination of the q_j derivatives results in a condition on the a_{ij} coefficients:

$$a_{2j}^2\frac{\partial}{\partial x_1}\left(\frac{a_{3j}}{a_{2j}}\right) + a_{3j}^2\frac{\partial}{\partial x_2}\left(\frac{a_{1j}}{a_{3j}}\right) + a_{1j}^2\frac{\partial}{\partial x_3}\left(\frac{a_{2j}}{a_{1j}}\right) = 0, \quad j = 1,2 \tag{6.64}$$

for the existence of a solution of Equations (6.63a to c). This equation, however, can be shown to hold for all j, including $j = 3$ (see Problem 6.8). Thus, a solution of Equations (6.63a to c) exists for the h_j scale factors.

Each of Equations (6.63a to c) can be viewed as a separate equation forq_j and, thus, solved independently of the other two. (Hence, the system of equations is overdetermined.) Each of these solutions will involve an arbitrary function of integration. There are no boundary or initial conditions that can be used to evaluate these functions of integration. These functions are chosen so that the resulting q_j is a solution of all three of Equations (6.63a to c). To obtain an explicit formulation for this solution, we use superscripts a, b, and c, respectively, to denote the solutions of Equations (6.63a to c). These equations are first-order PDEs and their general solution is obtained by utilizing a version of the method of characteristics that is described in Appendix E. With this approach, we obtain from Equation (6.63a) the characteristic equations [Equations (E.8) in Appendix E]:

$$\frac{dx_1}{a_{2j}} = -\frac{dx_2}{a_{1j}} = \frac{dx_3}{0} = \frac{dq_j^{(a)}}{\dfrac{\partial a_{2j}}{\partial x_1} - \dfrac{\partial a_{1j}}{\partial x_2}} \tag{6.65}$$

For a fixed j, let $u_{jk}^{(a)}(x_i,t)$, $k = 1,2,3$ denote the functional form of the unique solutions to these three ODEs. To avoid an infinity, the dx_3 term is made indeterminant by setting x_3 equal to a constant. The solution of the two leftmost ODEs can be written as

$$u_{j1}^{(a)} = x_3 = c_{j1}^{(a)}, \quad u_{j2}^{(a)} = c_{j2}^{(a)} \tag{6.66a,b}$$

where the $c_{jk}^{(a)}$ are integration constants. The relation $u_{j2}^{(a)} = c_{j2}^{(a)}$ is the functional form for the solution of the leftmost of Equations (6.65); x_3 and t are held fixed in a_{1j} and a_{2j} when obtaining this solution. There are two equivalent possibilities for $q_j^{(a)}$; for purposes of brevity only one is presented. The solution of the dx_1, $dq_j^{(a)}$ equation can be written as

$$u_{j3}^{(a)}(x_i,q_j^{(a)},t) = q_j^{(a)} - \int\left(\frac{\partial a_{2j}}{\partial x_1} - \frac{\partial a_{1j}}{\partial x_2}\right)\frac{dx_1}{a_{2j}} = c_{j3}^{(a)} \tag{6.66c}$$

where, if necessary, x_2 and x_3 are replaced in the integrand with the aid of Equations (6.66a,b). After the integration is performed, the constants $c_{jk}^{(a)}$, $k = 1,2$, are then replaced by x_3, which equals $u_{j1}^{(a)}$, and by $u_{j2}^{(a)}$.

Although theoretically equivalent, the quadrature that results from using dx_2, $dq_j^{(a)}$ may be simpler or more complicated than the one stemming from dx_1, $dq_j^{(a)}$. In either case, the general solution of Equation (6.63a) can be written as

$$u_{j3}^{(a)} = \ell n g_j^{(a)}(u_{j1}^{(a)}, u_{j2}^{(a)})$$

in accordance with Equation (E.9) in Appendix E, where $g_j^{(a)}$ is an arbitrary function of its two arguments. Equation (6.66c) with $q_{j3}^{(a)} = \ell n h_j^{(a)}$ then yields

$$\ell n h_j^{(a)} = \int\left(\frac{\partial a_{2j}}{\partial x_1} - \frac{\partial a_{1j}}{\partial x_2}\right)\frac{dx_1}{a_{2j}} + \ell n g_j^{(a)}(x_3, u_{j2}^{(a)}(x_i,t))$$

or finally

$$h_j^{(a)} = g_j^{(a)}(x_3,u_{j2}^{(a)}) \exp\left[\int\left(\frac{\partial a_{2j}}{\partial x_1} - \frac{\partial a_{1j}}{\partial x_2}\right)\frac{dx_1}{a_{2j}}\right] \tag{6.67a}$$

The same procedure, when applied to Equations (6.63b,c), results in

$$h_j^{(b)} = g_j^{(b)}(x_2,u_{j2}^{(b)}) \exp\left[\int\left(\frac{\partial a_{3j}}{\partial x_1} - \frac{\partial a_{1j}}{\partial x_3}\right)\frac{dx_1}{a_{3j}}\right] \tag{6.67b}$$

$$h_j^{(c)} = g_j^{(c)}(x_1,u_{j2}^{(c)}) \exp\left[-\int\left(\frac{\partial a_{3j}}{\partial x_2} - \frac{\partial a_{2j}}{\partial x_3}\right)\frac{dx_3}{a_{2j}}\right] \tag{6.67c}$$

where $u_{j2}^{(b)}$ and $u_{j2}^{(c)}$ are solutions of

$$\frac{dx_1}{a_{3j}} = -\frac{dx_3}{a_{1j}}, \qquad \frac{dx_2}{a_{3j}} = -\frac{dx_3}{a_{2j}}$$

respectively. The various g_j coefficients are chosen by inspection so that

$$h_j = h_j^{(a)} = h_j^{(b)} = h_j^{(c)}, \qquad j = 1, 2$$

The choicc for the g_j is not unique. Since $h_j^{(a)} = h_j^{(b)} = h_j^{(c)}$, their selection must satisfy the constraint

$$g_j^{(a)} \exp\left(\int\left(\frac{\partial a_{2j}}{\partial x_1} - \frac{\partial a_{1j}}{\partial x_2}\right)\frac{dx_1}{a_{2j}}\right) = g_j^{(b)} \exp\left(\int\left(\frac{\partial a_{3j}}{\partial x_1} - \frac{\partial a_{1j}}{\partial x_3}\right)\frac{dx_1}{a_{3j}}\right)$$
$$= g_j^{(c)} \exp\left(-\int\left(\frac{\partial a_{3j}}{\partial x_2} - \frac{\partial a_{2j}}{\partial x_3}\right)\frac{dx_3}{a_{2j}}\right)$$

For each g_j choice, there is a different h_j and, through Equation (6.60), a different ξ_j.

Once the scalc factors are known, Equation (6.60) is integrated in a stepwise fashion for ξ_1 and ξ_2. This integration is assured, since the h_j satisfy Equation (6.63). The final result, Equation (6.55a), and the associated scale factors hold globally, both upstream and downstream of the shock, although we will use the transformation only just downstream of the shock. Problems 6.18 and 6.19 illustrate the use of this approach.

Application to the Theory in Section 6.3

A steady, two-dimensional or axisymmetric flow of a perfect gas is assumed. The velocity in the uniform freestream is written as

$$\vec{w}_1 = w_1 \hat{\imath}_1$$

where w_1 is a constant. It is convenient to introduce a transverse radial position vector

$$\vec{R} = x_2 \hat{\imath}_2 + \sigma x_3 \hat{\imath}_3$$

where

$$R = \left(x_2^2 + \sigma x_3^2\right)^{1/2}, \qquad \frac{\partial R}{\partial x_2} = \frac{x_2}{R}, \quad \frac{\partial R}{\partial x_3} = \frac{\sigma x_3}{R}$$

and its normalized form

$$\hat{e}_R = \frac{\vec{R}}{R} = \frac{x_2}{R}\hat{\imath}_2 + \frac{\sigma x_3}{R}\hat{\imath}_3$$

The arbitrary shape of a shock can be written as

$$F = f(x_1) - R = 0$$

In the axisymmetric case, $f(0) = 0$. The gradient of F and its magnitude are

$$\nabla F = \frac{df}{dx_1}\hat{\imath}_1 - \frac{x_2}{R}\hat{\imath}_2 + \frac{\sigma x_3}{E}\hat{\imath}_3 = f'\hat{\imath}_1 - \hat{\varepsilon}_R$$

$$|\nabla F| = \left(f'^2 + \frac{x_2^2}{R^2} + \frac{\sigma x_3^2}{R^2}\right)^{1/2} = (1 + f'^2)^{1/2}$$

where $f' = (df/dx_1)$.

We now determine the $\hat{n}$, $\hat{b}$, and $\hat{t}$ vectors, where

$$\hat{n} = \frac{\nabla F}{|\nabla F|} = \frac{f'\hat{\imath}_1 - \hat{\varepsilon}_R}{(1 + f'^2)^{1/2}}$$

Write $\vec{b}$ as

$$\vec{b} = \hat{w}_1 \times \hat{n} = -\frac{w_1}{(1 + f'^2)^{1/2}}\hat{\imath}_1 \times \hat{\varepsilon}_R$$

where

$$\hat{\imath}_1 \times \hat{\varepsilon}_R = \frac{x_2}{R}\hat{\imath}_1 \times \hat{\imath}_2 + \frac{\sigma x_3}{R}\hat{\imath}_1 \times \hat{\imath}_3 = -\frac{1}{R}(\sigma x_3\hat{\imath}_2 - x_2\hat{\imath}_3)$$

This yields

$$\vec{b} = \frac{w_1}{R(1 + f'^2)^{1/2}}(\sigma x_3\hat{\imath}_2 - x_2\hat{\imath}_3)$$

and

$$\vec{b} = \frac{1}{R}(\sigma x_3\hat{\imath}_2 - x_2\hat{\imath}_3)$$

With

$$\hat{t} = \hat{n} \times \hat{b}$$

we obtain

$$\hat{t} = \frac{1}{(1 + f'^2)^{1/2}} (\hat{\imath}_1 + f'\hat{\varepsilon}_R)$$

The orthonormal basis, $\hat{n}, \hat{b}, \hat{t}$, is only defined at the shock. We extend it to the rest of space by simply writing

$$\hat{e}_1 = \hat{t} = \frac{1}{(1 + f'^2)^{1/2}} (\hat{\imath}_1 + f'\hat{\varepsilon}_R)$$

$$\hat{e}_2 = \hat{b} = \frac{1}{R}(\sigma x_3 \hat{\imath}_2 - x_2 \hat{\imath}_3)$$

$$\hat{e}_3 = \hat{n} = \frac{1}{(1 + f'^2)^{1/2}} (f'\hat{\imath}_1 - \hat{\varepsilon}_R)$$

With the above, the $a_{ij} = \hat{\imath}_i \cdot \hat{e}_j$ are readily evaluated as

$$a_{11} = \frac{1}{(1 + f'^2)^{1/2}}, \quad a_{12} = 0, \quad a_{13} = \frac{f'}{(1 + f'^2)^{1/2}}$$

$$a_{21} = \frac{x_2 f'}{R(1 + f'^2)^{1/2}}, \quad a_{22} = \frac{\sigma x_3}{R}, \quad a_{23} = -\frac{x_2}{R(1 + f'^2)^{1/2}}$$

$$a_{31} = \frac{\sigma x_3 f'}{R(1 + f'^2)^{1/2}}, \quad a_{32} = -\frac{x_2}{R}, \quad a_{33} = -\frac{\sigma x_3}{R(1 + f'^2)^{1/2}}$$

We next develop Equation (6.63a) for $j = 1$; i.e.,

$$a_{21}\frac{\partial q_1}{\partial x_1} - a_{11}\frac{\partial q_1}{\partial x_2} = \frac{\partial a_{21}}{\partial x_1} - \frac{\partial a_{11}}{\partial x_2}$$

where

$$\frac{\partial a_{11}}{\partial x_2} = 0$$

$$\frac{\partial a_{21}}{\partial x_1} = \frac{x_2}{R}\frac{d}{dx_1}\left[\frac{f'}{(1 + f'^2)^{1/2}}\right] = \frac{x_2 f''}{R(1 + f'^2)^{3/2}}$$

The equation simplifies to

$$f'\frac{\partial q_1}{\partial x_1} - \frac{R}{x_2}\frac{\partial q_1}{\partial x_2} - \frac{f''}{1 + f'^2} = 0$$

The characteristic equations are

$$\frac{dx_1}{f'} = -\frac{x_2 dx_2}{R} = \frac{dx_3}{0} = \frac{1+f'^2}{f''} dq_1$$

which yield

$$x_3 = c_1$$

$$\frac{dx_1}{f'} + \frac{x_2 dx_2}{(x_2^2 + \sigma c_1^2)^{1/2}} = 0$$

$$\frac{dq_1}{dx_1} - \frac{1}{f'(1+f'^2)} \frac{df'}{dx_1} = 0$$

The second of these equations integrates to

$$\int \frac{x_2 dx_2}{(x_2^2 + \sigma c_1^2)^{1/2}} = (x_2^2 + \sigma c_1^2)^{1/2} = (x_2^2 + \sigma x_3^2)^{1/2} = R$$

with the result

$$R + \int \frac{dx_1}{f'} = c_2$$

The third characteristic equation integrates to

$$\int \frac{df'}{f'(1+f'^2)} = \ell n \frac{f'}{(1+f'^2)^{1/2}}$$

and

$$q_1 - \ell n \frac{f'}{(1+f')^{1/2}} = c_3$$

We thus obtain

$$q_1 = \ell n h_1^{(a)} = \ell n \frac{f'}{(1+f'^2)^{1/2}} + \ell n g_1^{(a)}\left(x_3, R + \int \frac{dx_1}{f'}\right)$$

or

$$h_1^{(a)} = \frac{f'}{(1+f'^2)^{1/2}} g_1^{(a)}\left(x_3, R + \int \frac{dx_1}{f'}\right)$$

where $g_1^{(a)}$ is an arbitrary function of its two arguments.

The same process for Equations (6.63b) and (6.63c) yields

$$h_1^{(b)} = \frac{f'}{(1+f'^2)^{1/2}} g_1^{(b)}\left(x_2, \sigma R + \int \frac{dx_1}{f'}\right)$$
$$h_1^{(c)} = g_1^{(c)}(x_1, R)$$

A simple (nonunique) choice is

$$g_1^{(a)} = 1, \quad g_1^{(b)} = 1, \quad g_1^{(c)} = \frac{f'}{(1+f'^2)^{1/2}}$$

since f' is a function only of x_1. Consequently, h_1 is

$$h_1 = \frac{f'}{(1+f'^2)^{1/2}}$$

and is only a function of x_1.

Equation (6.63a) for $j = 2$ is

$$a_{22}\frac{\partial q_2}{\partial x_1} - a_{12}\frac{\partial q_2}{\partial x_2} = \frac{\partial a_{22}}{\partial x_1} - \frac{\partial a_{12}}{\partial x_2}$$

where

$$a_{12} = 0, \quad \frac{\partial a_{22}}{\partial x_1} = 0, \quad \frac{\partial a_{12}}{\partial x_2} = 0$$

This yields

$$\frac{\sigma x_3}{R}\frac{\partial q_2}{\partial x_1} = 0$$

with the result

$$h_2^{(a)} = \begin{cases} g_2^{(a)}(x_1, x_2, x_3), & \sigma = 0 \\ g_2^{(a)}(x_2, x_3), & \sigma = 1 \end{cases}$$

Similarly, the (b) and (c) equations produce

$$h_2^{(a)} = g_2^{(b)}(x_2, x_3)$$

$$h_2^{(c)} = \begin{cases} g_2^{(c)}(x_1, x_2), & \sigma = 0 \\ x_2 g_2^{(a)}\left(x_1, \dfrac{x_2}{x_3}\right), & \sigma = 1 \end{cases}$$

A common solution for h_2 is $h_2 = x_2^{\sigma}$.

With the aid of Equation (6.61), we summarize the results for the scale factors as

$$h_1 = \frac{f'}{(1+f'^2)^{1/2}}, \qquad h_2 = x_2^{\sigma}, \qquad h_3 = \frac{1}{(1+f'^2)^{1/2}}$$

Since $f' = \tan\beta$, we have

$$h_1 = \sin\beta, \qquad h_3 = \cos\beta$$

at the shock wave.

Our next task is to use Equation (6.60) to obtain the transformation equations. For $j = 1$, we write

$$\frac{\partial \xi_1}{\partial x_1} = \frac{a_{11}}{h_1} = \frac{1}{f'}$$

which integrates to

$$\xi_1 = \phi_1(x_2, x_3) + \int \frac{dx_1}{f'}$$

where ϕ_1 is a function of integration. To evaluate this function, we use

$$\frac{\partial \xi_1}{\partial x_2} = \frac{\partial \phi_1}{\partial x_2}$$

$$\frac{\partial \xi_1}{\partial x_2} = \frac{a_{21}}{h_1} = \frac{x_2}{R}$$

or

$$\frac{\partial \phi_1}{\partial x_2} = \frac{x_2}{R}$$

$$\phi = \phi_2(x_3) + \int \frac{x_2 dx_2}{(x_2^2 + \sigma x_3^2)^{1/2}}$$

In the integrand, x_3 is held constant, and ϕ_2 is a second function of integration. Upon evaluation of the integral, we have

$$\phi_1 = \phi_2 + (x_2^2 + \sigma x_3^2)^{1/2} = \phi_2 + R$$

To evaluate ϕ_2, we utilize

$$\frac{\partial \xi_1}{\partial x_3} = \frac{d\phi_2}{dx_3} + \frac{\sigma x_3}{R}$$

$$\frac{\partial \xi_1}{\partial x_3} = \frac{a_{31}}{h_1} = \frac{\sigma x_3}{R}$$

As a consequence, we obtain

$$\frac{d\phi_2}{dx_3} = 0$$

or

$$\phi_2 = 0$$

Thus, the ξ_1 transformation equation is

$$\xi_1 = R + \int \frac{dx_1}{f'}$$

A similar process for ξ_2 results in

$$\xi_2 = -\left[(1-\sigma)x_3 + \sigma \ell n\left(\frac{x_2}{R - x_3}\right)\right]$$

With the aid of Equation (6.56), we have

$$\xi_1 = R + \int \frac{dx_1}{f'}, \quad \xi_2 = -\left[(1-\sigma)x_3 + \sigma \ell n\left(\frac{x_2}{R - x_3}\right)\right], \quad \xi_3 = f - R \qquad \textbf{(6.68)}$$

The transformation is not fully explicit until $f(x_1)$ is prescribed. As various checks, one can show that Equations (6.63) are satisfied for $j = 1, 2, 3$, the $\hat{e}_i$ basis is tangent to the ξ_i, and that

$$(ds)^2 = h_j^2 (d\xi_j)^2 = (dx_1)^2 + (dx_2)^2 + (dx_3)^2$$

Problem 6.18, using experimental data for several blunt body flows, illustrates the application of the foregoing analysis.

The steady Euler equations can be written as

$$\nabla \cdot (\rho \vec{w}) = 0$$

$$\rho \frac{D\vec{w}}{Dt} + \nabla p = 0$$

$$\frac{D}{Dt}\left(\frac{p}{\rho^r}\right) = 0$$

where

$$\vec{w} = v_i \hat{e}_i, \qquad v_2 = 0, \qquad \frac{\partial p}{\partial \xi_2} = \frac{\partial \rho}{\partial \xi_2} = 0$$

We shall need (Appendix A, Table 4)

$$\frac{\partial \hat{e}_j}{\partial \xi_i} = \sum_{k \neq j} \left(\frac{\delta_{ik}}{h_j} \frac{\partial h_k}{\partial \xi_j} - \frac{\delta_{ij}}{h_k} \frac{\partial h_j}{\partial \xi_k} \right) \hat{e}_k$$

which yields

$$\frac{\partial \hat{e}_1}{\partial \xi_1} = -\frac{1}{h_2} \frac{\partial h_1}{\partial \xi_2} \hat{e}_2 - \frac{1}{h_3} \frac{\partial h_1}{\partial \xi_3} \hat{e}_3, \ldots$$

Note that the h_i and ξ_i are explicitly known in terms of x_j. Nevertheless, the Euler equations reduce to

$$\frac{\partial}{\partial \xi_1}(h_2 h_3 \rho v_1) + \frac{\partial}{\partial \xi_3}(h_1 h_2 \rho v_3) = 0$$

$$\frac{v_1}{h_1} \frac{\partial v_1}{\partial \xi_1} + \frac{v_3}{h_3} \frac{\partial v_1}{\partial \xi_3} + \frac{v_3}{h_1 h_3} \left(v_1 \frac{\partial h_1}{\partial \xi_3} - v_3 \frac{\partial h_3}{\partial \xi_1} \right) + \frac{1}{h_1 \rho} \frac{\partial p}{\partial \xi_1} = 0$$

$$\frac{v_1^2}{h_1} \frac{\partial h_1}{\partial \xi_2} + \frac{v_3^2}{h_3} \frac{\partial h_3}{\partial \xi_2} = 0$$

$$\frac{v_1}{h_1} \frac{\partial v_3}{\partial \xi_1} + \frac{v_3}{h_3} \frac{\partial v_3}{\partial \xi_3} + \frac{v_1}{h_1 h_3} \left(v_3 \frac{\partial h_3}{\partial \xi_1} - v_1 \frac{\partial h_1}{\partial \xi_3} \right) + \frac{1}{h_3 \rho} \frac{\partial p}{\partial \xi_3} = 0$$

$$\frac{v_1}{h_1} \frac{\partial}{\partial \xi_1} \left(\frac{p}{\rho^{\gamma}} \right) + \frac{v_3}{h_3} \frac{\partial}{\partial \xi_3} \left(\frac{p}{\rho^{\gamma}} \right) = 0$$

The form of the continuity equation means that a stream function can be introduced.

The h_1 and h_3 scale factors and their derivatives with respect to ξ_1 and ξ_3 need to be written in terms of the ξ_i. This is neatly accomplished with Jacobian theory (see Appendix B). For this, we need

$$\frac{\partial h_1}{\partial x_1} = \frac{f''}{(1 + f'^2)^{3/2}}, \qquad \frac{\partial h_1}{\partial x_2} = 0, \qquad \frac{\partial h_1}{\partial x_3} = 0$$

$$\frac{\partial h_2}{\partial x_1} = 0, \qquad \frac{\partial h_2}{\partial x_2} = \sigma, \qquad \frac{\partial h_2}{\partial x_3} = 0$$

$$\frac{\partial h_3}{\partial x_1} = -\frac{f' f''}{(1 + f'^2)^{3/2}}, \qquad \frac{\partial h_3}{\partial x_2} = 0, \qquad \frac{\partial h_3}{\partial x_3} = 0$$

and

$$\frac{\partial \xi_1}{\partial x_1} = \frac{1}{f'}, \quad \frac{\partial \xi_1}{\partial x_2} = \frac{x_2}{R}, \quad \frac{\partial \xi_1}{\partial x_3} = \frac{\sigma x_3}{R}$$

$$\frac{\partial \xi_2}{\partial x_1} = 0, \quad \frac{\partial \xi_2}{\partial x_2} = \frac{\sigma x_3}{x_2 R}, \quad \frac{\partial \xi_2}{\partial x_3} = -\frac{x_2^{1-\sigma}}{R}$$

$$\frac{\partial \xi_3}{\partial x_1} = f', \quad \frac{\partial \xi_3}{\partial x_2} = \frac{x_2}{R}, \quad \frac{\partial \xi_3}{\partial x_3} = -\frac{\sigma x_3}{R}$$

The Jacobian of the transformation reduces to

$$J = \frac{\partial(\xi_1, \xi_2, \xi_3)}{(x_1, x_2, x_3)} = -\frac{1 + f'^2}{x_2^\sigma f'}$$

The $\partial h_1/\partial \xi_1$ derivative is then given by

$$\frac{\partial h_1}{\partial \xi_1} = \frac{\partial(h_1, \xi_2, \xi_3)}{\partial(\xi_1, \xi_2, \xi_3)} = \frac{\dfrac{\partial(h_1, \xi_2, \xi_3)}{\partial(x_1, x_2, x_3)}}{\dfrac{\partial(\xi_1, \xi_2, \xi_3)}{\partial(x_1, x_2, x_3)}}$$

$$= \frac{1}{J}\begin{vmatrix} \dfrac{\partial h_1}{\partial x_1} & 0 & 0 \\ \dfrac{\partial \xi_2}{\partial x_1} & \dfrac{\partial \xi_2}{\partial x_2} & \dfrac{\partial \xi_2}{\partial x_3} \\ \dfrac{\partial \xi_3}{\partial x_1} & \dfrac{\partial \xi_3}{\partial x_2} & \dfrac{\partial \xi_3}{\partial x_3} \end{vmatrix} = \frac{1}{J}\frac{\partial h_1}{\partial x_1}\left(\frac{\partial \xi_2}{\partial x_2}\frac{\partial \xi_3}{\partial x_3} - \frac{\partial \xi_2}{\partial x_3}\frac{\partial \xi_3}{\partial x_2}\right)$$

$$= \frac{f' f''}{(1 + f'^2)^{5/2}}$$

We thus obtain

$$\frac{\partial h_1}{\partial \xi_1} = \frac{f' f''}{(1 + f'^2)^{5/2}}, \quad \frac{\partial h_1}{\partial \xi_2} = 0, \quad \frac{\partial h_1}{\partial \xi_3} = \frac{f' f''}{(1 + f'^2)^{5/2}}$$

$$\frac{\partial h_2}{\partial \xi_1} = \frac{\sigma x f'^2}{R(1 + f'^2)}, \quad \frac{\partial h_2}{\partial \xi_2} = \frac{\sigma x_2 x_3}{R}, \quad \frac{\partial h_2}{\partial \xi_3} = \frac{\sigma x_2}{R(1 + f'^2)}$$

$$\frac{\partial h_3}{\partial x_1} = -\frac{f'^2 f''}{(1 + f'^2)^{5/2}}, \quad \frac{\partial h_3}{\partial \xi_2} = 0, \quad \frac{\partial h_3}{\partial \xi_3} = -\frac{f'^2 f''}{(1 + f'^2)^{5/2}}$$

With this result, the Euler equations become

$$\frac{f'}{1 + f'^2}\left(\frac{f''}{1 + f'^2} - \frac{\sigma}{R}\right)\rho(v_3 - f' v_1) + \frac{\partial(\rho v_1)}{\partial \xi_1} + f'\frac{\partial(\rho v_3)}{\partial \xi_3} = 0 \qquad \textbf{(6.69a)}$$

$$v_1 \frac{\partial v_1}{\partial \xi_1} + f' v_3 \frac{\partial v_1}{\partial \xi_3} + \frac{f' f''}{(1 + f'^2)^{3/2}} v_3 (v_1 + f' v_3) + \frac{1}{\rho} \frac{\partial p}{\partial \xi_1} = 0 \quad \textbf{(6.69b)}$$

$$v_1 \frac{\partial v_3}{\partial \xi_1} + f' v_3 \frac{\partial v_3}{\partial \xi_3} - \frac{f' f''}{(1 + f'^2)^2} v_1 (v_1 + f' v_3) + \frac{f'}{\rho} \frac{\partial p}{\partial \xi_3} = 0 \quad \textbf{(6.69c)}$$

$$v_1 \frac{\partial}{\partial \xi_1}\left(\frac{p}{\rho^r}\right) + f' v_3 \frac{\partial}{\partial \xi_3}\left(\frac{p}{\rho^r}\right) = 0 \quad \textbf{(6.69d)}$$

The momentum equation in the ξ_2 direction does not appear because it is identically satisfied. In the above equations, f, f', f'', and R must still be written in terms of the ξ_i. This is possible only after a specific $f(x_1)$ is given.

For example, suppose the shock surface is a parabola ($\sigma = 0$) or a paraboloid of revolution ($\sigma = 1$) with a radius of curvature of unity at the origin (Van Dyke, 1965). In this case,

$$f = (2x_1)^{1/2}$$

and the inverse transformation for $(2x_1)^{1/2}$ is a cubic equation

$$(2x_1)^{3/2} + 3(2x_1)^{1/2} - 3(\xi_1 + \xi_3) = 0$$

6.5 TANGENTIAL DERIVATIVES

With Equation (6.55a) for the ξ_j coordinates established, we turn our attention to finding the tangential derivatives. In contrast to the analysis in Section 6.3, we have two sets, one along ξ_1 and the other along ξ_2. We assume that the enthalpy function, $h(p,\rho)$, and the shock surface, Equation (6.1), are known. In addition, β and all unity subscripted quantities are presumed to be known functions of ξ_1, ξ_2, and τ. Consequently, Equations (6.22) to (6.27) provide w_2^*, ρ_2, p_2, h_2, and θ.

We first differentiate Equation (6.22) with respect to ξ_j, $j = 1,2$, to obtain

$$\frac{\partial w_2^*}{\partial \xi_j} = \frac{\cos\beta}{\cos(\beta - \theta)} \frac{\partial w_1^*}{\partial \xi_j} - \frac{w_1^* \sin\beta}{\cos(\beta - \theta)} \frac{\partial \beta}{\partial \xi_j} + \frac{w_1^* \cos\beta \tan(\beta - \theta)}{\cos(\beta - \theta)} \left(\frac{\partial \beta}{\partial \xi_j} - \frac{\partial \theta}{\partial \xi_j}\right)$$

which rearranges to

$$\frac{\partial w_2^*}{\partial \xi_j} + \frac{w_1^* \cos\beta \sin(\beta - \theta)}{\cos^2(\beta - \theta)} \frac{\partial \theta}{\partial \xi_j} = \frac{\cos\beta}{\cos(\beta - \theta)} \frac{\partial w_1^*}{\partial \xi_j} - \frac{w_1^* \sin\theta}{\cos(\beta - \theta)} \frac{\partial \beta}{\partial \xi_j} = A_1$$

where $\partial\theta/\partial\xi_j$ is one of the derivatives to be found. A similar procedure is used for Equations (6.23) to (6.25); the result is summarized in Appendix F. Part F1 provides the system of equations for $\partial w_2^*/\partial \xi_j$, $\partial\theta/\partial\xi_j$, $\partial \rho_2/\partial \xi_j$, and $\partial p_2/\partial \xi_j$ when $j = 1,2$, where

$$h_\rho = \left(\frac{\partial h}{\partial \rho}\right)_p, \quad h_p = \left(\frac{\partial h}{\partial p}\right)_\rho$$

and the A_n are the known inhomogeneous terms. The h_ρ and h_p derivatives are based on Equation (6.26) and are related to the speed of sound by means of

$$a^2 = \left(\frac{\partial p}{\partial \rho}\right)_s = \frac{\rho h_\rho}{1 - \rho h_p} \tag{6.70}$$

where s is the entropy. Part F2 provides the explicit solution to the linear equations of Part F1. (Note that $h_{\rho,2}$ and $h_{p,2}$ are known quantities.) Once these $j = 1{,}2$ derivatives are known, the tangential derivatives in any other direction can be found, as illustrated by the example near the end of Section 6.3.

6.6 NORMAL DERIVATIVES

We now consider the normal derivatives that require the Euler equations. These are written in vectorial form as

$$\frac{D\rho}{Dt} + \rho \nabla \cdot \vec{w} = 0 \tag{6.71a}$$

$$\rho \frac{Dh}{Dt} - \frac{Dp}{Dt} = 0 \tag{6.71b}$$

$$\rho \frac{D\vec{w}}{Dt} + \nabla p = 0 \tag{6.71c}$$

They hold in the ξ_j orthogonal coordinate system, which is particularly convenient, since ξ_3 is normal to the shock and the ξ_1 and ξ_2 tangential derivatives are provided by Appendix F. Since the coordinates may be time dependent, the velocity of the coordinate system is introduced as

$$\vec{W} = \frac{\partial \xi_j}{\partial t} \hat{e}_j = W^j \hat{e}_j \tag{6.72}$$

where $W^3 = \partial F / \partial t$ and W^1 and W^2 also can be nonzero because the upstream flow may be unsteady. By comparison with Equation (6.5), we see that $-W^3$, at $\xi_3 = 0$, does not equal the shock speed w_s, except when $|\nabla F| = 1$.

The x_i,t coordinates are for an inertial system, whereas the ξ_j,τ coordinates are noninertial. Since these latter variables do not provide a rigid body motion, the theory of Section 2.5 cannot be used. Instead, we start with Equations (6.55) from which we obtain

$$\frac{\partial}{\partial t} = \frac{\partial \tau}{\partial t}\frac{\partial}{\partial \tau} + \frac{\partial \xi_j}{\partial t}\frac{\partial}{\partial \xi_j} = \frac{\partial}{\partial \tau} + \vec{W} \cdot \nabla_\xi \tag{6.73a}$$

$$\frac{\partial}{\partial x_i} = \frac{\partial \tau}{\partial x_i}\frac{\partial}{\partial \tau} + \frac{\partial \xi_j}{\partial x_i}\frac{\partial}{\partial \xi_j} = \frac{a_{ij}}{h_j}\frac{\partial}{\partial \xi_j} \tag{6.73b}$$

where

$$\nabla_\xi = \frac{\hat{e}_i}{h_i}\frac{\partial}{\partial \xi_i} \tag{6.74}$$

The differential operator ∇_ξ is just the del operator for the ξ_j system.

We write the fluid velocity in the ξ_j system as

$$\vec{w} = u^j \hat{e}_j \tag{6.75}$$

where the $\hat{e}_j$ are given by Equations (6.58b). By comparison with Equation (6.8b), we have

$$u_2^1 = w_1^* \cos(\beta - \theta) \tag{6.76a}$$

$$u_2^2 = 0 \tag{6.76b}$$

$$u_2^3 = w_2^* \sin(\beta - \theta) + w_s \tag{6.76c}$$

just downstream of the shock. (Remember that the $\hat{n}, \hat{t}$ system is defined so that u_2^2 is identically zero just downstream of the shock.) Because the substantial derivative follows a fluid particle, it is invariant with respect to the coordinate system. In Cartesian coordinates, we have

$$\frac{D}{Dt} = \frac{\partial}{\partial t} + \vec{w} \cdot \nabla_x$$

where ∇_x is the operator

$$\nabla_x = \hat{\imath}_i \frac{\partial}{\partial x_i}$$

Since $\nabla_x = \nabla_\xi$, we now obtain, with the aid of Equations (6.73a), (6.74), and (6.75),

$$\frac{D}{Dt} = \frac{\partial}{\partial \tau} + \frac{W^j}{h_j}\frac{\partial}{\partial \xi_j} + \vec{w} \cdot \nabla_\xi = \frac{\partial}{\partial \tau} + \left(\frac{u^j + W^j}{h_j}\right)\frac{\partial}{\partial \xi_j} \tag{6.77}$$

for the substantial derivative in the ξ_j, τ system.

In the ξ_j coordinate system, the divergence of the velocity is

$$\begin{aligned} \nabla \cdot \vec{w} &= \frac{1}{h_1 h_2 h_3}\frac{\partial}{\partial \xi_j}\left(\frac{h_1 h_2 h_3}{h_j} u^j\right) \\ &= \frac{1}{h_1 h_2 h_3}\left[\frac{\partial}{\partial \xi_1}(h_1 h_2 u^1) - \frac{\partial}{\partial \xi_3}(h_1 h_2 u^3)\right] \end{aligned} \tag{6.78}$$

just downstream of the shock where u^2 and $\partial u^2/\partial\xi_2$ are both zero. For notational convenience, we again omit the subscript 2, although most of the subsequent formulas hold only just downstream of the shock. Equation (6.71a) for continuity therefore becomes

$$\frac{\partial\rho}{\partial\tau}+\left(\frac{u^j+W^j}{h_j}\right)\frac{\partial\rho}{\partial\xi_j}+\frac{\rho}{h_1h_2h_3}\left[\frac{\partial}{\partial\xi_1}(h_2h_3u^1)+\frac{\partial}{\partial\xi_3}(h_1h_2u^3)\right]=0 \tag{6.79}$$

The density ρ_2 is provided by Equation (6.23), while $(\partial\rho/\partial\tau)_2$ is obtained by differentiating this equation with respect to time. To obtain $(\partial u^1/\partial\xi_1)_2$, Equation (6.76a) is differentiated with respect to ξ_1 and the equations in Appendix F2 are then utilized. The density derivatives $(\partial\rho/\partial\xi_1)_2$ and $(\partial\rho/\partial\xi_2)_2$ are also found with the aid of Appendix F2. The ξ_1 and ξ_2 derivatives of the scale factors are provided by

$$\frac{\partial h_k}{\partial\xi_j}=\frac{1}{\frac{\partial\xi_j}{\partial t}}\frac{\partial h_k}{\partial t}+\frac{1}{\frac{\partial\xi_j}{\partial x_i}}\frac{\partial h_k}{\partial x_i}=\frac{1}{\frac{\partial\xi_j}{\partial t}}\frac{\partial h_k}{\partial t}+\frac{h_j}{a_{ij}}\frac{\partial h_k}{\partial x_i}\qquad(\text{no } j \text{ sum})$$

where the derivatives of h_k with respect to t and x_i come from Equation (6.55a) and where Equation (6.50) is utilized. The rightmost term is set equal to zero if $a_{ij}=0$, since, in this case, the $\partial h_k/\partial x_i$ are also zero. Hence, the only unknowns in Equation (6.79) are the normal derivatives $(\partial\rho/\partial\xi_3)_2$ and $(\partial u^3/\partial\xi_3)_2$.

Equation (6.71b) can be written as

$$\rho\left(h_p\frac{Dp}{Dt}+h_p\frac{D\rho}{Dt}\right)-\frac{Dp}{Dt}=0$$

With Equation (6.70), this becomes

$$\frac{Dp}{Dt}-a^2\frac{D\rho}{D_t}=0$$

or

$$\frac{\partial p}{\partial\tau}-a^2\frac{\partial\rho}{\partial\tau}+\left(\frac{u^j+W^j}{h_j}\right)\left(\frac{\partial p}{\partial\xi_j}-a^2\frac{\partial\rho}{\partial\xi_j}\right)=0 \tag{6.80}$$

When evaluated behind the shock, the unknowns in this relation are $(\partial p/\partial\xi_3)_2$ and $(\partial\rho/\partial\xi_3)_2$.

For the two differential terms in the momentum equation, we have

$$\nabla p=\frac{1}{h_i}\frac{\partial p}{\partial\xi_j}\hat{e}_i \tag{6.81a}$$

$$\frac{D\vec{w}}{Dt}=\frac{\partial u^i}{\partial\tau}\hat{e}_i+u^i\frac{\partial\hat{e}_i}{\partial\tau}+\frac{u^i+W^i}{h_i}\left(\frac{\partial u^j}{\partial\xi_i}\hat{e}_i+u^i\frac{\partial\hat{e}_j}{\partial\xi_i}\right) \tag{6.81b}$$

where (Appendix A, Table 4)

$$\frac{\partial \hat{e}_j}{\partial \xi_i} = \sum_{k \neq j} \left(\frac{\delta_{ik}}{h_j} \frac{\partial h_i}{\partial \xi_j} - \frac{\delta_{ij}}{h_k} \frac{\partial h_j}{\partial \xi_k} \right) \hat{e}_k \tag{6.82}$$

Evaluation of the $(\partial u^i/\partial \tau)_2$ term in Equation (6.81b) requires the use of Equations (6.76) and differentiation, with respect to t, of Equations (6.11), (6.22), and (6.27). The $(\partial \hat{e}_i/\partial \tau)_2$ derivatives require differentiation of Equation (6.4) and $\hat{b} = (\vec{w}_1 \times \hat{n})/(w_1 \cos \beta)$, after which the Cartesian basis $\hat{\imath}_i$ is replaced with the $\hat{e}_j$ basis. This last step is expedited with the use of Equations (6.59).

In view of its complexity, it is not feasible to explicitly write the momentum equation in scalar form, partly because of the $(\partial \hat{e}_i/\partial \tau)_2$ term, nor is it feasible to explicitly solve for its unknowns, which are $(\partial p/\partial \xi_3)_2$ and $(\partial u^i/\partial \xi_3)_2$, $i = 1,2,3$. In spite of Equation (6.76b), the $(\partial u^2/\partial \xi_3)_2$ derivative is generally not zero. Nevertheless, the foregoing Euler equations and Equation (6.26) represent five linear inhomogeneous equations for $(\partial p/\partial \xi_3)_2$, $(\partial \rho/\partial \xi_3)_2$, and $(\partial u^i/\partial \xi_3)_2$. The normal derivative of other quantities can be found once these are available. For instance, $(\partial w/\partial \xi_3)_2$ and $(\partial \vec{w}/\partial \xi_3)_2$ are given by

$$\left(\frac{\partial w}{\partial \xi_3}\right)_2 = \left\{ \frac{u^1 \frac{\partial u^1}{\partial \xi_3} + u^3 \frac{\partial u^3}{\partial \xi_3}}{[(u^1)^2 + (u^3)^2]^{1/2}} \right\}_2$$

$$\left(\frac{\partial \vec{w}}{\partial \xi_3}\right)_2 = \left(\frac{\partial u^i}{\partial \xi_3} \hat{e}_i + u^i \frac{\partial \hat{\imath}}{\partial \xi_3} + u^3 \frac{\partial \hat{n}}{\partial \xi_3}\right)_2$$

Thus, the ξ_3 derivative of the basis enters into the last derivative but not the first.

Vorticity Just Downstream of the Shock

As an illustration of the theory, we derive $\vec{\omega}$. With Equation (6.75) for $\vec{w}$, Table 4 in Appendix A yields

$$\vec{\omega}_2 = \frac{|\nabla F|}{h_1 h_2} h_1 \left[\frac{\partial}{\partial \xi_2}\left(\frac{u^3}{|\nabla F|}\right) - h_2 \frac{\partial u^2}{\partial \xi_3} \right] \hat{\imath} + h_2 \left[\frac{\partial}{\partial \xi_3}(h_1 u^1) - \frac{\partial}{\partial \xi_1}\left(\frac{u^3}{|\nabla F|}\right) \right] \hat{b} - \frac{1}{|\nabla F|} \frac{\partial}{\partial \xi_2}(h_1 u^1) \hat{n}_2 \tag{6.83}$$

where $u_2^2 = 0$, but $(\partial u^2/\partial \xi_3)_2$ is not zero, and Equation (6.61) is used for h_3. Some further algebraic development is feasible by replacing the scale factor derivatives and utilizing Appendix F for the tangential derivatives of the u^i. The jump in the tangential component of the vorticity is zero (Hayes, 1957), since one can show that

$$\left[\frac{\partial}{\partial \xi_2}(h_1 w_t^*)\right]_1 = \left[\frac{\partial}{\partial \xi_2}(h_1 w_t^*)\right]_2$$

where w_t^* is the magnitude of the tangential component of $\vec{w}^*$.

For a two-dimensional or axisymmetric flow (without swirl), both $\partial(\ )/\partial\xi_2$ and $\partial u^2/\partial\xi_3$ are zero. (Remember that u^2 is the velocity component along $\hat{b}$.) Equation (6.83) therefore becomes

$$\vec{\omega}_2 = \frac{|\nabla F|}{h_1}\left[\frac{\partial}{\partial\xi_3}(h_1 u^1) - \frac{\partial}{\partial\xi_1}\left(\frac{u^3}{|\nabla F|}\right)\right]\hat{b} \tag{6.84}$$

and the vorticity is tangential to the shock. Both vorticity equations hold for steady or unsteady flows in which the upstream flow need not be uniform [Equation (6.84) requires a two-dimensional or axisymmetric upstream flow], and the gas need not be perfect. Of course, the evaluation of derivatives, such as $(\partial u^1/\partial\xi_3)_2$ and $(\partial u^3/\partial\xi_1)_2$, requires an enthalpy function, Equation (6.26). As shown in Problem 6.3, Equation (6.84) can be further simplified by introducing the balance of the Section 6.3 assumptions.

REFERENCES

Ames Research Staff, *Equations, Tables, and Charts for Compressible Flow*, NACA Report 1135, 1953.

Billig, F.S., "Shock-Wave Shapes around Spherical- and Cylindrical-Nosed Bodies," *J. Spacecraft and Rockets* 4, 822 (1967).

Emanuel, G., *Gasdynamics: Theory and Applications*, AIAA Education Series, Washington, D.C., 1986.

Emanuel, G. and Liu, M.-S., "Shock Wave Derivatives," *Phys. Fluids* 31, 3625 (1988).

Hayes, W.D., "The Vorticity Jump Across a Gasdynamic Discontinuity," *J. Fluid Mech.* 2, 595 (1957).

Moretti, G., "Computation of Flows with Shocks," *Annu. Rev. Fluid Mech.* 19, 313 (1987).

Rand, R.H., *Computer Algebra in Applied Mathematics, an Introduction to MACSYMA*, Research Notes in Mathematics, Vol. 94, Pitman, London, 1984.

Stoker, J.J., *Differential Geometry*, John Wiley, New York, 1969, 392.

Truesdell, C., "On Curved Shocks in Steady Plane Flow of an Ideal Fluid," *J. Aeron. Sci.* 19, 826 (1952).

Van Dyke, M., "Hypersonic Flow behind a Paraboloidal Shock Wave," *J. Mécanique* 4, 477 (1965).

Van Dyke, M., *An Album of Fluid Motion*, The Parabolic Press, Stanford, CA, 1982.

Vincenti, W.G. and Kruger, C.H., Jr., *Introduction to Physical Gas Dynamics*, John Wiley, New York, 1965.

Zel'dovich, Ya. B. and Raizer, Yu. P., *Physics of Shock Waves and High-Temperature Phenomena*, Vols. I and II, Academic Press, New York, 1966.

Zucrow, M.J. and Hoffman, J.D., *Gas Dynamics*, Vol. I, John Wiley, New York, 1976, Sections 7-5 and 7-8.

PROBLEMS

6.1 Start with Equation (6.27) and derive Equation (6.28).

6.2 Prove the statement made in Section 6.3 that β/y equals the shock wave curvature on the symmetry axis for a detached shock.

6.3 With the assumptions used in Section 6.3, Emanuel (1986), p. 246, shows that the vorticity can be written as

$$\vec{\omega} = \left[\frac{1}{h_1}\frac{\partial v_2}{\partial\xi_1} - \frac{1}{h_2}\frac{\partial v_1}{\partial\xi_2} + v_2\kappa_2 - v_1\kappa_1\right]\hat{e}_3$$

(a) Derive a form for $\vec{\omega}_2$ consistent with Appendix D1.

(b) Use the result of part (a) to derive a form for $\vec{\omega}_2/(\beta' w_1)$ that depends only on the θ and β angles (Truesdell, 1952).

6.4 Utilize the thermodynamics of a van der Waals gas provided in Problem 5.18. For this gas, determine the counterpart to Equation (6.28) in the form

$$F(\theta, \beta; m, c_v/R, p_{r1}, \rho_{r1}) = 0$$

where

$$m^2 = \frac{w_1^{*2}}{RT_c}$$

6.5 Continue with Problem 6.4; use the assumptions (except for perfect gas) of Section 6.3 and introduce

$$U_i = \frac{v_i^*}{(RT_c)^{1/2}}, \qquad i = 1, 2$$

where the v_i^* are the upstream and downstream velocity components that are normal to the shock. Determine a cubic equation for U_2/U_1 whose other parameters are c_v/R, ρ_{r1}, ρ_{r1}/U_1^2, and $p_{r1}/(\rho_{r1}U_1^2)$.

6.6 Continue with Problems 6.4 and 6.5. Determine $(a_{r2}/a_{r1})^2$ and $s_{r2} - s_{r1}$ in terms of U_2/U_1 and the other parameters used in Problem 6.5.

6.7 Consider a triple point in the steady flow of a perfect gas. The I and R shocks are weak solution shocks while the shock M is a strong solution shock. Emanating from the triple point is a slipstream, SS, which satisfies the conditions that $p_4 = p_3$ and $\vec{w}_4$ is parallel to $\vec{w}_3$. Let $M_1 = 3$, $\gamma = 1.4$, and $\theta_i = 30°$ and determine M_2, M_3, and M_4.

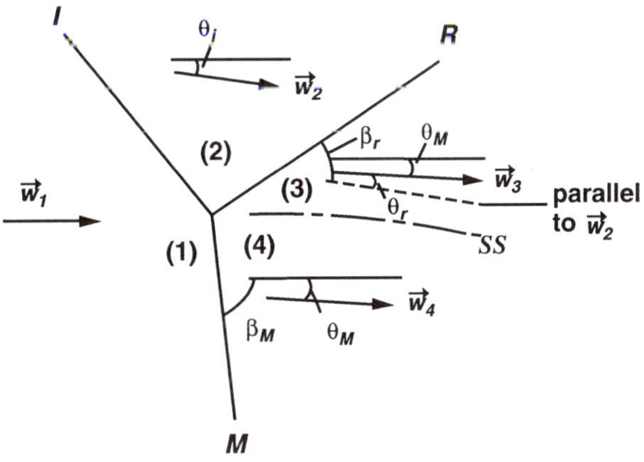

6.8 **(a)** Show that Equation (6.64) is identically satisfied for $j = 3$.

(b) Show that Equation (6.64) is also satisfied for $j = 1,2$ at any arbitrary point on the shock.

6.9 Develop a code to compute $d(\mu - \theta)/d\beta$ as a function of γ, M_1, and β. Tabulate the derivative vs. β for several M_1 values using $\gamma = 1.4$.

6.10 A normal shock has speed $w_s(t)$. The upstream flow is uniform with a constant velocity, $w\hat{\imath}_3$, that is perpendicular to the shock.

(a) Determine $F(x_3,t)$ and verify that all tangential derivatives are zero.

(b) Use the theory of Section 6.6 to determine equations for $(\partial p/\partial\xi_3)_2$, $(\partial\rho/\partial\xi_3)_2$, and $(\partial w_2^*/\partial\xi_3)$. Do not assume a specific form for the enthalpy function, Equation (6.26).

6.11 Continue with Problem 6.10 by assuming a van der Waals gas; see Problem 5.18. Determine w_2^* and $\partial w_2/\partial\xi_3$ in terms of γ, p_1/ρ_1, and $w_1 - w_s$. Assume a nearly perfect gas in order to obtain an explicit equation for w_2^*.

6.12 A spherically symmetric flow is caused by an intense point explosion in a uniform atmosphere. Assume air to be a perfect gas and ignore the analysis in Sections 6.4 to 6.6. Use spherical coordinates for the Euler equations where all scalar variables depend only on r and t (see Problem 5.3a). Eliminate h_o in favor or p, ρ, and w. Use a coordinate system fixed at the shock by introducing

$$R(r, t) = r_s(t) - r, \qquad \tau = t, \qquad W = W_s - w$$

where r_s is the radius of the spherical shock and the shock speed w_s is dr_s/dt.

(a) Determine the Euler equations in terms of these variables.

(b) Define the shock Mach number $M_s = (w_s/a_1)$, where a_1 is the speed of sound ahead of the shock, and develop jump conditions for p, ρ, and W assuming a strong shock; i.e., $M_s >> 1$.

(c) With this assumption, determine $\rho_{2R}[= (\partial\rho/\partial R)_2]$, p_{2R}, and W_{2R} in terms of r_s and its derivatives just behind the shock. Obtain simplified results for these derivatives when $\gamma = 1.4$ and

$$r_s = ct^{2/5}$$

where c is a constant.

6.13 Consider the steady, homenergetic, two-dimensional or axisymmetric flow, without swirl, of a perfect gas. These same assumptions are used in Problem 5.17.

(a) Show that the vorticity ω_2 just downstream of a shock can be related to the upstream vorticity ω_1 by means of

$$\omega_2 = \frac{2}{\gamma+1}\left(\gamma w - \frac{\gamma-1}{2}\right)\omega_1 + \frac{\gamma}{\gamma+1}x_2^\sigma p_1 \frac{(w-1)^2}{w\left(1+\frac{\gamma-1}{2}w\right)}\frac{dw}{d\psi}$$

where $w = M_1^2 \sin^2\beta$ and β is the shock wave angle.

(b) Assume M_1 is constant and derive a relation between ω_1 and $d\beta/d\psi$ such that ω_2 is identically zero. Thus, for a given curved shock shape, $\beta = \beta(\psi)$, the upstream vorticity just cancels the shock-produced vorticity.

(c) Determine the entropy $s_1 = s_1(w)$ when M_1 is constant and $\omega_2 = 0$. Is s_2 constant?

(d) Use Crocco's equation to show that $\vec{\omega} = 0$ everywhere downstream of a curved shock when M_1 is constant and $\omega_2 = 0$.

6.14 Start with Equation (6.28) and derive $(\partial\theta/\partial s)_2$ as given in Appendix D2.

6.15 Use the theory in Section 6.3 to determine

$$\frac{dp}{d\eta}, \quad \frac{d\rho}{d\eta}, \quad \frac{dM^2}{d\eta}$$

just downstream of the conical shock in a Taylor–Maccoll flow. See Problem 8.2 for nomenclature and a brief discussion of this type of flow. Note that the η derivative is based on the n derivative used in Appendix D. Your final result should be in terms of conventional nomenclature, not m, w, X, Y, etc.

6.16 Consider steady, two-dimensional or axisymmetric flow of a perfect gas. Assume that all flow conditions just downstream of a curved shock, including θ, are known. Conditions upstream of the shock are not necessarily uniform or parallel. A local analysis, at an arbitrary shock point, is to be performed. Remember that both θ and β are measured relative to the velocity $\vec{w}_1$ just upstream of the shock at the point in question.

(a) Derive a cubic polynomial for $\cos^2\beta$ whose coefficients depend only on γ, θ, and M_2. To check your result, use $\gamma = 1.4$, $M_1 = 3$, and $\theta = 30°$.

(b) Provide equations for M_1^2, p_2/p_1, and ρ_2/ρ_1 in terms of γ, M_2, θ, and β. (Although p_1 and ρ_1 are the unknowns, it is still convenient to write p_2/p_1 and ρ_2/ρ_1.)

6.17 Consider a steady, two-dimensional or axisymmetric flow of a perfect gas. Let s and n be shock-fixed coordinates, as sketched in Figure 6.1. Let $\tilde{s}$ and $\tilde{n}$ be natural coordinates, where $\tilde{s}$ and $\tilde{n}$ are along and normal to the streamlines (see Figure 5.4).

(a) Derive an equation for $(\partial p/\partial\tilde{n})_2$ that has the form of the Appendix D3 equations. For this, utilize

$$\frac{\partial}{\partial\tilde{n}} = \frac{\partial s}{\partial\tilde{n}}\frac{\partial}{\partial s} + \frac{\partial n}{\partial\tilde{n}}\frac{\partial}{\partial n}$$

and the analogous derivations in Section 6.3.

(b) Obtain an equation for $(\partial\theta/\partial\tilde{s})_2$.

(c) The curvature of the shock κ_s and curvature of a streamline κ_o, just downstream of the shock, are given by

$$\kappa_s = -\beta', \qquad \kappa_o = \left(\frac{\partial\theta}{\partial\tilde{s}}\right)_2$$

The minus signs mean that both curvatures are positive when the shock appears as shown in Figure 6.5. Use the part (b) result to write, in conventional notation, a relation between the curvatures.

(d) Evaluate this curvature relation when

$$\sigma = 0, \qquad \gamma = 1.4, \qquad M_1 = 3, \qquad \theta_2 = 30°$$

(e) For extra credit, plot κ_o/κ_s vs. β, in degrees, when $\sigma = 0$, $\gamma = 1.4, 5/3$, $M_1 = 2,4,6,8$, and $\mu \le \beta \le (\pi/2)$.

6.18 A cylinder-wedge ($\sigma = 0$) or sphere-cone ($\sigma = 1$) is at zero incidence, as shown in the sketch. The sphere and cylinder each has a radius R_b, while the nose radius and stand-off distance for the detached bow shock are R_s and Δ, respectively. The wedge and cone have half angles of θ_b. Far downstream, the shock angle β_∞ is for a sharp cone or wedge with the same θ_b

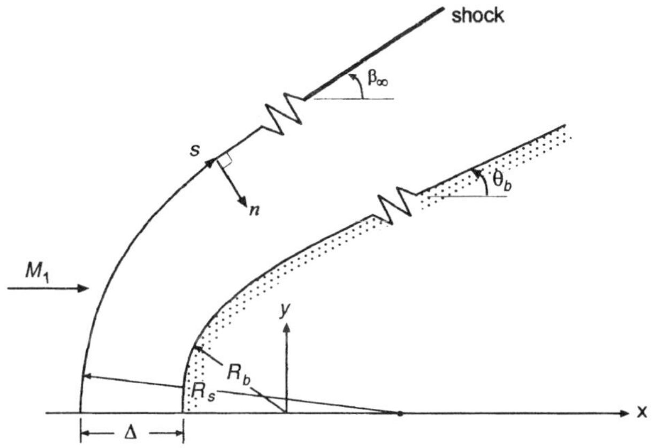

half angle. If θ_b is zero, β_∞ then equals the freestream Mach angle. Introduce the nondimensional notation

$$\bar{x} = \frac{x}{R_b}, \quad \bar{y} = \frac{y}{R_b}, \quad \bar{\Delta} = \frac{\Delta}{R_b}, \quad \bar{r} = \frac{R_s}{R_b}, \quad S = \frac{s}{R_b}, \quad N = \frac{n}{R_b}$$

where s is arc length along the shock measured from $y = 0$ and n is normal to the shock. An empirical representation for the shock is (Billig, 1967)

$$\bar{x} = 1 - \bar{\Delta} + \bar{r}\,\frac{\left(1 + \bar{y}^2\,\dfrac{\tan^2\beta_\infty}{\bar{r}^2}\right)^{1/2} - 1}{\tan^2\beta_\infty}$$

Curve fits to experimental data yield

$$\bar{\Delta} = \begin{cases} 0.143\ \exp(3.24/M_1^2), & \sigma = 1 \\ 0.386\ \exp(4.67/M_1^2), & \sigma = 0 \end{cases}$$

$$\bar{r} = \begin{cases} 1.143\ \exp[(0.54/(M_1 - 1))^{1/2}], & \sigma = 1 \\ 1.386\ \exp[1.8/(M_1 - 1)^{0.75}], & \sigma = 0 \end{cases}$$

for $\bar{\Delta}$ and $\bar{r}$. In normalized form, the shock shape depends only on M_1 and θ_b when $\gamma = 1.4$.

(a) Determine equations for $\bar{y}$, β, and S in terms of $\bar{x}$.

(b) With the parameters

$$\gamma = 1.4, \quad M_1 = 4, \quad \theta_b = 15°$$

numerically evaluate β^* and $\bar{x}^*$ for both geometries, where the asterisk indicates that $M_2 = 1$. [Use Appendix C for β_∞ when $\sigma = 0$. When $\sigma = 1$, the Ames Research Staff (1953) report yields $\beta_\infty = 21.9°$.]

(c) Develop a computer code to evaluate

$$\bar{y}, \quad \beta, \quad M_2, \quad \frac{1}{p_1}\left(\frac{\partial p}{\partial S}\right)_2, \quad \frac{1}{p_1}\left(\frac{\partial p}{\partial N}\right)_2$$

at the five shock locations

$$\bar{x} = -1 - \bar{\Delta}, \quad \frac{1}{2}(-1 - \bar{\Delta} + \bar{x}^*), \quad \bar{x}^*, \quad 1 + \bar{\Delta} + 2\bar{x}^*, \quad 2 + 2\bar{\Delta} + 3\bar{x}^*$$

for both geometries using the γ, M_1, and θ_b values of part (b). Present your results in tabular form. (Note the comment about $\cos\beta/\bar{y}$ in Problem 6.2, which applies when $\sigma = 1$ and $\bar{x} = -1 - \bar{\Delta}$.)

(d) Apply the results at the end of Section 6.4 to obtain $h_i = h_i(\bar{x}_j)$ and $\xi_i = \xi_i(\bar{x}_j)$ equations, where $\bar{x}_j = x_j/R_b$ and the $\bar{x}_1$ and R are respectively $\bar{x}$ and $\bar{y}$. Simplify all answers as much as possible; e.g., evaluate the integral that appears in ξ_1.

6.19 Consider a steady, axisymmetric shock wave

$$F = x - \frac{1}{2}e^2(y^2 + z^2) = 0$$

where e is a constant and the upstream velocity is

$$\vec{w}_1 = w_1\hat{\imath}_1$$

Determine the $\hat{n}$, $\hat{b}$, $\hat{t}$, the h_i, and the ξ_i in terms of x, y, and z.

7 The Hodograph Transformation and Limit Lines

7.1 PRELIMINARY REMARKS

The hodograph transformation was first investigated by Molenbroek in 1890. Most simply, the transformation interchanges the dependent and independent variables; the result is called the hodograph equation or equations. The name specifically refers to an interchange for a partial differential equation or system of equations; its use for solving ODEs long predates 1890.

As we know, the Euler equations are nonlinear. They can be linearized, for example, by assuming a slightly perturbed flow about a known reference state, such as its freestream state. An alternate route for obtaining linear equations, without approximating the nonlinear terms, is by means of the hodograph transformation. As we shall see, a linear system is obtained only when the flow is assumed to be steady, two-dimensional, homentropic, and irrotational. (Starting with a potential equation, Oyibo, 1990, provides a three-dimensional extension.) Hence, the flow is also barotropic, and by Crocco's equation, it is homenergetic. In this case, the price of linearity is rather steep.

The transformation is not restricted to a steady flow but can be used with the unsteady, one-dimensional equations of gas dynamics (see Problem 13.13 of Emanuel, 1986). Once again, the resulting equations are linear. In this chapter, however, we will focus on steady flows.

An important step was taken in 1904 by Chaplygin who provided a general solution to the hodograph equations and also solved the problem of a subsonic free jet that emanates from a two-dimensional slit. We will examine his general solution in Section 7.5. In 1940, Ringleb provided a solution that first revealed the presence of a limit line. The solution and the accompanying limit line are the subject of Section 7.3.

Approximate analytical or numerical methods for solving the Euler equations are commonly encountered. For subsonic flow, e.g., there is the Rayleigh–Janzen method where an expansion in powers of M^2 is used. For supersonic flow, there is the numerical method of characteristics. The situation for transonic flow, however, is more difficult. In this circumstance, part of the flow is subsonic and part is supersonic. From a mathematical viewpoint, the equations are sometimes elliptic and sometimes hyperbolic; this is referred to as a mixed system. The transonic equations, without the use of the hodograph transformation, are still nonlinear even when a small perturbation assumption is utilized. The hodograph equations, however, are linear for subsonic, transonic, and supersonic flows.

Aside from two-dimensional jets (Chaplygin, 1944; Chang, 1952), the primary use of the hodograph equations is in analyzing transonic flow. In particular, three types of flows have received attention. These are nozzle flow (Cherry, 1950; Libby and Reiss, 1951; Lighthill, 1947), flow about an airfoil (Nieuwland and Spee, 1973), and flow about a symmetric wedge that is aligned with the freestream (Cole, 1951; Mackie and Pack, 1952; Vincenti and Wagoner, 1951). (It is convenient to view the flow about a wedge as distinct from the more complicated flow about a smoothly contoured airfoil.)

Let us briefly discuss the nature of the transonic flow field about a symmetric double wedge that is aligned with the freestream (Figure 7.1). The AE line is a detached bow shock, BCE is a sonic line, the dashed Mach lines from the shoulder at B represent an expansion fan, while the solid Mach lines that originate along BCE represent a compression wave. The compression wave reflects from the wedge as a compression; the gradual coalescence of the Mach lines becomes a

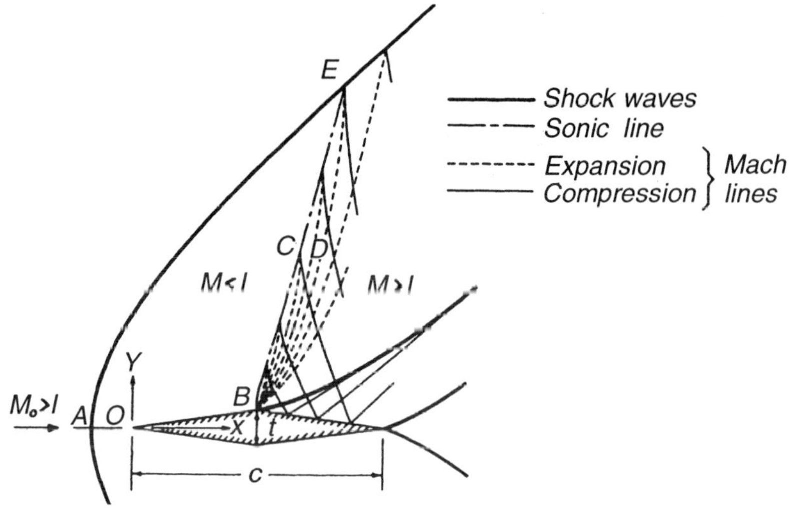

FIGURE 7.1 Transonic flow about a symmetric wedge at zero angle of attack. (After Vincenti and Wagoner, 1951.)

weak shock wave. There are also shock waves that start at the trailing edge of the wedge. The BDE Mach line is special in that any small disturbance upstream of it is felt throughout the region bordered by the bow shock, wedge, and BDE Mach line. (Of course, this runs counter to the usual rule that disturbances do not propagate in the upstream direction in a supersonic flow.) Under the assumption that the freestream Mach number is near unity, the detached bow shock is weak and the rotation introduced by its curvature can be neglected. In Vincenti and Wagoner (1951), the irrotational hodograph equations in the region upstream of BDE are numerically solved in a plane whose coordinates are u and v. Downstream of BDE, the supersonic flow is numerically solved using the method of characteristics. The objective is to obtain the inviscid transonic wave drag of the wedge.

Although the hodograph equations are linear and superposition of solutions can be used, they are, nevertheless, mathematically complicated. Section 7.2 derives these equations and the associated transformation in which we emphasize a compressible flow. Ringleb's solution and the nature of the limit lines are discussed in Section 7.3. Section 7.4 provides a general discussion of limit lines, while Section 7.5 discusses the solution of Chaplygin and an alternative class of solutions. As a coordinate transformation, the hodograph method does not require that the flow be two-dimensional, irrotational, isoenergetic, or homentropic. Section 7.6 derives the pertinent equations when these assumptions are not invoked. Of course, the resulting equations are nonlinear; their solution is discussed in the next chapter.

Some of the material presented in Sections 7.2, 7.3, and 7.5 is based on Pai (1959) and Shapiro (1954), while Section 7.6 is based on Rodriguez and Emanuel (1989). More mathematically oriented discussions of the hodograph transformation are provided by Lighthill (1953) and Manwell (1971), while Chang (1952) contains, at the time of its publication, a comprehensive bibliography.

7.2 TWO-DIMENSIONAL, IRROTATIONAL FLOW

As discussed in the above section, a steady, homentropic and two-dimensional, irrotational flow is assumed. We will use x,y for the Cartesian coordinates and u,v for the corresponding velocity components, where

$$u = w\cos\theta, \qquad v = w\sin\theta \tag{7.1}$$

and θ is the angle a streamline has with respect to the x-axis. From Equation (5.69) with $\vec{\omega} = 0$, we have

$$\frac{\partial v}{\partial x} = \frac{\partial u}{\partial y} \tag{7.2}$$

and a velocity potential can be introduced as

$$u = \phi_x, \qquad v = \phi_y \tag{7.3}$$

where a subscript indicates a partial derivative. A stream function ψ that satisfies continuity is given by [see Equations (5.70)]

$$u = \frac{\rho_o}{\rho}\psi_y, \qquad v = -\frac{\rho_o}{\rho}\psi_x \tag{7.4}$$

where it is dimensionally convenient to introduce the constant stagnation density ρ_o.

The compressible flow Bernoulli equation can be written as

$$\frac{1}{2}w^2 + \int \frac{dp}{\rho} = h_o \tag{7.5a}$$

or as

$$w\,dw + \frac{dp}{\rho} = 0 \tag{7.5b}$$

where both forms are used shortly and the stagnation enthalpy h_o is a constant.

Observe that for an incompressible flow, $\rho = \rho_o$, Equations (7.3) and (7.4) yield the Cauchy–Riemann equations

$$\phi_x = \psi_y, \qquad \phi_y = -\psi_x$$

In this circumstance, both ϕ and ψ satisfy Laplace's equation, while Equation (7.5a) becomes Bernoulli's equation for the pressure. We thus obtain the standard equations for steady, incompressible, two-dimensional, potential flow. The use of complex variable theory is especially convenient in this case. For a compressible flow, however, complex variable theory is of less value and we will not need it.

The Hodograph Transformation

The hodograph transformation utilizes w and θ, or u and v, as the independent variables. In Sections 7.2 to 7.5 we primarily use w and θ, while Section 7.6 uses u and v. One exception to the use of w and θ occurs when graphically showing streamline patterns, where u and v coordinates are preferable. For the dependent variables, we initially use both ϕ and ψ. At some point in the analysis, we also need to determine how ρ, p, x, and y depend on w and θ. (We note that much of the following derivation was foreshadowed by Problem 5.9.) Some of the ensuing analysis can be slightly shortened by using a Legendre transformation of the form (Pai, 1959; Lighthill, 1953)

$$\Phi = \phi - ux - vy$$

where Φ is a new dependent variable. However, in the interest of clarity, we will pursue a more straightforward analytical approach.

We first consider ϕ and ψ to be functions of x and y so that Equations (7.3) and (7.4) yield

$$d\phi = \phi_x dx + \phi_y dy = w(\cos\theta\, dx + \sin\theta\, dy)$$

$$d\psi = \psi_x dx + \psi_y dy = w\frac{\rho}{\rho_o}(-\sin\theta\, dx + \cos\theta\, dy)$$

These relations are solved for dx and dy, with the results

$$dx = \frac{1}{w}\left(\cos\theta\, d\phi - \frac{\rho_o}{\rho}\sin\theta\, d\psi\right) \tag{7.6a}$$

$$dy = \frac{1}{w}\left(\sin\theta\, d\phi + \frac{\rho_o}{\rho}\cos\theta\, d\psi\right) \tag{7.6b}$$

We next write

$$\phi = \phi(w, \theta), \qquad \psi = \psi(w, \theta) \tag{7.7}$$

to obtain

$$d\phi = \phi_w dw + \phi_\theta d\theta$$

$$d\psi = \psi_w dw + \psi_\theta d\theta$$

These relations are substituted into Equation (7.6), thereby yielding

$$dx = \frac{1}{w}\left[\left(\cos\theta\phi_w - \frac{\rho_o}{\rho}\sin\theta\,\psi_w\right)dw + \left(\cos\theta\phi_\theta - \frac{\rho_o}{\rho}\sin\theta\,\psi_\theta\right)d\theta\right] \tag{7.8a}$$

$$dy = \frac{1}{w}\left[\left(\sin\theta\phi_w + \frac{\rho_o}{\rho}\cos\theta\,\psi_w\right)dw + \left(\sin\theta\phi_\theta + \frac{\rho_o}{\rho}\sin\theta\,\psi_\theta\right)d\theta\right] \tag{7.8b}$$

We now write

$$x = x(w, \theta), \qquad y = y(w, \theta) \tag{7.9}$$

and

$$dx = x_w dw + x_\theta d\theta$$

$$dy = y_w dw + y_\theta d\theta$$

By comparison with Equation (7.8), we have

$$x_w = \frac{1}{w}\left(\cos\theta\phi_w - \frac{\rho_o}{\rho}\sin\theta\,\psi_w\right) \tag{7.10a}$$

$$x_\theta = \frac{1}{w}\left(\cos\theta\phi_\theta - \frac{\rho_o}{\rho}\sin\theta\,\psi_\theta\right) \tag{7.10b}$$

$$y_w = \frac{1}{w}\left(\sin\theta\phi_w + \frac{\rho_o}{\rho}\cos\theta\,\psi_w\right) \tag{7.10c}$$

$$y_\theta = \frac{1}{w}\left(\sin\theta\phi_\theta + \frac{\rho_o}{\rho}\cos\theta\,\psi_\theta\right) \tag{7.10d}$$

We can eliminate x from Equations (7.10a,b) by setting $x_{w\theta} = x_{\theta w}$. In the process, the ρ_θ and ρ_w derivatives are encountered. To evaluate these derivatives, we differentiate Equation (7.5a) with respect to θ, keeping w fixed, to obtain

$$\frac{\partial}{\partial\theta}\int\frac{dp}{\rho} = 0$$

or

$$\int\frac{\rho_\theta}{\rho^2}\,dp = 0$$

Hence, we have

$$\rho_\theta = 0$$

and ρ depends only on w; i.e.,

$$\rho_w = \frac{d\rho}{dw}$$

By performing the indicated x-coordinate cross-derivative operation, we obtain

$$w\tan\theta\phi_w - \phi_\theta = -w\frac{\rho_o}{\rho}\psi_w + w^2\tan\theta\frac{d}{dw}\left(\frac{\rho_o}{\rho w}\right)\psi_\theta \tag{7.11a}$$

after simplification. Similarly, y is eliminated by means of $y_{w\theta} = y_{\theta w}$, with the result

$$w\phi_w + \tan\theta\phi_\theta = w\frac{\rho_o}{\rho}\tan\theta\,\psi_w + w^2\frac{d}{dw}\left(\frac{\rho_o}{\rho w}\right)\psi_\theta \tag{7.11b}$$

These equations are solved for ϕ_w and ϕ_θ thereby yielding

$$\phi_w = w\frac{d}{dw}\left(\frac{\rho_o}{\rho w}\right)\psi_\theta \tag{7.12a}$$

$$\phi_\theta = w\frac{\rho_o}{\rho}\psi_w \tag{7.12b}$$

These are the hodograph equations in their most compact form. With the barotropic relation, $p = p(\rho)$, Equation (7.5a) provides ρ as a known function of w. The above are two first-order, coupled, linear equations for ϕ and ψ.

By cross-differentiation, we easily eliminate ϕ, to obtain a single hodograph equation

$$w\frac{d}{dw}\left(\frac{\rho_o}{\rho w}\right)\psi_{\theta\theta} = \frac{d}{dw}\left(w\frac{\rho_o}{\rho}\right)\psi_w + w\frac{\rho_o}{\rho}\psi_{ww} \tag{7.13}$$

Alternatively, ψ could be eliminated in favor of ϕ (see Problem 7.2). In view of streamline boundary conditions, a ψ equation is more useful than one for ϕ; we therefore do not bother to obtain the second-order PDE for ϕ.

It is also convenient to eliminate ρ_o/ρ in favor of the Mach number, which can be done without assuming a perfect gas. Recall that the entropy is a constant; consequently, we have

$$M^2 = \frac{w^2}{a^2} = \frac{w^2}{\dfrac{dp}{d\rho}} = \frac{d\rho}{\rho}\frac{w^2}{\dfrac{dp}{\rho}} = -\frac{w}{\rho}\frac{d\rho}{dw} = -w\frac{\rho_o}{\rho}\frac{d(\rho/\rho_o)}{dw}$$

where Equation (7.5b) is utilized. Since

$$\frac{d(\rho/\rho_o)}{dw} = -\left(\frac{\rho}{\rho_o}\right)^2\frac{d(\rho_o/\rho)}{dw}$$

we obtain

$$M^2 = w\frac{\rho}{\rho_o}\frac{d(\rho_o/\rho)}{dw} \tag{7.14}$$

The two speed derivatives that appear in Equation (7.13) then become

$$\frac{d}{dw}\left(\frac{\rho_o}{dw}\right) = -\frac{\rho_o}{\rho}\frac{1}{w^2} + \frac{1}{w}\frac{d(\rho_o/\rho)}{dw} = -\frac{\rho_o}{\rho}\frac{1}{w^2} + \frac{\rho_o}{\rho}\frac{M^2}{w^2} = \frac{1}{w^2}\frac{\rho_o}{\rho}(M^2-1)$$

$$\frac{d}{dw}\left(w\frac{\rho_o}{\rho}\right) = \frac{\rho_o}{\rho} + w\frac{d(\rho_o/\rho)}{dw} = \frac{\rho_o}{\rho} + M^2\frac{\rho_o}{\rho} = \frac{\rho_o}{\rho}(1+M^2)$$

Thus, Equation (7.13) can be written as

$$(M^2-1)\psi_{\theta\theta} = (1+M^2)w\psi_w + w^2\psi_{ww} \tag{7.15}$$

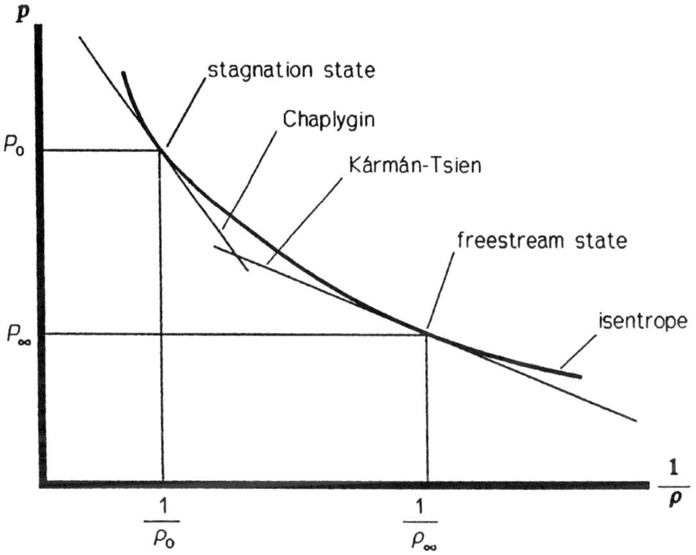

FIGURE 7.2 Two possible choices for Equation (7.16). (After Pai, 1959.)

This equation is still linear inasmuch as M^2 is a function only of w. Since the equation is homogeneous, with only derivatives of ψ, we see that if ψ is a solution, then

$$\tilde{\psi} = a\psi + b$$

is also a solution, where a and b are constants. Consequently, ψ can always be set equal to zero along a wall or along a symmetry streamline. Furthermore, ψ can be taken as nondimensional.

Thermodynamic Equation of State

An explicit algebraic relation between M and w, or ρ_o/ρ and w, requires a thermal equation of state, most conveniently of the form $p = p(\rho,s)$. Although Equation (7.15) is linear, the coefficients are variable and their complexity depends on the choice of a state equation. The two simplest forms for Equation (7.15) do not result from the assumption of a perfect gas. For an incompressible flow, we simply set $M = 0$ and use Bernoulli's equation to determine the pressure. A second simple form, which is for a compressible flow, stems from the Kármán–Tsien approximation

$$p = A - \frac{B}{\rho}$$

where A and B are constants. With the entropy constant, the $p = p(\rho,s)$ state equation provides the curved line shown in Figure 7.2 as an isentrope. The parameters A and B are chosen so that Equation (7.16), at some point, is tangent to the isentrope. As indicated in the figure there are two common choices. In the first, introduced by Chaplygin (1944), Equation (7.16) is chosen to be tangent to the isentrope at the stagnation state. The common choice for transonic flow analysis, due to Kármán and Tsien, is to choose Equation (7.16) to be tangent to the isentrope at the freestream state (see Problem 7.1). Either choice can be referred to as a first-order approximation. There are also second- and third-order approximations (Chang, 1952).

Although much effort has been put into Equation (7.16) and its higher-order approximations (Chang, 1952; Lighthill, 1953), we will not pursue this line of analysis. Our first reason is that Equation (7.16) yields for the speed of sound

$$a^2 = \frac{dp}{d\rho} = \frac{B}{\rho^2}$$

With $B > 0$, as is always the case, the speed of sound increases as ρ decreases, which is unrealistic for the homentropic flow of most real gases as evident in a one-dimensional nozzle flow. Second, we will ultimately utilize the theory in the next chapter, which requires a perfect gas.

We thus assume a perfect gas, which means the speed of sound is given by

$$\frac{a^2}{\gamma - 1} + \frac{1}{2}w^2 = \frac{a_o^2}{\gamma - 1}$$

or

$$a^2 = a_o^2 - \frac{\gamma - 1}{2}w^2$$

and the Mach number can be written as

$$M^2 = \frac{w^2}{a^2} = \frac{\dfrac{w^2}{a_o^2}}{1 - \dfrac{\gamma - 1}{2}\dfrac{w^2}{a_o^2}}$$

It is traditional and convenient to introduce a ratio τ defined by

$$\tau = \frac{w^2}{w_{max}^2} = \frac{\gamma - 1}{2}\frac{w^2}{a_o^2} \tag{7.17}$$

We thus obtain

$$M^2 = \frac{2}{\gamma - 1}\frac{\tau}{1 - \tau} \tag{7.18a}$$

and, conversely,

$$\tau = \frac{(\gamma - 1)M^2/2}{1 + (\gamma - 1)M^2/2} = \begin{cases} 0, & M = 0 \\ \dfrac{\gamma - 1}{\gamma + 1}, & M = 1 \\ 1, & M = \infty \end{cases} \tag{7.18b}$$

where the sonic value of $(\gamma - 1)/(\gamma + 1)$ will hereafter be written as τ^*.

The replacement of M^2 in Equation (7.15) is expedited with the aid of

$$M^2 - 1 = \frac{\frac{\tau}{\tau^*} - 1}{1 - \tau}, \qquad M^2 + 1 = \frac{1 + \frac{3-\gamma}{\gamma - 1}\tau}{1 - \tau}, \qquad 1 + \frac{\gamma - 1}{2}M^2 = \frac{1}{1 - \tau} \tag{7.19}$$

and w with the aid of

$$\frac{\partial}{\partial w} = \frac{d\tau}{dw}\frac{\partial}{\partial \tau} = (\gamma - 1)\frac{w}{a_o^2}\frac{\partial}{\partial \tau} = \frac{2\tau}{w}\frac{\partial}{\partial \tau}$$

$$\frac{\partial^2}{\partial w^2} = \frac{\partial}{\partial w}\left[(\gamma - 1)\frac{w}{a_o^2}\frac{\partial}{\partial \tau}\right] = \frac{\gamma - 1}{a_o^2}\frac{\partial}{\partial \tau} + (\gamma - 1)^2\frac{w^2}{a_o^4}\frac{\partial^2}{\partial \tau^2}$$

$$= \frac{\gamma - 1}{a_o^2}\left(\frac{\partial}{\partial \tau} + 2\tau\frac{\partial^2}{\partial \tau^2}\right)$$

With Equation (7.19), the homentropic relations for the density and pressure are

$$\frac{\rho}{\rho_o} = (1 - \tau)^{1/(\gamma - 1)} \tag{7.20a}$$

$$\frac{p}{p_o} = (1 - \tau)^{\gamma/(\gamma - 1)} \tag{7.20b}$$

and Equations (7.1) become

$$\frac{u}{a_o} = \left(\frac{2}{\gamma - 1}\right)^{1/2}\tau^{1/2}\cos\theta \tag{7.20c}$$

$$\frac{v}{a_o} = \left(\frac{2}{\gamma - 1}\right)^{1/2}\tau^{1/2}\sin\theta \tag{7.20d}$$

while Equation (7.15) transforms to

$$\tau^2(1 - \tau)\psi_{\tau\tau} + \tau\left(1 + \frac{2-\gamma}{\gamma - 1}\tau\right)\psi_\tau + \frac{1}{4}\left(1 - \frac{\tau}{\tau^*}\right)\psi_{\theta\theta} = 0 \tag{7.21}$$

Thus, τ and θ are the independent variables where this relation, in contrast to Equation (7.15), assumes a perfect gas. A solution $\psi(\tau,\theta)$ of Equation (7.21) is referred to as a solution in the hodograph plane; the same terminology is used for a solution $\psi(w,\theta)$ of Equation (7.15).

Transformation to the Physical Plane

It is, of course, essential to be able to transform the solution of Equation (7.21) back to the physical plane in which the coordinates are x and y. The equations for the inverse transformation are based on Equations (7.10). We first eliminate ϕ with the aid of Equation (7.12), to obtain

$$x_w = \cos\theta \frac{d}{dw}\left(\frac{\rho_o}{\rho w}\right)\psi_\theta - \frac{\rho_o}{\rho}\frac{\sin\theta}{w}\psi_w \tag{7.22a}$$

$$x_\theta = \frac{\rho_o}{\rho}\left(-\frac{\sin\theta}{w}\psi_\theta + \cos\theta\psi_w\right) \tag{7.22b}$$

$$y_w = \sin\theta \frac{d}{dw}\left(\frac{\rho_o}{\rho w}\right)\psi_\theta - \frac{\rho_o}{\rho}\frac{\cos\theta}{w}\psi_w \tag{7.22c}$$

$$y_\theta = \frac{\rho_o}{\rho}\left(\frac{\cos\theta}{w}\psi_\theta + \sin\theta\psi_w\right) \tag{7.22d}$$

With the use of Equations (7.14), (7.17), (7.18a), and (7.20a), we eliminate ρ and w, with the result

$$x_\tau = \left(\frac{\gamma-1}{2}\right)^{1/2}\frac{1}{a_o}\left[\frac{1}{2}\cos\theta\frac{\frac{\tau}{\tau^*}-1}{\tau^{3/2}(1-\tau)^{\gamma/(\gamma-1)}}\psi_\theta - \frac{\sin\theta}{\tau^{1/2}(1-\tau)^{1/(\gamma-1)}}\psi_\tau\right] \tag{7.23a}$$

$$x_\theta = \left(\frac{\gamma-1}{2}\right)^{1/2}\frac{1}{a_o}\left[-\frac{\sin\theta}{\tau^{1/2}(1-\tau)^{1/(\gamma-1)}}\psi_\theta + 2\cos\theta\frac{\tau^{1/2}}{(1-\tau)^{1/(\gamma-1)}}\psi_\tau\right] \tag{7.23b}$$

$$y_\tau = \left(\frac{\gamma-1}{2}\right)^{1/2}\frac{1}{a_o}\left[\frac{1}{2}\sin\theta\frac{\frac{\tau}{\tau^*}-1}{\tau^{3/2}(1-\tau)^{\gamma/(\gamma-1)}}\psi_\theta + \frac{\cos\theta}{\tau^{1/2}(1-\tau)^{1/(\gamma-1)}}\psi_\tau\right] \tag{7.23c}$$

$$y_\theta = \left(\frac{\gamma-1}{2}\right)^{1/2}\frac{1}{a_o}\left[\frac{\cos\theta}{\tau^{1/2}(1-\tau)^{1/(\gamma-1)}}\psi_\theta + 2\sin\theta\frac{\tau^{1/2}}{(1-\tau)^{1/(\gamma-1)}}\psi_\tau\right] \tag{7.23d}$$

Integration of these equations would provide x and y once $\psi(\tau,\theta)$ is known, as will be demonstrated in the subsequent sections and in several problems.

Uniqueness of the Transformation

The transformation between x,y and τ,θ is well-defined only if it is one-to-one; i.e., to every x,y point there corresponds a unique τ,θ point and vice versa. This will occur when the Jacobian of the transformation

$$J = \frac{\partial(x,y)}{\partial(\tau,\theta)} = \begin{vmatrix} x_\tau & x_\theta \\ y_\tau & y_\theta \end{vmatrix} = x_\tau y_\theta - x_\theta y_\tau \tag{7.24}$$

is neither zero nor infinite. With the aid of Equations (7.23) this becomes

$$J = -\frac{\gamma-1}{a_o^2}\frac{1}{(1-\tau)^{1/(\gamma-1)}}\left[-\frac{1}{4}\frac{\frac{\tau}{\tau^*}-1}{\tau^2(1-\tau)}\psi_\theta^2+\psi_\tau^2\right] \tag{7.25a}$$

$$= -\frac{\gamma-1}{a_o^2}\frac{1}{(1-\tau)^{1/(\gamma-1)}}\left(\frac{1-M^2}{4\tau^2}\psi_\theta^2+\psi_\tau^2\right) \tag{7.25b}$$

If we assume M is finite, then J is zero only when

$$\frac{1-M^2}{4\tau^2}\psi_\theta^2+\psi_\tau^2 = 0 \tag{7.26}$$

This relation cannot be satisfied if $M < 1$. Thus, J can be zero only in a sonic or supersonic flow.

A limit line occurs when Equation (7.26) is satisfied and $J = 0$. The name stems from the fact that a streamline doubles back on itself when it encounters a limit line. Similarly, for the inverse transformation to be one-to-one, J^{-1} cannot be zero; i.e., J cannot be infinite. It is also worth noting that a one-to-one transformation of the independent variables does not imply a unique value for the stream function in the physical plane. In fact, we will encounter a multivalued $\psi(x,y)$ when discussing the compressible Ringleb solution in the next section.

Let us examine one important implication of Equation (7.26). To do this, we evaluate the magnitude of the acceleration

$$\frac{Dw}{Dt} = u\frac{\partial w}{\partial x}+v\frac{\partial w}{\partial y} = w\left(\cos\theta\frac{\partial w}{\partial x}+\sin\theta\frac{\partial w}{\partial y}\right)$$

where

$$\frac{\partial w}{\partial x} = \frac{\partial \tau}{\partial x}\frac{\partial w}{\partial \tau}+\frac{\partial \theta}{\partial x}\frac{\partial w}{\partial \theta} = \frac{1}{2}\left(\frac{2}{\gamma-1}\right)^{1/2}\frac{a_o}{\tau^{1/2}}\frac{\partial \tau}{\partial x}$$

$$\frac{\partial w}{\partial y} = \frac{\partial \tau}{\partial y}\frac{\partial w}{\partial \tau}+\frac{\partial \theta}{\partial y}\frac{\partial w}{\partial \theta} = \frac{1}{2}\left(\frac{2}{\gamma-1}\right)^{1/2}\frac{a_o}{\tau^{1/2}}\frac{\partial \tau}{\partial y}$$

to obtain

$$\frac{Dw}{Dt} = \frac{a_o^2}{\gamma-1}\left(\cos\theta\frac{\partial \tau}{\partial x}+\sin\theta\frac{\partial \tau}{\partial y}\right)$$

To determine the partial derivatives of τ, it is convenient to use Jacobian theory (see Appendix B) as follows:

$$\left.\frac{\partial \tau}{\partial x}\right|_y = \frac{\partial(\tau,x)}{\partial(x,y)} = \frac{\dfrac{\partial(\tau,x)}{\partial(\tau,\theta)}}{\dfrac{\partial(x,y)}{\partial(\tau,\theta)}} = \frac{y_\theta}{J} \tag{7.27a}$$

$$\left.\frac{\partial \tau}{\partial y}\right|_x = \frac{\partial(\tau,x)}{\partial(y,x)} = \frac{\dfrac{\partial(\tau,x)}{\partial(\tau,\theta)}}{\dfrac{\partial(y,x)}{\partial(\tau,\theta)}} = -\frac{x_\theta}{J} \tag{7.27b}$$

We thereby obtain

$$\frac{Dw}{Dt} = \frac{a_o^2}{\gamma - 1}\frac{1}{J}(\cos\theta\, y_\theta - \sin\theta\, x_\theta)$$

With Equations (7.23b), (7.23d), and (7.25b), we have

$$\frac{Dw}{Dt} = -\frac{a_o^3}{(2\tau)^{1/2}(\gamma-1)^{3/2}}\frac{\psi_\theta}{\left(\frac{1-M^2}{4\tau^2}\psi_\theta^2 + \psi_\tau^2\right)} \tag{7.28}$$

Equation (7.26) shows that the acceleration is infinite when $J = 0$ unless $\psi_\theta = 0$. The presence of an infinite acceleration can be interpreted as a limit line, thereby extending the concept to flows that do not stem from the hodograph transformation.

7.3 RINGLEB'S SOLUTION

Ringleb (1940) found an elementary, but informative, particular solution of Equation (7.21):

$$\psi = \frac{\cos\theta}{\tau^{1/2}} \tag{7.29}$$

where ψ is a nonnegative stream function and $\sin\theta$ can be used in place of the $\cos\theta$. For convenience in the later compressible flow analysis, ψ is already normalized by ℓa_o, per unit depth, where ℓ is a characteristic length and a_o is the stagnation speed of sound. As we will see, the above solution is simple in the u,v hodograph plane but quite complicated in the physical plane. (By solution, we typically mean the streamline pattern in the hodograph and physical planes. We thereby see the advantage of using ψ instead of ϕ.) Because of this complexity, we deviate from our usual compressible-flow emphasis to first examine the incompressible streamline pattern in both the hodograph and physical planes. These patterns will serve as a rough guide to the corresponding compressible flow patterns.

Incompressible Solution

Equation (7.29) is written as

$$\psi_i = \frac{\cos\theta}{w} \tag{7.30}$$

where an i subscript indicates an incompressible flow and both w and ψ are nondimensional. This is readily seen to be a solution of Equation (7.15) with $M = 0$. This relation is used to eliminate θ

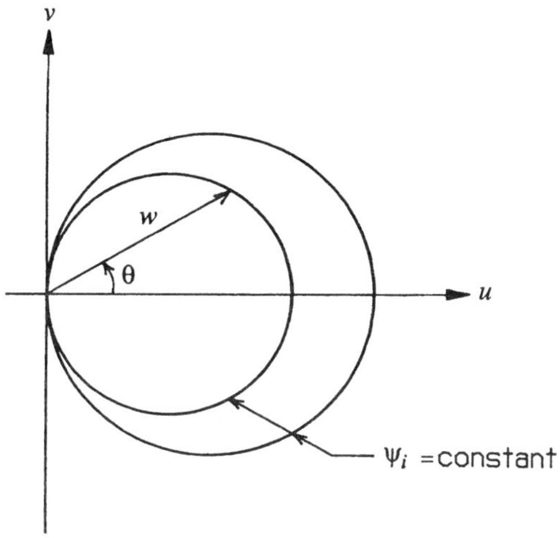

FIGURE 7.3 Incompressible Ringleb solution in the hodograph plane. All streamlines are circles with $u \geq 0$ when $\psi_i \geq 0$.

from Equations (7.1), with the result

$$u = w\cos\theta = w^2\psi_i$$
$$v = w(1-\cos^2\theta)^{1/2} = w(1-\psi_i^2 w^2)^{1/2}$$

We next eliminate w, to obtain

$$\left(u - \frac{1}{2\psi_i}\right)^2 + v^2 = \frac{1}{4\psi_i^2} \tag{7.31}$$

Thus, streamlines are circles to the right of the v-axis in the hodograph plane, as shown in Figure 7.3. These circles have their center on the positive u-axis and are tangent to the v-axis at the origin, which is the stagnation point for each streamline. Since the streamlines are circles, each one has its maximum speed at the point where it crosses the u-axis with $u > 0$. From the above equation, we see that the maximum speed, on a given streamline, is

$$w_{max} = \frac{1}{\psi_i} \tag{7.32}$$

which is in accordance with Equation (7.30) with $\theta = 0°$. Observe that a large ψ_i value corresponds to a small circle and a slowly moving fluid particle. On the other hand, the fluid will move with a relatively high speed over part of its trajectory when ψ_i is small.

To determine the solution in the physical plane, we utilize Equations (7.22), with $\rho = \rho_o$, and

$$\psi_{i\theta} = -\frac{\sin\theta}{w}, \qquad \psi_{iw} = -\frac{\cos\theta}{w^2}$$

to obtain

$$x_w = \frac{2\sin\theta\cos\theta}{w^3} = \frac{\sin 2\theta}{w^3} \tag{7.33a}$$

$$x_\theta = \frac{\sin^2\theta - \cos^2\theta}{w^2} = -\frac{\cos 2\theta}{w^2} \tag{7.33b}$$

$$y_w = -\frac{\cos 2\theta}{w^3} \tag{7.33c}$$

$$y_\theta = -\frac{\sin 2\theta}{w^2} \tag{7.33d}$$

where x and y are normalized by ℓ. Equations (7.33b,d) are readily integrated, to yield

$$x = -\frac{\sin 2\theta}{2w^2} + f(w)$$

$$y = \frac{\cos 2\theta}{2w^2} + g(w)$$

where f and g are functions of integration. By differentiating these relations with respect to w and comparing the result with Equations (7.33a,c), we obtain

$$\frac{df}{dw} = 0, \qquad \frac{dg}{dw} = 0$$

We take the constants of integration to be zero; hence, the hodograph transformation for Equation (7.30) is

$$x = -\frac{\sin 2\theta}{2w^2}, \qquad y = \frac{\cos 2\theta}{2w^2} \tag{7.34}$$

The streamlines in the physical plane are determined by Equations (7.30) and (7.34). Equation (7.30) is first used to eliminate θ from the other two equations, with the result

$$x = -\frac{\psi_i}{w}(1 - w^2\psi_i^2)^{1/2}$$

$$y = \psi_i^2 - \frac{1}{2w^2}$$

Next, w is eliminated, to yield

$$x^2 + 2\psi_i^2 y = \psi_i^4 \tag{7.35}$$

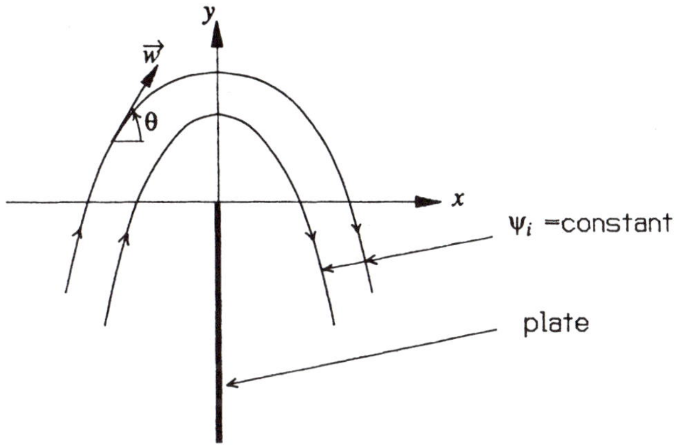

FIGURE 7.4 Streamlines of the incompressible Ringleb solution in the physical plane.

Thus, the streamlines are parabolas in the physical plane, as seen in Figure 7.4. As shown, the angle θ is positive for a counterclockwise rotation of the velocity; hence, θ is positive when x is negative. If a fluid particle moves in a clockwise fashion along a streamline in Figure 7.3, then the corresponding direction of flow is shown by the arrows on the streamlines in Figure 7.4. The flow represents a 180° turn about a semi-infinite flat plate, which is located along the negative y-axis. Of course, any two streamlines can also serve as the bounding walls for a flow.

As indicated by Equations (7.34), the flow stagnates when $y \to -\infty$ at both ends of each streamline. As was the case in the hodograph plane, the maximum flow speed on a given streamline occurs when $\theta = 0°$ or $x = 0$ and $y \geq 0$. From Equation (7.34), we see that the flow speed is infinite when $x = y = 0$, which is the tip of the flat plate. Hence, $\psi_i = 0$ on the plate, which is also evident from Equation (7.32). The value of the stream function thus increases for the slower-moving flow that occurs away from the plate.

Compressible Solution in the Hodograph Plane

Our analysis is patterned after the incompressible solution; hence, we shortly obtain the streamlines in the hodograph plane. From Equations (7.20c,d), the velocity components are

$$u = \left(\frac{2}{\gamma - 1}\right)^{1/2} \tau^{1/2} \cos\theta \tag{7.36a}$$

$$v = \left(\frac{2}{\gamma - 1}\right)^{1/2} \tau^{1/2} \sin\theta \tag{7.36b}$$

where u and v are normalized by a_o. As in Figure 7.3, the hodograph plane origin corresponds to a stagnation condition for all streamlines. In Figure 7.5, the sonic line is a dashed circle, labeled $\tau = \tau^*$, and is centered at the origin. There is a larger circle, labeled $\tau = 1$ on which the Mach number is infinite. All curves shown in Figures 7.5 and 7.6 are to scale for $\gamma = 7/5$. Thus, for the $\tau^* = (\gamma - 1)/(\gamma + 1) = (1/6)$ sonic circle, the radius is $[2/(\gamma + 1)]^{1/2} = 0.9192$, while for the $\tau = 1$ circle the radius is $[2/(\gamma - 1)]^{1/2} = 2.236$.

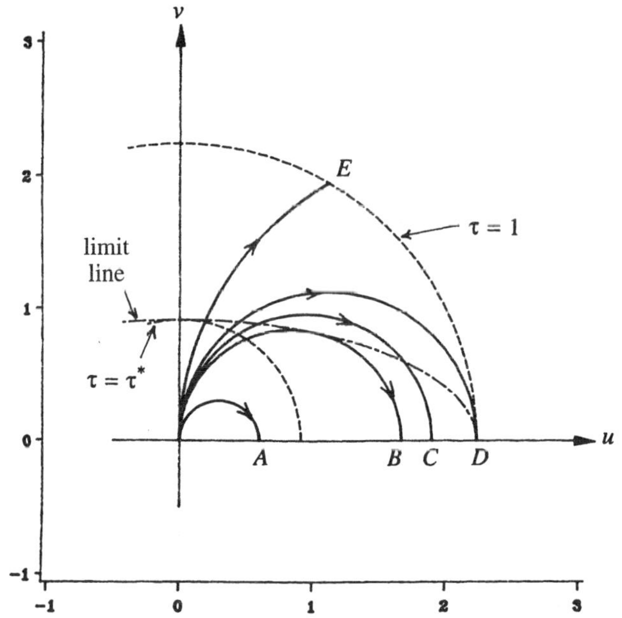

FIGURE 7.5 Compressible Ringleb solution in the hodograph plane, $\gamma = 7/5$.

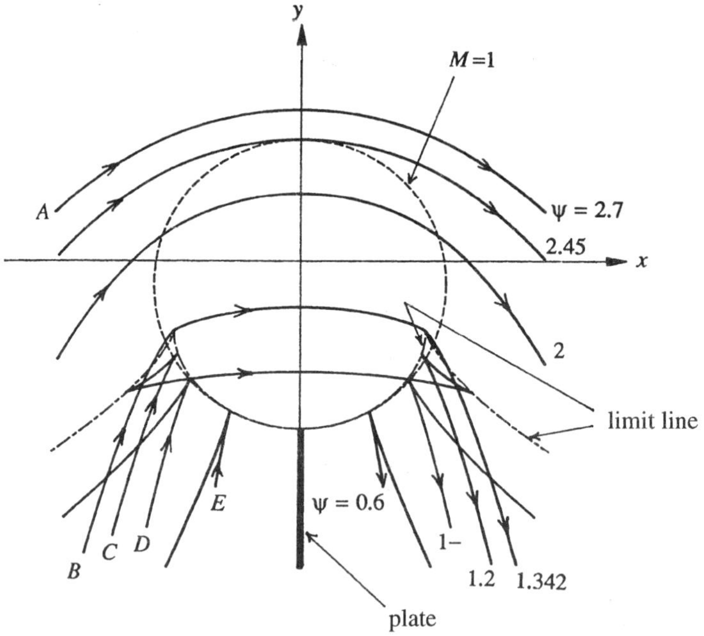

FIGURE 7.6 Streamlines, sonic line, and limit line of the compressible Ringleb solution, $\gamma = 7/5$.

We eliminate both τ and θ from Equations (7.29) and (7.36), to obtain

$$\left\{u - \frac{1}{[2(\gamma-1)]^{1/2}\psi}\right\}^2 + v^2 = \frac{1}{2(\gamma-1)}\frac{1}{\psi^2} \tag{7.37}$$

for the streamlines in the hodograph plane. With $\psi \geq 0$, the streamlines are circles on the right side of the plane as in the incompressible case. For example, curve A represents a subsonic streamline that has a maximum Mach number where it crosses the positive u-axis. Because its radius is small, we see from the above equation that ψ is relatively large for this streamline. Streamlines B to E have progressively smaller ψ values; these streamlines all become supersonic. Again, all features are symmetric about the u-axis.

To determine the limit line, we use Equations (7.1) and (7.29), to obtain

$$\psi_\theta = -\frac{\sin\theta}{\tau^{1/2}} = -\frac{v}{w\tau^{1/2}}$$

$$\psi_\tau = -\frac{\cos\theta}{2\tau^{3/2}} = -\frac{u}{2w\tau^{3/2}}$$

From Equation (7.26), the limit line condition is then

$$\frac{1}{4}\frac{1-\frac{\tau}{\tau^*}}{\tau^2(1-\tau)}\frac{v^2}{w^2\tau} + \frac{y^2}{4w^2\tau^3} = 0$$

which simplifies to

$$\left(1 - \frac{\gamma+1}{\gamma-1}\tau\right)v^2 + (1-\tau)u^2 = 0$$

With w normalized by a_o in Equation (7.17), we replace τ and obtain the final result

$$\frac{\gamma-1}{2}u_\ell^2 + \frac{\gamma+1}{2}v_\ell^2 = 1 \tag{7.38}$$

where an ℓ subscript denotes the limit line; thus, the hodograph plane limit line is an ellipse. As evident in Figure 7.5, the limit line is tangent to the $\tau = \tau^*$ circle at $u = 0$ and tangent to the $\tau = 1$ circle at $v = 0$.

The streamline that is tangent to the sonic line at $v = 0$ has a nondimensional ψ value of $[(\gamma+1)/(\gamma-1)]^{1/2}$, which equals 2.450 when $\gamma = 7/5$ (see Problem 7.3). This streamline, which is not shown in Figure 7.5, does not intersect the limit line. The B streamline is tangent to the limit line and has a ψ value of $(\gamma+1)/[8(\gamma-1)]^{1/2} = 1.342$ (see Problem 7.3). For in-between values, $1.342 < \psi < 2.450$, the streamlines become supersonic but do not intersect the limit line. For a smaller value of ψ, e.g., 1.2, the C streamline intersects the limit line twice. A special streamline is D for which $\psi = 1$. This streamline also intersects the limit line twice with the second time at $v = 0$. All streamlines in the range $0 < \psi < 1$ intersect the limit line only once. These streamlines terminate at the $\tau = 1$ circle where M is infinite. When $\psi = 0$, the E streamline would become coincident with the v-axis.

Transformation to the Physical Plane

Equations (7.23) are now simplified with the aid of Equation (7.29). In these equations, x and y are normalized with ℓ, whereas ψ is normalized with ℓa_o, as before. We thus obtain

$$x_\tau = \frac{1}{2}\left(\frac{\gamma-1}{2}\right)^{1/2}\sin 2\theta\,\frac{1-\frac{\gamma}{\gamma-1}\tau}{\tau^2(1-\tau)^{\gamma/(\gamma-1)}} \tag{7.39a}$$

$$x_\theta = -\left(\frac{\gamma-1}{2}\right)^{1/2}\frac{\cos 2\theta}{\tau(1-\tau)^{1/(\gamma-1)}} \tag{7.39b}$$

$$y_\tau = -\frac{1}{2}\left(\frac{\gamma-1}{2}\right)^{1/2}\cos 2\theta\,\frac{1-\frac{\gamma}{\gamma-1}\tau}{\tau^2(1-\tau)^{\gamma/(\gamma-1)}} - \frac{1}{[8(\gamma-1)]^{1/2}\tau(1-\tau)^{\gamma/(\gamma-1)}} \tag{7.39c}$$

$$y_\theta = -\left(\frac{\gamma-1}{2}\right)^{1/2}\frac{\sin 2\theta}{\tau(1-\tau)^{1/(\gamma-1)}} \tag{7.39d}$$

Equations (7.39b,d) integrate to

$$x = -\frac{1}{2}\left(\frac{\gamma-1}{2}\right)^{1/2}\frac{\sin 2\theta}{\tau(1-\tau)^{1/(\gamma-1)}} + f(\tau) \tag{7.40a}$$

$$y = \frac{1}{2}\left(\frac{\gamma-1}{2}\right)^{1/2}\frac{\cos 2\theta}{\tau(1-\tau)^{1/(\gamma-1)}} + g(\tau) \tag{7.40b}$$

where f and g are again functions of integration. Since

$$\frac{d}{d\tau}\left[\frac{1}{\tau(1-\tau)^{1/(\gamma-1)}}\right] = -\frac{1-\frac{\gamma}{\gamma-1}\tau}{\tau^2(1-\tau)^{\gamma/(\gamma-1)}}$$

differentiation of Equation (7.40a) with respect to τ yields

$$x_\tau = \frac{1}{2}\left(\frac{\gamma-1}{2}\right)^{1/2}\sin 2\theta\,\frac{1-\frac{\gamma}{\gamma-1}\tau}{\tau^2(1-\tau)^{\gamma/(\gamma-1)}} + \frac{df}{d\tau}$$

By comparing this result with Equation (7.39a), we see that x is given by Equation (7.40a) with $f = 0$. Similarly, differentiation of Equation (7.40b) with respect to τ results in

$$y_\tau = -\frac{1}{2}\left(\frac{\gamma-1}{2}\right)^{1/2}\cos 2\theta\,\frac{1-\frac{\gamma}{\gamma-1}\tau}{\tau^2(1-\tau)^{\gamma/(\gamma-1)}} + \frac{dg}{d\tau}$$

Comparing this relation with Equation (7.39c) produces

$$\frac{dg}{d\tau} = -\frac{1}{[8(\gamma-1)]^{1/2}\tau(1-\tau)^{\gamma/(\gamma-1)}}$$

which integrates to

$$g = -\frac{1}{[8(\gamma-1)]^{1/2}}\int\frac{d\tau}{\tau(1-\tau)^{\gamma/(\gamma-1)}} \tag{7.41a}$$

where the constant of integration is set equal to zero. For certain values of γ, the integral can be performed; for instance, for $\gamma = 7/5$ we have

$$g = 1.118\left[\tanh^{-1}(1-\tau)^{1/2} - \frac{1}{5(1-\tau)^{5/2}} - \frac{1}{3(1-\tau)^{3/2}} - \frac{1}{(1-\tau)^{1/2}}\right] \tag{7.41b}$$

Thus, y is given by Equation (7.40b) with g provided by Equation (7.41a) or (7.41b).

Compressible Solution in the Physical Plane

The angle θ is eliminated between Equations (7.29) and (7.40), with the result

$$x^2 + (y-g)^2 = R^2 \tag{7.42a}$$

where the normalized radius is

$$R = \frac{1}{2}\left(\frac{\gamma-1}{2}\right)^{1/2}\frac{1}{\tau(1-\tau)^{1/(\gamma-1)}} \tag{7.42b}$$

Since g and R are functions only of τ, constant Mach number lines are circles with centers on the y-axis. Consequently, the sonic line, $\tau = \tau^*$, is a circle in the physical plane. As in the incompressible case, the maximum Mach number on a streamline occurs when $\theta = 0°$, which in turn implies $x = 0$. Hence, the maximum Mach number on a streamline can be obtained from

$$y = \frac{1}{2}\left(\frac{\gamma-1}{2}\right)^{1/2}\frac{1}{\tau(1-\tau)^{1/(\gamma-1)}} + g(\tau)$$

and Equation (7.18a).

As in the incompressible case, the wall occurs where $\psi = 0$ or $\theta = \pm(\pi/2)$. In accordance with the clockwise flow direction indicated by the arrows in Figure 7.5, we have $\theta = \pi/2$ on the left side of the wall, where x is 0 and $\theta = -\pi/2$ on the right side. Again, the wall is a vertically oriented, semi-finite, flat plate. The tip of the plate is located at the point where y in Equation (7.40b), with $\theta = -\pi/2$, has a maximum value. From this equation, we find that $(dy/d\tau) = 0$ yields $\tau = \tau^*$. Thus, the tip of the plate is coincident with the $x = 0$ point on the sonic line.

Figure 7.6 shows the streamlines, sonic line, and limit line in the physical plane. The $A,B,\ldots,E$ labeling and arrows on the streamlines coincide with that shown in Figure 7.5. In contrast to the incompressible case, the tip of the plate is at a negative y value and so also is the center of the sonic line circle. Except for θ, there is complete symmetry about the y-axis. Only the flow direction, shown by the streamline arrows, is opposite on the two sides of the y-axis.

When $\tau << 1$ we expect the streamline pattern in Figure 7.6 to resemble the incompressible pattern in Figure 7.4. Streamline A, with $\psi = 2.7$, is such a streamline. With a ψ value of 2.450, the streamline is tangent to the sonic line at $x = 0$, where all streamlines attain their maximum speed. With a further reduction in ψ to 2, the streamline accelerates to a supersonic speed inside the sonic circle. For x-positive, the flow on this streamline smoothly decelerates to a stagnation condition as x and $-y$ approach infinity.

We can view the $\psi = 2$ and 2.7 streamlines as representing two walls that would bound an inviscid flow. In this circumstance, there is a supersonic "bubble" adjacent to the lower wall in which the flow smoothly decelerates without the presence of a shock wave. This bubble is similar to that which can occur on the upper surface of an airfoil in transonic flow (Nieuwland and Spee, 1973).

Location of the Limit Line

Before resuming the discussion of the B, …, E streamlines, we first discuss the limit line shown as a short/long dashed curve in Figure 7.6. To obtain the limit line equation in the physical plane, we again substitute Equation (7.29) into (7.26), with the result

$$1 - M_\ell^2 \sin^2\theta_\ell = 0$$

By replacing the Mach number with τ_ℓ and eliminating θ_ℓ, with the aid of Equation (7.29), we obtain

$$\psi_\ell^2 \tau_\ell^2 - \frac{\gamma+1}{2}\tau_\ell + \frac{\gamma-1}{2} = 0 \tag{7.43}$$

By setting ψ_ℓ equal to zero, we have $\tau_\ell = \tau^*$, where this point is the tip of the plate. Another result is obtained by solving this equation for τ_ℓ:

$$\tau_\ell = \frac{\gamma+1}{4\psi_\ell^2}\left\{1 \pm \left[1 - \frac{8(\gamma-1)}{(\gamma+1)^2}\psi_\ell^2\right]^{1/2}\right\} \tag{7.44}$$

Hence, for a real solution, ψ_ℓ is restricted to the range

$$0 \le \psi_\ell \le \frac{\gamma+1}{[8(\gamma-1)]^{1/2}} \tag{7.45}$$

where the upper limit equals 1.342 when γ is 7/5.

To place the limit line discussion in perspective, we briefly return to Figure 7.5. In this figure, the E streamline becomes coincident with the v-axis when $\psi = 0$. We therefore have a family of streamlines that cross the limit line once and end on the $\tau = 1$ circle. At one end of the family is streamline D for which $\psi = 1$; hence, this family occurs when ψ is in the range

$$0 \le \psi \le 1$$

Consequently, the limit line, from its tangency points with the $\tau = \tau^*$ and $\tau = 1$ circles, is traversed only once whenever ψ_ℓ ranges from 0 to 1. This traverse requires the minus sign before the square root in Equation (7.44).

The B through D streamlines, however, intersect the limit line twice. Since the B streamline has $\psi = (\gamma + 1)/[8(\gamma - 1)]^{1/2}$, we now have

$$1 \leq \psi_\ell \leq \frac{\gamma + 1}{[8(\gamma - 1)]^{1/2}}$$

and the plus sign is used in Equation (7.44). Each ψ_ℓ value in this region corresponds to two points on the limit line. Both points are supersonic, with the lower speed point occurring to the left of where the B streamline is tangent to the limit line. Because of this behavior, there is a variety of different streamline families in the physical plane that intersect the limit line.

We still need the actual equations for the limit line in the physical plane. Equation (7.43) is solved for ψ_ℓ, to obtain

$$\psi_\ell = \left(\frac{\gamma + 1}{2}\right)^{1/2} \frac{(\tau_\ell - \tau^*)^{1/2}}{\tau_\ell}$$

We eliminate ψ_ℓ between this relation and Equation (7.29) to obtain the following trigonometric results:

$$\cos\theta_\ell = \left(\frac{\gamma + 1}{2} \frac{\tau_\ell - \tau^*}{\tau_\ell}\right)^{1/2}, \qquad \sin\theta_\ell = \left(\frac{\gamma - 1}{2} \frac{1 - \tau_\ell}{\tau_\ell}\right)^{1/2}$$

$$\cos 2\theta_\ell = \frac{\gamma\tau_\ell - (\gamma - 1)}{\tau_\ell}, \qquad \sin 2\theta_\ell = \left[\frac{(\gamma^2 - 1)(\tau_\ell - \tau^*)(1 - \tau_\ell)}{\tau_\ell^2}\right]^{1/2}$$

These relations are substituted into Equations (7.40), to yield (with $f = 0$)

$$x_\ell = -\left(\frac{\gamma + 1}{2}\right)^{1/2} \frac{\gamma - 1}{2} \frac{(\tau_\ell - \tau^*)^{1/2}}{\tau_\ell^2(1 - \tau_\ell)^{(3-\gamma)/[2(\gamma-1)]}} \tag{7.46a}$$

$$y_\ell = \frac{1}{2}\left(\frac{\gamma + 1}{2}\right)^{1/2} \frac{\gamma\tau_\ell - (\gamma - 1)}{\tau_\ell^2(1 - \tau_\ell)^{1/(\gamma-1)}} + g(\tau_\ell) \tag{7.46b}$$

Equation (7.46a) provides x_ℓ when $x_\ell < 0$; we simply delete the minus sign for the $x_\ell > 0$ curve. These equations provide the limit line shown in Figure 7.6. At $x_\ell = 0$ this curve is tangent to the sonic line. It then remains quite close to the sonic line, finally deviating from it near where streamline D intersects the limit line.

For purposes of clarity, we, henceforth, discuss the limit line in Figure 7.6 when $x_\ell < 0$. The curve has a sharp peak where it intersects streamline B (see Problem 7.4). Between this peak and where streamline C first intersects it, x_ℓ has a local minimum value which is not discernible in Figure 7.6. At the peak itself, y_ℓ has a maximum value while x_ℓ has a local maximum value. Consequently, both $dx_\ell/d\tau_\ell$ and dy_ℓ/dx_ℓ are zero and dy_ℓ/dx_ℓ is indeterminate at the peak. This indeterminacy is the reason a sharp peak exists in the first place. Outside of the peak, the limit line persists as $-x_\ell$ and $-y_\ell$ go to infinity.

Streamline Discussion in the Physical Plane

Streamline B, which has a ψ value of $(\gamma + 1)/[8(\gamma - 1)]^{1/2}$ (= 1.342), is the last streamline to be able to make a smooth transition across the physical plane. Any streamline with a smaller ψ value will intersect the limit line once or twice when $x < 0$. Thus, curve C with a ψ value of 1.2 intersects the limit line both to the left and to the right of the sharp peak. These two intersections are also evident in Figure 7.5. Anytime a streamline intersects a point on the limit line, θ changes by 180° (Shapiro, 1954); hence, the streamline has a cusp at the point of intersection and doubles back on itself.

The change a streamline experiences when it intersects a limit line is discontinuous. This difficulty occurs in the physical plane; the streamlines do not exhibit a discontinuity in the hodograph plane. Unlike a shock wave, a limit line is a purely mathematical discontinuity in the physical plane. In the real world, a streamline cannot suddenly change its direction by 180°.

Although complex variable theory has not been used, the concept of a multi-sheeted Riemann surface now becomes indispensable. Each time a streamline reflects from the limit line, the solution moves onto a different Riemann surface or sheet. Streamline C, before it first intersects the oval-shaped portion of the limit line, is said to be on sheet I. This sheet also includes all streamlines for which $\psi \geq (\gamma + 1)/[8(\gamma - 1)]^{1/2}$, such as streamline A. The part of streamline C that is between its two limit line intersections is on sheet II. Sheet III consists of those streamlines between the two outermost branches of the limit line; these streamlines smoothly cross the y-axis. Sheets I and II are separated by a branch cut, which is the oval part of the limit line. Sheets II and III are also separated by a branch cut, namely, the non-oval part of the limit line. Below streamline B and below the outermost portion of the limit line, after it crosses B, we have either a two- or three-sheeted solution. For instance, an x,y point that lies on curve C might be on any one of the three sheets. On the other hand, a point above the outermost parts of the limit line and above the central part of streamline B can only be on sheet I.

Streamline E, as is evident from Figure 7.5, approaches the limit line on sheet I with a positive slope near 90°. Just before intersection with the limit line, this streamline crosses the sonic line. After intersection, it heads toward infinity (i.e., $x \rightarrow -\infty$, $y \rightarrow -\infty$) on sheet II as τ tends to unity. Streamline D with a ψ value of 1- is similar to E; it does not intersect the outermost portion of the limit line.

For ψ values in the range of 1 to just below $(\gamma + 1)/[8(\gamma - 1)]^{1/2}$, such as streamline C, there are two limit line intersections and three separate sheets. It is important to note that the third sheet does not contain the oval-shaped portion of the limit line, the sonic line, or the plate. The supersonic streamlines on the third sheet are nearly horizontal; they smoothly accelerate from the leftmost outer section of the limit line to the y-axis after which they smoothly decelerate as they approach the rightmost portion of the limit line. For a ψ value just above unity, not shown in Figure 7.6, sheet III streamlines cross the y-axis where the sheet I plate would occur.

The solution on each of the three sheets satisfies the steady Euler equations and represents a legitimate flow field; thus, the solution associated with a given sheet is not necessarily any more realistic than that of the other two sheets. Furthermore, the solution on each sheet is distinct; the solution on one sheet is not an analytic continuation from either of the other sheets.

To our knowledge, the foregoing theory has yet to be experimentally verified. For such a verification, consider a duct whose walls replace two streamlines. Moreover, the subsequent discussion is limited to a duct whose inlet flow is supersonic. Because there is no upstream influence in a supersonic flow, a supersonic inlet is easier to generate than a prescribed subsonic inlet flow. This simplicity is realistic, of course, only when shock waves are not present.

The flows on sheets II and III and the flow above streamline B within the sonic circle on sheet I are supersonic and thus relevant to the discussion. Consider a sheet I duct whose inlet is located on the y-axis of Figure 7.6 where the velocity vector is parallel to the x-axis. To be specific, the values of 2.0 and 2.4 are chosen for ψ and represent the bounding walls of the duct, which is to the right of the y-axis. We therefore have a nonuniform but parallel, supersonic flow at the inlet of the duct. For the design approach alluded to in the next paragraph, the supersonic restriction is essential, whereas the parallel flow restriction is not, although it simplifies the analysis. Inside the duct, the fluid gradually decelerates toward a stagnation condition. Along the way the flow smoothly passes through a circular arc sonic line.

To generate this supersonic inlet flow, a contoured nozzle with an asymmetric diverging section would be used that is based on the theory of characteristics. A known solution on the duct's inlet data line is sufficient to design the upstream nozzle. A general design approach is discussed in Emanuel (1986, Chapter 18), where it is used to design an asymmetric nozzle that generates a supersonic potential vortex for the flow downstream of the nozzle.

The foregoing approach might be envisioned for a sheet III duct whose supersonic inlet is also on the y-axis. It would be interesting to experimentally determine what transpires when the flow encounters the outermost limit line. Of course, the duct can terminate before the limit line is encountered. This termination condition is readily imposed because there is no upstream influence in a supersonic flow.

7.4 LIMIT LINES

Within the context of the hodograph transformation, experimental results that focus on limit lines appear to be nonexistent. The one exception is for transonic flow over airfoils (Nieuwland and Spee, 1973). In this circumstance, experiments demonstrate that a shock-free transition from a supersonic flow to a subsonic flow is possible. The maximum value for the Mach number inside the supersonic region, however, is near unity and a limit line is not present. The occurrence of a smooth transition is really not surprising, since it can appear in a Taylor–Maccoll flow. In this section, limit lines are discussed in a more general manner, with emphasis on unresolved issues.

A simple observation is that a uniform flow in the physical plane corresponds to a single point in the hodograph plane. Clearly, the x,y to w,θ transformation cannot be one-to-one in this case. Furthermore, any nonuniform flow that depends on a single similarity variable will transform into a line in the hodograph plane. This is the case for Prandtl–Meyer flow, which depends typically on an angle rather than x and y separately, and with unsteady, one-dimensional rarefaction waves. These flows all have zero Jacobian values; they are referred to as missing solutions when using the hodograph equations. Hence, nonexistence of a hodograph solution does not mean nonexistence in terms of the original variables.

A second observation is that a limit line is not a shock wave, or vice versa. Like a missing solution, it is a mathematical consequence of the theory. As we will discuss shortly, limit lines are not just restricted to the hodograph transformation, but may occur more generally. Furthermore, while a limit line is not a shock wave, we suggest that it may *induce* a shock wave, or a shock wave system, in the flow somewhere upstream of where the shock-free limit line otherwise would occur. For instance, supersonic flow over a transonic airfoil normally terminates with a shock. A second illustration would be supersonic diffusers, where relatively strong shock waves are thought to be unavoidable.

As mentioned earlier, the extension to a general flow is via the infinite acceleration condition. In a spiral flow (Emanuel, 1986, Section 15.2), there is a minimum radius, with a slightly supersonic Mach number, where the acceleration is infinite. In the limiting case of a vortex flow, there is no limit line. Alternatively, when the flow is purely a source or sink flow, the minimum radius limit line is also a sonic line. This latter solution is used, e.g., in the design of a two-dimensional minimum length nozzle with a curved (circular) sonic line at the throat (Emanuel, 1986, Section 17.2). In this case, there is no difficulty in starting a flow field solution at a limit line. The flow field upstream of the limit line is not an analytic continuation of the downstream source flow. In fact, there is no known suitable design procedure for the upstream flow (Argrow and Emanuel, 1991). The situation is different for a minimum-length nozzle with a straight sonic line, where an upstream design procedure is known (Ho and Emanuel, 2000).

Both Cherry (1950) and Lighthill (1947) have examined the transonic flow field in a smoothly contoured nozzle using the hodograph transformation (see Problem 17.15). The parabolically shaped sonic line is concave in the upstream direction, with respect to the direction of flow on the symmetry axis. The parabolically shaped limit line is convex with respect to this direction. The two curves are tangent to each other where they meet on the symmetry axis. Cherry provides an analytically continued solution, downstream of the limit line, although it is not clear what happens to the infinite acceleration. Actual nozzle design does not utilize these studies, and a troublesome limit line is not encountered. As evident from the concept of missing solutions, a parabolically shaped sonic line, e.g., is not a required feature of the flow.

Lens Analogy

A steady, inviscid flow can be transformed from one uniform, supersonic state to another uniform, supersonic state using the lens analogy (Emanuel, 1986, Section 17.2). (The name stems from the similarity with what several lenses can do with an optical beam.) The flow field may be two-dimensional or axisymmetric, and is homentropic, i.e., shock-free. The design procedure for the enclosing duct has features in common with a minimum-length nozzle, and in neither case is a hodograph transformation utilized. Let the upstream and downstream Mach numbers be M_1 and M_2, respectively. If $M_1 > M_2 \geq 1$, we have a supersonic diffuser; otherwise, with $M_2 > M_1 \geq 1$, we have a supersonic nozzle.

A'Rafat (1994) computationally investigated the diffuser case using the lens analogy. For a given value of γ and M_1, a downstream limit line occurs if $M_2 - 1$ becomes too small. At this line, the acceleration is infinite and the streamlines go through a 180° reversal in direction (Emanuel and Rodriguez, 1998). In contrast to a theorem by Friedrichs (1948), this limit line starts at the wall of the diffuser. Actual supersonic diffusers are not designed using the lens analogy, but their converging shape is often reminiscent of a lens analogy diffuser. Aside from problems associated with an adverse pressure gradient, a limit line may force a shock system to develop upstream of the theoretical limit line location.

Since a lens analogy flow is homentropic, it is reversible. If a diffuser has a limit line, then, by interchanging M_1 and M_2, a supersonic nozzle is obtained with a limit line near the inlet. In either case, an explicit equation can be written for the Mach number along the limit line (Emanuel and Rodriguez, 1998). As a consequence, the lens analogy design procedure can be modified to avoid the presence of the line. For example, A'Rafat (1994) uses two diffusers in series, thereby avoiding a limit line.

In view of the importance and conventional difficulties associated with supersonic diffusers, it would be interesting to test the lens analogy approach, both with and without the presence of a limit line. There would be a major improvement in diffuser efficiency, especially at high inlet Mach numbers, if a nearly homentropic flow were obtained. A weak normal shock, or oblique shock system, near the exit of the supersonic diffuser would have little impact on its performance. The magnitude of the adverse pressure gradient can be limited (Emanuel and Rodriguez, 1998) so that (turbulent) boundary-layer separation is avoided.

7.5 GENERAL SOLUTION

We commence by finding a general solution to Equation (7.21) by means of a method devised by Chaplygin (1944). It is worth recalling that a perfect gas has been assumed. Observe that θ only appears in a partial derivative; hence,

$$\psi = A + B\theta \tag{7.47}$$

is a particular solution, where A and B are constants. The constant A merely adjusts the level of ψ and can be set equal to zero. The $B\theta$ term represents a compressible line source or sink flow.

In view of the dependence on θ, it is natural to seek a nondimensional separation of variable solution of the form

$$\psi = \tau^m \chi_m(\tau) \sin n\theta$$

or

$$\psi = \tau^m \chi_m(\tau) \cos n\theta$$

Upon substituting either of these into Equation (7.21), we obtain

$$\tau^2(1-\tau)\chi_m'' + \tau\left[2m+1-\left(2m+1-\frac{1}{\gamma-1}\right)\tau\right]\chi_m'$$
$$+\left\{m^2-\left(\frac{n}{2}\right)^2+\tau\left[\left(\frac{n}{2}\right)^2\frac{\gamma+1}{\gamma-1}+m\frac{2-\gamma}{\gamma-1}-m(m-1)\right]\right\}\chi_m = 0$$

where

$$\chi_m' = \frac{d\chi_m}{d\tau}, \qquad \chi_m'' = \frac{d^2\chi_m}{d\tau^2}$$

Observe that a τ factor can be canceled by choosing $m = n/2$. With this simplification, we obtain

$$\tau(1-\tau)\chi_n'' + \left[n+1-\left(n+1-\frac{1}{\gamma-1}\right)\tau\right]\chi_n' + \frac{n(n+1)}{2(\gamma-1)}\chi_n = 0 \qquad \textbf{(7.48)}$$

where $\chi_{n/2}$ is relabeled as χ_n for notational convenience. A general solution of Equation (7.21) is

$$\psi = A + B\theta + \sum_{n=0}^{\infty} \tau^{n/2}\chi_n(\tau)(A_n \cos n\theta + B_n \sin n\theta) \qquad \textbf{(7.49)}$$

where each χ_n satisfies Equation (7.48). Observe that the form of this solution does not change if θ is replaced with $\theta + \pi$ or with $-\theta$. Thus, the direction of flow is reversible. For instance, if Equation (7.49) represents an accelerating flow, it also represents the corresponding decelerating flow by simply increasing θ by π.

A solution of Equation (7.48) is the hypergeometric function (Spanier and Oldham, 1987; *Handbook of Mathematical Functions*, 1964), written as

$$\chi_n = F(a_n, b_n; c_n; \tau) \qquad \textbf{(7.50a)}$$

where F is the Gauss or hypergeometric series

$$F(a,b;c;\tau) = \frac{\Gamma(c)}{\Gamma(a)\Gamma(b)}\sum_{j=0}^{\infty}\frac{\Gamma(a+j)\Gamma(b+j)}{\Gamma(c+j)}\frac{\tau^j}{j!} \tag{7.50b}$$

and where Γ is the gamma function. (The Γ function generalizes $j!$, while the hypergeometric series is a generalization of the geometric series, $1+x+x^2+\ldots$.) The a_n, b_n, and c_n parameters are connected to the coefficients of Equation (7.48) by means of

$$a_n + b_n = n - \frac{1}{\gamma-1}, \quad a_n b_n = -\frac{n(n+1)}{2(\gamma-1)}, \quad c_n = n+1$$

An explicit solution for these parameters is then

$$a_n = \frac{(\gamma-1)n - 1 - [1+(\gamma^2-1)n^2]^{1/2}}{2(\gamma-1)} \tag{7.51a}$$

$$b_n = \frac{(\gamma-1)n - 1 + [1+(\gamma^2-1)n^2]^{1/2}}{2(\gamma-1)} \tag{7.51b}$$

$$c_n = n+1 \tag{7.51c}$$

We will need a number of properties of the gamma function as follows:

$$\begin{aligned}\Gamma(1) &= 0! = 1 \\ \Gamma(j) &= (j-1)!, \quad j = 1, 2, 3, \ldots \\ |\Gamma(j)| &= \infty, \quad j = 0, -1, -2, \ldots \\ \Gamma(z+1) &= z\Gamma(z)\end{aligned}$$

Since $\Gamma(c)$ is infinite when $c = 0, -1, -2, \ldots$, the Gauss series is not defined for these c values, providing $a - c$ or $b - c$ is not equal to 0, 1, 2, … . For instance, if $n = -2$ and $\gamma = 5/3$, then neither $a - c$ nor $b - c$ equals 0 or a positive integer. The ratio of two infinite gamma function values, however, is finite and is given by (Spanier and Oldham, 1987)

$$\frac{\Gamma(-m)}{\Gamma(-N)} = (-1)^{N-m}\frac{N!}{m!} \tag{7.52}$$

when m and N are nonnegative integers. As a consequence of the restriction on the c value, the Gauss series generally does not exist when n is a negative integer, which is the reason the summation in Equation (7.50b) excludes negative integer values for j.

Nevertheless, Equation (7.48) possesses solutions for negative integer n values as demonstrated for $n = -1$ in Problem 7.12. The situation is conceptually similar to the expansion

$$\frac{1}{1-x} = 1 + x + x^2 + x^3 + \cdots$$

where the right side is divergent when $|x| \geq 1$, but the left side possesses a finite value. Solutions to Equation (7.48) therefore exist even though the Gauss series does not.

Since Equation (7.48) is second order, it has two linearly independent solutions. From Spanier and Oldham (1987, p. 600), these can be written as

$$c_1 F(a_n, b_n; n+1; \tau) + c_2 \tau^{-n} F(a_n - n, b_n - n; 1 - n; \tau)$$

where the c_i are constant. For $n = 0$, the c_1 and c_2 terms are redundant, while for $n = 1, 2, 3, \ldots$, the Gauss series that is multiplied by c_2 does not exist. We thus obtain the form of the solution to Equation (7.21) that is given by Equations (7.49) and (7.50).

For the hodograph transformation, Equations (7.23) can be integrated in closed form. We illustrate the procedure with a typical term in Equation (7.49), which we write as

$$\psi = \tau^{n/2} \chi(\tau) \cos n\theta \tag{7.53}$$

where χ is the hypergeometric function, χ_n. Equations (7.23a,b), with x now normalized with $[(\gamma - 1)/2]^{1/2}(1/a_o)$ for notational convenience, become

$$x_\tau = -\frac{n}{2} \frac{\tau^{(n-3)/2}}{(1-\tau)^{\gamma/(\gamma-1)}} \left[\left(\frac{\tau}{\tau^*} - 1\right)\chi \cos\theta \sin n\theta + (1-\tau)\left(\chi + \frac{2}{n}\tau\chi'\right)\sin\theta \cos n\theta\right] \tag{7.54a}$$

$$x_\theta = \frac{n\tau^{(n-1)/2}}{(1-\tau)^{1/(\gamma-1)}} \left[\chi \sin\theta \sin n\theta + \left(\chi + \frac{2}{n}\tau\chi'\right)\cos\theta \cos n\theta\right] \tag{7.54b}$$

with similar equations for y_τ and y_θ (see Problem 7.6). The integration of x_θ is expedited with the assistance of

$$\int_0^\theta \sin\theta \sin n\theta \, d\theta = \frac{\cos\theta \sin n\theta - n \sin\theta \cos n\theta}{n^2 - 1}$$

$$\int_0^\theta \cos\theta \cos n\theta \, d\theta = \frac{n \cos\theta \sin n\theta - \sin\theta \cos n\theta}{n^2 - 1}$$

which requires that $|n| \neq 1$. (The special cases when $n = \pm 1$ are easily evaluated.) Upon performing the θ integration, we obtain

$$x = \frac{1}{n^2 - 1}(X(\tau; n)\cos\theta \sin n\theta - X(\tau; 1)\sin\theta \cos n\theta) + f(\tau)$$

where

$$X(\tau; \tilde{n}) \equiv \frac{(n^2 + n)\tau^{(n-1)/2}\chi + 2\tilde{n}\tau^{(n+1)/2}\chi'}{(1-\tau)^{1/(\gamma-1)}}$$

and $\tilde{n}$ is either n or 1. To determine the function of integration f, we differentiate x with respect to τ and compare the result with Equation (7.54a). This computation utilizes

$$\frac{dX}{d\tau} = \frac{(n+1)\tau^{(n-3)/2}}{(1-\tau)^{\gamma/(\gamma-1)}} \left\{\frac{n}{2}\left[(n-1)(1-\tau) - (\tilde{n}-1)\frac{2\tau}{\gamma - 1}\right]\chi + (n - \tilde{n})\tau(1-\tau)\chi'\right\}$$

where the second derivative, χ'', has been eliminated with the aid of Equation (7.48). We thereby obtain

$$\frac{df}{d\tau} = 0$$

A similar computation for y (see Problem 7.6) results in

$$\frac{dg}{d\tau} = \frac{\tau^{(n-3)/2}}{(1-\tau)^{1/(\gamma-1)}}\left(\tau\chi' + \frac{n}{2}\chi\right)$$

With the integration constants set equal to zero, we finally obtain

$$x = \frac{n}{n^2-1}\frac{\tau^{(n-1)/2}}{(1-\tau)^{1/(\gamma-1)}}$$
$$\times\left\{[(n+1)\chi + 2\tau\chi']\cos\theta\sin n\theta - \left[(n+1)\chi + \frac{2}{n}\tau\chi'\right]\sin\theta\cos n\theta\right\} \tag{7.55a}$$

$$y = \frac{n}{n^2-1}\frac{\tau^{(n-1)/2}}{(1-\tau)^{1/(\gamma-1)}}$$
$$\times\left\{\left[(n+1)\chi + \frac{2}{n}\tau\chi'\right]\cos\theta\cos n\theta + [(n+1)\chi + 2\tau\chi']\sin\theta\sin n\theta\right\} + g(\tau) \tag{7.55b}$$

In view of Equation (7.49), we see that the method is an inverse one in which we hope to find suitable boundary conditions that satisfy the solution. Walls and a symmetry axis, if there is any, are specified by prescribing constant values for ψ for them and similarly by specifying θ. The stream function is also constant on the surface of a free jet. By Bernoulli's equation, τ is constant on this surface, if the pressure external to the jet is also constant. Satisfying all prescribed boundary conditions in the hodograph plane is usually difficult.

Solutions Based on a Noninteger *n* Value

The n parameter in Equation (7.48) is not restricted to a positive integer value (Lighthill, 1953). Since n determines a_n, b_n, and c_n by means of Equations (7.51), a noninteger n value provides a new class of solutions. We illustrate the general approach by considering a particular solution, Equation (7.53), with a noninteger n value.

We write χ as

$$\chi = F(a, b; n+1; \tau)$$

where subscripts on a, b, and c are no longer necessary. An especially simple form for the solution is obtained by setting

$$a = -m \tag{7.56}$$

where m is a positive integer. Although not essential for the analysis, this choice greatly simplifies the algebra. Equation (7.50b) for the hypergeometric series now becomes a polynomial of degree m:

$$
\begin{aligned}
F(-m, b;c;\tau) &= \frac{\Gamma(c)}{\Gamma(-m)(\Gamma b)} \sum_{j=0}^{m} \frac{\Gamma(-m+j)\Gamma(b+j)}{\Gamma(c+j)} \frac{\tau!}{j!} \\
&= 1 + \frac{\Gamma(-m+1)\Gamma(b+1)\Gamma(c)}{\Gamma(-m)\Gamma(b)\Gamma(c+1)}\tau + \frac{\Gamma(-m+2)\Gamma(b+2)\Gamma(c)}{2\Gamma(-m)\Gamma(b)\Gamma(c+2)}\tau^2 + \cdots \\
&= 1 - \frac{m!b!(c-1)!}{(m-1)!(b-1)!c!}\tau + \frac{m!(b+1)!(c-1)!}{2(m-2)!(b-1)!(c+1)!}\tau^2 - \cdots \\
&= 1 - \frac{mb}{c}\tau + \frac{m(m-1)b(b+1)}{2c(c+1)}\tau^2 - \cdots + (-1)^m \frac{\Gamma(b+m)}{\Gamma b}\frac{\Gamma(c)}{\Gamma(c+m)}\tau^m \qquad (7.57)
\end{aligned}
$$

where Equation (7.52) is used for those gamma functions with a zero or negative integer in their argument. With Equations (7.51) and (7.56), n, b, and c become

$$
\begin{aligned}
n_{\pm} &= (\gamma-1)m - \frac{1}{2} \pm \left[(\gamma^2-1)m^2 - (\gamma+1)m + \frac{1}{4}\right]^{1/2} \\
b_{\pm} &= n_{\pm} + m - \frac{1}{\gamma-1} \\
c_{\pm} &= n_{\pm} + 1
\end{aligned}
$$

In order for $n_{\pm}$ to be real, the quantity within the square root must be nonnegative; i.e.,

$$(\gamma^2-1)m^2 - (\gamma+1)m + \frac{1}{4} \ge 0$$

This results in

$$
\begin{aligned}
&\gamma \ge \frac{1}{2} + 2^{1/2}, && m = 1 \\
&\gamma \ge \frac{1}{4} + \frac{6^{1/2}}{2}, && m = 2 \\
&\gamma \ge \frac{7}{6}, && m = 3 \\
&\vdots && \vdots
\end{aligned}
$$

and consequently the $m = 1$ case is not of practical interest for a perfect gas, since γ cannot exceed 5/3.

To illustrate the method, we choose

$$m = 2, \qquad \gamma = 5/3$$

with the result that $a = -2$ and

$$
\begin{aligned}
n_+ &= 2.257, & n_- &= -0.5907 \\
b_+ &= 2.757, & b_- &= -0.09067 \\
c_+ &= 3.257, & c_- &= 0.4093
\end{aligned}
$$

where the ± subscripts stem from the two values for n. With $m = 2$, the hypergeometric function simplifies to a quadratic:

$$F(-2, b;c;\tau) = 1 - \frac{2b}{c}\tau + \frac{b(b+1)}{c(c+1)}\tau^2$$

where b and c are either b_+ and c_+ or b_- and c_-. With the aid of Equations (7.36), (7.53), and (7.55) and a computer, solutions for both n values have been generated in the hodograph and physical planes.* These are established by holding ψ constant and letting τ range from near zero to near unity. The range of τ is further limited by the requirement that $|\cos n\theta| \le 1$, or

$$\left|\frac{\psi}{\tau^{n/2}\chi(\tau)}\right| \le 1$$

A streamline terminates when this inequality fails to hold. For each ψ,τ pair, the left side of Equation (7.26) is evaluated. If this quantity changes sign along a streamline, then a limit line is encountered that cannot be crossed by the streamline.

It is also essential that the solutions in the hodograph and physical planes be consistent with each other. For example, if u and v are both positive, then x and y should increase with τ. Typically, if the solution is consistent when $\tau \gtrless \tau_\ell$, it is inconsistent when $\tau \lessgtr \tau_\ell$. Because of this, it is convenient to take advantage of the even property of the cosine function. We therefore write Equation (7.53) as

$$\theta = \frac{\delta}{n}\cos^{-1}\left(\frac{\psi}{\tau^{n/2}F}\right)$$

where δ is ±1. This form is convenient for computer analysis, since both ψ and τ are specified.

Figures 7.7 and 7.8 show the hodograph and physical planes when $n = -0.5907$. In each of the figures, the $\psi = 0$, ±0.5, ±1 normalized streamlines are shown. Whether in the physical or hodograph plane, the $\psi = 0$ curve is a straight line whose slope is given by $n\theta = \pm(\pi/2)$. The other ψ curves are then symmetric about this curve. [In the two hodograph planes, u and v are normalized by $[2/(\pi - 1)]^{1/2}a_o$.]

Figures 7.7(a, b) show that the largely subsonic solution requires that $\delta = -1$. [In Figures 7.8(a, b), the consistent supersonic solution requires $\delta = 1$.] In the hodograph plane in Figure 7.7(a), the flow originates near the origin of the u,v coordinates and accelerates outward. The asterisk on each streamline indicates where the flow is sonic. The streamlines terminate at the limit line, which is near the sonic line. For the $\psi = 0$ streamline, the sonic and limit lines coincide. This is not the case for the $\psi = \pm0.5$ streamlines, although the distance separating the two lines is quite small. A similar behavior is evident in Figure 7.5 in the region where the sonic and limit lines are nearly tangent to each other.

The arrows in Figures 7.7(a, b) show the direction of flow. As expected, in the physical plane the subsonic flow converges. In accordance with Figure 7.6, the asterisks on the streamlines are even closer to the limit line than in the hodograph plane. Recall that any two streamlines can be used as walls. For instance, we have flow in an asymmetric duct if $\psi = -1$ and $\psi = 0.5$ are used as the bounding walls.

* I am indebted to Dr. H.-K. Park for the computations and the associated figures.

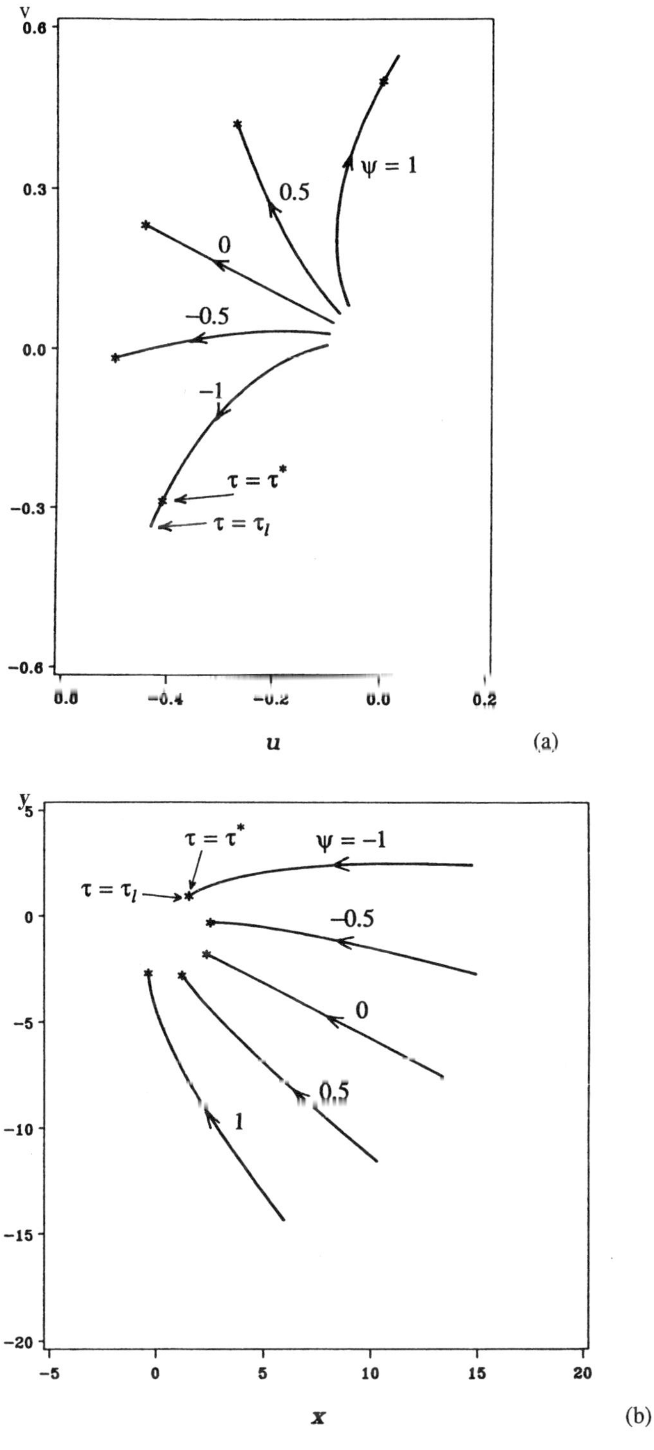

FIGURE 7.7 Hodograph (a) and physical (b) planes when $m = 2$, $\gamma = 5/3$, $n = -0.5907$, and $\delta = -1$. The flow is largely subsonic; the asterisks mark where $\tau = \tau^*$.

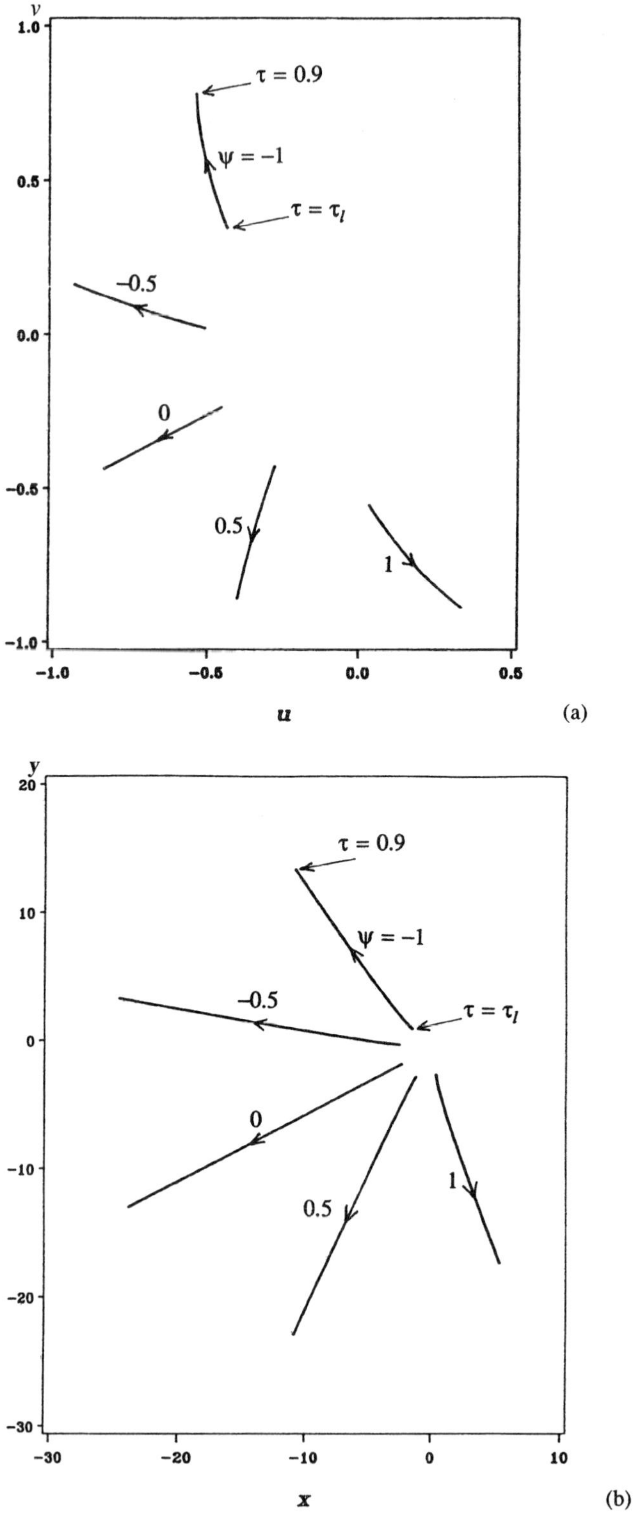

FIGURE 7.8 Hodograph (a) and physical (b) planes when $m = 2$, $\gamma = 5/3$, $n = -0.5907$, and $\delta = 1$. The flow is supersonic.

Figures 7.8(a,b) provide the corresponding $\delta = 1$ supersonic solution. The streamlines now originate at the limit line. In the physical plane, the $\psi = \pm 0.5, \pm 1$ streamlines are slightly curved. The flow is nearly a source flow.

Some care must be exercised when this procedure is used. For example, the $\delta = -1$, $\psi = -1.36$ streamline results in a consistent solution that is not cut by a limit line for τ in the range $0 < \tau \le 0.75$. The solution terminates before $\tau = 0.8$ is reached because the $|\cos n\theta| \le 1$ condition is violated. On the other hand, the adjacent $\psi = -1.35$ streamline is cut twice by a limit line.

Other solutions with noninteger n values can be constructed; e.g.,

$$\psi = A + B\theta + A_+\tau^{n_+/2}F(-2, b_+;c_+;\tau)\cos n_+\theta + A_-\tau^{n_-/2}F(-2, b_-;c_-;\tau)\cos n_-\theta \quad \textbf{(7.58a)}$$

Similarly, different integer m values can be used, such as

$$\psi = A_1\tau^{n_1/2}F(-m_1, b_1;c_1;\tau)\cos n_1\theta + A_2\tau^{n_2/2}F(-m_2, b_2;c_2;\tau)\cos n_2\theta + \cdots \quad \textbf{(7.58b)}$$

where each n_i represents an n_+ or an n_- value. Based on our experience in generating the solutions shown in Figures 7.7 and 7.8, this approach is especially amenable to computer analysis. As previously noted, it is an inverse method in which we start with a solution and hope to find a useful flow field by choosing wall and/or symmetry plane values for ψ.

7.6 ROTATIONAL FLOW

Section 7.1 pointed out that the hodograph transformation does not necessarily require irrotationality. To significantly broaden the scope of the theory, we do not assume an irrotational, homenergetic, or homentropic flow. A steady, two-dimensional or (for the first time in this chapter) axisymmetric flow of a perfect gas is assumed. In this circumstance, the Euler equations can be written as

$$\frac{D\rho}{Dt} + \rho(u_x + v_y) + \frac{\sigma\rho v}{y} = 0 \quad \textbf{(7.59a)}$$

$$\rho\frac{Du}{Dt} + p_x = 0 \quad \textbf{(7.59b)}$$

$$\rho\frac{Dv}{Dt} + p_y = 0 \quad \textbf{(7.59c)}$$

$$\frac{D}{Dt}\left(\frac{p}{\rho^\gamma}\right) = 0 \quad \textbf{(7.59d)}$$

where y is the radial coordinate when the flow is axisymmetric ($\sigma = 1$) and

$$\frac{D}{Dt} = u\frac{\partial}{\partial x} + v\frac{\partial}{\partial y}$$

Because h_o and s are now only constant along streamlines, the homenergetic relation, Equation (7.5a), is replaced with that for a constant entropy along streamlines. For a perfect gas, this relation is written as Equation (7.59d).

Rotational Hodograph Transformation

Instead of w and θ, it is simpler to use u and v as the independent coordinates. For the dependent coordinates, we will ultimately utilize x,y, and the speed of sound a. Consequently, the final system of hodograph equations will consist of three PDEs. As in Section 7.2, it is convenient to use Jacobian theory for the transformation. Hence, we have

$$J = \frac{\partial(x, y)}{\partial(u, v)} = \begin{vmatrix} x_u & x_v \\ y_u & y_v \end{vmatrix} = x_u y_v - x_v y_u \tag{7.60}$$

$$p_x = \frac{\partial p}{\partial x} = \frac{\partial(p, y)}{\partial(x, y)} = \frac{\dfrac{\partial(p, y)}{\partial(u, v)}}{\dfrac{\partial(x, y)}{\partial(u, v)}} = \frac{1}{J}(y_v p_u - y_u p_v)$$

$$v_y = \frac{\dfrac{\partial(v, x)}{\partial(u, v)}}{\dfrac{\partial(y, x)}{\partial(u, v)}} = \frac{x_u}{J}$$

$$\frac{D(\ )}{Dt} = u\frac{\partial(\ , y)}{\partial(x, y)} + v\frac{\partial(\ , x)}{\partial(y, x)} = u\frac{\dfrac{\partial(\ , y)}{\partial(u, v)}}{\dfrac{\partial(x, y)}{\partial(u, v)}} + v\frac{\dfrac{\partial(\ , x)}{\partial(u, v)}}{\dfrac{\partial(y, x)}{\partial(u, v)}}$$

$$= \frac{u}{J}\begin{vmatrix} \dfrac{\partial}{\partial u} & \dfrac{\partial}{\partial v} \\ y_u & y_v \end{vmatrix} - \frac{v}{J}\begin{vmatrix} \dfrac{\partial}{\partial u} & \dfrac{\partial}{\partial v} \\ x_u & x_v \end{vmatrix} = \frac{1}{J}\left(A\frac{\partial}{\partial u} - B\frac{\partial}{\partial v}\right) \tag{7.61}$$

with similar equations for the other derivatives, and where

$$A = uy_v - vx_v \tag{7.62a}$$

$$B = uy_u - vx_u \tag{7.62b}$$

We thus obtain

$$A\frac{\rho_u}{\rho} - B\frac{\rho_v}{\rho} + x_u + y_v + \frac{\sigma v J}{y} = 0 \tag{7.63a}$$

$$\rho A + y_v p_u - y_u p_v = 0 \tag{7.63b}$$

$$\rho B + x_v p_u - x_u p_v = 0 \tag{7.63c}$$

$$A\frac{p_u}{p} - B\frac{p_v}{p} - \gamma\left(A\frac{\rho_u}{\rho} - B\frac{\rho_v}{\rho}\right) = 0 \tag{7.63d}$$

These equations are in the hodograph form but use p, ρ, x, and y as the dependent variables instead of the desired choice.

Equations (7.63b,c) for momentum are solved for p_u and p_v, with the result

$$p_u = -\frac{\rho}{J}(Ax_u - By_u) \tag{7.64a}$$

$$p_v = -\frac{\rho}{J}(Ax_v - By_v) \tag{7.64b}$$

Along with $a^2 = (\gamma p/\rho)$, these are substituted into the entropy equation, to obtain

$$A^2 x_u - AB(x_v + y_u) + B^2 y_v + a^2 J\left(A\frac{\rho_u}{\rho} - B\frac{\rho_v}{\rho}\right) = 0$$

The rightmost term is replaced with the aid of Equation (7.36a) and A^2 is replaced with

$$A^2 = u^2 y_v^2 - 2uv y_v x_v + v^2 x_v^2$$

with a similar replacement for AB and B^2. After rearrangement and cancellation, we have

$$\begin{aligned}(x_u y_v^2 - x_v y_u y_v)u^2 + (-x_u x_v y_v + x_v y_u^2 + y_v^2 y_u - x_u y_u y_v)uv \\ + (-x_u x_v y_u + x_u^2 y_v)v^2 \quad a^2 J(x_u + y_v) - \frac{\sigma v a^2 J^2}{y} = 0\end{aligned}$$

A J factor can be canceled, leaving

$$(a^2 - v^2)x_u + uv(x_v + y_u) + (a^2 - u^2)y_v + \frac{\sigma v a^2 J}{y} = 0 \tag{7.65}$$

as the first of the x,y,a hodograph equations. This relation is nonlinear because of the σ term and the a^2 terms, since a is a dependent variable.

The next equation stems from the substantial derivative of the stagnation enthalpy:

$$\frac{Dh_o}{Dt} = \frac{1}{J}\left(A\frac{\partial}{\partial u} - B\frac{\partial}{\partial v}\right)\left[\frac{a^2}{\gamma - 1} + \frac{1}{2}(u^2 + v^2)\right] = 0$$

where h_o may vary from streamline to streamline. This relation yields

$$A\frac{a_u}{a} - B\frac{a_v}{a} + \frac{\gamma - 1}{2a^2}(Au - Bv) = 0$$

With the aid of Equation (7.65), we obtain

$$Au - Bv = u^2 y_v - uvx_v - uvy_u + v^2 x_u = a^2\left(x_u + y_v + \frac{\sigma v J}{y}\right)$$

or, more simply,

$$A\frac{a_u}{a} - B\frac{a_v}{a} + \frac{\gamma-1}{2}\left(x_u + y_v + \frac{\sigma v J}{y}\right) = 0 \tag{7.66}$$

for the second hodograph equation.

We next cross-differentiate Equations (7.64), to obtain

$$p_{vu} = \frac{\rho}{J}\left[(By_v - Ax_v)\left(\frac{\rho_u}{\rho} - \frac{J_u}{J}\right) + (B_u y_v - A_u x_v) + (By_{vu} - Ax_{uv})\right]$$

$$p_{uv} = \frac{\rho}{J}\left[(By_u - Ax_u)\left(\frac{\rho_v}{\rho} - \frac{J_v}{J}\right) + (B_v y_u - A_v x_u) + (By_{uv} - Ax_{vu})\right]$$

We equate p_{vu} with p_{uv}, with the result

$$(By_v - Ax_v)\left(\frac{\rho_u}{\rho} - \frac{J_u}{J}\right) - (By_u - Ax_u)\left(\frac{\rho_v}{\rho} - \frac{J_v}{J}\right) + (B_u y_v - A_u x_v) - (B_v y_u - A_v x_u) = 0$$

or

$$\left(J_v - J\frac{\rho_v}{\rho}\right)(Ax_u - By_u) - \left(J_u - J\frac{\rho_u}{\rho}\right)(Ax_v - By_v) = D \tag{7.67}$$

where

$$\begin{aligned} D &= J(A_v x_u - B_v y_u - A_u x_v + B_u y_v) \\ &= -(x_u + y_v)(x_v + y_u)J + \{[x_u y_{vv} - (x_v + y_u)y_{uv} + y_v y_{uu}]u - [x_u x_{vv} - (x_v + y_u)x_{uv} + y_v x_{uu}]v\}J \end{aligned}$$

To eliminate the density, we first differentiate the equation for the speed of sound with respect to u and v:

$$p_u = \frac{a^2}{\gamma}\rho_u + \frac{2}{\gamma}\rho a a_u$$

$$p_v = \frac{a^2}{\gamma}\rho_v + \frac{2}{\gamma}\rho a a_v$$

These are combined with Equations (7.64), to obtain

$$-J\frac{\rho_u}{\rho} = 2J\frac{a_u}{a} + \frac{\gamma}{a^2}(Ax_u - By_u)$$

$$-J\frac{\rho_v}{\rho} = 2J\frac{a_v}{a} + \frac{\gamma}{a^2}(Ax_v - By_v)$$

which are used to eliminate the density terms in Equation (7.67). In addition, Equation (7.60) is differentiated with respect to u and v in order to eliminate J_u and J_v. We thus obtain for the third hodograph equation

$$\begin{aligned} &J\omega[x_u(Ay_{uv} - By_{vv} - J) + y_v(Ax_{uu} - Bx_{uv} - J)] \\ &+ (y_u^2 - x_u y_v)(Ax_{uv} - Bx_{vv}) - (x_v^2 - u_u y_v)(Ay_{uu} - By_{uv}) \\ &+ 2J\left[(Ax_v - By_v)\frac{a_u}{a} - (Ax_u - By_u)\frac{a_v}{a}\right] = 0 \end{aligned} \tag{7.68}$$

where the vorticity is

$$\omega = v_x - u_y = \frac{1}{J}(x_v - y_u) \tag{7.69}$$

Irrotational Limit

Equations (7.65), (7.66), and (7.68) are coupled, nonlinear equations for x, y, and a. These three equations, or Equations (7.63), are simply alternate versions, within the context of their assumptions, of the Euler equations. A method for obtaining solutions for either system is discussed in Section 8.5. The equations, however, should reduce to a simpler form if the flow is irrotational. From Equation (7.69) this occurs when

$$x_v = y_u \tag{7.70}$$

Equations (7.65) and (7.70), in conjunction with

$$J\frac{Dh_o}{Dt} = Ah_{ou} - Bh_{ov} = 0 \tag{7.71}$$

yield Equation (7.66) (see Problem 7.8). With Equation (7.70), Equation (7.68) simplifies to

$$(Ay_u - By_v)a_u - (Ax_u - By_u)a_v = 0$$

which further reduces to

$$va_u - ua_v = 0$$

The general solution of this PDE is provided by Appendix E as

$$a = \tilde{a}(u^2 + v^2) = a(w)$$

With the definition

$$h_o = \frac{a^2}{\gamma - 1} + \frac{1}{2}(u^2 + v^2) = \frac{a^2}{\gamma - 1} + \frac{1}{2}w^2 \tag{7.72}$$

and Equation (7.71), we readily obtain

$$\frac{Dw}{Dt}\left(a\frac{da}{dw} + \frac{\gamma - 1}{2}w\right) = 0$$

for $a(w)$.

Consequently, there are two possibilities for the flow to be irrotational. First, if $Dw/Dt = 0$ one can show that the flow is a parallel flow or a potential vortex flow (Nemenyi and Prim, 1948). Neither of these flows needs to be homenergetic or homentropic. Alternatively, if $Dw/Dt \neq 0$ then

$$a\frac{da}{dw} + \frac{\gamma - 1}{2}w = 0 \tag{7.73}$$

which implies that $\nabla h_o = 0$ (see Problem 7.8). Crocco's equation for a steady, irrotational flow then yields $\nabla s = 0$. Hence, if the flow is irrotational and $Dw/Dt \neq 0$, it is also homentropic and homenergetic. In this circumstance, we can introduce a potential function by means of

$$x = \phi_u, \qquad y = \phi_v$$

and obtain the gas dynamic equation in the hodograph plane (see Problem 7.8)

$$(a^2 - v^2)\phi_{uu} + 2uv\phi_{uv} + (a^2 - u^2)\phi_{vv} + \frac{\sigma v a^2 J}{\phi_v} = 0 \tag{7.74}$$

where the Jacobian is

$$J = \phi_{uu}\phi_{vv} - \phi_{uv}^2$$

In Equation (7.74), a^2 is provided by Equation (7.72) with h_o as a constant. Thus, a single hodograph equation is sufficient when the flow is irrotational. This equation is nonlinear when $\sigma = 1$ but is linear when the flow is two dimensional and $\sigma = 0$.

REFERENCES

Abramowitz, M. and Stegun, I.A., eds., *Handbook of Mathematical Functions*, NBS Applied Mathematical Series 55, 1972.

A'Rafat, S., "Numerical Analysis of the Viscous Flow in a Supersonic Diffuser," M.S. Thesis, Embry-Riddle Aeronautical University, Daytona Beach, FL., 1994.

Argrow, B.M. and Emanuel, G., "Computational Analysis of the Transonic Flow Field of Two-Dimensional Minimum Length Nozzles," *J. Fluids Eng.* 113, 479 (1991).

Chang, C.-C., "General Consideration of Problems in Compressible Flow Using the Hodograph Method," NACA TN 2582 (January 1952).

Chaplygin, S.A., "On Gas Jets," available as NACA TM 1063 (1944).

Cherry, T.M., "Exact Solutions for Flow of a Perfect Gas in a Two-Dimensional Laval Nozzle," *Proc. R. Soc. London A* 203, 551 (1950).

Cole, J.D., "Drag of a Finite Wedge at High Subsonic Speeds," *J. Math. Phys.* 30, 79 (1951).

Emanuel, G., *Gasdynamics: Theory and Applications*, AIAA Education Series, Washington, D.C., 1986.

Emanuel, G. and Rodriguez, J., "Lense Analogy for Diffusers and Nozzles," University of Oklahoma, AME Report 98-1, 1998.

Friedrichs, K.O., "On the Non-Occurrence of a Limiting Line in Transonic Flow," *Commun. Appl. Math.* I, 287 (1948).

Ho, T.-L. and Emanuel, G., "Design of a Nozzle Contraction for Uniform Sonic Throat Flow," *AIAA J.* 38, 720 (2000).

Libby, P.A. and Reiss, H.R., "The Design of Two-Dimensional Contraction Sections," *Q. Appl. Math.* 9, 95 (1951).

Lighthill, M.J., "The Hodograph Transformation in Trans-Sonic Flow. I. Symmetrical Channels," *Proc. R. Soc. London* A 191, 323 (1947).

Lighthill, M.J., "The Hodograph Transformation," in *Modern Developments in Fluid Dynamics*, edited by L. Howarth, Vol. I, Clarendon Press, Oxford, 1953, Chapter VII.

Mackie, A.G. and Pack, D.C., "Transonic Flow Past Finite Wedges," *Proc. Camb. Philos. Soc.* 48, 178 (1952).

Manwell, A.R., *The Hodograph Equations*, Hafner Publ. Co., Darien, CT, 1971.

Molenbroek, P.,"Uber einige Bewegungen eines Gases mit Annahme eines Geschwindigkeitspotentials," *Arch. Math. Phys.* 9, 157 (1890).

Nemenyi, P. and Prim, R., "Some Geometric Properties of Plane Gas Flow," *Stud. Appl. Math.* 27, 130 (1948).

Nieuwland, G.Y. and Spee, B.M., "Transonic Airfoils: Recent Developments in Theory, Experiment, and Design," *Annu. Rev. Fluid Mech.* 5, 119 (1973).

Oyibo, G., "Formulation of Three-Dimensional Hodograph Method and Separable Solutions for Nonlinear Transonic Flows," *AIAA J.* 28, 1745 (1990).

Pai, S.-I., *Introduction to the Theory of Compressible Flow*, D. Van Nostrand Co., New York, 1959.

Ringleb, F., "Exakte Loesungen der Differentialgleichungen einer Adiabatischen Gasstroemung," *ZAMM* 20, 185 (1940).

Rodriguez Azara, J.L. and Emanuel, G., "Compressible Rotational Flows Generated by the Substitution Principle, II," *Phys. Fluids A* 1, 600 (1989).

Shapiro, A.H., *The Dynamics and Thermodynamics of Compressible Fluid Flow*, Vol. II, John Wiley, New York, 1954, Chapter 20.

Spanier, J. and Oldham, K.B., *An Atlas of Functions*, Hemisphere Publ. Co., New York, 1987.

Vincenti, W.G. and Wagoner, C.B., "Transonic Flow Past a Wedge Profile with Detached Bow Wave — General Analytical Method and Final Calculated Results," NACA TN 2339 (April 1951).

PROBLEMS

7.1 **(a)** Assume a perfect gas and determine A and B in Equation (7.16) for the Kármán–Tsien approximation.

(b) Assume a van der Waals gas with constant c_v and repeat part (a). Utilize reduced variables and the relations in Problem 5.18.

7.2 Derive the equations equivalent to Equations (7.15) and (7.21) using a potential function ϕ instead of ψ with w and θ as the independent variables.

7.3 For the nondimensional compressible Ringleb solution, derive the result that the normalized ψ equals $[(\gamma + 1)/(\gamma - 1)]^{1/2}$ for the streamline tangent to the sonic line (on the u-axis), $(\gamma + 1)[8(\gamma - 1)]^{-1/2}$ for the streamline tangent to the limit line, and 1 for the streamline tangent to the $\tau = 1$ circle (on the u-axis).

7.4 With $\gamma = 7/5$, determine for the nondimensional compressible Ringleb solution the y value where the sonic and limit lines are tangent to each other. Determine the x (with $x > 0$) and y values for the peak point on the limit line. What is the value of M at this point? What is the maximum value of the Mach number on streamline B in Figure 7.6?

7.5 Consider Equation (7.53) with $n = 1$ and $\gamma = 7/5$.

(a) Determine a, b, c, and the first four terms in the Equation (7.50b) expansion for $\chi(\tau)$.

(b) Obtain the hodograph equations that relate x and y to τ and θ. Note that Equations (7.55) do not hold when $n = \pm 1$.

7.6 **(a)** Derive the y_τ and y_θ counterparts to Equations (7.54) for the Equation (7.53) stream function.
(b) Utilize Equation (7.56), $\gamma = 2$, $m = 1$, and the n_+ root to determine x and y as normalized functions of τ and θ.
(c) Start with the results of part (b) and determine θ_ℓ and ψ_ℓ as functions of τ_ℓ for the limit line.

7.7 **(a)** Determine the solution to Equation (7.48) when $n = 0$.
(b) Compare this result to Equations (7.50) and (7.51).
(c) For the rest of this problem, use the part (a) result with $\gamma = 3/2$. Write ψ as

$$\psi = B\theta + \chi_o$$

and determine x and y in terms of τ and θ. Call the constants of integration x_o and y_o.
(d) Determine the equation for the limit line in terms of τ_ℓ, θ_ℓ variables. Determine the equation for the limit line in the physical plane.

7.8 **(a)** Show that Equations (7.65) and (7.70) yield Equation (7.66).
(b) Show that Equation (7.73) implies $\nabla h_o = 0$.
(c) Derive Equation (7.74).

7.9 Under what conditions can the method of characteristics be used to solve Equation (7.74)?

7.10 Define a nondimensional potential function and a nondimensional stream function by means of

$$\phi_{x_i} = w_i, \qquad \psi_{x_1} = -\frac{\rho}{\rho_o} w_2 x_2^\sigma, \qquad \psi_{x_2} = \frac{\rho}{\rho_o} w_1 x_2^\sigma$$

where the x_i are normalized with a characteristic length and the flow speed with the stagnation speed of sound. Assume a perfect gas, use τ and θ as the independent variables, and obtain the counterpart of Equations (7.12) that holds for both two-dimensional and axisymmetric flows.

7.11 **(a)** Show that

$$\frac{y_\tau \psi_\theta - y_\theta \psi_\tau}{y_\tau \psi_\theta - x_\theta \psi_\tau} = \tan\theta$$

(b) Verify that Equations (7.23) satisfy this relation.

7.12 **(a)** Determine the general solution to Equation (7.48) when $n = -1$, where this solution contains two constants of integration.
(b) Combine the part (a) solution, using the $\cos\theta$ form, and Equation (7.47) to obtain an equation for the stream function ψ that contains four arbitrary constants. Demonstrate that ψ satisfies Equation (7.21).
(c) Utilize Equation (7.23) to obtain x and y as functions of τ and θ.

7.13 **(a)** Determine the equation for the Mach number on the limit line, $M_{c\ell}$, when the flow in Problem 5.21 has a limit line, where M_c is defined in Problem 5.8.
(b) Show that $M_{c\ell}$ is finite and is greater than unity.

7.14 **(a)** Utilize the theory in Section 7.2 to determine a polynomial equation for the Mach number, M_ℓ^2 on the limit line in terms of ψ_θ and ψ_τ. Check this result by seeing if it is consistent with the compressible, limit line, Ringleb result.
(b) Use the relation given in Problem 7.11(a) to obtain an equation for the streamline angle θ_ℓ at the limit line, in terms of x_τ, x_θ, y_τ, y_θ, and τ_ℓ.
(c) With $\gamma = 1.4$ and $\psi = 0.9$, determine the Ringleb solution limit line values for M, x, y, θ, and the Mach angle μ.

7.15 Consider steady, irrotational, two-dimensional flow of a perfect gas in a symmetric converging/diverging nozzle. Let $\bar{x}$ and $\bar{y}$ be dimensional Cartesian coordinates, where $\bar{x}$ is along the symmetry axis and is measured from the point where the velocity is sonic. Also let $\overline{\psi}$ be zero on the symmetry axis and note that $\overline{\psi}$ is an antisymmetric function of θ. Consider only the upper half of the nozzle. Normalize $\bar{p}$ and $\bar{\rho}$ with their stagnation values, all speeds with a_o, lengths with $\bar{s}^*$ [which is the (unknown) arc length of the sonic line measured from the symmetry axis to the wall] and the stream function with $\bar{a}_0\bar{s}^*\,[2/(\gamma+1)]^{(\gamma+1)/[2(\gamma-1)]}$.

(a) Simplify Equation (7.49) to account for the symmetry and centerline conditions. Sketch typical streamlines in a nondimensional u,v hodograph plane for the upper half of a nozzle whose inlet and exit conditions are uniform.

(b) To simplify the algebra, replace the infinity in Equation (7.49) with unity and assume $\gamma = 1.5$. Determine the equations for the inverse transformation.

(c) Determine the equation for the limit line. Determine $B_1, f(\tau)$, and $g(\tau)$ such that the sonic line and limit line pass through the origin of the x,y coordinate system.

8 The Substitution Principle

8.1 PRELIMINARY REMARKS

Previous to the application of the substitution principle, exact rotational solutions of the Euler equations were rare (or nonexistent). Such solutions are therefore of intrinsic interest, particularly when they are simple and represent extensions of well-known irrotational flows. These solutions are obtained by means of the substitution principle, which holds for the three-dimensional, steady Euler equations with no body force and where the gas is perfect. Moreover, the flow field may be subsonic, transonic, or supersonic and may contain shock waves, contact surfaces, and slipstreams. In brief, the principle enables us to transform an irrotational, homenergetic, and homentropic flow field into one that is rotational, nonhomenergetic, and nonhomentropic. Both flows are exact solutions of the steady Euler equations. Generating rotational flows from irrotational flows is but one use of the principle. A number of homework problems illustrate alternate applications.

Our focus is on compressible flows where rotationality is often important. The principle, however, has been used in atmospheric studies (Yih, 1960) and in studies of incompressible stratified flows. A stratified flow occurs, for example, when fresh water mixes with the higher-density salty ocean water. It is also utilized in Chapter 23, which deals with second-order boundary-layer theory.

The principle was first discovered by Munk and Prim (1947) and Prim (1952) and later rediscovered by Yih (1960). Our presentation, however, is primarily based on Rodriguez and Emanuel (1988, 1989), since these authors are the first to systematically apply the substitution principle to a compressible flow.

Aside from the substitution principle, a unifying feature in the analysis is the ability to have an arbitrary variation, transverse to the streamlines, of the stagnation enthalpy in the transformed flow. Such a variation, for example, may stem from nonuniform combustion or heat addition. This type of heat addition process occurs in jet engines, ramjets, scramjets, and chemical or gas dynamic lasers. The heat addition process in these devices is invariably nonuniform, and the downstream flow is not homentropic or homenergetic and is rotational. The analysis of the downstream flow is of interest, since it may be passing through a gas turbine or a propulsion nozzle.

In the next section, we provide the transformation, discuss invariance under it, and establish the reason for the perfect gas restriction. The section also provides the corresponding transformation for the vorticity. The next two sections deal with applications, first to a parallel flow and then to Prandtl–Meyer flow. Other applications are briefly mentioned in the Prandtl–Meyer flow section. The final section applies the principle to flow in the hodograph plane; it represents a continuation of the analysis in Section 7.6.

8.2 TRANSFORMATION EQUATIONS

We start with the steady Euler equations in the form

$$w_j \frac{\partial \rho}{\partial x_j} + \rho \frac{\partial w_i}{\partial x_i} = 0 \tag{8.1a}$$

$$\rho w_j \frac{\partial w_i}{\partial x_j} + \frac{\partial p}{\partial x_i} = 0, \qquad i = 1, 2, 3 \tag{8.1b}$$

$$\rho w_j \frac{\partial h}{\partial x_j} - w_i \frac{\partial p}{\partial x_i} = 0 \tag{8.1c}$$

where Cartesian coordinates are used and Equation (8.1c) stems from Equation (2.39). This particular form of the energy equation is being used because a perfect gas is not yet assumed. We presume there are two distinct solutions of Equations (8.1), one of which is referred to as the baseline solution and is denoted with a subscript b. No special symbol denotes the second solution; it will be referred to as the transformed solution. In this section, either solution may be rotational or irrotational.

The two solutions are presumed to be connected by means of a transformation of the dependent variables given by

$$p = \lambda^{c_1} p_b, \qquad \rho = \lambda^{c_2} \rho_b, \qquad h = \lambda^{c_3} h_b, \qquad w_i = \lambda^{c_4} w_{bi} \tag{8.2}$$

where the c_i exponents are constants that are to be determined and λ is an arbitrary function. Upon substitution of Equations (8.2) into (8.1a), we obtain

$$\lambda^{c_4} w_{bj}\left(\lambda^{c_2} \frac{\partial \rho_b}{\partial x_j} + c_2 \lambda^{c_2 - 1} \rho_b \frac{\partial \lambda}{\partial x_j}\right) + \lambda^{c_2} \rho_b \left(\lambda^{c_4} \frac{\partial w_{bj}}{\partial x_j} + c_4 \lambda^{c_4 - 1} w_{bj} \frac{\partial \lambda}{\partial x_j}\right) = 0$$

or, after simplification,

$$\left(w_{bj} \frac{\partial \rho_b}{\partial x_j} + \rho_b \frac{\partial w_{bj}}{\partial x_j}\right) + \frac{c_2 + c_4}{\lambda} \rho_b \left(w_{bj} \frac{\partial \lambda}{\partial x_j}\right) = 0$$

The terms within the leftmost parentheses sum to zero, since these terms satisfy continuity for the baseline flow. Thus, for the transformed variables to represent a solution, we require that

$$\frac{D\lambda}{Dt} = w_j \frac{\partial \lambda}{\partial x_j} = \lambda^{c_4} \left(w_{bj} \frac{\partial \lambda}{\partial x_j}\right) = 0 \tag{8.3}$$

Hence, λ is a constant along streamlines in both flows, although we anticipate that it will vary from streamline to streamline in the transformed flow. The possibility of

$$c_2 + c_4 = 0$$

is ruled out by the analysis in the next paragraph.

The same procedure is applied to Equation (8.1b), with the result

$$\left(\lambda^{-c_1 + c_2 + 2c_4} \rho_b w_{bj} \frac{\partial w_{bi}}{\partial x_j} + \frac{\partial p_b}{\partial x_i}\right) + c_4 \lambda^{-c_1 + c_2 + 2c_4 - 1} \rho_b w_{bi} \left(w_{bj} \frac{\partial \lambda}{\partial x_j}\right) + \frac{c_1}{\lambda} p_b \frac{\partial \lambda}{\partial x_i} = 0$$

The terms within the first set of parentheses sum to zero, providing

$$-c_1 + c_2 + 2c_4 = 0 \tag{8.4a}$$

The term within the second set of parentheses is zero in view of Equation (8.3), while the last term requires that

$$c_1 = 0 \tag{8.4b}$$

With $c_1 = 0$, Equation (8.1c) becomes

$$\lambda^{c_2+c_3}\rho_b w_{bj}\frac{\partial h_b}{\partial x_j} - w_{bj}\frac{\partial p_b}{\partial x_j} + c_3\lambda^{c_2+c_3-1}\rho_b h_b\left(w_{bj}\frac{\partial \lambda}{\partial x_j}\right) = 0$$

which now utilizes Equation (8.3) and requires that

$$c_2 + c_3 = 0 \tag{8.4c}$$

Equations (8.4) summarize as

$$c_1 = 0, \quad c_2 = -2c_4, \quad c_3 = 2c_4$$

The exponent c_4 is arbitrary; a convenient (nonzero) choice is ½. We thus obtain

$$p = p_b, \quad \rho = \lambda^{-1}\rho_b, \quad h = \lambda h_b, \quad w_i = \lambda^{1/2}w_{bi} \tag{8.5}$$

Hence, any baseline solution can be used to generate a new, one-parameter family of solutions of the steady Euler equations providing λ satisfies Equation (8.3). The transformation provided by Equations (8.5) is referred to as the substitution principle.

We can show that this transformation satisfies the properties of a mathematical group (Carmichael, 1927), where λ is the continuous parameter of the group. For instance, the identity transformation is provided by $\lambda = 1$, while the inverse transformation is obtained by replacing λ with λ^{-1}. Clearly, this transformation differs from the hodograph transformation, which involves the independent variables and is usually intended to obtain linear equations. The two transformations, in fact, are quite compatible with each other, as will become apparent in Section 8.5.

Equations (8.3) and (8.5) leave the steady Euler equations invariant. It is a simple matter to show that shock wave jump conditions are also invariant. For example, since h_0 is constant across a shock, we have

$$h_{b2} + \frac{1}{2}w_{b2}^2 = h_{b1} + \frac{1}{2}w_{b1}^2$$

for the baseline flow, which transforms to the jump condition

$$h_2 + \frac{1}{2}w_2^2 = h_1 + \frac{1}{2}w_1^2$$

in the new flow.

Gas Model

We need to determine the thermodynamic gas model that is consistent with Equations (8.5). We note that the enthalpy can be considered as a function of the pressure and density. Hence, we can write

$$h(p,\rho) = \lambda h_b = \lambda h(p_b,\rho_b) = \lambda h(p,\lambda\rho)$$

or

$$h(p,\lambda\rho) = \lambda^{-1}h(p,\rho) \tag{8.6}$$

Thus, h is homogeneous (Emanuel, 1987) of degree −1 with respect to ρ. This means that it satisfies an Euler equation [which is a mathematical equation that has nothing to do with Equations (8.1)] of the form (Emanuel, 1987)

$$\rho\left(\frac{\partial h}{\partial \rho}\right) = -h$$

where the subscript indicates the state variable held fixed when performing the partial derivative. [Euler's equation is obtained by differentiating Equation (8.6) with respect to λ and then setting $\lambda = 1$.] This relation is integrated, to yield

$$\ln h = -\ln\rho + \ln\bar{P}(p)$$

or

$$h = \frac{\bar{P}(p)}{\rho} \tag{8.7}$$

where $\bar{P}$ is an arbitrary function of integration.

When the enthalpy is a function of the entropy and pressure, it is called a potential function (see Section 3.3). In this circumstance, we can write

$$h = h(s,p)$$
$$dh = \left(\frac{\partial h}{\partial s}\right)_p ds + \left(\frac{\partial h}{\partial p}\right)_s dp \tag{8.8}$$

By comparison with Equation (5.25a), which is a general thermodynamic relation, two state relations are obtained:

$$T = \left(\frac{\partial h}{\partial s}\right)_p, \quad \frac{1}{\rho} = \left(\frac{\partial h}{\partial p}\right)_s \tag{8.9a,b}$$

The density is eliminated from Equation (8.7), with the result

$$h = \bar{P}(p)\left(\frac{\partial h}{\partial p}\right)_s$$

This relation is also integrated, to yield

$$\int \frac{dp}{\bar{P}(p)} = \ln h - \ln S(s)$$

where $S(s)$ is another arbitrary function of integration. We thus obtain

$$h = S \exp\left(\int \frac{dp}{\bar{P}}\right) = P(p)S(s) \tag{8.10}$$

where $P(p)$ replaces the exponential factor. Observe that $h = \lambda h_b$ requires the homogeneity condition, which in turn requires that the potential function for h be a separable product, in contrast to the more general relation, Equation (8.8). The two state equations now have the form

$$T = PS', \quad \frac{1}{\rho} = P'S \tag{8.11a,b}$$

where a prime denotes differentiation with respect to the argument of the differentiated variable. Equation (8.11a) is a caloric state equation, while the other is a thermal state equation.

Equation (8.10) has a useful implication for the specific heat c_p, which is given by

$$c_p = \left(\frac{\partial h}{\partial T}\right)_p = \frac{\partial(h, p)}{\partial(T, p)} = \frac{\dfrac{\partial(h, p)}{\partial(s, p)}}{\dfrac{\partial(T, p)}{\partial(s, p)}} = \frac{\left(\dfrac{\partial h}{\partial s}\right)_p}{\left(\dfrac{\partial T}{\partial s}\right)_p} = \frac{S'}{S''} \tag{8.12}$$

where Equations (8.10) and (8.11a) are utilized, and c_p is a function only of entropy.

We next examine the possibility that a thermally perfect gas, $p = \rho RT$, can satisfy Equations (8.10) and (8.12). For a thermally perfect gas, we know that the specific enthalpy is a function only of T. Consequently, c_p is also a function only of T. By Equation (8.11a), however, T is a function of both pressure and entropy. On the other hand, Equation (8.12) requires c_p to be a function only of s. This dilemma is resolved by requiring c_p to be a constant; i.e., we also have a calorically perfect gas. In other words, the assumption of a thermally perfect gas in conjunction with Equation (8.10) implies a calorically perfect gas. [From Equations (8.11), we might presume $P = 1$, but this results in $\rho = \infty$.]

We must still show that a thermally and calorically perfect gas is consistent with Equation (8.10). For this type of gas, we know that

$$\frac{s - s_r}{R} = \ln\left[\left(\frac{T}{T_r}\right)^{\gamma/(\gamma-1)} \frac{p_r}{p}\right] \tag{8.13}$$

and

$$h = c_p T$$

where the r subscript denotes a reference condition. By eliminating T, we obtain

$$\frac{h}{h_r} = \left(\frac{p}{p_r}\right)^{(\gamma-1)/\gamma} \exp\left[\left(\frac{\gamma-1}{\gamma}\right) \frac{s-s_r}{R}\right] \tag{8.14}$$

which indeed is consistent with Equation (8.10).

Equation (8.10) and, consequently, Equations (8.11) are required by the substitution principle when applied to the Euler equations and the shock jump conditions. However, virtually all well-known and not-so-well-known thermal equations of state, such as a virial equation, van der Waals equation, or the Clausius-II equation, do not fit the form of Equation (8.11b); the one notable exception is the equation of a thermally perfect gas. For all practical purposes, the substitution principle for a compressible flow is therefore restricted to this gas. In view of this and Equations (8.5), we see that the gas is also calorically perfect and has constant values for its specific heats.

Discussion

The speed of sound can be evaluated, with the aid of Equation (8.11b), as

$$a^2 = \left(\frac{\partial p}{\partial \rho}\right)_s = -\frac{1}{\rho^2 S P''} = -\frac{P'}{\rho P''} = -\lambda\frac{P'(p_b)}{\rho_b P''(p_b)} = \lambda a_b^2$$

As a consequence, the Mach number transforms as

$$M^2 = \frac{w^2}{a^2} = \frac{\lambda w_b^2}{\lambda a_b^2} = M_b^2 \tag{8.15}$$

This result is also obtained by utilizing the perfect gas equation, $a^2 = (\gamma p/\rho)$, for the speed of sound. Both the pressure and Mach number therefore are invariant under the transformation. Additionally, the stagnation pressure, given by

$$p_o = p\left(1 + \frac{\gamma - 1}{2}M^2\right)^{\gamma/(\gamma-1)}$$

is invariant. Hence, if p_{ob} is a constant in the baseline flow, then p_o is the same constant in the transformed flow.

Since λ is a constant along streamlines of a steady Euler flow, it may be a function of s, h_o, and any other stagnation quantity, such as ρ_o. If the baseline flow is homentropic, then p_o is not a suitable choice for λ, since it is the same constant in both flows. Aside from s, h_o, ρ_o, …, λ can also be a function of a stream function when the flow is two dimensional or axisymmetric.

For our purposes, the most convenient choice for λ is

$$\lambda = \frac{h_o}{h_{ob}} \tag{8.16}$$

where h_{ob} and h_o are constant along streamlines in the baseline and transformed flows, respectively. This form for λ is chosen because h_o is constant across shock waves, λ is positive and nondimensional, and it can be used with three-dimensional flows.

With λ given by the above equation, other quantities that are constant along streamlines can be found. Although T is not constant along streamlines, we have

$$T = \frac{p}{R\rho} = \lambda\frac{p_b}{R\rho_b} = \lambda T_b$$

With Equations (8.13), (8.16), and the above, we obtain for the entropy, which is constant along streamlines,

$$\frac{s - s_r}{R} = \ln\left[\lambda^{\gamma/(\gamma-1)}\left(\frac{T_b}{T_r}\right)^{\gamma/(\gamma-1)}\frac{p_r}{p_b}\right] = \frac{\gamma}{\gamma-1}\ln\lambda + \frac{s_b - s_r}{R}$$

or

$$s = s_b + \frac{\gamma R}{\gamma - 1}\ln\lambda \tag{8.17}$$

Even if the baseline entropy is a constant, the value of the transformed entropy changes from streamline to streamline. The transformation for a stream function, ψ, assuming a two-dimensional or axisymmetric flow, is obtained as follows:

$$\psi_x = -\rho v y^\sigma, \qquad \psi_y = \rho u y^\sigma \tag{8.18a,b}$$

$$\begin{aligned} d\psi &= \psi_x dx + \psi_y dy = -\rho v y^\sigma dx + \rho u y^\sigma dy \\ &= \lambda^{-1/2}(-\rho_b v_b y_b^\sigma dx_b + \rho_b u_b y_b^\sigma dy_b) = \lambda^{-1/2} d\psi_b \end{aligned} \tag{8.19}$$

where

$$x = x_b, \qquad y = y_b$$

as discussed in the next paragraph.

Since streamlines are determined by Equation (1.21b), we see that they are invariant under the transformation. Because Mach lines are at the Mach angle μ, with respect to the streamlines, they too are invariant. The substitution principle preserves the geometry of the flow field, and wall locations, shock shapes, and slipstream locations are unaltered. For instance, if the baseline flow involves Mach waves that (do not) coalesce to form a shock wave, then the new flow field will similarly involve Mach waves that (do not) coalesce. Since the pressure and wall geometry are unaltered, so are all inviscid forces and moments. As we will see in later sections, the principal difference between the two flows is in their rotationality.

Since the inviscid forces and moments are invariant, the substitution principle would appear to be of no utility. This is not the case, because the skin friction and wall heat transfer are altered by nonzero values for the vorticity and the gradient of the stagnation enthalpy, both evaluated at the wall. Chapter 23 discusses the reasons for this and how to determine the heat transfer and skin friction.

The physical notion behind the substitution principle is that there is a degree of freedom between adjacent streamlines in a steady Euler flow. This freedom is evident in the inviscid flow about a cylinder (Section 1.2), or any body shape, and is especially evident in Problem 5.22. Thus, a steady Euler flow is not unique unless upstream (or downstream) conditions are prescribed. This freedom, e.g., is not present in a viscous flow.

Vorticity Transformation

Because of its central role, it is appropriate to derive several transformation equations for the vorticity. We thus have

$$\vec{\omega} = \nabla \times \vec{w} = \nabla \times (\lambda^{1/2}\vec{w}_b) = (\nabla\lambda^{1/2}) \times \vec{w}_b + \lambda^{1/2}\nabla \times \vec{w}_b$$
$$= \lambda^{1/2}\vec{\omega}_b + \frac{1}{2\lambda^{1/2}}\nabla\lambda \times \vec{w}_b \tag{8.20}$$

where $\nabla\lambda \times \vec{w}_b$ represents a vector that is perpendicular to the streamlines. The component of $\vec{\omega}$ along the streamlines is obtained by multiplying Equation (8.20) with $\vec{w}$, with the result

$$\vec{w}_b \cdot \vec{\omega} = \lambda^{1/2}\vec{w}_b \cdot \vec{\omega}_b$$

Thus, the component of $\vec{\omega}$ along the streamlines equals that of $\vec{\omega}_b$ when $\vec{\omega}_b$ is multiplied by $\lambda^{1/2}$.

A particularly elegant formula can be obtained when the flow is two-dimensional or axisymmetric. We start with Equation (7.69)

$$\omega = \frac{\partial v}{\partial x} - \frac{\partial u}{\partial y}$$

which becomes

$$\omega = \frac{\partial(\lambda^{1/2}v_b)}{\partial x} - \frac{\partial(\lambda^{1/2}u_b)}{\partial y} = \lambda^{1/2}\left(\frac{\partial v_b}{\partial x} - \frac{\partial u_b}{\partial y}\right) + \frac{1}{2\lambda^{1/2}}\left(v_b\frac{\partial\lambda}{\partial x} - u_b\frac{\partial\lambda}{\partial y}\right)$$
$$= \lambda^{1/2}\omega_b + \frac{1}{2\lambda^{1/2}}\left(v_b\frac{\partial\lambda}{\partial x} - u_b\frac{\partial\lambda}{\partial y}\right)$$

With the aid of Equations (8.18), we have

$$\frac{\partial\lambda}{\partial x} = \frac{\partial\lambda}{\partial\psi}\frac{\partial\psi}{\partial x} = -\rho v y^\sigma\frac{\partial\lambda}{\partial\psi} = -\frac{1}{\lambda^{1/2}}\rho_b v_b y^\sigma\frac{\partial\lambda}{\partial\psi}$$
$$\frac{\partial\lambda}{\partial y} = \frac{\partial\lambda}{\partial\psi}\frac{\partial\psi}{\partial y} = -\rho u y^\sigma\frac{\partial\lambda}{\partial\psi} = \frac{1}{\lambda^{1/2}}\rho_b u_b y^\sigma\frac{\partial\lambda}{\partial\psi}$$

Consequently, we obtain

$$\omega = \lambda^{1/2}\omega_b + \frac{1}{2\lambda}\rho_b(-v_b^2 - u_b^2)y^\sigma\frac{\partial\lambda}{\partial\psi} = \lambda^{1/2}\omega_b - \frac{1}{2\lambda}\rho_b w_b^2 y^\sigma\frac{\partial\lambda}{\partial\psi}$$
$$= \lambda^{1/2}\omega_b - \frac{\lambda}{2}pM^2\frac{y^\sigma}{\lambda}\frac{\partial\lambda}{\partial\psi} \tag{8.21}$$

It is worth recalling that pM^2y^σ is invariant under the substitution principle. Hence, it is a simple matter to evaluate the vorticity of the transformed flow for a known two-dimensional or axisymmetric baseline flow. This vorticity may have a baseline contribution plus a contribution associated with the transverse gradient of λ.

8.3 PARALLEL FLOW

A parallel flow is one in which the x_1 coordinate can be aligned with the velocity. This type of flow may or may not be uniform. In any case, we have

$$w_2 = w_3 = 0$$

and Equations (8.1) become

$$w_1 \frac{\partial \rho}{\partial x_1} + \rho \frac{\partial w_1}{\partial x_1} = 0$$

$$\rho w_1 \frac{\partial w_1}{\partial x_1} + \frac{\partial p}{\partial x_1} = 0, \qquad \frac{\partial p}{\partial x_2} = \frac{\partial p}{\partial x_3} = 0$$

$$\rho \frac{\partial h}{\partial x_1} - \frac{\partial p}{\partial x_1} = 0$$

As shown by Problem 5.4, the general solution of these equations can be written as

$$\rho w_1 = f_1(x_2, x_3)$$

$$p = p(x_1)$$

$$p + \rho w_1^2 = f_2(x_2, x_3)$$

where f_1, f_2, and $p(x_1)$ are arbitrary functions of integration. (Integration of the energy equation yields a result that is not needed in the analysis.)

Before proceeding, we discuss why we are interested in a parallel flow. This type of flow represents the *uniform* upstream flow for most of the well-known irrotational solutions of supersonic gas dynamics. These include Prandtl–Meyer flow, flow with an oblique shock wave, and flow about a cone (Taylor–Maccoll flow). Thus, for a vortical flow, the counterpart of a baseline uniform flow is a *nonuniform* parallel flow. In the cases to be discussed, the uniform baseline flow field is either two dimensional or axisymmetric. Hence, this and the next section assume a parallel flow that is adjacent to a planar wall or that has an axial symmetry axis.

We therefore assume a baseline parallel flow that is uniform; i.e.,

$$p_b = p_\infty, \qquad \rho_b = \rho_\infty, \qquad M_b = M_\infty, \qquad u_b = w_\infty, \qquad v_b = 0$$

where the infinity subscript denotes a constant freestream parameter. For λ, we use Equation (8.16), where

$$h_{ob} = \frac{\gamma}{\gamma - 1}\frac{p_\infty}{\rho_\infty} + \frac{1}{2} w_\infty^2 = \frac{\gamma}{\gamma - 1}\frac{p_\infty}{\rho_\infty} X_\infty \tag{8.22a}$$

$$h_o = h_o(y) = \frac{\gamma}{\gamma - 1}\frac{p_\infty}{\rho(y)} X_\infty \tag{8.22b}$$

and X is defined by Equation (5.42). Because of their invariance, p_∞ and M_∞ replace p and M in h_o. Once $\lambda(y)$ is prescribed, $h_o(y)$ and therefore $\rho(y)$ are known. This density result, of course, agrees with that provided by Equations (8.5), which can be used to determine $T(y)$ and $w(y)$. Consequently, $\lambda(y)$ provides a one-parameter family of two-dimensional or axisymmetric parallel flow rotational solutions of the Euler equations.

Stream Function, Stagnation Enthalpy Relation

While we have considered h_o as a function of y, it is essential for the subsequent applications to evaluate h_o in terms of ψ, or vice versa. The reason for this is that both h_o and ψ are constants along streamlines. A relationship such as

$$h_o = h_o(\psi) \qquad \text{or} \qquad \psi = \psi(h_o) \tag{8.23}$$

then holds throughout the flow field, including across shock waves. We thus evaluate one of the equations in (8.23) in the upstream parallel flow but continue to use this equation throughout the nonparallel, downstream flow field.

We start with Equations (8.18), with the parallel flow result

$$\psi_x = 0$$

$$\psi_y = \rho w y^\sigma = \lambda^{-1/2}\rho_b w_b y^\sigma = \left(\frac{h_{o\infty}}{h_o}\right)^{1/2} \rho_\infty w_\infty y^\sigma$$

where

$$h_{o\infty}^{1/2}\rho_\infty w_\infty = \left[\frac{\gamma}{\gamma-1}\frac{p_\infty}{\rho_\infty}\left(1+\frac{\gamma-1}{2}M_\infty^2\right)\right]^{1/2}\rho_\infty\left(\gamma\frac{p_\infty}{\rho_\infty}\right)^{1/2} M_\infty = \gamma p_\infty M_\infty\left(\frac{X_\infty}{\gamma-1}\right)^{1/2}$$

We thus have

$$\frac{d\psi}{dy} = \gamma p_\infty M_\infty\left(\frac{X_\infty}{\gamma-1}\right)^{1/2}\frac{y^\sigma}{h_o^{1/2}} \tag{8.24a}$$

or

$$\psi = \gamma p_\infty M_\infty\left(\frac{X_\infty}{\gamma-1}\right)^{1/2}\int_0^y \frac{y^\sigma dy}{h_o^{1/2}(y)} \tag{8.24b}$$

where y is measured from the planar wall or the symmetry axis, where ψ is zero. While these relations require a parallel flow, once the integration is performed, y can be eliminated to yield one of the equations in (8.23), which then holds throughout the flow field.

Example

As an example, suppose we have a parallel flow with

$$h_o = h_o(0)(1+y)^2 \tag{8.25}$$

where $h_o(0)$ is the value of h_o when $y = 0$. Equation (8.24b) then yields

$$\psi = \gamma p_\infty M_\infty \left(\frac{X_\infty}{\gamma - 1} \frac{1}{h_o(0)} \right)^{1/2} \ln(1 + y), \quad \sigma = 0$$

$$= \gamma p_\infty M_\infty \left(\frac{X_\infty}{\gamma - 1} \frac{1}{h_o(0)} \right)^{1/2} [y - \ln(1 + y)], \quad \sigma = 1$$

We eliminate y by using Equation (8.25), to obtain the relations

$$\frac{h_o(\psi)}{h_o(0)} = \exp\left\{ \frac{2}{\gamma p_\infty M_\infty} \left[\frac{(\gamma - 1) h_o(0)}{X_\infty} \right]^{1/2} \psi \right\}, \quad \sigma = 0 \tag{8.26a}$$

$$\psi(h_o) = \gamma p_\infty M_\infty \left(\frac{X_\infty}{\gamma - 1} \frac{1}{h_o(0)} \right)^{1/2} \left[\left(\frac{h_o}{h_o(0)} \right)^{1/2} - 1 - \frac{1}{2} \ln\left(\frac{h_o}{h_o(0)} \right) \right], \quad \sigma = 1 \tag{8.26b}$$

that hold throughout the flow field. Since the baseline flow is irrotational, Equation (8.21) yields

$$\omega = -\frac{\gamma}{2} p_\infty M_\infty^2 \frac{y^\sigma}{h_o} \frac{dh_o}{d\psi} \tag{8.27}$$

With the aid of Equations (8.26), this becomes

$$\omega = -\left[\frac{(\gamma - 1) M_\infty^2 h_o(0)}{X_\infty} \right]^{1/2}$$

for the vorticity of the transformed, two-dimensional or axisymmetric parallel flow. In this case, ω is a constant. Generally, the stagnation enthalpy distribution is arbitrary since h_o can increase, decrease, or vary in a nonmonotonic manner with y. Consequently, ω can be positive or negative, depending on the sign of $dh_o/d\psi$.

8.4 PRANDTL–MEYER FLOW

For simplicity, the discussion of Prandtl–Meyer flow is restricted to a sharp expansive wall turn, as shown in Figure 8.1. Nevertheless, the theory applies directly to smoothly varying expansive or compressive wall turns. For the baseline case, the upstream flow is uniform with a supersonic Mach number M_∞. The leading edge of the expansion is a straight, left-running Mach line that has an acute angle μ_∞, which is the Mach angle, relative to the upstream wall. As shown in Figure 8.1, the expansion terminates at its trailing edge, which is a left-running characteristic, at the Mach angle μ_2, relative to the downstream wall. A subscript 2 denotes the uniform flow downstream of the trailing edge. The Mach number M_2 is given by the gas dynamic relation

$$\nu(M_2) = \nu(M_\infty) + \theta_w = \nu_\infty + \theta_w \tag{8.28}$$

where θ_w is the positive wall turn angle and ν is the Prandtl–Meyer function

$$\nu(M) = \left(\frac{\gamma + 1}{\gamma - 1} \right)^{1/2} \tan^{-1}\left[\left(\frac{\gamma - 1}{\gamma + 1} \right) (M^2 - 1) \right]^{1/2} - \tan^{-1}(M^2 - 1)^{1/2} \tag{8.29}$$

With homentropic relations, conditions downstream of the expansion are readily established.

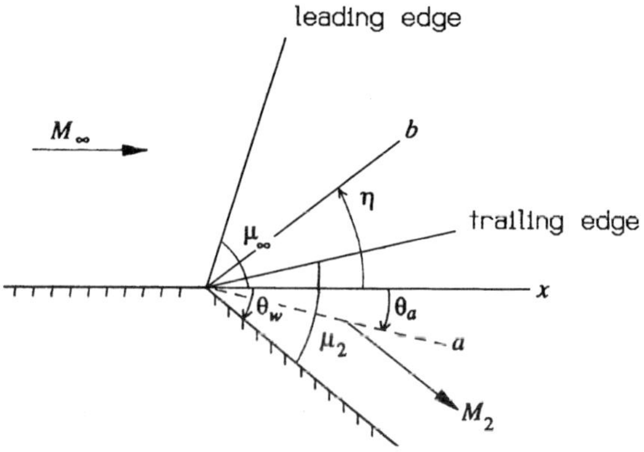

FIGURE 8.1 Nomenclature for a Prandtl–Meyer flow.

Irrotational Prandtl–Meyer Flow

For the rotational analysis, we will first need the baseline flow solution inside the expansion. Since the form obtained for the solution is not provided by other compressible flow texts, we proceed to derive it. This irrotational flow has a similarity solution that can be found in terms of an angle η (see Figure 8.1). Thus, all flow properties are constant along ray b, which is an arbitrary Mach line inside the expansion.

As θ_w increases, the expansion fan broadens. This broadening is achieved by the trailing edge rotating in a clockwise direction. The previously existing expansion, before θ_w increased, is unaltered by the further broadening of the fan. We find the solution on ray b by considering an imaginary wall at an angle θ_a, where $\theta_w > \theta_a$. For the flow with wall a, ray b is the trailing edge Mach line. Hence, we have

$$\nu(M_b) = \nu_\infty + \theta_a$$

and

$$\sin \mu_b = \frac{1}{M_b}$$

However, μ_b is also given by

$$\mu_b = \theta_b + \eta$$

with the result

$$\nu(M_b) = \nu_\infty + \mu_b - \eta = \nu_\infty + \sin^{-1}\frac{1}{M_b} - \eta$$

For the actual flow, where the wall is at an angle θ_w, the Mach number on ray b is provided by the same relation

$$\nu(M) = \nu_\infty + \sin^{-1}\frac{1}{M} - \eta$$

where η is the angle the ray has with respect to the x-axis.

This is an implicit equation for M; a useful explicit version can be found by noting that

$$\mu = \sin^{-1}\frac{1}{M} = \cot^{-1}(M^2 - 1)^{1/2}$$

and the trigonometric identity

$$\tan^{-1}\phi + \cot^{-1}\phi = \frac{\pi}{2}$$

which holds for any angle ϕ. We thereby obtain

$$\begin{aligned} \nu(M) &= \left(\frac{\gamma+1}{\gamma-1}\right)^{1/2} \tan^{-1}\left[\left(\frac{\gamma-1}{\gamma+1}\right)(M^2 - 1)\right]^{1/2} - \tan^{-1}(M^2 - 1)^{1/2} \\ &= \nu_\infty + \cot^{-1}(M^2 - 1)^{1/2} - \eta \end{aligned}$$

or

$$\tan^{-1}\left[\frac{\gamma-1}{\gamma+1}(M^2 - 1)\right]^{1/2} = \left(\frac{\gamma-1}{\gamma+1}\right)^{1/2}\left(\nu_\infty + \frac{\pi}{2} - \eta\right) \tag{8.30}$$

It is convenient to replace η with a scaled angular coordinate

$$z = \left(\frac{\gamma-1}{\gamma+1}\right)^{1/2}\left(\nu_\infty + \frac{\pi}{2} - \eta\right)$$

On the leading edge of the expansion we have $z = 0$ when $M_\infty = 1$, whereas $z = \pi/2$ on the trailing edge when the trailing edge Mach number is infinite, regardless of the value of M_∞. While η may be positive or negative, z is confined to the 0 to $\pi/2$ range.

With z, Equation (8.30) becomes

$$M^2 = 1 + \frac{\gamma+1}{\gamma-1}\tan^2 z \tag{8.31}$$

which is the sought-after relation for the Mach number inside the expansion. Other variables are now readily obtained, since the baseline flow is homentropic. For this, we write

$$X = 1 + \frac{\gamma-1}{2}M^2 = \frac{\gamma+1}{2}(1 + \tan^2 z) = \frac{\gamma+1}{2}\sec^2 z$$

Consequently, the pressure inside the expansion is

$$p = p_o X^{-\gamma/(\gamma-1)} = p_o\left(\frac{\gamma+1}{2}\sec^2 z\right)^{-\gamma/(\gamma-1)}$$

(A subscript b is not used on p, p_o, or M, since these are invariant under the transformation.) On the other hand, p_o can be evaluated in the freestream as

$$p_o = p_\infty X_\infty^{\gamma/(\gamma-1)}$$

so that

$$p = p_\infty\left(\frac{2}{\gamma+1}X_\infty \cos^2 z\right)^{\gamma/(\gamma-1)} \tag{8.32a}$$

Other baseline variables of interest are

$$\rho_b = \rho_{b\infty}\left(\frac{2}{\gamma+1}X_\infty \cos^2 z\right)^{1/(\gamma-1)} \tag{8.32b}$$

$$\theta = \mu - \eta = \cot^{-1}\left[\left(\frac{\gamma+1}{\gamma-1}\right)^{1/2} \tan z\right] - \eta \tag{8.32c}$$

$$u_b = w_b \cos\theta = \left(\gamma \frac{p}{\rho_b}\right) M \cos\theta$$

$$= \left(\frac{2}{\gamma+1}X_\infty\right)^{1/2} a_{b\infty} M \cos\theta \cos z \tag{8.32d}$$

$$v_b = -\left(\frac{2}{\gamma+1}X_\infty\right)^{1/2} a_{b\infty} M \sin\theta \cos z \tag{8.32e}$$

Observe that a downward-pointing streamline has a positive (clockwise) angle θ relative to the x-axis; this definition is consistent with θ_w being positive. In view of Equation (8.31), all variables given by the above equations only depend on η.

Rotational Prandtl–Meyer Flow

With the use of Equations (8.5) and (8.16), Equations (8.32) are easily transformed, noting that p, M, θ, η, and z are invariant. Thus, ρ is given by

$$\rho = \lambda^{-1}\rho_b = \frac{h_{ob}}{h_o(\psi)}\rho_b = \frac{\gamma}{\gamma-1}\frac{p_\infty X_\infty}{h_o}\left(\frac{2}{\gamma+1}X_\infty \cos^2 z\right)^{1/(\gamma-1)}$$

or

$$\frac{\rho}{\rho_{b\infty}} = \frac{1}{\gamma-1}\left(\frac{2}{\gamma+1}\right)^{1/(\gamma-1)} X_\infty^{\gamma/(\gamma-1)}(\cos^{2/(\gamma-1)} z)\frac{a_{b\infty}^2}{h_o(\psi)} \tag{8.33a}$$

where h_{ob} and ρ_b are replaced by Equations (8.22a) and (8.32b), respectively. We similarly obtain for u and v

$$u = \left[2\frac{\gamma-1}{\gamma+1}h_o(\psi)\right]^{1/2} M\cos\theta\cos z \tag{8.33b}$$

$$v = -\left[2\frac{\gamma-1}{\gamma+1}h_o(\psi)\right]^{1/2} M\sin\theta\cos z \tag{8.33c}$$

As discussed in the preceding section, $h_o(\psi)$ is established in the upstream parallel flow. In contrast to the previous irrotational solution, this solution is not a similar one, since ρ, u, and v depend on both η and ψ. Note that by replacing h_o with h_{ob}, we recover the irrotational solution.

Vorticity

We denote the variable vorticity of the upstream rotational flow as ω_∞; it is given by Equation (8.27). Since the vorticity of the baseline flow is zero, the vorticity inside the expansion is provided by Equation (8.21) as

$$\omega = -\frac{\gamma}{2}pM^2\frac{1}{h_o}\frac{dh_o}{d\psi}$$

which becomes

$$\omega = -\frac{\gamma}{2}p_\infty\left(\frac{2}{\gamma+1}X_\infty\cos^2 z\right)^{\gamma/(\gamma-1)}\left(1+\frac{\gamma+1}{\gamma-1}\tan^2 z\right)\frac{1}{h_o}\frac{dh_o}{d\psi} \tag{8.34a}$$

with the aid of Equations (8.31) and (8.32a). A normalized vorticity is thus given by

$$\frac{\omega}{\omega_\infty} = \frac{1}{M_\infty^2}\left(\frac{2}{\gamma+1}X_\infty\cos^2 z\right)^{\gamma/(\gamma-1)}\left(1+\frac{\gamma+1}{\gamma-1}\tan^2 z\right) \tag{8.34b}$$

and is displayed in Figure 8.2 for $\gamma = 7/5$. This form is especially convenient, since it does not depend on $h_o(\psi)$, and basically represents the vorticity along a streamline inside the expansion. The leftmost point of each curve corresponds to the leading edge of the expansion, where $\omega = \omega_\infty$. When the trailing-edge Mach number is infinite, we have $\omega = 0$ at $z = 90°$. The curves have a maximum value that is greater than unity when $M_\infty < 2^{1/2}$, which stems from a multiplicative factor

$$\cos^2 z - \frac{\gamma+1}{2\gamma} \tag{8.35}$$

that appears in the equation for $D\omega/Dt$ (see Problem 8.1). In the narrow region between the curves labeled $M_\infty = 1$ and $2^{1/2}$, ω/ω_∞ is double-valued. One value corresponds to a solution where $1 \le M_\infty \le 2^{1/2}$, the other to a solution where M_∞ is slightly greater than $2^{1/2}$.

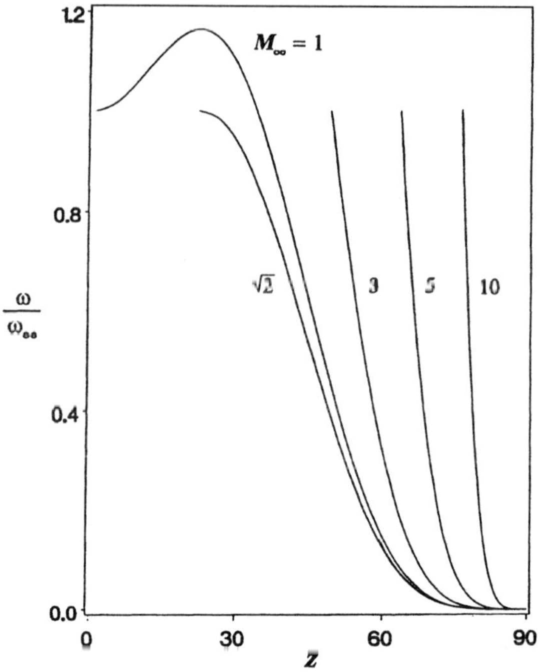

FIGURE 8.2 Normalized vorticity distribution for a rotational Prandtl–Meyer flow, $\gamma = 7/5$.

The vorticity of a fluid particle, as it traverses the expansion, may increase or decrease and may exhibit a minimum or a maximum value. These trends depend on the values of M_∞, the trailing-edge Mach number, factor (8.35), and the sign of $dh_o/d\psi$.

Other Solutions

Aside from a parallel flow and a Prandtl–Meyer flow, Rodriguez and Emanuel (1988) also consider two other well-known gas dynamic flows. One of these is the Taylor–Maccoll flow about a cone, which is further discussed in Problems 8.2 and 8.3. The other is a flow with a planar, oblique shock wave. When the upstream parallel flow is rotational, the vorticity goes through a jump at the shock, and the downstream flow is again a parallel, rotational flow. The ratio ω_2/ω_∞ across a planar shock is also independent of h_o; this ratio is exhibited in Figure 8.3 for $\gamma = 7/5$. The leftmost point on each of the curves corresponds to the shock becoming a Mach wave, while at the rightmost point, where the shock angle β is 90°, the shock is normal to the flow. For relatively small shock angles, the freestream vorticity is amplified by the shock. At large M_∞ values this amplification is quite significant. At large shock angles, however, the vorticity is attenuated by the shock.

Suppose a parallel flow upstream of a curved shock is rotational. As we know, a curved shock generates vorticity, and the two effects act in combination. For a vorticity jump condition, we can utilize Problem 6.13.

It is natural to expect that a transverse gradient in the vorticity upstream of a shock would alter the shape of the shock. This is not the case, however, if the upstream flow remains parallel. When there is a change in the shape of the shock, the upstream flow is no longer parallel. To the author's knowledge, experimental verification of the supersonic substitution principle has yet to be performed. It should not be too difficult to demonstrate the principle with a Prandtl–Meyer or planar oblique shock experiment.

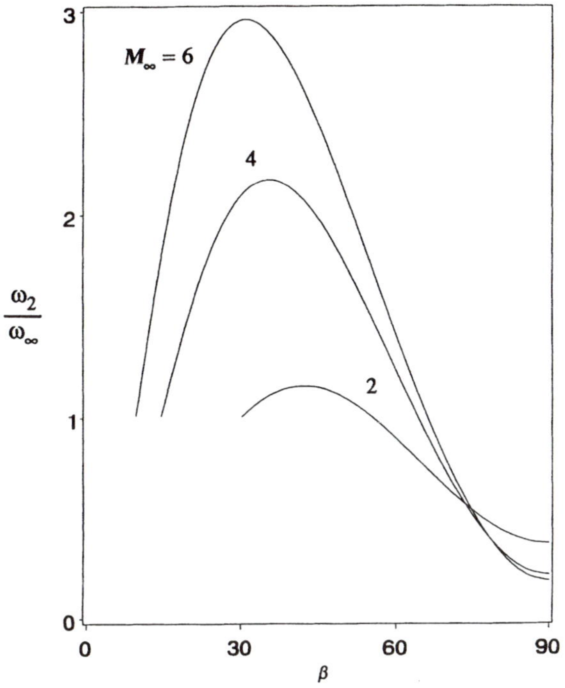

FIGURE 8.3 Normalized vorticity distribution behind a planar shock, $\gamma = 7/5$.

8.5 ROTATIONAL SOLUTIONS IN THE HODOGRAPH PLANE

Suppose we have a solution, such as Equation (7.29) or Equation (7.49), of the linear hodograph equations. As indicated in Section 7.6, an irrotational flow is a necessary condition for the linear hodograph equations. We now transform this irrotational baseline solution by means of the substitution principle

$$x = x_b, \quad y = y_b, \quad a = \lambda^{1/2} a_b, \quad u = \lambda^{1/2} u_b, \quad v = \lambda^{1/2} v_b \tag{8.36}$$

as applied in the hodograph plane. As before, the parameter λ must satisfy

$$\frac{D\lambda}{Dt} = 0$$

or, in the hodograph plane,

$$A\lambda_u - B\lambda_v = 0 \tag{8.37}$$

where A and B are provided by Equations (7.62). The transformed result is an exact solution of Equations (7.65), (7.66), and (7.68), which are the nonlinear, rotational, hodograph equations for a two-dimensional or axisymmetric flow. If the baseline flow is in terms of w, τ, and θ, then Equations (8.36) are supplemented with

$$w = \lambda^{1/2} w_b, \quad \tau = \tau_b, \quad \theta = \theta_b \tag{8.38}$$

Irrotational Source Flow

To illustrate the method, let us consider two-dimensional supersonic nozzle flow between planar, diverging walls (Rodriguez, 1988). This is a cylindrical source flow in which fluid properties vary only with the distance r from the (apparent) source. This type of flow has a number of applications, including use in the design of minimum-length nozzles with a curved sonic line (see Section 7.4).

The irrotational solution for a cylindrical source flow can be based on the area, Mach number relation of gas dynamics

$$\frac{r}{r^*} = \frac{1}{M}\left[\left(\frac{2}{\gamma+1}\right)X\right]^{(\gamma+1)/[2(\gamma-1)]} \tag{8.39}$$

where r^* is the radius of the sonic line. This relation is to be used for the baseline flow in the hodograph plane, with x,y as the dependent variables and u,v as the independent variables. We begin by setting

$$x = r\cos\theta = \frac{r^*}{M}\left[\left(\frac{2}{\gamma+1}\right)X\right]^{(\gamma+1)/[2(\gamma-1)]}\cos\theta \tag{8.40a}$$

$$y = r\sin\theta = \frac{r^*}{M}\left[\left(\frac{2}{\gamma+1}\right)X\right]^{(\gamma+1)/[2(\gamma-1)]}\sin\theta \tag{8.40b}$$

We obtain from Equation (7.71)

$$a = (\gamma-1)^{1/2}\left(h_o - \frac{1}{2}w^2\right)^{1/2} \tag{8.41}$$

so that

$$M = \frac{w}{a} = \frac{1}{(\gamma-1)^{1/2}}\frac{w}{\left(h_o - \frac{1}{2}w^2\right)^{1/2}}$$

and

$$X = \frac{h_o}{h_o - \frac{1}{2}w^2}$$

With these relations and Equations (7.1), Equations (8.40) become

$$x_b = Ch_{ob}^{(\gamma+1)/[2(\gamma-1)]}\left[h_{ob} - \frac{1}{2}(u_b^2 + v_b^2)\right]^{-1/(\gamma-1)}\frac{u_b}{u_b^2 + v_b^2} \tag{8.42a}$$

$$y_b = Ch_{ob}^{(\gamma+1)/[2(\gamma-1)]}\left[h_{ob} - \frac{1}{2}(u_b^2 + v_b^2)\right]^{-1/(\gamma-1)}\frac{v_b}{u_b^2 + v_b^2} \tag{8.42b}$$

for the baseline flow, where

$$C = (\gamma-1)^{1/2}\left(\frac{2}{\gamma+1}\right)^{(\gamma+1)/[2(\gamma-1)]} r^*$$

is a constant. In these hodograph equations, there is no dependence on the speed of sound, since h_{ob} is a constant.

Rotational Source Flow

With Equations (8.16) and (8.36), the rotational form of Equations (8.42) is obtained. The rotational equations, however, are identical to Equations (8.42) but with the subscript b deleted. The parameter λ does not appear because x and y are invariant under the substitution principle. Nevertheless, in the rotational solution, h_o is not a constant but is a function of the stream function ψ. In a rotational source flow, ψ is proportional to θ, which equals $\tan^{-1}(v/u)$; hence, h_o is an arbitrary function of v/u. As seen from Section 7.6, the rotational solution may involve the speed of sound, which is given by Equation (8.41), with $u^2 + v^2$ replacing w^2 and $h_o = h_o(v/u)$.

To evaluate the Jacobian of the transformation and the vorticity, we will need the derivatives of the coordinates. From Equations (8.42), with the subscript b deleted, we have

$$x_v = x\left\{\frac{\gamma+1}{2(\gamma-1)}\frac{h_{ov}}{h_o} - \frac{h_{ov}-v}{(\gamma-1)\left[h_o - \frac{1}{2}(u^2+v^2)\right]} - \frac{2v}{u^2+v^2}\right\}$$

$$y_v = y\left\{\frac{\gamma+1}{2(\gamma-1)}\frac{h_{ou}}{h_o} - \frac{h_{ov}-u}{(\gamma-1)\left[h_o - \frac{1}{2}(u^2+v^2)\right]} - \frac{2u}{u^2+v^2}\right\}$$

with similar relations for x_u and y_v. Equation (7.60) then yields for the Jacobian (Rodriguez, 1988)

$$J = -\frac{xy}{uv}\,\frac{h_o - \frac{\gamma+1}{2(\gamma-1)}(u^2+v^2)}{h_o - \frac{1}{2}(u^2+v^2)} \tag{8.43}$$

Equation (7.69) provides the vorticity

$$\omega = \frac{C h_o^{(3-\gamma)/[2(\gamma-1)]}}{2(u^2+v^2)}\,\frac{h_o - \frac{\gamma+1}{2(\gamma-1)}(u^2+v^2)}{\left[h_o - \frac{1}{2}(u^2+v^2)\right]^{\gamma/(\gamma-1)}}(uh_{ov} - vh_{ou}) \tag{8.44}$$

where the substantial derivative of h_o

$$uh_{ou} + vh_{ov} = 0$$

is used to simplify this result. The vorticity can also be derived from Equation (8.21).

Observe that when h_o is a constant, ω is zero, as expected. From Equation (8.43), we see that $J \neq 0$ for either the transformed or baseline flows, except when x or y is zero or

$$h_o - \frac{\gamma+1}{2(\gamma-1)}(u^2+v^2)$$

which occurs when $M = 1$. In addition, J is infinite when u or v is zero or M is infinite. On the sonic line one can show that the acceleration is infinite (see Problem 8.5); hence, it is a limit line. The other zeros or infinities of J cause no difficulties.

Although some variables, such as p, ρ, and M, depend only on r, the irrotational solution is not a similarity solution in the hodograph plane, since x and y depend on both u and v. Consequently, the irrotational source-flow solution is not a missing solution in the hodograph plane. Similarly, the rotational solution is also not a missing solution.

REFERENCES

Ames Research Staff, *Equations, Tables, and Charts for Compressible Flow,* NACA Report 1135 (1953).

Carmichael, R.D., "Transformations Leaving Invariant Certain Partial Differential Equations of Physics," *Am. J. Math.* 49, 97 (1927).

Emanuel, G., *Advanced Classical Thermodynamics*, AIAA Education Series, Washington, D.C., 1987.

Munk, M. and Prim, R., "On the Multiplicity of Steady Gas Flows Having the Same Streamline Pattern," *Proc. Natl. Acad. Sci. U.S.A.* 33, 127 (1947).

Prim, R.C., III, "Steady Rotational Flow of Ideal Gases," *J. Rat. Mech. Anal.* 1, 425 (1952).

Rodriguez Azara, J.L., "Substitution Principle Theory for Compressible Flows," Ph.D. dissertation, University of Oklahoma, 1988.

Rodriguez Azara, J.L. and Emanuel, G., "Compressible Rotational Flows Generated by the Substitution Principle," *Phys. Fluids.* 31, 1058 (1988).

Rodriguez Azara, J.L. and Emanuel, G., "Compressible Rotational Flows Generated by the Substitution Principle, II," *Phys. Fluids.* A 1, 600 (1989).

Yih, C.-S., "A Transformation for Non-Homentropic Flows, with an Application to Large-Amplitude Motion in the Atmosphere," *J. Fluid Mech.* 9, 68 (1960).

PROBLEMS

8.1 **(a)** Assume a two-dimensional or axisymmetric flow and derive the vorticity relation

$$\frac{D\omega}{Dt} = \lambda^{1/2}\frac{D\omega_b}{Dt} - \left[y^{\sigma}\frac{D}{Dt}\left(\frac{1}{2}\rho w^2\right) + \frac{1}{2}\rho w^2(\sigma v)\right]\frac{1}{\lambda}\frac{d\lambda}{d\psi}$$

The σv term represents the vortex stretching (or compression) by the radial component of the velocity in order to keep the strength of a vortex tube constant as its cross-sectional area changes.

(b) Apply this result to the Prandtl–Meyer flow, thereby obtaining the factor (8.35).

8.2 Taylor–Maccoll flow is a steady, supersonic flow about a cone at zero angle of attack with an attached conical shock. For the baseline flow, the upstream flow is uniform and the flow downstream of the shock is homentropic, homenergetic, and only depends on the angle η, as shown in the sketch. Assume the baseline flow field is fully known. In order to apply the substitution principle to the flow, assume $h_o(y)$ is known in the freestream flow.

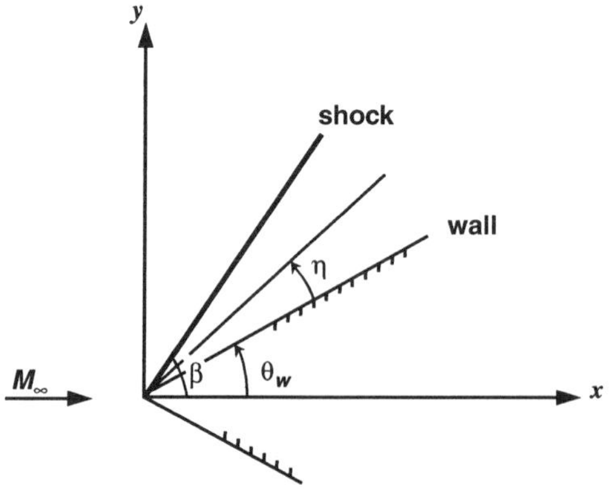

(a) Determine the vorticity $\omega_\infty(y)$ and the $d\psi/dy$ equation in the upstream flow.
(b) Determine the equation for y/y_s for a streamline, in terms of a quadrature, for the flow downstream of the shock. The ordinate y_s is the y value at the shock, while y is downstream of the shock; both values are for the same streamline.
(c) Derive the equation for $\omega\,(\eta,\psi)$ for the flow downstream of the shock.

8.3 **(a)** Utilize the material in Problem 8.2 and determine the vorticity $\omega(y)$ on the surface of the cone. The solution $\omega\,(\eta,\psi)$ of Problem 8.2 is singular on the surface of the cone, where both η and ψ are zero and cannot be used.
(b) Use conical flow tables or figures, such as in Ames Research Staff (1953), and the conditions

$$\gamma = 1.4, \qquad M_\infty = 3, \qquad p_\infty = 1 \text{ atm}, \qquad \beta = 35.6°$$

to evaluate ω/x on the surface of the cone when the nondimensional variation of h_o in the freestream is

$$h_o = 1 + y^2$$

8.4 Use the results of Section 8.3 to consider a parallel, rotational baseline flow that has gradients only in the y direction, where y is perpendicular to a planar wall; i.e.,

$$\rho u = f_1(y), \qquad p + \rho u^2 = f_2(y)$$

In addition, assume homenergetic flow of a perfect gas.
(a) Find f_1 as a function of f_2 and determine p, ρ, u, and ω in terms of f_2 and $f_2' = (df_2/dy)$.
(b) Assume $u(0) = 0$, $u(\infty) = U$ = constant and determine $f_2(0)$ and $f_2(\infty)$ in terms of U and other constants. Determine conditions on f_2 such that the following inequalities

$$\frac{du}{dy} \geq 0, \qquad \frac{d^2u}{dy^2} \leq 0, \qquad 0 \leq y \leq \infty$$

hold.
(c) Consider the results of parts (a) and (b) as providing a rotational baseline flow. Use a subscript b to denote the baseline flow in this part of the problem. Apply the substitution principle in the form

$$u(y) = [\lambda(y)]^{1/2}\, u_b(y)$$

and derive the conditions on λ such that

$$u(0) = u_b(0), \qquad u(\infty) = U, \qquad \frac{d^2u}{dy^2}(y^*) = 0$$

where y^* is a positive y value. The objective of this problem is to develop an inviscid profile with an inflection point for use in the stability analysis of a laminar boundary layer.

8.5 **(a)** Show that a sonic line has an infinite value for the magnitude of the acceleration in a cylindrical source flow of a perfect gas.

(b) Assume $h_o = h_o(\theta)$, with $h_o(0) = h_o(2\pi)$, and determine the vorticity $\vec{\omega}(M,\theta)$.

8.6 Consider the flow in Problem 5.21 under the substitution principle. This is a spiral flow with a constant velocity component normal to the plane containing the original spiral. Use Equation (8.16) and write the rotational solution for w_θ, w_r, w_x, ρ, and p. Use Appendix E to explain how h_o is determined when $h_{oa} = h_o(\theta_a,\eta_a)$ is prescribed in the $x = x_a$ plane. Finally, determine the vorticity $\vec{\omega}$.

8.7 Consider a baseline flow that is rotational and is not homenergetic or homentropic.

(a) Derive the condition such that this flow can be transformed into (i) a homenergetic flow, (ii) a homentropic flow, or (iii) an irrotational flow. Each of these conditions should be independent of λ; e.g., the case (iii) condition is simply $\nabla \times \vec{\omega} = 0$.

(b) For each of these cases, derive algebraic equations for λ in terms of only baseline flow variables plus constants associated with the transformed flow; hence, λ is no longer an arbitrary function. In case (iii) use Cartesian coordinates and the theory in Section 6.4 to obtain an answer for λ that depends on the w_{bi} velocity components.

8.8 Consider a baseline flow that is rotational and homenergetic but not homentropic. We shall use the substitution principle to transform this flow into a rotational and homentropic flow that is not homenergetic.

(a) Determine h_o in terms of the entropy and baseline flow variables.

(b) Assume a two dimensional or axisymmetric flow without swirl. Starting with the Euler equations, derive two PDEs whose dependent variables are ρ and the stream function ψ for the transformed flow.

(c) If $h_o(\psi)$ is known, what is an integral of these two PDEs?

8.9 The lower wall of a two-dimensional duct has a sharp turn as shown in the adjoining sketch. The upper wall is contoured so that the Prandtl–Meyer expansion does not reflect from it; i.e., between the leading and trailing edges of the expansion, the wall is a streamline of a Prandtl–Meyer flow.

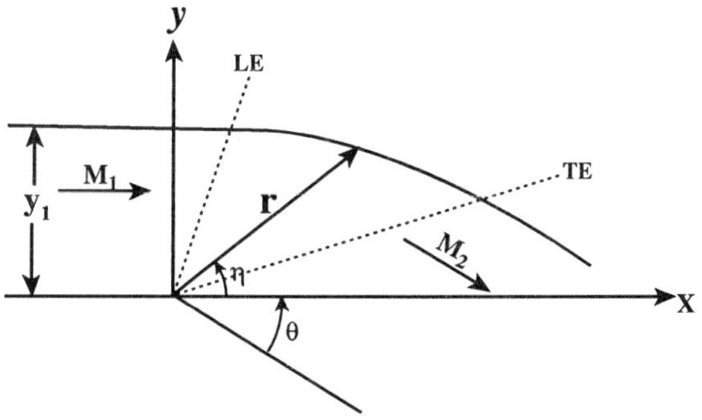

(a) Determine the shape, x/y_1 and y/y_1, of the upper wall between the leading and trailing edges of the expansion as a function of η. Use continuity and assume $\rho_1(y)$ is a constant.

(b) Determine analytical results for the force components per unit depth, F_x and F_y, acting on the upper surface between the leading and trailing edges. These components can be found by performing an integration that involves the pressure along the section of wall of interest. Alternatively, the momentum theorem (see Problem 5.16)

$$\int_{CS} [\rho\vec{w}(\vec{w}\cdot\hat{n}) + p\hat{n}]\,ds = 0$$

can be used where part of the control surface, CS, is adjacent to the curved wall.

(c) Evaluate F_x and F_y when

$$\gamma = 1.4, \quad M_1 = 2.2, \quad \theta = 15^\circ$$

$$p_1 = 10^5\text{Pa}, \quad y_1 = 10\text{ cm}, \quad \Delta z = 30\text{ cm}$$

where Δz is the depth of the duct.

(d) Do your results for parts (a) to (c) hold if the parallel flow in region 1 is rotational, and why?

8.10 Consider a rotating system, such as that associated with a compressor or turbine blade, where $\vec{R} = 0$ and the angular velocity, $\vec{\omega}_{rot}$, is a constant vector. With respect to the rotating system (see Section 2.5), assume a steady inviscid, compressible flow of a perfect gas. Does a nontrivial form of the substitution principle exist for the rotating system?

8.11 Continue with the analysis in Problem 8.10 and use a Cartesian coordinate system with $\vec{\omega}_{rot}$ aligned with the x_3-axis.

(a) Assume a steady flow in the noninertial system and derive a nondimensional condition so that the flow in the x_1, x_3 plane is approximately two dimensional.

(b) Suppose the inertial flow is irrotational and homenergetic. Is the two-dimensional flow in the x_1, x_3 noninertial plane irrotational and homenergetic?

(c) For the (approximately) two-dimensional flow in the x_1,x_3 plane, apply the hodograph transformation to the continuity equation. Use Jacobian theory to interchange x_1,x_3 with w_1,w_3, where the overbar has been deleted from the w_i. Is the resulting equation linear or nonlinear? Can you explain this answer?

8.12 **(a)** Use Jacobian theory and derive the relation for the specific heat $c_v[=(\partial e/\partial T)_v]$ that is the counterpart to Equation (8.12).

(b) Similarly, derive a relation for the Joule–Thomson coefficient, which is defined as

$$\mu_{JT} = \frac{\partial T}{\partial p_h}$$

8.13 Consider a centered, rotational Prandtl–Meyer expansion.

(a) Use Jacobian theory to determine $D\phi/Dt$, where ϕ is an arbitrary scalar, for a streamline located a distance y_∞ above the upstream wall. Assume $h_o(y_\infty)$ is known and utilize the $r = r(\eta, y_\infty)$ relation from Problem 8.9.

(b) Evaluate Dw/Dt.

8.14 Consider a centered Prandtl–Meyer expansion in which the parallel upstream flow is rotational. Denote conditions just upstream and downstream of the leading edge (LE) of the expansion with – and + subscripts, respectively, and assume $h_o(\psi)$ is continuous.

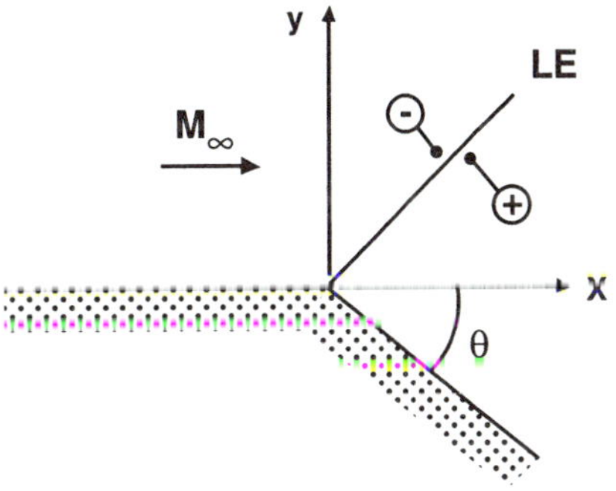

(a) Evaluate $(\partial u/\partial y)_\pm$ and $(\partial v/\partial x)_\pm$, thereby demonstrating that (i) derivatives normal to the LE characteristic are discontinuous, and (ii) the vorticity

$$\omega = \frac{\partial v}{\partial x} - \frac{\partial u}{\partial y}$$

is continuous at the LE.

(b) Under what conditions are $\partial v/\partial x$ and $\partial u/\partial y$ continuous across the LE?

8.15 Consider a steady, three-dimensional flow.

(a) Determine a relation, similar to Equations (8.5), for how the helicity density, $H_d = \vec{w} \cdot \vec{\omega}$, transforms under the substitution principle.

(b) Consider an irrotational baseline flow that is transformed under the substitution principle to a rotational one. Determine H_d and an equation for the angle α between the transformed velocity and vorticity vectors.

(c) For a two-dimensional or axisymmetric, rotational baseline flow, we have the following expansions:

$$w_b = \sum_{i=0} w_{bi}(x)y^i$$

$$\omega_b = \sum_{i=0} \omega_{bi}(x)y^i$$

$$\cos \alpha_b = \sum_{i=0} \alpha_{bi}(x)y^i$$

where α_b is the angle between $\vec{w}_b$ and $\vec{\omega}_b$, and

$$\lambda = \sum_{i=0} \lambda_{bi}(x)y^i$$

Determine the expansion, in the order specified, for w, H_{db}, H_d, and ω to $O(y^2)$ in terms of the known coefficients w_{bi}, ω_{bi}, α_{bi}, and λ_i, where $i = 0, 1, 2$. Be sure to simplify your answers.

9 Calorically Imperfect Flows

9.1 PRELIMINARY REMARKS

There are many engineering situations where real-gas effects cannot be overlooked. They range from the flow in a cryogenic wind tunnel to the aerodynamics of high-speed reentry. Real-gas effects are important in the vicinity of the coexistence curve, especially for a polyatomic molecule with large specific heats. Our interest in this chapter, however, is primarily with air when its temperature is well above room temperature.

Throughout the early part of this book, a thermally and calorically perfect gas is sometimes assumed. This assumption, however, is not appropriate for air at cryogenic temperatures or when the temperature is well above room temperature. In the latter case, the perfect gas assumption should be replaced with one that incorporates real-gas phenomena. For air, and many other gases, a thermally perfect, calorically imperfect gas model is useful, since it is applicable over a broad range of pressures and temperatures. In thermodynamics, this model is called an ideal gas. Further revision is required at still higher temperatures when molecules begin to dissociate or react.

At all temperatures, except near absolute zero, simple polyatomic molecules in the gaseous state are rotationally fully excited. At room temperature, however, the vibrational mode, or modes, are not necessarily excited. For instance, the single vibrational mode of N_2 and of O_2 is not excited; hence, N_2 and O_2, or a mixture of the two, has a specific heat ratio of 1.4. At temperatures above about 600 K, however, the vibrational mode of O_2 gradually starts to become excited with a similar occurrence for N_2 at a higher temperature. The reason is evident from the characteristic vibrational temperature, T_v, of O_2, which is equal to 2219 K, and of N_2, which equals 3352 K.

Tsien (1947a) performed the first gas dynamic study of imperfect gas flows. A van der Waals state equation is used, but the paper is marred by many errors (Tsien, 1947b; Donaldson, 1948). A more systematic treatment is provided by Eggers (1950) and the Ames Research Staff (1953) in which a Berthelot thermal equation of state is used in conjunction with a harmonic oscillator model for vibrational excitation. The Ames report contains several charts for evaluating imperfect gas effects for a streamtube flow, a uniform flow containing a planar shock wave, and a Prandtl–Meyer flow.

The discussion in this chapter assumes a diatomic gas consisting of a single species that is thermally perfect but whose vibrational excitation is provided by the harmonic oscillator model of Section 3.4. This is the simplest possible self-consistent thermodynamic model for an imperfect gas. Nevertheless, our approach is quite accurate at temperatures below where dissociation begins. It is less accurate for a mixture of diatomics, such as air. In this circumstance, the Ames report suggests 3056 K as a compromise air value for T_v. In addition, the gas is assumed to be in thermodynamic equilibrium; vibrational or chemical nonequilibrium effects are beyond the scope of this book.

Emphasis is on the method of formulation, with due care given to mathematical and computational subtleties. Trends are discussed, and the overall approach is suitable for approximate, back-of-the-envelope estimates (see Problem 9.1). The numerical accuracy, however, is adequate for checking more elaborate CFD computations. For instance, a streamtube flow analysis revealed an error in a rocket nozzle thrust study (Lentini, 1992; Christy and Emanuel, 1994; Lentini, 1994). We hope this material will enable the reader to develop an intuitive feel for the behavior of equilibrium imperfect gas flows.

Although the presentation somewhat resembles the Ames report, there are significant differences. A simpler, nondimensional, computer-oriented approach is used and a broader range of flows and phenomena are examined. A concise description of the thermodynamic model is provided in the next section. Four M.S. theses are the basis of the material discussed in the subsequent sections. We are pleased to acknowledge the assistance of these former students, who generously provided the many figures that appear in this chapter. Isentropic streamtube flow (Christy, 1993) is discussed in Section 9.3, while Section 9.4 discusses flows with normal or oblique planar shock waves (Bultman, 1994). Sections 9.5 and 9.6, respectively, cover Prandtl–Meyer flow (Ismail, 1994) and supersonic flow over a cone (Lampe, 1994). A duct with heat transfer and skin friction, i.e., Rayleigh/Fanno flow, is the subject of Problem 9.2. These flows constitute the building blocks of gas dynamics.

9.2 THERMODYNAMICS

HARMONIC OSCILLATOR MODEL

We start with Equation (3.53) and write

$$c_{vv} = Rz(\theta_v) \tag{9.1}$$

where

$$\theta_v = \frac{T_v}{2T}, \qquad z(\theta_v) = \left(\frac{\theta_v}{\sinh\theta_v}\right)^2 \tag{9.2a,b}$$

For the gas as a whole, the constant volume specific heat is

$$\frac{c_v(T)}{R} = \frac{5}{2} + \delta z(\theta_v) \tag{9.3}$$

where the 5/2 represents the translational and rotational contribution, and

$$\delta = \begin{cases} 0, & \text{diatomic molecule with no vibrational excitation} \\ 1, & \text{diatomic molecule with vibrational excitation} \end{cases} \tag{9.4}$$

The parameter δ should not be confused with the one defined by Equation (3.49). It conveniently allows simultaneous consideration of any $\gamma = 1.4$ calorically perfect gas in conjunction with the more general case. Indeed, whenever the $\delta = 0$ case is under discussion, we presume $\gamma = 1.4$ and use the corresponding compressible flow relations.

THERMODYNAMIC MODEL

With the aid of standard thermodynamic procedures (see Chapter 3), the following self-consistent model is obtained:

$$p = \rho RT \tag{9.5}$$

$$c_p = c_v + R \tag{9.6}$$

$$\gamma(T) = \frac{c_p(T)}{c_v(T)} \tag{9.7}$$

$$a^2 = \gamma RT \tag{9.8}$$

$$e(T) = e_r + \int_{T_r}^{T} c_v(T')dT' \tag{9.9}$$

$$h(T) = e(T) + RT \tag{9.10}$$

$$s(T, \rho) = s_r + R\ln\frac{\rho_r}{\rho} + \int_{T_r}^{T} c_v(T')\frac{dT'}{T'} \tag{9.11}$$

In the above, p is the pressure, ρ is the density, c_p is the specific heat at constant pressure, a is the speed of sound, e is the specific internal energy, h is the specific enthalpy, and s is the specific entropy. The r subscript denotes an arbitrary reference state while a prime denotes a dummy integration variable. The two integrals are provided by Equations (3.60) and (3.62):

$$\int_{T_r}^{T} c_v(T')dT' = \frac{5}{2}RT_r\left(\frac{\theta_{vr}}{\theta_v} - 1\right) + \delta RT_r\theta_{vr}(\coth\theta_v - \coth\theta_{vr}) \tag{9.12}$$

$$\int_{T_r}^{T} c_v(T')\frac{dT'}{T'} = \frac{5}{2}R\ \ln\frac{\theta_{vr}}{\theta_v} + \delta R(\theta_v\coth\theta_v - \theta_{vr}\coth\theta_{vr}) + \delta R\ \ln\frac{\sinh\theta_{vr}}{\sinh\theta_v} \tag{9.13}$$

where

$$\theta_{vr} = \frac{T_v}{2T_r} \tag{9.14}$$

By combining the above, we obtain

$$e = e_r + \frac{5}{2}RT_r\left(\frac{\theta_{vr}}{\theta_v} - 1\right) + \frac{1}{2}\delta RT_v(\coth\theta_v - \coth\theta_{vr}) \tag{9.15a}$$

$$h = e_r + \frac{RT_v}{2\theta_v} + \frac{5}{2}RT_r\left(\frac{\theta_{vr}}{\theta_v} - 1\right) + \frac{1}{2}\delta RT_v(\coth\theta_v - \coth\theta_{vr}) \tag{9.15b}$$

$$s = s_r + R\ \ln\left[\frac{\rho_r}{\rho}\left(\frac{\theta_{vr}}{\theta_v}\right)^{5/2}\left(\frac{\sinh\theta_{vr}}{\sinh\theta_v}\right)^{\delta}\right] + \delta R(\theta_v\coth\theta_v - \theta_{vr}\coth\theta_{vr}) \tag{9.15c}$$

Equations (9.2) through (9.15) fully represent the thermodynamics of the gas. The only free parameters are δ, R, T_v, and the reference state (T_r, e_r, s_r, ρ_r).

For the reference state, we use

$$(T_r)_{\delta=0} = (T_r)_{\delta=1} = T_o$$
$$(\rho_r)_{\delta=0} = (\rho_r)_{\delta=1} = \rho_o$$
$$(e_r)_{\delta=0} = (e_r)_{\delta=1} = \frac{5}{2}RT_o$$
$$(s_r)_{\delta=0} = (s_r)_{\delta=1} = 0$$

where a zero subscript denotes a stagnation value. The stagnation state is the same for both $\delta = 0$ and $\delta = 1$ and, moreover, represents an upstream state. Equations (9.15) thus become

$$\frac{e}{RT_v} = \frac{5}{4\theta_v} + \frac{1}{2}\delta(\coth\theta_v - \coth\theta_{vo}) \tag{9.16a}$$

$$\frac{h}{RT_v} = \frac{7}{4\theta_v} + \frac{1}{2}\delta(\coth\theta_v - \coth\theta_{vo}) \tag{9.16b}$$

$$\frac{s}{R} = \ln\left[\frac{\rho_o}{\rho}\left(\frac{\theta_{vr}}{\theta_v}\right)^{5/2}\left(\frac{\sinh\theta_{vr}}{\sinh\theta_v}\right)^{\delta}\right] + \delta(\theta_v\coth\theta_v - \theta_{vo}\coth\theta_{vo}) \tag{9.16c}$$

where

$$\theta_{vo} = \frac{T_v}{2T_o}$$

Aside from T_o and ρ_o, the other thermodynamic variables have the following upstream stagnation values:

$$p_o = \rho_o RT_o, \quad \frac{e_o}{RT_v} = \frac{5}{4\theta_{vo}}, \quad \frac{h_o}{RT_v} = \frac{7}{4\theta_{vo}}, \quad \frac{s_o}{R} = 0 \tag{9.17}$$

for both $\delta = 0$ and $\delta = 1$. Moreover, e_o and h_o are both proportional to T_o.

Discussion

Throughout the subsequent analysis, a key parameter is the upstream value of θ_{vo}. Three values are assigned to it when graphical results are presented, namely, 0.5, 1.5, and 5.0. The value 0.5 represents a large stagnation temperature and a substantially altered flow caused by equilibrium changes in the vibrational energy. At the other extreme, the value 5 represents a flow with a relatively small T_o value and a negligible change in flow properties caused by the vibrational energy; i.e., the vibrational energy is nearly zero throughout the flow. A θ_{vo} value of 1.5 yields changes in between these two cases. At any rate, all results are nondimensional and are *not* restricted to specific T_v and T_o values. Graphical results, however, are limited to the above three θ_{vo} values.

The downstream state is fixed by prescribing the Mach number, M, which is the universally agreed upon choice when $\delta = 0$. The streamtube equations with vibrational excitation, however, are most simply expressed in terms of θ_v, i.e., the inverse static temperature. With θ_v fixed, the two flows have slightly different Mach numbers. The choice of θ_v as the independent variable, however, complicates the analysis of flows with shock waves. Hence, the Mach number is consistently used as the independent variable for the comparisons; i.e., the two flows are compared for the same value of M. There are exceptions, such as those quantities that depend on only a single variable, as is the case with γ. As we shall see, this Mach number choice comes at a price; the equations typically require a more complicated iterative computer solution.

9.3 ISENTROPIC STREAMTUBE FLOW

INTRODUCTORY DISCUSSION

Because of its simplicity, a steady, quasi-one-dimensional, isentropic flow has many engineering applications, such as providing performance estimates for nozzle flows. This section discusses a wide variety of comparisons in which the state of a calorically perfect gas ($\delta = 0$) is contrasted with one possessing equilibrium vibrational energy, or excitation ($\delta = 1$). The comparison will take the form of a ratio, referred to as a δ comparison. For instance, the first comparison will be for γ and is written as

$$\frac{\gamma_{\delta=1}}{\gamma_{\delta=0}} = \frac{\gamma(\theta_v)}{1.4}$$

This ratio is a unique function of θ_v, as shown in the next subsection. In this subsection, the Mach number is introduced and the governing equations are written in nondimensional form. Subsequent subsections then examine a range of δ comparisons in a systematic fashion.

GOVERNING EQUATIONS

Isentropic streamtube flows utilizes Equation (9.5), conservation of the mass flow rate:

$$\rho w A = \dot{m} \tag{9.18}$$

and

$$h + \frac{1}{2}w^2 = h_o \tag{9.19}$$

$$s = s_o \tag{9.20}$$

Note that the $\dot{m}$, h_o, and s_o parameters are constants. The cross-sectional area, A, of the streamtube is viewed as a known function of distance in the flow direction.

With Equations (9.17) for h_o and s_o, we obtain

$$\ln\left[\frac{\rho_o}{\rho}\left(\frac{\theta_{vo}}{\theta_v}\right)^{5/2}\left(\frac{\sinh\theta_{vo}}{\sinh\theta_v}\right)^{\delta}\right] = \delta(\theta_{vo}\coth\theta_{vo} - \theta_v\coth\theta_v)$$

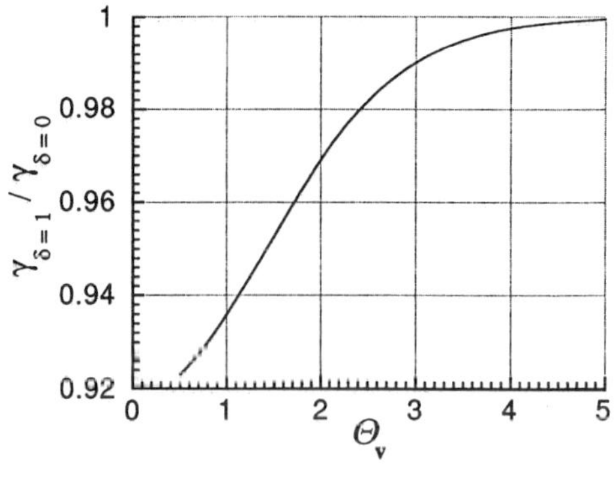

FIGURE 9.1 Comparison of specific heat ratios.

which simplifies to

$$\frac{\rho}{\rho_o} = \left(\frac{\theta_{vo}}{\theta_v}\right)^{5/2}\left(\frac{\sinh\theta_{vo}}{\sinh\theta_v}\right)^{\delta}\exp[\delta(\theta_v\coth\theta_v - \theta_{vo}\coth\theta_{vo})] \tag{9.21}$$

This relation replaces Equation (9.20). When $\delta = 0$, the usual result

$$\frac{\rho}{\rho_o} = \left(\frac{T}{T_o}\right)^{5/2} \tag{9.22}$$

is obtained for a gas with $\gamma = 1.4$. (The reduction to the $\delta = 0$ formulation provides a check on the more general analysis.)

The equations in the preceding section are combined to yield, for the ratio of specific heats,

$$\gamma = \frac{c_p}{c_v} = \frac{7 + 2\delta z(\theta_v)}{5 + 2\delta z(\theta_v)} \tag{9.23}$$

The first δ comparison

$$\frac{\gamma_{\delta=1}}{\gamma_{\delta=0}} = \frac{\gamma}{1.4} = \frac{1 + \frac{2}{7}z(\theta_v)}{1 + \frac{2}{5}z(\theta_v)} \tag{9.24}$$

is shown in Figure 9.1. At low temperatures, when θ_v is large, the vibrational mode has little or no excitation and the $\gamma/1.4$ ratio is near unity. At higher temperatures, the excitation increases with a consequent decrease in $\gamma/1.4$. When $\theta_v \le 2.5$, the reduction in $\gamma_{\delta=1}$ starts to become significant. For air, $\theta_v \cong 2.5$ corresponds to a temperature of about 600 K.

As usual, the Mach number is defined as

$$M = \frac{w}{a} \tag{9.25}$$

where the speed of sound is

$$a^2 = \gamma \frac{RT_v}{2\theta_v} \tag{9.26}$$

The flow speed is replaced with

$$w = \left(\frac{\gamma RT_v}{2\theta_v}\right)^{1/2} M \tag{9.27}$$

With the aid of Equation (9.16b), Equation (9.19) becomes

$$\gamma M^2 = 7\left(\frac{\theta_v}{\theta_{vo}} - 1\right) + 2\delta\theta_v(\coth\theta_{vo} - \coth\theta_v) \tag{9.28}$$

As expected, this relation reduces to

$$\frac{\theta_v}{\theta_{vo}} = \frac{T_o}{T} = 1 + \frac{1}{5}M^2 \tag{9.29}$$

when $\delta = 0$.

Equation (9.28) enables θ_v to be compared with fixed values for the Mach number and θ_{vo}. This comparison is equivalent to one for $T_{\delta=0}/T_{\delta=1}$. We thus write

$$M^2 = 5\left(\frac{(\theta_v)_{\delta=0}}{\theta_{vo}} - 1\right) \tag{9.30}$$

$$M^2 = \frac{1}{\gamma_{\delta=1}}\left\{7\left[\frac{(\theta_v)_{\delta=1}}{\theta_{vo}} - 1\right] + 2(\theta_v)_{\delta=1}[\coth\theta_{vo} - \coth(\theta_v)_{\delta=1}]\right\} \tag{9.31}$$

for the respective $\delta = 0$ and 1 cases. Although the first equation is easily solved for $(\theta_v)_{\delta=0}$, an iterative solution is required with the second one when M is prescribed.

The nondimensionalization of the governing equations is completed by writing Equation (9.18) as

$$\frac{(RT_o)^{1/2}\dot{m}}{p_o A} = \left(\frac{\theta_{vo}}{\theta_v}\right)^3\left(\frac{\sinh\theta_{vo}}{\sinh\theta_v}\right)^\delta \{\exp[\delta(\theta_v\coth\theta_v - \theta_{vo}\coth\theta_{vo})]\}\gamma^{1/2}M \tag{9.32}$$

and Equation (9.5) as

$$\frac{p}{p_o} = \left(\frac{\theta_{vo}}{\theta_v}\right)^{7/2}\left(\frac{\sinh\theta_{vo}}{\sinh\theta_v}\right)^\delta \exp[\delta(\theta_v\coth\theta_v - \theta_{vo}\coth\theta_{vo})] \tag{9.33}$$

Temperature, Density, and Pressure Comparisons

The θ_v, or T, δ comparison is shown in Figure 9.2. The curves level off when M is large, since the vibrational energy decreases to zero as T decreases. Each curve has a minimum value, although it is not discernible when $\theta_{vo} = 5$ and 1.5. The Mach number at the minimums is 1.3, 2.2, and 3.8 for $\theta_{vo} = 5$, 1.5, and 0.5, respectively. The minimum is caused by $(\theta_v)_{\delta=0}$ increasing with M more rapidly than $(\theta_v)_{\delta=1}$ at subsonic and low supersonic Mach numbers. Physically, $T_{\delta=1}$ is larger than would be predicted by a calorically perfect gas model because of the transfer of vibrational energy to the translational and rotational modes. The effect is significant at low θ_{vo}, or large T_o, values and at supersonic speeds. The energy transfer is negligible when $\theta_{vo} \gtrsim 5$, and the imperfect gas is effectively perfect. As noted earlier, this conclusion is a general one; it will uniformly hold throughout the rest of the analysis in this chapter.

The streamtube flows $\delta = 0$ and 1 start from the same stagnation state. Consequently, a δ comparison may be unity when $M = 0$. (As discussed later, not all comparisons are unity when $M = 0$.) For example, this is the case for the density in Figure 9.3. As with the temperature ratio, there is a substantial decrease in the density ratio for small θ_{vo} and supersonic Mach numbers. The same trend will hold in the later shock wave and Prandtl–Meyer comparisons. Overall, the deviation from unity seen in the figure is among the largest observed. It will influence later comparisons, such as the one for the cross-sectional area of a nozzle.

In Figure 9.4, the pressure ratio first rises above unity before decreasing. In view of the similarity of Equations (9.21) and (9.33), this behavior is nonintuitive and unexpected. As with the density comparison, later δ comparisons for the pressure will show the same behavior.

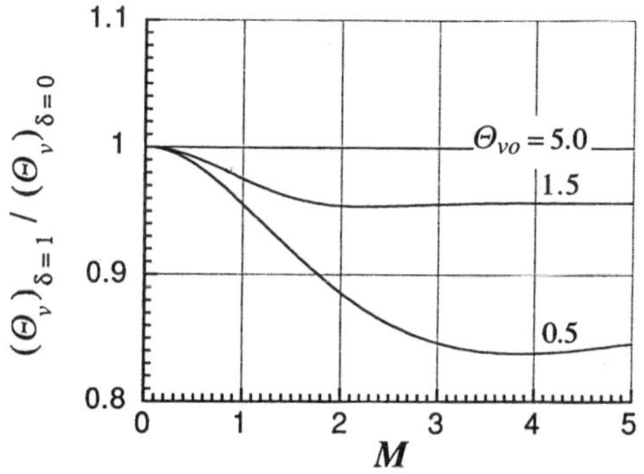

FIGURE 9.2 Isentropic comparison for θ_v or the temperature.

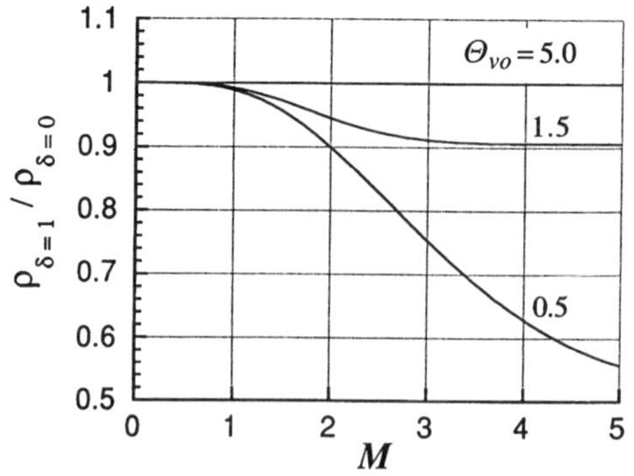

FIGURE 9.3 Isentropic comparison for the density.

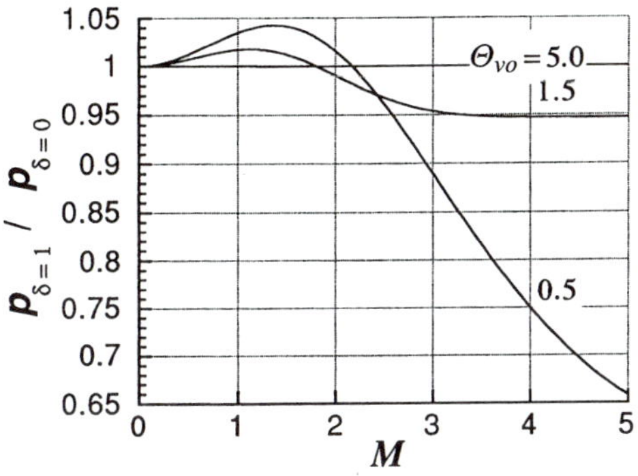

FIGURE 9.4 Isentropic comparison for the pressure.

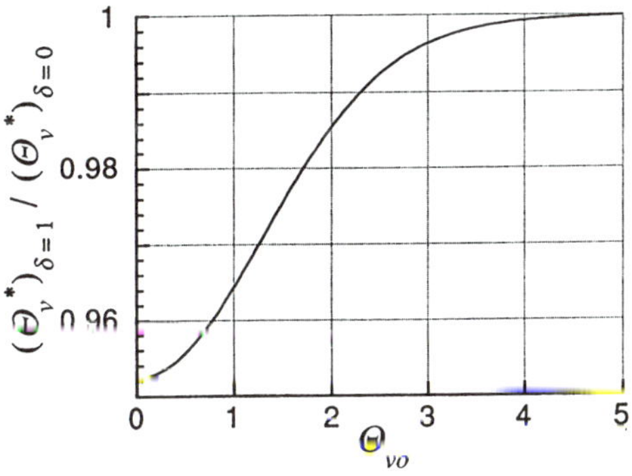

FIGURE 9.5 Isentropic comparison for θ_v^* or the temperature when the flow is sonic.

An explanation is provided by the perfect gas equation of state. At low Mach numbers, the variation in $(\theta_v)_{\delta=1}/(\theta_v)_{\delta=0}$, or the equivalent temperature ratio, dominates, while the density ratio dominates at larger Mach numbers.

Trends seen in Figures 9.2–9.4 also can be understood by examining x_2/x_1, where x represents T/T_o, ρ/ρ_o, or p/p_o, and γ_2 is chosen as 1.3 while γ_1 is 1.4. Two calorically perfect gases are compared with different specific heat ratios. A plot of x_2/x_1 vs. M, where x is T/T_o, would show a slow monotonic increase starting from unity. On the other hand, the x_2/x_1 ratio for the density decreases monotonically with M. The rate of decrease is quite rapid when the flow is supersonic. The ratio for the pressure, therefore, first increases with M before decreasing.

Cross-Sectional Area Comparison

Sonic conditions, denoted with an asterisk, are isentropically related to the stagnation state. This is done by setting $M = 1$ in Equation (9.28) and using Equation (9.23), to obtain

$$\frac{21 + 8\delta z(\theta_v^*)}{5 + 8\delta z(\theta_v^*)} = \frac{\theta_v^*}{\theta_{vo}}\left[\frac{7}{2} + \delta\theta_{vo}(\coth\theta_{vo} - \coth\theta_v^*)\right] \tag{9.34}$$

for the δ comparison shown in Figure 9.5.

Observe that the $T^*_{\delta=0}/T^*_{\delta=1}$ ratio does not significantly deviate from unity, in accord with the $M = 1$ values in Figure 9.2. The θ_v^* solution is obtained iteratively, as is generally the case with the $\delta = 1$ equations. Since $T_o > T^*$, we have $\theta_v^* > \theta_{vo}^*$. When T_o goes to zero, the parameters θ_{vo} and θ_v^* approach infinity, with the asymptotic results

$$\sinh\theta_{vo} \sim \frac{1}{2}e^{\theta_{vo}}, \qquad \coth\theta_{vo} \sim 1 \tag{9.35}$$

which leads to

$$\frac{\theta_v^*}{\theta_{vo}} \sim \frac{6}{5}$$

Hence, $\theta_{vo} < \theta_v^* < (6\theta_{vo}/5)$, and these bounds on θ_v^* can be used to simplify the numerical procedure. We readily obtain

$$\frac{\theta_v^*}{\theta_{vo}} = \frac{6}{5} \tag{9.36}$$

when $\delta = 0$.

The Mach number is next eliminated from Equations (9.28) and (9.32), with the result

$$\frac{(RT_o)^{1/2}\dot{m}}{p_oA} = \left(\frac{\theta_{vo}}{\theta_v}\right)^3\left(\frac{\sinh\theta_{vo}}{\sinh\theta_v}\right)^\delta\{\exp[\delta(\theta_v\coth\theta_v - \theta_{vo}\coth\theta_{vo})]\}$$
$$\times\left[7\left(\frac{\theta_v}{\theta_{vo}} - 1\right) + 2\delta\theta_v(\coth\theta_{vo} - \coth\theta_v)\right]^{1/2} \tag{9.37}$$

With $(RT_o)^{1/2}\dot{m}/p_o$ fixed, the streamtube area ratio is given by

$$\frac{A}{A^*} = \left(\frac{\theta_v}{\theta_v^*}\right)^3\left(\frac{\sinh\theta_v}{\sinh\theta_v^*}\right)^\delta\{\exp[\delta(\theta_v^*\coth\theta_v^* - \theta_v\coth\theta_v)]\}$$
$$\times\left[\frac{7\left(\frac{\theta_v^*}{\theta_{vo}} - 1\right) + 2\delta\theta_v^*(\coth\theta_{vo} - \coth\theta_v^*)}{7\left(\frac{\theta_v}{\theta_{vo}} - 1\right) + 2\delta\theta_v(\coth\theta_{vo} - \coth\theta_v)}\right]^{1/2} \tag{9.38}$$

Observe that θ_v^* is a function only of θ_{vo}, while θ_v is a function of θ_{vo} and M. Hence, A/A^* is a function of δ, M, and θ_v^*. The area ratio, of course, is a function only of M when $\delta = 0$.

With $\delta = 0$, Equation (9.29) is

$$\frac{(\theta_v)_{\delta=0}}{\theta_{vo}} = 1 + \frac{1}{5}M^2 \tag{9.39}$$

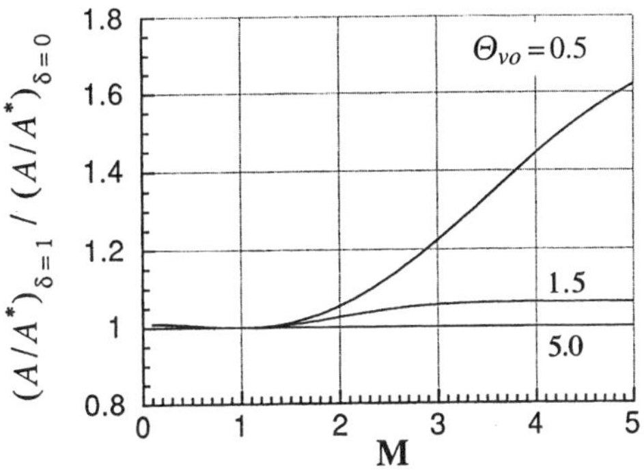

FIGURE 9.6 Isentropic area ratio comparison.

and the corresponding streamtube area ratio becomes

$$\left(\frac{A}{A^*}\right)_{\delta=0} = \frac{\left[\frac{5(\theta_v)_{\delta=0}}{6\theta_{vo}}\right]^3}{\left\{5\left[\frac{(\theta_v)_{\delta=0}}{\theta_{vo}} - 1\right]\right\}^{1/2}} = \left(\frac{5}{6}\right)^3 \frac{\left(1 + \frac{1}{5}M^2\right)^3}{M} \tag{9.40}$$

The above equations provide the comparison

$$\frac{\left(\frac{A}{A^*}\right)_{\delta=1}}{\left(\frac{A}{A^*}\right)_{\delta=0}} = \frac{M}{\left(1 + \frac{1}{5}M^2\right)^3}\left(\frac{6\theta_v}{5\theta_v^*}\right)^3 \frac{\sinh\theta_v}{\sinh\theta_v^*}[\exp(\theta_v^*\coth\theta_v^* - \theta_v\coth\theta_v)]$$
$$\times\left[\frac{7\left(\frac{\theta_v^*}{\theta_{vo}} - 1\right) + 2\theta_v^*(\coth\theta_{vo} - \coth\theta_v^*)}{7\left(\frac{\theta_v}{\theta_{vo}} - 1\right) + 2\theta_v(\coth\theta_{vo} - \coth\theta_v)}\right]^{1/2} \tag{9.41}$$

where (for notational brevity) $\theta_v = (\theta_v)_{\delta=1}$, which is an implicit function of θ_{vo} and M.

The δ comparison is shown in Figure 9.6. Remember that the flows $\delta = 0$ and $\delta = 1$ have the same stagnation state. Consequently, if $\dot{m}_{\delta=1} = \dot{m}_{\delta=0}$ in this comparison, then the $A^*_{\delta=1}$ and $A^*_{\delta=0}$ throat areas are different. Alternatively, if we assume the throat areas are equal, then the mass flow rates differ. The comparison in the figure holds for both cases. At high supersonic speeds, the large deviation from unity, when θ_{vo} is small, is primarily due to the density; i.e., ρA is roughly constant. When M goes to zero, both (A/A^*) ratios become infinite. This is evident from Equation (9.40) for $\delta = 0$. When $\delta = 1$, θ_v becomes θ_{vo} and the denominator within the square root in Equation (9.38) goes to zero. Thus, the comparison in the figure is indeterminant, but finite, when M goes to zero.

In accord with Figure 9.6, when air expands in a wind tunnel nozzle from a high stagnation temperature to a large exit Mach number, the nozzle area ratio must be substantially larger than would be estimated by using $\gamma = 1.4$. This is certainly the case if molecular collisions can maintain the vibrational modes of O_2 and N_2 in equilibrium. In actuality, however, this does not occur. Shortly downstream of the throat, the reduced density and temperature result in a sharply reduced molecular collision frequency and collision energy. The rate of transfer of vibrational energy to the translational and rotational modes is then unable to keep pace with the increasing flow speed (Vincenti and Kruger, 1965), and the vibrational energy "freezes" at a nonequilibrium value. A vibrational relaxation model is then required for an accurate assessment of the expansion process.

By way of summary, the computational situation is reviewed. For purposes of discussion, known values are presumed for

$$\theta_{vo}, \quad (RT_o)^{1/2}\dot{m}/p_o, \quad M$$

Equation (9.34) then provides $(\theta_v^*)_{\delta=0}$ and $(\theta_v^*)_{\delta=1}$, while Equations (9.30) and (9.31) provide $(\theta_v)_{\delta=0}$ and $(\theta_v)_{\delta=1}$. With M known, Equations (9.38) and (9.40) respectively provide $(A/A^*)_{\delta=1}$ and $(A/A^*)_{\delta=0}$, while Equation (9.37) can be used to determine the areas $A_{\delta=1}$ and $A_{\delta=0}$. Other $\delta = 0$ or $\delta = 1$ parameters of interest can then be found. For example, the density and pressure are given by Equations (9.21) and (9.33), respectively.

Mass Flow Rate Comparison

For a choked flow, it is instructive to compare $\dot{m}_{\delta=1}$ with $\dot{m}_{\delta=0}$ for the same throat area. Equations (9.32) and (9.36) yield

$$\frac{(RT_o)^{1/2}\dot{m}_{\delta=0}}{p_o A^*} = \left(\frac{7}{5}\right)^{1/2}\left(\frac{5}{6}\right)^3 = 0.6847 \tag{9.42}$$

when $\delta = 0$ and $M = 1$. Equations (9.32) and (9.42) are utilized, with $\delta = 1$ and $M = 1$, to obtain

$$\frac{\dot{m}_{\delta=1}}{\dot{m}_{\delta=0}} = \left(\frac{5}{7}\right)^{1/2}\left(\frac{6\theta_{vo}}{5\theta_v^*}\right)^3 \frac{\sinh\theta_{vo}}{\sinh\theta_v^*}[\exp(\theta_v^*\coth\theta_v^* - \theta_{vo}\coth\theta_{vo})] \times \left[\frac{7 + 2\left(\dfrac{\theta_v^*}{\sinh\theta_v^*}\right)^2}{5 + 2\left(\dfrac{\theta_v^*}{\sinh\theta_v^*}\right)^2}\right]^{1/2} \tag{9.43}$$

where (for notational brevity) $\theta_v^* = (\theta_v^*)_{\delta=1}$, which is given in terms of θ_{vo} by Equation (9.34). This result is shown as Figure 9.7. Observe that the $\delta = 1$ mass flow rate is slightly less than the calorically perfect mass flow rate. Although the deviation from unity is not large, it nevertheless is of interest for thrust nozzles, which are discussed shortly.

Maximum Flow Speed Comparison

Another parameter of interest is the maximum flow speed, w_{max}. As before, this comparison is performed with fixed stagnation conditions. From Equation (9.19), observe that w is a maximum when the temperature is zero. Hence, set

$$\theta_v \to \infty, \quad h = \frac{1}{2}\delta RT_v(1 - \coth\theta_{vo}), \quad h_o = \frac{7RT_v}{4\theta_{vo}}$$

in this equation, with the result

$$\frac{1}{2}w_{max}^2 = \frac{7RT_v}{4\theta_{vo}} + \frac{1}{2}\delta RT_v(\coth\theta_{vo} - 1)$$

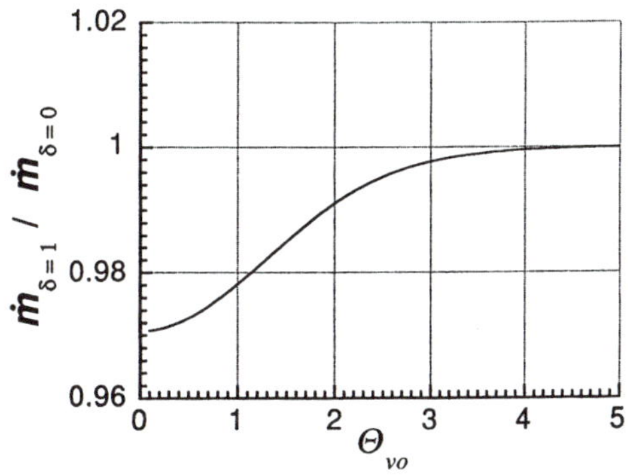

FIGURE 9.7 Isentropic mass flow rate comparison.

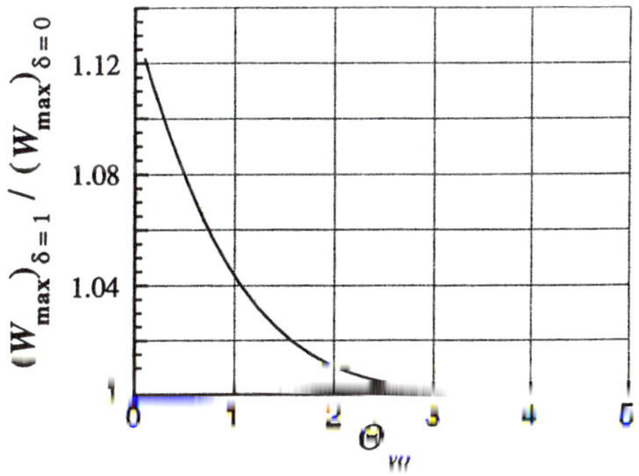

FIGURE 9.8 Isentropic maximum flow speed comparison.

or

$$w_{\max} = \left\{ \frac{7RT_v}{2\theta_{vo}} \left[1 + \frac{2}{7}\delta\theta_{vo}(\coth\theta_{vo} - 1) \right] \right\}^{1/2} \tag{9.44}$$

Consequently, the speed ratio comparison, shown in Figure 9.8, is given by

$$\left(\frac{w_{\delta=1}}{w_{\delta=0}}\right)_{\max} = \left[1 + \frac{2}{7}\theta_{vo}(\coth\theta_{vo} - 1) \right]^{1/2} \tag{9.45}$$

Since $\coth x \geq 1$, the ratio exceeds unity. This is a general result caused by vibrational energy becoming translational and rotational energy as the gas cools while accelerating along a streamline. The effect is significant when θ_{vo} is below 2.

Nozzle Thrust Comparison

The trends exhibited in Figures 9.7 and 9.8 tend to counter each other when the thrust of a rocket nozzle is evaluated. In other words, real-gas behavior reduces $\dot{m}$ but increases the flow speed, w_e, at the exit plane of the nozzle (see Problem 9.1). To examine this trade-off, the thrust, $\Im$, is evaluated for a nozzle operating with an ambient, or back, pressure p_a.

The analysis has several limitations. An actual thrust nozzle generally has a gas with a specific heat ratio different from 1.4 and $\gamma_{\delta=1}$. Another limitation is the equilibrium assumption for the vibrational excitation. As discussed earlier, this assumption breaks down for flow in a nozzle at a low supersonic Mach number. The equilibrium assumption, nevertheless, is useful, since it provides a maximum value for the thrust with only a modest computational effort. At the other extreme, a minimum thrust value is obtained by using a constant value for γ, starting with its value in the plenum. In terms of real-gas behavior, the actual thrust is bounded by these two values.

Impulse Function

Before discussing the thrust, the impulse function, defined as

$$F = pA + \dot{m}w \tag{9.46}$$

is evaluated. This parameter is used when the thrust or drag of a quasi one-dimensional flow is evaluated. With our gas model, this transforms to

$$\frac{F}{p_o A^*} = \frac{p}{p_o}\frac{A}{A^*}(1 + \gamma M^2) \tag{9.47}$$

and with the aid of Equation (9.28), we have

$$1 + \gamma M^2 = 7\frac{\theta_v}{\theta_{vo}} - 6 + 2\delta\theta_v(\coth\theta_{vo} - \coth\theta_v) \tag{9.48}$$

Along with Equations (9.33) and (9.38), this is substituted into Equation (9.47), with the result

$$\frac{F}{p_o A^*} = \left(\frac{\theta_{vo}}{\theta_v}\right)^{1/2}\left(\frac{\theta_{vo}}{\theta_v^*}\right)^3\left(\frac{\sinh\theta_{vo}}{\sinh\theta_v^*}\right)^{\delta}\{\exp[\delta(\theta_v^*\coth\theta_v^* - \theta_{vo}\coth\theta_{vo})]\}$$
$$\times\left[\frac{7\left(\frac{\theta_v^*}{\theta_{vo}} - 1\right) + 2\delta\theta_v^*(\coth\theta_{vo} - \coth\theta_v^*)}{7\left(\frac{\theta_v}{\theta_{vo}} - 1\right) + 2\delta\theta_v(\coth\theta_{vo} - \coth\theta_v)}\right]^{1/2}\left[7\frac{\theta_v}{\theta_{vo}} - 6 + 2\delta\theta_v(\coth\theta_{vo} - \coth\theta_v)\right] \tag{9.49}$$

With the use of Equation (9.39), this reduces to

$$\left(\frac{F}{p_o A^*}\right)_{\delta=0} = \left(\frac{5}{6}\right)^3\frac{1 + \frac{7}{5}M^2}{M\left(1 + \frac{1}{5}M^2\right)^{1/2}} \tag{9.50}$$

which is the $\gamma = 1.4$ impulse function equation. The parameters $p_o A^*$ and θ_{vo} are kept fixed, so that $\dot{m}_{\delta=1} \neq \dot{m}_{\delta=0}$, to obtain the δ comparison

$$\frac{F_{\delta=1}}{F_{\delta=0}} = \frac{\left(\frac{F}{p_o A^*}\right)_{\delta=1}}{\left(\frac{F}{p_o A^*}\right)_{\delta=0}} \tag{9.51}$$

shown in Figure 9.9.

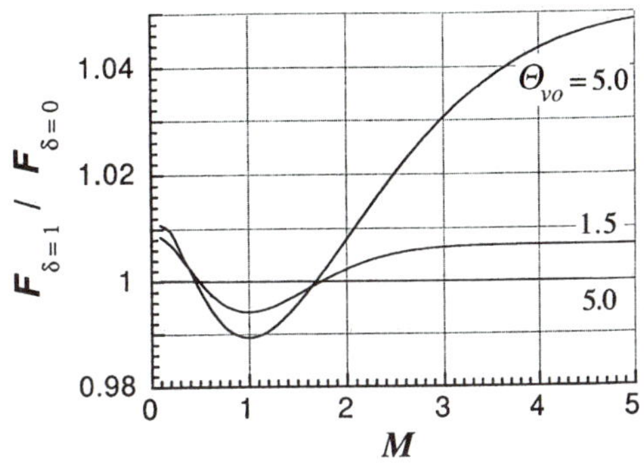

FIGURE 9.9 Isentropic comparison for the impulse function.

As evident in Equation (9.46), F consists of two terms. From Figures 9.4 and 9.6, observe that p and A have opposite trends with M. Similarly, w and $\dot{m}$ have opposite trends. These opposing trends account for the variation seen in Figure 9.9 and the relatively modest deviation from unity.

Thrust

For a rocket nozzle, the thrust is given by (Shapiro, 1953)

$$\Im = F_e - p_a A_e \tag{9.52a}$$

or nondimensionally as

$$\frac{\Im}{p_o A^*} = \frac{F_e}{p_o A^*} - \frac{p_a}{p_e}\frac{p_e}{p_o}\frac{A_e}{A^*} \tag{9.52b}$$

where Equation (9.49) provides F_e/p_oA^* by setting $\theta_v = \theta_{ve}$. The ambient pressure, p_a, need not equal the nozzle exit pressure, p_e, when $M_e \geq 1$. In view of Equations (9.47) and (9.48), we can write

$$\frac{\Im}{p_o A^*} = \frac{p_e}{p_o}\frac{A_e}{A^*}\left[7\frac{\theta_{ve}}{\theta_{vo}} - 6 + 2\delta\theta_{ve}(\coth\theta_{vo} - \coth\theta_{ve}) - \frac{p_a}{p_e}\right] \tag{9.53}$$

where

$$\frac{p_e}{p_o}\frac{A_e}{A^*} = \left(\frac{\theta_{vo}}{\theta_{ve}}\right)^{1/2}\left(\frac{\theta_{vo}}{\theta_v^*}\right)^3\left(\frac{\sinh\theta_{vo}}{\sinh\theta_v^*}\right)^{\delta}\{\exp[\delta(\theta_v^*\coth\theta_v^* - \theta_{vo}\coth\theta_{vo})]\}$$
$$\times\left[\frac{7\left(\frac{\theta_v^*}{\theta_{vo}} - 1\right) + 2\delta\theta_v^*(\coth\theta_{vo} - \coth\theta_v^*)}{7\left(\frac{\theta_{ve}}{\theta_{vo}} - 1\right) + 2\delta\theta_{ve}(\coth\theta_{vo} - \coth\theta_{ve})}\right]^{1/2} \tag{9.54}$$

This pressure area relation is useful for analyzing an internal duct flow containing a normal shock wave. [Note that p_oA^* is not constant across a normal shock when $\delta = 1$ because of the change in γ, whereas p_oA^* is constant when $\delta = 0$. The difference, however, in the value of $(p_oA)_{\delta=1}$ across a normal shock is expected to be fairly small.] When $\delta = 0$, the above equations reduce to

$$\left(\frac{\Im}{p_oA^*}\right)_{\delta=0} = \left(\frac{5}{6}\right)^3 \frac{1+\frac{7}{5}M^2 - \frac{p_a}{p_e}}{M_e\left(1+\frac{1}{5}M_e^2\right)^{1/2}} \tag{9.55}$$

The desired comparison is given by

$$\frac{\Im_{\delta=1}}{\Im_{\delta=0}} = \frac{\left(\frac{\Im}{p_oA^*}\right)_{\delta=1}}{\left(\frac{\Im}{p_oA^*}\right)_{\delta=0}} \tag{9.56}$$

which is shown in Figure 9.10 for different p_a/p_e ratios. (For simplicity, the $M_e \geq 1$ restriction is ignored.) As in the impulse function comparison, p_oA^* and θ_{vo} are fixed, and the comparison is for different $\dot{m}$ values. Note that the abscissa is actually the nozzle exit Mach number M_e.

The ratio $(p_a/p_e) = 0$ provides the vacuum thrust, and Figure 9.10(a) is identical to Figure 9.9. The curves have a minimum when $M = 1$. At low M values, the effect of $\dot{m}$ is dominant; its influence is overcome by the pressure and flow speed at larger M values. Above an exit Mach number of about 1.7, the imperfect gas thrust exceeds that for a perfect gas. This increase stems from the transfer of vibrational energy to rotational and translational energy, which ultimately increases the speed of the exit flow. The allowable payload weight of a rocket is sensitive to small changes in thrust. Although the deviation from unity in Figure 9.10 is usually small, on the order of a few percent, the effect on payload weight can be significant.

The unity value for p_a/p_e in panel (b) corresponds to a maximum thrust for a given nozzle configuration with fixed plenum conditions. Above an exit Mach number of about 2 (equilibrium), real-gas effects improve performance.

A $(p_a/p_e) = 2$ ratio represents an overexpanded nozzle. This ratio is sufficiently small that boundary-layer separation, for either a laminar or turbulent layer, inside the nozzle should not occur. As evident from Equation (9.52a), a sufficiently subsonic nozzle, with a large ambient pressure, has a negative thrust. The $\theta_{vo} = 1.5$ and 0.5 curves in panel (c) become negative at different nozzle exit subsonic Mach numbers. For the perfect gas case, this occurs when $M_e = (5/7)^{1/2}$. Of course, thrust nozzles do not have subsonic exit conditions. Consequently, a lower bound of unity is used for the abscissa in this figure.

Dynamic Pressure Comparison

The dynamic pressure

$$q = \frac{1}{2}\rho w^2 \tag{9.57}$$

is often encountered in fluid dynamics and aerodynamics. With the aid of earlier equations, q can be written nondimensionally as

$$\frac{q}{p_o} = \frac{1}{2}\left(\frac{\theta_{vo}}{\theta_v}\right)^{7/2}\left(\frac{\sinh\theta_{vo}}{\sinh\theta_v^*}\right)^{\delta}\{\exp[\delta(\theta_v\coth\theta_v - \theta_{vo}\coth\theta_{vo})]\}$$
$$\times\left[7\left(\frac{\theta_v}{\theta_{vo}} - 1\right) + 2\delta\theta_v(\coth\theta_{vo} - \coth\theta_v)\right] \tag{9.58}$$

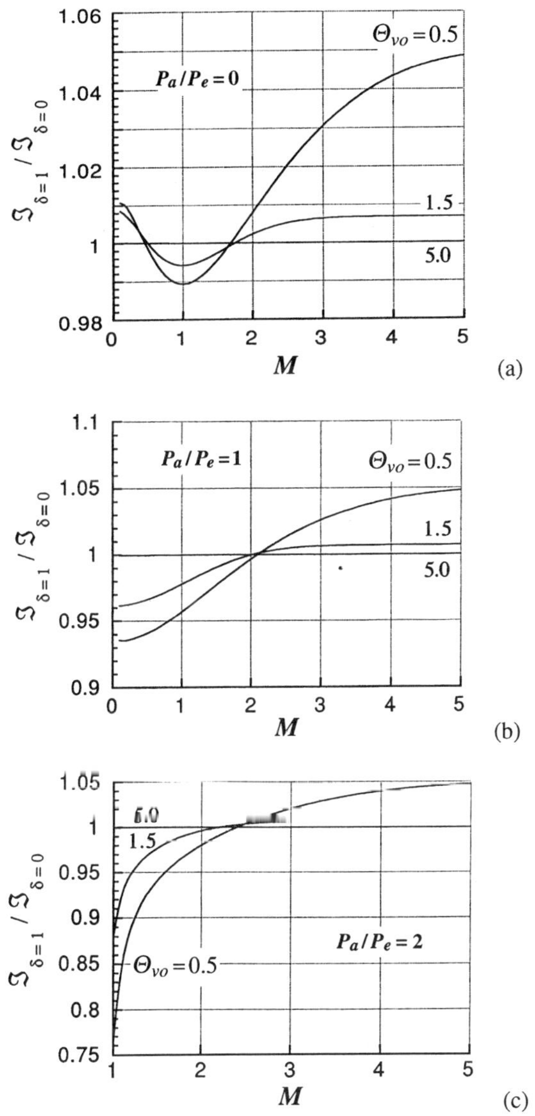

FIGURE 9.10 Isentropic comparison for the thrust vs. nozzle exit Mach number; (a) $p_a/p_o = 0$, (b) $p_a/p_e = 1$, (c) $p_a/p_e = 2$.

The $\delta = 0$ result is given by

$$\frac{q_{\delta=0}}{p_o} = \frac{7}{10}M^2\left(1 + \frac{1}{5}M^2\right)^{-7/2} \tag{9.59}$$

and the comparison, $q_{\delta=1}/q_{\delta=0}$, is shown as Figure 9.11.

The dynamic pressure of an imperfect gas is less than its perfect gas counterpart. Aside from this, the curves are reminiscent of the pressure curves in Figure 9.4. As in several other figures, the q ratio is not unity when M is zero. To explain this, the ratio is written as

$$\frac{q_{\delta=1}}{q_{\delta=0}} = \frac{\rho_{\delta=1}}{\rho_{\delta=0}}\left(\frac{a_{\delta=1}}{a_{\delta=0}}\right)^2 = \frac{(\rho\gamma T)_{\delta=1}}{(\rho\gamma T)_{\delta=0}}$$

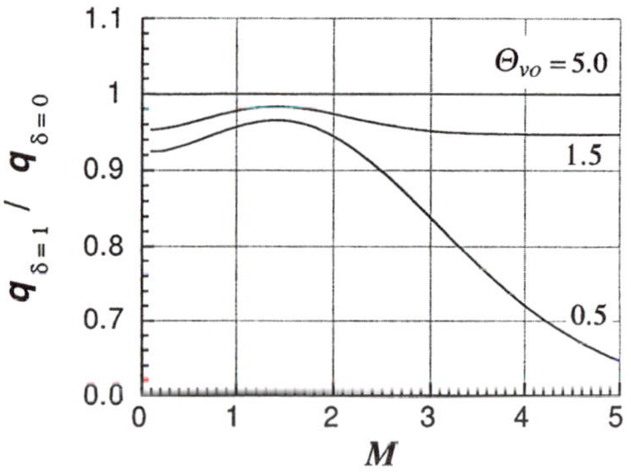

FIGURE 9.11 Isentropic comparison for the dynamic pressure.

when $M_{\delta=1} = M_{\delta=0}$. This becomes, when M is zero,

$$\frac{q_{\delta=1}}{q_{\delta=0}} = \frac{\gamma_{\delta=1}}{1.4}$$

since

$$\frac{\rho_{\delta=1}}{\rho_{\delta=0}} = 1, \qquad \frac{T_{\delta=1}}{T_{\delta=0}} = \frac{T_o}{T_o} = 1$$

With $\theta_v = \theta_{vo}$ as the abscissas, the q ratio is in accord with Figure 9.1 when M is zero.

9.4 PLANAR SHOCK FLOW

Steady flow across a planar shock wave is discussed. The upstream flow, denoted with a unity subscript, and the downstream flow, denoted with a subscript 2, are uniform and in thermodynamic equilibrium. The next subsection treats a normal shock while the last subsection considers an oblique shock.

As before, the upstream comparison flows possess identical stagnation values. The comparison is made unique for a nominal shock by choosing

$$(M_1)_{\delta=0} = (M_1)_{\delta=1} = M_1 \tag{9.60}$$

for the upstream Mach number. Thus, $(\theta_{v1})_{\delta=1}$ does not equal $(\theta_{v1})_{\delta=0}$. These θ_{v1} values are provided by Equations (9.30) and (9.31) and their δ comparison is shown in Figure 9.2. Similarly, $(p_1)_{\delta=1}$ does not equal $(p_1)_{\delta=0}$. For the comparison, it is therefore advisable to use $(p_2/p_1)_{\delta=1}/(p_2/p_1)_{\delta=0}$ rather than $(p_2)_{\delta=1}/(p_2)_{\delta=0}$. These remarks also apply to other δ comparisons.

Normal Shock Comparison

The conventional shock jump conditions are

$$(\rho w)_1 = (\rho w)_2 \tag{9.61a}$$

$$(p + \rho w^2)_1 = (p + \rho w^2)_2 \tag{9.61b}$$

$$\left(h + \frac{1}{2}w^2\right)_1 = \left(h + \frac{1}{2}w^2\right)_2 \tag{9.61c}$$

Note that h_{o2} equals h_{o1} for $\delta = 0$ and 1. Consequently, we have

$$(h_{o1})_{\delta=0} = (h_{o2})_{\delta=0} = (h_{o1})_{\delta=1} = (h_{o2})_{\delta=0} = h_o$$

In view of Equation (9.17), a similar set of equations holds for θ_{vo}. Our goal is to determine the state of the gas in region 2, in terms of that in region 1, in as simple a form as possible. Toward this end, continuity is revised to

$$\frac{\rho_2}{\rho_1} = \frac{w_1}{w_2} \tag{9.62a}$$

Similarly, momentum becomes

$$p_2 = p_1 + \rho_1 w_1^2\left(1 - \frac{w_2}{w_1}\right)$$

$$\rho_2 R T_2 = \rho_1 R T_1 + \rho_1 \gamma_1 R T_1 M_1^2\left(1 - \frac{w_2}{w_1}\right)$$

$$\frac{w_1}{w_2}\frac{T_2}{T_1} = 1 + \gamma_1 M_1^2\left(1 - \frac{w_2}{w_1}\right)$$

$$\frac{\theta_{v1}}{\theta_{v2}} = (1 + \gamma_1 M_1^2)\frac{w_2}{w_1} - \gamma_1 M_1^2\left(\frac{w_2}{w_1}\right)^2 \tag{9.62b}$$

Equation (9.27) is used to eliminate w_2/w_1. With the use of Equations (9.5) and (9.62), we have

$$\frac{\rho_2}{\rho_1} = \frac{w_1}{w_2} = \frac{\gamma_1}{\gamma_2}\frac{M_1^2}{M_2^2}\left(\frac{1 + \gamma_2 M_2^2}{1 + \gamma_1 M_1^2}\right) \tag{9.63a}$$

$$\frac{\theta_{v2}}{\theta_{v1}} = \frac{\gamma_1}{\gamma_2}\frac{M_1^2}{M_2^2}\left(\frac{1 + \gamma_2 M_2^2}{1 + \gamma_1 M_1^2}\right)^2 \tag{9.63b}$$

$$\frac{p_2}{p_1} = \frac{1 + \gamma_1 M_1^2}{1 + \gamma_2 M_2^2} \tag{9.63c}$$

for the jump conditions in terms of γ_1, γ_2, M_1, and M_2. These relations hold for $\delta = 0$ and 1.

With M_1 and θ_{vo} known, Equation (9.29) is used to determine θ_{v1} when $\delta = 0$. Similarly, Equations (9.23) and (9.28) determine γ_1 and θ_{v1} when $\delta = 1$. For downstream conditions, we utilize Equations (9.16b), (9.17), and (9.27) in conjunction with

$$h_o = h_2 + \frac{1}{2}w_2^2 \tag{9.64}$$

to obtain

$$\gamma_2 M_2^2 = 7\left(\frac{\theta_{v2}}{\theta_{vo}} - 1\right) + 2\delta\theta_{v2}(\coth\theta_{vo} - \coth\theta_{v2}) \tag{9.65}$$

When $\delta = 0$, this relation in conjunction with Equation (9.63b) yields θ_{v2} and M_2. When $\delta = 1$, we also need Equation (9.23) to determine θ_{v2} and M_2. After θ_{v2} and M_2 are known, other parameters are easily evaluated for both $\delta = 0$ and $\delta = 1$. Note that M_1 must be subsonic.

With the aid of Equation (9.16c), the entropy jump across the shock is

$$\frac{s_2 - s_1}{R} = \ln\left[\left(\frac{\gamma_2 M_2^2}{\gamma_1 M_1^2}\right)^{7/2}\left(\frac{1 + \gamma_1 M_1^2}{1 + \gamma_2 M_2^2}\right)^6\left(\frac{\sinh\theta_{v1}}{\sinh\theta_{v2}}\right)^\delta\right] + \delta(\theta_{v2}\coth\theta_{v2} - \theta_{v1}\coth\theta_{v1}) \tag{9.66}$$

It is also useful to derive $(p_{o2}/p_{o1})_{\delta=1}$ and its δ comparison. Equation (9.33) is used to eliminate p_2/p_1 from Equation (9.63c), with the result

$$\left(\frac{p_{o2}}{p_{o1}}\right)_{\delta=1} = \frac{1 + \gamma_1 M_1^2}{1 + \gamma_2 M_2^2}\left(\frac{\theta_{v2}}{\theta_{v1}}\right)^{7/2}\frac{\sinh\theta_{v2}}{\sinh\theta_{v1}}\exp(\theta_{v1}\coth\theta_{v1} - \theta_{v2}\coth\theta_{v2}) \tag{9.67}$$

Equation (9.66) is rewritten as

$$\exp(\theta_{v1}\coth\theta_{v1} - \theta_{v2}\coth\theta_{v2}) = \left(\frac{\gamma_2 M_2^2}{\gamma_1 M_1^2}\right)^{7/2}\left(\frac{1 + \gamma_1 M_1^2}{1 + \gamma_2 M_2^2}\right)^6\frac{\sinh\theta_{v1}}{\sinh\theta_{v2}}\exp\left[-\frac{(s_2 - s_1)_{\delta=1}}{R}\right]$$

which yields

$$\left(\frac{p_{o2}}{p_{o1}}\right)_{\delta=1} = \left(\frac{\theta_{v2}}{\theta_{v1}}\right)^{7/2}\left(\frac{\gamma_2 M_2^2}{\gamma_1 M_1^2}\right)^{7/2}\left(\frac{1 + \gamma_1 M_1^2}{1 + \gamma_2 M_2^2}\right)^7\exp\left[-\frac{(s_2 - s_1)_{\delta=1}}{R}\right]$$

Equation (9.63b) is used to eliminate θ_v, with the simple result

$$\left(\frac{p_{o2}}{p_{o1}}\right)_{\delta=1} = \exp\left[-\frac{(s_2 - s_1)_{\delta=1}}{R}\right]$$

The same equation holds when $\delta = 0$:

$$\left(\frac{p_{o2}}{p_{o1}}\right)_{\delta=0} = \exp\left[-\frac{(s_2 - s_1)_{\delta=0}}{R}\right]$$

Hence, the stagnation pressure ratio comparison is

$$\frac{\left(\frac{p_{o2}}{p_{o1}}\right)_{\delta=1}}{\left(\frac{p_{o2}}{p_{o1}}\right)_{\delta=0}} = \frac{\exp\left[-\frac{(s_2 - s_1)_{\delta=1}}{R}\right]}{\exp\left[-\frac{(s_2 - s_1)_{\delta=0}}{R}\right]} \tag{9.68}$$

The formulation concludes with the Rayleigh pitot tube pressure ratio comparison for a supersonic flow. When $\delta = 0$, this ratio is

$$\left(\frac{p_{o2}}{p_1}\right)_{\delta=0} = 5^{1/2}\left(\frac{6}{5}\right)^6 \frac{M_1^7}{(7M_1^2 - 1)^{5/2}} \tag{9.69}$$

For $\delta = 1$, the pitot tube pressure ratio is written as

$$\left(\frac{p_{o2}}{p_1}\right)_{\delta=1} = \left(\frac{p_{o2}}{p_2}\right)_{\delta=1}\left(\frac{p_2}{p_1}\right)_{\delta=1} \tag{9.70}$$

where the p_2/p_1 ratio is given by Equation (9.63c). The p_{o2}/p_2 ratio is provided by Equation (9.33), written as

$$\left(\frac{p_2}{p_{o2}}\right)_{\delta=1} = \left(\frac{\theta_{vo}}{\theta_{v2}}\right)^{1/2} \frac{\sinh\theta_{vo}}{\sinh\theta_{v2}} \exp(\theta_{v2}\coth\theta_{v2} - \theta_{vo}\coth\theta_{vo}) \tag{9.71}$$

where θ_{v2} requires $\delta = 1$. The above equations provide the supersonic pitot tube comparison.

Figures 9.12 through 9.18 provide the following comparisons:

$$\frac{(M_2)_{\delta=1}}{(M_2)_{\delta=0}}, \quad \frac{\left(\frac{\rho_2}{\rho_1}\right)_{\delta=1}}{\left(\frac{\rho_2}{\rho_1}\right)_{\delta=0}}, \quad \frac{\left(\frac{p_2}{p_1}\right)_{\delta=1}}{\left(\frac{p_2}{p_1}\right)_{\delta=0}}, \quad \frac{\left(\frac{\theta_{v2}}{\theta_{v1}}\right)_{\delta=1}}{\left(\frac{\theta_{v2}}{\theta_{v1}}\right)_{\delta=0}}, \quad \frac{(s_2 - s_1)_{\delta=1}}{(s_2 - s_1)_{\delta=0}}, \quad \frac{\left(\frac{p_{o2}}{p_{o1}}\right)_{\delta=1}}{\left(\frac{p_{o2}}{p_{o1}}\right)_{\delta=0}}, \quad \frac{\left(\frac{p_{o2}}{p_1}\right)_{\delta=1}}{\left(\frac{p_{o2}}{p_1}\right)_{\delta=0}}$$

For a perfect gas, the limiting value

$$M_2 = \left(\frac{\gamma - 1}{2\gamma}\right)^{1/2}$$

occurs when $M_1 \to \infty$. As γ decreases, the limiting M_2 value decreases. This behavior is observed in Figure 9.12, where the limiting values for M_2, with $\delta = 0$ and 1, are nearly obtained by the time M_1 is 5. There is a discernible minimum in the curves because a and w have opposite trends with M_1.

As in isentropic flow, there is a substantial change in the density ratio, as seen in Figure 9.13. This figure also holds for the flow speed ratio, $(w_1/w_2)_{\delta=1}/(w_1/w_2)_{\delta=0}$. As with M_2, the large asymptotic

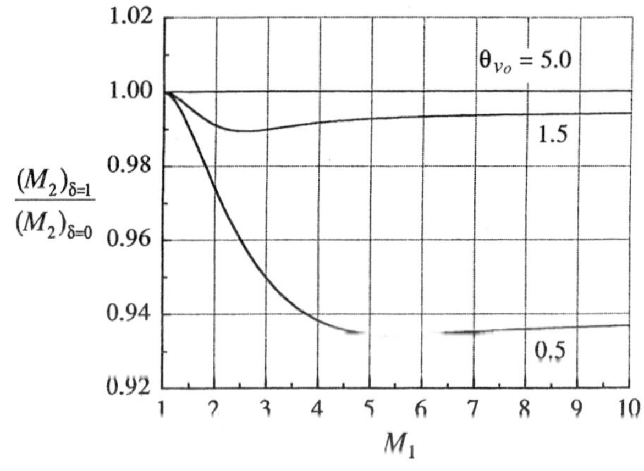

FIGURE 9.12 Normal shock comparison for the Mach number.

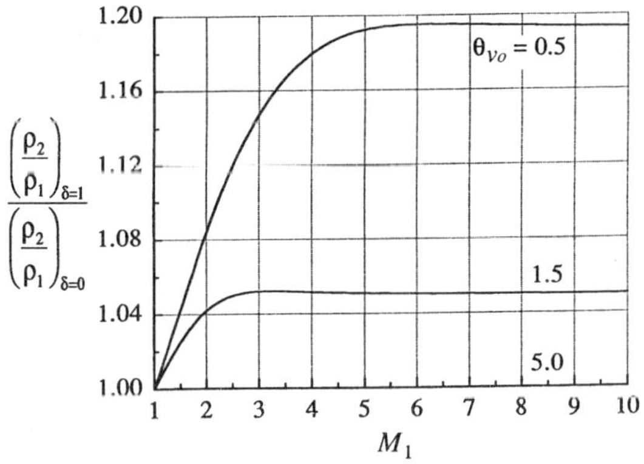

FIGURE 9.13 Normal shock comparison for the density.

value for M_1 is essentially achieved by the time M_1 is 5. For a perfect gas, the density ratio has a limiting value

$$\frac{\rho_2}{\rho_1} = \frac{\gamma + 1}{\gamma - 1}$$

when $M_1 \to \infty$. This ratio increases when γ decreases, thus yielding the imperfect to perfect gas trend seen in the figure.

Again, as in isentropic flow, the pressure ratio, shown in Figure 9.14, has a rather complicated behavior. This is caused by γ_1, γ_2, and M_2 varying with M_1 at different rates when $\delta = 1$. The deviation from unity, however, is not large. The δ comparison for θ_v is shown in Figure 9.15. The ordinate also represents $(T_1/T_2)_{\delta=1}/(T_1/T_2)_{\delta=0}$. As with other comparisons, the maximum is caused by different rates of change for γ and T with increasing M_1.

Figure 9.16 shows the entropy difference comparison. The peculiar behavior near $M_1 = 1$ is caused by $s_2 - s_1$ going to zero when M_1 goes to unity for both $\delta = 0$ and 1. The rate at which $s_2 - s_1$ goes to zero, however, is different for the imperfect and perfect gases. This difference accounts for the nonunity values at $M_1 = 1$ and the crossover in the $\theta_{vo} = 0.5$ and 1.5 curves.

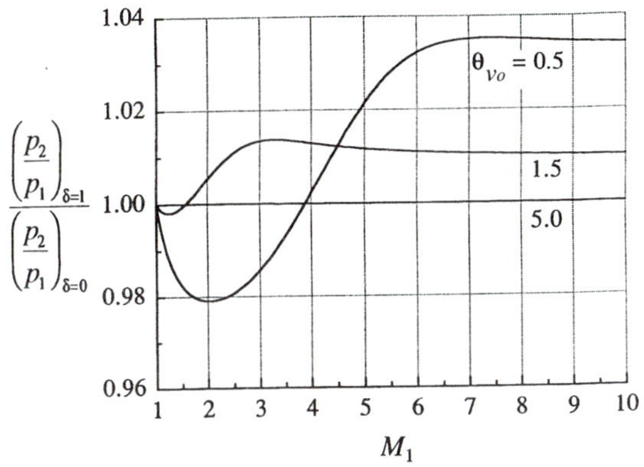

FIGURE 9.14 Normal shock comparison for the pressure.

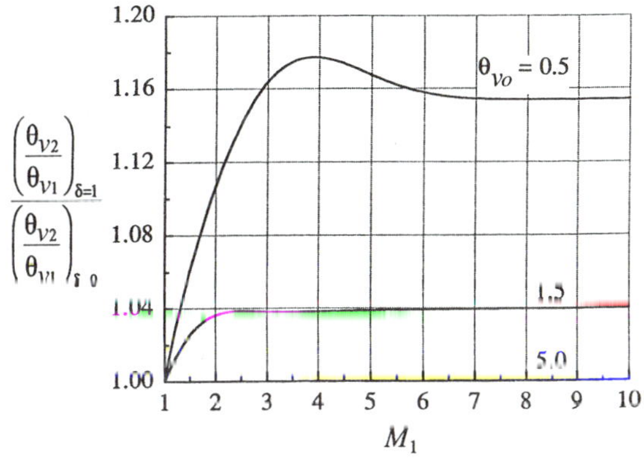

FIGURE 9.15 Normal shock comparison for θ_v or the temperature.

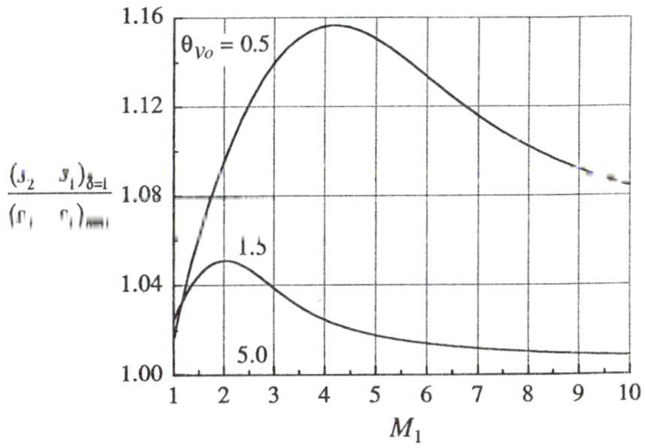

FIGURE 9.16 Normal shock comparison for the entropy jump.

Although the stagnation pressure and entropy jumps are closely related, the stagnation pressure comparison (see Figure 9.17) is quite different. Of some practical importance is the significant reduction in the loss of stagnation pressure for an imperfect gas, relative to a perfect gas, at large M_1 and T_o values. In a recirculating wind tunnel, e.g., this loss determines the pressure ratio for the wind tunnel's compressors.

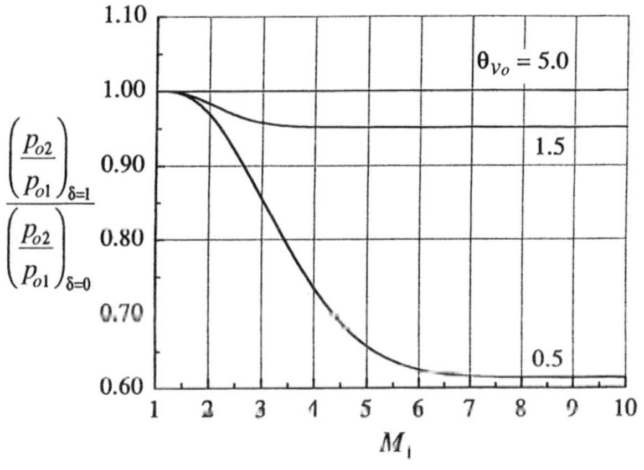

FIGURE 9.17 Normal shock comparison for the stagnation pressure.

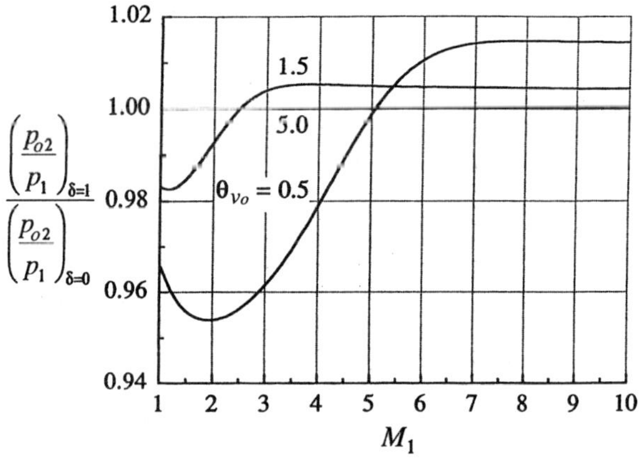

FIGURE 9.18 Normal shock comparison for the Rayleigh pitot tube pressure ratio.

A pitot pressure comparison is shown in Figure 9.18. The deviation from unity is modest and the trends are similar to those for the pressure in Figure 9.14.

Oblique Shock Comparison

The angles θ and β, respectively, designate the wall, or velocity, turn angle and the shock angle (see Figure 9.19). As with a normal shock, upstream stagnation conditions and M_1 are kept fixed. For a unique comparison, either θ or β must also be fixed. It is physically more meaningful to fix the wall turn angle; i.e.,

$$\theta_{\delta=0} = \theta_{\delta=1} = \theta$$

As a consequence, $\beta_{\delta=1}$ does not equal $\beta_{\delta=0}$. (It is important not to confuse the turn angle θ with θ_v.)

As indicated in Figure 9.19, a t subscript denotes the velocity component that is tangential to the shock, while an n subscript denotes the normal component. We thus have

$$w_{1n} = w_1 \sin\beta = w_{1t} \sin\beta, \quad w_{2n} = w_2 \sin(\beta - \theta) = w_{2t} \sin(\beta - \theta) \tag{9.72a,b}$$

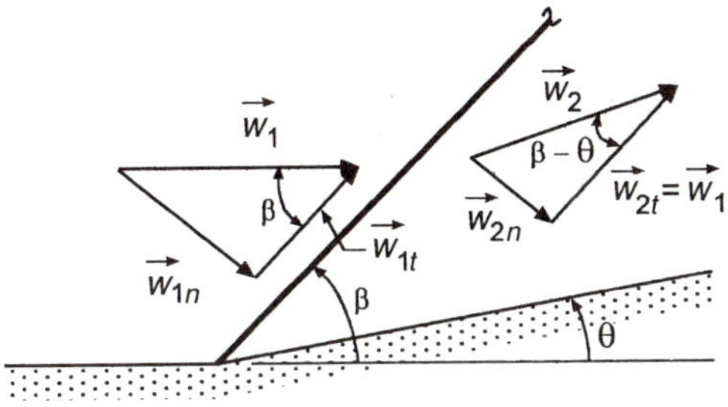

FIGURE 9.19 Schematic for an oblique shock wave.

and recall that $w_{2t} = w_{1t}$. The normal component Mach numbers

$$M_{1n} = \frac{w_{1n}}{a_1}, \quad M_{2n} = \frac{w_{2n}}{a_2} \tag{9.73}$$

are introduced, and note that

$$M_{1n} = M_1 \sin\beta, \quad M_{2n} = M_2 \sin(\beta - \theta) \tag{9.74a,b}$$

The above are the usual relations for an oblique shock and hold for both $\delta = 0$ and 1.

When $\delta = 0$, the standard oblique shock relations can be used. Among these is the β,θ equation

$$\tan\theta = \frac{\cot\beta(M_1^2 \sin^2\beta - 1)}{1 + \left(\frac{\gamma + 1}{2} - \sin^2\beta\right)M_1^2} \tag{9.75}$$

The $\delta = 1$ counterpart stems from Equations (9.63a), (9.72), and (9.74), with the result

$$\frac{\gamma_1 M_1^2 \sin^2\beta}{\gamma_2 M_2^2 \sin^2(\beta - \theta)} \frac{1 + \gamma_2 M_2^2 \sin^2(\beta - \theta)}{1 + \gamma_1 M_1^2 \sin^2\beta} = \frac{\tan\beta}{\tan(\beta - \theta)} \tag{9.76}$$

An explicit solution for θ can be found in terms of β, M_1, M_2, θ_{v1}, and θ_{v2}. This complicated result, however, has no utility because of the explicit dependence on M_2 and θ_{v2}. For instance, there is no simple counterpart to the $\delta = 0$ equation

$$M_2^2 \sin^2(\beta - \theta) = \frac{1 + \frac{\gamma - 1}{2} M_1^2 \sin^2\beta}{\gamma M_1^2 \sin^2\beta - \frac{\gamma - 1}{2}} \tag{9.77}$$

In view of the foregoing situation, an iterative procedure is outlined for the $\delta = 1$ case. First, note that θ_{v1} is determined in the preceding subsection. In the following, it is notationally convenient to retain M_{2n} in place of $M_2 \sin(\beta - \theta)$. Equations (9.63b) and (9.65) are written as

$$\frac{\theta_{v2}}{\theta_{v1}} = \frac{\gamma_1 M_1^2 \sin^2\beta}{\gamma_2 M_{2n}^2} \left(\frac{1 + \gamma_2 M_{2n}^2}{1 + \gamma_1 M_1^2 \sin^2\beta}\right)^2 \tag{9.78}$$

$$\gamma_2 M_{2n}^2 = 7\left(\frac{\theta_{v2}}{\theta_{vo}} - 1\right) + 2\theta_{v2}(\coth\theta_{vo} - \coth\theta_{v2}) \tag{9.79}$$

With $\gamma_2 M_{2n}^2$ eliminated, the unknowns are β and θ_{v2}. Although θ, not β, is prescribed, it is nevertheless convenient to assume a value for β and thus determine θ_{v2}. Subsequently, $\gamma_2 M_{2n}^2$ and, therefore, γ_2 and M_{2n} are found. Equation (9.76) is written with M_{2n} in place of $M_2 \sin(\beta - \theta)$ and solved for θ:

$$\theta = \beta - \tan^{-1}\left[\frac{\gamma_2 M_{2n}^2(1 + \gamma_1 M_1^2 \sin^2\beta)}{\gamma_1 M_1^2 \sin\beta \cos\beta(1 + \gamma_2 M_{2n}^2)}\right] \tag{9.80}$$

The foregoing procedure, starting with Equation (9.78), is iteratively solved by varying β until the specified θ value is obtained.

Once θ_{v2}, β, and $\gamma_2 M_{2n}^2$ are known for specified values of M_1, θ, and θ_{vo}, all other quantities are easily determined. For instance, the pressure ratio across the shock is

$$\frac{p_2}{p_1} = \frac{1 + \gamma_1 M_1^2 \sin^2\beta}{1 + \gamma_2 M_{2n}^2} \tag{9.81}$$

Care must be exercised to distinguish between the weak and strong shock solutions. This is readily done by fixing M_1 and θ_{vo}, and varying β from the Mach angle

$$\mu = \sin^{-1}\frac{1}{M_1} \tag{9.82}$$

to $\pi/2$. This approach generates an inverted U curve for θ whose maximum value represents the demarcation between the weak and strong solutions. A value for θ_{vo} needs to be specified to obtain a β,θ plot when $\delta = 1$. Hence, each θ_{vo} value requires its own β,θ figure; see Figure 9.20(a,b) for plots when θ_{vo} equals 1.5 and 0.5, respectively. A plot is not shown for $\theta_{vo} = 5$, since the perfect and imperfect curves overlay each other. (See Figure 9.28a for a composite figure.)

Figure 9.20 also shows the β,θ curves as dashed lines when $\gamma = 1.4$. These curves are most easily obtained using the inversion formulas of Appendix C.

The comparisons shown in Figures 9.12 through 9.17 also hold for an oblique shock by simply replacing M_1 and M_2 with M_{1n} and M_{2n}, respectively. Note that this replacement does not hold for the pitot pressure result, since this figure is valid only for a normal shock.

The imperfect gas curves in Figures 9.20 have a larger detachment value for θ than the corresponding perfect gas curves. For example, with $M_1 = 3$ and $\theta_{vo} = 0.5$, the difference in the detachment angles is about 3°. For a fixed θ, the weak (strong) solution imperfect gas shock is weaker (stronger) than the corresponding perfect gas shock.

9.5 PRANDTL–MEYER FLOW

Our first task is to verify Equation (9.82). Equation (9.76) is written as

$$\frac{\gamma_1 M_{1n}^2}{\gamma_2 M_{2n}^2} \frac{1 + \gamma_2 M_{2n}^2}{1 + \gamma_1 M_{1n}^2} = \frac{\tan\beta}{\tan(\beta - \theta)}$$

which holds for both the $\delta = 0$ and 1 cases. In the limit of $\theta \to 0$, we obtain $M_{2n} = M_{1n}$, which is possible only when

$$M_{2n} = M_{1n} = 1$$

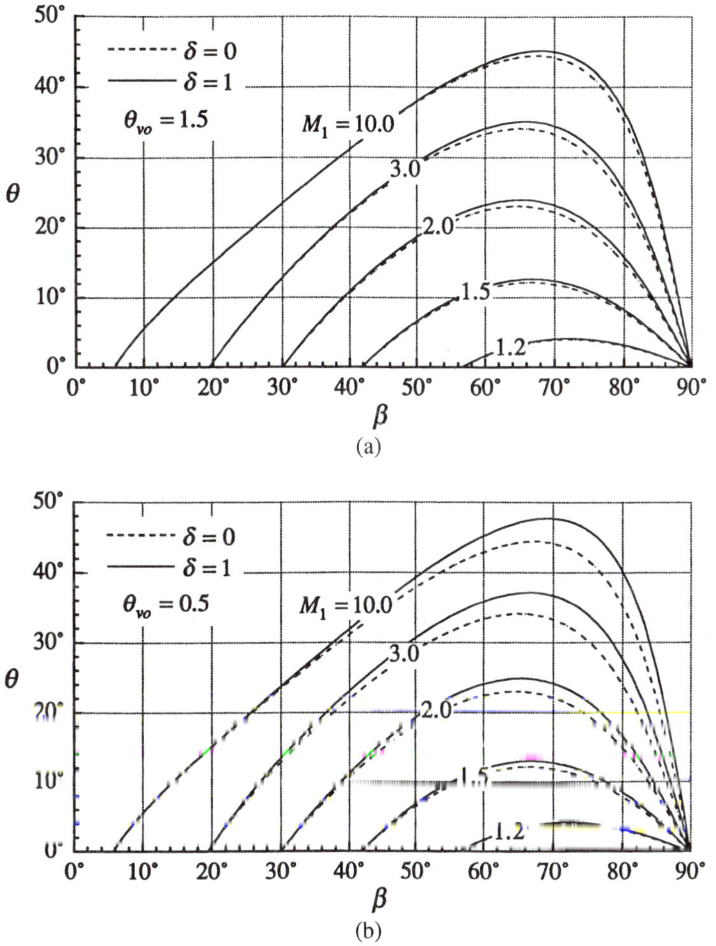

FIGURE 9.20 θ vs. β curves for (a) $\theta_{vo} = 1.5$ and (b) $\theta_{vo} = 0.5$. The solid curves are for $\delta = 1$, the dashed curves for $\delta = 0$.

Consequently, Equation (9.74a) yields

$$\beta = \sin^{-1}\frac{1}{M} = \mu \qquad (9.83)$$

as expected. Thus, the equation for the Mach angle is unaltered by equilibrium real-gas effects.

Prandtl–Meyer flow is isentropic; hence, Section 9.3 is directly applicable. An expansive wall turn with a positive angle θ is considered. The turn may be sharp, thereby yielding a centered expansion, or it may be gradual. A gradual compressive turn, with a negative θ value, is also appropriate for the region of homentropic flow that is adjacent to the wall, providing the downstream Mach number is still supersonic. As was done for an oblique shock, the wall turn angle θ is the same for both the $\delta = 0$ and 1 cases.

The Mach numbers upstream and downstream of the expansion are denoted as M_1 and M_2, respectively. With stagnation conditions fixed, we retain Equation (9.60), and note that

$$(M_2)_{\delta=0} \neq (M_2)_{\delta=1}$$

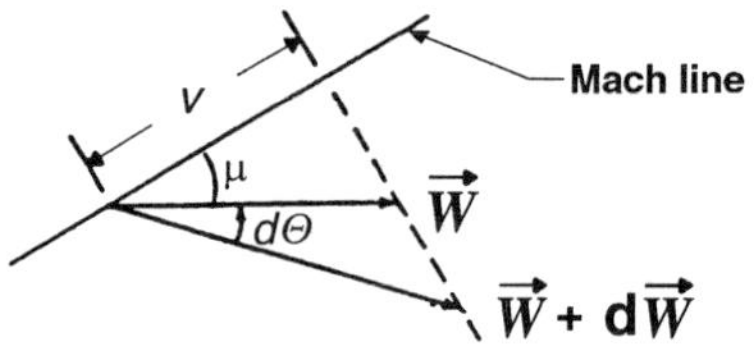

FIGURE 9.21 Schematic for a two-dimensional supersonic expansion.

Consequently, Figures 9.2 to 9.4, e.g., cannot be used at station 2, since these figures require $M_{\delta=0} - M_{\delta=1}$.

When $\delta = 0$, we have the well-known relation

$$\nu(M_2) = \nu(M_1) + \theta \tag{9.84}$$

where the Prandtl–Meyer function is

$$\nu = \left(\frac{\gamma+1}{\gamma-1}\right)^{1/2} \tan^{-1}\left[\left(\frac{\gamma-1}{\gamma+1}\right)^{1/2}(M^2-1)^{1/2}\right] - \tan^{-1}(M^2-1)^{1/2} \tag{9.85}$$

From the sketch in Figure 9.21, we obtain

$$\frac{v}{w} = \cos\mu = \frac{(M^2-1)^{1/2}}{M}$$

and

$$\frac{v}{w+dw} = \cos(\mu + d\theta)$$

which results in the equation

$$\frac{dw}{w} = \frac{d\theta}{(M^2-1)^{1/2}} \tag{9.86}$$

This relation requires a two-dimensional flow but otherwise holds for $\delta = 0$ and 1.

In view of the above, the subsequent discussion is limited to the $\delta = 1$ case. From equations (9.27) and (9.28), we have

$$\frac{w^2}{RT_v} = \frac{7}{2}\left(\frac{1}{\theta_{vo}} - \frac{1}{\theta_v}\right) + \coth\theta_{vo} - \coth\theta_v \tag{9.87}$$

By differentiation, this becomes

$$\frac{w^2}{RT_v}\frac{dw}{w} = \frac{1}{4}[7 + 2z(\theta_v)]\frac{d\theta_v}{\theta_v^2}$$

Since

$$\frac{w^2}{RT_v} = \frac{\gamma M^2}{2\theta_v}$$

we write

$$\frac{dw}{w} = \frac{[7 + 2z(\theta_v)]}{2\gamma M^2} \frac{d\theta_v}{\theta_v}$$

where γM^2 is given by Equation (9.28). Equation (9.86) thus becomes

$$d\theta = \frac{1}{2} \frac{(M^2 - 1)^{1/2}}{M^2} [5 + 2z(\theta_v)] \frac{d\theta_v}{\theta_v} \tag{9.88}$$

where

$$M^2 = \frac{1}{\gamma}\left[7\left(\frac{\theta_v}{\theta_{vo}} - 1\right) + 2\theta_v(\coth\theta_{vo} - \coth\theta_v)\right] \tag{9.89}$$

and γ is given by Equation (9.23). The above relation is integrated:

$$\int_0^{\nu} d\theta = \frac{1}{2}\int_{\theta_v^*}^{\theta_v} \frac{(M^2 - 1)^{1/2}}{M^2} [5 + 2z(\theta_v)] \frac{d\theta_v}{\theta_v}$$

to obtain

$$\nu(M) = N(\theta_v(M), \theta_{vo}) = \frac{1}{2}\int_{\theta_v^*}^{\theta_v} \frac{(M^2 - 1)^{1/2}}{M^2} [5 + 2z(\theta_v)] \frac{d\theta_v}{\theta_v} \tag{9.90}$$

Remember that θ_v^* depends only on θ_{vo} and that M and θ_v are related by Equation (9.89). Figure 9.22 provides the Prandtl–Meyer function. As expected, the curves for a perfect gas and $\theta_{vo} = 5$ overlay each other. The difference between $\theta_{vo} = 1.5$ and a perfect gas is small, whereas the $\theta_{vo} = 0.5$ difference is substantial.

Let us review the computational procedure for the $\delta = 1$ case when stagnation conditions, M_1, and the wall turn angle θ are known. Equation (9.34) is utilized for θ_v^*, after which Equation (9.89) is used to determine θ_{v1}. With θ_v^* and θ_{v1} established, Equations (9.89) and (9.90) provide the upstream value, $N(\theta_{v1}, \theta_{vo})$. Next, the integral equation

$$N(\theta_{v2}, \theta_{vo}) = N(\theta_{v1}, \theta_{vo}) + \theta \tag{9.91}$$

is iteratively solved for θ_{v2}, which appears as the upper limit on the integral in Equation (9.90). This iteration again requires Equations (9.23) and (9.89) for γ and M. The above equation is the

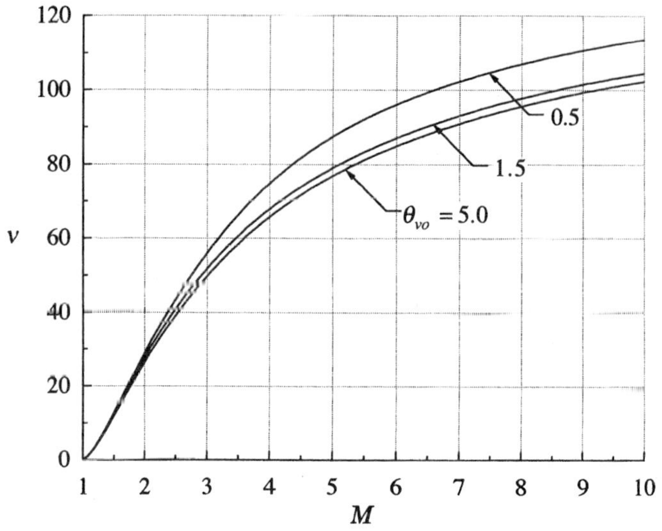

FIGURE 9.22 Prandtl–Meyer function ν for various θ_{vo} values.

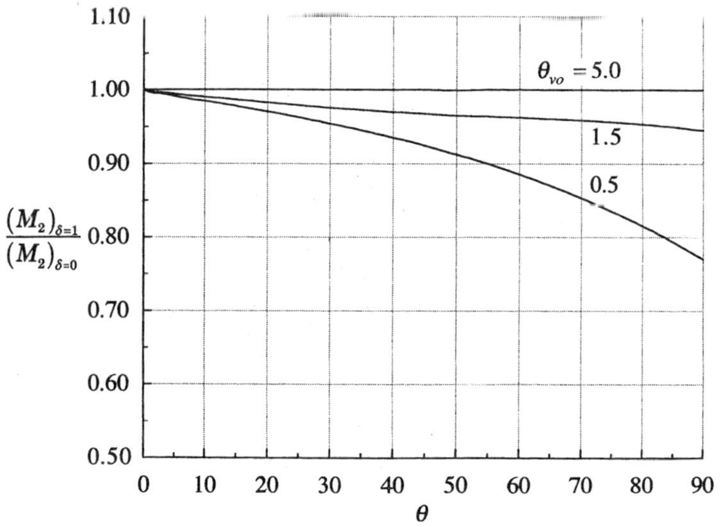

FIGURE 9.23 Prandtl–Meyer comparison for the downstream Mach number when $M_1 = 1$.

$\delta = 1$ counterpart of Equation (9.84). As an assist, it is advisable to tabulate $N(\theta_{v1}, \theta_{vo})$ vs. θ_v for a variety of θ_{vo} values. At any rate, θ_{v2} then provides M_2 via Equation (9.89). Other flow variables for station 2 are readily obtained once θ_{v2} and M_2 are known.

Figures 9.23 through 9.26, respectively, show δ comparisons for M, θ_v, ρ, and p at station 2 when $M_1 = 1$. As with a shock wave, upstream state conditions are used in the comparisons whenever their $\delta = 0$ and 1 values differ.

In view of the trends in Figure 9.22, we expect the imperfect gas downstream Mach number to be less than its perfect gas counterpart, as is shown in Figure 9.23. For a large wall turn angle, the decrease in M can be substantial. This Mach number decrease is in accord with the temperature,

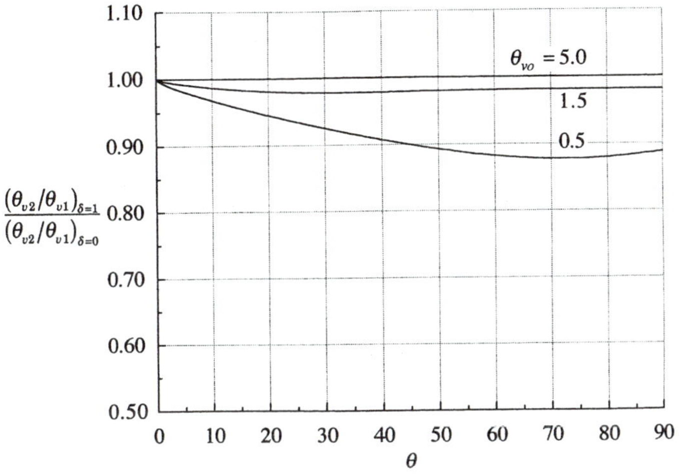

FIGURE 9.24 Prandtl–Meyer comparison for the downstream θ_v value when $M_1 = 1$.

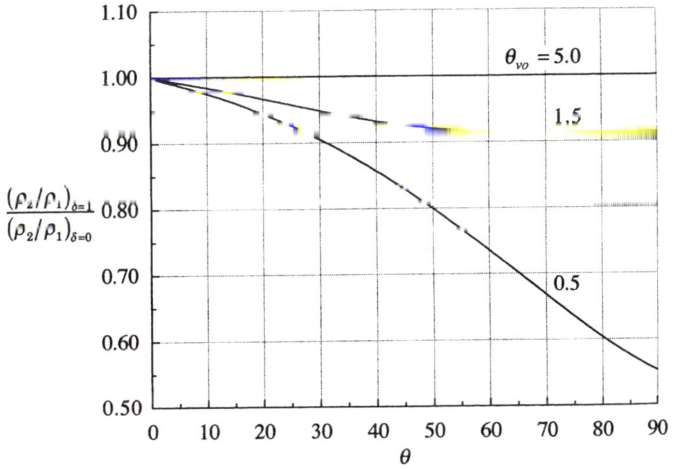

FIGURE 9.25 Prandtl–Meyer comparison for the downstream density when $M_1 = 1$.

or θ_v, and density deviations seen in Figures 9.24 and 9.25. The minimum values seen in Figure 9.24 are again caused by different rates of change, with θ or M, of γ and θ_v when $\delta = 1$. As in the previous flows, the density deviation is quite substantial. Again, as in the previous flows, the pressure deviation in Figure 9.26 is nonmonotonic.

9.6 TAYLOR–MACCOLL FLOW

A uniform, supersonic freestream flow over a cone at zero incidence is considered. When the bow shock is attached to the vertex of the cone, the disturbed flow is called a Taylor–Maccoll flow. In this situation, the shock is also conical (see Figure 9.27), as is the flow field between the shock and body.

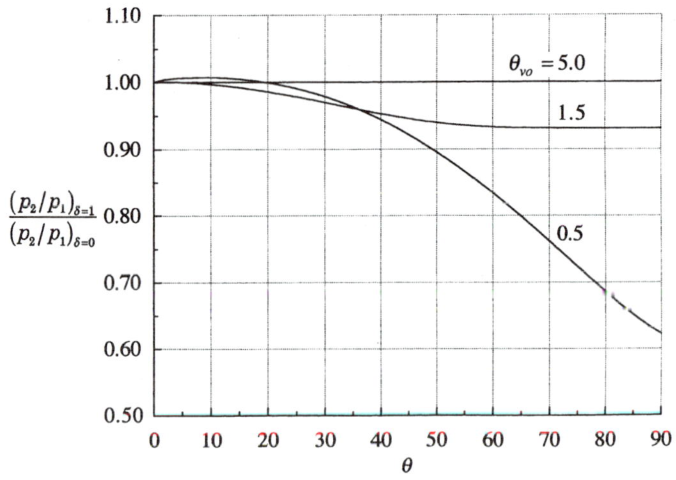

FIGURE 9.26 Prandtl–Meyer comparison for the downstream pressure when $M_1 = 1$.

We follow the same guidelines previously used. Freestream stagnation conditions, M_1, and the cone's semi-vertex angle, θ_b, are prescribed. Our objective is to determine and compare the flow field, including the unknown shock wave angle, β, when $\delta = 0$ and 1. Once this is done, δ comparisons can be provided for the parameters of interest. These are β, the Mach number M_2 just downstream of the shock, the Mach number M_b on the surface of the body, and the surface pressure coefficient

$$C_{pb} = \frac{p_b - p_1}{q_1} = \frac{2\left(\frac{p_b}{p_1} - 1\right)}{\gamma_1 M_1^2} \tag{9.92}$$

which represents the wave drag of the cone.

Once β and M_2 are known, other $\delta = 0$ or 1 parameters, just downstream of the shock, are found by virtue of the analysis in Section 9.4. Although this section is for a planar shock, the theory is nevertheless applicable, since a ray along a conical shock is straight when it passes through the vertex of the cone. Consequently, h_o and θ_{vo} are the same constants throughout both the $\delta = 0$ and 1 flow fields. The streamline angle ϕ_2, just downstream of the shock, therefore equals θ in Equation (9.75) when $\delta = 0$. Flow conditions between the shock and cone, at the cone's vertex, however, are singular. For instance, the streamline angle is θ_b at the vertex, whereas elsewhere along the shock it is ϕ_2 with $\phi_2 < \theta_b$.

General Formulation

Although this is a well-known flow when $\delta = 0$, a relatively complete treatment is provided, since our approach differs from the conventional one. One aspect, which is not different, is that the equations are integrated from the shock inward. Thus, β is iterated on until the desired semi-vertex angle θ_b is found. For a given β value, however, conditions at station 2, just downstream of the shock, are known from Section 9.4 for both the $\delta = 0$ and 1 cases.

Figure 9.27 shows a sketch of the x,r coordinate system. The solution for both the $\delta = 0$ and 1 cases depends only on the angle η, which is measured from the shock. The governing inviscid

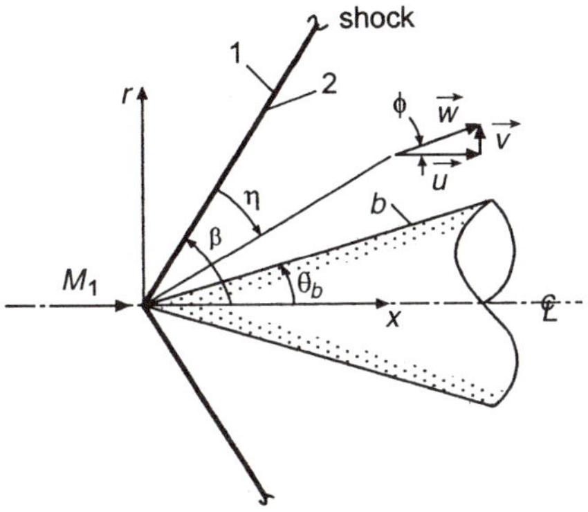

FIGURE 9.27 Schematic for Taylor–Maccoll flow.

equations can be written as

$$\frac{\partial(r\rho u)}{\partial x} + \frac{\partial(r\rho v)}{\partial r} = 0 \tag{9.93a}$$

$$\frac{Du}{Dt} + \frac{1}{\rho}\frac{\partial p}{\partial x} = 0 \tag{9.93b}$$

$$\omega = \frac{\partial v}{\partial x} - \frac{\partial u}{\partial r} = 0 \tag{9.93c}$$

$$s = s_2 \tag{9.93d}$$

where ω is the scalar vorticity and the substantial derivative is

$$\frac{D}{Dt} = u\frac{\partial}{\partial x} + v\frac{\partial}{\partial r} \tag{9.94}$$

Equation (9.93a) represents continuity; Equation (9.93b) is an x-momentum equation, while the remaining two relations, respectively, express irrotationality and the homentropic nature of the flow. The foregoing relations hold for both $\delta = 0$ and 1.

New independent variables are introduced (Emanuel, 1970) as

$$\xi = x, \qquad \eta = \beta - \tan^{-1}\frac{r}{x} \tag{9.95}$$

such that

$$\frac{\partial}{\partial x} = \frac{\partial}{\partial \xi} + \frac{\sin(\beta-\eta)\cos(\beta-\eta)}{\xi}\frac{\partial}{\partial \eta}, \qquad \frac{\partial}{\partial r} = -\frac{\cos^2(\beta-\eta)}{\xi}\frac{\partial}{\partial \eta}$$

A similarity solution, dependent only on η, is assumed. Toward this end, we introduce (Emanuel, 1970)

$$p = p_2 P(\eta), \qquad \rho = \rho_2 R(\eta)$$
$$u = w_2 Q(\eta)\cos\phi(\eta), \qquad v = w_2 Q(\eta)\cos\phi(\eta) \tag{9.96}$$

where R is not to be confused with the gas constant. Initial conditions at the shock are

$$P(0) = Q(0) = R(0) = 1, \quad \phi(0) = \phi_2 \tag{9.97}$$

while at the body, where $\eta_b = \beta - \theta$, we have

$$\phi(\beta - \theta_b) = \theta_b \tag{9.98}$$

Satisfaction of this latter condition requires a β iteration. In view of the above, the substantial derivative becomes

$$\frac{D}{Dt} = \frac{w_2 Q}{\xi}\cos(\beta - \eta)\sin(\beta - \eta - \phi)(\ )' \tag{9.99}$$

where

$$w_2 = a_2 M_2 = \left(\gamma\frac{p}{\rho}\right)_2^{1/2} M_2 \tag{9.100}$$

and $d(\ )/d\eta$ is written as $(\ )'$.

With the foregoing, continuity, vorticity, and the x-momentum equation become

$$\sin(\beta - \eta)\tan(\beta - \eta)[RQ\cos\phi]' - \cos(\beta - \eta)[\tan(\beta - \eta)RQ\sin\phi]' = 0$$

$$\sin(\beta - \eta)(Q\sin\phi)' + \cos(\beta - \eta)(Q\cos\phi)' = 0$$

$$\frac{\rho_2 w_2^2}{p_2}RQ\sin(\beta - \eta - \phi)(Q\cos\phi)' + \sin(\beta - \eta)P' = 0$$

These equations, which hold for $\delta = 0$ and 1, can be rearranged to

$$\frac{Q'}{Q} = -\tan(\beta - \eta - \phi)\phi' \tag{9.101a}$$

$$\frac{R'}{R} = \frac{\sin\phi}{\sin(\beta - \eta)\sin(\beta - \eta - \phi)} + \frac{\phi'}{\sin(\beta - \eta - \phi)\cos(\beta - \eta - \phi)} \tag{9.101b}$$

$$\frac{P'}{P} = -\gamma M^2\frac{Q'}{Q} \tag{9.101c}$$

where

$$M^2 = \frac{\gamma_2}{\gamma}M_2^2\frac{RQ^2}{P} \tag{9.102}$$

Perfect Gas Formulation

Equation (9.93) is replaced with

$$P = R^{\gamma} \tag{9.103}$$

With the aid of Equations (9.102) and (9.103), Equation (9.101c) can be integrated to yield the homentropic relation

$$P = \left[1 + \frac{\gamma - 1}{2} M_2^2 (1 - Q^2)\right]^{\gamma/(\gamma-1)} \tag{9.104a}$$

Hence, we obtain for R

$$R = \left[1 + \frac{\gamma - 1}{2} M_2^2 (1 - Q^2)\right]^{1/(\gamma-1)} \tag{9.104b}$$

and the differential equations

$$\phi' = \frac{\cos(\beta - \eta - \phi)\sin\phi}{\sin(\beta - \eta)\left[1 - \dfrac{M_2^2 Q^2 \sin^2(\beta - \eta - \phi)}{1 + \dfrac{\gamma - 1}{2} M_2^2 (1 - Q^2)}\right]} \tag{9.105}$$

$$\frac{Q'}{Q} = -\tan(\beta - \eta - \phi)\phi' \tag{9.106}$$

Shock and body conditions for these differential equations are provided by Equations (9.97) and (9.98). Other variables are readily found; e.g., the temperature is given by Equation (9.5).

Imperfect Gas Formulation

Equation (9.93) is replaced with Equation (9.21), which is written as

$$R = \left(\frac{\theta_{v2}}{\theta_v}\right)^{5/2} \frac{\sinh\theta_{v2}}{\sinh\theta_v} \exp(\theta_v \coth\theta_v - \theta_{v2}\coth\theta_{v2}) \tag{9.107a}$$

Since the gas is thermally perfect, we also have

$$P = R\frac{\theta_{v2}}{\theta_v} = \left(\frac{\theta_{v2}}{\theta_v}\right)^{7/2} \frac{\sinh\theta_{v2}}{\sinh\theta_v} \exp(\theta_v \coth\theta_v - \theta_{v2}\coth\theta_{v2}) \tag{9.107b}$$

By differentiation, this results in

$$\frac{R'}{R} = -\frac{1}{2}[5 + z(\theta_v)]\frac{\theta_v'}{\theta_v} \tag{9.108a}$$

$$\frac{P'}{P} = -\frac{1}{2}[7 + z(\theta_v)]\frac{\theta_v'}{\theta_v} \tag{9.108b}$$

The above equations are used to eliminate P, Q, and R from Equations (9.101b,c), with the result

$$\frac{\theta_v'}{\theta_v} = \frac{2M^2 \sin(\beta - \eta - \phi)\sin\phi}{[5 + z(\theta_v)]\sin(\beta - \eta)[M^2 \sin^2(\beta - \eta - \phi) - 1]} \tag{9.109a}$$

$$\phi' = \frac{7 + 2z(\theta_v)}{2\gamma M^2 \tan(\beta - \eta - \phi)} \frac{\theta_v'}{\theta_v} \tag{9.109b}$$

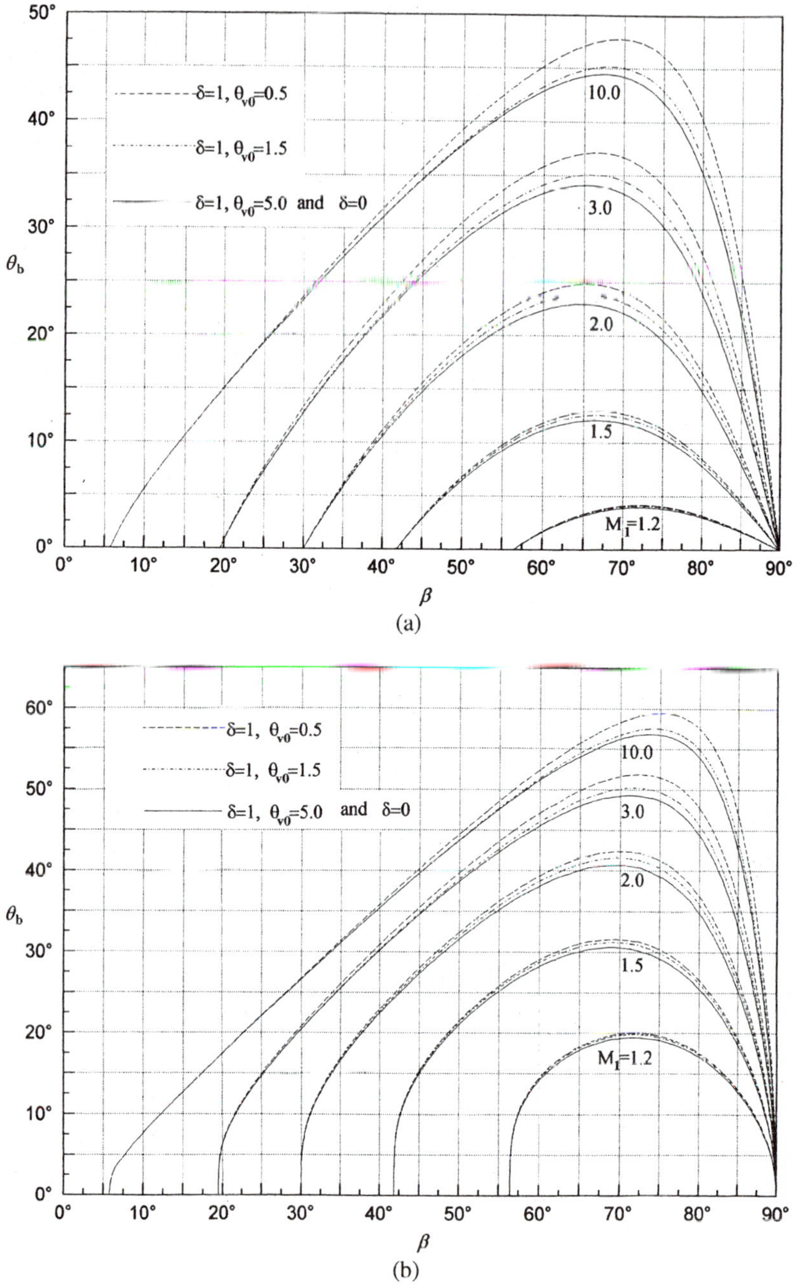

FIGURE 9.28 θ_b vs. β; (a) wedge, (b) cone.

where β is a known constant, η is the independent variable, and z, γ, and M_2 are given by Equations (9.2b), (9.23), and (9.28), respectively. Equations (9.101a) and (9.109) are numerically integrated subject to Equations (9.97) and (9.98) and $\theta_v(0) = \theta_{v2}$.

Figure 9.28 shows the shock wave angle for a wedge in panel (a) and a cone in panel (b), where the wedge panel can be viewed as a composite of Figure 9.20, with $\theta_b = \theta$. For a given upstream Mach number, the different θ_{vo} curves have the same relative relationship to each other

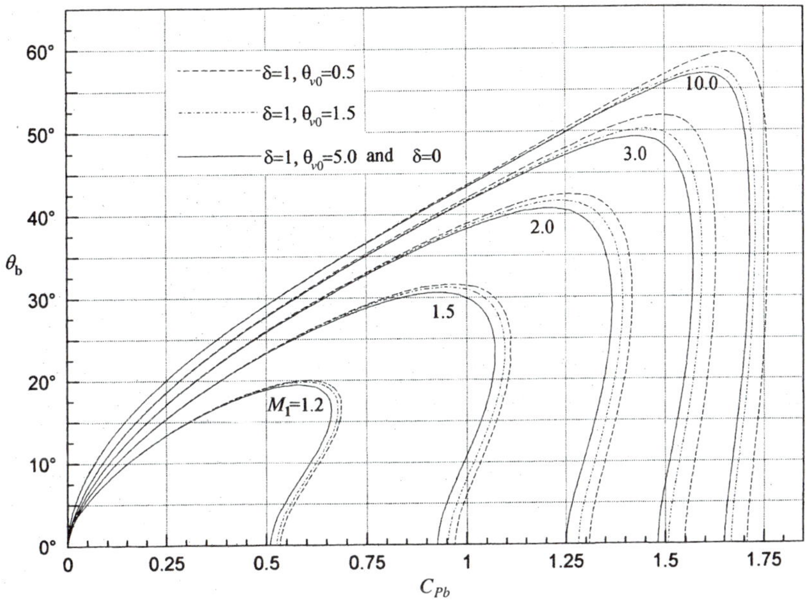

FIGURE 9.29 Pressure coefficient on the body vs. body half-angle.

in the two panels. The detachment angle is larger for the cone, very significantly so at low M_1 values. In contrast to the wedge, the curves are vertical when θ_b goes to zero for both the weak and strong solutions. This effect and the increased detachment angle are a consequence of the three-dimensional relief phenomenon that occurs for a flow about an axisymmetric body.

On the abscissa, β equals the Mach angle, in both panels, for the weak solution. Of course, when β is 90°, the shock is a normal one. In the wedge case, both weak and strong solutions physically occur, e.g., in supersonic flow about a two-dimensional blunt body. On the other hand, only the weak solution occurs for a cone. Nevertheless, the strong solution is shown for the cone in Figures 9.28 and 9.29 for academic reasons (i.e., to see what the solution looks like). For example, the cone's strong shock solution would be of interest in stability studies.

The surface pressure coefficient of the cone is shown in Figure 9.29. The weak solution values are to the left of the maximum. For a given θ_b, an imperfect gas has little effect on C_{pb}, except near detachment, where the effect rapidly increases. In accord with Figure 9.28(b), an imperfect gas has a slightly reduced value for C_{pb} and therefore a slightly reduced wave drag. This phenomenon of wave drag reduction appears to be more important in flows where chemical or ionizational effects are pronounced.

Figure 9.30 has four panels that show δ comparisons for β, M_2, M_b, and C_{pb} vs. M_1 for the weak solution of a cone with a 10° semi-vertex angle. This is a fairly representative angle for cones used in supersonic and hypersonic experiments. Calculations determine that the detachment Mach number for the cone is roughly 1.06 for all θ_{vo} values, in accord with Figure 9.28b. The detachment Mach number, however, does increase slightly with decreasing θ_{vo}. This effect is responsible for the rapid change in direction of the curves in Figure 9.30 as M_1 approaches 1.06 from above. This change does not occur when $\theta_{vo} = 5$, since this case and the $\delta = 0$ result are essentially identical.

Aside from the variation near $M_1 = 1.06$, the curves show various minimum or maximum values. More important, however, is the small deviation from unity in all panels. Consequently, this real-gas effect represents a small perturbation when $\theta_b = 10°$. The magnitude of the perturbation would increase with θ_b.

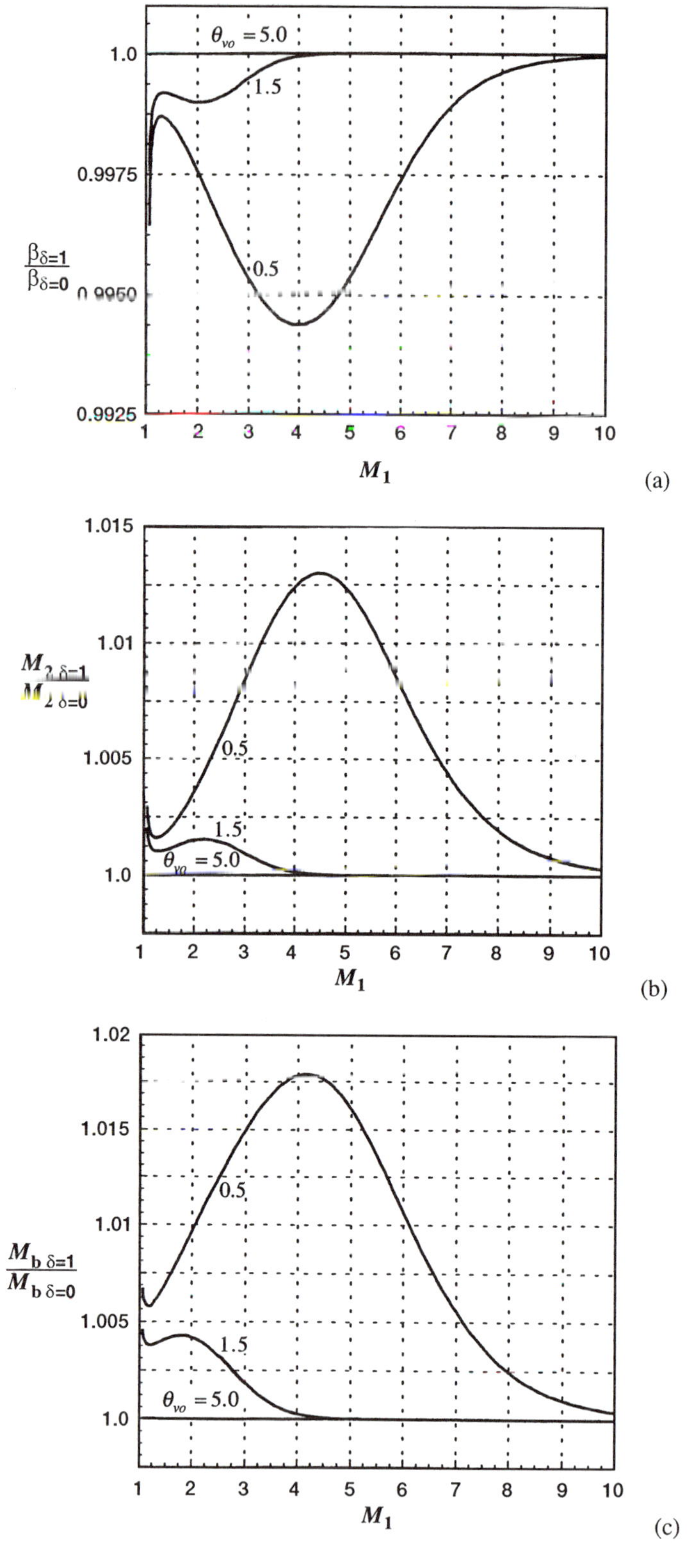

FIGURE 9.30 Comparisons for the weak solution of a $\theta_b = 10°$ cone; (a) β, (b) M_2, (c) M_b, (d) C_{pb}.

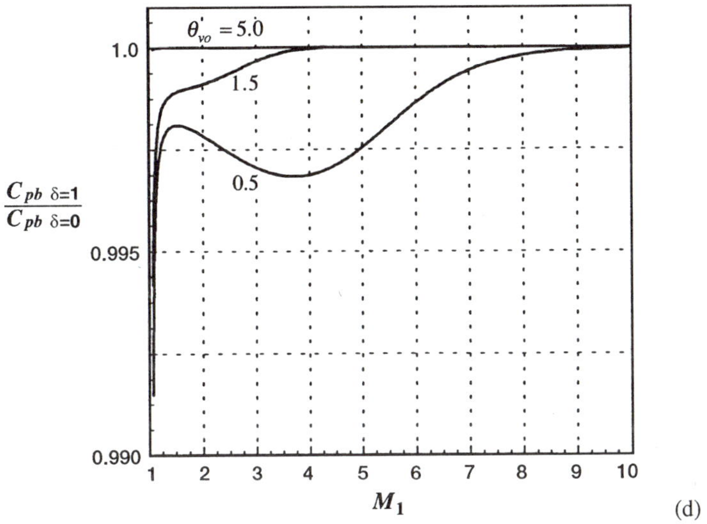

FIGURE 9.30 (*continued*).

REFERENCES

Ames Research Staff, *Equations, Tables, and Charts for Compressible Flow*, NACA Report 1135 (1953).

Bultman, M.L., "Thermally Perfect, Calorically Imperfect Planar Shock Flow," M.S. thesis, University of Oklahoma, 1994.

Christy, G.T., "Calorically Imperfect Isentropic Flow," M.S. thesis, University of Oklahoma, 1993.

Christy, G. and Emanuel, G., "Comment on 'Extension of the λ Formulation to Imperfect Gas Flows,'" *AIAA J.* 32, 1554 (1994).

Donaldson, C. duP., "Note on the Importance of Imperfect Gas Effects and Variation of Heat Capacities on the Isentropic Flow of Gases," NACA RM L8J14 (1948).

Eggers, A.J., Jr., "One-Dimensional Flows of an Imperfect Diatomic Gas," NACA Report 959 (1950).

Emanuel, G., "Blowing from a Porous Cone with an Embedded Shock Wave," *AIAA J.* 8, 283 (1970).

Ismail, M., "Prandtl–Meyer Flow of a Calorically Imperfect Gas," M.S. thesis, University of Oklahoma, 1994.

Lampe, D.R., "Thermally Perfect, Calorically Imperfect Taylor–Maccoll Flow," M.S. thesis, University of Oklahoma, 1994.

Lentini, D., "Extension of the λ Formulation to Imperfect Gas Flows," *AIAA J.* 30, 2785 (1992).

Lentini, D., "Reply by the Author to G. Christy and G. Emanuel," *AIAA J.* 32, 1554 (1994).

Shapiro, A.H., *The Dynamics and Thermodynamics of Compressible Fluid Flow*, Vol. I, The Ronald Press Co., New York, 1953.

Tsien, H.-S., "One-Dimensional Flows of a Gas Characterized by van der Waal's Equation of State," *J. Math. and Phys.* 25, 301 (1947a).

Tsien, H.-S., "Corrections on the Paper 'One-Dimensional Flows of a Gas Characterized by van der Waal's Equation of State,'" *J. Math. Phys.* 26, 76 (1947b).

Vincenti, W.G. and Kruger, C.H., Jr., *Introduction to Physical Gas Dynamics*, John Wiley, New York, 1965.

PROBLEMS

9.1 Pure oxygen ($T_v = 2240$ K) is flowing in a thrust nozzle where $p_o = 10^7$ Pa and $T_o = 2240$ K. The nozzle has an area ratio of 20.075 and a throat area of 10^2 cm^2. Evaluate M_e, p_e, T_e, ρ_e, $\dot{m}$, and the thrust assuming (a) $\gamma = 1.4$ and (b) a calorically imperfect gas. For the thrust, assume a vacuum for the ambient pressure, and tabulate your answers. [Hint: Note that $M_{e,\delta=1} \neq M_{e,\delta=0}$.]

9.2 Consider steady subsonic or supersonic flow in a duct with a constant cross-sectional area A. Both heat addition and skin friction are present; i.e., this is a Rayleigh/Fanno flow to be analyzed with the influence coefficient method. (This type of flow is of interest for subsonic and supersonic mixing lasers, jet engine combustors, and molecular beam devices. For a perfect gas, this topic is the subject of Section 13.5.) Introduce the parameters

$$D = \frac{4A}{c}, \qquad c_f = \frac{2\tau}{\rho w^2}, \qquad \Gamma(\theta_v) = \frac{8z(\theta_v \coth\theta_v - 1)}{(5+2z)(7+2z)}$$

where c is the perimeter, D is the hydraulic diameter, and τ is the wall shear stress. Start with the governing conservation equations, and note that stagnation conditions, such as h_o, θ_{vo}, p_o, and s_o, are functions of the axial distance x.

(a) Develop three linear equations for $d\rho/\rho$, dM/M, and $d\theta_v/\theta_v$. In these equations the heat transfer, per unit mass, $q(x)$ should be replaced with $\theta_{vo}(x)$.

(b) Use Cramer's rule to solve these equations, thereby obtaining explicit equations for $d\rho/dx$, dM/dx, and $d\theta_v/dx$.

(c) Determine the condition for a choked flow. Is the Mach number M^*, when the flow is choked, smaller or larger than unity when $\delta = 1$?

(d) Provide formulas for the net heat transfer q and the net thrust, which is negative.

9.3 Consider a normal shock whose overall pressure ratio is p_2/p_1. Aside from p_2/p_1, assume M_1 and θ_{vo} are known.

(a) Decompose the entropy change, $\Delta s = s_2 - s_1$, into a part, $s_a - s_1$, associated with a constant γ shock, and a part, $s_2 - s_a$, associated with a downstream vibrational relaxation process. Plot

$$f_s = \frac{s_a - s_1}{s_2 - s_1}$$

vs. M_1 when $\theta_{vo} = 0.5$. Note that $f_s \to 1$ when $\theta_{vo} \to \infty$ and when $M_1 = 1$.

(b) Repeat the analysis for the stagnation pressure. Plot

$$f_{p_o} = \frac{p_{oa}/p_{o1}}{p_{o2}/p_{o1}}$$

vs. M_1 when $\theta_{vo} = 0.5$.

10 Sweep

10.1 PRELIMINARY REMARKS

Wings and missile fins with sweep are a common occurrence in transonic and supersonic aerodynamics. The leading edge shock wave impinges on the surface of the vehicle and causes shock-wave/boundary-layer interaction (Settles and Dolling, 1992). In this chapter, however, we focus on the fluid dynamics of sweep, but without the interaction. Admittedly, this is a vastly simpler topic. Nevertheless, it has received scant attention, and, as we shall see, contains several non-intuitive features.

For purposes of simplicity, a steady, inviscid flow of a perfect gas is considered in which the upstream flow is supersonic and uniform. Only two wall configurations are discussed. In the first, a wedge generates an attached, weak solution shock. In the second, a centered Prandtl–Meyer expansion is caused by an expansive wall turn. In both cases, the straight leading edge has an arbitrary sweep angle relative to the freestream velocity. Our topic is, thus, a natural extension of Chapters 6 and 8.

The concept of using sweep with a wing, to either delay transonic drag or reduce supersonic wave drag, goes back to the 1930s. The first analytical formulation, however, is due to Poritsky (1946). The basic idea is simple. Consider a steady, inviscid, two-dimensional flow, which may be subsonic or supersonic. A constant velocity vector may be added that is normal to the plane of the original two-dimensional flow. The resulting three-dimensional flow field is easily shown to satisfy the steady Euler equations.

Poritsky's analysis does not require that the upstream flow be uniform, supersonic, or irrotational. Moreover, the vorticity is invariant; e.g., if one flow is irrotational, then so is the other. The addition of a constant transverse velocity component leaves all static thermodynamic variables, such as the pressure and speed of sound a, unchanged. Stagnation conditions (when a shock occurs), velocities, Mach numbers, and the shape of streamlines do change. The subsequent analysis will focus on these quantities.

The strength of a shock wave depends only on the component $M_{\infty n}$ of the freestream Mach number that is normal to the wave. Sweep reduces this normal component, thereby reducing the wave drag or, alternatively, the loss in stagnation pressure. Sweep might also be used profitably in the design of an otherwise two-dimensional supersonic inlet.

Although the concept of sweep is simple enough, a detailed analysis is not trivial. We shall have to deal with issues such as the minimum value of the upstream Mach number for an attached oblique shock and various detachment conditions. Moreover, several results are unexpected. It is also worth noting that some of the previous restrictions imposed on the freestream flow can be relaxed, since the shock and Prandtl–Meyer flow fields satisfy the substitution principle. Thus, the upstream flow may be a parallel, vortical flow that is nonhomenergetic.

10.2 OBLIQUE SHOCK FLOW

Introductory Discussion

A wedge is envisioned whose lower surface is aligned with the freestream. A Cartesian coordinate system is introduced, where x and z are in the lower surface, x is aligned with the freestream velocity $\vec{w}_\infty$, and y and z are in a plane perpendicular to $\vec{w}_\infty$. In the x,y plane, the upper wedge

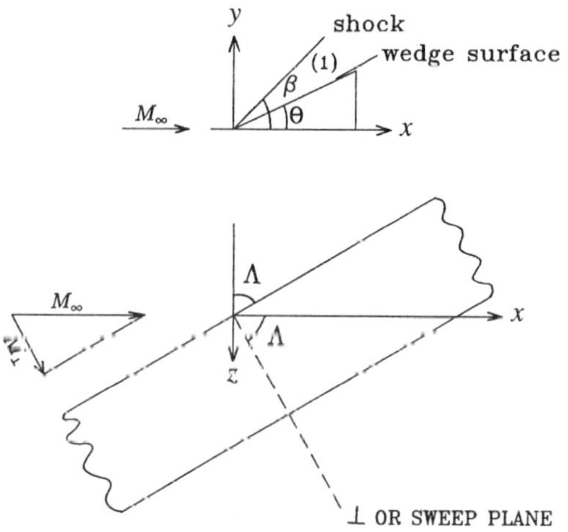

FIGURE 10.1 Schematic depicting a wedge, oblique shock, and the sweep angle.

surface and planar shock have included angles θ and β, respectively, with respect to $\vec{w}_\infty$; see Figure 10.1. Now let the sharp leading edge of the wedge have a sweep angle Λ. A plane that is perpendicular to the leading edge is called the sweep plane; it is denoted with a subscript $\perp$. The sweep angle is in the x,z plane between the leading edge and the z axis. Downstream of the attached shock, the uniform flow is denoted with a unity subscript.

Angles

In the sweep plane, the shock and wedge angles are $\beta_\perp$ and $\theta_\perp$; see Figure 10.2. These angles, not β and θ are associated with an oblique shock in a two-dimensional flow. From this figure, we obtain

$$\tan\theta = \cos\Lambda \tan\theta_\perp, \qquad \tan\beta = \cos\Lambda \tan\beta_\perp \tag{10.1a,b}$$

With $0 \le \Lambda \le 90°$, observe that $\theta_\perp \ge \theta$ and $\beta_\perp \ge \beta$. When there is no sweep, $\Lambda = 0$, then $\theta_\perp = \theta$, $\beta_\perp = \beta$, and conventional results for an oblique shock hold. We presume γ, M_∞, and Λ are prescribed. It is still necessary to fix a wedge angle, either θ or $\theta_\perp$. The analysis with θ fixed is found in Emanuel (1992a). It is more interesting, however, to fix $\theta_\perp$ (Emanuel, 1992b). In this case, Λ is a free parameter and changes in Λ correspond to a solid body rotation about the y-axis of the wedge. Only this flow will be discussed.

Standard oblique shock relations hold for the two-dimensional flow in the sweep plane. From Figure 10.1, the upstream Mach number in this plane is

$$M_\perp = M_\infty \cos\Lambda \tag{10.2}$$

where it is notationally convenient not to write $M_\perp$ as $M_{\perp\infty}$. Moreover, $\theta_\perp$ and $\beta_\perp$ are related by

$$\tan\theta_\perp = \frac{\cot\beta_\perp(M_\perp^2 \sin^2\beta_\perp - 1)}{1 + \left(\frac{\gamma+1}{2} - \sin^2\beta_\perp\right)M_\perp^2} \tag{10.3}$$

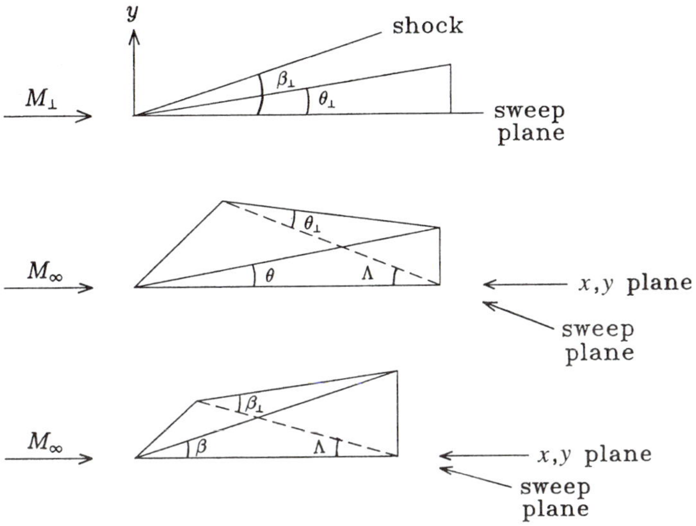

FIGURE 10.2 Schematic relating various angles; the planes respectively containing θ, β, $\theta_\perp$, and $\beta_\perp$ are perpendicular to the plane containing the sweep angle Λ.

Since $\theta_\perp$ is to be prescribed, the inversion of this relation is more useful. It is provided by Appendix C, where β, θ, and M are replaced with $\beta_\perp$, $\theta_\perp$, and $M_\perp$, respectively. Only the weak solution is of interest in this discussion; hence, we set $\delta = 1$. In this circumstance, χ (in Appendix C) is generally near unity, except when the weak solution shock approaches detachment, and then χ rapidly decreases toward its detachment value of -1. With γ, M_∞, Λ, and $\theta_\perp$ prescribed, the foregoing relations explicitly determine $M_\perp$, θ, $\beta_\perp$, and β.

The shock wave angle for detachment in the sweep plane is (Emanuel, 1986)

$$\beta_{\perp d} = \sin^{-1}\left(\frac{\gamma+1}{4\gamma M_\perp^2}\left\{M_\perp^2 - \frac{4}{\gamma+1} + \left[M_\perp^4 + 8\,\frac{\gamma-1}{\gamma+1}\,M_\perp^2 + \frac{16}{\gamma+1}\right]^{1/2}\right\}\right)^{1/2} \qquad \textbf{(10.4)}$$

while the corresponding wedge angle, $\theta_{\perp d}$, is provided by Equation (10.3). Results for $\theta_{\perp d}$ are shown in Figure 10.3 for various M_∞ and γ values. Here, $\beta_{\perp d}$ and $\theta_{\perp d}$ only depend on γ and $M_\infty \cos\Lambda$. When $\Lambda = 0$, $\theta_{\perp d}$ equals the conventional detachment angle, with θ_d increasing as γ decreases. When $\theta_{\perp d}$ is zero, the component of the upstream Mach number normal to the shock is unity and the disturbance is a Mach wave. At low freestream Mach numbers, $\theta_{\perp d}$ smoothly decreases with Λ, whereas at hypersonic Mach numbers the decrease is rapid only at large values for Λ. For instance, if $M_\infty = 9$, $\gamma = 1.4$, and $\Lambda = 60°$, detachment requires that $\theta_\perp$ exceed about 40°. At $M_\infty = 3$, however, detachment occurs at about 12.5°. At this freestream Mach number, an attached solution does not exist if $\theta_\perp$ is above 12.5°.

Mach Numbers and Velocity Components

The normal component of the upstream Mach number in the sweep plane is

$$M_{\perp n} = M_\perp \sin\beta_\perp = M_\infty \cos\Lambda \sin\beta_\perp \qquad \textbf{(10.5)}$$

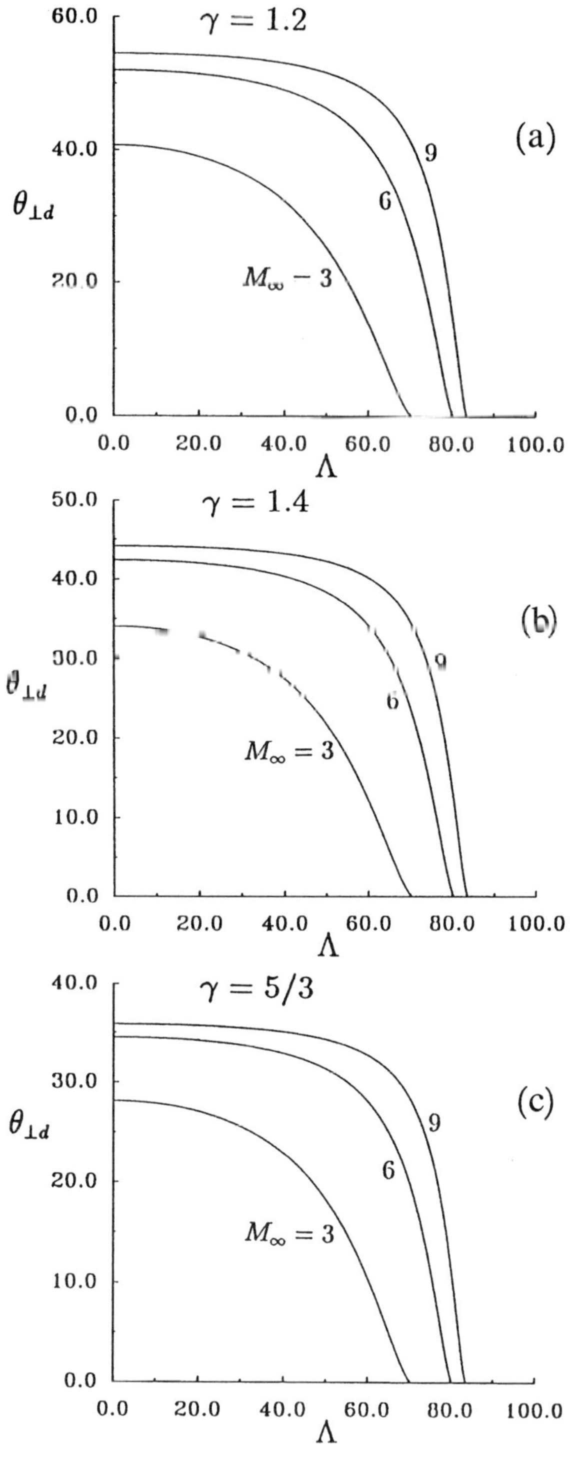

FIGURE 10.3 Detachment wedge angle in the sweep plane $\theta_{\perp d}$ vs. Λ for various M_∞ and γ values.

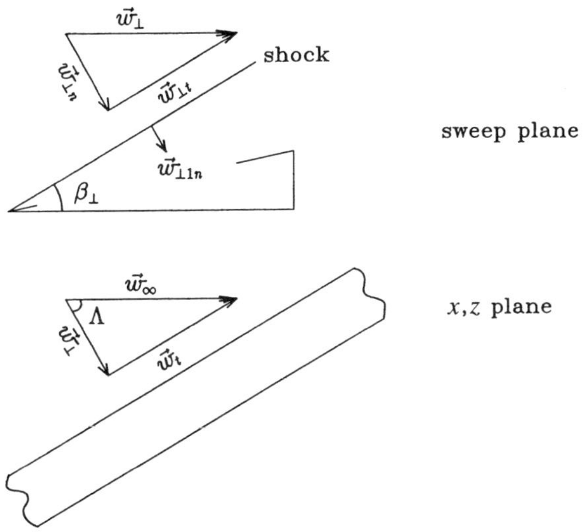

FIGURE 10.4 Schematic illustrating various velocity components.

where $M_{\perp n} > 1$ in order for a shock wave of finite strength to exist. Its downstream counterpart is

$$M_{\perp 1n} = \left[\frac{1 + \frac{\gamma - 1}{2} M_{\perp n}^2}{\gamma M_{\perp n}^2 - \frac{\gamma - 1}{2}}\right]^{1/2} \tag{10.6}$$

The Mach number component, downstream of the shock, in the sweep plane is given by

$$M_{\perp 1} = \frac{M_{\perp 1n}}{\sin(\beta_\perp - \theta_\perp)} \tag{10.7}$$

Shock wave jump ratios, such as p_1/p_∞, T_1/T_∞, $p_{01}/p_{0\infty}$, …, only depend on γ and $M_{\perp n}$. For instance, we write

$$\frac{p_1}{p_\infty} = \frac{2}{\gamma + 1}\left(\gamma M_{\perp n}^2 - \frac{\gamma - 1}{2}\right) \tag{10.8}$$

$$\frac{\rho_1}{\rho_\infty} = \frac{\gamma + 1}{2} \frac{M_{\perp n}^2}{1 + \frac{\gamma - 1}{2} M_{\perp n}^2} \tag{10.9}$$

for the pressure and density jumps.

Notice that the full downstream Mach number M_1 has yet to be established. For this, we shall need to evaluate $\vec{w}_1$. To do this, $\vec{w}_\infty$ is first decomposed into a component $\vec{w}_t$ that is parallel to the leading edge of the wedge and a component $\vec{w}_\perp$ that is perpendicular; see Figure 10.4. The $\vec{w}_t$ component is the constant velocity that is added to the two-dimensional flow in the sweep plane. The sweep plane velocity $\vec{w}_\perp$ is further decomposed into $\vec{w}_{\perp t}$ and $\vec{w}_{\perp n}$, which, respectively, are the tangential and normal components of the upstream velocity in the sweep plane. The tangential components, $\vec{w}_t$ and $\vec{w}_{\perp t}$, are perpendicular to each other, but each is the same on both sides of the shock.

To obtain $\vec{w}_1$, we write

$$\vec{w}_1 = \vec{w}_t + \vec{w}_{\perp t} + \vec{w}_{\perp 1n} \tag{10.10}$$

The velocities on the right-hand side can be written as

$$\vec{w}_t = \vec{w}_\infty - \vec{w}_\perp, \qquad \vec{w}_{\perp t} = \vec{w}_\perp - \vec{w}_{\perp n}, \qquad \vec{w}_{\perp 1n} = \frac{2}{\gamma+1}\frac{1+\frac{\gamma-1}{2}M^2_{\perp n}}{M^2_{\perp n}}\vec{w}_{\perp n} \tag{10.11a,b,c}$$

with the result

$$\vec{w}_1 = \vec{w}_\infty - \frac{2}{\gamma+1}\frac{M^2_{\perp n}-1}{M^2_{\perp n}}\vec{w}_{\perp n} \tag{10.12}$$

The change in magnitude and orientation from $\vec{w}_\infty$ to $\vec{w}_1$ thus depends on the rightmost term. However, we have

$$0 \le \frac{M^2_{\perp n}-1}{M^2_{\perp n}} < 1$$

In fact, $M^2_{\perp n} - 1$ is positive but near zero for large sweep angles. In this circumstance, the upstream Mach number $M_{\perp n}$ that governs the strength of the shock is transonic. We thus have the result that the change between $\vec{w}_\infty$ and $\vec{w}_1$ is rather small for large sweep angles. This result, and the corresponding one for a Prandtl–Meyer flow, stems from the weakening of the disturbance for a solid body rotation with increasing sweep.

Further progress requires the introduction of the orthonormal basis $\hat{1}_x$, $\hat{1}_y$, and $\hat{1}_z$, where $\hat{1}_x$ is parallel to $\vec{w}_\infty$. In terms of this basis, the unit normal vector to the shock, in the downstream direction, is (see Problem 10.1)

$$\hat{n} = \frac{\hat{1}_x - \cot\beta\hat{1}_y + \tan\Lambda\,\hat{1}_z}{(\csc^2\beta + \tan^2\Lambda)^{1/2}} \tag{10.13a}$$

or, with the aid of Equation (10.1b),

$$\hat{n} = \cos\Lambda\sin\beta_\perp\hat{1}_x - \cos\beta_1\hat{1}_y + \sin\Lambda\sin\beta_\perp\hat{1}_z \tag{10.13b}$$

The velocities on the right side of Equation (10.12) become

$$\vec{w}_\infty = w_\infty\hat{1}_x \tag{10.14}$$

$$\vec{w}_{\perp n} = a_\infty M_{\perp n}\hat{n} = w_\infty\cos\Lambda\sin\beta_\perp\hat{n} \tag{10.15}$$

with the result

$$\vec{w}_1 = w_\infty\left[\left(1 - \frac{2}{\gamma+1}\frac{M_{\perp n}^2 - 1}{M_\infty^2}\right)\hat{\imath}_x + \frac{2}{\gamma+1}\frac{M_{\perp n}^2 - 1}{M_\infty^2 \cos\Lambda \tan\beta_\perp}\hat{\imath}_y - \frac{2}{\gamma+1}\frac{M_{\perp n}^2 - 1}{M_\infty^2}\tan\Lambda\,\hat{\imath}_z\right] \tag{10.16}$$

The magnitude of $\vec{w}_1$ simplifies to

$$\frac{w_1}{w_\infty} = \left[1 - \frac{4}{\gamma+1}\frac{M_{\perp n}^2 - 1}{M_\infty^2} + \frac{4}{(\gamma+1)^2}\frac{(M_{\perp n}^2 - 1)^2}{M_\infty^2 M_{\perp n}^2}\right]^{1/2} \tag{10.17}$$

With this relation, one can show that w_1/w_∞ is below unity for any sweep angle. The speed of sound, downstream of the shock, can be written as

$$\frac{a_1}{a_\infty} = \left(\frac{T_1}{T_\infty}\right)^{1/2} = \frac{2}{\gamma+1}\frac{\left(1 + \frac{\gamma-1}{2}M_{\perp n}^2\right)^{1/2}\left(\gamma M_{\perp n}^2 - \frac{\gamma-1}{2}\right)^{1/2}}{M_{\perp n}} \tag{10.18}$$

Consequently, the downstream Mach number is

$$M_1 = \frac{w_1}{a_1} = M_\infty \frac{\frac{w_1}{w_\infty}}{\frac{a_1}{a_\infty}} \tag{10.19}$$

where Equations (10.17) and (10.18) provide the flow speed and speed of sound ratios.

With the above formulas, the orientation of $\vec{w}_1$, with respect to the x,y,z coordinate system, is obtained as direction cosines

$$\cos\alpha_x = \frac{\vec{w}_1}{w_1}\cdot\hat{\imath}_x, \qquad \cos\alpha_y = \frac{\vec{w}_1}{w_1}\cdot\hat{\imath}_y, \qquad \cos\alpha_z = \frac{\vec{w}_1}{w_1}\cdot\hat{\imath}_z \tag{10.20}$$

where α_i is the angle between $\vec{w}_\infty$ and $\vec{w}_1$. Table 10.1 provides results with the nonsweep cases serving as references. Even though Λ is 60°, the change in orientation of $\vec{w}_1$ is small relative to a corresponding nonsweep case, as mentioned earlier. Observe that α_x may increase or decrease from its nonsweep value, while the α_z change represents a small lateral turn toward the sweep plane.

TABLE 10.1
Angles, in Degrees, between $\vec{w}_1$ and the x, y, and z Axes for an Oblique Shock when $\gamma = 1.4$

M_∞	$\theta_\perp$	Λ	α_x	α_y	α_z
3	5	0	5.00	85.0	90.0
3	5	60	9.26	87.7	87.8
9	15	0	15.0	75.0	90.0
9	15	60	10.5	83.1	87.2

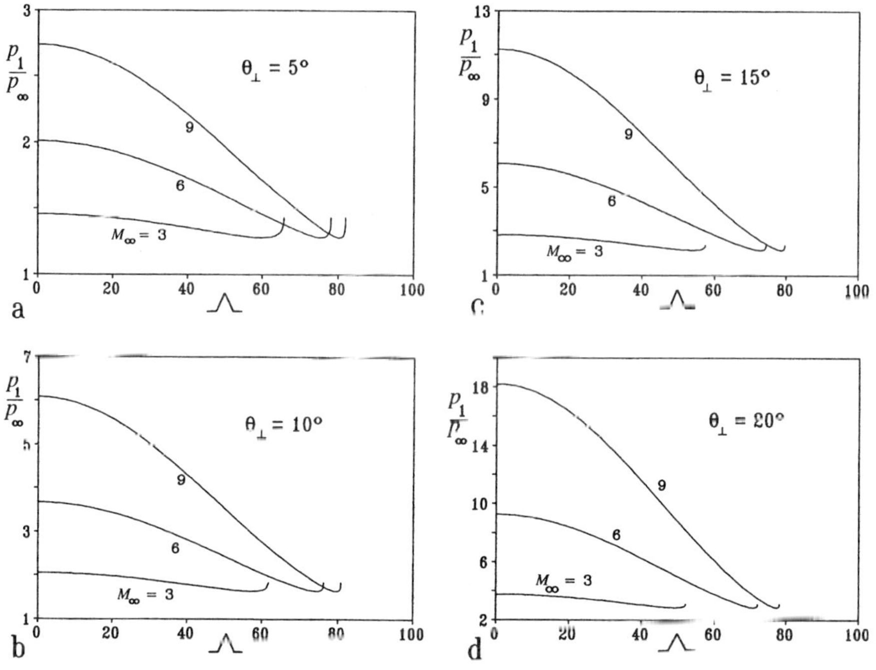

FIGURE 10.5 Pressure ratio vs. Λ for $\gamma = 1.4$ for various M_∞ and $\theta_\perp$ values.

Flow Results

Results are presented with a view toward illustrating trends that are not especially obvious. Figures 10.5 to 10.7 show p_1/p_∞, $p_{01}/p_{0\infty}$, and M_1 vs. Λ for $\gamma = 1.4$, $M_\infty = 3$, 6, 9, and $\theta_\perp = 5°$, 10°, 15°, 20°. The equation for the stagnation pressure ratio is given by

$$\frac{p_{01}}{p_{0\infty}} = \frac{p_{01}}{p_1}\frac{p_1}{p_\infty}\frac{p_\infty}{p_{0\infty}} = \frac{2}{\gamma+1}\left(\gamma M_{\perp n}^2 - \frac{\gamma-1}{2}\right)\left(\frac{1+\frac{\gamma-1}{2}M_1^2}{1+\frac{\gamma-1}{2}M_\infty^2}\right)^{\gamma/(\gamma-1)} \quad \textbf{(10.21)}$$

When Λ is zero, the figures agree with nonsweep results.

The curves terminate at a large Λ value when detachment first occurs. In this situation, $\beta_{\perp d}$ is given by Equation (10.4) and the upstream and downstream Mach numbers $M_{\perp n}$ and $M_{\perp 1}$, respectively, are transonic. The curves in the three figures exhibit a rapid change in slope near detachment. In this region, the above Mach numbers are transonic and χ (in the inversion formula) experiences a rapid decrease.

Except near detachment, the static pressure variation (see Figure 10.5) is slight at small M_∞ values but quite large when M_∞ is hypersonic. All the curves exhibit a minimum in p_1/p_∞ near detachment. This minimum corresponds to the maximum value seen in the $p_{01}/p_{0\infty}$ curves of Figure 10.6. The extremum stems from a minimum value for $M_{\perp n}$ with respect to Λ; i.e.,

$$\frac{\partial M_{\perp n}}{\partial \Lambda} = \frac{\partial M_{\perp n}}{\partial M_\perp}\frac{\partial M_\perp}{\partial \Lambda} = -M_\infty \sin\Lambda \, \frac{\partial M_{\perp n}}{\partial M_\perp} = 0 \quad \textbf{(10.22)}$$

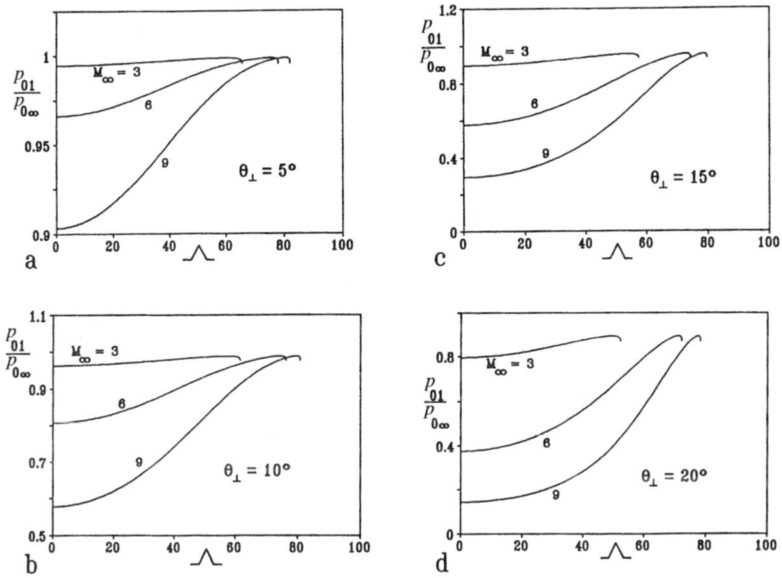

FIGURE 10.6 Stagnation pressure ratio vs. Λ for $\gamma = 1.4$ for various M_∞ and $\theta_\perp$ values.

where $\theta_\perp$ is held fixed. With the aid of Equation (10.5), the extremum condition becomes

$$\frac{\partial M_{\perp n}}{\partial M_\perp} = \sin\beta_\perp + M_\perp \cos\beta_\perp \frac{\partial \beta_\perp}{\partial M_\perp} = 0 \tag{10.23}$$

Equation (10.3), with $\theta_\perp$ again held fixed, results in

$$\frac{\partial \beta_\perp}{\partial M_\perp} = \frac{2M_\perp \cos^2\beta_\perp\left[\sin^2\beta_\perp - \tan\beta_\perp\left(\frac{\gamma+1}{2} - \sin^2\beta_\perp\right)\tan\theta_\perp\right]}{\left[1 + \left(\frac{\gamma+1}{2} - 3\sin^2\beta_\perp + 2\sin^4\beta_\perp\right)M_\perp^2\right]\tan\theta_\perp - 2\sin\beta_\perp + \cos^3\beta_\perp M_\perp^2} \tag{10.24}$$

Hence Equation (10.23) simplifies to

$$\tilde{M}_\perp^2 = \left(\frac{\gamma+1}{2} - \gamma\sin^2\tilde{\beta}_\perp\right)^{-1} \tag{10.25}$$

where a tilde denotes conditions when $M_{\perp n}$ is a minimum, and the static and stagnation pressure ratios have an extremum. With the above, Equation (10.3) provides $\tilde{\beta}_\perp$ as

$$\tilde{\beta}_\perp = \frac{\pi}{2} - \frac{1}{2}\tan^{-1}\left(\frac{1}{\tan\theta_\perp}\right) \tag{10.26}$$

where a more convenient form is (see Problem 10.2)

$$\tilde{\beta}_\perp = \frac{\pi}{4} + \frac{1}{2}\theta_\perp \tag{10.27a}$$

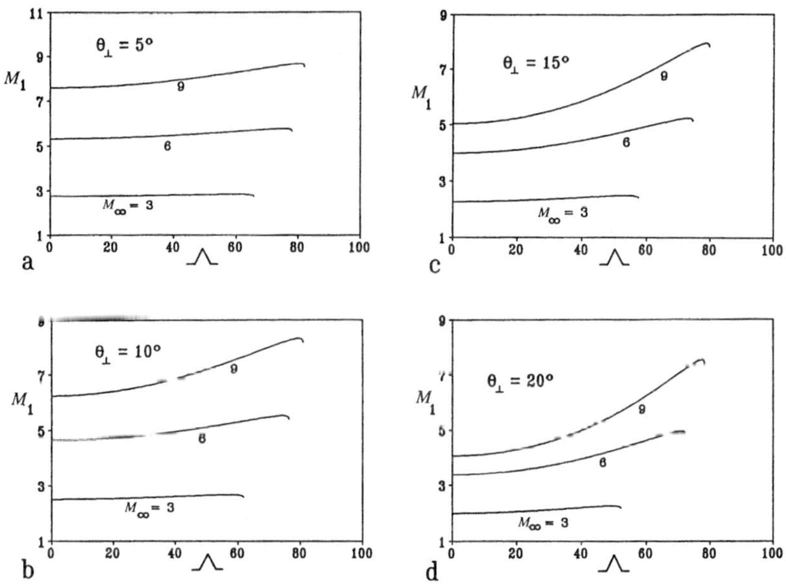

FIGURE 10.7 Downstream Mach number vs. Λ for $\gamma = 1.4$ for various M_∞ and $\theta_\perp$ values.

This last relation, in turn, yields

$$\tilde{M}_\perp = \left[\frac{1}{2}(1-\gamma\sin\theta_\perp)\right]^{-1/2} \tag{10.27b}$$

$$\frac{\tilde{p}_1}{p_\infty} = \frac{1+\gamma\sin\theta_\perp}{1-\gamma\sin\theta_\perp} \tag{10.27c}$$

$$\frac{\tilde{p}_{01}}{p_{0\infty}} = \left(\frac{1+\sin\theta_\perp}{1-\sin\theta_\perp}\right)^{\gamma/(\gamma-1)}\left(\frac{1-\gamma\sin\theta_\perp}{1+\gamma\sin\theta_\perp}\right)^{1/(\gamma-1)} \tag{10.27d}$$

$$\tilde{\Lambda} = \cos^{-1}\left[\frac{1}{M_\infty}\left(\frac{2}{1-\gamma\sin\theta_\perp}\right)^{1/2}\right] \tag{10.27e}$$

As is evident in Figures 10.5 and 10.6, the extremum values for p_1/p_∞ and $p_{01}/p_{0\infty}$ are independent of M_∞. Moreover, for given values of M_∞ and $\theta_\perp$, the p_1/p_∞ and $p_{01}/p_{0\infty}$ extremum values occur at the same sweep angle. Similarly, $M_{\perp n}$ has a minimum value and M_1 has a maximum value (see Figure 10.7) at this sweep angle. By adjusting Λ, a given wedge can thus attain a fixed value for p_1/p_∞ and for $p_{01}/p_{0\infty}$, independent of M_∞. (The design of supersonic inlets might benefit from this insensitivity to M_∞. A condition at a Λ value somewhat smaller than $\tilde{\Lambda}$ is necessary, however, since the extremum state is close to detachment.) This independence, e.g., extends to other jump relations, such as T_1/T_∞ and ρ_1/ρ_∞.

Conditions at the extremum are provided by Equations (10.27). The loss in stagnation pressure (see Figure 10.6) is modest at this condition; i.e., the shock is relatively weak. For instance, for $\gamma = 1.4$, $M_\infty = 9$, and $\theta_\perp = 20°$, we have

$$\tilde{\Lambda} = 77.4°, \quad \frac{\tilde{p}_1}{p_\infty} = 2.84, \quad \frac{\tilde{p}_{01}}{p_{0\infty}} = 0.893, \quad \tilde{M}_1 = 7.54$$

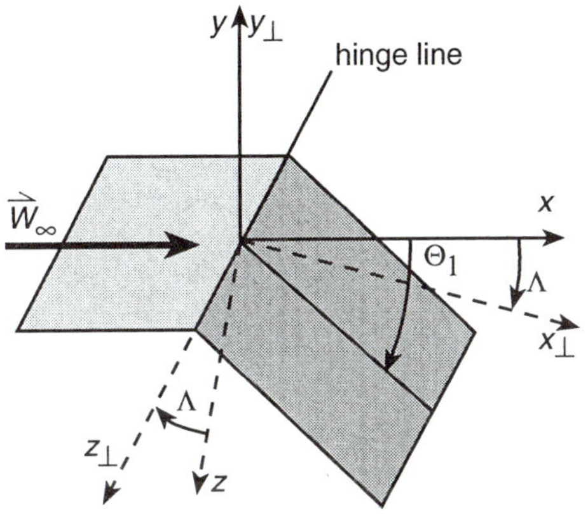

FIGURE 10.8 Schematic and coordinate systems for a Prandtl–Meyer expansion.

The weakness of the shock is also evident in the large $\tilde{M}_1$ value. On the other hand, without sweep, we obtain

$$\Lambda = 0, \qquad \frac{p_1}{p_\infty} = 18.2, \qquad \frac{p_{01}}{p_{0\infty}} = 0.143, \qquad M_1 = 4.06$$

for a much stronger shock. There is thus a trade-off between introducing sweep or reducing a ramp or wedge angle in order to weaken the shock. (Of course, both sweep and low angle ramps could be utilized; this might reduce the overall size of the inlet.) This weakening is evident in Figure 10.7, where M_1 increases toward M_∞ as Λ increases. This trend again does not hold near detachment, nor is it effective at small M_∞ values, where the nonswept shock is already relatively weak.

10.3 PRANDTL–MEYER FLOW

INTRODUCTORY DISCUSSION

Two Cartesian coordinate systems are utilized (see Figure 10.8) in which x is aligned with the freestream velocity $\vec{w}_\infty$. The origin of both systems is on the hinge line, where the wall has a sharp expansive turn. The coordinate y is normal to the upstream wall and the z coordinate completes the right-handed system. The freestream velocity has a sweep angle Λ with respect to the hinge line. A coordinate system, associated with the sweep plane, is introduced in which $z_\perp$ is along the hinge line, $z_\perp = y$, and $x_\perp$ is rotated about the y-axis by the angle Λ from the $\vec{w}_\infty$ direction. As with the shock wave case, a change in Λ corresponds to a solid body rotation. For purposes of brevity, a centered Prandtl–Meyer expansion is considered. The extension to a noncentered expansion is straightforward, and much of the subsequent analysis would still apply. As in the preceding section, thermodynamic static variables are unchanged. Since the flow is homentropic, any stagnation quantity is a constant and is unchanged from its upstream value by sweep. As mentioned in the first section, nonhomenergetic solutions are available by means of the substitution principle. Our discussion is largely based on the thesis by Vahrenkamp (1992).

As usual, flow conditions downstream of the expansion are denoted with a unity subscript, while sweep plane conditions are denoted with a subscript $\perp$. A conventional Prandtl–Meyer flow

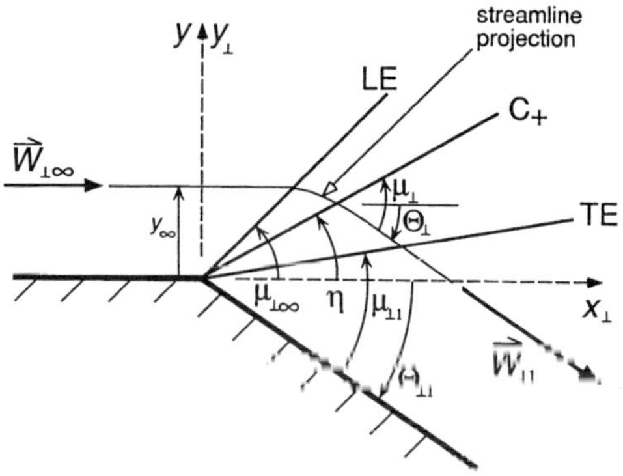

FIGURE 10.9 Schematic of Prandtl–Meyer flow in the sweep plane.

(see Figure 10.9) occurs in the sweep plane. The governing Mach number is

$$M_{\perp\infty} = M_\infty \cos\Lambda = \frac{w_\infty}{a_\infty}\cos\Lambda \tag{10.28}$$

and must exceed unity if an expansion is to exist. Consequently, the sweep angle has a maximum value

$$\Lambda_{max} = \cos^{-1}\frac{1}{M_\infty} \tag{10.29}$$

when the transonic $M_{\perp\infty} = 1$ condition holds. If Λ nevertheless exceeds Λ_{max} and $M_\infty > 1$, there is a supersonic flow in which the solution in the sweep plane, however, is subsonic and elliptic.

As indicated in Figure 10.9, the overall wall turn angle, in the sweep plane, is $\theta_{\perp 1}$. In parallel with the shock analysis, γ, M_∞, $\theta_{\perp 1}$, and Λ are prescribed, where Λ is treated as a free parameter. Ahead of the expansion, a streamline is straight and parallel to the upstream wall, while after the expansion it is straight and parallel to the downstream wall. When $\Lambda > 0$, however, the streamlines are not tangent, at any location, to the sweep plane. Figure 10.9 shows the projection of a streamline, initially at $y = y_\infty$, onto the sweep plane. The various Mach angles shown are measured relative to a projected streamline; e.g.,

$$\sin\mu_{\perp\infty} = \frac{1}{M_{\perp\infty}} \tag{10.30}$$

Figure 10.10 is a view in a plane where y is constant. The locations are indicated where the leading edge (LE) and trailing edge (TE) of the expansion cross this plane. A streamline projection is also shown. Downstream of the expansion, the streamlines are not parallel to the $x_\perp$ coordinate when $\Lambda > 0$. The freestream velocity is decomposed as

$$\vec{w}_\infty = \vec{w}_{\perp\infty} + \vec{w}_{1t} \tag{10.31}$$

where

$$\vec{w}_{\perp\infty} = \vec{w}_\infty \cos\Lambda, \qquad \vec{w}_{1t} = -\vec{w}_\infty \sin\Lambda \tag{10.32a,b}$$

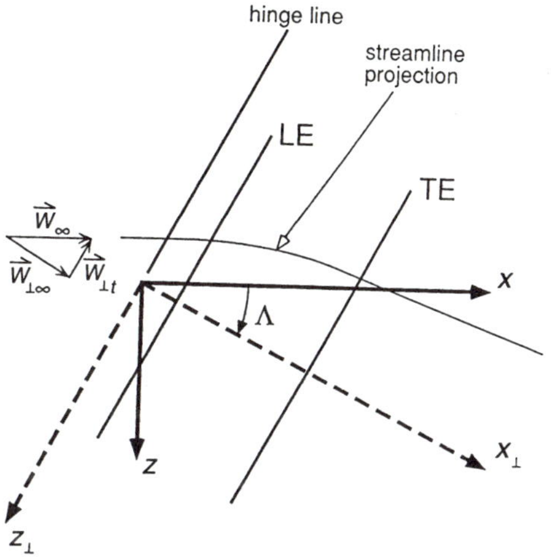

FIGURE 10.10 Sketch of the flow in a constant y plane.

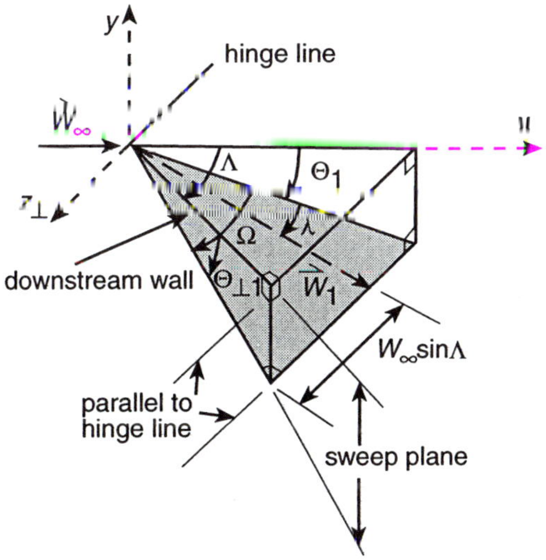

FIGURE 10.11 The shaded side of the pyramid is on the downstream wall. This construction defines the angles θ_1, Ω, and λ.

and the need for the minus sign in Equation (10.32b) is evident from Figure 10.10. The component $\vec{w}_{\perp t}$ is a constant throughout the flow field.

A number of angles are introduced in Figure 10.11, where the shaded triangle is located on the downstream wall. It is a right triangle that is perpendicular to the sweep plane. Its size is determined by the magnitude of the downstream velocity $\vec{w}_1$, which is not normal to the base of the triangle. By definition, Λ is in the x,z plane; Ω is its counterpart in the plane of the downstream wall. As evident from the pyramid, the orientation of $\vec{w}_1$ is fixed by θ_1 and λ, where θ_1 is in the x,y plane. This figure is useful in establishing various trigonometric relations. For instance,

we obtain

$$\tan\theta_1 = \cos\Lambda \tan\theta_{\perp 1} \tag{10.33}$$

$$\sin\Omega = \sin\Lambda \cos\theta_1 \tag{10.34}$$

$$\lambda = \Omega - \sin^{-1}\left(\frac{w_\infty}{w_1}\sin\Lambda\right) \tag{10.35}$$

Although not required by fluid dynamics, these trigonometric relations limit $\theta_{\perp 1}$ to a maximum value of 90°. With $\theta_{\perp 1}$ and Λ limited to a 0° to 90° range, we observe that θ_1 and Ω are also restricted to this range. As we shall see, λ is not similarly restricted. We observe that when there is no sweep, $\theta_1 = \theta_{\perp 1}$ and $\lambda = \Omega = 0$.

General Features of the Solution

At any location, the velocity can be written as

$$\vec{w} = \vec{w}_\perp + \vec{w}_{\perp t} = -aM_\perp \frac{\vec{w}_\perp}{w_\perp} - w_\infty \sin\Lambda \frac{\vec{w}_{\perp t}}{w_{\perp t}} \tag{10.36}$$

where $a_\perp = a$, $\vec{w}_\perp / w_\perp$ and $\vec{w}_{\perp t}/w_{\perp t}$ are unit vectors, and the minus sign stems from $\vec{w}_{\perp t}$ and positive $z_\perp$ being in opposite directions. In turn, this yields

$$M^2 = \frac{w^2}{a^2} = M_\perp^2 + M_\infty^2\left(\frac{a_\infty}{a}\right)^2 \sin^2\Lambda \tag{10.37}$$

Since the flow is homentropic, we have

$$\left(\frac{a}{a_\infty}\right)^2 = \frac{X_\infty}{X} \tag{10.38}$$

where

$$X = 1 + \frac{\gamma - 1}{2} M^2 \tag{10.39}$$

Consequently, the Mach number, anywhere in the flow field, can be written as

$$M^2 = \frac{X_\infty M_\perp^2 + M_\infty^2 \sin^2\Lambda}{X_{\perp\infty}} \tag{10.40}$$

where $M_{\perp\infty}$, which appears in $X_{\perp\infty}$, is given by Equation (10.28). With this relation, the elegant result is obtained as

$$X = \frac{X_\infty X_\perp}{X_{\perp\infty}} \tag{10.41}$$

Thus, the static and stagnation pressures are, respectively,

$$\frac{p}{p_\infty} = \left(\frac{X_\infty}{X}\right)^{\gamma/(\gamma-1)}, \qquad \frac{p_0}{p_\infty} = X^{\gamma/(\gamma-1)} \tag{10.42a,b}$$

with similar homentropic relations for other variables. In view of Equation (10.41), the pressure ratio can also be written as

$$\frac{p}{p_\infty} = \left(\frac{X_{\perp\infty}}{X_\perp}\right)^{\gamma/(\gamma-1)}$$

Consequently, any thermodynamic quantity can be evaluated from just the flow in the sweep plane.

Solution Inside the Expansion

As one would expect, the analysis is performed in the sweep plane (see Figure 10.9) where C_+ denotes an arbitrary left-running characteristic within the expansion. To some extent, the analysis here parallels that in Section 8.4. The constant Mach number $M_\perp$ on this characteristic is associated with the Mach angle

$$\mu_\perp = \mu(M_\perp) = \cot^{-1}(M_\perp^2 - 1)^{1/2} = \frac{\pi}{2} - \tan^{-1}(M_\perp^2 - 1)^{1/2} \tag{10.43}$$

For this characteristic, which has an angle η relative to the $x_\perp$ axis,

$$\nu(M_\perp) = \nu(M_{\perp\infty}) + \theta_\perp = \nu(M_{\perp\infty}) + \mu_\perp - \eta \tag{10.44}$$

where ν is the Prandtl–Meyer function. With the aid of the above relations, $M_\perp$ is explicitly given by

$$M_\perp^2 = 1 + b^2 \tan^2 \hat{z} \tag{10.45}$$

where

$$\hat{z} = \frac{1}{b}\left[\nu(M_{\perp\infty}) + \frac{\pi}{2} - \eta\right], \qquad b = \left(\frac{\gamma+1}{\gamma-1}\right)^{1/2} \tag{10.46}$$

The variable angle $\theta_\perp$ in Equation (10.44) can be shown to be given by

$$\theta_\perp = \cot^{-1}(b \tan \hat{z}) - \eta \tag{10.47}$$

In Figure 10.9, η is constrained to lie between the leading and trailing edges of the expansion

$$\eta_{LE} = \frac{\pi}{2} - \tan^{-1}(M_{\perp\infty}^2 - 1)^{1/2}, \qquad \eta_{TE} = \mu(M_{\perp 1}) - \theta_{\perp 1} \tag{10.48}$$

where Equation (10.28) provides $M_{\perp\infty}$, while $M_{\perp 1}$ is given in the next subsection. The corresponding $\hat{z}$ values are

$$\hat{z}_{LE} = \tan^{-1}\left[\frac{(M_{\perp\infty}^2 - 1)^{1/2}}{b}\right], \qquad \hat{z}_{TE} = \tan^{-1}\left[\frac{(M_{\perp 1}^2 - 1)^{1/2}}{b}\right] \tag{10.49}$$

With the foregoing relations, we obtain

$$X_\perp = \frac{\gamma+1}{2\cos^2\hat{z}}, \qquad \frac{X_\infty}{X} = \frac{2}{\gamma+1} X_{\perp\infty} \cos^2\hat{z} \tag{10.50a,b}$$

where Equation (10.50b) is useful with homentropic equations, and the flow speed can be written as

$$\frac{w_\perp}{a_\infty} = \frac{aM_1}{a_\infty} = \left(\frac{2}{\gamma+1} X_{\perp\infty}\right)^{1/2} (1 + b^2 \tan^2\hat{z})^{1/2} \cos\hat{z} \tag{10.51}$$

The solution inside the expansion is thus given in terms of $\hat{z}$, which, in turn, is linear with η.

Downstream Condition

The downstream Mach number $M_{\perp 1}$, in the sweep plane, is given by

$$\nu(M_{\perp 1}) = \nu(M_{\perp\infty}) + \theta_{\perp 1} \tag{10.52}$$

Once $M_{\perp 1}$ is known, Equation (10.40) provides M_1. Other parameters, such as a_1 and p_1, are, respectively, given by Equations (10.38) and (10.42a). The downstream flow speed is

$$\frac{w_1}{w_\infty} = \frac{M_1 a_1}{w_\infty} = \frac{M_1}{M_\infty}\left(\frac{X_\infty}{X_1}\right)^{1/2} \tag{10.53}$$

The condition when the TE of the expansion is coincident with the downstream wall is also referred to as detachment, and is denoted with a d subscript. In this circumstance, $M_{\perp 1d} \to \infty$, $p_{1d} = 0$, and the wall turn angle is

$$\theta_{\perp 1d} = \frac{\pi}{2}(b-1) - \nu(M_{\perp\infty}) \tag{10.54}$$

where $\pi(b-1)/2$ equals 130.5° when $\gamma = 7/5$. Should $\theta_{\perp 1} > \theta_{\perp 1d}$, there is a void between the wall and the TE of the expansion. For purposes of simplicity, the subsequent discussion assumes $\theta_{\perp 1d} \geq \theta_{\perp 1}$, which results in a lower bound for Λ:

$$\nu(M_\infty \cos\Lambda_{min}) \leq \frac{\pi}{2}(b-1) - \theta_{\perp 1} \tag{10.55}$$

When the inequality sign holds, $\Lambda_{min} = 0$, while the equality sign yields $\Lambda_{min} > 0$, which occurs when M_∞ and $\theta_{\perp 1}$ are large. Hence, with sufficient sweep, a void can be avoided. Consequently, Λ is bounded by Λ_{min} and Λ_{max}. The lower bound avoids detachment, while the upper bound stems from the $M_{\perp\infty} \geq 1$ condition.

Figure 10.12 shows $\theta_{\perp 1d}$ increasing with Λ for a given M_∞ value. Since $\gamma = 1.4$, the maximum value for $\theta_{\perp 1d}$ is 130.5°, and the curves terminate when $\Lambda = \Lambda_{max}$. As expected, the detachment angle decreases with increasing M_∞.

Figures 10.13 and 10.14, respectively, show M_1 and p_1/p_∞ when $\gamma = 1.4$, $M_\infty = 3$, 6, 9, and $\theta_{\perp 1} = 20°$, 40°, 60°. The curves terminate when $\Lambda = \Lambda_{max}$, while curves based on large $\theta_{\perp 1}$ and M_∞ values also terminate when $\Lambda = \Lambda_{min}$. This occurs at detachment when M_1 becomes infinite.

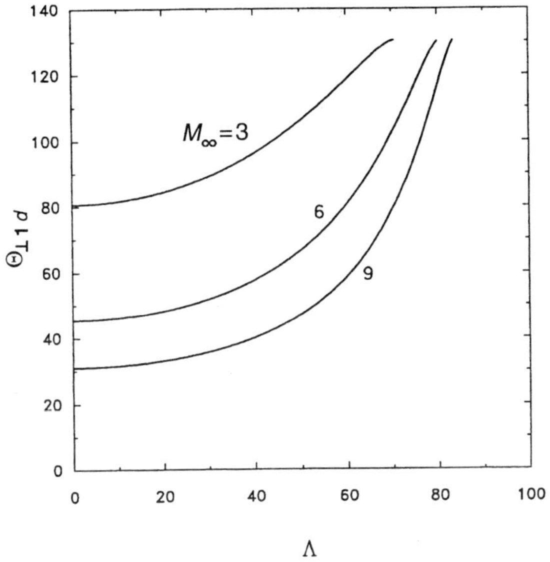

FIGURE 10.12 Detachment angle vs. Λ when $\gamma = 1.4$.

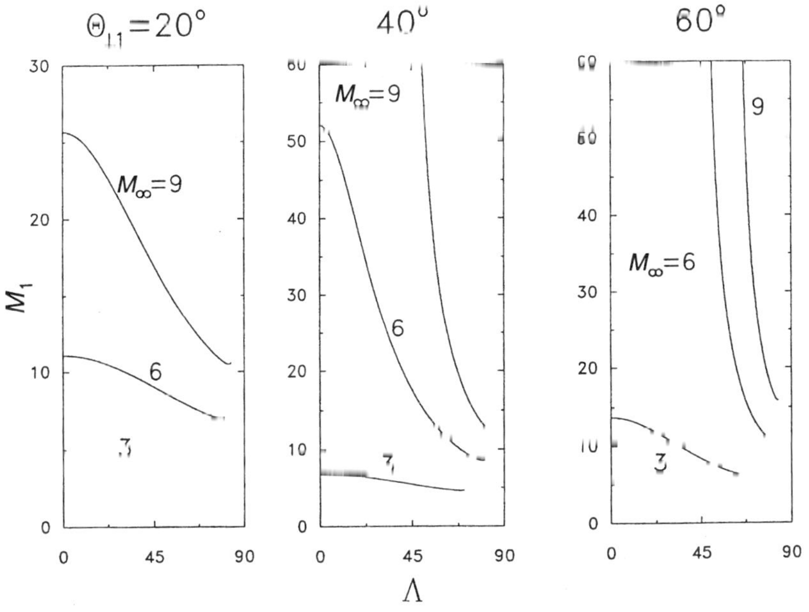

FIGURE 10.13 Downstream Mach number vs. Λ for various $\theta_{\perp 1}$ and M_∞ when $\gamma = 1.4$.

Although barely discernible, each M_1 curve has a minimum value, denoted with a tilde, when Λ is near Λ_{max}. This minimum corresponds to (see Problem 10.3)

$$\frac{\tilde{M}_{\perp 1}^4}{\tilde{M}_{\perp 1}^2 - 1} = \frac{\tilde{M}_{\perp \infty}^4}{\tilde{M}_{\perp \infty}^2 - 1} \tag{10.56}$$

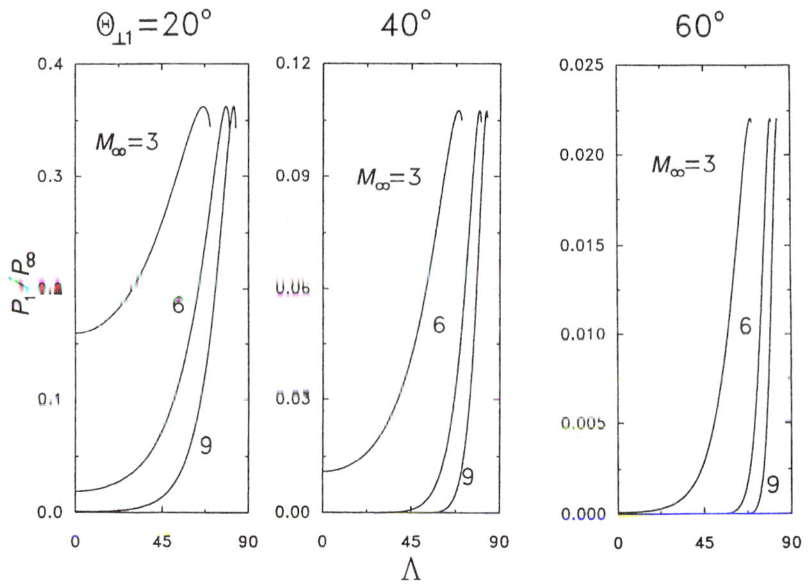

FIGURE 10.14 The pressure ratio p_1/p_∞ vs. Λ for various $\theta_{\perp 1}$ and M_∞ when $\gamma = 1.4$.

which is a quadratic equation for either $\tilde{M}_{\perp\infty}^2$ or $\tilde{M}_{\perp 1}^2$. By symmetry, one solution is $\tilde{M}_{\perp 1} = \tilde{M}_{\perp\infty}$, which occurs when $\theta_{\perp 1} = 0$. The other solution is

$$\tilde{M}_{\perp 1} = \frac{\tilde{M}_{\perp\infty}}{(\tilde{M}_{\perp\infty}^2 - 1)^{1/2}} \quad \text{or} \quad \tilde{M}_{\perp\infty} = \frac{\tilde{M}_{\perp 1}}{(\tilde{M}_{\perp 1}^2 - 1)^{1/2}} \tag{10.57}$$

Either of these equations, in conjunction with Equation (10.52), yields $\tilde{M}_{\perp 1}$ and $\tilde{M}_{\perp\infty}$, and therefore $\tilde{\Lambda}$. Since $\tilde{\Lambda}$ is close to Λ_{max}, the value for $\tilde{M}_{\perp\infty}$ is always close to unity. Aside from a small region near $\tilde{\Lambda}$, M_1 decreases with sweep. As with a shock wave, sweep reduces the overall strength of the disturbance.

The pressure ratio is shown in Figure 10.14, where the curves are subject to the same Λ_{min} and Λ_{max} constraints as in Figure 10.13. Moreover, the maximum p_1/p_∞ value corresponds to the minimum M_1 value. For a given $\theta_{\perp 1}$ value, the maximum p_1/p_∞ value is independent of M_∞. The similarity with the earlier shock wave analysis is evident.

Streamline Equations

Inside the expansion, the velocity can be written as

$$\vec{w} = w_{\perp x}\,\hat{\imath}_{\perp x} + w_{\perp y}\,\hat{\imath}_{\perp y} + w_{\perp z}\,\hat{\imath}_{\perp z} = w_\perp \cos\theta_\perp\,\hat{\imath}_{\perp x} - w_\perp \sin\theta_\perp\,\hat{\imath}_y - w_\infty \sin\Lambda\,\hat{\imath}_{\perp z} \tag{10.58}$$

where the $\hat{\imath}_\perp$ represent an orthonormal basis associated with the sweep plane. The streamline equations then are

$$\frac{dx_\perp}{dt} = w_\perp \cos\theta_\perp, \qquad \frac{dy}{dt} = -w_\perp \sin\theta_\perp, \qquad \frac{dz_\perp}{dt} = -w_\infty \sin\Lambda \tag{10.59}$$

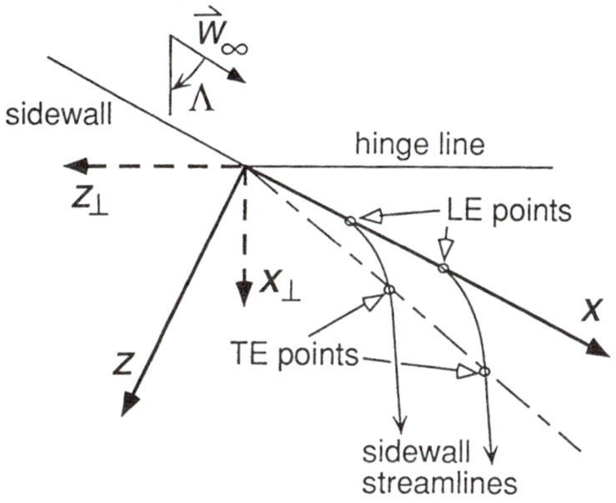

FIGURE 10.15 Depiction of two sidewall streamlines inside and downstream of the expansion.

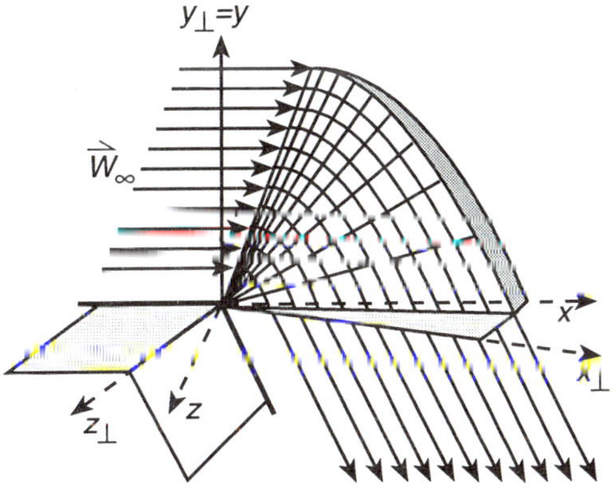

FIGURE 10.16 Depiction of a sidewall when the sweep angle is 45°.

Each streamline originates at the LE of the expansion at time $t = 0$. For convenience, the streamlines are required to pass directly above the origin. This construction thereby provides a sidewall, upstream of the expansion, that is planar and contains the origin of the coordinate systems (see Figure 10.8). The planar sidewall is in the x,y plane and terminates at the LE. The streamlines that are tangent to it become tangent to a curved wall between the LE and TE. This approach would expedite the design of a duct for generating the swept flow field.

Figure 10.15 depicts two streamlines with different y_∞ values. Upstream of the LE, the streamlines are straight and in the $z = 0$ plane. Inside the expansion, they curve with their x and z values increasing, while y decreases. Downstream of the expansion, the streamlines are again straight and parallel to the downstream wall. Figure 10.16 depicts the curved sidewall, adjacent to the expansion, that stems from the planar upstream sidewall when $\gamma = 1.4$, $M_\infty = 3$, $\theta_{\perp 1} = 20°$, and $\Lambda = 45°$. As the figure indicates, the TE is nearly coincident with the x,y and x,z planes. Moreover, as seen from Figure 10.15, the sidewall, downstream of the expansion, is also planar.

Initial conditions for Equations (10.59) are

$$x_{\perp LE} = y_\infty (M_{\perp\infty}^2 - 1)^{1/2}, \quad y_{\perp LE} = y_\infty, \quad z_{\perp LE} = -y_\infty (M_{\perp\infty}^2 - 1)^{1/2} \tan \Lambda \tag{10.60}$$

where y_∞ is the arbitrary height of a streamline above the upstream wall. From the last of Equations (10.59), we obtain

$$\frac{z_\perp}{y_\infty} = -M_\infty(\sin\Lambda)\tau - (M_{\perp\infty}^2 - 1)^{1/2}\tan\Lambda \tag{10.61}$$

where $\tau = (a_\infty t)/y_\infty$ is a nondimensional time measured from when a fluid particle crosses the LE.

The $dx_\perp$ and dy equations are integrated to obtain $x_\perp$ and y in terms of τ, as was just done for $z_\perp$. Toward this end, we start with the transformation (see Figure 10.9)

$$x_\perp = r_\perp\cos\eta, \qquad y = r_\perp\sin\eta \tag{10.62}$$

These equations are differentiated to obtain, with the aid of Equations (10.59),

$$\frac{dr_\perp}{dt} = a_\infty b\left(\frac{2}{\gamma+1}X_{\perp\infty}\right)^{1/2}\sin\hat{z}, \qquad \frac{d\eta}{dt} = -a_\infty\left(\frac{2}{\gamma+1}X_{\perp\infty}\right)^{1/2}\frac{\cos\hat{z}}{r_\perp} \tag{10.63}$$

They are combined, to yield

$$\frac{dr_\perp}{d\hat{z}} = b^2 r_\perp\tan\hat{z} \tag{10.64}$$

where $d\eta = bd\hat{z}$ is used. The initial condition is $r_{\perp LE} = y_\infty M_{\perp\infty}$, with the result

$$\frac{r_\perp}{y_\infty} = \frac{M_{\perp\infty}}{\left(\frac{2}{\gamma+1}X_{\perp\infty}\right)^{b^2/2}(\cos\hat{z})^{b^2}} \tag{10.65}$$

Equations (10.63) and (10.65) relate τ to $\hat{z}$:

$$\frac{d\hat{z}}{d\tau} = \frac{1}{b}\left(\frac{2}{\gamma+1}X_{\perp\infty}\right)^{\gamma/(\gamma-1)}\frac{(\cos\hat{z})^{2\gamma/(\gamma-1)}}{M_{\perp\infty}} \tag{10.66}$$

Upon integration, we have

$$\tau = bM_{\perp\infty}\left(\frac{2}{\gamma+1}X_{\perp\infty}\right)^{-\gamma/(\gamma-1)} I(\hat{z};\hat{z}_{LE},\gamma) \tag{10.67}$$

where

$$I = \int_{\hat{z}_{LE}}^{\hat{z}} (\cos x)^{-2\gamma/(\gamma-1)}\,dx \tag{10.68}$$

Thus, the time τ it takes a fluid particle to go from the LE to the η characteristic only depends on γ, $M_{\perp\infty}$, and η. The integral can be evaluated in closed form whenever $\gamma = (n+2)/n$, $n = 3, 4, \ldots$. For $\gamma = 7/5$ (or $n = 5$), this yields

$$\int\frac{dx}{(\cos x)^7} = \frac{\sin x}{48\cos^6 x}(8 + 10\cos^2 x + 15\cos^4 x) + \frac{5}{32}\ell n\left(\frac{1+\sin x}{1-\sin x}\right) + \text{constant} \tag{10.69}$$

The desired relation, with $\gamma = 7/5$, is then

$$\tau = 6^{1/2} M_{\perp\infty}\left(\frac{5}{6}X_{\perp\infty}\right)^{-7/2} I(\hat{z};\hat{z}_{LE}, 7/5) \tag{10.70}$$

The equations for the streamlines are given in the x,y,z coordinate system by a solid body rotation about y:

$$x = x_\perp \cos\Lambda - z_\perp \sin\Lambda, \qquad z = x_\perp \sin\Lambda + z_\perp \cos\Lambda \tag{10.71}$$

With the aid of Equations (10.61), (10.62), and (10.65), these become

$$\frac{x}{y_\infty} = \frac{M_{\perp\infty}\cos\eta\cos\Lambda}{\left(\frac{2}{\gamma+1}X_{\perp\infty}\right)^{b^2/2}(\cos\hat{z})^{b^2}} + M_\perp\tau\sin^2\Lambda + (M_{\perp\infty}^2 - 1)^{1/2}\sin\Lambda\tan\Lambda \tag{10.72a}$$

$$\frac{y}{y_\infty} = \frac{M_{\perp\infty}\sin\eta}{\left(\frac{2}{\gamma+1}X_{\perp\infty}\right)^{b^2/2}(\cos\hat{z})^{b^2}} \tag{10.72b}$$

$$\frac{z}{y_\infty} = \sin\Lambda\left[\frac{M_{\perp\infty}\cos\eta}{\left(\frac{2}{\gamma+1}X_{\perp\infty}\right)^{b^2/2}(\cos\hat{z})^{b^2}} - M_\perp\tau\cos\Lambda - (M_{\perp\infty}^2 - 1)^{1/2}\right] \tag{10.72c}$$

where η and $\hat{z}$ are related by Equation (10.46) and τ and $\hat{z}$ by Equation (10.67).

The angle λ, shown in Figure 10.11, is the minimum angle between $\vec{w}_1$ and the x,y plane. It thus represents the overall turn made by a streamline when projected onto the downstream wall. This angle is given by Equation (10.35), and can be written as

$$\lambda = \sin^{-1}\left[\frac{\sin\Lambda}{(1+\cos^2\Lambda\tan^2\theta_{\perp 1})^{1/2}}\right] - \sin^{-1}\left[\frac{M_\infty}{M_1}\left(\frac{X_1}{X_\infty}\right)^{1/2}\sin\Lambda\right] \tag{10.73}$$

Figure 10.17 shows λ, in degrees, for the same γ, M_∞, and $\theta_{\perp 1}$ values used in Figures 10.13 and 10.14. The magnitude of λ is relatively small, since it is given by the difference of two angles. Thus, the change in orientation between $\vec{w}_1$ and $\vec{w}_\infty$ is not especially large, even when Λ is large. In this regard, the behavior is similar to the shock case. For instance, when $\gamma = 1.4$, $M_\infty = 6$, $\theta_{\perp 1} = 20°$, and $\Lambda = 60°$, λ is only about 2°. A small λ value is also apparent in Figure 10.16.

Moreover, λ is often negative, especially when M_∞ and $\theta_{\perp 1}$ are large. Figure 10.18 helps clarify this result by repeating Figure 10.11, but for a negative λ value. Assume all prescribed parameters are fixed and that Λ gradually increases from zero. When $\Lambda = 0$, the sweep and x,y planes coincide and $\lambda = 0$. When Λ is small and detachment does not occur, the angle, $\sin^{-1}[(w_\infty/w_1)\sin\Lambda]$, often increases slightly faster than Ω, thereby yielding Figure 10.18 and a negative λ. Alternatively, when M_∞ is large, w_∞/w_1 is close to unity, and λ is approximately given by

$$\lambda \cong \Omega - \Lambda \tag{10.74}$$

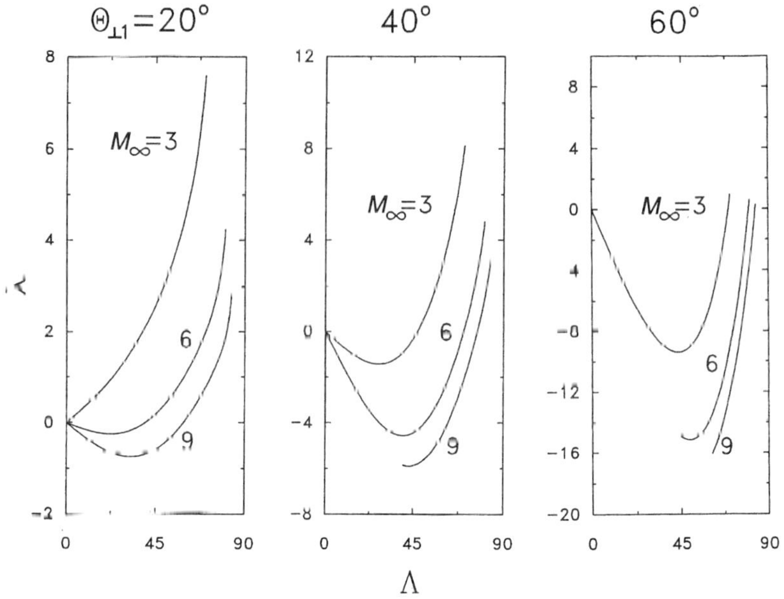

FIGURE 10.17 The angle λ vs. Λ for various $\theta_{\perp 1}$ and M_∞ when $\gamma = 1.4$.

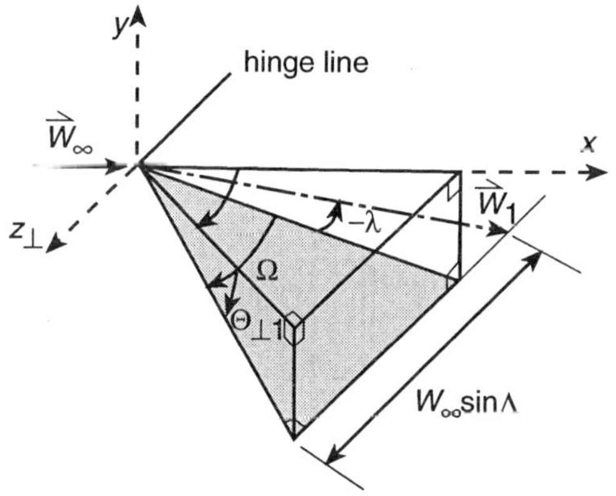

FIGURE 10.18 A repeat of Figure 10.11 to illustrate the situation when λ is negative.

From Equation (10.34), we observe that Ω is smaller than Λ when $\theta_{\perp 1} > 0$ and, consequently, λ is again negative.

REFERENCES

Emanuel, G., *Gasdynamics: Theory and Applications*, AIAA Education Series, Washington, D.C., 1986, 356.

Emanuel, G., "Oblique Shock Wave with Sweep," *Shock Waves* 2, 13 (1992a).

Emanuel, G., "Oblique Shock Wave with Sweep II," *Shock Waves* 2, 273 (1992b).

Poritsky, H., "Compressible Flows Obtainable from Two-Dimensional Flows through the Addition of a Constant Normal Velocity," *ASME Trans.* 68, A-61 (1946).

Settles, G.S. and Dolling, D.S., "Swept Shock-Wave/Boundary-Layer Interactions," in *Tactical Missile Aerodynamics: General Topics*, edited by M. J. Hemsch, Progress in Astronautics and Aeronautics, Vol. 141, 505 (1992).
Vahrenkamp, M., "Prandtl–Meyer Flow with Sweep," M.S. thesis, University of Oklahoma, 1992.

PROBLEMS

10.1 Derive Equations (10.13a,b).
10.2 Derive Equations (10.27).
10.3 Derive Equations (10.57).
10.4 A wall has an expansive turn of 30°, and the gas is helium at $M_\infty = 4$.
  (a) For a 40° sweep angle, determine Λ_{max}, p_1/p_∞, w_1/w_∞, M_1, and λ.
  (b) What would M_1 be if there was no sweep?
  (c) Determine p/p_∞ as an explicit function of η for the flow inside the expansion.
  (d) Write the equations for a streamline inside the expansion. Be sure to include the necessary auxiliary equations, such as Equation (10.67).

11 Interaction of an Expansion Wave with a Shock Wave

11.1 PRELIMINARY REMARKS

The interaction of a centered Prandtl–Meyer expansion with a weak-solution, planar shock wave is discussed. A perfect gas flow is assumed that is steady, inviscid, and two dimensional. Figure 11.1 illustrates the configuration in which the flow on the underside of the wedge is not sketched. Upstream, the flow is assumed to be uniform and supersonic. The shoulder, where the wall has a turn angle ϕ, generates a centered Prandtl–Meyer expansion that causes the shock to curve and weaken. Although not indicated in the figure, the flow downstream of the curved portion of the shock is rotational, or vortical.

Problem 11.1 demonstrates that the leading edge (LE) of the expansion and the planar shock intersect at a finite point. A small angle ϕ is shown in Figure 11.1 in order that the trailing edge (TE) of the expansion can also intersect the shock at a finite point. By the time ϕ equals the wedge angle θ_w, the TE no longer intersects the shock. This intersection is not required for the subsequent analysis. It is, however, pedagogically and possibly computationally convenient for this to be the case, as we generally assume.

The flow behind the planar part of the shock is also considered to be supersonic. For a given freestream Mach number M_∞, this is the usual situation, since the flow is subsonic for only a very narrow range of θ_w values. If the flow is subsonic, a planar shock does not occur. This is illustrated in Figure 11.2, which shows that the subsonic region is bounded, on the downstream side, by a curved sonic line that starts at the shoulder. A small increase in θ_w, or a small decrease in M_∞, would cause the shock in the figure to detach.

The flow sketched in the first figure is fundamental to gas dynamics. Consequently, it and similar flows have attracted the attention of a number of authors; e.g., see Shapiro (1953), Marshall and Plohr (1984), Saad (1993), or Schreier (1982). Shapiro and the last two references use an approximate "wavelet" approach to model the expansion and its reflection from the shock. Other basic studies that deal with the interaction of an expansion wave with a shock are by Friedrichs (1948), Lighthill (1949), and Chu (1952). The article by Friedrichs deals with a broad range of topics, including unsteady, one-dimensional flow and the process of shock formation. Lighthill develops a linearized approach, while Chu investigates the effect of a wedge whose face is slightly perturbed. Other authors, e.g., Munk and Prim (1948) and Pai (1952), have examined the flow of a related but different problem, namely, that due to a plane ogive with an attached shock. [Eggers et al. (1955) point out that the analyses by Lighthill (1949), Chu (1952), and Pai (1952) are not free of error.] The configuration in Figure 11.1, but with $\phi = \theta_w$, has been used in CFD studies by Nicola et al. (1996), and by Nasuti and Onofri (1996). This configuration has also been used in an engine inlet experiment by Wang et al. (1995). Li and Ben-Dor (1996) consider the interaction when the expansion is upstream of the shock, i.e., the shock runs into the expansion.

Rand (1950) examines the interesting possibility, shown in Figure 11.3, where upstream vorticity is used to cancel shock-produced vorticity. (A related, but different, vorticity-canceling shock is the subject of Problem 6.13.) In the figure, regions I, III, and IV are uniform, supersonic flows. Region II is a supersonic, parallel, vortical flow. From the substitution principle, in Chapter 8, we know that this type of flow is possible. With the presumption that the vorticity just downstream of

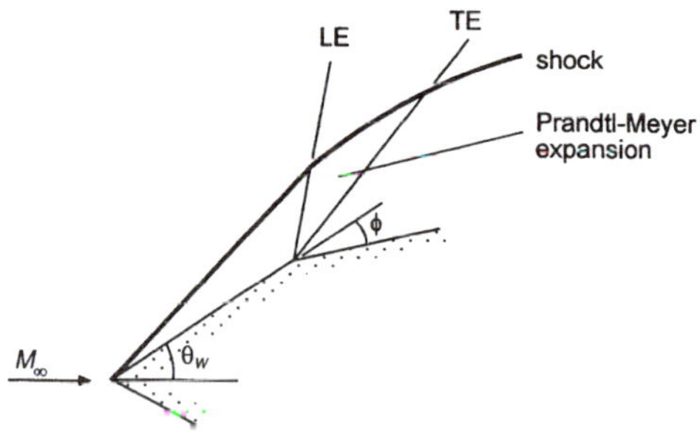

FIGURE 11.1 Supersonic flow about a wedge with an expansive turn.

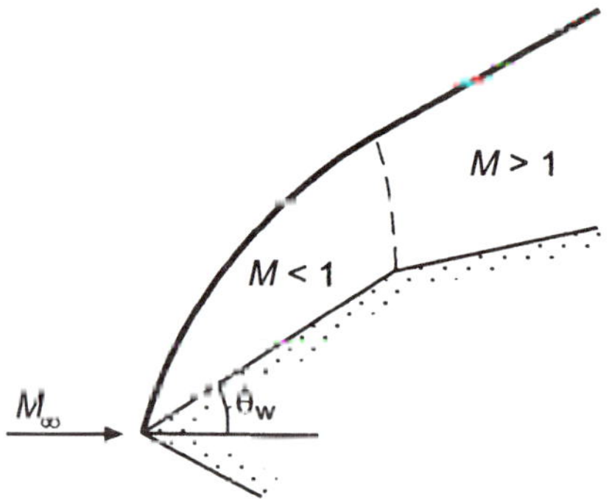

FIGURE 11.2 Attached bow shock with an embedded subsonic region.

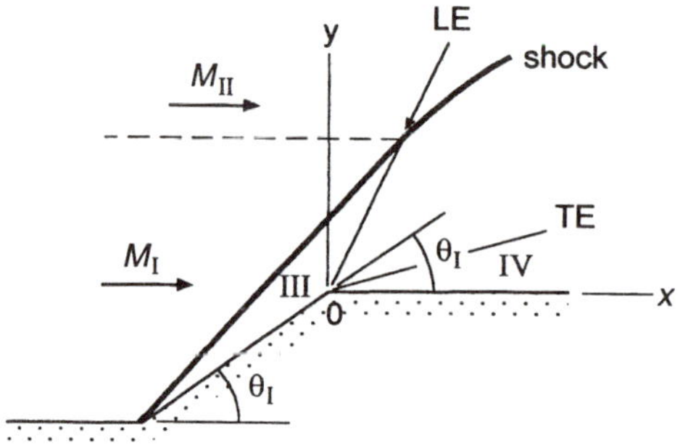

FIGURE 11.3 Schematic in which the Prandtl–Meyer flow is unperturbed.

the curved shock is zero, the vorticity then remains zero along the streamlines that pass through the curved shock. Consequently, the centered Prandtl–Meyer expansion is unperturbed as it propagates toward the downstream side of the shock. In accord with the substitution principle, the pressure and Mach number in region II are constants and respectively equal their region I values.

Rand's conjecture appears plausible in that the magnitude of the pressure jump across the curved shock decreases with increasing distance in accord with the decreasing pressure inside the expansion. Flow properties along any left-running characteristic, inside the unperturbed expansion, are constant. These include, e.g., the pressure and the stagnation enthalpy. Hence, the stagnation enthalpy on the downstream side of the shock is a constant and equals its value in region I. This, however, is inconsistent with region II being a parallel vortical flow, which requires a nonzero stagnation enthalpy gradient transverse to the streamlines. This point was established in Section 8.3. Consequently, the flow sketched in Figure 11.3 is not physically realizable.

On a conceptual basis, our analysis is reminiscent of shock-expansion theory, although ϕ may not be as large as this theory would require for a finite chord length. In its simplest form, shock-expansion theory ignores any interaction between the wall-generated expansion wave and the upstream, attached shock wave. The theory provides an inviscid estimate for the wall pressure that can be used to obtain lift and wave drag coefficients; see Section 14.10. More precise results sometimes can be obtained with the method-of-characteristics (MOC). In the days before computers, however, the MOC calculations were tedious, and, as a practical matter, required a relatively large grid. Moreover, a wide-ranging parametric study was out of the question. Hence, a number of approximate analytical (with some required computation) treatments were developed by Hayes and Probstein (1959), Eggers et al. (1955), Waldman and Probstein (1957), and Mahoney (1955) to improve upon the original theory or assess its accuracy. The report by Waldman and Probstein provides a review of the theory current at the time of publication. (Little in the way of analysis has been done since, because numerical methods became the topic of interest.) Even the more refined versions of the theory usually neglect the right-running reflected wave, since it is generally weak or because it may not impinge on the downstream airfoil surface. The studies by Eggers et al. (1955) and by Mahoney (1955) note that shock-expansion theory generally yields a relatively accurate prediction for the surface pressure.

Our motivation for accurately analyzing the flow field sketched in Figure 11.1 does not come from shock-expansion theory, although selected results will bear on this topic. Rather, it is pedagogical in that we wish to show how analysis, numerical methods, and gas dynamic concepts are combined to yield both insight as well as trends and specific results. (At the time of writing, actual numerical results were not available. This would be a good thesis or dissertation topic.) The material in this chapter can be viewed as the prototype for this type of wave interaction. We do not assume a weak shock, linearize the equations, or utilize a wavelet approach. Rather, a global solution is obtained that depends on only four nondimensional parameters, namely, the ratio of specific heats γ, M_∞, and the two angles, θ_w and ϕ. Although this type of flow configuration has been of interest for many years, our approach is believed to be new. A major reason for this is the recognition that the shock wave's curvature can be discontinuous, and whose treatment requires a novel MOC unit process. The next section discusses the topology of the flow field, which is much more complicated than Figure 11.1 indicates. Specific objectives are more clearly outlined at the end of this section. Section 11.3 provides the analytical solution for selected flow regions. The next section discusses the shock curvature singularity. The chapter concludes with a presentation of the unit processes and the MOC scheme.

11.2 FLOW TOPOLOGY

General Remarks

Figure 11.4 illustrates the many regions that constitute only part of the flow field downstream of the 0-2-6-10 shock wave. (The angle ϕ is drawn larger than it should be for purposes of clarity.) This type of patchwork is typical of a supersonic flow containing one, or more, shocks. The flow is governed by a wave equation that allows a discontinuous change in some of the primary variables, such as the entropy, when a shock wave or a slipstream is present, and in the first derivative of the primary variables across selected Mach lines and streamlines. The lines drawn in the figure show the location of these discontinuities. Eleven regions are pictured; a solution for the first ten is to be obtained. In other words, ten distinct solutions, one per region, are to be found. For some regions, an analytical solution is possible, while others require a computational approach. It is convenient, of course, to perform all of the work, analytical and computational, with a computer, thereby making a wide-ranging parametric study possible.

In an irrotational region, we have the left-running C_+, and right-running C_-, Mach lines or characteristics. When the flow is rotational, the C_o streamlines constitute a third family of characteristics, as sketched in Figure 11.5. In the figure, $\vec{w}$ is the velocity, μ is the Mach angle, and θ is the streamline

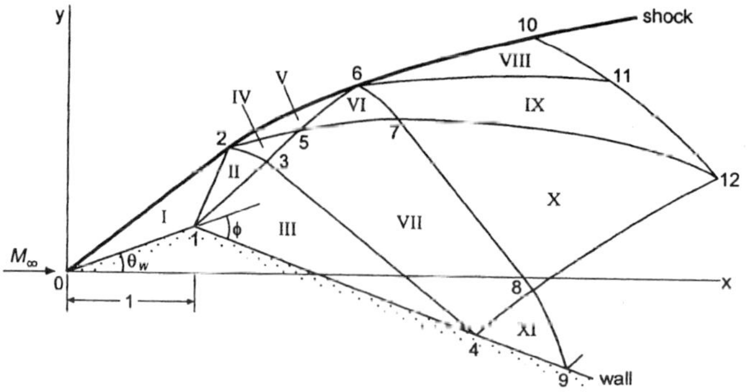

FIGURE 11.4 Schematic of flow regions.

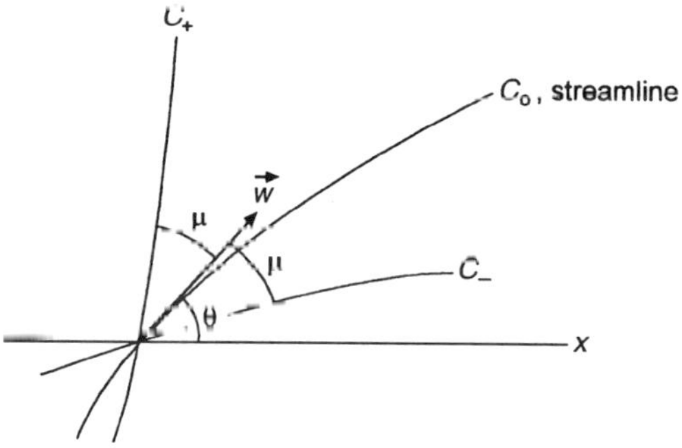

FIGURE 11.5 Characteristic curves.

angle relative to $\vec{w}_\infty$ or the x-axis. A uniform flow region is irrotational with just straight $C_\pm$ characteristics. Its rotational counterpart is a parallel flow region with three families of characteristics. In the irrotational case, a region can be described as uniform, simple, or nonsimple. In a uniform flow region, such as I, the $C_\pm$ characteristics are straight, as are the streamlines. In a simple wave region, such as II, one characteristic family is straight, the other is curved, as are the streamlines. In a nonsimple wave region, such as IV, both $C_\pm$ families and the streamlines are curved. The characterization of some of the regions is simplified by recalling the theorem for a steady, two-dimensional, irrotational, supersonic flow. It states that uniform and nonsimple wave regions border simple wave regions. In addition, a simple or nonsimple irrotational region can be bordered by a shock or a rotational flow region. Lines between different types of flow regions, including rotational regions, are always shock waves, slipstreams, or $C_\pm$ and C_o characteristics. These demarcation lines involve some sort of discontinuity. Moreover, these lines may possess properties associated with both adjacent regions. For example, the Mach line 1-2 has the properties of the uniform flow in region I and the Prandtl–Meyer flow in region II. These comments are important in the discussion of the flow field.

Another limitation of the analysis is that the wave that reflects from the downstream wall is ignored. The solution thus terminates on the 4-8-12-11-10 characteristic line segments. The reflected wave that begins at the wall in region XI is therefore not discussed. Later, we observe that regions VIII, IX, and X can be truncated on their downstream side without sacrificing any objectives.

Flow Field Description

A Cartesian coordinate system (see Figure 11.4) is utilized, in which all lengths are normalized by the distance, in the freestream direction, between the apex and shoulder of the body. Tables 11.1 and 11.2, respectively, characterize the different regions and their demarcation lines.

TABLE 11.1
Description of Regions Shown in Figure 11.4

Region	Description
I	Uniform, irrotational
II	Centered Prandtl–Meyer irrotational expansion
III	Uniform, irrotational
IV	Nonsimple, irrotational
V	Rotational
VI	Rotational
VII	Simple, irrotational
VIII	Rotational
IX	Rotational
X	Nonsimple, irrotational
XI	Nonsimple, irrotational

TABLE 11.2
Demarcation Lines in Figure 11.4

Lines	Description
0-1, 1-4-9	Straight wall segments
0-2	Straight shock
2-6-10	Curved shock
1-2, 1-3	Straight C_+ characteristics
3-5-6, 4-8-12	Curved C_+ characteristics
3-4, 7-8	Straight C_- characteristics
2-3, 6-7, 8-9, 10-11-12	Curved C_- characteristics
2-5-7-12, 6-11	Curved C_o characteristics

The Prandtl–Meyer expansion reflects from the shock between points 2 and 6. (If ϕ is too large, point 6 is at infinity.) There are actually two reflected waves. The first consists of the C_o characteristics that constitute a vortical layer, located in regions V, VI, and IX. The second wave consists of the C_- characteristics in region IV, V, VI, VII, and XI. The Prandtl–Meyer expansion interacts with both waves in the regions where there is overlap.

In the subsequent discussion, frequent reference is made to the shock wave angle β, which is measured relative to the x-axis. It is viewed as a function of the arc length s along the shock. Its first derivative, with respect to s, is written as β'. Along the 0-2 shock, β is a constant, equal to β_I, whereas beyond point 2, β decreases in magnitude. Although β is continuous at point 2, β' is not, going from zero to a finite (negative) value, as discussed later. Consequently, β is nonanalytic (i.e., β does not possess a Taylor series expansion with respect to s) at point 2.

The 2-5-7-12 and 6-11 streamlines separate regions of different rotationality and, thus, are C_o characteristics. Another possibility is that these boundaries are actually slipstreams. It is essential for the subsequent MOC discussion to resolve this question. Slipstreams normally start at a triple point, where three shock waves intersect. The magnitude of the velocity is then different on the two sides of a slipstream. For this to be the case, the strength of the shock at points 2 and 6 would have to change discontinuously. All primary variables, however, change continuously in the Prandtl–Meyer expansion. In turn, this means the strength of the shock is continuous. In short, the bounding streamlines are not slipstreams, and β is a continuous function of s, including at points 2 and 6.

The first derivative of quantities, such as the pressure and Mach number, is discontinuous along the leading and trailing edges of the Prandtl–Meyer expansion in a direction normal to these edges (see Problem 8.14). Thus, the derivative, $\partial M/\partial s$, is discontinuous at points 2 and 6. Consequently, β' is similarly discontinuous at these two points. Since the curvature of the shock equals $-\beta'$, it also is discontinuous at these points.

The overall strength of the reflected C_- wave is frequently weak in comparison to the Prandtl–Meyer expansion. This occurs when the expansion significantly weakens the bow shock. In turn, the weakened shock has a relatively large slope change; hence, the vortical layer is relatively intense. In the literature (e.g., Lighthill, 1949), a parameter is defined that represents, at the shock, the local ratio of the reflected C_- pressure wave to the incoming C_+ pressure wave. A more convenient parameter for our purposes is $d(\mu - \theta)/d\beta$, given by Equation (6.51). The sign of this derivative locally determines if the C_- wave is compressive or expansive. As the Section 6.3 analysis shows, both types of waves are possible. In particular, the C_- wave is generally compressive when $38° \geq \beta \geq \mu_\infty$. Figure 11.4 is drawn as if the C_- wave is compressive; i.e., it converges in the flow direction. When this wave is compressive, it is theoretically possible for an embedded shock wave to form somewhere inside regions IV, V, VI, or VII when C_- characteristics attempt to overlap. If this occurs, the shock would be quite weak, especially where it first forms. It should be noted that embedded shocks form in this manner inside the jet from an underexpanded nozzle (Emanuel, 1986, Section 19.4).

One might anticipate that the shock beyond point 6 would be planar. In this circumstance, VIII would be a uniform flow region and 6-11 would be a straight streamline with a constant pressure. To be consistent with this picture, X would also be a uniform flow region and region IX would contain a parallel, rotational flow. Such a region is permissible; however, the streamline angle θ would then have to be constant along the 6-7 characteristics. The last condition is quite unlikely; hence, we consider the shock, beyond point 6, to be curved. More importantly, the left-running C_+ characteristics that originate at the wall in region III, but pass through and interact with the vortical flow, cause the shock to curve. Since the C_+ wave entering region VIII initially is from the uniform flow in region III, the curvature of the shock should be quite small compared to that along 2-6.

Goals

A key objective would be to establish a numerical procedure that properly represents the Figure 11.4 flow field. This task is not trivial, in view of the nonanalytic nature of points 2 and 6. Thus, a major goal is to verify the previous description, or to modify it, based on accurate computational results. The overall strength of the two reflected waves, relative to the Prandtl–Meyer expansion, should be evaluated parametrically. Other objectives include locating points 4 and 6, obtaining an estimate for the maximum value of ϕ when the coordinates of point 6 are still finite, and finding the location of an embedded shock, if there is one.

A final objective would be to develop suitable flow conditions for experimental verification. For instance, a wind tunnel experiment might answer the question as to whether or not the shock is curved beyond point 6. If a weak expansion is used, i.e., ϕ is only a few degrees in magnitude, the amount of shock curvature may not be detectable. On the other hand, as ϕ increases, point 6 may move out of the field of view. There would be similar experimental difficulties in detecting the possible presence of a weak embedded shock.

11.3 SOLUTION FOR REGIONS I, II, AND III

With γ, M_∞, θ_w, and ϕ known, the solutions for regions I and III are routinely obtained. We thus focus on obtaining the coordinates of points 2, 3, and 4, the shape of the 2-3 characteristic, and the pressure, Mach number, and flow angle along this characteristic. Along the 3-4 part of this characteristic, these parameters have their constant region III values; i.e.,

$$p_{34} = p_{III}, \qquad M_{34} = M_{III}, \qquad \theta_{34} = -\theta_w + \phi \tag{11.1}$$

Conditions along 2-3-4 are needed for the MOC computation of the outer regions.

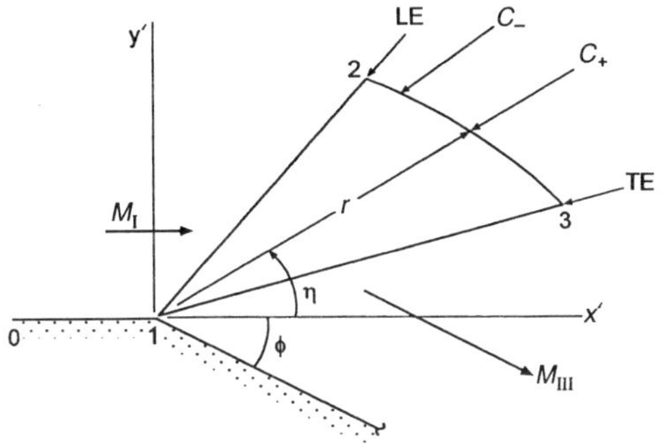

FIGURE 11.6 Rotated coordinate system centered at point 1 of Figure 11.4.

The coordinates of the first three points in Figure 11.4 are

$$x_0 = 0, \qquad y_0 = 0 \tag{11.2a}$$

$$x_1 = 1, \qquad y_1 = \tan \theta_w \tag{11.2b}$$

$$x_2 = \frac{\tan(\mu_I + \theta_w) - \tan \theta_w}{\tan(\mu_I + \theta_w) - \tan \beta_I}, \qquad y_2 = x_2 \tan \beta_I \tag{11.2c}$$

where

$$\mu_I = \sin^{-1}\left(\frac{1}{M_I}\right) \tag{11.3}$$

For the expansion, the angular coordinate η, shown in Figure 11.6, is convenient. Its leading and trailing edge values are

$$\mu_{LE} = \mu_I, \qquad \eta_{TE} = \mu_{III} - \phi \tag{11.4}$$

Even more convenient is the scaled angle

$$z(\eta) = \left(\frac{\gamma-1}{\gamma+1}\right)^{1/2}\left(\nu_I + \frac{\pi}{2} - \eta\right) \tag{11.5}$$

previously defined in Section 8.4. From the Prandtl–Meyer solution in that section, we can write

$$M = \left(1 + \frac{\gamma+1}{\gamma-1}\tan^2 z\right)^{1/2} \tag{11.6a}$$

$$\frac{p}{p_I} = \left[\frac{2}{\gamma+1}\left(1 + \frac{\gamma-1}{2}M_I^2\right)\cos^2 z\right]^{\gamma/(\gamma-1)} \tag{11.6b}$$

$$\theta = -\mu(M) + \eta + \theta_w \tag{11.6c}$$

where the parameters on the left-hand side are viewed as functions of η. The above relations hold throughout the Prandtl–Meyer flow region. In particular, they hold along the 2-3 characteristic.

The shape of this characteristic (see Problem 11.2) is provided by

$$z_{\mathrm{LE}} = \tan^{-1}\left[\frac{\gamma-1}{\gamma+1}(M_{\mathrm{I}}^2-1)\right]^{1/2} \tag{11.7a}$$

$$r_{\mathrm{LE}} = [(x_2-1)^2+(y_2-\tan\theta_w)^2]^{1/2} \tag{11.7b}$$

$$\frac{r}{r_{\mathrm{LE}}} = \left[\frac{\sin z_{\mathrm{LE}}}{\sin z}\left(\frac{\cos z_{\mathrm{LE}}}{\cos z}\right)^{(\gamma+1)/(\gamma-1)}\right]^{1/2} \tag{11.8a}$$

$$x = 1 + r_{\mathrm{LE}}\frac{r}{r_{\mathrm{LE}}}\cos(\eta+\theta_w) \tag{11.8b}$$

$$y = \tan\theta_w + r_{\mathrm{LE}}\frac{r}{r_{\mathrm{LE}}}\sin(\eta+\theta_w) \tag{11.8c}$$

Point 3 is obtained when

$$z_{\mathrm{TE}} = \left(\frac{\gamma-1}{\gamma+1}\right)^{1/2}\left(\nu_{\mathrm{I}}+\frac{\pi}{2}-\mu_{\mathrm{III}}+\phi\right) \tag{11.9}$$

is substituted into Equations (11.8). Finally, we note that point 4 is given by

$$x_4 = \frac{y_3 + x_3\tan(\phi-\theta_w+\mu_{\mathrm{III}}) - \tan\theta_w - \tan(\phi-\theta_w)}{\tan(\phi-\theta_w+\mu_{\mathrm{III}}) - \tan(\phi-\theta_w)} \tag{11.10a}$$

$$y_4 = \tan\theta_w + (1-x_4)\tan(\phi-\theta_w) \tag{11.10b}$$

11.4 CURVATURE SINGULARITY

The curvature of a shock wave is discontinuous whenever a dispersed wave starts, or finishes, interacting with the shock. Ferri (1954), e.g., presents a derivation, quite different from the following one, that relates the streamline and shock curvatures (see Problem 6.17). His analysis, however, assumes analyticity. A brief examination of the CFD literature indicates that the curvature discontinuity is not considered (see, e.g., Nasuti and Onofri, 1996; Nicola et al., 1996). In the book by Zucrow and Hoffmann (1977), which extensively treats the MOC, it is not mentioned.

In the subsequent analysis, a formula is obtained for the shock curvature just above point 2. This type of result is particularly useful when developing a numerical scheme, such as an MOC code, that is expected to accurately represent the shock.

Let point 2′ in Figure 11.7 be a point on the shock above point 2. With point 2 fixed, our objective is to determine the (negative of the) curvature, β'_2 of the 2 – 2′ circular arc in the limit of point 2′ approaching point 2. In this limit, neither reflected wave has any influence on β'_2. The 2 subscript hereafter denotes the curved part of the shock just above point 2.

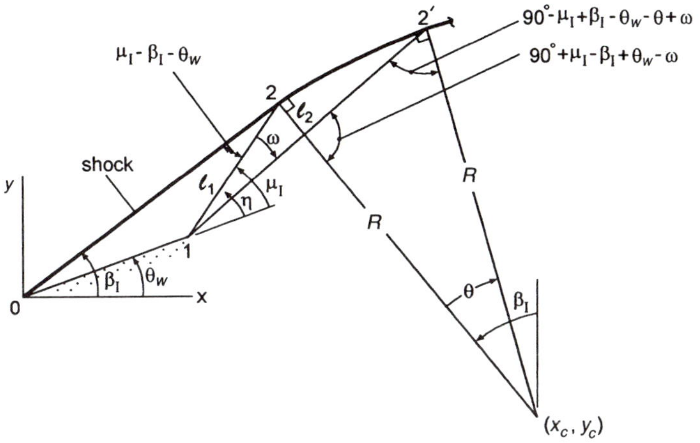

FIGURE 11.7 Shock wave curvature just above point 2.

The radius of curvature, R, which equals $-(\beta'_2)^{-1}$, is normal to the $2 - 2'$ arc. The arc is tangent to the planar shock at point 2, since β is continuous. To begin with, a number of trigonometric results are established. From the law of sines for the 0-1-2 triangle, we have

$$l_1 = \frac{\sin(\beta_I - \theta_w)}{\cos\theta_w \sin(\mu_I - \beta_I + \theta_w)} \tag{11.11}$$

where $x_1 = 1$ is utilized. The law of sines for the ω triangle yields

$$\frac{l_2}{\sin\omega} = \frac{l_1}{\sin\left(\frac{\pi}{2} + \mu_I - \beta_I + \theta_w - \omega\right)} \tag{11.12a}$$

which becomes, after l_1 is eliminated,

$$l_2 = \frac{\sin\omega \sin(\beta_I - \theta_w)}{\cos\theta_w \sin(\mu_I - \beta_I + \theta_w)\cos(\mu_I - \beta_I + \theta_w - \omega)} \tag{11.12b}$$

The law of sines for the θ triangle is

$$\frac{R}{\sin\left(\frac{\pi}{2} + \mu_I - \beta_I + \theta_w - \omega\right)} = \frac{R - l_2}{\sin\left(\frac{\pi}{2} - \mu_I + \beta_I - \theta_w + \omega\right)} \tag{11.13a}$$

which becomes

$$\cos(\mu_I - \beta_I + \theta_w - \omega) - \cos(\mu_I - \beta_I + \theta_w + \theta - \omega) = \frac{\sin\omega \sin(\beta_I - \theta_w)}{\cos\theta_w \sin(\mu_I - \beta_I + \theta_w)}\frac{1}{R} \tag{11.13b}$$

The purpose of Equations (11.11) and (11.12b) is to eliminate l_1 and l_2. With the replacement

$$\omega \rightarrow -d\eta, \qquad \theta \rightarrow d\theta \tag{11.14}$$

and expanding the left side of Equation (11.13b), we have

$$\sin(\mu_{\mathrm{I}} - \beta_{\mathrm{I}} + \theta_w)d\theta = \frac{d\eta \sin(\beta - \theta_w)}{\cos\theta_w \sin(\mu_{\mathrm{I}} - \beta_{\mathrm{I}} + \theta_w)}\frac{1}{R} \tag{11.13c}$$

Since $Rd\theta = ds$, we finally obtain

$$\left(\frac{ds}{d\eta}\right)_2 = \frac{\sin(\beta_{\mathrm{I}} - \theta_w)}{\cos\theta_w \sin^2(\mu_{\mathrm{I}} - \beta_{\mathrm{I}} + \theta_w)} \tag{11.15}$$

This relation provides the differential arc length along the shock, just above point 2, with respect to the negative of the differential angle in the Prandtl–Meyer expansion, at the LE of the expansion.

Along the shock, Appendix D provides

$$\left(\frac{dp}{ds}\right)_2 = \frac{2\gamma}{\gamma+1} p_\infty M_\infty^2 \beta_2' \sin 2\beta_{\mathrm{I}} \tag{11.16}$$

With Equation (11.15), this yields

$$\beta_2' = -\frac{\gamma+1}{2}\frac{\cos\theta_w \sin^2(\mu_{\mathrm{I}} - \beta_{\mathrm{I}} + \theta_w)}{\sin 2\beta_{\mathrm{I}} \sin(\beta_{\mathrm{I}} - \theta_w)}\frac{1}{p_\infty M_\infty^2}\left(\frac{dp}{d\eta}\right)_2 \tag{11.17}$$

where the pressure derivative is evaluated at the LE of the expansion. Equation (11.6b) provides the pressure; its derivative is

$$\frac{dp}{d\eta} = \frac{2}{(\gamma^2-1)^{1/2}} p_{\mathrm{I}} \left[\frac{2}{\gamma+1}\left(1 + \frac{\gamma-1}{2}M_{\mathrm{I}}^2\right)\right]^{\gamma/(\gamma-1)} \sin z(\cos z)^{(\gamma+1)/(\gamma-1)} \tag{11.18}$$

The value of z on the LE is given by Equation (11.7a), or

$$\tan^2 z_{\mathrm{LE}} = \left(\frac{\gamma-1}{\gamma+1}\right)(M_{\mathrm{I}}^2 - 1) \tag{11.19a}$$

which can be written as

$$\sin^2 z_{\mathrm{LE}} = 1 - \cos^2 z_{\mathrm{LE}} = \frac{\gamma-1}{2}\frac{M_{\mathrm{I}}^2 - 1}{1 + \frac{\gamma-1}{2}M_{\mathrm{I}}^2} \tag{11.19b}$$

Equation (11.18) thus becomes

$$\left(\frac{dp}{d\eta}\right)_2 = \frac{2\gamma}{\gamma+1} p_1 (M_{\mathrm{I}}^2 - 1)^{1/2} \tag{11.20}$$

Our final result is obtained by combining this with Equation (11.17):

$$\beta_2' = -\frac{2\gamma}{\gamma+1} \frac{\gamma M_\infty^2 \sin^2\beta_{\mathrm{I}} - \frac{\gamma-1}{2}}{M_\infty^2} (M_{\mathrm{I}}^2 - 1)^{1/2} \frac{\cos\theta_w \sin^2(\mu_{\mathrm{I}} - \beta_{\mathrm{I}} + \theta_w)}{\sin 2\beta_{\mathrm{I}} \sin(\beta_{\mathrm{I}} - \theta_w)} \tag{11.21}$$

With the aid of a software program, Appendix D can be used to evaluate quantities such as $(M_{\mathrm{I}}^2 - 1)^{1/2}$ and $\sin(\beta_{\mathrm{I}} - \theta_w)$. Note that β_2' is continuous at point 2 only when M_I is unity; otherwise, it is negative. In fact, the above relation verifies that β' is discontinuous at point 2. Observe that β_2' only depends on γ, M_∞, and θ_w, and not on ϕ. As shown by Problem 11.3, $-\beta_2'$ rapidly increases with θ_w; the increase with M_∞ is slower. From Appendix D, we observe that the tangential derivatives are proportional to β', while the normal derivatives also contain a β' term. These derivatives are therefore discontinuous at point 2. Similarly, the vorticity is also discontinuous.

Although a global analysis is utilized to derive Equation (11.21), the result is actually local to point 2. With the replacement,

$$\theta_w \to \theta, \qquad \beta_{\mathrm{I}} \to \beta, \qquad \mu_{\mathrm{I}} \to \mu, \qquad M_{\mathrm{I}} \to M \tag{11.22}$$

where $\theta, \ldots, M$ refer to the flow in region I just adjacent to point 2, the planar shock is now of infinitesimal length. In further support of the local argument, note that β_2' does not depend on the strength of the expansion, i.e., the wall turn angle ϕ. However, in order to have a shock curvature discontinuity, the wall curvature at point 1 must be discontinuous.

11.5 NUMERICAL PROCEDURE

General Remarks

If a conventional finite difference scheme is utilized, then shock capturing rather than shock fitting is the customary approach. With this approach, the shock and the characteristics that border different regions are not well resolved. This is particularly detrimental at points 2 and 6, where the solution is nonanalytic. What is desired, for an accurate solution, is a fitting procedure in which the shock and all bordering characteristics become part of the computational grid! This is best done with an MOC procedure.

As previously noted, the solution only depends on γ, M_∞, θ_w, and ϕ. The solution in regions I, II, and III and the location of the 2-3-4 characteristic can be performed analytically. With data along 3-5-7, the simple wave solution for region VII can be found. The other regions require a computational approach, which here is done with the MOC. Note that the solution for regions IV, V, and VI must be found before the region VII solution is analytically or numerically attempted. By examining the C_- characteristics for these regions, we can establish where the flow is expansive or compressive. If an embedded shock is present, the location where it starts can be estimated.

Once data along 6-7-8 is established, a numerical solution for regions VIII, IX, and X can be found. In this regard, the solution can be truncated starting at an arbitrary point on the 7-8 characteristics. All that is required is to evaluate enough of these regions to establish their nature. This truncation procedure should enable solutions to be obtained even if there is an embedded shock.

The solution procedure has the following (approximate) sequential pattern. Regions I, II, and III are analytically determined. Next, regions IV, V, and VI are computationally established. Region VII is computationally evaluated with an analytical solution serving as a check and as a way of assessing numerical accuracy. Finally, all or part of regions VIII, IX, and X are computationally established.

As previously indicated, points 2 and 6 are special. Standard MOC procedures (e.g., see Zucrow and Hoffman, 1977) encounter a serious difficulty at these points. Because the shock is nonanalytical at these points, we end up with one unknown parameter in excess of the number of equations. For point 2, this difficulty is resolved by analytically establishing the curvature of the shock just above point 2. This result cannot be used for point 6, however, because the strength of the expansion, where it intersects the shock near this point, is analytically unknown. (This is because the expansion has been altered when passing through both reflected waves.) At this location, the supposition can be used that the curvature of the shock, beyond point 6, is small. An iterative procedure is utilized, which terminates when a self-consistent solution is obtained. The rotational MOC equations are used even when the region is irrotational or when a unit process straddles the boundary between irrotational and rotational regions. For instance, this situation occurs along line 2-5. Remarkably, only four unit processes are needed. To avoid confusion with earlier figures, grid points in the unit processes are labeled alphabetically.

Shock Equations

At a shock point, the unknowns are x, y, and β. With $K = M_\infty^2 \sin^2\beta$, the jump conditions are

$$\tan\theta = \cot\beta \frac{K-1}{1-K+\frac{\gamma+1}{2}M_\infty^2} \tag{11.23a}$$

$$M = \frac{1}{\sin(\beta-\theta)}\left(\frac{1+\frac{\gamma-1}{2}K}{\gamma K-\frac{\gamma-1}{2}}\right)^{1/2} \tag{11.23b}$$

$$\frac{p}{p_\infty} = \frac{2}{\gamma+1}\left(\gamma K-\frac{\gamma-1}{2}\right) \tag{11.23c}$$

$$\frac{s-s_\infty}{R} = \frac{1}{\gamma-1}\ln\left[\frac{\left(1+\frac{\gamma-1}{2}K\right)^{\gamma}\left(\gamma K-\frac{\gamma-1}{2}\right)}{\left(\frac{\gamma+1}{2}\right)^{\gamma+1}K^{\gamma}}\right] \tag{11.23d}$$

Of course, γ and M_∞ are known. Prescribing β then yields values for θ, M, p/p_∞, and $(s-s_\infty)/R$ just downstream of the shock. The shock shape is determined by

$$\frac{dy}{dx} = \tan\beta \tag{11.24}$$

MOC Equations

The rotational MOC equations utilize P, M, θ, x, and y as the unknowns, where

$$P = \ln \frac{p}{p_\infty} \tag{11.25}$$

These equations can be written as

$$\left.\begin{aligned} dP + \frac{\gamma M^2}{\alpha} d\theta &= 0 \\ \frac{dx}{dy} &= a^+ \end{aligned}\right\} C_+ \tag{11.26a}$$

$$\left.\begin{aligned} P + \frac{\gamma}{\gamma - 1} \ln\left(1 + \frac{\gamma - 1}{2} M^2\right) &= a^o(\beta) \\ \frac{dy}{dx} &= \tan\theta \end{aligned}\right\} C_o \tag{11.26b}$$

$$\left.\begin{aligned} dP - \frac{\gamma M^2}{\alpha} d\theta &= 0 \\ \frac{dy}{dx} &= a^- \end{aligned}\right\} C_- \tag{11.26c}$$

where

$$\alpha = (M_2 - 1)^{1/2}, \qquad a^+ = \frac{\alpha \tan\theta + 1}{\alpha - \tan\theta}, \qquad a^- = \frac{\alpha \tan\theta - 1}{\alpha + \tan\theta} \tag{11.27a}$$

$$a^o(\beta) = \frac{1}{\gamma - 1} \ln\left[\frac{\left(\frac{\gamma + 1}{2}\right)^{\gamma + 1}\left(1 + \frac{\gamma - 1}{2} M_\infty^2\right)^\gamma K^\gamma}{\left(1 + \frac{\gamma - 1}{2} K\right)^\gamma \left(\gamma K - \frac{\gamma - 1}{2}\right)}\right] \tag{11.27b}$$

The first $C_\pm$ and C_o equations are the compatibility equations; the second are the characteristic equations. The C_o compatibility equation can only be used for grid points just downstream of the shock. This relation is replaced with

$$P_n + \frac{\gamma}{\gamma - 1} \ln\left(1 + \frac{\gamma - 1}{2} M_n^2\right) = P_{n+1} + \frac{\gamma}{\gamma - 1} \ln\left(1 + \frac{\gamma - 1}{2} M_{n+1}^2\right) \tag{11.28}$$

for internal points, where n and $n + 1$ denote consecutive grid points along a streamline. Required auxiliary equations are

$$\frac{p}{p_\infty} = e^P \tag{11.29}$$

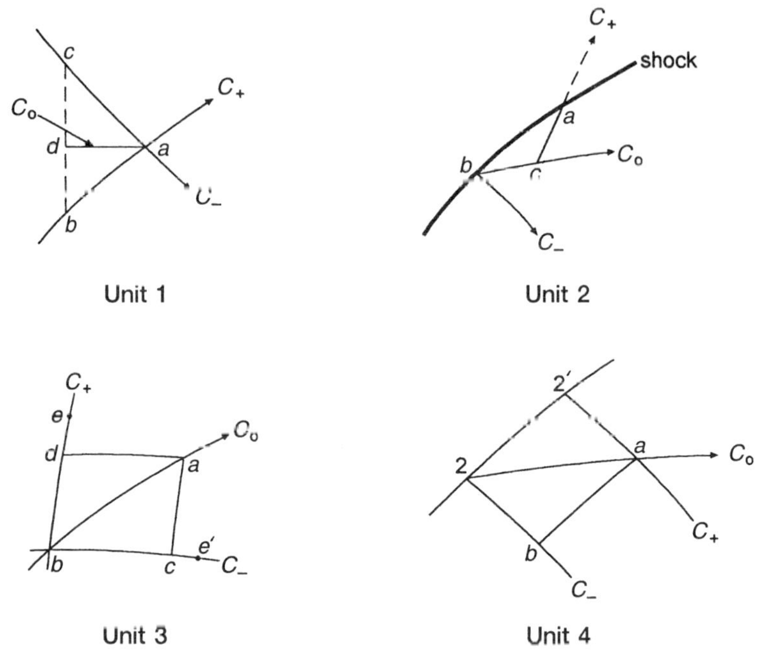

FIGURE 11.8 Sketches of the four unit processes. Unit 1 is for an internal point, unit 2 is for an ordinary shock point, unit 3 is for the streamlines that bound the vortical layer, and unit 4 is for points 2 and 6.

and along a streamline

$$\frac{s - s_\infty}{R} = \frac{\gamma}{\gamma - 1}\ln\left(1 + \frac{\gamma - 1}{2}M_\infty^2\right) - a^o(\beta) \qquad \textbf{(11.30)}$$

Unit Process 1

This process (Zucrow and Hoffman, 1977) is sketched in Figure 11.8, unit 1. It is used for any internal, nonspecial point in an irrotational or rotational region. There are five unknowns (P, M, θ, x, y) at point a. Line a-d is a straight line estimate of the C_o characteristic, or streamline, that intersects point a. Conditions at point d are provided by linear interpolation between known values at points b and c. The interpolation parameter is a sixth unknown. There are six MOC (three compatibility and three characteristic) equations for these unknowns.

The straight line estimate can lead to significant errors, as pointed out by Powers and O'Neill (1963) and confirmed by Liu (1988). These publications provide an alternative unit process that is more accurate.

Unit Process 2

This is used for a shock point (see Figure 11.8, unit 2), except when point b is at point 2 or at point 6. Conditions are known at points b and c, while (x, y, β) are the unknowns at point a. A solution is obtained by guessing a value β_a, where β_a is slightly less than β_b. The shock jump conditions then yield a first estimate for $(P, M, \theta)_a$. The C_+ equation

$$P_a - P_c + \frac{\gamma M_{ac}^2}{\alpha_{ac}}(\theta_a - \theta_c) = \in \qquad \textbf{(11.31)}$$

is evaluated for $\in$, where M_{ac} and α_{ac} are average values using data at points a and c. (Any systematic averaging procedure should suffice.) We iterate on β_a until $\in$ is zero. Once this occurs, the location of point a is established by solving

$$\left(\frac{dy}{dx}\right)_{ab} = \tan\beta_{ab}, \quad \left(\frac{dy}{dx}\right)_{ac} = a^{+} \tag{11.32}$$

for $(x,y)_a$. Note that the β_a iteration is completed before the location of point a is established.

Unit Process 3

This process is for the 2-5-7-12 and 6-11 streamlines. The unit process shown in Figure 11.8, unit 3, has known data at points b, c, and e. We seek a solution for the five unknowns at point a. The a-d straight line is an estimate for the C_- characteristic through point a. Point e, which may lie on either side of point d, provides point d data by linear interpolation or extrapolation. It may be advantageous to reverse the above procedure and use data at point e' to evaluate conditions at point c, where data at point d are now known. Whichever version provides the most compact grid is probably best. Thus, two calculations, one with each version, can be performed and compared. For convenience, we only discuss the version where data at point e are known.

The unknown are $(P, M, \theta, x, y)_a$ and the interpolation/extrapolation parameter for point d. These unknowns are determined by the six MOC equations that involve point a.

Unit Process 4

This process is for the first grid point on the streamline downstream of point 2; see Figure 11.8, unit 4. Use is made of Equation (11.21) and the shock shape equation

$$(x_{2'} - x_2)^2 + (y_{2'} - y_2)^2 + \frac{1}{\beta_2'}[(x_{2'} - x_2)\sin\beta_{\mathrm{I}} - (y_{2'} - y_2)\cos\beta_{\mathrm{I}}] = 0 \tag{11.33}$$

This is the equation for a circular arc that passes through points 2 and $2'$ that is tangent to the 0-2 shock at point 2, and that has a curvature $-\beta'_2$. We also use the slope of the shock at point $2'$:

$$\tan\beta_{2'} = \frac{(x_2 - x_{2'})\beta_2' - \sin\beta_{\mathrm{I}}}{(y_{2'} - y_2)\beta_2' - \cos\beta_{\mathrm{I}}} \tag{11.34}$$

Of course, this formulation is accurate only if point $2'$ is close to point 2, which emphasizes the need for grid compression.

For point 6, the same unit process can be used, but with β'_6 set equal to zero or a small negative value. The curvature of the shock should be apparent from shock grid points beyond point $6'$.

REFERENCES

Chu, B.-T., "On Weak Interaction of Strong Shock and Mach Waves Generated Downstream of the Shock," *J. Aeron. Sci.* 19, 443 (1952).

de Nicola, C., Iaccarino, G., and Tognaccini, R., "Rotating Dissipation for Accurate Shock Capture," *AIAA J.* 34, 1289 (1996).

Eggers, A.J., Jr., Savin, R.C., and Syvertson, C.A., "The Generalized Shock-Expansion Method and Its Application to Bodies Traveling at High Supersonic Speeds," *J. Aeron. Sci.* 22, 231, 248 (1955).

Emanuel, G., *Gasdynamics: Theory and Applications*, AIAA Education Series, Washington, D.C., 1986.
Ferri, A., "Supersonic Flows with Shock Waves," in *General Theory of High Speed Aerodynamics*, High Speed Aerodynamics and Jet Propulsion, Vol. VI, Princeton, NJ, 1954, 678.
Friedrichs, K.O., "Formation and Decay of Shock Waves," *Commun. Appl. Math.* 1, 211 (1948).
Hayes, W.D. and Probstein, R.F., *Hypersonic Flow Theory*, Academic Press, New York, 1959, 265.
Li, H. and Ben-Dor, G., "Oblique-Shock/Expansion-Fan Interaction-Analytical Solution," *AIAA J.* 34, 418 (1996).
Lighthill, M.J., "The Flow Behind a Stationary Shock," *Philos. Mag.* 40, 214 (1949).
Liu, M.-S., "Method of Characteristic Studies for Rotational Flow Downstream of a Curved Shock Wave," Ph.D. dissertation, University of Oklahoma, Norman, OK, 1988.
Mahoney, J.J., "A Critique of Shock-Expansion Theory," *J. Aeron. Sci.* 22, 673, 720 (1955).
Marshall, G. and Plohr, B., "A Random Choice Method for Two-Dimensional Steady Supersonic Shock Wave Diffraction Problems," *J. Comp. Phys.* 56, 410 (1984).
Munk, M.M. and Prim, R.C., "Surface-Pressure Gradient and Shock-Front Curvature at the Edge of a Plane Ogive with Attached Shock Front," *J. Aeron. Sci.* 15, 691 (1948).
Nasuti, F. and Onofri, M., "Analysis of Unsteady Supersonic Viscous Flows by a Shock-Fitting Technique," *AIAA J.* 34, 1428 (1996).
Pai, S.I., "On the Flow behind an Attached Curved Shock," *J. Aeron. Sci.* 19, 734 (1952).
Powers, S.A. and O'Neill, J.B., "Determination of Hypersonic Flow Fields by the Method of Characteristics," *AIAA J.* 1, 1693 (1963).
Rand, R.C., "Prandtl–Meyer Flow behind a Curved Shock Wave," *J. Math. Phys.* 29, 124 (1950).
Saad, M.A., *Compressible Fluid Flow*, 2nd ed., Prentice-Hall, Englewood Cliffs, NJ, 1993, 494.
Schreier, S., *Compressible Flow*, John Wiley, New York, 1982, 177.
Shapiro, A.H., *Compressible Fluid Flow*, Vol. I, The Ronald Press Co., New York, 1953, 559.
Waldman, G.D. and Probstein, R. F., "An Analytic Extension of the Shock-Expansion Method," WADC TN 57-214, ASTIA Doc. No. AD-130751, May 1957.
Wang, K.C., Smith, O.I., and Karagozian, A.R., "In-Flight Imaging of Transverse Gas Jets Injected into Compressible Crossflows," *AIAA J.* 33, 2259 (1995).
Zucrow, M.J. and Hoffmann, J.D., *Gas Dynamics*, Vol. II, John Wiley, New York, 1977, 187.

PROBLEMS

11.1 Verify that the LE of the expansion always intersects the planar shock. Do this by assuming the two lines don't intersect; i.e., they are parallel:

$$\hat{\mu}_{\mathrm{I}} + \hat{\theta}_w = \hat{\beta}_{\mathrm{I}}$$

where a caret denotes the parallel assumption. Derive an explicit quadratic equation for $\sin^2\hat{\beta}_{\mathrm{I}}$. Show that $\sin^2\hat{\beta}_{\mathrm{I}}$ either exceeds unity or is complex.

11.2 Derive Equations (11.7) and (11.8).

11.3 With $\gamma = 1.4$, $M_\infty = 2, 4, 6$, and $\theta_w = 0°(2°)30°$, determine β_{I}, M_{I}, x_1, y_1, x_2, y_2, and $-\beta'_2$. Tabulate your results.

12 Unsteady, One-Dimensional Flow

12.1 PRELIMINARY REMARKS

In this chapter, the flow is assumed to be adiabatic and inviscid with a constant cross-sectional area, and the gas is thermally and calorically perfect. This type of flow occurs in shock tubes, internal ballistics, and generally inside ducts, such as the flexible tubes of a stethoscope. Our presentation emphasizes physical and mathematical aspects, although applications such as internal ballistics are treated. Other applications of unsteady, one-dimensional flow, such as pressure exchangers, pulse combustors, and ejectors, can be fgound in the books by Azoury (1992), Kentfield (1993), and Weber (1995).

As is customary, an Eulerian formulation is utilized in which space and time are the independent variables. A Lagrangian formulation, however, is sometimes encountered in the literature. In this circumstance, x_0 and t are the independent variables, where x_0 is the initial position of a fluid particle. An introduction to Lagrangian coordinates can be found in the book by Karamcheti (1980) and on page 5 of Landau and Lifshitz (1987). These authors observe that Euler should be credited with originating both the Eulerian and Lagrangian formulations. The articles by Ludford and Martin (1954), Steketee (1972), and by Sharma et al. (1987) can be consulted for application of the Lagrangian approach to some of the flows of interest in this chapter.

The scope of the subsequent sections ranges from elementary to sophisticated, from introductory to relatively comprehensive, and from engineering applications to essentially applied mathematics. The next two sections treat incident and reflected normal shock waves. The fourth section develops characteristic theory, while the fifth and sixth sections apply the theory to unsteady rarefaction and compression waves, respectively. Sections 12.2 through 12.6 thus provide the basic elements for the analysis of compressible, unsteady, one-dimensional flows. The remaining two sections are more advanced and can be bypassed. The earlier material, however, is utilized in Section 12.7, which contains an introductory presentation of interior ballistics, i.e., what happens inside a gun barrel up until the time the projectile reaches the muzzle. The last section introduces Riemann function theory, which is used to establish a general solution for a nonsimple wave region.

12.2 INCIDENT NORMAL SHOCK WAVES

Shock waves were previously discussed where their position is fixed relative to a bounding wall. Moving shocks, however, are a frequent occurrence in compressible flows. Unsteady shock waves are present in shock tube flows, ballistics, pressure exchangers, and explosions. They also appear in naturally occurring phenomena, such as lightning discharges or when a volcano violently erupts. For instance, Mt. St. Helens, in the state of Washington, generated an unsteady shock (see Problem 12.2) when it erupted in 1980. If the surface area of a shock increases with time, as happens with diverging shocks, the strength of the shock—measured by the pressure ratio across it—rapidly attenuates. Conversely, the strength of a converging shock increases. In between, the strength of a planar, normal shock remains constant with time. This is the simplest case, and its dynamics are studied in this section. We begin with an incident shock that is moving into a quiescent gas. The next section considers what happens after the shock reflects from a planar endwall.

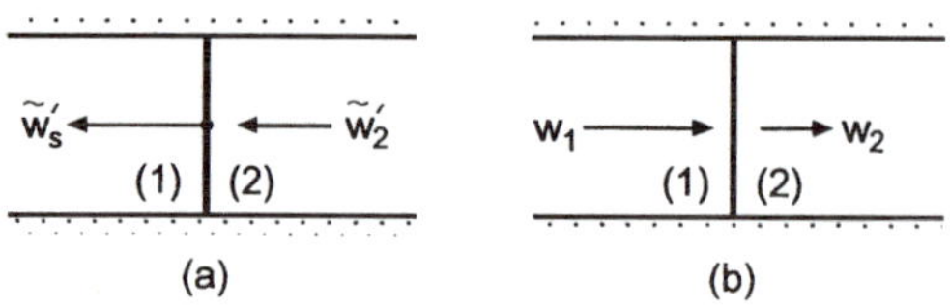

FIGURE 12.1 Moving (a) and fixed (b) shock waves.

Figure 12.1(a) shows a normal shock traveling with speed $\tilde{w}_s'$ into a motionless gas. The shock is moving to the left and causes the flow downstream of it to move in the same direction with a speed $\tilde{w}_2'$. The shock is therefore moving with respect to the duct; i.e., Figure 12.1(a) represents an unsteady flow in a laboratory, or fixed, coordinate system. A prime is used for flow conditions in this system, especially for velocities or flow speeds (The reason for the tilde will become apparent shortly.) Figure 12.1(b) represents the same flow, but with the shock brought to rest; i.e., a shock-fixed coordinate system is used. This is done by adding to the flow speeds in regions 1 and 2 the magnitude, $-\tilde{w}_s'$, of the shock speed:

$$w_1 = -\tilde{w}_s' + \tilde{w}_1' = -\tilde{w}_s', \qquad w_2 = -\tilde{w}_s' + \tilde{w}_2' \tag{12.1a,b}$$

where $\tilde{w}_1' = 0$ and the shock speed in Figure 12.1(b) is zero. A leftward (rightward) directed flow speed is taken as negative (positive); thus, $\tilde{w}_s'$ and w_2' are negative, whereas w_1 and w_2 are positive.

This velocity transformation is similar to the one used when an oblique shock wave is analyzed. In both situations, static thermodynamic conditions are unaltered, so that

$$p_1' = p_1, \qquad \rho_1 = \rho_1', \qquad s_2 = s_2', \qquad a_2 = a_2', \qquad \ldots \tag{12.2}$$

On the other hand, stagnation conditions change; e.g., the stagnation enthalpies are related by

$$h_{01} = h_1 + \frac{1}{2}w_1^2 = h_{01}' + \frac{1}{2}\tilde{w}_s'^2 \tag{12.3}$$

since $h_{01}' = h_1' = h_1$ and $h_{01} \neq h_{01}'$.

As usual, our goal is to determine region 2 conditions in terms of those in region 1, which are presumed to be known. Since the gas is perfect, we also introduce Mach numbers and they become the focus of the analysis. Moreover, the steady flow in Figure 12.1(b) has already been deciphered. We, therefore, introduce the procedure of transforming the unsteady flow to a shock-fixed coordinate system, solve the problem in this system, and then transform back to the unsteady system. This procedure is only conceptual, since the subsequent algebra will enable us to go directly to the desired result.

In line with this scenario, we introduce

$$M_1 = \frac{w_1}{a_1} = -\frac{\tilde{w}_s'}{a_1} = M_s \tag{12.4a}$$

where M_s is the shock Mach number and equals the magnitude of the shock speed divided by the speed of sound just upstream of the shock. By convention, a shock Mach number determines the strength of the shock and is associated with standing as well as moving shocks. It is notationally convenient, therefore, not to encumber M_s with a prime. Generally, steady and unsteady Mach numbers are defined as nonnegative; hence, a minus sign appears in the above equation. As will

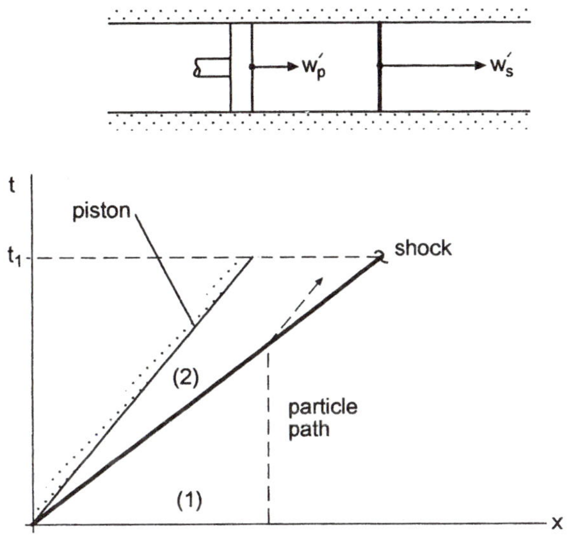

FIGURE 12.2 An x,t diagram with a constant-speed piston that generates a shock.

become apparent, M_s is the key parameter in this section, with results usually given in terms of it. For M_2, the standard normal shock relation is utilized:

$$M_2 = \frac{w_2}{a_2} = \left(\frac{X_s}{Y_s}\right)^{1/2} \tag{12.4b}$$

where the convenient shorthand

$$X = 1 + \frac{\gamma - 1}{2}M^2, \qquad Y = \gamma M^2 - \frac{\gamma - 1}{2}, \qquad Z = M^2 - 1 \tag{12.5}$$

is again introduced, and Z is utilized shortly.

Piston Speed

Before discussing unsteady Mach numbers, it is necessary to obtain a relation for the speed, w_p', of a hypothetical piston that could generate the shock. Imagine in Figure 12.2 a piston starting impulsively at $t = 0$ and moving into the gas with a straight trajectory (i.e., a constant speed)

$$x_p = w_p' t \tag{12.6a}$$

The piston compresses the gas, thereby producing a constant speed shock, which has the trajectory

$$x_s = w_s' t \tag{12.6b}$$

It is convenient to have the piston moving to the right; hence, w_s' is positive and equals $-\tilde{w}_s'$ of Figure 12.1(a). (This change in direction is the reason why a tilde appears in the earlier figure.) Figure 12.2 is an x,t diagram, which is the conventional way to represent any kind of unsteady, one-dimensional flow. In view of the above equations, the piston and shock wave paths are straight lines through the origin. Observe that the shock moves faster than the piston; i.e., $w_s' > w_p' > 0$. Region 1 contains a quiescent gas, with $w_1' = 0$, while region 2 contains a uniform flow with a constant speed. The dashed line represents the path of a fluid particle. This path is vertical in region 1, where $w_1' = 0$, and parallels the piston in region 2. The piston/cylinder sketch above the x,t diagram depicts the location of the piston's face and the shock wave at time t_1.

In actuality, a piston cannot start impulsively. The simple picture embodied in the figure, however, expedites the analysis, and this sketch is fairly realistic for part of a shock tube flow field.

From Equation (12.1b), we have

$$w_p' = w_s' - w_2$$

or

$$\frac{w_p'}{a_1} = \frac{w_s'}{a_1'} - \frac{a_2}{a_1}\frac{w_2}{a_2} = M_s - \left(\frac{T_2}{T_1}\right)^{1/2} M_2 \tag{12.7a}$$

With the normal shock relation

$$\frac{T_2}{T_1} = \left(\frac{2}{\gamma+1}\right)^2 \frac{X_s Y_s}{M_s^2} \tag{12.8}$$

the desired result is obtained as

$$\frac{w_p'}{a_1} = \frac{2}{\gamma+1}\frac{Z_s}{M_s} \tag{12.7b}$$

where w_p'/a_1 can be viewed as a piston Mach number. Since $w_p' \geq 0$, we have $M_s \geq 1$. Equation (12.7b) can be inverted:

$$M_s = \frac{(\gamma+1)w_p'}{4a_1} + \left\{1 + \left[\frac{(\gamma+1)w_p'}{4a_1}\right]^2\right\}^{1/2} \tag{12.9}$$

which provides M_s in terms of $(\gamma+1)w_p'/(4a_1)$. Note that the shock becomes a Mach wave, with $M_s = 1$, when $w_p' = 0$. This condition actually determines the sign in front of the square root.

Flow Parameters

We return to the analysis of the flow sketched in Figure 12.1(a). The unsteady Mach numbers are

$$M_1' = \frac{\tilde{w}_1'}{a_1} = 0 \tag{12.10a}$$

$$M_2' = -\frac{\tilde{w}_2'}{a_1} = \frac{a_1}{a_2}\frac{w_p'}{a_1} = \left(\frac{T_1}{T_2}\right)^{1/2}\frac{2}{\gamma+1}\frac{Z_s}{M_s} = \frac{Z_s}{(X_s Y_s)^{1/2}} \tag{12.10b}$$

With these and Equations (12.4), the Mach numbers for regions 1 and 2 are given in terms of M_s. In turn, M_s is related to the piston speed through Equations (12.7b) or (12.9).

We are now in a position to determine various pressures, with the undisturbed pressure p_1 as a reference value. We thus write

$$\frac{p_2'}{p_1} = \frac{p_2}{p_1} = \frac{2}{\gamma+1}Y_s \tag{12.11a}$$

$$\frac{p_{01}'}{p_1} = 1, \qquad \frac{p_{01}}{p_1} = X_s^{\gamma/(\gamma-1)} \tag{12.11b}$$

$$\frac{p_{02}}{p_1} = \frac{p_{02}}{p_2}\frac{p_2}{p_1} = \left(\frac{\gamma+1}{2}\right)^{\gamma+1/(\gamma-1)}\frac{M_s^{2\gamma/(\gamma-1)}}{Y_s^{1/(\gamma-1)}} \tag{12.11c}$$

$$\frac{p_{02}'}{p_1} = \frac{p_2}{p_1}\frac{p_{02}'}{p_2'} = \left(\frac{\gamma+1}{2Y_s}\right)^{1/(\gamma-1)}\left\{\frac{M_s^2}{X_s}[(\gamma-1)M_s^2+(3-\gamma)/2]\right\}^{\gamma/(\gamma-1)} \tag{12.11d}$$

Equations (12.11c,d) are a combination of the normal shock relation, Equation (12.11a), and homentropic point relations.

Table 12.1 contains a compendium of incident shock equations, including temperature relations. For the temperature, note that

$$\frac{T_2'}{T_1'} = \frac{T_2}{T_1}, \qquad \frac{T_{02}}{T_{01}} = 1, \qquad \frac{T_{02}'}{T_{01}} \neq 1 \tag{12.12}$$

where T_2/T_1 is also given by Equation (12.8). Density relations are obtained by utilizing

$$\rho = \frac{p}{RT}, \qquad \rho_0 = \frac{p_0}{RT_0} \tag{12.13}$$

as in the ratio

$$\frac{\rho_{02}'}{\rho_1} = \frac{p_{02}'}{p_1}\frac{T_1}{T_{02}'} \tag{12.14}$$

Example

Consider an intense normal shock. In this circumstance, we let M_s become infinite and determine various relations in this limit. The strong shock asymptotic relations

$$X \sim \frac{\gamma-1}{2}M^2, \qquad Y \sim \gamma M^2, \qquad Z \sim M^2, \qquad (\gamma-1)M_s^2 + \frac{3-\gamma}{2} \sim (\gamma-1)M_s^2 \tag{12.15}$$

are used, where the tilde means that $Y/(\gamma M^2) \to 1$, for example, as $M \to \infty$. These relations yield

$$\frac{w_p'}{a_1} \sim \frac{2}{\gamma+1}M_s, \qquad M_2 = \left(\frac{\gamma-1}{2\gamma}\right)^{1/2}, \qquad M_2' = \left[\frac{2}{\gamma(\gamma-1)}\right]^{1/2} \tag{12.16a}$$

TABLE 12.1
Incident Normal Shock Wave Formulas for a Perfect Gas

$$w_1 = w_s', \quad w_2 = w_s' - w_2'$$

$$\frac{w_p'}{a_1} = \frac{2}{\gamma+1}\frac{Z_s}{M_s}$$

$$M_s = \frac{(\gamma+1)w_p'}{4a_1}\left(1+\left\{1+\left[\frac{4a_1}{(\gamma+1)w_p'}\right]^2\right\}^{1/2}\right)$$

$$M_1 = M_s, \quad M_2 = \left(\frac{X_s}{Y_s}\right)^{1/2}$$

$$M_1' = 0, \quad M_2' = \frac{Z_s}{(X_s Y_s)^{1/2}}$$

$$\frac{p_2'}{p_1'} = \frac{p_2}{p_1} = \frac{2}{\gamma-1}Y_s$$

$$\frac{p_{01}'}{p_1} = 1, \quad \frac{p_{02}'}{p_1} = \left(\frac{\gamma+1}{2Y_s}\right)^{1/(\gamma-1)}\left\{\frac{M_s^2}{X_s}[(\gamma-1)M_s^2+(3-\gamma)/2]\right\}^{\gamma/(\gamma-1)}$$

$$\frac{T_2'}{T_1'} = \frac{T_2}{T_1} = \left(\frac{2}{\gamma+1}\right)^2\frac{X_sY_s}{M_s^2}$$

$$\frac{T_{01}'}{T_1} = 1, \quad \frac{T_{02}'}{T_1} = \frac{2}{\gamma+1}[(\gamma-1)M_s^2+3-\gamma/2]$$

and the stagnation relations

$$\frac{p_{02}'}{p_1} \sim 2\left(\frac{\gamma+1}{\gamma}\right)^{1/2}M_s^2, \qquad \frac{T_{02}'}{T_1} \sim 2\frac{\gamma-1}{\gamma+1}M_s^2, \qquad \frac{\rho_{02}'}{\rho_1} = \frac{\gamma+1}{\gamma-1}\left(\frac{\gamma+1}{\gamma}\right)^{1/(\gamma-1)} \tag{12.16b}$$

For a fixed shock, p_{02}/p_1 and T_{02}/T_1 are asymptotic to M_1^2, which equals M_s^2. For a moving shock, $p_{02}'/p_{01}' = p_{02}'/p_1$ and $T_{02}'/T_{01}' = T_{02}'/T_1$, and these ratios increase as M_s^2, whereas $p_{02}/p_{01} \to 0$ and $T_{02}/T_{01} = 0$. This sharp difference in behavior is evident from the energy equation

$$\frac{T_{02}'}{T_1} = \frac{T_2}{T_1} + \frac{\gamma-1}{2}\left(\frac{w_2'}{a_1}\right)^2 \tag{12.17}$$

which shows that T_{02}'/T_1 is the sum of a static temperature increase plus an increase due to the shock-induced flow speed w_2'. Another way of viewing this is to realize that the piston that generates the shock imparts energy and momentum to the flow, thereby increasing T_{02}' and p_{02}'.

Discussion

Equations (12.16a) include an equation for M_2 in order to compare this fixed shock Mach number with M_2'. Observe that

$$0 \le M_2' \le \left[\frac{2}{\gamma(\gamma-1)}\right]^{1/2} = 1.889 \tag{12.18}$$

where the numerical value is for $\gamma = 1.4$. In contrast with M_2 which cannot exceed unity, M_2' can be subsonic or supersonic. This is a consequence of the different reference frames for the two Mach numbers. Starting with Equation (12.10b), one can show that

$$M_s = \left\{ \frac{1}{4(2-\gamma)} [7 - \gamma + (\gamma^2 + 2\gamma + 17)^{1/2}] \right\}^{1/2} \tag{12.19}$$

when $M_2' = 1$. Thus, when $\gamma = 1.4$ and $M_s \geq 2.068$, M_2' is supersonic. As discussed in the fourth section, any unsteady inviscid flow is hyperbolic, regardless of the value of M_2'. Whether or not M_2' is supersonic or subsonic, in an unsteady flow, is therefore not of fundamental importance. Notice that both p_{02}'/p_1 and T_{02}'/T_1 become infinite as $M_s^2 \to \infty$. However, ρ_{02}'/ρ_1 remains finite although it can be quite large; e.g., it equals 23.09 when $\gamma = 1.4$.

In the unsteady shock wave applications discussed by Weber (1995), the airflow is generally considered as subsonic. This means that M_s is below 2.068. Pressure changes nevertheless can be significant. For instance, with $M_s = 2$ the static pressure increase across the shock is 4.5, which is adequate for a wave engine or supercharger. This increase is incurred with a relatively modest loss in stagnation pressure, since the stagnation pressure ratio is 0.7209 across the shock. (The stagnation pressure loss is for a stationary shock.) Thus, by limiting M_2' to subsonic values, a significant static pressure increase is obtained with only a modest loss of stagnation pressure. At a large value for M_s, the loss of stagnation pressure becomes exceedingly large.

As with a steady, or fixed, shock, the theory here is purely algebraic. It still applies if the upstream and downstream flows themselves are unsteady. In this circumstance, all parameters, such as M_s, p_1, M_2', ..., parametrically depend on time. In other words, the flow field that contains the shock is evaluated at a given instant of time, but can change with time.

Similarly, an unsteady shock can be oblique. This can occur for a variety of reasons, such as unsteady boundary-layer detachment, or when a normal shock encounters an obstacle, such as a wedge. In addition, if the flow downstream of the shock is subsonic and unsteady, the shock will also be unsteady. As in the steady case, an oblique shock is established by adding a parallel velocity component, which may be unsteady, to both sides of the shock (Emanuel and Yi, 2000). The relations in this subsection then govern the velocity components that are normal to the shock. Stagnation properties, such as and p_{02}/p_1, p_{02}'/p_1, and T_{02}'/T_1, however, are an exception. For instance, relations (12.11c,d) utilize homentropic point relations that depend on the full Mach number, not just its normal component.

12.3 REFLECTED NORMAL SHOCK WAVES

An incident normal shock is considered that propagates down a tube until it reflects from an endwall, as sketched in Figure 12.3. An incident shock reflects in a nonnormal manner from an arbitrarily shaped obstacle, although the simple reflection process depicted in the figure does occur in a shock tube flow. The flow in regions 1 and 2 was analyzed in the preceding subsection; hence, the reflected shock propagates to the left into a uniform flow whose speed is w_2'. The gas in region 3 has the same speed as the wall; i.e., $w_3' = 0$. A particle path, shown as a dashed line, has two sharp turns, with the path parallel to the endwall in regions 1 and 3. It is convenient to continue to normalize thermodynamic variables with their region 1 values. We also continue to use M_s as the principal independent parameter.

A slight, but useful, generalization of the flow sketched in the figure is considered by allowing w_3' to be nonzero; see Figure 12.4. By adding $-w_r'$ to the flow, the shock becomes fixed, with the result

$$\hat{w}_2 = w_2' - w_r', \qquad w_3 = w_3' - w_r' \tag{12.20}$$

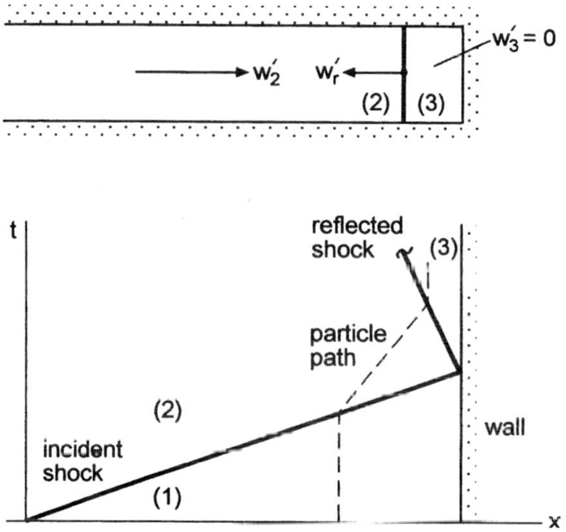

FIGURE 12.3 An x,t diagram with an incident shock that reflects off an endwall.

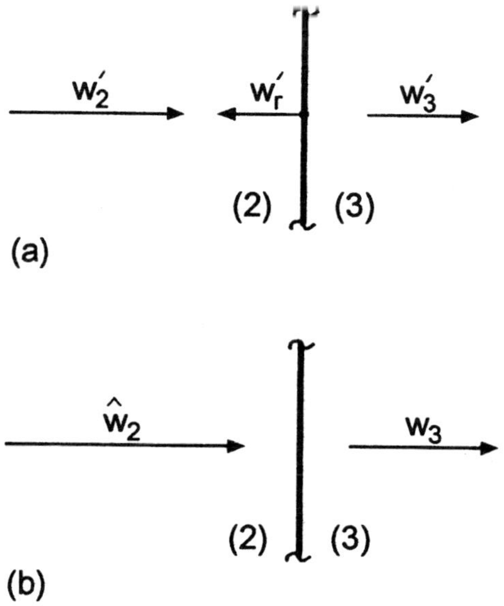

FIGURE 12.4 Reflected shock schematic when (a) it is moving and (b) fixed.

The flow speed $\hat{w}_2$ is associated with a fixed reflected shock and should not be confused with w_2, which is associated with a fixed incident shock. We next define the Mach numbers

$$M_r = \frac{\hat{w}_2}{a_2}, \qquad M_2 = \frac{w_2}{a_2}, \qquad M_r' = -\frac{w_r'}{a_2},$$

$$M_2' = \frac{w_2'}{a_2}, \qquad M_3' = \frac{w_3'}{a_3}, \qquad M_3 = \frac{w_3}{a_3} \tag{12.21}$$

where M_r (not M_r') is the reflected shock Mach number, since the strength of the shock depends on this Mach number. Note that M_2 and M_2' were introduced in the preceding section. Equations (12.20) become

$$M_r = M_2' + M_r', \qquad M_3 = M_3' + \left(\frac{T_2}{T_3}\right)^{1/2} M_r' \tag{12.22a,b}$$

where the temperature ratio is [see Equation (12.8)]

$$\left(\frac{T_2}{T_3}\right)^{1/2} = \frac{\gamma+1}{2}\frac{M_r}{(X_r Y_r)^{1/2}} \tag{12.23}$$

The steady shock Mach numbers, M_r and M_3, are related by

$$M_3 = \left(\frac{X_r}{Y_r}\right)^{1/2} \tag{12.24}$$

Hence, Equation (12.22b) becomes

$$\left(\frac{X_r}{Y_r}\right)^{1/2} = M_3' + \frac{\gamma+1}{2}\frac{M_r M_r'}{(X_r Y_r)^{1/2}}$$

which can be written as

$$M_3' = \frac{1 + \frac{\gamma-1}{2}M_r^2 - \frac{\gamma+1}{2}M_r M_r'}{\left[\left(1+\frac{\gamma-1}{2}M_r^2\right)\left(\gamma M_r^2 - \frac{\gamma-1}{2}\right)\right]^{1/2}} \tag{12.25a}$$

Equation (12.22a) is used to replace M_r, with the fundamental result

$$M_3' = \frac{1 - \frac{\gamma+1}{2}M_r'(M_2'+M_r') + \frac{\gamma-1}{2}(M_2'+M_r')^2}{\left\{\left[1+\frac{\gamma-1}{2}(M_2'+M_r')^2\right]\left[\gamma(M_2'+M_r')^2 - \frac{\gamma+1}{2}\right]\right\}^{1/2}} \tag{12.25b}$$

that establishes M_3' in terms of M_2' and M_r'.

This equation covers three special cases, two of which were previously dealt with. In the first, Equation (12.10b) is recovered for the unsteady Mach number downstream of an incident shock, by setting

$$M_2' \to 0, \qquad M_r' \to M_s, \qquad M_3' \to M_2' \tag{12.26}$$

In the second, Equation (12.4b) for a fixed normal shock is recovered, by setting

$$M_2' \to M_s, \qquad M_r' \to 0, \qquad M_3' \to M_2 \tag{12.27}$$

The last case corresponds to the flow sketched in Figure 12.3. We set $M_3' = 0$ and use Equation (12.22a) to eliminate M_r', with the result

$$M_r^2 - \frac{\gamma+1}{2} M_2' M_r - 1 = 0 \tag{12.28a}$$

which yields

$$M_r = \frac{\gamma+1}{4} M_2' + \left[1 + \left(\frac{\gamma+1}{4} M_2'\right)^2\right]^{1/2} \tag{12.28b}$$

An elegant relation is obtained by using Equation (12.10b) to eliminate M_2':

$$M_r = \left(\frac{Y_s}{X_s}\right)^{1/2} \tag{12.29}$$

that directly provides the reflected shock Mach number in terms of M_s. In view of Equation (12.4b), we have $M_2 M_r = 1$, and M_2 and M_r are inverse to each other. The reflected shock Mach number is bounded as

$$1 \le M_r \le \left(\frac{2\gamma}{\gamma-1}\right)^{1/2} = 2.646 \tag{12.30}$$

where $\gamma = 1.4$ for the numerical value. The lower bound occurs when $M_s = 1$ and the upper one when $M_s \to \infty$. One can show that $M_r \le M_s$ (see Problem 12.4), where the equal sign only holds when the shocks are Mach waves or $\gamma = 1$. The disparity in the strength of the two shocks is particularly large when M_s is large. In this case, the strong incident shock in the x,t diagram in Figure 12.3 will be close to the x-axis, while the relatively weak reflected shock will be much more vertically oriented.

The subsequent analysis, which continues to use M_s as the independent parameter, is expedited by noting that

$$X_r = \frac{\gamma+1}{2} \frac{(\gamma-1)M_s^2 + (3-\gamma)/2}{X_s} \tag{12.31a}$$

$$Y_r = \frac{\gamma+1}{2} \frac{(3\gamma-1)M_s^2/2 - (\gamma-1)}{X_s} \tag{12.31b}$$

The pressure p_3, which equals p_3' and p_{03}', is given by

$$\frac{p_3}{p_1} = \frac{p_{03}'}{p_1} = \frac{p_3}{p_2}\frac{p_2}{p_1} = \left(\frac{2}{\gamma+1}\right)^2 Y_r Y_s = \frac{2}{\gamma+1}\frac{Y_s}{X_s}\left[\frac{3\gamma-1}{2} M_s^2 - (\gamma-1)\right] \tag{12.32}$$

Table 12.2 summarizes key reflected shock relations, including one for T_3/T_1. Problem 12.5 deals with the entropy change $(s_3 - s_1)/R$.

TABLE 12.2
Reflected Normal Shock Wave Formulas for a Perfect Gas

$$\hat{w}_2 = w_2' - w_r', \quad w_3 = -w_r', \quad w_3 = 0$$

$$M_r = \frac{\hat{w}_2}{a_2} = \left(\frac{Y_s}{X_s}\right)^{1/2}$$

$$\frac{p_3}{p_1} = \frac{p'_{03}}{p_1} = \frac{2}{\gamma+1}\frac{Y_s}{X_s}\left[\frac{3\gamma-1}{2}M_s^2-(\gamma-1)\right]$$

$$\frac{T_3}{T_1} = \frac{T'_{03}}{T_1} = \left(\frac{2}{\gamma+1}\right)^2\frac{1}{M_s^2}\left[(\gamma-1)M_s^2+\frac{3-\gamma}{2}\right]\left[\frac{3\gamma-1}{2}M_s^2-(\gamma-1)\right]$$

Example

The $M_s \to \infty$ example of the preceding section is continued. We now obtain

$$M_r = \left(\frac{2\gamma}{\gamma-1}\right)^{1/2} \tag{12.33a}$$

$$\frac{p'_{03}}{p_1} = \frac{p_3}{p_1} \sim \frac{2\gamma(3\gamma-1)}{\gamma^2-1}M_s^2 \tag{12.33b}$$

$$\frac{T'_{03}}{T_1} = \frac{T_3}{T_1} \sim \frac{2(\gamma-1)(3\gamma-1)}{(\gamma+1)^2}M_s^2 \tag{12.33c}$$

$$\frac{\rho'_{03}}{\rho_1} = \frac{\rho_3}{\rho_1} \sim \frac{\gamma(\gamma+1)}{(\gamma-1)^2} \tag{12.33d}$$

With $\gamma = 1.4$ and, e.g., $M_s = 10$, the $M_s \to \infty$ formulas yield the approximate but reasonably accurate results

$$\frac{p_2}{p_1} = 116.5, \quad \frac{T_2}{T_1} = 20.39, \quad \frac{\rho_2}{\rho_1} = 5.714$$

$$\frac{p'_{02}}{p_1} \cong 769.6, \quad \frac{T'_{02}}{T_1} \cong 33.33, \quad \frac{\rho'_{02}}{\rho_1} \cong 23.09$$

$$\frac{p'_{03}}{p_1} \cong 933.3, \quad \frac{T'_{03}}{T_1} \cong 44.44, \quad \frac{\rho'_{03}}{\rho_1} \cong 21.00$$

where the static ratios in the first row are exact. Thus, the pressure p'_{03} and temperature T'_{03} behind the reflected shock exceed their counterparts, static and stagnation, behind the incident shock. Surprisingly, the density behind the reflected shock is slightly less than the stagnation density behind the incident shock. The magnitudes of the pressure and temperature behind the reflected shock are quite large. For instance, if $p_1 = 1$ atm and $T_1 = 300$ K, we then have $p_3 = 933$ atm and $T_3 = 1.33 \times 10^4$ K. These region 3 values are overestimates, since air experiences significant real-gas effects (dissociation and ionization) well before a temperature of 1.3×10^4 K is achieved. Nevertheless, these crude estimates illustrate that a high-enthalpy gas can be obtained behind a reflected shock. (For instance, shock tunnels require a high-enthalpy gas.) Still larger pressure and temperature values occur with a monatomic gas, such as helium or argon. With a shock tube, temperatures of about 4×10^4 K have been obtained behind the reflected shock in argon. Dissociation now does not occur and a higher temperature is required for ionization than with air. Consequently, region 3 estimates, when $M_s = 10$, are more reliable. With sufficient ionization, region 3 in a shock tube flow is a high-temperature plasma that cools primarily by intense radiative heat transfer to the walls.

Discussion

As noted earlier, an explosion generates a shock wave that reflects from any obstacle it might encounter. Structural and thermal damage is caused by the transient, but large, pressure and temperature that occur behind reflected shock waves. In short, it is the reflection process that destroys buildings and causes fires. (Strong rarefaction waves also occur and cause further structural damage.)

12.4 CHARACTERISTIC THEORY

This theory is generally associated with a second-order, hyperbolic PDE, as was the case in Chapter 11. We therefore initially show, with appropriate assumptions, that the system of governing first-order PDEs for a one-dimensional, inviscid, unsteady flow can be transformed into a single, second-order PDE. Secondly, the resulting PDE is shown to be hyperbolic. Finally, a solution based on characteristic theory, called the method-of-characteristics (MOC), is obtained. This approach is especially well suited for the analysis, or computation, of unsteady, one-dimensional flows or steady, supersonic, two-dimensional flows.

Basic Equations

The pertinent conservation equations can be written as

$$\frac{\partial \rho}{\partial t} + w\frac{\partial \rho}{\partial x} + \rho\frac{\partial w}{\partial x} = 0 \tag{12.34a}$$

$$\frac{\partial w}{\partial t} + w\frac{\partial w}{\partial x} + \frac{1}{\rho}\frac{\partial p}{\partial x} = 0 \tag{12.34b}$$

$$\frac{Dh_0}{Dt} = \frac{1}{\rho}\frac{\partial p}{\partial t} \tag{12.34c}$$

Equation (12.34c) indicates that the stagnation enthalpy is not a constant. Consequently, other stagnation parameters are also variable. There is, however, one important exception. Since the flow is adiabatic and inviscid, the entropy is a constant

$$\frac{Ds}{Dt} = 0 \tag{12.35}$$

along particle paths, such as the dashed lines in Figures 12.2 and 12.3. When a particle path originates in a quiescent flow — region 1 in the figures — then $s(x,t) = s_0 = s_1$ can be used in place of the above equation. A region of flow where s is a constant is referred to as homentropic, whereas a region in which s changes its value from one particle path to the next is referred to as isentropic. (This terminology applies to both steady and unsteady flows, where a streamline is equivalent to a particle path in an unsteady flow.) As in a steady flow, the entropy goes through a jump discontinuity across a shock wave. Since the shocks in these figures are of constant strength, regions 2 and 3 remain homentropic, although $s_3 > s_2 > s_1$. If, for example, the piston path in Figure 12.2 is curved, then region 2 becomes isentropic, while region 1 is still homentropic.

In the balance of this section, only homentropic flows of a perfect gas are considered. In this case, the relation $p = p\,(\rho,s)$ can be written as

$$\frac{p}{p_1} = \left(\frac{\rho}{\rho_1}\right)^{\gamma} \tag{12.36}$$

where a unity subscript denotes a reference condition, such as a quiescent state. This relation replaces the energy equation and can be used to remove the pressure as a variable with

$$\frac{\partial p}{\partial x} = \left(\frac{\partial p}{\partial \rho}\right)_s \frac{\partial \rho}{\partial x} = a^2 \frac{\partial \rho}{\partial x} \tag{12.37}$$

It is convenient to also replace ρ with the speed of sound by introducing

$$\frac{\rho}{\rho_1} = \left(\frac{a}{a_1}\right)^{2/(\gamma-1)} \tag{12.38}$$

These manipulations result in the basic equations for an unsteady, one-dimensional flow:

$$\frac{\partial a}{\partial t} + w\frac{\partial a}{\partial x} + \frac{\gamma-1}{2}a\frac{\partial w}{\partial x} = 0 \tag{12.39a}$$

$$\frac{\partial w}{\partial t} + w\frac{\partial w}{\partial x} + \frac{2}{\gamma-1}a\frac{\partial a}{\partial x} = 0 \tag{12.39b}$$

where both dependent variables are speeds. We thus have two first-order, coupled PDEs. In the literature, these equations are often taken as the starting point for an unsteady analysis. They are restricted, however, to a homentropic flow, since Equation (12.36) is essential for their derivation. For a steady flow, the counterpart to Equations (12.34a,b) and (12.35) is the Euler equations, whereas the counterpart to Equations (12.39) would be a (homentropic) potential flow equation for a perfect gas.

Second-Order PDE

We would like to replace Equations (12.39) with a single, second-order PDE. This is not as straightforward as it may sound, since both equations are nonlinear. Nevertheless, this task will be accomplished. Because of their similarity in appearance, we are motivated to multiply the first of the equations by a constant, λ, and add the result to the second, to obtain

$$\frac{\partial}{\partial t}(w+\lambda a) + w\frac{\partial}{\partial x}(w+\lambda a) + \frac{\gamma-1}{2}\lambda a\frac{\partial}{\partial x}\left[w + \left(\frac{2}{\gamma-1}\right)^2\frac{a}{\lambda}\right] = 0 \tag{12.40}$$

For the rightmost term to conform to the other two terms, we set

$$\left(\frac{2}{\gamma-1}\right)^2\frac{1}{\lambda} = \lambda$$

which yields

$$\lambda = \pm\frac{2}{\gamma-1} \tag{12.41}$$

These two λ values result in

$$\frac{\partial}{\partial t}\left(w+\frac{2}{\gamma-1}a\right)+(w+a)\frac{\partial}{\partial x}\left(w+\frac{2}{\gamma-1}a\right) = 0 \tag{12.42a}$$

$$\frac{\partial}{\partial t}\left(w-\frac{2}{\gamma-1}a\right)+(w-a)\frac{\partial}{\partial x}\left(w-\frac{2}{\gamma-1}a\right) = 0 \tag{12.42b}$$

In view of the form of these equations, w and a are replaced with new dependent variables:

$$J_{\pm} = w \pm \frac{2a}{\gamma-1} \tag{12.43}$$

The inverse transformation is

$$w = \frac{1}{2}(J_+ + J_-), \qquad a = \frac{\gamma-1}{4}(J_+ - J_-) \tag{12.44}$$

Equations (12.42) thus become

$$\frac{\partial J_+}{\partial t}+(w+a)\frac{\partial J_+}{\partial x} = 0, \qquad \frac{\partial J_-}{\partial t}+(w-a)\frac{\partial J_-}{\partial x} = 0 \tag{12.45a,b}$$

or

$$\frac{\partial J_+}{\partial t}+\left(\frac{\gamma+1}{4}J_+ + \frac{3-\gamma}{4}J_-\right)\frac{\partial J_+}{\partial x} = 0, \qquad \frac{\partial J_-}{\partial t}+\left(\frac{3-\gamma}{4}J_+ + \frac{\gamma+1}{4}J_-\right)\frac{\partial J_-}{\partial x} = 0 \tag{12.46a,b}$$

Note that only J_-, not its derivatives, appears in Equation (12.46a), with a similar remark for J_+ in Equation (12.46b). As shown by Problem 12.8, one can use Equation (12.46a) to replace J_- in Equation (12.46b). [Alternatively, J_+ can be eliminated in Equation (12.46a).] We thereby obtain

$$AJ_{tt} + 2BJ_{tx} + CJ_{xx} = \Phi \tag{12.47}$$

where

$$J = J_+, \qquad J_x = \frac{\partial J_+}{\partial x}, \qquad \ldots$$

and

$$A = J_x^2 \tag{12.48a}$$

$$B = -\frac{2}{3-\gamma} J_x \left(J_t + \frac{\gamma - 1}{2} J J_x \right) \tag{12.48b}$$

$$C = \frac{\gamma + 1}{3 - \gamma} J_t \left(J_t + 2 \frac{\gamma - 1}{\gamma + 1} J J_x \right) \tag{12.48c}$$

$$\Phi = \frac{\gamma^2 - 1}{2(3 - \gamma)} J_x^3 (J_t + J J_x) \tag{12.48d}$$

Equation (12.47) is the sought-after second-order PDE.

Mathematical Properties

We momentarily digress from further discussion of this equation in order to consider the mathematical properties of a second-order PDE with the form

$$A u_{xx} + 2B u_{xy} + C u_{yy} = \Phi(x, y, u, u_x, u_y) \tag{12.49}$$

where A, B, and C are not given by Equations (12.48) and can depend only on the arguments shown for Φ. The dependence restriction means the equation is quasilinear; i.e., it is linear in its highest-order derivatives, which here are u_{xx}, u_{xy}, and u_{yy}. Thus, A, B, C, and Φ do not contain any of these second-order derivatives. If, for example, Φ happened to contain a term that is linear in u_{xy}, this term would be transferred to the B term. Note that Equation (12.47) is quasilinear. The quasilinear restriction is a necessary condition for the method-of-characteristics. In other words, this method cannot be used with an equation that contains a u_{xx}^2 term.

Equation (12.49) is linear if A, B, and C only depend on x and y and Φ is linear with respect to u, u_x, and u_y. In this case, superposition of solutions applies. Superposition also requires linear initial and/or boundary conditions. By definition, a linear PDE is also quasilinear. If the equation is quasilinear but nonlinear, then superposition is no longer applicable. This is the case for Equation (12.47).

Equation (12.49) comes in three general types:

$$B^2 - AC < 0, \qquad \text{elliptic} \tag{12.50a}$$

$$B^2 - AC = 0, \qquad \text{parabolic} \tag{12.50b}$$

$$B^2 - AC > 0, \qquad \text{hyperbolic} \tag{12.50c}$$

The nature of the flow depends strongly on the type. Steady, inviscid, subsonic, two- or three-dimensional flow is elliptic. This type of flow occurs behind the detached shock, near the nose, of a blunt reentry vehicle, such as the shuttle. Waves, such as a shock wave or a Prandtl–Meyer expansion, cannot occur inside an elliptic flow region. The viscous boundary-layer equations are an important example of a parabolic flow. The most important examples of a hyperbolic flow are steady, inviscid, two- or three-dimensional, supersonic flow, and unsteady, inviscid flow with any Mach number value and with any number of dimensions. In general, fluid dynamics involves all three types, although this chapter is solely concerned with hyperbolic flows. Only a hyperbolic equation, or system of equations, can admit wave behavior.

Equations (12.48a to c) yield (see Problem 12.8)

$$B^2 - AC = \left[\frac{\gamma-1}{3-\gamma} J_x (J_t + JJ_x)\right]^2 > 0 \tag{12.51}$$

and Equation (12.47) is therefore hyperbolic. (One exception is the limiting case of γ equaling unity, when the equation is parabolic.) A region of flow either can have waves or cannot, and this physical property is invariant regardless of the number and form of the governing equations. Thus, Equations (12.34), (12.39), (12.42), (12.46), and (12.47) are all hyperbolic. The necessary and sufficient condition that the MOC be applicable to Equation (12.49) is that the equation is both quasilinear and hyperbolic.

Method-of-Characteristics

Wave behavior, in part, means there may be lines, or paths, in the flow field along which certain combinations of the dependent variables may be constant. Finding these lines is our next objective. The derivation used to obtain Equation (12.47) greatly expedites the subsequent analysis, since J_+ and J_- are these combinations. We are to find a set of lines in the x,t plane along which J_+ is a constant; i.e., J_+ is constant along a given line but may have a different value on any adjacent line. We thus write

$$J_+ = J_+(x,t) \tag{12.52}$$

and by differentiation

$$dJ_+ = \frac{\partial J_+}{\partial x}dx + \frac{\partial J_+}{\partial t}dt \tag{12.53}$$

We eliminate $\partial J_+/\partial t$ using Equation (12.45a), to obtain

$$dJ_+ = \frac{\partial J_+}{\partial x}\left[\frac{dx}{dt} - (w+a)\right]dt \tag{12.54}$$

Thus, $dJ_+ = 0$ means that J_+ is constant on lines generated by

$$\frac{dx}{dt} = w + a \tag{12.55a}$$

This is not the equation for a particle path, which is given by $dx/dt = w$. Rather, it is an ODE whose solution provides a special set of paths in the x,t plane along which J_+ is a constant. When this equation is integrated, either analytically or numerically, the constant of integration may select a particular member of the family of lines along which J_+ is a constant. A similar derivation results in J_- being constant along lines given by

$$\frac{dx}{dt} = w - a \tag{12.55b}$$

The integration of these ODEs provides a nonorthogonal grid in the x,t plane. This grid represents a characteristic net and these equations, or their solution, are referred to as characteristic equations. These equations, in conjunction with Equations (12.43), which are called compatibility equations, provide the MOC solution. The compatibility equations provide values for $J_\pm$, which are known as the Riemann invariants.

We concisely summarize this approach as follows. The J_+ Riemann invariant

$$J_+ = w + \frac{2}{\gamma - 1}a \tag{12.56a}$$

is constant along the C_+ characteristic lines, given by

$$\frac{dx}{dt} = w + a \tag{12.56b}$$

Similarly, the J_- Riemann invariant

$$J_- = w - \frac{2}{\gamma - 1}a \tag{12.56c}$$

is constant along the C_- characteristic lines, given by

$$\frac{dx}{dt} = w - a \tag{12.56d}$$

These relations represent an *exact* solution for unsteady, one-dimensional, homentropic flow of a perfect gas. The next two sections are devoted to illustrating this approach.

Restrictions

There are several limitations underlying the above MOC. As mentioned, the flow must be homentropic. With an isentropic flow, either Equation (12.34c) or (12.35) is used in place of Equation (12.36). An MOC formulation can be developed, but now there are three, first-order PDEs. A wave-diagram (Rudinger, 1969) solution is required. (This is a numerical or graphical solution of the MOC equations. Several extensions not dealt with here, such as a nonconstant cross-sectional area $A(x,t)$ and wall skin friction, can be included.)

A second restriction is that the flow is planar. There are also cylindrically and spherically symmetric unsteady, one-dimensional flows. In this circumstance, x is a radial coordinate measured from the axis, or origin, of symmetry. Let us assume a homentropic flow, in which case only the continuity equation is altered. We can start by writing a general, one-dimensional form for this equation:

$$\frac{1}{\rho}\frac{D\rho}{Dt} + \frac{1}{A}\frac{DA}{Dt} + \frac{\partial w}{\partial x} = 0$$

where A is the cross-sectional area of the flow. This area can be written as

$$A = \pi(2x)^{\sigma}$$

where σ equals 0, 1, or 2 for planar, cylindrically symmetric, or spherically symmetric geometries, respectively. With this alteration, Equation (12.34a) becomes

$$\frac{\partial \rho}{\partial t} + w\frac{\partial \rho}{\partial x} + \rho\frac{\partial w}{\partial x} = -\frac{\sigma \rho w}{x}$$

while Equation (12.39a) changes to

$$\frac{\partial a}{\partial t} + w\frac{\partial a}{\partial x} + \frac{\gamma-1}{2}a\frac{\partial w}{\partial x} = -\frac{\gamma-1}{2}\frac{\sigma a w}{x} \tag{12.57}$$

Equation (12.39b) is unaltered and is coupled to Equation (12.57), where again a and w are the dependent variables. When σ is one or two, the inhomogeneous term results in nonconstant Reimann invariants along the characteristics. Within the context of characteristic theory, a numerical solution is required. Nevertheless, Problem 12.26 shows that Equations (12.39b) and (12.57) can be reduced to ODEs with a similarity transformation.

12.5 RAREFACTION WAVES

CENTERED WAVES

In this subsection, the simplest nontrivial case is considered, namely, a centered expansion, or rarefaction, wave. The wave is generated by an impulsively moving piston; see Figure 12.5. The piston travels to the right with a constant speed w_p. Section 12.2 was concerned with the opposite situation of a constant speed piston moving into a gas. As with an incident shock, there is no length scale. The pressure can therefore adjust by means of a discontinuity or, as is the case here, a continuous centered wave.

Region 1 in the figure is quiescent, and static and stagnation conditions are the same. Region 2 is a uniform flow region, where $w_2 = w_p$. Between these two regions is a centered expansion wave whose leading and trailing edges are denoted as LE and TE, respectively. A dashed line particle path is shown in the figure; inside the expansion the path curves to the right; inside region 2 it is a

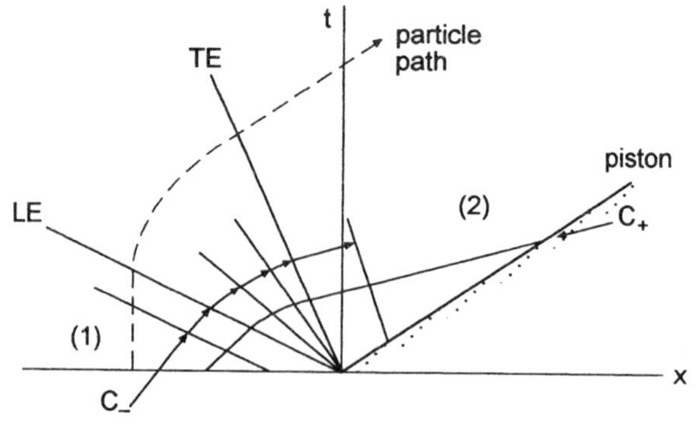

FIGURE 12.5 Centered rarefaction wave caused by an impulsively moving piston.

straight line that is parallel to the path of the piston. As with a shock wave, the slope of a particle path is discontinuous at the leading and trailing edges [see Problem. 12.9(c)]. As you may have noticed, a prime is not used in this section to denote unsteady variables. With a spreading rarefaction wave, a simple velocity transformation — as was used with shock waves — cannot yield a steady flow. The distinction made earlier between steady and unsteady coordinate systems is now unnecessary.

With t as the ordinate in any x,t diagram, it is convenient to write Equations (12.56b) and (12.56d) as

$$\frac{dt}{dx} = \frac{1}{a}\frac{1}{M+1}, \qquad C_+ \tag{12.58a}$$

$$\frac{dt}{dx} = \frac{1}{a}\frac{1}{M-1}, \qquad C_- \tag{12.58b}$$

Figure 12.5 indicates that $w \geq 0$ and $M \geq 0$. Thus, any C_+ characteristic has a positive slope, while a C_- characteristic has a negative (positive) slope when $M < 1$ ($M \geq 1$). When $M = 1$, the C_- characteristic is vertical.

In region 1, $a = a_1$ and $M = 0$, so that

$$\frac{dt}{dx} = \frac{1}{a_1}, \qquad C_+ \tag{12.59a}$$

$$\frac{dt}{dx} = -\frac{1}{a_1}, \qquad C_- \tag{12.59b}$$

Hence, the two characteristic families are straight with the same slopes but of opposite sign. In region 2, Equations (12.58) reduce to

$$\frac{dt}{dx} = \frac{1}{a_2}\frac{1}{M_2+1}, \qquad C_+ \tag{12.60a}$$

$$\frac{dt}{dx} = \frac{1}{a_2}\frac{1}{M_2-1}, \qquad C_- \tag{12.60b}$$

where a_2 and M_2 are still to be evaluated.

The key to establishing a_2 and M_2 is to note that a C_+ characteristic crosses the rarefaction wave. We thus have a "shooting method" type of solution in which the known value for J_+ in region 1 is used to determine unknown values in region 2. From Equation (12.56a), we write

$$J_{+1} = w_1 + \frac{2}{\gamma-1}a_1 = \frac{2}{\gamma-1}a_1 = J_{+2} = w_2 + \frac{2}{\gamma-1}a_2 \tag{12.61}$$

with the result

$$\frac{a_2}{a_1} = 1 - \frac{\gamma-1}{2}\frac{w_p}{a_1} \tag{12.62}$$

since $w_2 = w_p$. The J_+ invariant is a constant even when it crosses the expansion. Consequently, this result is a special case of the more general result

$$\frac{a}{a_1} = 1 - \frac{\gamma - 1}{2}\frac{w}{a_1} \tag{12.63}$$

that holds throughout the entire flow field. Moreover, this relation also holds for noncentered waves, discussed later in this section. We generally assume γ, conditions in region 1, and w_p are known. Equation (12.62) then establishes a_2, while

$$M_2 = \frac{w_p}{a_2} = \frac{\frac{w_p}{a_2}}{1 - \frac{\gamma - 1}{2}\frac{w_p}{a_1}} \tag{12.64}$$

determines M_2.

The leading edge of the expansion propagates into the quiescent gas with the speed of sound. Along the leading edge, we can write

$$a_{LE} = a_1, \qquad M_{LE} = 0, \qquad x_{LE} = -a_1 t_{LE} \tag{12.65a}$$

Similarly, along the trailing edge, we have

$$a_{TE} = a_2, \qquad M_{TE} = M_2, \qquad x_{TE} = a_2(M_2 - 1)t_{TE} \tag{12.65b}$$

If $M_2 < 1$, then the trailing edge is in the second quadrant (as shown in Figure 12.5), while it is in the first quadrant when $M_2 > 1$. Thus, the trailing edge should be inside the first quadrant in the figure if $M_2 > 1$. For this to be the case, the slope of the piston's trajectory would be shallower. The leading and trailing edges are members of the C_- characteristic family. Characteristic C_- lines in region 1 are parallel to the leading edge, while they are parallel to the trailing edge in region 2. Inside the expansion, they form a centered fan of straight lines that pass through the origin.

Consider an arbitrary point inside the expansion. At this point, we have

$$J_- = w - \frac{2}{\gamma - 1}a = \text{constant} \tag{12.66}$$

and Equation (12.63), which stems from $J_+ = $ constant. We eliminate w from these two equations, with the result

$$\frac{a}{a_1} = \frac{1}{2}\left(1 - \frac{\gamma - 1}{2}\frac{J_-}{a_1}\right) \tag{12.67}$$

Since J_- is constant along a C_- characteristic, a is constant, as well. In turn, other variables, such as w and p, are also constants along these centered C_- characteristics. As we shall see, a and the other variables do vary along any path that crosses the expansion. The similarity with Prandtl–Meyer flow is quite evident.

The foregoing discussion is not limited to the C_- characteristics inside a centered expansion. For example, suppose a characteristic of either family is straight. If this characteristic is C_+, then

dx/dt in Equation (12.56b) is a constant, and $w + a$ along this straight characteristic is constant. Since J_+ is also constant along this characteristic, both a and w are constants. Consequently, other variables, such as p and ρ, are also constants along this straight C_+ characteristic.

There are two methods for evaluating $a, w, p, \ldots$ along a straight characteristic. When the characteristic starts, crosses, or ends in a region with known values, these values can be used. Alternatively, a Riemann invariant of the opposite family that intersects a straight characteristic can be utilized. This is the procedure to be used with the straight C_- characteristics that form a centered or noncentered wave. It will yield the structure of the wave. Moreover, the wave may be an expansion wave, as in this section, or a compression wave, which is discussed in Section 12.6.

Solution for Static Variables

Our next goal is to obtain a solution for the flow inside the centered expansion in terms of x and t. Note that w and a are constants along a J_- characteristic. Hence, Equation (12.56d) readily integrates to

$$x = (w - a)t \tag{12.68}$$

where the constant of integration is zero, since the line passes through the origin. (The constant of integration is not zero for lines that do not pass through the origin.) Equation (12.63) is used to eliminate a and the result is solved for w:

$$\frac{w}{a_1} = \frac{2}{\gamma + 1}(1 + \eta) \tag{12.69}$$

where

$$\eta = \frac{x}{a_1 t} \tag{12.70}$$

Equation (12.69) does *not* hold outside the expansion, unlike Equation (12.63). As we have seen, $w = 0$ in region 1 and $w = w_p$ in region 2. Thus, the derivative of w that is in a transverse direction to a C_- characteristic is discontinuous at the leading and trailing edges of the expansion. This discontinuity of a first derivative, in a transverse direction, on the leading and trailing edges of the rarefaction extends to the other variables.

Hyperbolic equations admit different types of discontinuities. Shock waves are the most familiar, in which the basic dependent variables themselves are discontinuous. As we see, first derivatives can also be discontinuous. In the case at hand, the first derivatives of the variables $a, w, p, \ldots,$ experience a discontinuity normal to characteristics that bound different flow regions. Here, these characteristics are the leading and trailing edges of an expansion wave. The same behavior occurs in steady, supersonic, two- or three-dimensional flow, where first derivatives are discontinuous on Mach lines or surfaces that separate different flow regions (uniform and simple wave regions, for example). As with shock waves, this behavior has a profound effect on numerical methods.

Hyperbolic equations admit discontinuous derivatives of higher order than the first derivative. For instance, suppose an otherwise smooth curved wall, with a steady supersonic flow, has a point where its curvature is discontinuous. Second derivatives, transverse to the Mach line emanating from this point, are also discontinuous.

By substituting Equation (12.69) into Equation (12.63), the solution for the speed of sound inside the wave

$$\frac{a}{a_1} = \frac{2}{\gamma+1}\left(1 - \frac{\gamma-1}{2}\eta\right) \qquad \textbf{(12.71a)}$$

is obtained. At a given instant of time, w and a each have a linear variation with x inside the centered expansion. When $\gamma > 1$, none of the other variables have this linear dependency. Since the flow is homentropic, we can readily write for other variables

$$\frac{T}{T_1} = \left(\frac{a}{a_1}\right)^2 \qquad \textbf{(12.71b)}$$

$$\frac{p}{p_1} = \left(\frac{a}{a_1}\right)^{2\gamma/(\gamma-1)} \qquad \textbf{(12.71c)}$$

$$\frac{\rho}{\rho_1} = \left(\frac{a}{a_1}\right)^{2/(\gamma-1)} \qquad \textbf{(12.71d)}$$

$$M = \frac{w}{a} = \frac{\frac{w}{a_1}}{\frac{a}{a_1}} = \frac{1+\eta}{1-\frac{\gamma-1}{2}\eta} \qquad \textbf{(12.71e)}$$

The last equation, of course, is consistent with Equation (12.64). Observe that $M = 0$ when $\eta = -1$, $M = 1$ when $\eta = 0$, and $M \to \infty$ when $\eta = 2/(\gamma - 1)$. For $\gamma = 1.4$, this last result corresponds to $\eta = 5$. At this η value, a, p, ρ, and T are zero. Figure 12.6 shows the normalized variation of T, ρ, and p for this γ value, where the abscissa is $-\eta$. Observe that p/p_1 changes more rapidly than ρ/ρ_1, which varies more rapidly than T/T_1. The reason stems from the exponents; e.g., the pressure ratio exponent is 7 when $\gamma = 1.4$.

Equations (12.69) and (12.71) can be viewed as the nondimensional solution for static variables. As evident from Figure 12.5, there is no length scale. Thus, the solution only depends on a single similarity variable η. We could have bypassed the MOC approach and directly transformed the Euler equations into several ODEs with η as the independent variable. As shown by Emanuel (1981), the result would be the above similarity solution. Similarity methods, while elegant, are not as flexible as the MOC. For instance, the flow field between a piston that is accelerating into a gas and the increasing shock strength it produces can be solved with the MOC, but this isentropic flow does not possess a similarity solution.

Explicit algebraic equations can be obtained for a particle path and the curved portion of a C_+ characteristic that are shown in Figure 12.5. This topic is the subject of Problem 12.9.

Solution for Stagnation Variables

It is of interest to obtain results for various stagnation quantities in terms of η. For instance, the stagnation temperature (or stagnation enthalpy) is given by

$$\frac{T_0}{T_1} = \frac{T_0}{T}\frac{T}{T_1} = X\left(\frac{2}{\gamma+1}\right)^2\left(1 - \frac{\gamma-1}{2}\eta\right)^2 \qquad \textbf{(12.72a)}$$

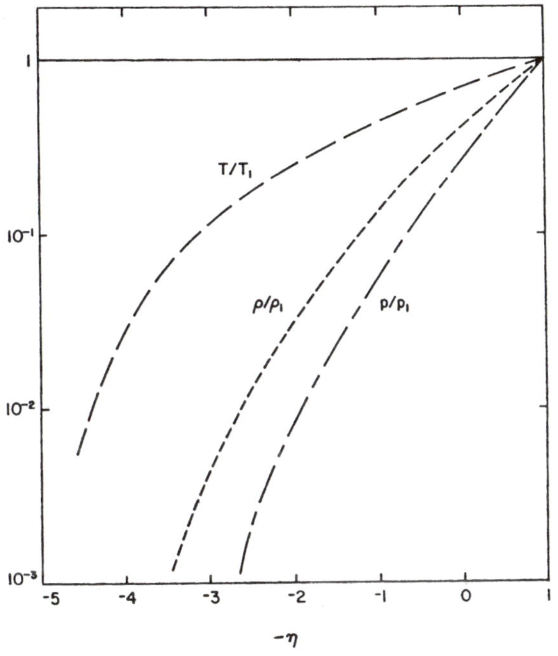

FIGURE 12.6 Variation of T, p, and ρ inside a centered rarefaction wave with $\gamma = 1.4$ (from Emanuel, 1981).

where a homentropic point relation is used for T_0/T. We next obtain

$$X = 1 + \frac{\gamma - 1}{2} M^2 = \frac{\gamma + 1}{2} \frac{1 + \frac{\gamma - 1}{2}\eta^2}{\left(1 - \frac{\gamma - 1}{2}\eta\right)^2} \tag{12.73}$$

so that

$$\frac{T_0}{T_1} = \frac{\gamma + 1}{2}\left(1 + \frac{\gamma - 1}{2}\eta^2\right) \tag{12.72b}$$

The stagnation pressure and density are then given by the homentropic relations

$$\frac{p_0}{p_1} = \left(\frac{T_0}{T_1}\right)^{\gamma/(\gamma - 1)}, \qquad \frac{\rho_0}{\rho_1} = \left(\frac{T_0}{T_1}\right)^{1/(\gamma - 1)} \tag{12.74a,b}$$

These stagnation quantities, along with M, are shown in Figure 12.7 for $\gamma = 1.4$. Note that T_0, p_0, and ρ_0 only depend on η^2, and therefore are symmetric about their minimum value, which occurs when $\eta = 0$. Exceptionally large stagnation values are possible when $\eta \le 2$. In particular, these values are large relative to their stagnation counterparts in the quiescent region.

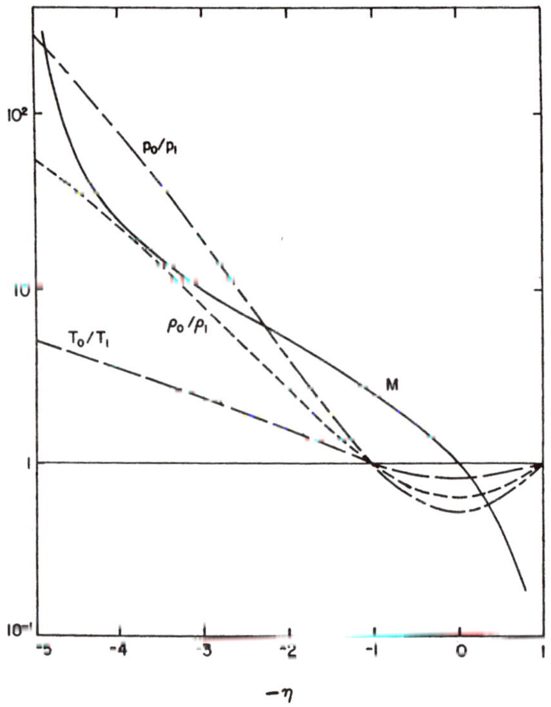

FIGURE 12.7 Variation of T_0, ρ_0, and p_0 inside a centered rarefaction wave with $\gamma = 1.4$ (from Emanuel, 1981).

Discussion

In Figure 12.5, the $\partial(\ )/\partial t$ derivative represents a change in a variable in the vertical direction, whereas $D(\ )/Dt$ represents a change along a particle path. Since $\partial(\ )/\partial t$ appears in the substantial derivative, this derivative has a component in the vertical direction. This observation underlies an important energy transfer mechanism that occurs when $\eta > 0$, and especially when $\eta \le 2$. It is this mechanism, which has no steady flow counterpart, that produces the large stagnation values seen on the left side of Figure 12.7.

As previously noted, the pressure is constant along the C_- characteristics in Figure 12.5. Inside the expansion, the pressure decreases in a clockwise direction, which corresponds to an increasing η value. As a result, $\partial p/\partial t$ is negative (positive) in the second (first) quadrant. We next examine the behavior of h_0, or T_0, for a fluid particle inside the expansion when x is negative, and the flow is locally subsonic. In this circumstance, Equation (12.34c) shows that h_0 *decreases* with increasing time. When $\eta > 0$, the rarefaction extends into the first quadrant, where the Mach number is supersonic. Hence, for a fluid particle in the first quadrant, $\partial p/\partial t$ is positive and h_0 *increases* with increasing time. Moreover, the rate at which h_0 increases with time is amplified by the $1/\rho$ coefficient in Equation (12.34c), which becomes large when $\eta > 2$. This h_0 variation means the wave motion transfers energy from the subsonic region to the supersonic region where the kinetic energy of the flow is large. In an unsteady, inviscid flow, this is a general phenomenon. This generality is directly evident from the energy equation, Equation (12.34c), which is not restricted to a one-dimensional flow. The mechanism is essential for a high-stagnation enthalpy wind tunnel, such as the expansion tube and, more recently, the superorbital expansion tube (Neely and Morgan, 1994). These experimental devices simulate high-speed flight in the upper atmosphere, as would occur on reentry back into Earth's atmosphere from a Mars or Venus mission.

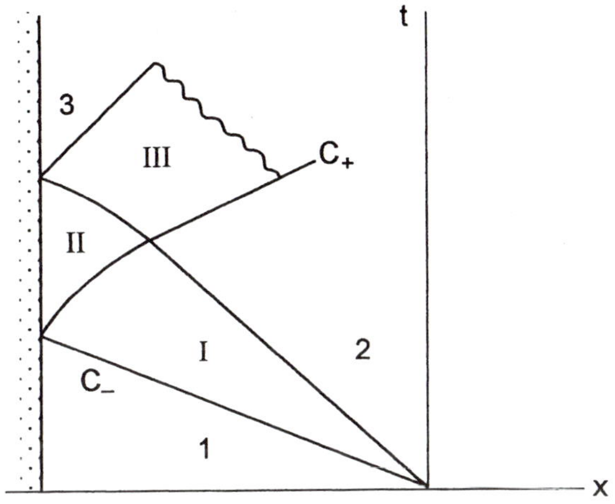

FIGURE 12.8 Reflection of a rarefaction wave from an endwall.

Figure 12.8 illustrates what happens when a rarefaction wave reflects from an endwall. (This sketch contains no piston; e.g., it is applicable to a shock tube flow.) The reflected wave in region III is also a rarefaction wave, since one can show that $p_2 > p_3$. (This is established in Section 12.8.) The C_- characteristics in regions I and II reflect from the endwall as C_+ characteristics. If region III were a compression, the C_+ characteristics would tend to converge instead of diverging. Regions 1, 2, and 3 are uniform flow regions, where 1 and 3 are quiescent. Both characteristic families are straight lines in these three uniform flow regions. Region I is a simple wave region, where the C_- characteristics are straight and centered, but the crossing C_+ characteristics are curved. (This characterization is slightly modified in Section 12.8.) Region III is a noncentered simple wave region. In region II, both families are curved and this region is referred to as nonsimple. Simple wave regions, in steady or unsteady flow, possess relatively elementary analytical solutions. As we shall see in Section 12.8, region II also possesses an analytical solution, but this solution is usually quite complicated. Because of this, nonsimple wave regions are evaluated numerically using the MOC.

All properties, such as the pressure, are constant along a given C_- characteristic in region I. Similarly, properties are constant along a given C_+ characteristic in region III. In a simple wave region, the straight characteristics represent acoustic waves. These are entirely analogous to the Mach lines or Mach waves in a steady flow, and, in fact, are sometimes called Mach lines in an unsteady flow. The disturbance propagates into the flow field along these Mach lines, whether the flow is steady or unsteady. In a nonsimple wave region, such as region II, disturbances propagate along both families of characteristics.

The vapor of many large molecules typically has specific heat ratios near unity. For instance, the specific heat ratios of SF_6 and UF_6 at 300 K are 1.093 and 1.069, respectively. As is true for all room-temperature polyatomic gases or vapors, the specific heat ratio decreases toward unity with increasing temperature. A useful approximation, therefore, for a large polyatomic gas is to set γ equal to unity. (The result of this limiting process is valid even though Equation (12.47) is now parabolic.) In this case, Equations (12.71) readily yield

$$\frac{a}{a_1} = 1, \qquad \frac{T}{T_1} = 1, \qquad M = 1 + \eta$$

Thus, M has a linear variation inside the expansion and the flow is isothermal. Because of their exponents, however, the equations for p/p_1 and ρ/ρ_1 are indeterminate. Their evaluation is based on L'Hospital's rule. We thus write

$$\frac{p}{p_1} = \left(\frac{a}{a_1}\right)^{2\gamma/\gamma-1} = \left(1 - \frac{\gamma-1}{2}\eta\right)^{2/(\gamma-1)}$$

where the $2/(\gamma + 1)$ in a/a_1 and the γ in the numerator of the p/p_1 exponent are both set equal to unity without altering the limit process. By taking the logarithm

$$\ln\frac{p}{p_1} = \frac{2\ln\left(1 - \frac{\gamma-1}{2}\eta\right)}{\gamma - 1}$$

the conventional "0/0" form that is needed for L'Hospital's rule is obtained when $\gamma \rightarrow 1$. The numerator and denominator are each individually differentiated with respect to γ, to yield

$$\ln\frac{p}{p_1} = \lim_{\gamma\rightarrow 1}\left(-\frac{\eta}{1 - \frac{\gamma-1}{2}\eta}\right) = -\eta$$

or

$$\frac{p}{p_1} = \frac{\rho}{\rho_1} = e^{-\eta}$$

when $\gamma = 1$. The density result stems from the replacement of $2\gamma/(\gamma - 1)$ with $2/(\gamma - 1)$ in the pressure exponent and is in accord with an isothermal flow.

A similar process for stagnation quantities yields

$$\frac{T_0}{T_1} = 1, \qquad \frac{p_0}{p_1} = \frac{\rho_0}{\rho_1} = e^{\eta^2/2}$$

This limit process can be applied to any compressible perfect gas flow. For example, Problem 12.6 considers a reflected shock flow. An isothermal result is quite general, holding for both homentropic and shock-containing flows. An exponential dependence for the pressure and density is also a general result for homentropic flows, as occurs, e.g., in nozzle flows.

Double Impulsive Acceleration

Suppose the piston in Figure 12.5 has a speed w_{p2}. At time t_c, its speed impulsively increases to w_{p3}. As indicated in Figure 12.9, point c becomes the origin of a second centered rarefaction. With $TE3$ pointing into the first quadrant, the flow to the right of x_c is supersonic. A rather large value for w_{p3} is required for this to be the case. For instance, suppose $M_3 = 2$, which is a modest supersonic Mach number for a steady flow. From Equation (12.71e), we obtain

$$\eta_{TE3} = \frac{M_3 - 1}{1 + \frac{\gamma-1}{2}M_3} = 0.714 \qquad \textbf{(12.75a)}$$

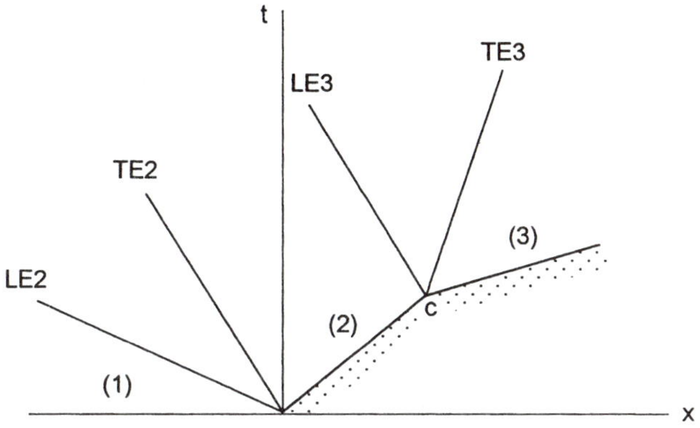

FIGURE 12.9 Schematic of a double impulsive acceleration.

when $\gamma = 1.4$. Equation (12.71a) yields $a_3 = 243$ m/s when $a_1 \cong 340$ m/s. Hence, we have

$$w_{p3} \cong a_3 M_3 = 243 \times 2 = 486 \text{ m/s} \tag{12.75b}$$

for the piston's speed, which is slightly faster than the muzzle velocity of a handgun.

Since region 2 in Figure 12.9 is a uniform flow, the C_- $TE2$ and $LE3$ characteristics are parallel to each other. Thus, flow conditions in region 3 are directly connected to conditions in region 1 as if there is only a single rarefaction wave generated by a piston with speed w_{p3}. In fact, this is the basis of the above calculation. This situation is entirely analogous to having a flow with two centered Prandtl–Meyer expansions.

Continuous Acceleration

Realistically, pistons do not start impulsively and travel with constant speeds. In this subsection, the flow field that is generated by a piston that accelerates away from an initially quiescent gas is discussed. In this circumstance, the flow field remains homentropic and an exact analytical solution is still possible.

For a rarefaction, two cases are distinguished. In the first, the initial piston $w_p(0)$ speed is zero. This is sketched in Figure 12.10(a), where the piston's trajectory is tangent to the time axis at the origin. In the second case, $w_p(0) > 0$ [see Figure 12.10(b)], and the first part of the expansion consists of a centered wave. [Since $w_p(0) > 0$, the initial piston motion is impulsive.]

First Case

For the flow pictured in Figure 12.10a, there is a single, noncentered expansion whose leading edge is given by Equations (12.65a). [This would be a centered wave if the piston's trajectory corresponded to a particle path of a centered wave, as analyzed in Problem 12.9(a).] The expansion is still a simple wave and the C_- characteristics that define it are straight. As with a centered expansion, flow conditions along a straight characteristic are constant. Thus, any C_- characteristic can be written as

$$x - x_p = (w - a)(t - t_p) \tag{12.76a}$$

which replaces Equation (12.68). Similarly, Equation (12.63) replaces Equation (12.71a), which requires a centered wave. (Any equation whose independent variable is the similarity variable η is

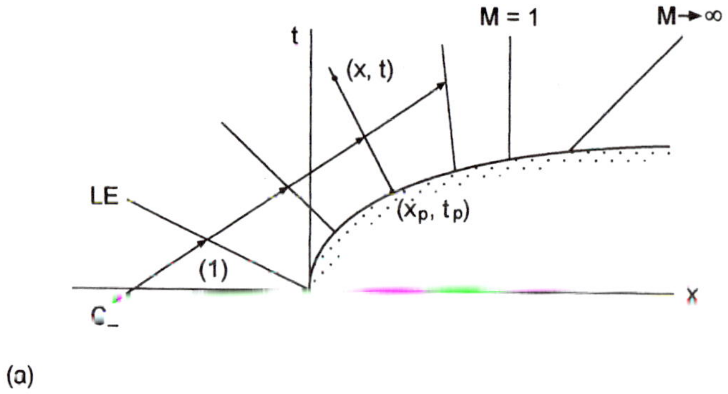

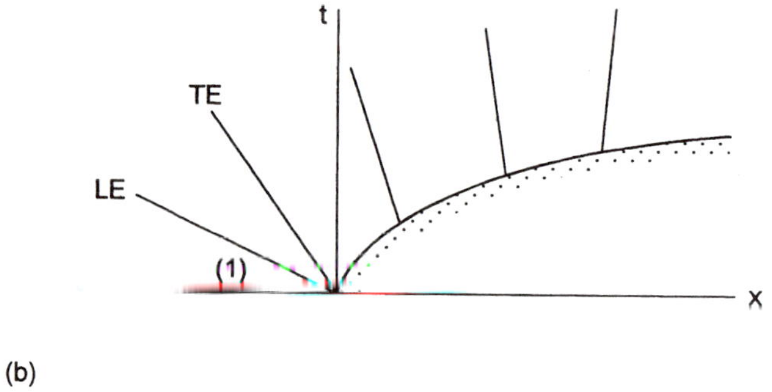

FIGURE 12.10 Schematic of an accelerating piston, where (*a*) $w_p\,(0) = 0$, and (b) $w_p\,(0) > 0$.

restricted to a centered wave.) Equation (12.71e) for the Mach number is replaced with

$$M = \frac{\frac{w}{a_1}}{\frac{a}{a_1}} = \frac{\frac{w}{a_1}}{1 - \frac{\gamma - 1}{2}\frac{w}{a_1}} \tag{12.77}$$

Along a given C_- characteristic in Figure 12.10a, we have

$$w(x, t) = w(x_p, t_p) = w_p = \frac{dx_p}{dt_p} \tag{12.78}$$

Thus, the w/a_1 factor in Equation (12.77) is given by w_p/a_1. Hence, the solution primarily hinges on connecting two points on a C_- characteristic, one of which is at the piston. This connection, however, is provided by Equation (12.76a) with w and a replaced with their piston values

$$x - x_p = (w_p - a_p)(t - t_p) \tag{12.76b}$$

This can be further simplified by using the C_+ compatibility equation

$$w_p + \frac{2}{\gamma - 1}a_p = \frac{2}{\gamma - 1}a_1 \tag{12.79}$$

to eliminate a_p, with the result

$$x - x_p = \left(\frac{\gamma+1}{2}w_p - a_1\right)(t - t_p) \tag{12.76c}$$

With a known piston trajectory

$$x_p = x_p(t_p) \tag{12.80}$$

we have $w_p = dx_p/dt_p$, and Equation (12.76c) can be viewed as providing

$$t_p = t_p(x, t) \tag{12.81}$$

Most often, this is an implicit equation for t_p.

Example

Consider the parabolic trajectory

$$x_p = \frac{1}{2}\alpha t_p^2 \tag{12.82a}$$

where α is a positive constant, and the corresponding piston speed is

$$w_p = \frac{dx_p}{dt_p} = \alpha t_p \tag{12.82b}$$

Our objective is to determine $p(x, t)$ and $M(x, t)$ for the flow inside the expansion.

Equation (12.76c) becomes

$$x - \frac{1}{2}\alpha t_p^2 = \left(\frac{\gamma+1}{2}\alpha t_p - a_1\right)(t - t_p) \tag{12.83a}$$

This is a quadratic equation for t_p that can be written nondimensionally as

$$\frac{\gamma}{2}\xi_p^2 - \left(1 + \frac{\gamma+1}{2}\xi\right)\xi_p + X + \xi = 0 \tag{12.83b}$$

where

$$\xi_p = \frac{\alpha t_p}{a_1}, \qquad \xi = \frac{\alpha t}{a_1}, \qquad X = \frac{\alpha x}{a_1^2} \tag{12.84}$$

Although the equation is linear in ξ, it is convenient, whenever possible, to solve it for ξ_p:

$$\xi_p = \frac{1}{\gamma}\left\{1 + \frac{\gamma+1}{2}\xi \pm \left[1 - (\gamma-1)\xi + \left(\frac{\gamma+1}{2}\right)^2\xi^2 - 2\gamma X\right]^{1/2}\right\} \tag{12.83c}$$

The piston's trajectory now is

$$X_p = \frac{1}{2}\xi_p^2 \tag{12.85}$$

and when the ξ, X point is on the piston, we see that the minus sign in front of the square root is correct. Equation (12.83c) represents $t_p = t_p(x, t)$ for two points on the same C_- characteristic. Once t_p is known in terms of x and t, Equations (12.82) provide x_p and w_p.

Equations (12.78) and (12.82b) yield

$$\frac{w(x, t)}{a_1} = \xi_p \tag{12.86}$$

Hence, we obtain

$$\frac{a}{a_1} = 1 - \frac{\gamma - 1}{2}\xi_p \tag{12.87a}$$

$$\frac{p}{p_1} = \left(1 - \frac{\gamma - 1}{2}\xi_p\right)^{2\gamma/(\gamma - 1)} \tag{12.87b}$$

$$M = \frac{\xi_p}{1 - \frac{\gamma - 1}{2}\xi_p} \tag{12.87c}$$

where the last two relations represent the desired solution. Because there is a length scale, there is no similarity solution in terms of η. Since variables are constant along straight characteristics, a similarity solution, however, does exist. Equations (12.86) and (12.87) represent this solution, where ξ_p is the appropriate, nondimensional, similarity variable.

From Equation (12.87c), we readily obtain

$$\xi_p^* = \frac{2}{\gamma + 1} \tag{12.88}$$

when $M = 1$. Thus, the flow to the right (left) of the vertical C_- characteristic in Figure 12.10a, labeled with $M = 1$, is supersonic (subsonic). The corresponding X value is obtained by substituting ξ_p^* into Equation (12.83b), which then yields $X_p^* = 2/(\gamma + 1)^2$. As shown in Problem 12.10, stagnation parameters, such as T_0, have a minimum on this characteristic. Similarly, the $M \rightarrow \infty$ characteristic in the figure starts at

$$\xi_{p\infty} = \frac{2}{\gamma - 1}, \qquad X_{p\infty} = \frac{2}{(\gamma - 1)^2} \tag{12.89}$$

To the right of this characteristic, there is a void; i.e., there is a vacuum between the piston and this characteristic. In this circumstance, the piston's motion no longer influences the flow of the gas. In theory, the density is zero along this characteristic. A continuum flow assumption, which underlies the analysis, however, is no longer valid in the limit when $(\rho/\rho_1) \rightarrow 0$.

Second Case

We now turn our attention to the flow depicted in Figure 12.10b. As is typical of hyperbolic flows, a solution is obtained by patching together several distinct solutions. Thus, the flow between the leading and trailing edges is given by a piston that impulsively starts with the $w_p(0)$ speed. The rest of the rarefaction is then treated as the one shown in Figure 12.10a. The detailed solution process is the subject of Problems 12.14 and 12.15.

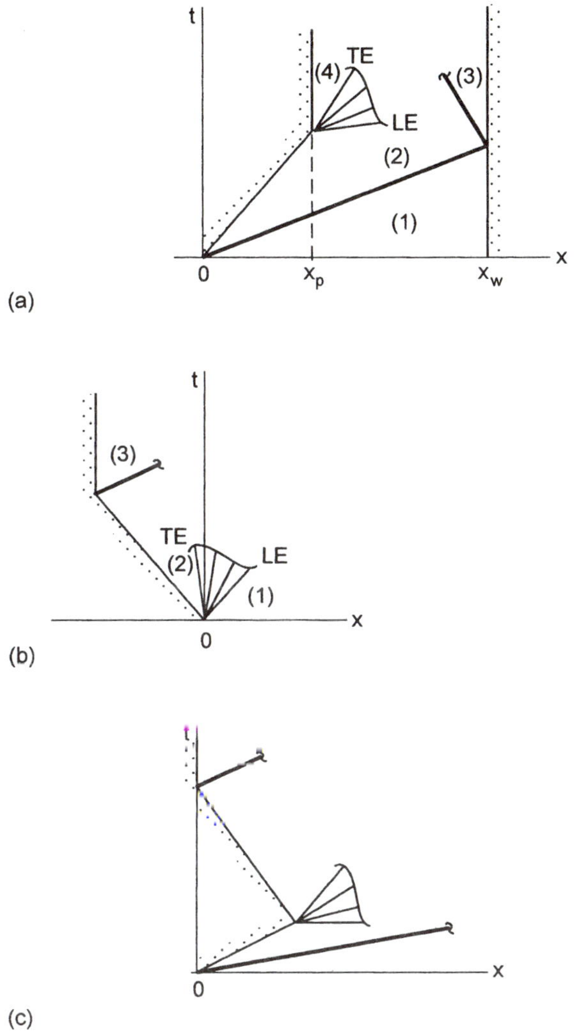

FIGURE 12.11 (a) Constant-speed piston moving into a quiescent gas and then stopping impulsively; (b) constant-speed piston withdrawing from a quiescent gas and then stopping impulsively; (c) piston motion for the generation of an N wave.

Flows Containing Shock and Rarefaction Waves

An introductory discussion is provided of several flows, each containing a constant strength shock, or shocks, and a centered rarefaction wave. The flows are sketched in Figures 12.11(a) and (b); Figure 12.11(c) is discussed at the end of this section. The constant speed piston in Figure 12.11(a) moves into a quiescent gas, thereby generating an incident shock that reflects off the endwall. When the piston abruptly stops, it generates a centered rarefaction wave. With increasing time the shock and expansion waves interact with each other and continually reflect from both endwalls. In this subsection, we shall be content to just discuss the flow before the interaction occurs; otherwise, the wave-diagram method (Rudinger, 1969) or a computer solution is required. It nevertheless is worth noting that, with the aid of viscous dissipation, the two waves ultimately cancel each other. At infinite time, the gas is quiescent. In Figure 12.11(b), the piston is withdrawn from a quiescent gas at a constant speed. When the piston abruptly stops, a normal shock is generated that will catch up and interact with the expansion. The solution for both flows is based on the analysis in Section 12.2 and in the earlier part of this section. These types of flows occur in devices such as pressure exchangers (Azoury, 1992; Kentfield, 1993; Weber, 1995), and it has been suggested (Emanuel, 1981) that studies involving both condensation and evaporation, in a single experiment, could be done with the configuration in Figure 12.11(b). (Glass et al., 1977, have performed condensation-only experiments that utilize the rarefaction wave in a shock tube.)

First Case

For the flow in Figure 12.11(a), the gas in regions (1), (3), and (4) is quiescent; i.e.,

$$w'_1 = w'_3 = w'_4 = 0 \tag{12.90}$$

where a prime again indicates the use of an unsteady x,t coordinate system. From Table 12.1, the incident shock Mach number M_s and the steady and unsteady Mach numbers for region 2 are respectively

$$M_s = \frac{(\gamma+1)w'_p}{4a_1}\left(1+\left\{1+\left[\frac{4a_1}{(\gamma+1)w'_p}\right]^2\right\}^{1/2}\right) \tag{12.9}$$

$$M_2 = \left(\frac{X_s}{Y_s}\right)^{1/2} \tag{12.4b}$$

$$M'_2 = \frac{Z_s}{(X_sY_s)^{1/2}} \tag{12.10b}$$

where w'_p is a positive piston speed. For the reflected shock Mach number, we have

$$M_r = \frac{1}{M_2} \tag{12.91}$$

Figure 12.12 more closely examines the expansion shown in Figure 12.11(a). In region 2, the flow speed w'_2, which equals w'_p, can be larger or smaller than a_2. In either case, the slope dt/dx of the C_+ characteristics is positive and relatively shallow. On the other hand, the slope of the C_- characteristics can be positive — as shown in the figure — if $M'_2 > 1$, or negative, if $M'_2 < 1$. In region 4, the C_- characteristics have a negative slope, since $w'_4 = 0$, while the C_+ characteristics still have a positive slope. The leading and trailing edges, and the centered rays within the expansion, are of the C_+ family.

Previously, we examined expansions that propagated into a quiescent gas. Here, the gas is in motion ahead of the expansion and quiescent behind it. Nevertheless, the method of analysis is the same. For instance, the key to determining the state of the gas in region 4 is to use the compatibility equation that crosses the expansion, which is

$$J_- = w'_2 - \frac{2}{\gamma-1}a_2 = -\frac{2}{\gamma-1}a_4 \tag{12.92a}$$

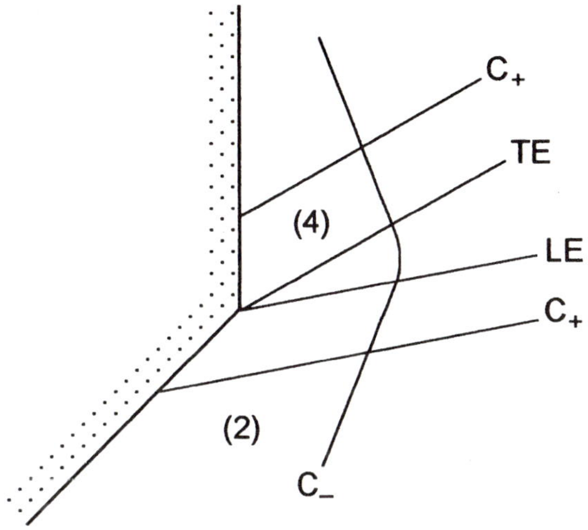

FIGURE 12.12 Characteristic lines in regions 2 and 4 of Figure 12.11(a).

This is written as

$$\frac{u_4}{a_1} = \frac{a_2}{a_1} - \frac{\gamma-1}{2}\frac{a_2}{a_1}\frac{w_2'}{a_2} = \frac{a_2}{a_1}\left(1 - \frac{\gamma-1}{2}M_2'\right) \tag{12.92b}$$

From Table 12.1

$$\frac{a_2}{a_1} = \left(\frac{T_2}{T_1}\right)^{1/2} = \frac{2}{\gamma+1}\frac{(X_s Y_s)^{1/2}}{M_s} \tag{12.93}$$

and with Equation (12.10b), we obtain

$$\frac{a_4}{a_1} = \frac{2}{\gamma+1}\frac{(X_s Y_s)^{1/2}}{M_s}\left[1 - \frac{\gamma-1}{2}\frac{Z_s}{(X_s Y_s)^{1/2}}\right] \tag{12.92c}$$

This relation, along with $w_4' = 0$, is sufficient to establish any other parameter in region 4. For instance, the temperature and pressure are

$$\frac{T_4}{T_1} = \left(\frac{a_4}{a_1}\right)^{1/2} \tag{12.94}$$

$$\frac{p_4}{P_1} = \frac{p_2}{p_1}\frac{p_4}{p_2} = \frac{2}{\gamma+1}Y_s\left(\frac{a_4}{a_2}\right)^{2\gamma/(\gamma-1)} = \frac{2}{\gamma+1}Y_s\frac{\left(\frac{a_4}{a_2}\right)^{2\gamma/(\gamma-1)}}{\left(\frac{a_2}{a_1}\right)^{2\gamma/(\gamma-1)}}$$

$$= \frac{2}{\gamma+1}Y_s\left[1 - \frac{\gamma-1}{2}\frac{Z_s}{(X_s Y_s)^{1/2}}\right]^{2\gamma/(\gamma-1)} \tag{12.95}$$

Example

Suppose we set $\gamma = 1.4$ and

$$\frac{w'_p}{a_1} = 1 \tag{12.96}$$

which represents a considerable piston speed. (For instance, it exceeds the maximum piston speed of a car engine.) We then obtain

$$M_s = 1.7662, \quad M_2 = 0.62425, \quad M'_2 = 0.81473, \quad M_r = 1.6019$$

$$\frac{p_2}{p_1} = 3.4727, \qquad \frac{p_3}{p_1} = 9.8178, \qquad \frac{p_4}{p_1} = 0.99987$$

$$\frac{T_2}{T_1} = 1.5065, \qquad \frac{T_3}{T_1} = 2.0929, \qquad \frac{T_4}{T_1} = 1.0136$$

Five significant digits are shown because p_4/p_1 and T_4/T_1 are quite close to unity. The expansion virtually cancels the effect of the incident shock. At early times, an exact cancellation is not possible since there is an entropy increase across the shock but no entropy change across the expansion. The reason for the near-unity values is that M_s is not very large; i.e., despite the substantial piston speed, the incident shock is relatively weak. (With a stronger shock, p_4/p_1 and T_4/T_1 start to significantly differ from unity; see Problem 12.17.) Since $M'_2 < 1$, the C_- characteristics in region 2 have a negative slope.

Final State Conditions for the First Case

Final state conditions in the chamber can be evaluated using elementary thermodynamics. At the start of the process, the mass of gas is given by

$$m = \rho_1 A x_w \tag{12.97}$$

where x_w is the initial chamber length (see Figure 12.11(a)), and A is the cross-sectional area. The density ρ_∞, at infinite time, is

$$\frac{\rho_\infty}{\rho_1} = \frac{x_w}{x_w - x_p} = \frac{1}{1 - \frac{x_p}{x_w}} \tag{12.98}$$

where we assume x_p/x_w is known. The work, per unit mass, done by the piston on the gas is

$$\tilde{w} = \frac{1}{m}\int_0^{x_p} pAdx = \frac{p_2 A x_p}{m} = RT_1 \frac{x_p}{x_w}\frac{p_2}{p_1} \tag{12.99}$$

where p_2/p_1 equals $2Y_s/(\gamma+1)$. Let e denote the specific internal energy. The first law of thermodynamics is

$$e_\infty - e_1 = \tilde{w} \tag{12.100a}$$

since the closed system is adiabatic and possesses no kinetic or potential energy initially or at infinite time. With Equation (12.99) and $RT/(\gamma-1)$ for e, we obtain

$$\frac{T_\infty}{T_1} = 1 + (\gamma-1)\frac{x_p}{x_w}\frac{p_2}{p_1} = 1 + 2\frac{\gamma-1}{\gamma+1}\frac{x_p}{x_w}Y_s \tag{12.100b}$$

where the Mach number M_s is determined by the piston speed via Equation (12.9). The pressure ratio p_∞/p_1 is then given by

$$\frac{p_\infty}{p_1} = \frac{\rho_\infty}{\rho_1}\frac{T_\infty}{T_1} \tag{12.101}$$

These results are for an irreversible process, since a shock wave is involved. This aspect finds its way into the analysis through the constant pressure p_2 that acts on the face of the moving piston. Different results would be obtained for p_∞ and T_∞, but not ρ_∞, had the piston slowly and reversibly compressed the gas. The pressure on the piston's face is then a variable, given by the standard $p \sim \rho^\gamma$ homentropic relation.

Second Case

We next turn our attention to the flow in Figure 12.11(b). The piston speed w_p', which equals w_2', is now negative. Equations (12.71) hold, providing Equations (12.69) and (12.70) have a sign change; i.e.,

$$\eta = \frac{x}{a_1 t} \tag{12.102}$$

$$\frac{w}{a_1} = -\frac{2}{\gamma+1}(1+\eta) \tag{12.103}$$

We thus obtain

$$\eta_{LE} = -1, \qquad \eta_{TE} = -1 - \frac{\gamma+1}{2}\frac{w_p'}{a_1} \tag{12.104a}$$

$$\frac{a_2}{a_1} = 1 + \frac{\gamma-1}{2}\frac{w_p'}{a_1} \tag{12.104b}$$

$$\frac{T_2}{T_1} = \left(\frac{a_2}{a_1}\right)^2 \tag{12.104c}$$

$$\frac{p_2}{p_1} = \left(\frac{a_2}{a_1}\right)^{2\gamma/(\gamma-1)} \tag{12.104d}$$

$$M_2' = -\frac{w_2'}{a_2} = -\frac{\dfrac{w_p'}{a_1}}{1 + \dfrac{\gamma-1}{2}\dfrac{w_p'}{a_1}} \tag{12.104e}$$

for the flow in region 2.

The positive shock speed w_s' is determined by the $w_3' = 0$ condition. In a shock-fixed coordinate system, the upstream $\hat{w}_2$ and downstream w_3 flow speeds are

$$\hat{w}_2 = w_2' - w_s', \quad w_3 = -w_s' \tag{12.105}$$

where w_2', $\hat{w}_2$, and w_3 are negative. (A caret is used to avoid confusion with previous definitions.) The upstream, shock-fixed Mach number that determines the strength of the shock is

$$\hat{M}_2 = -\frac{\hat{w}_2}{a_2} = -\frac{w_2'}{a_2} + \frac{w_s'}{a_2} = M_2' + \hat{M}_s \tag{12.106}$$

where

$$\hat{M}_s = \frac{w_s'}{a_2} \tag{12.107}$$

The equation for the ratio of flow speeds across a fixed normal shock yields

$$\frac{w_3}{\hat{w}_2} = \frac{2}{\gamma+1}\,\frac{1 + \frac{\gamma-1}{2}\hat{M}_2^2}{\hat{M}_2^2} \tag{12.108}$$

The left side of this equation is replaced with

$$\frac{w_3}{\hat{w}_2} = -\frac{\frac{w_s'}{a_2}}{\frac{w_2'}{a_2} - \frac{w_s'}{a_2}} = \frac{\hat{M}_s}{M_2' + \hat{M}_s} \tag{12.109}$$

and on the right side $\hat{M}_2$ is replaced with Equation (12.106), with the result

$$\hat{M}_s^2 + \frac{3-\gamma}{2}M_2'\hat{M}_s - \frac{\gamma-1}{2}M_2'^2 = 0 \tag{12.110a}$$

This differs, e.g., with Equation (12.28a), since flow conditions are different. We readily obtain

$$\hat{M}_s = \frac{\gamma-1}{2}M_2' \tag{12.110b}$$

In conjunction with Equations (12.104e) and (12.106), this yields

$$\hat{M}_2 = -\frac{\frac{\gamma+1}{2}\frac{w_p'}{a_1}}{1 + \frac{\gamma-1}{2}\frac{w_p'}{a_1}} \tag{12.111}$$

for the shock Mach number. This Mach number must equal or exceed unity, which means that $(-w_p'/a_1) \geq (1/\gamma)$ if a shock is to occur. In this situation, the magnitude of the expansion-produced speed, $-w_2'$, is added to the shock speed. Thus, the shock is a Mach wave when the region 2 Mach number, M_2', equals $2/(\gamma+1)$ and the flow in this region is subsonic. At the other extreme, $\hat{M}_2 \rightarrow \infty$ when $(w_p'/a_1) = -2/(\gamma - 1)$.

The evaluation of region 3 properties is assisted with

$$\hat{X}_2 = 1 + \frac{\gamma-1}{2}\hat{M}_2^2 = \frac{1 + (\gamma-1)\dfrac{w_p'}{a_1} + \dfrac{\gamma-1}{8}(\gamma^2 + 4\gamma - 1)\left(\dfrac{w_p'}{a_1}\right)^2}{\left(1 + \dfrac{\gamma-1}{2}\dfrac{w_p'}{a_1}\right)^2} \tag{12.112a}$$

$$\hat{Y}_2 = \gamma\hat{M}_2^2 - \frac{\gamma-1}{2} = \frac{-\dfrac{\gamma-1}{2} - \dfrac{(\gamma-1)^2}{2}\dfrac{w_p'}{a_1} + \dfrac{1}{8}(\gamma^3 + 7\gamma^2 - \gamma + 1)\left(\dfrac{w_p'}{a_1}\right)^2}{\left(1 + \dfrac{\gamma-1}{2}\dfrac{w_p'}{a_1}\right)^2} \tag{12.112b}$$

The pressure and temperature are then given by

$$\begin{aligned}\frac{p_3}{p_1} &= \frac{p_2}{p_1}\frac{p_3}{p_2} = \left(1 + \frac{\gamma-1}{2}\frac{w_p'}{a_1}\right)^{2\gamma/(\gamma-1)} \frac{2}{\gamma+1}\hat{Y}_2 \\ &= \frac{2}{\gamma+1}\left(1 + \frac{\gamma-1}{2}\frac{w_p'}{a_1}\right)^{2/(\gamma-1)}\left[-\frac{\gamma-1}{2} - \frac{(\gamma-1)^2}{2}\frac{w_p'}{a_1} + \frac{1}{8}(\gamma^3 + 7\gamma^2 - \gamma + 1)\left(\frac{w_p'}{a_1}\right)^2\right]\end{aligned} \tag{12.113}$$

$$\begin{aligned}\frac{T_3}{T_1} &= \frac{T_2}{T_1}\frac{T_3}{T_2} = \left(1 + \frac{\gamma-1}{2}\frac{w_p'}{a_1}\right)^2\left(\frac{2}{\gamma+1}\right)^2\frac{\hat{X}_2\hat{Y}_2}{\hat{M}_2^2} \\ &= \left(\frac{2}{\gamma+1}\right)^2\frac{1}{\left(\dfrac{w_p'}{a_1}\right)^2}\left[1 + (\gamma-1)\frac{w_p'}{a_1} + \frac{(\gamma-1)}{8}(\gamma^2 + 4\gamma - 1)\left(\frac{w_p'}{a_1}\right)^2\right] \\ &\quad \times\left[-\frac{\gamma-1}{2} - \frac{(\gamma-1)^2}{2}\frac{w_p'}{a_1} + \frac{1}{8}(\gamma^3 + 7\gamma^2 - \gamma + 1)\left(\frac{w_p'}{a_1}\right)\right]\end{aligned} \tag{12.114}$$

Discussion

Problems 12.17 and 12.18, respectively, deal with the flows in Figures 12.11(a) and (b). In these problems, the magnitude of the piston's speed and conditions in region 1 are the same. Nevertheless, conditions in the regions that experience both an expansion and a compression, although in opposite order, are quite different. Several factors account for this difference. For instance, the shock Mach number is larger in Problem 12.18 than it is in Problem 12.17, with a correspondingly larger increase in entropy. The nature of the work done by the piston also differs. In Problem 12.17, the piston does a significant amount of work on the gas, since p_2 is relatively large. On the other hand, in Problem 12.18 the gas does a minute amount of work on the piston, since p_2 is small.

Figure 12.11(c) is similar to 12.11(a), except that the end wall at x_w has been removed and the piston reverses direction and returns to its initial position. As indicated in the sketch, the magnitude of the return speed need not equal the piston's incident speed. Upon stopping at $x = 0$, a second shock propagates into the gas. In due time, the expansion will overtake both shocks and fill the region between them. As a result of the interaction, both shocks weaken and the distance between them increases (Friedrichs, 1948). In this circumstance, the overall wave system is called an N-wave, since the pressure disturbance has this shape. An N-wave is most often associated with a steady, supersonic flow. For instance, a two-dimensional airfoil generates a bow shock, a trailing edge shock, and an expansion wave between the shocks. On the ground, we are familiar with N-waves as sonic booms.

12.6 COMPRESSION WAVES

If a piston gradually accelerates from $w_p'(0) = 0$ into a quiescent gas, a compression wave forms. Initially, the wave is a simple wave with converging C_- characteristics, as sketched in Figure 12.13. As usual, region 1 is quiescent and the dashed line is a particle path. The converging C_- characteristics start to overlap with the straight leading edge characteristic at the point where the indicated $C_+^\dagger$ and C_- paths and the particle path all cross. This location is referred to as the start-of-the-shock and conditions on the C_+ characteristic that passes through this point are denoted with a † superscript. Above this point, a shock forms whose strength gradually increases as more C_- characteristics run into it. As the shock strengthens, its speed and Mach number increase. Its path in the x,t plane is thus curved and concave downward. Other shocks, internal to the simple or nonsimple wave regions, may form as the result of characteristics of the same family attempting to overlap. This can occur, e.g., if the piston's motion is jerky.

The C_+ characteristics in region 2 originate in a quiescent gas region. Region 2 is thus a simple wave region in which the C_- characteristics are straight. The C_+ characteristics in region 3, however, originate just downstream of a curved shock, and consequently region 3 is a nonsimple wave region. The $C_+^\dagger$ characteristic represents a boundary between simple and nonsimple wave regions. Moreover, in the narrow region between the shock and the indicated particle path, the flow is isentropic, not homentropic. The flow is homentropic outside this region. Remember that the theory established earlier requires a homentropic flow and does not apply to the flow in the isentropic region.

The flow speed in the figure is either zero, as in region 1, or negative. With this in mind, Equation (12.63) states that the speed of sound at the piston gradually increases with time from its initial value of a_1. Thus, weak disturbance signals generated at the face of the piston gradually converge, ultimately forming a shock. The analytical reason for converging C_- characteristics is evident from Equation (12.56d) when written as

$$\frac{dt}{dx} = \frac{1}{w - a} \tag{12.115}$$

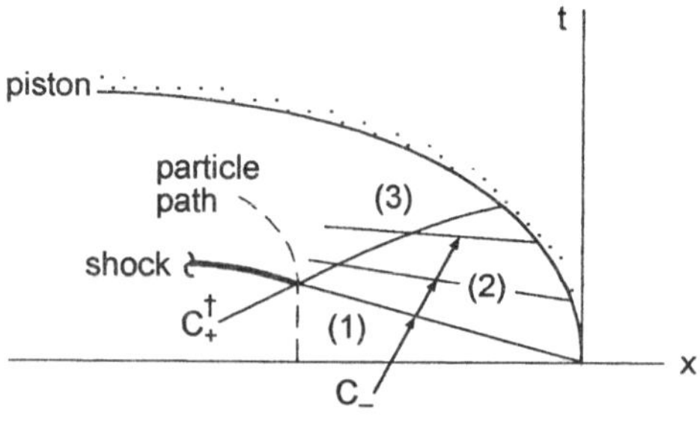

FIGURE 12.13 Compression wave generated by a gradually accelerating piston.

As we move along the trajectory of the piston, a and $|w|$ increase, but with w negative. Thus, dt/dx is negative and tending toward zero with increasing time.

Example

For region 2, the preceding (rarefaction) theory still applies. For instance, consider the trajectory and speed of a piston

$$x_p = -\frac{1}{2}\alpha t_p^2, \qquad w_p = -\alpha t_p \tag{12.116}$$

that is the compressive counterpart of Equations (12.82). With α replaced by $-\alpha$ and Equation (12.84) unaltered, we have, for a C_+ characteristic and the piston path

$$\frac{\gamma}{2}\xi_p^2 + \left(1 - \frac{\gamma+1}{2}\xi\right)\xi_p - X - \xi = 0 \tag{12.117a}$$

$$X_p = -\frac{1}{2}\xi_p^2 \tag{12.117b}$$

while Equations (12.86) and (12.87a) are

$$\frac{w}{a_1} = -\xi_p \tag{12.118a}$$

$$\frac{a}{a_1} = 1 - \frac{\gamma-1}{2}\frac{w}{a_1} = 1 + \frac{\gamma-1}{2}\xi_p \tag{12.118b}$$

Consequently, the (positive) Mach number inside region 2 is

$$M = -\frac{\frac{w}{a_1}}{\frac{a}{a_1}} = \frac{\xi_p}{1 + \frac{\gamma-1}{2}\xi_p} \tag{12.118c}$$

where the equation for an arbitrary C_+ characteristic

$$\xi_p = -\frac{1}{\gamma} + \frac{\gamma+1}{2\gamma}\xi + \frac{1}{\gamma}\left[1 + (\gamma-1)\xi + \left(\frac{\gamma+1}{2}\xi\right)^2 + 2\gamma X\right]^{1/2} \tag{12.119}$$

stems from Equation (12.117a).

The point where the shock first forms can be found by noting that it occurs on the leading edge of the compression, a C_- characteristic,

$$X^\dagger = -\xi^\dagger \tag{12.120}$$

This relation is substituted into Equation (12.117a), to obtain

$$\frac{\gamma}{2}\xi_p^\dagger + 1 - \frac{\gamma+1}{2}\xi^\dagger = 0 \tag{12.121}$$

However, $\xi_p^\dagger$ equals zero on the leading edge; hence,

$$\xi^\dagger = \frac{2}{\gamma+1}, \quad X^\dagger = -\frac{2}{\gamma+1} \tag{12.122}$$

At the start-of-the-shock point, as well as along the straight leading edge of region 2, the Mach number is zero.

The shape of two of the three borders of the simple wave region is elementary. We thus determine the equation for the $C_+^\dagger$ characteristic that borders the nonsimple wave region. We start with Equation (12.56b) and with aid of Equations (12.118) obtain

$$\frac{dX}{d\xi} = 1 - \frac{3-\gamma}{2}\xi_p \tag{12.123}$$

If ξ_p is replaced with Equation (12.119), the resulting equation cannot be analytically integrated in any obvious fashion. This difficulty is avoided by writing Equation (12.117a) as

$$X = \frac{\gamma}{2}\xi_p^2 + \left(1 - \frac{\gamma+1}{2}\xi\right)\xi_p - \xi \tag{12.124}$$

and by differentiation

$$dX = -\left(1 + \frac{\gamma+1}{2}\xi_p\right)d\xi + \left(\gamma\xi_p + 1 - \frac{\gamma+1}{2}\xi\right)d\xi_p \tag{12.125}$$

Along a C_- characteristic, ξ_p is a constant. We are dealing with a C_+ characteristic, however, and ξ_p is not a constant along it. Eliminate dX from Equation (12.23) and the above, with the result

$$\frac{d\xi_p}{d\xi} = \frac{2 + (\gamma-1)\xi_p}{1 - \frac{\gamma+1}{2}\xi + \gamma\xi_p} \tag{12.126}$$

The initial condition at the start-of-the-shock is

$$\xi^\dagger = \frac{2}{\gamma+1}, \quad \xi_p^\dagger = 0 \tag{12.127}$$

where ξ_p increases along the $C_+^\dagger$ characteristic until the face of the piston is reached.

Equation (12.126) is put in a standard form by a linear transformation

$$\xi_p = \frac{\gamma-1}{\gamma}\bar{\xi}_p - \frac{2}{\gamma-1}, \quad \xi = \bar{\xi} - \frac{2}{\gamma-1} \tag{12.128}$$

which yields

$$\frac{d\bar{\xi}_p}{d\bar{\xi}} = \frac{\bar{\xi}_p}{\bar{\xi}_p + \lambda\bar{\xi}} \tag{12.129}$$

where λ is a negative constant

$$\lambda = -\frac{1}{2}\frac{\gamma+1}{\gamma-1} \tag{12.130}$$

The initial condition at the start-of-the-shock point now is

$$\bar{\xi}^{\dagger} = \frac{4\gamma}{\gamma^2-1}, \qquad \bar{\xi}_p^{\dagger} = \frac{4\gamma}{(\gamma-1)^2} \tag{12.131a, b}$$

Equation (12.129) is singular when $\bar{\xi} = \bar{\xi}_p = 0$. The eigenvalues associated with this equation are unity and λ; hence, the singular point is a saddle point (Hurewicz, 1958). With Equations (12.128), it is easy to show that the saddle point is located well outside the region of interest and, therefore, is of no concern.

Equation (12.129) is homogeneous of degree one, and is solved by the standard substitution

$$\bar{\xi}_p = v(\bar{\xi})\bar{\xi} \tag{12.132}$$

which yields

$$\frac{d\bar{\xi}_p}{d\bar{\xi}} = v + \bar{\xi}\frac{dv}{d\bar{\xi}} \tag{12.133}$$

This substitution results in a separable equation. Equating this relation to Equation (12.29) results in

$$\bar{\xi}\frac{dv}{d\bar{\xi}} = \frac{v(-v+1-\lambda)}{v+\lambda} \tag{12.134}$$

where one initial value is provided by Equation (12.131a) and the other is

$$v^{\dagger} = \frac{\bar{\xi}_p^{\dagger}}{\bar{\xi}^{\dagger}} = -\lambda \tag{12.135}$$

Variables can be separated, to yield

$$\int_{\bar{\xi}^{\dagger}}^{\bar{\xi}} \frac{d\bar{\xi}}{\bar{\xi}} = \int_{-\lambda}^{v} \frac{dv}{-v+1-\lambda} + \lambda\int_{-\lambda}^{v} \frac{dv}{-v^2+(1-\lambda)v} \tag{12.136}$$

which ultimately results in

$$\left(\frac{\bar{\xi}}{\bar{\xi}^{\dagger}}\right)^{1-\lambda} = \left(-\frac{v}{\lambda}\right)^{\lambda}\frac{1}{-v+1-\lambda} \tag{12.137}$$

We return to the original variables and simplify, to obtain

$$\xi = \frac{2\gamma}{3\gamma-1}\xi_p - \frac{2}{3\gamma-1} + \frac{8\gamma}{(\gamma+1)(3\gamma-1)}\left(1+\frac{\gamma-1}{2}\xi_p\right)^{-(\gamma+1)/[2(\gamma-1)]} \tag{12.138}$$

We next utilize Equation (12.119) to eliminate ξ_p, to arrive at the desired implicit X, ξ equation

$$\xi = -\frac{2}{\gamma-1} + \frac{1}{\gamma-1}\left[1 + (\gamma-1)\xi + \left(\frac{\gamma+1}{2}\xi\right)^2 + 2\gamma X\right]^{1/2}$$
$$+ \frac{4\gamma}{\gamma^2-1}\left\{\frac{\gamma+1}{2\gamma} + \frac{\gamma^2-1}{4\gamma}\xi + \frac{\gamma-1}{2\gamma}\left[1 + (\gamma-1)\xi + \left(\frac{\gamma+1}{2}\xi\right)^2 + 2\gamma X\right]^{1/2}\right\}^{-(\gamma+1)/[2(\gamma-1)]} \quad \textbf{(12.139)}$$

for the $C_+^\dagger$ border characteristic.

At the piston, this relation yields

$$\xi_p^\dagger = \frac{2}{\gamma-1}\left[\left(\frac{4\gamma}{\gamma+1}\right)^{2(\gamma-1)/(3\gamma-1)} - 1\right] \quad \textbf{(12.140a)}$$

since

$$X_p = -\frac{1}{2}\xi_p^2 \quad \textbf{(12.140b)}$$

We thus have

$$\xi_p^\dagger = 1.18, \quad X_p^\dagger = -0.696 \quad \textbf{(12.141)}$$

when $\gamma = 1.4$. At the other end of this characteristic, Equations (12.122) provide

$$\xi^\dagger = 0.833, \quad X^\dagger = -0.833 \quad \textbf{(12.142)}$$

As expected, the $C_+^\dagger$ characteristic has a positive slope. Figure 12.13, however, is deceptive in showing the simple wave region as much broader than it really is. Finally, we note that $M^\dagger = 0$ and that $M_p^\dagger = 0.995$. Thus, the simple wave region is subsonic.

12.7 INTERNAL BALLISTICS

INTRODUCTORY DISCUSSION

Ballistics can be subdivided into internal, intermediate, external, and terminal regimes. For internal ballistics, the bullet or projectile is inside the gun barrel. In the intermediate regime (Merlen and Dyment, 1991; Jiang et al., 1998), the projectile is near the muzzle and the gas dynamics of the flow field caused by the gas discharge from the barrel are of primary interest. External ballistics deals with the projectile in free flight, while terminal ballistics involves the interaction with a target. A gas dynamically oriented introduction to internal ballistics is provided in this section. Other presentations, which include additional references, can be found in the book by Farrar and Leeming (1983) and in the articles by Krier and Adams (1979) and by Freedman (1988). Our objective is to illustrate how unsteady waves can be utilized to understand the dynamics involved in internal ballistics. To avoid undue length and complexity, a number of assumptions and approximations are introduced. Sufficient physical content, however, is retained in order that the presentation should still be representative of the actual situation.

A breech chamber contains the gun powder, or propellant, and the projectile. The diameter of this chamber is slightly larger than that of the barrel. Once the powder is ignited and starts to burn, the pressure in the chamber rapidly increases. At an early time, when this pressure is still relatively small compared to its subsequent peak value, the chamber pressure forces the projectile into the barrel and it starts to accelerate. The soft rotating band around the projectile and near its base is further squeezed into the rifling of the barrel, thereby causing the projectile to rotate. This spin provides aerodynamic stability for the projectile during its free flight. When the projectile is at the gun's muzzle, its rotational energy, however, is only about 0.3% of its translational energy (Krier and Adams, 1979, page 9). The subsequent discussion thus neglects the spin of the projectile.

Gun powder is often in the form of perforated grains. Before ignition, the breech chamber contains a mixture of powder and air, where the initial mass $m_g(0)$ of the powder (a g subscript denotes a grain property) greatly exceeds the mass of the trapped air. The volume of the two constituents, however, is roughly comparable. As we will observe in the next subsection, the fraction of volume devoted to air is an important parameter. The powder is engineered to burn smoothly and not to detonate. Detonation occurs when an unsteady normal shock wave is immediately followed by intense, nearly instantaneous combustion. In this situation, the combustion process is complete before the projectile has had time to move. The extreme breech pressure caused by the rapid combustion, which is amplified by the reflected shock, can rupture the wall of the chamber. Typically, the powder is not completely burned until the projectile is well down the barrel. The rate of burning for a given powder chemical composition is largely determined by the initial surface area, i.e., the number of perforations in a grain, per grain volume, and by the pressure. As the burning surface area increases (decreases) the rate of gas production increases (decreases). The rate at which the grains burn, i.e., the rate at which their surface recesses, is also pressure dependent, with more rapid recession occurring at a high pressure. Incidentally, these remarks also apply to the combustion process in a solid propellant rocket engine.

Figure 12.14(a) is a rough schematic of a gun barrel at an angle θ relative to gravity with the base of the projectile a distance x_p down the barrel, where a p subscript denotes the projectile. As

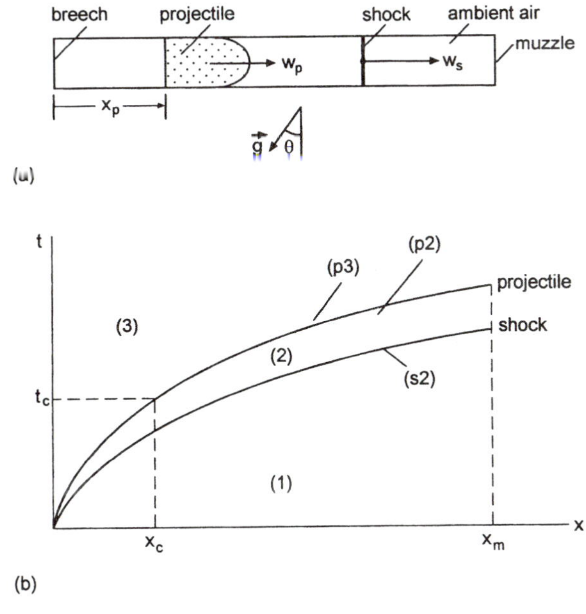

FIGURE 12.14 (a) Schematic of a gun barrel. (b) x,t diagram showing the labeling utilized.

indicated, we ignore the length of the breech chamber and that of the projectile. (The volume of air trapped in the breech chamber and the mass of the projectile, however, cannot be ignored.) At its location, the projectile has a speed w_p. For notational convenience, the prime notation, which we have sometimes used with an unsteady flow, is disregarded.

Nonplanar compression waves are generated by the accelerating projectile in the air ahead of it. These waves quickly become planar and coalesce into a normal shock. At early times, when w_p is small, the strength of this shock is quite negligible. In intermediate ballistics, this shock is important. In internal ballistics, however, it is usually overlooked or neglected (Farrar and Leeming, 1983). This effect, however, becomes significant when the muzzle speed of the projectile, w_{pm}, is large, where an m subscript denotes the muzzle. For instance, a large, but attainable, muzzle speed would be for w_{pm} to be six times the ambient speed of sound. (One way to increase the muzzle speed is to minimize the mass of the projectile.) Equation (12.9) then yields a shock Mach number of 7.34 and the pressure rise p_2/p_1 across the shock is 62.7. This estimate is for a conventional gun. There are hypervelocity launch devices, used in impact and penetration research, in which the strength of the normal shock can be a limiting factor for attaining a given muzzle speed. These projectile-in-a-tube devices go under various names, e.g., blast-wave accelerator, ram/scram accelerator, or a two-stage light gas gun (Wilson et al., 1996). For instance, with an 8 km/s muzzle speed, the shock Mach number is 28.8 and the p_2/p_1 ratio is now 970. When the ambient pressure is 1 atm, a retarding force occurs that is equivalent to about 10^3 atm pressure when the projectile is close to the muzzle. One approach for offsetting this effect is to use a diaphragm across the muzzle with the initial air pressure in the barrel reduced to a vacuum. Either the relatively weak shock or the pointed nose of the projectile can be used to rupture the diaphragm.

Figure 12.14(b) is an x,t diagram that schematically shows the trajectories for the shock and projectile. Combustion in the breech chamber is indicated as being completed when $t = t_c$ and $x_p = x_c$. Region 1 is quiescent, while $s2$ denotes the state just downstream of the shock. States $p2$ and $p3$ are respectively just ahead of and behind the projectile. Once the shock passes the muzzle, with the projectile still inside the barrel, a rarefaction wave starts to propagate down the barrel. This aspect is not indicated in the figure; it is part of intermediate ballistics. Almost immediately, however, the projectile overtakes this rarefaction wave.

In the Figure 12.14(a) sketch, the ambient air pressure is p_1. This pressure increases to p_{s2} just behind the shock and further slightly increases to p_{p2} at the nose of the projectile, where the flow has an unsteady stagnation point. The pressure at the base of the projectile is p_{p3}, which greatly exceeds p_{p2} at early and intermediate times. The breech (denoted with a b subscript) pressure is p_b, and it significantly exceeds p_{p3} at later times, as will be discussed. The ratio of specific heats for the combustion gases is denoted as γ_c and its value is usually near 1.2. Unsteady viscous boundary layers exist along the wall between the shock and projectile and between the projectile and the breech. These layers are typically thin compared to the diameter of the barrel, and they are neglected. Heat transfer to the walls, however, may represent a modest energy loss (Krier and Adams, 1979); we neglect this aspect. The combustion gas is modeled as thermally and calorically perfect, in line with the rest of this chapter. Neither assumption is entirely appropriate, since the hot gas has temperature-dependent specific heats, and the covolume should not be neglected in the thermal state equation, especially when p_b is near its maximum value.

Figure 12.15 shows normalized curves (loosely sketched from a figure in Krier and Adams, 1979) for a large-diameter (175 mm) military gun. The maximum breech pressure is about 3×10^3 atm, which determines — with a safety factor — the structural design of the breech mechanism and chamber. This pressure maximum typically occurs slightly before combustion of the powder terminates. By the time the projectile reaches the muzzle, the pressure in the breech p_{bm} is considerably reduced, with the pressure p_{p3m} on the base of the projectile further reduced from p_{bm}. The barrel length is about 8.7 m and the muzzle speed is about 10^3 m/s, or nearly three times the speed of sound in ambient air.

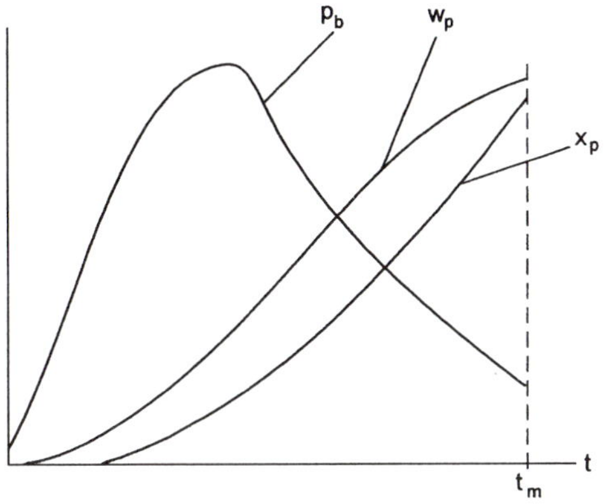

FIGURE 12.15 Schematic showing normalized values for the breech pressure and the position and speed of the projectile.

Simplified Model

Our goal is to formulate a computationally suitable form for Newton's second law for the projectile

$$m_p \frac{dw_p}{dt} = p_{p3}A - p_{p2}A - F - m_p g \sin\theta \qquad (12.143a)$$

where m_p, A, and F are the projectile mass, barrel cross-sectional area, and frictional force, respectively. We further simplify this relation by assuming $F = 0$ and $\theta = 0$:

$$\frac{m_p}{A}\frac{dw_p}{dt} = p_{p3} - p_{p2} \qquad (12.143b)$$

All that remains is to obtain p_{p3} and p_{p2} in terms of w_p, t, and known constants. First, several nondimensional variables and parameters are introduced:

$$X = \frac{x}{x_c}, \qquad \tau = \frac{t}{t_c}, \qquad W = \frac{w}{(\gamma_c R_c T_c)^{1/2}}$$

$$P = \frac{p}{\rho_g R_c T_c}, \qquad \Re = \frac{\rho}{\rho_g}, \qquad \hat{T} = \frac{T}{T_c}, \qquad \alpha_1 = \frac{A\rho_g t_c}{m_p}\left(\frac{R_c T_c}{\gamma_c}\right)^{1/2} \qquad (12.144)$$

where ρ_g is the density of the solid grain, and R_c and T_c are the gas constant and adiabatic flame temperature of the hot combustion gas. These three parameters are known constants. Newton's second law now has the form

$$\frac{dW_p}{d\tau} = \alpha_1(P_{p3} - P_{p2}) \qquad (12.143c)$$

The initial condition for this ODE is simply $W_p(0) = 0$.

Region 3

During combustion, the equation of state is

$$p_b = \rho_b R_c T_c \qquad (12.145)$$

for the air plus propellant gas in the breech chamber. Note that p_b is proportional to ρ_b, since $R_c T_c$ is a constant. Because of the very large mass disparity, the effect of the trapped air on R_c and T_c can be neglected. For purposes of simplicity, a linear burn rate for the powder is assumed:

$$\frac{m_g(\tau)}{m_g(0)} = 1 - \tau \tag{12.146}$$

where, for the initial powder mass,

$$m_g(0) = \rho_g \ell_g A \tag{12.147}$$

and $\ell_g A$ represents the grain volume as if it were a solid cylinder. Neither a linear burn rate nor a cylindrically shaped grain are realistic, but will suffice for this discussion. Actually, the length ℓ_g is only used as a way to represent the initial grain volume, not its actual configuration. The density of the gas in the breech then is

$$\rho_b(\tau) = \frac{m_g(0) - m_g(\tau)}{A\ell_b - \frac{m_g(\tau)}{\rho_g}} \tag{12.148}$$

where $A\ell_b$ is the initial volume of the trapped air plus the solid grain. With the foregoing, a relation for the gas density and pressure in the breech, during combustion, is obtained:

$$\Re_b = P_b = \frac{\tau}{\ell_1 - 1 + \tau} \tag{12.149}$$

where $\ell_1 = \ell_b/\ell_g$ and $\ell_1 \geq 1$. After combustion is completed, which occurs when $\tau = 1$, these relations require replacement. Hence, Equation (12.143) is solved in two stages, first when $0 \leq \tau \leq 1$ and then when $1 \leq \tau \leq \tau_m$. (Of course, if the barrel is short enough, there is only one stage and $\tau_m \leq 1$.) The maximum nondimensional pressure and density ($=1/\ell_1$) occurs when $\tau = 1$, and ℓ_1, which accounts for the comparable air and grain volumes, has a strong impact on the maximum breech pressure.

Region 3 is a nonsimple wave region. At early times, when the projectile is moving slowly, the expansion wave emanating from the base of the projectile has more than enough time to have numerous reflections between the base and breech endwall. These reflections smooth out the pressure distribution, and P_{p3} is effectively equal to P_b. However, as the projectile moves down the barrel, its speed increases and the generated expansion wave is more intense. Because the projectile's speed is increasing, the expansion resembles the one pictured in Figure 12.10(a), rather than the one in Figure 12.5. The expansion thus starts at the base of the projectile. Moreover, the gas in the vicinity of the base cools adiabatically with a significant reduction in the local speed of sound. In short, a large pressure gradient develops between the breech and the projectile. To model this flow, a considerable simplification is required if an iterative, complex numerical approach is to be avoided. Remember that the boundary location associated with the projectile base is unspecified, since it depends on w_p, which is unknown. Moreover, w_p also depends on the solution for the flow in region 2. To circumvent this complexity, an approximation is introduced that, at any given instant, the region between the breech and the projectile is a simple, centered rarefaction. This is a local, in time, approximation, whose justification would be that it yields appropriate physical trends. Local approximations, it should be noted, are common in fluid dynamics; an example is local similarity theory for a laminar boundary layer.

With this approach, we can write

$$W = \frac{W_p}{X_p} X \tag{12.150}$$

which directly stems from Equation (12.69), which is a linear w vs. x relation for a centered expansion. State 1 in Equation (12.69) is at the breech, where $w_b = 0$. The above equation is used when the grain is burning. The adiabatic breech temperature T_b is then fixed at T_c, while afterward T_b decreases with time. In other words, $\hat{T}_b = 1$ when $\tau \le 1$ and $\hat{T}_b < 1$ when $\tau > 1$. The speed and location of the projectile are related by

$$w_p = \frac{dx_p}{dt} \tag{12.151a}$$

or

$$\frac{dX_p}{d\tau} = \alpha_2 W_p \tag{12.151b}$$

with α_2 defined as

$$\alpha_2 = \frac{t_c}{x_c}(\gamma_c R_c T_c)^{1/2} \tag{12.152}$$

Equation (12.150) is used in conjunction with Equation (12.151b). In order for Equations (12.69) and (12.150) to be consistent, we require

$$\eta = \frac{\gamma_c + 1}{2}\hat{T}_b^{-1/2} W - 1 \tag{12.153a}$$

where a_b/a_{bc} equals $\hat{T}_b^{1/2}$, and

$$a_{bc} = (\gamma_c R_c T_c)^{1/2} \tag{12.154}$$

Consistency with the rarefaction wave solution of Section 12.5 also requires

$$\hat{T}_b^{-1/2}\frac{a}{a_{bc}} = 1 - \frac{\gamma_c - 1}{2}\hat{T}_b^{-1/2} W \tag{12.153b}$$

$$P = P_b\left(1 - \frac{\gamma_c - 1}{2}\hat{T}_b^{-1/2} W\right)^{2\gamma_c/(\gamma_c - 1)} \tag{12.153c}$$

$$\Re = \Re_b\left(1 - \frac{\gamma_c - 1}{2}\hat{T}_b^{-1/2} W\right)^{2/(\gamma_c - 1)} \tag{12.153d}$$

where P_b and $\Re_b$, which are time dependent, are given by Equation (12.149) when $\tau \leq 1$. The pressure on the base of the projectile thus becomes

$$P_{p3} = P_b\left(1 - \frac{\gamma_c - 1}{2}W_p\right)^{2\gamma_c/(\gamma_c - 1)} \tag{12.155}$$

when $\tau \leq 1$. This relation is directly used in Newton's second law when $\tau \leq 1$. We next determine its $\tau > 1$ counterpart.

Unknown quantities, such as P_b and $\Re_b$ when $\tau \geq 1$, are determined, in part, by conservation of mass at a given instant. (This approach is inappropriate when $\tau < 1$, since there is a source of gas in the breech chamber.) We start with

$$m_g(0) = A\int_0^{x_p} \rho dx \tag{12.156a}$$

which becomes

$$\ell_2 = \frac{\ell_g}{x_c} = \int_0^{Y_p} \Re dx \tag{12.156b}$$

where ℓ_2 is a positive constant. With the aid of Equations (12.150) and (12.153d), we have

$$\ell_2 = \Re_b\int_0^{X_p}\left(1 - \frac{\gamma_c - 1}{2}\hat{T}_b^{-1/2}\frac{W_p}{X_p}X\right)^{2/(\gamma_c - 1)} dX \tag{12.156c}$$

where $\Re_b$, $\hat{T}_b$, W_p, and X_p are only time dependent. The substitution

$$z = 1 - \frac{\gamma_c - 1}{2}\hat{T}_b^{-1/2}\frac{W_p}{X_p}X, \qquad dz = -\frac{\gamma_c - 1}{2}\hat{T}_b^{-1/2}\frac{W_p}{X_p}dX \tag{12.157}$$

results in

$$-\frac{\gamma_c - 1}{2}\ell_2\frac{W_p}{X_p} = \Re_b\hat{T}_b^{-1/2}\int_1^{z_p} z_p^{2/(\gamma_c - 1)} dz \tag{12.156d}$$

which integrates to

$$\frac{\gamma_c + 1}{2}\ell_2\frac{W_p}{X_p} = \Re_b\hat{T}_b^{1/2}\left[1 - \left(1 - \frac{\gamma_c - 1}{2}\hat{T}_b^{-1/2}W_p\right)^{(\gamma_c + 1)/(\gamma_c - 1)}\right] \tag{12.158}$$

Aside from X_p and W_p, $\Re_b$ and $\hat{T}_b$ are time-dependent parameters that require evaluation.

For this evaluation, the gas in the breech chamber is assumed to expand adiabatically:

$$\frac{p_b}{p_{bc}} = \left(\frac{\rho_b}{\rho_{bc}}\right)^{\gamma_c} = \left(\frac{T_b}{T_{bc}}\right)^{\gamma_c/(\gamma_c - 1)} \tag{12.159a}$$

when $\tau \geq 1$, where

$$p_{bc} = \frac{\rho_g}{\ell_1} R_c T_c, \qquad \rho_{bc} = \frac{\rho_g}{\ell_1}, \qquad T_{bc} = T_c$$

In nondimensional form, we have

$$\ell_1 P_b = (\ell_1 \mathfrak{R}_b)^{\gamma_c} = \hat{T}_b^{\gamma_c/(\gamma_c - 1)} \tag{12.159b}$$

The parameters $\mathfrak{R}_b$ and $\hat{T}_b$ are eliminated in favor of P_b in Equation (12.158), with the result

$$\frac{\gamma_c + 1}{2} \ell_2 \frac{W_p}{X_p} = \frac{P_b^{(\gamma_c + 1)/2\gamma_c}}{\ell_1^{(\gamma_c - 1)/2\gamma_c}} \left\{ 1 - \left[1 - \frac{\gamma_c - 1}{2} (\ell_1 P_b)^{-(\gamma_c - 1)/2\gamma_c} W_p \right]^{(\gamma_c + 1)/(\gamma_c - 1)} \right\} \tag{12.160a}$$

This is an implicit equation for P_b of the form

$$P_b = P_b(X_p, W_p;\ \gamma_c, \ell_1, \ell_2) \tag{12.160b}$$

where X_p and W_p are related by Equation (12.151b). This equation, which provides P_b, is used in conjunction with Equation (12.153c), when evaluated at the projectile:

$$P_{p3} = P_b \left[1 - \frac{\gamma_c - 1}{2} (\ell_1 P_b)^{-(\gamma_c - 1)/2\gamma_c} W_p \right]^{2\gamma_c/(\gamma_c - 1)} \tag{12.161}$$

This is the $\tau \geq 1$ counterpart of Equation (12.155), which is used in Newton's second law.

In view of the approximations already made, we may as well introduce one more approximation in order to simplify Equation (12.160a). We observe that

$$\frac{\gamma_c - 1}{2} (\ell_1 P_b)^{-(\gamma_c - 1)/2\gamma_c} W_p = \frac{\gamma_c - 1}{2} \hat{T}_b^{-1/2} W_p = \frac{\gamma_c - 1}{2} \frac{w_p}{(\gamma_c R_c T_b)^{1/2}} \tag{12.162}$$

is typically well below unity, even when the projectile is at the muzzle, where this term is a maximum. One reason for this is that γ_c is usually near 1.2 and, thus, $(\gamma_c - 1)/2 \cong 0.1$. Consequently, the square-bracketed term in Equation (12.160a) yields

$$\left[1 - \frac{\gamma_c - 1}{2} (\ell_1 P_b)^{-(\gamma_c - 1)/2\gamma_c} W_p \right]^{(\gamma_c + 1)/(\gamma_c - 1)} \cong 1 - \frac{\gamma_c + 1}{2} (\ell_1 P_b)^{-(\gamma_c - 1)/2\gamma_c} W_p \tag{12.163}$$

and this equation simplifies to

$$P_b \cong \left(\frac{\ell_2}{X_p} \right)^{\gamma_c} \ell_1^{\gamma_c - 1} \tag{12.164}$$

which can be used for P_b in Equation (12.161).

Region 2

We now turn our attention to region 2 and the shock wave denoted with an s subscript. We then have

$$w_s = \frac{dx_s}{dt}, \qquad M_s = \frac{w_s}{a_1}, \qquad \frac{p_{2s}}{p_1} = \frac{2}{\gamma+1}Y_s \tag{12.165a,b,c}$$

where γ is for air and equals 7/5. Region 2 is a compression, whose physical extent, $(x_s - x_p)/x_m$, is small. We thus assume, at any instant, a time dependent but approximately uniform pressure. This assumption readily yields

$$p_{2p} \cong p_{2s}, \qquad w_p \cong w_{2s} \tag{12.166}$$

where Equation (12.9) relates w_p and M_s. With this relation, P_{p2} becomes

$$P_{p2} = \alpha_3\left(1 + \frac{\gamma(\gamma+1)}{4}(\alpha_4 W_p)^2 + \gamma\alpha_4 W_p\left\{1 + \left[\left(\frac{\gamma+1}{4}\right)\alpha_4 W_p\right]^2\right\}^{1/2}\right) \tag{12.167}$$

where

$$\alpha_3 = \frac{p_1}{\rho_g R_c T_c}, \qquad \alpha_4 = \frac{(\gamma_c R_c T_c)^{1/2}}{a_1} \tag{12.168}$$

With the P_{p2} equation, the model is complete. In nondimensional terms, it depends on the four α_i and the two ℓ_i parameters. Problem 12.19 calls for a numerical solution.

12.8 NONSIMPLE WAVE REGION

Introductory Discussion

The focus in this section is to derive a closed-form solution for a nonsimple wave region when the gas is perfect and the flow is homentropic. Hence, steady flow behind a curved shock or an unsteady flow with a shock of variable strength is not considered. Here, the region of interest occurs when a rarefaction wave reflects from a solid wall. Figure 12.16 is an $\hat{x},\hat{t}$ diagram, where $\hat{x}$ and $\hat{t}$ are the dimensional position and time coordinates, that shows the nonsimple wave region, *abca*.

A general approach for obtaining an analytical solution of a single, second-order, hyperbolic PDE is pursued. Emphasis, however, is also placed on deriving an explicit solution for the reflection problem. General approaches typically assume known data on a single noncharacteristic curve. This is not the case for the problem at hand, which is why the general approach is not always utilized. Nevertheless, sufficient theory in combination with the nontrivial reflection problem should enable the reader to apply this method to similar problems. (A pertinent example would be flows governed by the linearized, supersonic, steady Euler equations.) The essential mathematical requirements for this transference is that the governing PDE be hyperbolic and linear.

The type of flow sketched in Figure 12.16 occurs in shock tube experiments. Region 1 in the figure is quiescent, region 2 is a uniform flow, while a centered, simple wave is situated between these two regions. As mentioned, the region of interest is *abca*, where *ab* is the wall, and *ac* and *bc* are, respectively, C_- and C_+ characteristics. The region above *bc* is also a simple, but noncentered, wave. The physical size is determined by the distance $\hat{\ell}$ between the diaphragm at the origin and the endwall. (The *PA* and *PB* lines in the figure are discussed later.) The best-known treatment for this flow is probably by Landau and Lifshitz (1987, Section 105). Their discussion, however,

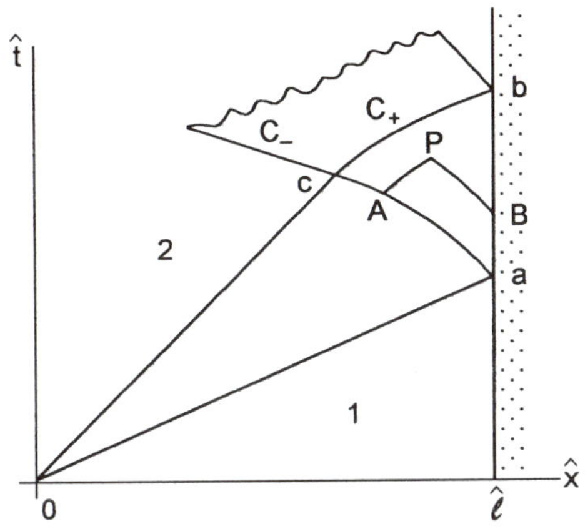

FIGURE 12.16 Nonsimple wave region caused by the reflection of a centered rarefaction wave from a wall.

is very concise, and the mathematical treatment, which differs from the one used here, is lacking in generality. They utilize a special method to obtain a solution that is limited to selected values for the ratio of specific heats.

A numerical solution for region *abca* can be obtained in a relatively straightforward manner using the method-of-characteristics. A central objective of this discussion, therefore, is to provide a systematic process for obtaining an analytical solution that does not lose sight of physical aspects of engineering interest. As such, the presentation can be viewed as a prototype for the analysis of other inviscid hyperbolic flows. Five steps are loosely followed:

i. Problem formulation
ii. Nondimensionalization
iii. Transformation of the PDE to a standard or canonical form
iv. Solution using the Riemann function method
v. Physical discussion and concluding remarks

The first step establishes an equation, or equations, and boundary and initial conditions. In terms of primitive variables, the governing equations are nonlinear, whereas the Riemann function method requires a linear equation. We therefore introduce a Legendre transformation that yields a single, second-order, linear PDE. In fluid dynamics, a linear PDE is the exception rather than the rule. Of course, a linear equation, when possible, is usually of major advantage. (Several reasons why this is *not* always the case will be discussed later.) Linear boundary conditions are established on *ac* and *ab*; they provide a unique solution for the *abca* region.

Coincident with establishing boundary conditions, the shape of the boundary must also be determined in both the original and transformed planes. Part of the formulation also establishes relations for quantities of engineering interest. These are the Mach number M, static pressure $\hat{p}$, and stagnation pressure $\hat{p}_0$. Other quantities are readily established once these three are known. As already evident, a caret denotes a dimensional quantity. Remember that stagnation quantities, such as $\hat{p}_0$, are not constant in an unsteady flow, even though the flow is homentropic.

Although relatively routine, the nondimensionalization step is important; it establishes a few key parameters and lends some generality to the analytical effort. This step is actually introduced as early as possible into the otherwise lengthy formulation.

The Riemann function method is based on a second-order PDE in a canonical form. Thus, a second transformation is introduced that performs this task with the linear equation aspect left unaltered. When the Riemann function method is appropriate, it yields a general solution in terms of quadratures.

The last step discusses several special cases and focuses on physical trends. These are established by examining the variation of M, $\hat{p}$, and $\hat{p}_0$ in *abca*. For expositional convenience, some of this material is interspersed in earlier subsections.

Formulation

As shown in Section 12.4, two first-order PDEs govern unsteady, one-dimensional flow. These equations are quasilinear and devoid of inhomogeneous terms. As a consequence, they are reducible (Courant and Friedrichs, 1948), which means they can be transformed into linear PDEs by means of an interchange of dependent and independent variables. At the end of Section 12.4, we demonstrated that the governing equations contain an inhomogeneous term when the flow possesses cylindrical or spherical symmetry. In this circumstance, the equations are irreducible, and an interchange of dependent and independent variables does not yield linear PDEs. (In fact, the transformed irreducible equations are quite nonlinear.) Moreover, the transformation does not require that the equations be hyperbolic; however, their hyperbolic nature is unchanged by the transformation. For the reducible equations of steady, two-dimensional flow, this interchange is referred to as a hodograph transformation (Chapter 7). As in Landau and Lifshitz (1987), upon which the next subsection is partly based, the interchange is expedited by utilizing a Legendre transformation (Emanuel, 1987). When completed, we shall end up with a single, second-order, linear PDE, which is the basis for the subsequent analysis.

Linear Equation

In place of a differential momentum equation, an unsteady Bernoulli equation is used:

$$\hat{\phi}_{\hat{t}} + \frac{1}{2}\hat{\phi}_{\hat{x}}^2 + \hat{h} = 0 \tag{12.169}$$

where $\hat{h}$ is enthalpy, a subscript denotes a partial derivative, $\hat{\phi}$ is a potential function

$$\hat{w} = \hat{\phi}_{\hat{x}} \tag{12.170}$$

and $\hat{w}$ is the flow speed. Any time-dependent function of integration in Bernoulli's equation is absorbed into the $\hat{\phi}_{\hat{t}}$ term. The flow is homentropic; i.e., the entropy $\hat{s}$ is a constant. Hence, we have from thermodynamics

$$\hat{p} = \hat{p}(\hat{\rho}) \tag{12.171}$$

$$\hat{h} = \int \frac{d\hat{p}}{\hat{\rho}} \tag{12.172}$$

for the pressure, density $\hat{\rho}$, and enthalpy. These relations replace the energy equation. Shortly, the continuity equation is utilized. First, however, a Legendre transformation is introduced.

We consider $\hat{\phi}$ as a function of $\hat{x}$ and $\hat{t}$:

$$\hat{\phi} = \hat{\phi}(\hat{x}, \hat{t}) \tag{12.173}$$

$$d\hat{\phi} = \hat{\phi}_{\hat{x}}\, d\hat{x} + \hat{\phi}_{\hat{t}} d\hat{t} = \hat{w} d\hat{x} - \left(\hat{h} + \frac{1}{2}\hat{w}^2\right) d\hat{t} \tag{12.174}$$

where Equations (12.169) and (12.170) are utilized. A new dependent variable $\hat{\psi}$ is introduced, which is not equivalent to a stream function, that depends on the dependent variables $\hat{w}$ and $\hat{h}$ instead of $\hat{x}$ and $\hat{t}$. This is done with

$$\hat{\psi}(\hat{w}, \hat{h}) = \hat{\phi}(\hat{x}, \hat{t}) - \hat{x}\hat{\phi}_{\hat{x}} - \hat{t}\hat{\phi}_{\hat{t}} = \hat{\phi} - \hat{x}\hat{w} + \hat{t}\left(\hat{h} + \frac{1}{2}\hat{w}^2\right) \tag{12.175}$$

where the leftmost equation is a Legendre transformation. This relation is differentiated and $d\hat{\phi}$ is eliminated, to yield

$$d\hat{\psi} = (\hat{t}\hat{w} - \hat{x})d\hat{w} + \hat{t}d\hat{h} \tag{12.176}$$

We also have

$$d\hat{\psi} = \hat{\psi}_{\hat{w}}d\hat{w} + \hat{\psi}_{\hat{h}}d\hat{h} \tag{12.177}$$

and by comparison

$$\hat{\psi}_{\hat{w}} = \hat{t}\hat{w} - \hat{x}, \qquad \hat{\psi}_{\hat{h}} = \hat{t} \tag{12.178a}$$

or

$$\hat{x} = \hat{w}\hat{\psi}_{\hat{h}} - \hat{\psi}_{\hat{w}}, \qquad \hat{t} = \hat{\psi}_{\hat{h}} \tag{12.178b}$$

Once $\hat{\psi}(\hat{w}, \hat{h})$ is known, these relations represent the transformation from $\hat{w}, \hat{h}$ coordinates to $\hat{x}, \hat{t}$ coordinates and vice versa.

To obtain an equation for $\hat{\psi}$, continuity is written as

$$\hat{\rho}_{\hat{t}} + \hat{w}\hat{\rho}_{\hat{x}} + \hat{\rho}\hat{w}_{\hat{x}} = 0 \tag{12.179a}$$

or in Jacobian form (Appendix B)

$$\frac{\partial(\hat{\rho}, \hat{x})}{\partial(\hat{t}, \hat{x})} - \hat{w}\frac{\partial(\hat{\rho}, \hat{t})}{\partial(\hat{t}, \hat{x})} - \hat{\rho}\frac{\partial(\hat{w}, \hat{t})}{\partial(\hat{t}, \hat{x})} = 0 \tag{12.179b}$$

Multiplication by the Jacobian

$$\hat{J} = \frac{\partial(\hat{t}, \hat{x})}{\partial(\hat{w}, \hat{h})} \tag{12.180}$$

of the transformation yields

$$\frac{\partial(\hat{\rho}, \hat{x})}{\partial(\hat{w}, \hat{h})} - \hat{w}\frac{\partial(\hat{\rho}, \hat{t})}{\partial(\hat{w}, \hat{h})} - \hat{\rho}\frac{\partial(\hat{w}, \hat{t})}{\partial(\hat{w}, \hat{h})} = 0 \tag{12.181}$$

We now see why the equations must be reducible. For a cylindrical or spherically symmetric flow, Equations (12.179) contain an inhomogeneous term, as noted at the end of Section 12.4. Multiplication by $\hat{J}$ then results in a complicated nonlinear equation.

In view of the $\hat{p}$ and $\hat{h}$ equations, we have

$$d\hat{h} = \frac{d\hat{p}}{\hat{\rho}}, \qquad \hat{p} = \hat{p}(\hat{h}) \tag{12.182}$$

and

$$\hat{\rho}_{\hat{h}} = \frac{d\hat{\rho}}{d\hat{h}} = \frac{d\hat{\rho}}{d\hat{p}}\frac{d\hat{p}}{d\hat{h}} = \frac{\hat{\rho}}{\left(\frac{d\hat{p}}{d\hat{\rho}}\right)_{\hat{s}}} = \frac{\hat{\rho}}{\hat{a}^2} \tag{12.183}$$

where $\hat{a}$ is the speed of sound and $\hat{\rho}_{\hat{w}} = 0$. Hence, the three determinants in Equation (12.181) become

$$\frac{\partial(\hat{\rho}, \hat{x})}{\partial(\hat{w}, \hat{h})} = \begin{vmatrix} \hat{\rho}_{\hat{w}} & \hat{\rho}_{\hat{h}} \\ \hat{x}_{\hat{w}} & \hat{x}_{\hat{h}} \end{vmatrix} = \begin{vmatrix} 0 & \frac{\hat{\rho}}{\hat{a}^2} \\ \hat{\psi}_{\hat{h}} + \hat{w}\hat{\psi}_{\hat{w}\hat{h}} - \hat{\psi}_{\hat{w}\hat{w}} & \hat{w}\hat{\psi}_{\hat{h}\hat{h}} - \hat{\psi}_{\hat{w}\hat{h}} \end{vmatrix} = -\frac{\hat{\rho}}{\hat{a}^2}(\hat{\psi}_{\hat{h}} + \hat{w}\hat{\psi}_{\hat{w}\hat{h}} - \hat{\psi}_{\hat{w}\hat{w}}) \tag{12.184a}$$

$$\frac{\partial(\hat{\rho}, \hat{t})}{\partial(\hat{w}, \hat{h})} = \begin{vmatrix} 0 & \frac{\hat{\rho}}{\hat{a}^2} \\ \hat{\psi}_{\hat{w}\hat{h}} & \hat{\psi}_{\hat{h}\hat{h}} \end{vmatrix} = -\frac{\hat{\rho}}{\hat{a}^2}\hat{\psi}_{\hat{w}\hat{h}} \tag{12.184b}$$

$$\frac{\partial(\hat{w}, \hat{t})}{\partial(\hat{w}, \hat{h})} = \begin{vmatrix} 1 & 0 \\ \hat{\psi}_{\hat{w}\hat{h}} & \hat{\psi}_{\hat{h}\hat{h}} \end{vmatrix} = \hat{\psi}_{\hat{h}\hat{h}} \tag{12.184c}$$

With these relations, continuity is given by

$$\hat{a}^2\hat{\psi}_{\hat{h}\hat{h}} - \hat{\psi}_{\hat{w}\hat{w}} + \hat{\psi}_{\hat{h}} = 0 \tag{12.185}$$

where $\hat{a} = \hat{a}(\hat{h})$. This is the previously mentioned linear equation that governs unsteady, one-dimensional flow. Moreover, it is clearly hyperbolic.

Nondimensionalization

A particularly convenient nondimensionalization is introduced as

$$X = \frac{\hat{x}}{\hat{\ell}}, \qquad \tau = \frac{\hat{a}_1\hat{t}}{\hat{\ell}}, \qquad \psi = \frac{1}{\hat{\ell}\hat{a}_1}\hat{\psi} \tag{12.186a}$$

$$w = \frac{\hat{w}}{\hat{a}_1}, \qquad h = \frac{\hat{h}}{\hat{a}_1^2}, \qquad a = \frac{\hat{a}}{\hat{a}_1}, \qquad p = \frac{\hat{p}}{\hat{p}_1}, \qquad \rho = \frac{\hat{\rho}}{\hat{\rho}_1} \tag{12.186b}$$

where conditions in the quiescent gas region provide the reference state. (Note that $\hat{h}_1$ does not equal $\hat{a}_1^2$.) This normalization leaves Equation (12.185) invariant; i.e., we simply delete the caret symbols. Similarly, the Mach number is invariant:

$$M = -\frac{\hat{w}}{\hat{a}} = -\frac{w}{a} \tag{12.187}$$

A minus sign is used because w is zero or negative for the flow of interest, and the Mach number is defined as nonnegative. This nondimensionalization will be extended whenever new variables or parameters are introduced. Quantities without a caret hereafter are nondimensional.

Perfect Gas Equation

For a simple thermodynamic fluid, the speed of sound can be considered a function of h and s. Since entropy is a constant, we have $a = a(h)$, as above. Any thermal state equation, such as a van der Waals equation, along with a specific heat equation, can be utilized to obtain $a(h)$. The simplest relevant choice, of course, is that of a perfect gas. The rest of the analysis is based on this choice; i.e.,

$$a^2 = (\gamma - 1)h \tag{12.188}$$

and Equation (12.185) becomes

$$(\gamma - 1)h\psi_{hh} - \psi_{ww} + \psi_w = 0 \tag{12.189}$$

Boundary Conditions

In line with characteristic theory (Courant and Friedrichs, 1948), a single condition on ab and on ac is sufficient to establish a unique solution for ψ in $abca$. On ab, we specify

$$w = 0 \tag{12.190a}$$

In view of Equation (12.178b), this becomes

$$\psi_w = -1 \tag{12.190b}$$

The solution for a centered rarefaction wave is required for the condition on ac, which also requires the shape of this boundary. From Section 12.6, this solution can be written as

$$w = -\frac{2}{\gamma + 1}(1 - \tilde{\eta}) \tag{12.191a}$$

$$a = \frac{2}{\gamma + 1}\left(1 + \frac{\gamma - 1}{2}\tilde{\eta}\right) \tag{12.191b}$$

$$h = \frac{4}{(\gamma - 1)(\gamma + 1)^2}\left(1 + \frac{\gamma - 1}{2}\tilde{\eta}\right)^2 \tag{12.191c}$$

$$M = -\frac{w}{a} = \frac{1 - \tilde{\eta}}{1 + \frac{\gamma - 1}{2}\tilde{\eta}} \tag{12.191d}$$

where the similarity variable is

$$\tilde{\eta} = \frac{\hat{x}}{\hat{a}_1 \hat{t}} = \frac{X}{\tau} \tag{12.192}$$

This solution holds on the ac characteristic.

Hereafter, it will sometimes be notationally convenient to utilize h instead of a and to replace γ with a new parameter n, where

$$n = \frac{1}{2}\frac{3-\gamma}{\gamma-1}, \qquad \gamma = \frac{2n+3}{2n+1} \tag{12.193}$$

Note that for $n = 0, 1, 2, \ldots$, we have $\gamma = 3, 5/3, 7/5, \ldots$. It is also convenient at this time to introduce the bc C_+ characteristic. Along the ac and bc characteristics, the compatibility equations

$$\xi = \frac{\hat{\xi}}{\hat{a}_1} = w + \frac{2}{\gamma-1}a = w + [2(2n+1)h]^{1/2}, \quad C_+ \tag{12.194a}$$

$$\eta = \frac{\hat{\eta}}{\hat{a}_1} = w - [2(2n+1)h]^{1/2}, \quad C_- \tag{12.194b}$$

hold, where ξ and η are the respective nondimensional Riemann invariants, and η should not be confused with the similarity variable $\tilde{\eta}$. For ac, η is evaluated at point a as

$$\eta_a = -(2n+1) \tag{12.195}$$

since the nondimensional enthalpy in the quiescent gas region is $h_1 = (\gamma-1)^{-1} = (2n+1)/2$. We thus have

$$w - [2(2n+1)h]^{1/2} = -(2n+1) \tag{12.196}$$

along ac. Shortly, the bc characteristic is dealt with.

We next, however, determine the value of ψ on ac and the shape of this characteristic. For the first item, we substitute Equations (12.191) into Equation (12.176), with the result

$$\begin{aligned} d\psi &= \left[-\frac{2\tau}{\gamma+1}(1-\tilde{\eta}) - X\right]\frac{2}{\gamma+1}d\tilde{\eta} + \frac{8\tau}{(\gamma-1)(\gamma+1)^2}\left(1+\frac{\gamma-1}{2}\tilde{\eta}\right)\frac{\gamma-1}{2}d\tilde{\eta} \\ &= \frac{2\tau}{\gamma+1}\left[-\frac{2}{\gamma+1}(1-\tilde{\eta}) - \tilde{\eta} + \frac{2}{\gamma+1}\left(1+\frac{\gamma-1}{2}\tilde{\eta}\right)\right]d\tilde{\eta} = 0 \end{aligned} \tag{12.197}$$

The solution ψ is continuous, including at point a, where the two boundary segments join. On ab, however, the Equation (12.190b) boundary condition is a derivative condition. Hence, we can set

$$\psi = 0 \tag{12.198}$$

on ac.

For the shape of *ac*, we start with the characteristic equation

$$\frac{dX}{d\tau} = w - a \tag{12.199a}$$

With the aid of Equations (12.191), this becomes

$$\frac{dX}{d\tau} = -\frac{4}{\gamma+1} + \frac{3-\gamma}{\gamma+1}\tilde{\eta} \tag{12.199b}$$

while from Equation (12.192)

$$\frac{dX}{d\tau} = \tilde{\eta} + \tau\frac{d\tilde{\eta}}{d\tau} \tag{12.200}$$

Elimination of $dX/d\tau$ yields

$$\frac{d\tau}{\tau} = -\frac{1}{2}\frac{\gamma+1}{\gamma-1}\frac{d\left(1+\frac{\gamma-1}{2}\tilde{\eta}\right)}{\left(1+\frac{\gamma-1}{2}\tilde{\eta}\right)} \tag{12.201}$$

which integrates to

$$\tau = \left(\frac{\frac{\gamma+1}{2}}{1+\frac{\gamma-1}{2}\tilde{\eta}}\right)^{(\gamma+1)/[2(\gamma-1)]} = \left(\frac{\frac{\gamma+1}{2}}{1+\frac{\gamma-1}{2}\frac{X}{\tau}}\right)^{(\gamma+1)/[2(\gamma-1)]}$$

A more convenient form is

$$X = 2(n+1)\tau^{n/(n+1)} - (2n+1)\tau \tag{12.202}$$

which represents the shape of the *ac* characteristic.

The location of point *c* is determined by the intersection of *ac* with the straight *oc* characteristic given by

$$X = \eta_c\tau \tag{12.203}$$

where $\tilde{\eta}_c$ stems from Equation (12.191d) with $M = M_2$; i.e.,

$$\tilde{\eta}_c = \frac{1-M_2}{1+\frac{\gamma-1}{2}M_2} = \frac{1-M_2}{1+\frac{M_2}{2n+1}} \tag{12.204}$$

From the above, we obtain

$$\tau_c = \left(1+\frac{M_2}{2n+1}\right)^{n+1}, \qquad X_c = (1-M_2)\left(1+\frac{M_2}{2n+1}\right)^{n} \tag{12.205a,b}$$

Observe that point c has $X_c = 1$ when $M_2 = 0$ (i.e., there is no rarefaction wave) and $X_c = 0$ when $M_2 = 1$. Thus, M_2 determines the strength of the centered rarefaction wave and the relative size of the *abca* region. For simplicity, we assume $X_c \geq 1$ by requiring $0 \leq M_2 \leq 1$. When X_c is negative, the interaction lasts for an infinite length of time (Landau and Lifshitz, 1987), since point b is now at infinity. Moreover, M_2 is supersonic and the portion of the *abca* region to the left of the τ-axis is also supersonic.

With Equation (12.204), we obtain from Equations (12.191a,c)

$$w_c = -\frac{M_2}{1 + \frac{M_2}{2n+1}}, \qquad h_c = \frac{2n+1}{2\left(1 + \frac{M_2}{2n+1}\right)^2} \tag{12.206}$$

which is the w,h counterpart of Equations (12.205). This yields for the *bc* Riemann invariant

$$\xi_c = (2n+1)\frac{1 - \frac{M_2}{2n+1}}{1 + \frac{M_2}{2n+1}} \tag{12.207}$$

and the *bc* characteristic is

$$w + [2(2n+1)h]^{1/2} = \xi_c \tag{12.208}$$

The shape of this characteristic is obtained by integrating

$$\frac{dX}{d\tau} = w + \left(\frac{2h}{2n+1}\right)^{1/2} \tag{12.209}$$

from point c to the wall, where $X = 1$ and $w = 0$. Either w or h can be eliminated by utilizing Equation (12.208), but a numerical integration is still required, except when $n = 0$. (This integration is further discussed near the end of this section.) Although there is no boundary condition along *bc* for the *abca* region, the subsequent physical discussion nevertheless utilizes this material, in part because *bc* is a boundary for the reflected wave.

By way of summary, the governing equation is written as

$$\frac{2h}{2n+1}\psi_{hh} - \psi_{ww} + \psi_h = 0 \tag{12.210}$$

The solution should satisfy Equation (12.190b) on $w = 0$ and Equation (12.198) on *ac*, which is given by Equation (12.196) or by Equation (12.202). In the nondimensional problem, the only prescribed parameters are n, or γ, and M_2.

Solution in the Physical Plane

Suppose $\psi(w, h)$ is a solution of Equation (12.210). Equations (12.178), in nondimensional form, provide the transformation to the physical X, τ plane. The Jacobian of the transformation, given by Equation (12.180), can be written as

$$J = \psi_{ww}\psi_{hh} - \psi_{wh}^2 - \psi_h\psi_{hh} \tag{12.211a}$$

With the aid of Equation (12.189), this becomes

$$J = a^2\psi_{hh}^2 - \psi_{wh}^2 \tag{12.211b}$$

In a simple wave region, $w = w(h)$ and Equation (12.180) directly yields $J = 0$. Thus, a uniform flow and a simple wave region are not encompassed by Equation (12.210); they are missing solutions. The analogy with a two-dimensional flow under the hodograph transformation is evident.

There are two reasons a linear equation, such as the one above, may not be advantageous. Missing solutions is the first. The second is the expected presence of a limit line when the flow is supersonic. This is a curve, not a region, along which the Jacobian of the transformation is zero (or infinite). Its presence is a serious mathematical difficulty with no physical counterpart, such as a shock wave. Limit lines are usually discussed in terms of the hodograph transformation for a steady, supersonic, two-dimensional flow, as in Chapter 7. We anticipate that a limit line will also occur in a supersonic, unsteady flow when the governing equations are linearized by a Legendre transformation. To our knowledge, this has not been demonstrated. Moreover, it will not be considered here, since M_2 is restricted to subsonic values.

In terms of w and h, the Mach number and pressure are given by

$$M = -w\left(\frac{2n+1}{2h}\right)^{1/2}, \qquad p = \left(\frac{2h}{2n+1}\right)^{(2n+3)/2} \tag{12.212}$$

The stagnation pressure is written as

$$p_0 = \frac{\hat{p}_0}{\hat{p}_1} = p\left(1+\frac{\gamma-1}{2}M^2\right)^{\gamma/(\gamma-1)} = \left(\frac{w^2+2h}{2n+1}\right)^{(2n+3)/2} \tag{12.213}$$

These quantities are functions of X and τ in *abca* through Equations (12.178). As mentioned, the subsequent physical discussion will focus on them.

Analysis

It is advisable to discuss several topics of a general or preliminary nature before embarking on the Riemann function method of solution. When appropriate, this practice will continue with later subsections. We first discuss canonical forms and subsequently the solution of Equation (12.210) when $n = 0$.

Canonical Forms

A linear or quasilinear PDE of the form (Sommerfeld, 1949)

$$Au_{xx} + 2Bu_{xy} + Cu_{yy} = \Phi(x,y,u,u_x,u_y) \tag{12.214}$$

is considered, where A, B, and C can depend only on x and y. This equation is hyperbolic providing

$$B^2 - AC > 0$$

which we presume. It is instructive to obtain the standard, or canonical, form for this relation. In the case of a second-order hyperbolic equation, there are actually two such forms (Courant and Hilbert, 1962). Both are obtained by a transformation of the independent variables written as

$$\xi = \tilde{\phi}(x, y), \qquad \eta = \tilde{\psi}(x, y) \tag{12.215}$$

where ξ and η are new coordinates and not (yet) the Riemann invariants. The transformation requires the chain rule derivatives

$$u_x = u_\xi \tilde{\phi}_x + u_\eta \tilde{\psi}_x \tag{12.216a}$$

$$u_y = u_\xi \tilde{\phi}_y + u_\eta \tilde{\psi}_y \tag{12.216b}$$

$$u_{xx} = u_{\xi\xi}\tilde{\phi}_x^2 + 2u_{\xi\eta}\tilde{\phi}_x\tilde{\psi}_x + u_{\eta\eta}\tilde{\psi}_x^2 + u_\xi\tilde{\phi}_{xx} + u_\eta\tilde{\psi}_{xx} \tag{12.216c}$$

with similar relations for u_{xy} and u_{yy}. We thus obtain

$$\begin{aligned}(A\tilde{\phi}_x^2 + 2B\tilde{\phi}_x\tilde{\phi}_y + C\tilde{\phi}_y^2)u_{\xi\xi} + 2[A\tilde{\phi}_x\tilde{\psi}_x + B(\tilde{\phi}_x\tilde{\psi}_y + \tilde{\phi}_y\tilde{\psi}_x) + C\tilde{\phi}_y\tilde{\psi}_y]u_{\xi\eta}\\ + (A\tilde{\psi}_x^2 + 2B\tilde{\psi}_x\tilde{\psi}_y + C\tilde{\psi}_y^2)u_{\eta\eta} + (A\tilde{\phi}_{xx} + 2B\tilde{\phi}_{xy} + C\tilde{\phi}_{yy})u_\xi\\ + (A\tilde{\psi}_{xx} + 2B\tilde{\psi}_{xy} + C\tilde{\psi}_{yy})u_\eta = \Phi\end{aligned} \tag{12.217}$$

in place of Equation (12.214). By setting the coefficients of $u_{\xi\xi}$ and $u_{\eta\eta}$ equal to zero, specific relations for the transformation and the first canonical form

$$u_{\xi\eta} = \frac{1}{2}\frac{\Phi - (A\tilde{\phi}_{xx} + 2B\tilde{\phi}_{xy} + C\tilde{\phi}_{yy})u_\xi - (A\tilde{\psi}_{xx} + 2B\tilde{\psi}_{xy} + C\tilde{\psi}_{yy})u_\eta}{A\tilde{\phi}_x\tilde{\psi}_x + B(\tilde{\phi}_x\tilde{\psi}_y + \tilde{\phi}_y\tilde{\psi}_x) + C\tilde{\phi}_y\tilde{\psi}_y} \tag{12.218}$$

are obtained.

By comparing Equations (12.210) and (12.214), we have

$$x \to h, \quad y \to w, \quad u \to \psi \tag{12.219a}$$

$$A = \frac{2h}{2n+1}, \quad B = 0, \quad C = -1, \quad \Phi = -\psi_h \tag{12.219b}$$

The coordinate transformation turns out to be

$$\xi = \tilde{\phi} = w + [2(2n+1)h]^{1/2}, \qquad \eta = \hat{\psi} = w - [2(2n+1)h]^{1/2} \tag{12.220a,b}$$

Thus, the ξ, η coordinates are just the Riemann invariants. In fact, the canonical form given by Equation (12.218) automatically utilizes the Riemann invariants as coordinates. Hence, any equation with the form

$$u_{\xi\eta} = f(\xi, \eta, u, u_\xi, u_\eta)$$

is in its hyperbolic canonical form with ξ and η as characteristic coordinates. Equation (12.210) reduces to

$$\psi_{\xi\eta} = \frac{n}{\eta - \xi}(\psi_\eta - \psi_\xi) \tag{12.221}$$

which illustrates the occasional advantage of n over γ as a parameter.

The second canonical form utilizes $u_{\xi^*\xi^*} - u_{\eta^*\eta^*}$ for the highest-order derivatives. An asterisk is used to distinguish these (noncharacteristic) coordinates from the preceding ones. For Equation (12.210), the transformation is given by

$$\xi^* = [2(2n+1)h]^{1/2}, \quad \eta^* = w \tag{12.222}$$

and the resulting PDE is

$$\psi_{\xi^*\xi^*} - \psi_{\eta^*\eta^*} = -\frac{2n}{\xi^*}\psi_{\xi^*} \tag{12.223}$$

When $n = 0$, the general solution is evident as

$$\psi_0 = f(\eta^* - \xi^*) + g(\eta^* + \xi^*) \tag{12.224a}$$

$$= f(\eta) + g(\xi) \tag{12.224b}$$

where f and g are arbitrary functions, and the zero subscript denotes the $n = 0$ case. (These functions are determined by boundary and initial conditions.) This result is easily obtained from Equation (12.221) or (12.223), but not from Equation (12.210). The solution of Equation (12.223) was first investigated by Riemann (see Copson, 1957), and it is his method of solution we follow. A simpler exposition, however, is obtained by utilizing Equation (12.221) in place of Equation (12.223).

Solution When $\gamma = 3$ or $n = 0$

For an ordinary gas, $5/3 \geq \gamma$ and $\gamma = 3$ is unrealistic. The high-density gas behind a detonation wave propagating through a condensed phase explosive, such as TNT, in fact, does have a large γ value near 3. This gas, however, is neither thermally nor calorically perfect. Nevertheless, this case is discussed in some detail because it is mathematically much simpler than any other choice for γ. It also demonstrates a quite unexpected feature of characteristic theory.

By choosing

$$f = -\eta - 1, \quad g = 0 \tag{12.225}$$

in Equation (12.224b), we obtain the $n = 0$ result

$$\psi_0 = -\eta - 1 = -w + (2h)^{1/2} - 1 \tag{12.226}$$

A systematic approach for determining ψ for arbitrary n is described later in this section.

The coordinate transformation, Equation (12.178b), can be written as

$$h = \frac{1}{2\tau^2}, \quad w = -\frac{1-X}{\tau} \tag{12.227}$$

Consequently, M, p, and p_0 are easily written in terms of X and τ, rather than h and w. We thus obtain

$$M = 1 - X, \quad p = \frac{1}{\tau^3}, \quad p_0 = \frac{(X^2 - 2X + 2)^{3/2}}{\tau^3} \tag{12.228a,b,c}$$

for the $n = 0$ solution in the *abca* region. Recall that M is constant on the straight characteristics with a positive or negative slope in the two simple wave regions that adjoin *abca*. Inside *abca*, however, M is constant along straight, vertical, noncharacteristic lines. Moreover, M is only a function of X, p of τ, and p_0 of both X and τ. We thus have a uniform pressure at any one instant, and we note that p and p_0 coincide at the wall where $M = 0$.

Deeper insight into the above trends can be obtained by examining the characteristic curves inside *abca*. On a general basis, these are given by

$$\frac{d\hat{x}}{d\hat{t}} = \hat{w} \pm \hat{a} \tag{12.229a}$$

where the minus (plus) sign is associated with the C_- (C_+) family. Nondimensionally, we have

$$\frac{dX}{d\tau} = w \pm \left(\frac{2h}{2n+1}\right)^{1/2} = w \pm (2h)^{1/2} \tag{12.229b}$$

where the rightmost result is for $n = 0$. With the aid of Equation (12.227), this becomes

$$\frac{dX}{d\tau} = \frac{X - 1 \pm 1}{\tau} \tag{12.230}$$

Generally, the ODEs for the characteristic curves in a nonsimple wave region cannot be analytically integrated, since their solution is coupled to the compatibility equations. Moreover, the right-hand side of the ODEs is such that the characteristics, in a nonsimple wave region, are curved. Because of its simplicity, however, this equation readily yields

$$X = c_-\tau + 2, \qquad X = c_+\tau \tag{12.231a,b}$$

where the $c_\pm$ integration constants are associated with the characteristics. Observe that Equation (12.202), which is the equation for a C_- characteristic, is consistent with Equation (12.231a) if $c_- = -1$.

Both families, given by Equations (12.231), are straight lines inside *abca*. Furthermore, the C_- family that crosses the centered simple wave region, situated between regions 1 and 2, also consists of straight characteristics. Actually, the C_+ family in this simple wave region continues unchanged into the *abca* region, as is evident from Equation (12.231b). Similarly, the reflected, straight C_- characteristics inside *abca* continue unaltered in shape and slope when they enter the adjoining simple wave region.

According to characteristic theory (Courant and Friedrichs, 1948), when both families are straight, we should have a uniform flow region. This conclusion, however, is not consistent with Equations (12.227) or (12.228a,b). Moreover, the characteristics that cross a simple wave region are supposed to be curved.

The $n = 0$ case clearly violates these well-established concepts. (They are not violated when $n \neq 0$.) This case is degenerate, caused by the simplicity of Equation (12.230), which admits straight-line solutions. In this situation, the $C_\pm$ families inside *abca* consist of straight but nonparallel characteristics. Similarly, the two simple wave regions consist of straight $C_\pm$ characteristics in which one family has parallel lines and the other does not. In this case, the characterization of uniform, simple, and nonsimple wave regions depends on whether or not the members of a given family are parallel.

In a simple wave region, a dependent variable, such as M, would depend on only one Riemann invariant, such as η. Although ψ_0 only depends on η, region *abca* is nevertheless nonsimple. This is also evident from Equations (12.228), since $1 - X$ or τ, when written in terms of the Riemann invariants, requires both η and ξ.

Transformed Boundary Conditions

As noted, the Riemann function method is somewhat simpler with the $u_{\xi\eta}$ canonical form; hence, we focus on Equation (12.221). With the Riemann invariants as coordinates, the transformed boundary conditions for ψ are determined as well as the location of the boundary of the *abca* region in the ξ, η plane. Since the overall transformation process goes from X, τ to w, a and then to ξ, η, it is convenient to first transform the *abca* region to the w, a coordinate system. One could also use w, h coordinates, but the first choice is simpler. It is important, however, not to confuse the speed of sound with point *a*.

From Equations (12.191) and (12.192), the *ac* characteristic is given by

$$w = -\frac{2n+1}{2(n+1)}\left(1 - \frac{X}{\tau}\right), \qquad a = \frac{2n+1}{2(n+1)}\left(1 + \frac{1}{2n+1}\frac{X}{\tau}\right) \tag{12.232}$$

By eliminating X/τ, the linear relation

$$a = 1 + \frac{w}{2n+1} \tag{12.233}$$

is obtained for *ac*. It is also useful to obtain point *c*. With Equation (12.206), we have

$$a_c = \frac{1}{1 + \frac{M_2}{2n+1}}, \qquad w_c = -\frac{M_2}{1 + \frac{M_2}{2n+1}} \tag{12.234}$$

which implies $0 \le -w_c \le a_c$ when $0 \le M_2 \le 1$.

Figure 12.17 is a sketch of the *abca* region, where *ab* represents the $w = 0$ condition. Point *b* is to the left of *a*, since $a_a > a_b > 0$. As indicated by Equation (12.233), the *ac* characteristic line is straight. (The shape of the *bc* characteristic is also straight. In fact, the triangle is isosceles with point *c* as the apex.)

The transformation to characteristic coordinates is based on Equations (12.194). Thus, the $w = 0$ condition becomes $\eta = -\xi$. The *ac* characteristic becomes

$$\xi = 2n + 1 + 2w, \qquad \eta = -(2n+1) \tag{12.235}$$

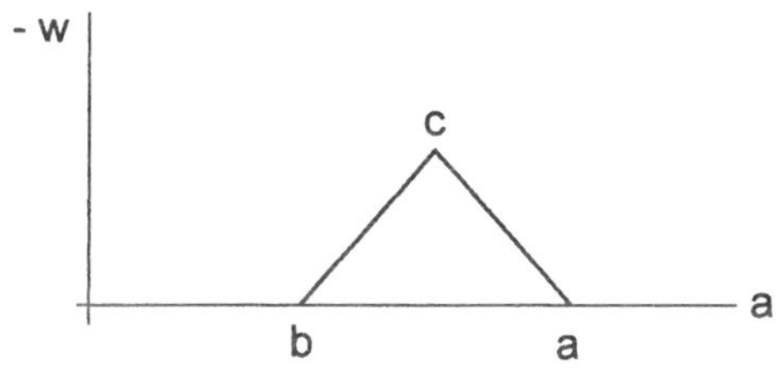

FIGURE 12.17 Nonsimple wave region when the coordinates are $-w$ and a.

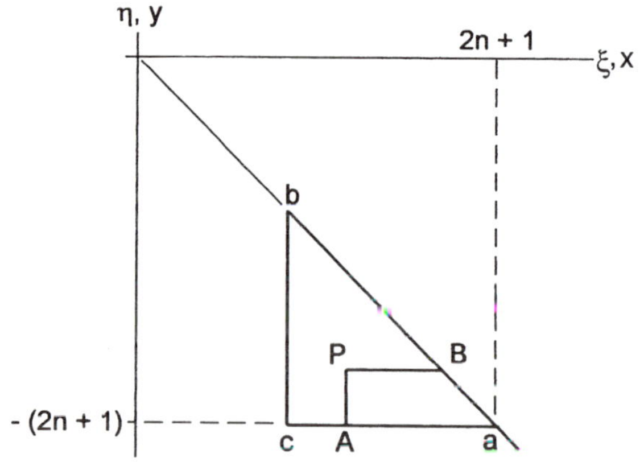

FIGURE 12.18 Nonsimple wave region using characteristic coordinates.

where $w_c \leq w \leq 0$, ξ_c is given by Equation (12.207), while ξ_a is

$$\xi_a = 2n + 1 \tag{12.236}$$

Hence, we have $\xi_a \geq \xi_c > 0$. The *ac* and *bc* characteristics transform into η and ξ constant value lines, respectively. We thus obtain the isosceles triangular region for *abca* that is sketched in Figure 12.18.

We still have $\psi = 0$ on *ac*. The *ab* boundary condition, Equation (12.190b), becomes

$$\psi_w = \xi_w \psi_\xi + \eta_w \psi_\eta = -1$$

or

$$\psi_\xi + \psi_\eta = -1 \tag{12.237}$$

since ξ_w and η_w are unity on *ab* (see Equations (12.220)).

Definition of the Riemann Function

As in the canonical form subsection, it is useful to first discuss the Riemann function (Courant and Hilbert, 1962) on a more general basis. We start with a linear, hyperbolic PDE in the canonical form

$$L(u) = u_{xy} + au_x + bu_y + cu = f \tag{12.238}$$

where *a*, *b*, *c*, and *f* are functions of the *x* and *y* characteristic coordinates. The adjoint L^* of the *L* operator is defined as

$$L^*(v) = v_{xy} - (av)_y - (bv)_y + cv \tag{12.239}$$

The important role of L^* will become apparent shortly.

Let the coordinates of point *P* in Figure 12.18 be denoted as ξ and η. The straight *PA* and *PB* lines are characteristics that intersect the *ac* and *ab* data lines. Figure 12.16 also shows these two characteristic lines, where they are curved. The solution at P thus depends only on the data along the aA and aB segments.

For $v(x,y)$, we choose the Riemann function $R(x,y;\ \xi,\eta)$, where x and y represent an arbi-trary point inside the region bounded by $AaBPA$, and ξ and η parametrically denote the fixed boundary point P. The Riemann function is fully specified by three conditions. These conditions yield a solution for $u(x, y)$ that is given in terms of one or more quadratures. Without these condi-tions, an integral equation for u would be obtained. The first condition is that R is a solution of the adjoint equation

$$L^*(R) = 0 \qquad \textbf{(12.240)}$$

Next, we require that

$$R_x = bR \quad \text{on} \quad y = \eta \qquad \textbf{(12.241a)}$$

$$R_y = aR \quad \text{on} \quad x = \xi \qquad \textbf{(12.241b)}$$

and finally that

$$R(\xi,\eta;\xi,\eta) = 1 \qquad \textbf{(12.242)}$$

Equations (12.241) are ODEs that integrate to

$$R(x,\eta;\xi,\eta) = \exp\left[\int_\xi^x b(\lambda,\ \eta)d\lambda\right], \qquad R(\xi,y;\xi,\eta) = \exp\left[\int_\eta^y a(\xi,\lambda)d\lambda\right] \qquad \textbf{(12.243a,b)}$$

where Equation (12.242) is utilized. We thus need to find a solution of Equation (12.240) that satis-fies Equations (12.243). This solution requires the a, b, c coefficients of Equation (12.238), but not f or the specific boundary (or initial) conditions for u. Because R does not depend on the bound-ary conditions for u, it represents the Riemann function for *any* unsteady, one-dimensional, homen-tropic flow. It should represent a simpler problem than the one for u, although still not trivial. In fact, the difficulty in finding R has been a major drawback for this method (Copson, 1957). It is worth noting that Equation (12.240) differs from $L(u) = 0$, unless the L operator is self-adjoint. As will become evident in the next subsection, this is not the case here.

The Riemann Function for One-Dimensional, Unsteady Flow

For the problem of interest, we use Equation (12.221); hence, the coefficients in Equation (12.238) become

$$a = \frac{n}{y-x}, \qquad b = -\frac{n}{y-x}, \qquad c = 0, \quad f = 0 \qquad \textbf{(12.244)}$$

Consequently, the adjoint equation is given by

$$L^*(R) = R_{xy} - n\left(\frac{R}{y-x}\right)_x + n\left(\frac{R}{y-x}\right)_y = R_{xy} + \frac{n}{y-x}(R_y - R_x) - \frac{2n}{(y-x)^2}R = 0 \qquad \textbf{(12.245)}$$

and Equations (12.243) reduce to

$$R(x, \eta; \xi, \eta) = \left(\frac{\eta - x}{\eta - \xi}\right)^n \quad \text{on} \quad y = \eta \tag{12.246a}$$

$$R(\xi, y; \xi, \eta) = \left(\frac{y - \xi}{\eta - \xi}\right)^n \quad \text{on} \quad x = \xi \tag{12.246b}$$

We need to find a Riemann function that satisfies these equations. In accord with Courant and Hilbert (1962), we attempt to reduce Equation (12.245) to an ODE whose solution is consistent with Equations (12.246). We try the form

$$R(x, y; \xi, \eta) = \left(\frac{y - x}{\eta - \xi}\right)^n F(z) \tag{12.247}$$

where

$$z = \frac{(x - \xi)(y - \eta)}{(\eta - \xi)(y - x)} \tag{12.248}$$

and

$$F(0) = 1 \tag{12.249}$$

Equation (12.247) thus satisfies both Equations (12.242) and (12.246). Its substitution into Equation (12.245) requires the following derivatives:

$$z_x = \frac{(y - \eta)(y - \xi)}{(\eta - \xi)(y - x)^2} \tag{12.250a}$$

$$z_y = \frac{(\eta - x)(x - \xi)}{(\eta - \xi)(y - x)^2} \tag{12.250b}$$

$$z_{xy} = \frac{1}{(\eta - \xi)(y - x)^3}[(y - \xi)(y - x) + (y - \eta)(y - x) - 2(y - \eta)(y - \xi)] \tag{12.250c}$$

and

$$R_x = \left(\frac{y - x}{\eta - \xi}\right)^n \left(z_x F' - \frac{n}{y - x} F\right) \tag{12.251a}$$

$$R_y = \left(\frac{y - x}{\eta - \xi}\right)^n \left(z_y F' + \frac{n}{y - x} F\right) \tag{12.251b}$$

$$R_{xy} = \left(\frac{y - x}{\eta - \xi}\right)^n \left[z_x z_y F'' + \left(z_{xy} + \frac{n}{y - x} z_x - \frac{n}{y - x} z_y\right) F' + \frac{n - n^2}{(y - x)^2} F\right] \tag{12.251c}$$

where a prime denotes differentiation with respect to z. We thus obtain

$$z(1-z)F' + (1-2z)F' + n(n-1)F = 0 \tag{12.252}$$

As with the hodograph transformation, an ODE is obtained whose solution is the hypergeometric function [see Equation (7.48)].

The general form (*Handbook of Mathematical Functions*, 1964; Spanier and Oldham, 1987) of the hypergeometric function, $F(a, b; c; z) = F(b, a; c; z)$, satisfies

$$z(1-z)F'' + [c-(a+b-1)z]F' - abF = 0 \tag{12.253}$$

where the a, b, and c constants should not be confused with earlier definitions. By comparing the above equations, we obtain

$$a+b = 1, \quad ab = -n(n-1), \quad c = 1 \tag{12.254a}$$

or

$$a = n, \quad b = 1-n, \quad c = 1 \tag{12.254b}$$

Thus, the Riemann function is given by

$$R_n = \left(\frac{y-x}{\eta-\xi}\right)^n F(n, 1-n;1;z) \tag{12.255}$$

where the n subscript on R is notationally convenient, and z is provided by Equation (12.248). This result is not restricted to the reflection problem, but holds for *any* unsteady, one-dimensional, homentropic flow. For arbitrary n, F is given by

$$F(n, 1-n;1;z) = \frac{1}{\Gamma(n)\Gamma(1-n)}\sum_{j=0}^{\infty}\Gamma(n+j)\Gamma(1-n+j)\frac{z^j}{(j!)^2} \tag{12.256a}$$

$$P_{-n}(1-2z) \tag{12.256b}$$

where Γ is the gamma function and P_{-n} is the Legendre function of the first kind. For nonintegral n values, P_{-n}, or F, is given by the infinite series. Note that when $z = 0$, we have

$$P_{-n}(1) = 1 \tag{12.257a}$$

and that

$$P_0(1-2z) = 1 \tag{12.257b}$$

when $n = 0$.

When n is a positive integer, F can be written in terms of a Legendre polynomial with a finite number of terms:

$$F(n, 1-n;1;z) = P_{n-1}(1-2z), \quad n = 1, 2, 3, \ldots \tag{12.256c}$$

rather than a Legendre function, which is an infinite series. We thus obtain

$$R_0(x,y;\xi,\eta) = 1 \tag{12.258a}$$

$$R_1(x,y;\xi,\eta) = \frac{y-x}{\eta-\xi} \tag{12.258b}$$

$$R_2(x,y;\xi,\eta) = \left(\frac{y-x}{\eta-\xi}\right)^2\left[1-2\frac{(x-\xi)(y-\eta)}{(\eta-\xi)(y-x)}\right] \tag{12.258c}$$

$$\vdots$$

for zero and integer n values. Noninteger n values require the infinite series Legendre function. Thus, the Riemann function is significantly simpler when $n = 0, 1, 2, \ldots$. As observed earlier, these n values respectively correspond to $\gamma = 3, 5/3, 7/5, \ldots$.

Stokes' Theorem

A particular form of Stokes' theorem is required in the next subsection. We start with a general form of the theorem, written as (see Appendix A)

$$\oint_C \vec{A}\cdot d\vec{r} = \int_S (\nabla\times\vec{A})\cdot\hat{n}ds \tag{12.259}$$

where $\vec{A}$ is an arbitrary vector, $\vec{r}$ is the position vector, S is a contiguous surface area, C is the simple, closed curve that borders S, and $\hat{n}$ is a unit normal vector to S in a right-handed sense relative to C. A Cartesian coordinate system is used in x, y, z space with $\hat{\imath}, \hat{\jmath}, \hat{k}$ the corresponding orthonormal basis. The S surface is taken to be in the $z = 0$ plane, and the corresponding A component, A^z, is zero. We now have

$$\vec{r} = x\hat{\imath} + y\hat{\jmath} + z\hat{k} \tag{12.260a}$$

$$\hat{n} = \hat{k} \tag{12.260b}$$

$$\vec{A} = A^x\hat{\imath} + A^y\hat{\jmath} \tag{12.260c}$$

$$\vec{A}\cdot d\vec{r} = A^x dx + A^y dy \tag{12.260d}$$

$$\nabla\times\vec{A} = \left(\frac{\partial A^y}{\partial x} - \frac{\partial A^x}{\partial y}\right)\hat{k} \tag{12.260e}$$

$$(\nabla\times\vec{A})\cdot\hat{n} = \frac{\partial A^y}{\partial x} - \frac{\partial A^x}{\partial y} \tag{12.260f}$$

and Equation (12.259) becomes

$$\oint_C (A^x dx + A^y dy) = \int_S \left(\frac{\partial A^y}{\partial x} - \frac{\partial A^x}{\partial y}\right)dxdy \tag{12.261}$$

in our formulation. Since $\vec{A}$ is arbitrary, so are its components A^x and A^y.

The Riemann Function Method

As previously mentioned, the standard Riemann method (Sommerfeld, 1949; Courant and Hilbert, 1962) is developed for problems where the data are specified along a single, noncharacteristic curve. The reflection problem has noncharacteristic data on *ab* but also has data on a second line segment, *ac*, which is a characteristic. For this reason, as well as for pedogogical purposes, the method is developed from scratch.

From Equations (12.238) and (12.239), we form

$$vL(u) - uL^*(v) = (vu_x + buv)_y - (uv_y - auv)_x \qquad \textbf{(12.262)}$$

This relation is integrated over S, with the result

$$-\int_S [vL(u) - uL^*(v)]dxdy = \int_S [(uv_y - auv)_x - (uv_x + buv)_y]dxdy \qquad \textbf{(12.263)}$$

Next, without loss of generality, we can set

$$A^x = vu_x + buv, \qquad A^y = uv_y - auv \qquad \textbf{(12.264)}$$

and use Stokes' theorem for the right side of Equation (12.263) to obtain

$$-\int_S [vL(u) - uL^*(v)]dxdy = \oint_C [(vu_x + buv)dx + (uv_y - auv)dy] \qquad \textbf{(12.265)}$$

This key result is not restricted to a particular Riemann function or to fluid dynamics. Rather, it represents a general solution to Equation (12.238).

Equation (12.265) is systematically applied to the reflection problem by setting

$$u \to \psi, \quad v \to R$$

and using Equations (12.244). The curve C is the $PAaBP$ closed curve in Figure 12.18, S is the enclosed region, and the n subscript on ψ and R is temporarily suppressed. Since

$$L(\psi) = 0, \quad L^*(R) = 0$$

the left side of Equation (12.265) is zero, and we have the line integral

$$\oint_C \left[R\left(\psi_x - \frac{n}{y-x}\psi\right)dx + \psi\left(R_y - \frac{n}{y-x}R\right)dy\right] = 0 \qquad \textbf{(12.266)}$$

where C is taken counterclockwise.

On various segments of C, the evaluation is assisted by an integration by parts. For instance, suppose we wish to evaluate the leftmost term on a segment where $dy = 0$. The simplest way to perform the integration by parts is to add and subtract the integral

$$\int_\Gamma \psi\left(R_x + \frac{n}{y-x}R\right)dx$$

to the one in Equation (12.266), where Γ represents the $dy = 0$ segment on C. For cancellation purposes, the R term has an opposite sign from the one in Equation (12.266). We thus obtain

$$\int_\Gamma R\left(\psi_x - \frac{n}{y-x}\psi\right)dx + \int_\Gamma \psi\left(R_x - \frac{n}{y-x}R\right)dx - \int_\Gamma \psi\left(R_x + \frac{n}{y-x}R\right)dx$$
$$= \int_\Gamma \frac{\partial(R\psi)}{\partial x}dx - \int_\Gamma \psi\left(R_x + \frac{n}{y-x}R\right)dx$$

or

$$\int_\Gamma R\left(\psi_x - \frac{n}{y-x}\psi\right)dx = R(\Gamma_2)\psi(\Gamma_2) - R(\Gamma_1)\psi(\Gamma_1) - \int_\Gamma \psi\left(R_x + \frac{n}{y-x}R\right)dx \tag{12.267a}$$

where Γ_2 and Γ_1 represent the end points on the Γ segment. Similarly, we have for the rightmost integral in Equation (12.266) on a $dx = 0$ segment

$$\int_\Gamma \psi\left(R_y - \frac{n}{y-x}R\right)dy = R(\Gamma_2)\psi(\Gamma_2) - R(\Gamma_1)\psi(\Gamma_1) - \int_\Gamma R\left(\psi_y + \frac{n}{y-x}\psi\right)dy \tag{12.267b}$$

If, for instance, ψ is known along Γ, then Equation (12.267a) is used.

We return to Equation (12.266) and first evaluate the BP contribution, where $dy = 0$:

$$\int_{BP} = \int_{BP} R\left(\psi_x - \frac{n}{y-x}\psi\right)dx \tag{12.268a}$$

Neither ψ nor ψ_x are known on BP; nevertheless, an integration by parts yields

$$\int_{BP} = R(P)\psi(P) - R(B)\psi(B) - \int_{BP} \psi\left(R_x - \frac{n}{y-x}R\right)dx \tag{12.268b}$$

where P represents the point $x = \xi$, $y = \eta$, and B represents the point $x = -\eta$, $y = \eta$. From Equations (12.241a) and (12.242), we have

$$R_x = -\frac{n}{y-x}R \quad \text{on} \quad y = \eta \quad \text{or} \quad BP \tag{12.269a}$$

$$R(P) = 1 \tag{12.269b}$$

which yields

$$\int_{BP} = \psi(\xi, \eta) - R(B)\psi(B) \tag{12.269c}$$

Along the PA segment, Equation (12.266) becomes

$$\int_{PA} = \int_{PA} \psi\left(R_y - \frac{n}{y-x}R\right)dy = 0 \tag{12.270}$$

since $dx = 0$, and by Equation (12.241b) the factor containing R_y and R is zero. For the Aa segment, we have

$$\int_{Aa} = \int_{Aa} R\left(\psi_x - \frac{n}{y-x}\psi\right)dx = \int_{Aa} R\,\psi_x dx = 0 \tag{12.271}$$

since $\psi = 0$ on ac, and hence $\psi_x = 0$ on ac. (This can be verified by integrating the $R\psi_x$ integral by parts.)

Along aB, $x = -y$, $dx = -dy$, and Equation (12.266) reduces to

$$\int_{aB} = \int_{aB} (-R\psi_x + \psi R_y)dy \tag{12.272a}$$

In the integrand, we subtract and add $R\psi_y$, with the result

$$\int_{aB} = \int_{aB} (-R\psi_x - R\psi_y + R\psi_y + R_y\psi)dy = -\int_{aB} R(\psi_x + \psi_y)dy + \int_{aB} \frac{\partial(R\psi)}{\partial y}dy \tag{12.272b}$$

Since $x = -y$, the rightmost integral is an ordinary line integral; hence,

$$\int_{aB} \frac{\partial(R\psi)}{\partial y}dy = R(B)\psi(B) - R(a)\psi(a)$$

With $\psi(a) = 0$ and $\psi_x + \psi_y = -1$ on aB, we obtain

$$\int_{aB} = R(B)\psi(B) + \int_{aB} R dy \tag{12.273}$$

By summing the above, Equation (12.266) becomes

$$\psi(\xi,\eta) = -\int_{y_a}^{y_B} R dy = -\int_{-(2n+1)}^{\eta} R(-y, y; \xi, \eta)dy \tag{12.274}$$

where y_a and y_B equal $-(2n+1)$ and η, respectively. The Riemann function in the integrand is given by Equation (12.255) in the form

$$R(-y, y; \xi, \eta) = \left(\frac{2y}{\eta - \xi}\right)^n F(n, 1-n; 1; z) \tag{12.275a}$$

where

$$z = -\frac{(y+\xi)(y-\eta)}{2(\eta-\xi)y} \tag{12.275b}$$

Thus, the solution of the reflection problem is relatively simple when $n = 0, 1, 2, \ldots$. For a noninteger n value, it is given by an integral over the Riemann function. In this case, the hypergeometric function in R is given by the infinite series.

When an infinite series is integrated, the order of summation and integration can be interchanged if the series converges uniformly (Courant, 1937). Moreover, the resulting series, obtained by term-by-term integration, also converges uniformly. If the infinite series in F, associated with a noninteger n and the above z, converges uniformly, then Equation (12.274) can be written as

$$\psi = -\left(\frac{2}{\eta-\xi}\right)^n \sum_{j=0}^{\infty}\left[-\frac{1}{2(\eta-\xi)}\right]^j \frac{\Gamma(n+j)\Gamma(1-n+j)}{\Gamma(n)\Gamma(1-n)(j!)^2}\int_{-(2n+1)}^{\eta}\left[\frac{(y+\xi)(y-\eta)}{y}\right]^j y^n dy \qquad \textbf{(12.276)}$$

where the integral is not singular, since η, in the upper limit, is negative.

Solution When $n = 0$

With Equation (12.258a), Equation (12.274) readily yields

$$\psi_0 = -\eta - 1$$

which agrees with Equation (12.226). The *abca* region is bordered by straight line segments, and the solution inside the region is given by Equations (12.228).

Solution when $n = 1$

Equation (12.258b) is utilized for R_1. From Equation (12.274), we obtain

$$\psi_1 = -\frac{2}{\eta-\xi}\int_{-3}^{\eta} y dy = \frac{9-\eta^2}{\eta-\xi} \qquad \textbf{(12.277a)}$$

With the aid of Equations (12.194), this becomes

$$\psi_1 = \frac{[w-(6h)^{1/2}]^2 - 9}{2(6h)^{1/2}} \qquad \textbf{(12.277b)}$$

Equations (12.178b), in nondimensional form, yield

$$\tau = \frac{3(9-w^2+6h)}{2(6h)^{3/2}}, \quad X = \frac{1}{2(6h)^{3/2}}[-3w^3 + 27w + w(6h) + 2(6h)^{3/2}] \qquad \textbf{(12.278a,b)}$$

for the w, h to X, τ transformation.

The *ac* characteristic is given by Equations (12.196) and (12.202) as

$$w - (6h)^{1/2} = -3, \quad X = 4\tau^{1/2} - 3\tau \qquad \textbf{(12.279a,b)}$$

and point c is given by Equations (12.205) as

$$\tau_c = \left(1 + \frac{M_2}{3}\right)^2, \quad X_c = (1 - M_2)\left(1 + \frac{M_2}{3}\right) \qquad \textbf{(12.280a,b)}$$

Since

$$M = -\frac{3w}{(6h)^{1/2}} \tag{12.281}$$

one can show that M, p, and p_0 are given by

$$M = \frac{3(\tau - X)}{3\tau + X}, \quad p = \left(1 + \frac{M}{3}\right)^{-5}, \quad p_0 = p\left(1 + \frac{M_2}{3}\right)^{5/2} \tag{12.282a,b,c}$$

along ac. Note that $M = 0$ and $p = p_0 = 1$ at point a, where $X = \tau = 1$.

The bc characteristic is provided by Equations (12.207) to (12.209). In particular, Equation (12.209) can be put in the form of an M,X ODE:

$$\frac{dM}{dX} = -\frac{M}{(1-X)^2 - \left(\frac{M}{3}\right)^4}\left[\frac{2^{1/2}}{9}\frac{3+M_2}{3-M_2}\frac{M^{3/2}(3-M)}{1-M}\left(1 - X - \frac{1}{6}M + \frac{1}{18}M^3\right)^{1/2} + (1 - X) + \left(\frac{M}{3}\right)^3\right] \tag{12.283}$$

which is integrated from point c to the wall, where $X = 1$ and $M = 0$. This ODE, however, is singular at the wall, where the right side of the equation is indeterminant. The nature of the singularity is further discussed in Problem 12.21.

From Equations (12.212) and (12.213), we have Equation (12.281) and

$$p = \frac{(6h)^{5/2}}{3^5}, \quad p_0 = \frac{(3w^2 + 6h)^{5/2}}{3^5} \tag{12.284}$$

for any point within $abca$ or on its border. The M, p, and p_0 quantities are given in terms of X and τ by means of Equations (12.278). Figures 12.19 and 12.20 show* constant value lines for M, p, and p_0 when $M_2 = 0.5$. Observe in Figure 12.19 that the lines for M are not vertical, ex-cept for $M = 0$, but slope to the left. The slope [illegible] decreases in magnitude as M increases. The [illegible] characteristics are slightly concave upward, except possibly for a small segment along bc near the wall.

The value for p_b is established in Problem 12.21 as 0.1862. Along the wall, $p_0 = p$, and both equal unity at point a. In Figure 12.20, lines for constant p and p_0 values are shown, where their wall values are given by $p = p_0 = 1 + \alpha(p_b - 1)$, with $\alpha = 0.2, 0.4, 0.6, 0.8$. The difference between the p and p_0 curves is not great, since $M < 0.5$. The isobar curves are very nearly horizontal, as in the $n = 0$ case. Consequently, an approximate solution, when M_2 is not too large, could be based on

$$p = p_1(\tau) + p_2(\tau, X)$$

where p_2 would represent a small perturbation.

---

* Figures 12.19 and 12.20 and the corresponding calculations were expertly done by Dr. M. Malik.

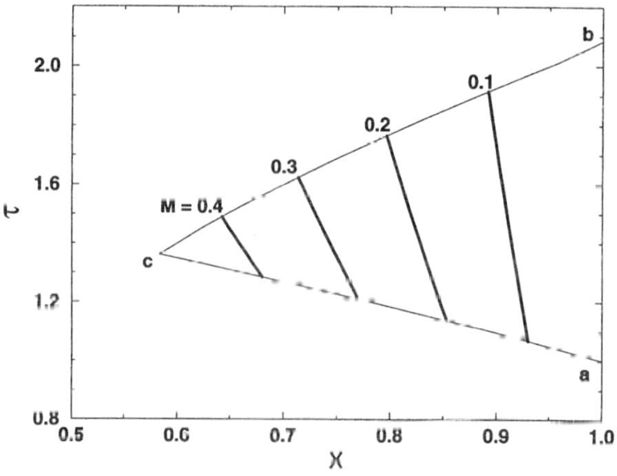

FIGURE 12.19 Constant Mach number lines in the nonsimple wave region when $M_2 = 0.5$ and $n = 1$ or $\gamma = 5/3$.

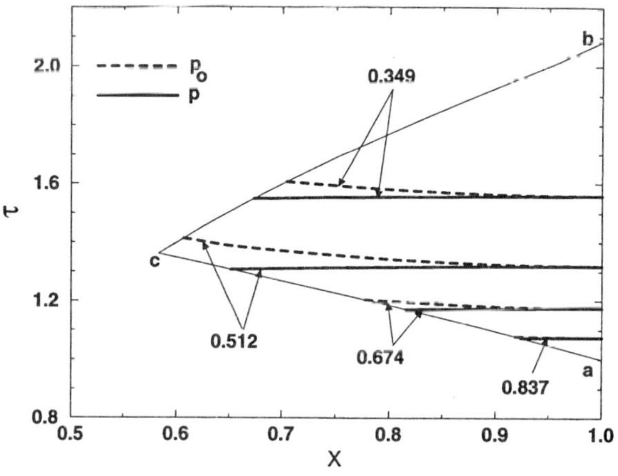

FIGURE 12.20 Constant pressure and stagnation pressure curves in the nonsimple wave region when $M_2 = 0.5$ and $n = 1$ or $\gamma = 5/3$.

Comparison with the Landau and Lifshitz Solution

Landau and Lifshitz give the solution to Equation (12.210), in dimensional form, as

$$\hat{\psi}_n = \left(\frac{\partial}{\partial \hat{h}}\right)^n [f(\hat{\eta}) + g(\hat{\xi})], \quad n = 0, 1, 2, \ldots \tag{12.285}$$

where f and g are arbitrary functions. For the reflection problem, $g \equiv 0$ and we can nondimensionally write

$$\psi_n = \frac{1}{2^{(2n+1)/2}(2n+1)^{(2n-1)/2} n!} \times \left(\frac{\partial}{\partial \hat{h}}\right)^{n-1} \left[\frac{1}{h^{1/2}}\left(\{w - [2(2n+1)h]^{1/2}\}^2 - (2n+1)^2\right)^n\right],$$

$$n = 1, 2, 3, \ldots \tag{12.286}$$

which does not hold for $n = 0$. With $n = 1$, this readily reduces to Equation (12.277b). Similarly, this reduces to the value obtained in Problem 12.22(a), which uses Equation (12.274). Problem 12.22 fully develops the equations for the $n = 2$ ($\gamma = 1.4$) case.

REFERENCES

Abramowitz, M. and Stegun, I.A., eds., *Handbook of Mathematical Functions,* NBS Applied Mathematics Series 55, 1964, Chapter 15.

Azoury, P.H., *Engineering Applications of Unsteady Fluid Flow,* John Wiley, New York, 1992.

Copson, E.T., "On the Riemann–Green Function," *Arch. Rat. Mech. Anal.* 1, 324 (1957–58).

Courant, R., *Differential and Integral Calculus,* Vol. I, Interscience Pub., New York, 1937, p. 540.

Courant, R. and Friedrichs, K.O., *Supersonic Flow and Shock Waves,* Interscience Pub., New York, 1948.

Courant, R. and Hilbert, D., *Methods of Mathematical Physics,* Vol. II, John Wiley, New York, 1962.

Emanuel, G., "Potential Applications of Piston Generated Unsteady Expansion Waves," *AIAA J.* 19, 1015 (1981).

Emanuel, G., *Advanced Classical Thermodynamics,* AIAA Education Series, Washington, D.C., 1987.

Emanuel, A. and Yi, T.H., "Unsteady Obligue Shock Waves," *Shock Waves* 10, 113 (2000).

Farrar, C.L. and Leeming, D.W., *Military Ballistics — A Basic Manual,* Brassey's Pub., Oxford, 1983.

Freedman, E., "Thermodynamic Properties of Military Gun Propellants," in *Gun Propulsion Technology,* edited by L. Stiefel, Progress in Astronautics and Aeronautics, 109, 103 (1988).

Friedrichs, K.O., "Formation and Decay of Shock Waves," *Comm. Appl. Math.* 1, 211 (1948).

Glass, I.I., Kalra, S.P., and Sislian, J.P., "Condensation Water Vapor in Rarefaction Waves III. Experimental Results," *AIAA J.* 15, 686 (1977).

Hurewicz, W., *Lectures on Ordinary Differential Equations,* John Wiley, New York, 1950, p. 75.

Jiang, Z., Takayama, K., and Skews, B.W., "Numerical Study on Blast Flowfields by Supersonic Projectiles Discharged from Shock Tubes," *Phys. Fluids* 10, 277 (1998).

Karamcheti, K., *Principles of Ideal-Fluid Aerodynamics,* R.E. Krieger Pub. Co., Malabar, FL, 1980, p. 158.

Kentfield, J.A.C., *Nonsteady, One Dimensional, Internal, Compressible Flows,* Oxford University Press, New York, 1993.

Krier, H. and Adams, M.J., "An Introduction to Gun Interior Ballistics and a Simplified Ballistic Code," in *Interior Ballistics of Guns,* edited by H. Krier and M. Summerfeld, Progress in Astronautics and Aeronautics, Vol. 66, 1 (1979).

Landau, L.D. and Lifshitz, E.M., *Fluid Mechanics,* 2nd ed., Pergamon Press, New York, 1987.

Ludford, G.S.S. and Martin, M.H., "One-Dimensional Anisentropic Flows," *Comm. Pure Appl. Math.* VII, 45 (1954).

Merlen, A. and Dyment, A., "Similarity and Asymptotic Analysis for Gun-Firing Aerodynamics," *J. Fluid Mech.* 225, 497 (1991).

Neely, A.J. and Morgan, R.G., "The Superorbital Expansion Tube Concept, Experiment and Analysis," *Aeronautical J.* 98, 97 (1994).

Rudinger, G., *Nonsteady Duct Flow: Wave Diagram Analysis,* Dover, New York, 1969.

Sharma, V.D., Ram, R., and Sachdev, P.L., "Uniformly Valid Analytical Solution to the Problem of a Decaying Shock Wave," *J. Fluid Mech.* 185, 153 (1987).

Sommerfeld, A., *Partial Differential Equations in Physics,* Academic Press, New York, 1949.

Spanier, J. and Oldham, K.B., *An Atlas of Functions,* Hemisphere Pub. Corp., New York, 1987.

Steketee, J.A., "An Expansion Wave in the Non-Homentropic Flows of Martin and Ludford," *Quart. of Appl. Math.* 30, 167 (1972).

Weber, H.E., *Shock Wave Engine Design,* John Wiley, New York, 1995.

Wilson, D., Tan, Z., and Varghese, P.L., "Numerical Simulation of the Blast-Wave Accelerator," *AIAA J.* 34, 1341 (1996).

PROBLEMS

12.1 Let the Mach number behind a moving normal shock be small compared to unity; i.e., set $M_2' = \varepsilon$. Derive perturbation formulas for w_p'/a, M_s, p_2'/p_1', and $(s_2' - s_1')/R$. It is necessary to retain terms to $O(\varepsilon^3)$ for the entropy change. Finally, eliminate ε to obtain

$(s_2' - s_1')/R$ in terms of $(p_2' - p_1')/p_1'$. (The pressure and entropy results hold for both fixed and moving normal shocks.)

12.2 Ground-based measurements indicated wind speeds as high as 600 mph occurred as a result of the Mount St. Helens 1980 volcanic eruption. Estimate the Mach number M_s for a normal incident shock that might have caused this wind speed.

12.3 Derive a formula for the work $\hat{w}$, per unit mass of shocked gas, done by a piston moving into a quiescent gas at constant speed. Your result for $\hat{w}$ should only depend on a_1, γ, and M_s. Compare your result with the change in internal energy, $\Delta e = e_2 - e_1$, and explain the difference. (Note that the system is adiabatic; hence, the first law of thermodynamics is $\hat{w} = \Delta e$.)

12.4 Prove the statement made below Equation (12.30) that $M_r \leq M_s$.

12.5 Derive an equation for $s_3 - s_1)/R$ that only depends on γ and M_s for the flow in Figure 12.3.

12.6 Determine equations for M_r, p_3/p_1, T_3/T_1, and $s_3 - s_1)/R$ in terms of M_s when $\gamma = 1$.

12.7 Argon gas is initially at 300 K. On the downstream side of the reflected shock, the temperature is 1500 K. Determine the incident w_s' and reflected w_r' shock speeds.

12.8 Start with Equations (12.46) and derive Equations (12.48) and (12.51).

12.9 **(a)** Determine equations, one for each of the three regions, for the particle path shown in Figure 12.5. The fluid particle initially is at x_1, and the three algebraic equations should be simplified and put in the form

$$\frac{x_p}{x_1} = f\left(\frac{x_1}{a_1 t}, \gamma, \eta_{TE}\right)$$

where x_p is the position of the particle.

(b) Determine an algebraic equation for the curved portion of the C_+ characteristic in Figure 12.5 that has the form

$$\frac{x}{a_1 t} = f\left(\frac{x_1}{a_1 t}, \gamma\right)$$

(c) Is the slope of a particle path continuous when it crosses the *LE* and *TE*? Is the slope of a C_+ characteristic continuous when it crosses the *LE* and *TE*?

12.10 For a rarefaction wave whose piston trajectory is given by Equation (12.82a), determine

$$\frac{T_0}{T_1} = f(\xi_p, \gamma)$$

for the flow inside the wave. Show that the stagnation temperature T_0 has an extremum on the characteristic where $M = 1$.

12.11 For the centered rarefaction wave sketched in Figure 12.5, determine $D(p/p_1)/Dt$ and $D(p_0/p_1)/Dt$ in terms of γ and η, and, if necessary, $(1/t)$.

12.12 For the flow in Figure 12.5, determine the piston speed $(w_p/a_1)_m$ that minimizes p_{02}/p_{01}. At this condition, what is M_2? If region 1 consists of air at 300 K, what is $(w_p)_m$? What is the piston speed if the trailing edge coincides with the piston path?

12.13 **(a)** Obtain $p(x, t)/p_1$ for the flow in the unsteady expansion caused by a piston whose speed is

$$w_p = \alpha t_p^{1/2}$$

where α is a positive constant. Your solution should be a function only of γ and $\xi_p = [(\alpha^2 t_p)/a_1^2]$.

(b) Determine the pressure on the piston's face and the time $t_{p\infty}$ when this pressure first becomes zero.

(c) With $\gamma = 7/5$, derive an algebraic equation for the work normalized by the force $p_1 A$, $W/(p_1 A)$, done on the piston's face by the gas during the time interval $0 \le t \le t_{p\infty}$.

12.14 A piston has the trajectory

$$x_p = \alpha e^{\beta t_p} - \alpha$$

where α and β are positive constants. Determine $p(x, t)$ for both the centered and noncentered expansions. Show that the location and pressure on the trailing edge of the centered expansion agrees with that for the noncentered expansion. Determine the time $t_{p\infty}$, location $x_{p\infty}$, and speed $w_{p\infty}$ when M_p first becomes infinite.

12.15 Derive $p(x,t)/p_1$ and $M(x, t)$ for the unsteady expansion caused by a piston whose motion is

$$x_p = x_0 \ln\left(1 - \frac{t_p}{t_0}\right)$$

where x_0 and t_0 are positive constants.

12.16 A piston trajectory, $x_p = f(t_p)$, results in an expansion, where $w_p(0) = 0$. The wall pressure, measured at $x = 0$, can be represented as

$$\frac{p(0, t)}{p_1} = \left(\frac{2}{\gamma + 1}\right)^{2\gamma/(\gamma-1)} + \left[1 - \left(\frac{2}{\gamma + 1}\right)^{2\gamma/(\gamma-1)}\right] e^{-\alpha^2 (t/t_p^*)}$$

where α is a constant and t_p^* is the time when $M_p = 1$. Establish a first-order ODE for f and the corresponding initial condition. The solution of the ODE provides the piston's trajectory during the interval $0 \le t_p \le t_p^*$. This is an inverse problem, in that $p(0,t)$ is used to determine the subsonic part of the piston's trajectory.

12.17 A constant-speed piston causes a normal shock in oxygen with a shock Mach number of 3. The temperature of the region 1 quiescent gas is 300 K.

(a) Utilize the notation in Figure 12.11(a) and determine the incident shock speed w_s' as well as w_p', p_2/p_1, and T_2/T_1.

(b) Determine the reflected shock speed w_r', p_3/p_1, and T_3/T_1.

(c) Determine p_4/p_1, and T_4/T_1.

(d) With $x_w = 1$ m and $x_p = 0.6$ m, determine the time t_c and location x_c for the intersection of the reflected shock wave with the leading edge of the expansion.

12.18 This problem is based on Figure 12.11(b), where the piston speed w_p' is −734.3 m/s and the piston travels a distance of 0.6 m. The gas in region 1 is oxygen at 300 K.

(a) Determine η_{TE}, M_2', $\hat{M}_2$, p_2/p_1, and T_2/T_1.

(b) Evaluate w_s', p_3/p_1, and T_3/T_1.

(c) Determine the time t_c and location x_c where the shock and trailing edge of the expansion intersect.

12.19 The following values are representative of a 175 mm tank gun:

$$d = 0.175 \text{ m}, \; m_p = 67 \text{ kg}, \; \ell_g = 1.5 \text{ m}, \; \ell_h = 3.75 \text{ m},$$

$$\gamma_c = 1.21, \; R_c = 300 \text{ J/kg} - \text{K}, \; T_c = 3800 \text{ K},$$

$$\rho_g = 650 \text{ kg/m}^3, \; t_c = 9.5 \times 10^{-3} \text{ s},$$

$$x_c = 1.8 \text{ m}, \; p_1 = 10^5 \text{ Pa}, \; a_1 = 340 \text{ m/s}$$

Use the nondimensional model with Equation (12.164) to obtain a solution for pertinent nondimensional parameters, such as X_p, P_b, etc. Utilize a fourth-order Runge-Kutta scheme for the integration. Terminate the solution when W_p is a maximum. What are the values for $P_{b\max}$, the time $\tau_{\max}$ when this occurs, and X_m and τ_m. Plot X_p and X_s vs. τ, and, in a separate figure, P_b, P_{p3}, and P_{p2} vs. τ.

12.20 Show that R_2 [see Equation (12.258c)] satisfies Equations (12.240) to (12.242).

12.21 Consider the *bc* characteristic in Figure 12.16 when $n = 1$.

(a) For this characteristic, derive the equation

$$\frac{d\tau}{dX} = \frac{3}{(1-M)(6h)^{1/2}}$$

(b) Show that

$$M \cong c(1-X)$$

where c is a positive constant, as the wall is approached along *bc*. A numerical solution of Equation (12.283) from point c to $X = 0.999$, followed by a linear extrapolation to the wall, shows that $c = 0.9825$ when $n = 1$ and $M_2 = 0.5$.

(c) Use this value for c to establish p_b as 0.1862.

12.22 **(a)** Use R_2 and Equation (12.274) to obtain $\psi_2(w,h)$.

(b) Determine the nondimensional transformation between w and h and X and τ when $n = 2$.

(c) Write the equation for the *ac* characteristic. Write the equations for X_c and τ_c in terms of M_2. Use this result to verify the = equation in part (b).

(d) Write the equations for M, p, and p_0 in terms of w and h.

(e) Show that ψ_2 from part (a) satisfies Equation (12.210).

12.23 As indicated in the sketch, a fast acting valve located at $X = 0$ fully opens at $\tau = 0$. The initial pressure $\hat{p}_1$ in the chamber, to the right of the closed valve, exceeds the ambient pressure $\hat{p}_{amb}$, which is constant and holds everywhere to the left of the valve. Region 2 thus has $\hat{p}_2 = \hat{p}_{amb}$, while region 3 is quiescent. The expansion between regions 2 and 3 partly reflects from the opening at $X = 0$ as a compression wave, sketched as short dashed lines. Part of the expansion wave transmits into the region to the left of the valve. The compression wave quickly develops a shock wave along the C_+ characteristic emanating from point d. For analytical simplicity, we assume $\gamma = 3$ and ignore the compression wave. The sketch is for $\gamma = 3$, where characteristic lines are correctly drawn as straight. We also assume that $0.125 < p_{amb} < 1$, where $p_{amb} = \hat{p}_{amb}/\hat{p}_1$, which implies that $0 < M_2 < 1$. The valve quickly closes when $\hat{w}$ first equals zero at $X = 0$. If the compression wave is considered, closure would occur at a point such as g. Under the simplifying assumption of no reflected wave, the valve instead closes at point e. In working this problem, remember that points d, g, and e are connected by straight C_- characteristic lines with points a, f, and b, respectively.

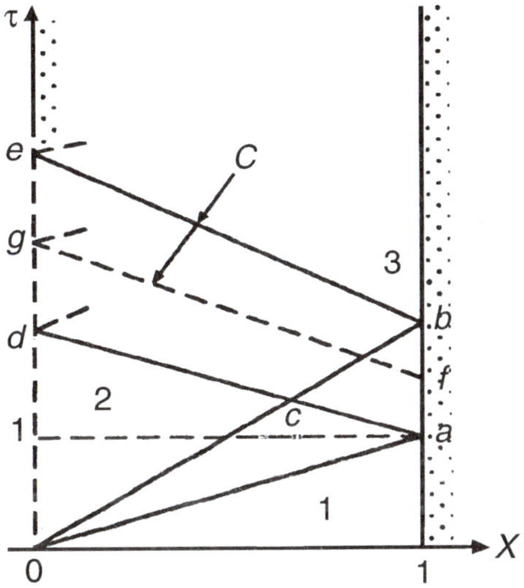

(a) Determine the nondimensional pressure $p(\tau)$ at $X = 0$ for $\tau_d \le \tau \le \tau_e$. Since p_e can be well below unity, this result illustrates the useful experimental observation (Azoury, 1992) that the originally high pressure chamber initially empties to a pressure below the ambient value.

(b) Compute the mass remaining in the chamber at the time the valve closes. Normalize this value to the initial mass in the chamber.

12.24 Start with the equation

$$L(u) = u_{\tilde{x}\tilde{x}} - \alpha_1^2 u_{\tilde{t}\tilde{t}} + \alpha_2 u_{\tilde{x}} + \alpha_3 u_{\tilde{t}} + \alpha_4 u = 0 \tag{1}$$

where the α_i are constants and $\alpha_1 > 0$. The solution is to satisfy the conditions

$$u(\tilde{x}, 0) = 0, \quad u(0, \tilde{t}) = g(\tilde{t}) \tag{2}$$

with $g(0) = 0$ and $\tilde{x} \ge 0$, $\tilde{t} \ge 0$.

(a) Use a linear transformation to obtain the u_{xy} canonical form for Equation (1), where x and y are characteristic coordinates. Establish equations for the constants a and b in terms of the α_i in order that the transformed version of Equation (1) has the form

$$L(u) = u_{xy} + au_x + bu_y + abu = 0 \tag{3}$$

and determine the value for α_4 if this relation is to hold. Transform conditions (2), and transfer the $u(\tilde{x}, 0) = 0$ condition to the leading edge of the disturbance, which occurs on the $\tilde{t} = \alpha_1 \tilde{x}$ line.

(b) Obtain the Riemann function $R(x,y; \xi,\eta)$ for Equation (3).

(c) Use the Riemann function method to obtain either

$$\oint_C R(u_x + bu)dx = 0 \quad \text{or} \quad \oint_C \frac{\partial(Ru)}{\partial x}dx = 0 \tag{4a,b}$$

(d) Obtain a solution to Equations (2) and (3) by noting that Equation (3) can be written as

$$(u_y + au)_x + b(u_y + au) = 0$$

(e) Transform this solution back to the original $\tilde{x}, \tilde{t}$ coordinates and the α_i constants. With the α_4 value from part (a), show that your $u(\tilde{x}, \tilde{t})$ result satisfies Equations (1) and (2), where the $u(\tilde{x}, 0) = 0$ condition is imposed on the leading edge of the disturbance. Observe that the amplitude of the wave, in the x direction, experiences amplification if $\alpha_3 > \alpha_1 \alpha_2$.

(f) Use the part (d) solution to show that Equation (4a) is satisfied.

12.25 Consider a form of the telegraph equation

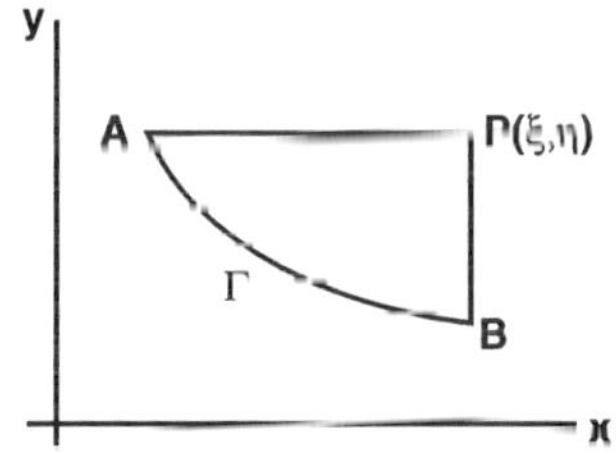

$$L(u) = u_{xy} + cu = g(x,y)$$

where c is a constant.

(a) Show that the Riemann function is

$$R(x,y;\xi,\eta) = J_o(z)$$

where J_o is the Bessel function of the first kind of zero order, and

$$z = [4c(x-\xi)(y-\eta)]^{1/2}$$

(b) Use the general solution

$$u(P) = \frac{1}{2}[u(A)R(A) + u(B)R(B)]$$

$$+ \int_A^B \left\{ \left[\frac{1}{2}Ru_x + \left(bR - \frac{1}{2}R_x\right)u\right]dx - \left[\frac{1}{2}Ru_y + \left(aR - \frac{1}{2}R_y\right)u\right]dy \right\}$$

$$+ \iint_S Rf\,dxdy$$

to the equation

$$L(u) = u_{xy} + au_x + bu_y + cu = f$$

where a, b, c, and f are known functions of x and y, to obtain the general solution to the telegraph equation. Some of the nomenclature is defined by the sketch where x and y are characteristic coordinates and the Γ segment is noncharacteristic. On Γ, u and either u_x or u_y are given functions. Simplify your result as much as possible.

(c) Set

$$A(0,1), \quad B(1,0), \quad \Gamma: y = -x+1, \quad g = 0$$

$$u_x = 1 \text{ on } \Gamma, \quad u(A) = u(B) = h = \text{constant}$$

Determine the solution for $u(1,1)$.

12.26 With

$$\eta = \frac{x}{a_r t}, \quad a = a_r A(\eta), \quad w = a_r W(\eta)$$

where a_r is a reference speed, show that Equations (12.39b) and (12.57) become ODEs. Use Cramer's rule to determine equations for A' and W'.

Part III

Viscous/Inviscid Fluid Dynamics

Outline of part III

Chapter 13 presents a more detailed discussion of the governing (Navier–Stokes) equations than was given in Chapter 2. A compressible, continuum fluid is assumed that ignores effects due to a gravitational body force, interfacial tension, multiphase constituents, chemical rections, electromagnetic fields, radiative heat transfer, molecular diffusion, and that of a non-Newtonian fluid. Hence, natural convection is not discussed.

The chapter beings by formulating the equations in orthogonal curvilinear coordinates and introducing the relevant nondimensional parameters, such as the Mach number and Reynolds number. One of the transport coefficients in the Navier–Stokes equations is the bulk viscosity, whose relevance has generated some controversy. Section 13.4 is devoted to a physical discussion of this parameter. The chapter concludes by discussing viscous flow in a heated duct. This is combined Rayleigh/Fanno flow using the influence coefficient method.

Chapter 14 discusses the forces and moments that can act on a body or a surface. The treatment centers around the momentum theorem, here generalized for steady or unsteady, incompressible or compressible, inviscid or viscous, and irrotational or rotational flows. A (possibly) new method is developed for treating a viscous force at a surface. A number of applications are presented, not all of which are based on the momentum theorem. Incompressible flows constitute the first group of applications, compressible flows the second group. Notable in the second group are the unique discussions of interference and of the forces and moments that act on a supersonic vehicle.

13 Coordinate Systems and Related Topics

13.1 PRELIMINARY REMARKS

The viscous, heat-conducting conservation equations utilizing Fourier's equation and a Newtonian fluid are reexamined. The equations are first formulated in orthogonal curvilinear coordinates. Subsequently, the equations are nondimensionalized using Cartesian coordinates. We thereby make explicit the dependence on a number of nondimensional parameters. Sections 13.2 and 13.3 are therefore particularly important for the later analysis in Parts IV and V. Section 13.4 is devoted to a brief discussion of the bulk viscosity, while the last section discusses flow in a duct with heat transfer and skin friction.

13.2 ORTHOGONAL COORDINATES

By way of introduction, the conservation equations are summarized:

$$\frac{1}{\rho}\frac{D\rho}{Dt} + \nabla\cdot\vec{w} = 0 \qquad \textbf{(2.2a)}$$

$$\rho\frac{D\vec{w}}{Dt} = -\nabla p + \nabla\cdot\overleftrightarrow{\tau} \qquad \textbf{(2.08)}$$

$$\rho\frac{Dh}{Dt} = \frac{Dp}{Dt} + \Phi - \nabla\cdot\vec{q} \qquad \textbf{(2.39)}$$

where

$$\frac{D\vec{w}}{Dt} = \frac{\partial\vec{w}}{\partial t} + \nabla\left(\frac{1}{2}w^2\right) + \vec{\omega}\times\vec{w} \qquad \textbf{(1.18)}$$

$$\overleftrightarrow{\tau} = 2\mu\overleftrightarrow{\varepsilon} + \lambda(\nabla\cdot\vec{w})\overleftrightarrow{I} \qquad \textbf{(1.54b)}$$

$$\Phi = \overleftrightarrow{\tau}:\overleftrightarrow{\varepsilon} \qquad \textbf{(2.30)}$$

$$\vec{q} = -\kappa\nabla T \qquad \textbf{(1.59)}$$

These relations are written in vector form and thus hold in any coordinate system, including nonorthogonal, curvilinear coordinates. Additional equations from thermodynamics are required in order to relate p, ρ, T, and h. From kinetic theory or experiments, relations are also needed for the transport coefficients μ, λ, and κ.

Let us consider two coordinate systems. The first, denoted by x_i, is Cartesian with the velocity written as

$$\vec{w} = w_i \hat{\imath}_i \qquad (1.7)$$

An orthogonal, curvilinear coordinate system, ξ_i, is the second one, with the velocity written as

$$\vec{w} = v_i \hat{e}_i \qquad \mathbf{(13.1)}$$

where the $\hat{e}_i$ is an orthonormal basis. With somewhat greater complexity, nonorthogonal coordinates can be considered by utilizing the relations in Appendix A. For the subsequent discussion, however, the simpler orthogonal system will suffice. Moreover, a relatively straightforward decomposition is available for $\nabla \cdot \overleftrightarrow{\tau}$ when the coordinates are orthogonal, as shown below. The two systems are related through a time-independent transformation:

$$x_j = x_j(\xi_1, \xi_2, \xi_3), \qquad j = 1,2,3 \qquad \mathbf{(13.2)}$$

The Jacobian of the transformation is assumed not to be zero or infinite, except possibly at isolated points. Hence, the inverse transformation exists if needed.

Appendix G

This appendix provides a systematic and straightforward method for obtaining the compressible conservation equations in any orthogonal curvilinear coordinate system. Let ϕ be an arbitrary scalar function and $\vec{A}$ an arbitrary vector; i.e.,

$$\vec{A} = A_i \hat{e}_i \qquad \mathbf{(13.3)}$$

The appendix then summarizes the relevant equations needed for the governing equations. The vector and tensor relations are based on those in Appendix A for an orthogonal coordinate system. The curl of $\vec{A}$ is provided, since it is useful for evaluating the vorticity. We write the rate of deformation tensor as

$$\overleftrightarrow{\varepsilon} = \varepsilon_{ki} \hat{e}_k \hat{e}_i \qquad \mathbf{(13.4)}$$

Appendix G provides explicit relations for the ε_{ki} components, which only depend on the v_m velocity components and the scale factors h_n.

In the momentum equation, with the aid of Equation (1.54b), the $\nabla \cdot \overleftrightarrow{\tau}$ term splits into two applied viscous force terms. The rightmost of these terms stems from

$$\nabla \cdot [\lambda(\nabla \cdot \vec{w})\overleftrightarrow{I}] = \nabla[\lambda(\nabla \cdot \vec{w})] \cdot \overleftrightarrow{I} + \lambda(\nabla \cdot \vec{w})\nabla \cdot \overleftrightarrow{I} = \nabla\lambda(\nabla \cdot \vec{w})$$

since

$$\nabla \cdot \overleftrightarrow{I} = 0, \qquad (\nabla\phi) \cdot \overleftrightarrow{I} = \nabla\phi$$

and where the del operator is

$$\nabla = \frac{\hat{e}_j}{h_j}\frac{\partial}{\partial \xi_j}$$

The other force term is written as

$$\vec{F}^s = F_i^s \hat{e}_i = 2\nabla \cdot (\mu \overleftrightarrow{\varepsilon}) \tag{13.5}$$

where a superscript s denotes a shear force. A somewhat different shear force is obtained if λ is replaced with Equation (1.57), and only the bulk viscosity term represents a nonshearing viscous force.

As shown in the appendix, a particularly elegant form is given for the F_i^s components that depends only on μ, ε_{ki}, and the scale factors. To derive this relation, we write the divergence term as follows:

$$\begin{aligned}
\nabla \cdot (\mu \overleftrightarrow{\varepsilon}) &= \frac{\hat{e}_j}{h_j}\frac{\partial}{\partial \xi_j} \cdot (\mu \varepsilon_{ki} \hat{e}_k \hat{e}_i) \\
&= \frac{\hat{e}_j}{h_j} \cdot \left[\frac{\partial}{\partial \xi_j}(\mu \varepsilon_{ki} \hat{e}_k)\hat{e}_i + \mu \varepsilon_{ki} \hat{e}_k \frac{\partial \hat{e}_i}{\partial \xi_j}\right] \\
&= \left[\frac{\hat{e}_j}{h_j}\frac{\partial}{\partial \xi_j} \cdot (\mu \varepsilon_{ki} \hat{e}_k)\right]\hat{e}_i + \frac{\mu \varepsilon_{ki}}{h_j}\delta_{jl} \sum_{m \neq i}\left(\frac{\delta_{jm}}{h_i}\frac{\partial h_m}{\partial \xi_i} - \frac{\delta_{ij}}{h_m}\frac{\partial h_i}{\partial \xi_m}\right)\hat{e}_m \\
&= [\nabla \cdot (\mu \varepsilon_{ki} \hat{e}_k)]\hat{e}_i + \frac{\mu \varepsilon_{jl}}{h_j} \sum_{m \neq i}\left(\frac{\delta_{jm}}{h_i}\frac{\partial h_m}{\partial \xi_i} - \frac{\delta_{ij}}{h_m}\frac{\partial h_i}{\partial \xi_m}\right)\hat{e}_m
\end{aligned}$$

where, in addition to m, the i, j, and k are summed over. While convenient, the $m \neq i$ designation is not essential, since the term within the brackets is zero whenever m equals i. The first term on the right involves the divergence of a vector instead of a tensor, and can be written as

$$\nabla \cdot (\mu \varepsilon_{ki} \hat{e}_k) = \frac{1}{h_1 h_2 h_3}\frac{\partial}{\partial \xi_j}\left(\frac{h_1 h_2 h_3}{h_j}\mu \varepsilon_{ij}\right)$$

in an orthogonal system. This result yields the first term on the right side of the F_i^s equation in Appendix G.

For F_i^s we also require the coefficient of $\hat{e}_i$ that stems from the above $m \neq i$ summation term. When $j \neq i$ and the j summation is explicitly shown, we observe that a nonzero value is obtained only if $m = j$. We thus have

$$\mu \sum_{i \neq j} \frac{\varepsilon_{ij}}{h_j}\frac{1}{h_i}\frac{\partial h_j}{\partial \xi_i}\hat{e}_i = \frac{\mu}{h_i}\sum_{j \neq i}\frac{\varepsilon_{ji}}{h_j}\frac{\partial h_i}{\partial \xi_j}\hat{e}_i$$

where the i and j indices are interchanged. This result is the first part of the $j \neq i$ term in F_i^s in the appendix.

Again, consider the $m \neq i$ summation term but now with $i = j$:

$$-\frac{\mu \varepsilon_{jj}}{h_j}\sum_{m \neq j}\frac{1}{h_m}\frac{\partial h_j}{\partial \xi_m}\hat{e}_m$$

Suppose we are evaluating F_1^s. This summation then contributes twice to F_i^s. For instance, when $j = 2$ we have

$$-\frac{\mu\varepsilon_{22}}{h_2}\left(\frac{1}{h_i}\frac{\partial h_2}{\partial \xi_1}\hat{e}_1 + \frac{1}{h_3}\frac{\partial h_2}{\partial \xi_3}\hat{e}_3\right)$$

which contributes an $\hat{e}_1$ term, and when $j = 3$ we have

$$-\frac{\mu\varepsilon_{33}}{h_3}\left(\frac{1}{h_1}\frac{\partial h_3}{\partial \xi_1}\hat{e}_1 + \frac{1}{h_2}\frac{\partial h_3}{\partial \xi_2}\hat{e}_2\right)$$

which contributes another $\hat{e}_1$ term. (The $\hat{e}_3$ and $\hat{e}_2$ terms, respectively, appear in F_3^s and F_2^s.) By synthesizing the foregoing relations, we obtain the final result for F_i^s, shown in Appendix G.

With Cartesian coordinates all h_i are unity, and the conservation equations simplify to

$$\frac{1}{\rho}\frac{D\rho}{Dt} + \frac{\partial w_i}{\partial x_i} = 0 \tag{13.6a}$$

$$\rho\frac{Dw_j}{Dt} = -\frac{\partial p}{\partial x_j} + \frac{\partial}{\partial x_i}\left[\mu\left(\frac{\partial w_j}{\partial x_i} + \frac{\partial w_i}{\partial x_j}\right)\right] + \frac{\partial}{\partial x_j}\left(\lambda\frac{\partial w_i}{\partial x_i}\right), \qquad j = 1, 2, 3 \tag{13.6b}$$

$$\rho\frac{Dh}{Dt} = \frac{Dp}{Dt} + \frac{\partial}{\partial x_i}\left(\kappa\frac{\partial T}{\partial x_i}\right) + \Phi \tag{13.6c}$$

where

$$\frac{D}{Dt} = \frac{\partial}{\partial t} + w_i\frac{\partial}{\partial x_i} \tag{13.7}$$

The viscous dissipation Φ is provided by Equation (2.32b), and the F_i^s components become

$$F_i^s = 2\frac{\partial}{\partial x_j}(\mu\varepsilon_{ij}), \qquad i = 1, 2, 3 \tag{13.8}$$

Cylindrical Polar Coordinates

As an illustration, we consider cylindrical polar coordinates for an axisymmetric flow, where

$$\begin{aligned}
&\xi_1 = r, && h_1 = h_r = 1, && v_1 = v_r \\
&\xi_2 = \theta, && h_2 = h_\theta = r, && v_2 = v_\theta \\
&\xi_3 = z, && h_3 = h_z = 1, && v_3 = v_z
\end{aligned}$$

and

$$\vec{w} = v_r\hat{e}_r + v_\theta\hat{e}_\theta + v_z\hat{e}_z$$

For scalar variables, the substantial derivative is

$$\frac{D}{Dt} = \frac{\partial}{\partial t} + v_r\frac{\partial}{\partial r} + \frac{v_\theta}{r}\frac{\partial}{\partial \theta} + v_z\frac{\partial}{\partial z}$$

while the acceleration can be written as

$$\frac{D\vec{w}}{Dt} = \frac{Dv_r}{Dt}\hat{e}_r + \frac{Dv_\theta}{Dt}\hat{e}_\theta + \frac{Dv_z}{Dt}\hat{e}_z + \frac{v_\theta v_r}{r}\hat{e}_\theta - \frac{v_\theta^2}{r}\hat{e}_r$$

The two rightmost terms stem from the term in the acceleration

$$\frac{v_i v_j}{h_i}\frac{\partial \hat{e}_j}{\partial \xi_i}$$

in which only

$$\frac{\partial \hat{e}_r}{\partial \theta} = \hat{e}_\theta, \qquad \frac{\partial \hat{e}_\theta}{\partial \theta} = -\hat{e}_r$$

are nonzero. From Appendix G, the rate-of-deformation tensor components are

$$\varepsilon_{rr} = \frac{1}{h_1}\left(\frac{\partial v_1}{\partial \xi_1} + \frac{v_2}{h_2}\frac{\partial h_1}{\partial \xi_2} + \frac{v_3}{h_3}\frac{\partial h_1}{\partial \xi_3}\right) = \frac{\partial v_r}{\partial r}$$

$$\varepsilon_{\theta\theta} = \frac{v_r}{r} + \frac{1}{r}\frac{\partial v_\theta}{\partial \theta}$$

$$\varepsilon_{zz} = \frac{\partial v_z}{\partial z}$$

$$\varepsilon_{r\theta} = \varepsilon_{\theta r} = \frac{1}{2}\left(\frac{1}{r}\frac{\partial v_r}{\partial \theta} + \frac{\partial v_\theta}{\partial r} - \frac{v_\theta}{r}\right)$$

$$\varepsilon_{rz} = \varepsilon_{zr} = \frac{1}{2}\left(\frac{\partial v_r}{\partial z} + \frac{\partial v_z}{\partial r}\right)$$

$$\varepsilon_{\theta z} = \varepsilon_{z\theta}\frac{1}{2}\left(\frac{\partial v_z}{\partial \theta} + \frac{\partial v_\theta}{\partial z}\right)$$

Hence, $\nabla \cdot \vec{w}$, $\vec{\omega}$, Φ, and the F_i^s are given by

$$\nabla \cdot \vec{w} = \frac{\partial v_r}{\partial r} + \frac{1}{r}\frac{\partial v_\theta}{\partial \theta} + \frac{\partial v_z}{\partial z} + \frac{v_r}{r}$$

$$\vec{\omega} = \nabla \times \vec{w} = \left(\frac{1}{r}\frac{\partial v_z}{\partial \theta} - \frac{\partial v_\theta}{\partial z}\right)\hat{e}_r + \left(\frac{\partial v_r}{\partial z} - \frac{\partial v_z}{\partial r}\right)\hat{e}_\theta + \left(\frac{\partial v_\theta}{\partial r} + \frac{v_\theta}{r} - \frac{1}{r}\frac{\partial v_r}{\partial \theta}\right)\hat{e}_z$$

$$\Phi = \mu\left\{2\left[\left(\frac{\partial v_r}{\partial r}\right)^2 + \frac{1}{r^2}\left(\frac{\partial v_\theta}{\partial \theta} + v_r\right)^2 + \left(\frac{\partial v_z}{\partial z}\right)^2\right] + \left(\frac{1}{r}\frac{\partial v_z}{\partial \theta} + \frac{\partial v_\theta}{\partial z}\right)^2 + \left(\frac{\partial v_r}{\partial z} + \frac{\partial v_z}{\partial r}\right)^2 + \left(\frac{1}{r}\frac{\partial v_r}{\partial \theta} + \frac{\partial v_\theta}{\partial r} - \frac{v_\theta}{r}\right)^2\right\} + \lambda(\nabla \cdot \vec{w})^2$$

$$F_r^s = \frac{2}{r}\frac{\partial}{\partial r}\left(r\mu\frac{\partial v_r}{\partial r}\right) + \frac{1}{r}\frac{\partial}{\partial \theta}\left[\mu\left(\frac{1}{r}\frac{\partial v_r}{\partial \theta} + \frac{\partial v_\theta}{\partial r} - \frac{v_\theta}{r}\right)\right] + \frac{\partial}{\partial z}\left[\mu\left(\frac{\partial v_r}{\partial z} + \frac{\partial v_z}{\partial r}\right)\right] - \frac{2\mu}{r^2}\left(\frac{\partial v_\theta}{\partial \theta} + v_r\right)$$

$$F_\theta^s = \frac{\partial}{\partial r}\left[\mu\left(\frac{1}{r}\frac{\partial v_r}{\partial \theta} + \frac{\partial v_\theta}{\partial r} - \frac{v_\theta}{r}\right)\right] + \frac{2}{r^2}\frac{\partial}{\partial \theta}\left[\mu\left(\frac{\partial v_\theta}{\partial \theta} + v_r\right)\right] + \frac{\partial}{\partial z}\left[\mu\left(\frac{1}{r}\frac{\partial v_z}{\partial z} + \frac{\partial v_z}{\partial r}\right)\right] + \frac{2\mu}{r}\left(\frac{1}{r}\frac{\partial v_r}{\partial \theta} + \frac{\partial v_\theta}{\partial r} - \frac{v_\theta}{r}\right)$$

$$F_z^s = \frac{1}{r}\frac{\partial}{\partial r}\left[r\mu\left(\frac{\partial v_r}{\partial z} + \frac{\partial v_z}{\partial r}\right)\right] + \frac{1}{r}\frac{\partial}{\partial \theta}\left[\mu\left(\frac{1}{r}\frac{\partial v_z}{\partial \theta} + \frac{\partial v_\theta}{\partial r}\right)\right] + 2\frac{\partial}{\partial z}\left(\mu\frac{\partial v_z}{\partial z}\right)$$

Problems 13.2 to 13.4 continue with cylindrical coordinates; e.g., the conservation equations are obtained in Problem 13.2(a). Natural coordinates are the subject of Problems 13.5 and 13.6, while Problems 13.12 and 13.13 consider parabolic coordinates. When dealing with a new coordinate system, the first step is to develop relations for h_i, $\partial\hat{e}_j/\partial\xi_i$, etc., as is done in the above illustrative example and in part (a) of Problem 13.12.

Vorticity

Let us continue with the above example and further assume the flow is radially oriented with only an r,t dependence. In this circumstance, we readily establish

$$\vec{w} = v_r(r,t)\hat{e}_r$$

$$F_r^s = 2\frac{\partial \mu}{\partial r}\frac{\partial v_r}{\partial r} + 2\mu\left(-\frac{v_r}{r^2} + \frac{1}{r}\frac{\partial v_r}{\partial r} + \frac{\partial^2 v_r}{\partial r^2}\right)$$

$$F_\theta^s = F_z^s = 0$$

$$\Phi = 2\mu\left[\left(\frac{\partial v_r}{\partial r}\right)^2 + \left(\frac{v_r}{r^2}\right)^2\right] + \lambda(\nabla \cdot \vec{w})^2$$

$$\vec{\omega} = 0$$

Although the flow is viscous with a nonzero viscous dissipation, the vorticity is, nevertheless, zero. Thus, a viscous flow can be irrotational. While a velocity potential function can be introduced, there is no point in doing so, since the velocity has only one nonzero component anyway.

Generally, the vorticity is a significant factor in a viscous flow. The reason for this stems from Equation (4.19), which holds for a viscous or inviscid flow. For purposes of simplicity, let us assume ρ and μ are constants. Then the acceleration is

$$\vec{a} = -\nabla \frac{p}{\rho} + \frac{1}{\rho}\nabla \cdot \overleftrightarrow{\tau} = -\nabla \frac{p}{\rho} + \frac{2}{\rho}\nabla \cdot (\mu\overleftrightarrow{\varepsilon}) = -\nabla \frac{p}{\rho} + \frac{1}{\rho}\vec{F}^s$$

and its curl becomes

$$\nabla \times \vec{a} = \frac{1}{\rho}\nabla \times \vec{F}^s$$

Hence, Equation (4.19) has the form

$$\frac{D\vec{\omega}}{Dt} = \vec{\omega} \cdot (\nabla\vec{w}) + \frac{1}{\rho}\nabla \times \vec{F}^s \tag{13.9}$$

where

$$\nabla \times \vec{F}^s = \mu\nabla^2\vec{\omega}$$

In an incompressible flow, the convection of vorticity depends on a stretching term and on a diffusion-like term.

13.3 SIMILARITY PARAMETERS

We return to the Cartesian form and assume a perfect gas in order to replace the enthalpy with c_pT. We nondimensionalize the equations using the following relations:

$$\begin{aligned}
x_i &= \ell x_i^*, & t &= \tau t^* & & \\
w_i &= U_\infty w_i^*, & \rho &= \rho_\infty \rho^*, & p &= p_\infty p^* \\
\mu &= \mu_\infty \mu^*, & \lambda &= \lambda_\infty \lambda^*, & c_p &= c_{p\infty} \\
T &= T_\infty T^*, & \kappa &= \kappa_\infty \kappa^* & &
\end{aligned} \tag{13.10}$$

Here ℓ and τ are a characteristic length and time in the flow, and the infinity subscript denotes a reference (or freestream) condition. In addition, we utilize

$$\begin{aligned}
c_{p\infty} &= \frac{\gamma}{\gamma - 1}R \\
p_\infty &= \rho_\infty R T_\infty \\
\Phi &= \frac{\mu_\infty U_\infty^2}{\ell^2}\Phi_\mu^* + \frac{\lambda_\infty U_\infty^2}{\ell^2}\Phi_\lambda^*
\end{aligned} \tag{13.11}$$

where

$$\Phi_\mu^* = \mu^*\left[2\sum_{i=1}^{3}\left(\frac{\partial w_i^*}{\partial x_i^*}\right)^2 + \left(\frac{\partial w_1^*}{\partial x_2^*} + \frac{\partial w_2^*}{\partial x_1^*}\right)^2 + \cdots\right]$$

$$\Phi_\lambda^* = \lambda^*\left(\sum_{i=1}^{3}\frac{\partial w_i^*}{\partial x_i^*}\right)^2$$

In the transformed equations we omit the asterisk, with the result

$$\frac{1}{S}\frac{\partial\rho}{\partial t} + \frac{\partial(\rho w_i)}{\partial x_i} = 0 \tag{13.12a}$$

$$\frac{1}{S}\frac{\partial w_j}{\partial t} + w_i\frac{\partial w_j}{\partial x_i} = -\frac{1}{\gamma M_\infty^2}\frac{1}{\rho}\frac{\partial p}{\partial x_j} + \frac{1}{\rho Re_\mu}\frac{\partial}{\partial x_i}\left[\mu\left(\frac{\partial w_j}{\partial x_i} + \frac{\partial w_i}{\partial x_j}\right)\right] + \frac{1}{\rho Re_\lambda}\frac{\partial}{\partial x_j}\left(\lambda\frac{\partial w_i}{\partial x_i}\right), \qquad j = 1, 2, 3 \tag{13.12b}$$

$$\frac{1}{S}\frac{\partial T}{\partial t} + w_i\frac{\partial T}{\partial x_i} = \frac{\gamma-1}{\gamma}\frac{1}{\rho S}\frac{\partial p}{\partial t} + \frac{\gamma-1}{\gamma}\frac{w_i}{\rho}\frac{\partial p}{\partial x_i} + \frac{1}{\rho Re_\mu Pr}\frac{\partial}{\partial x_i}\left(\kappa\frac{\partial T}{\partial x_i}\right) + \frac{(\gamma-1)M_\infty^2}{\rho}\left(\frac{\Phi_\mu}{Re_\mu} + \frac{\Phi_\lambda}{Re_\lambda}\right) \tag{13.12c}$$

Several of the six nondimensional parameters in the above equations

$$S = \text{Strouhal number} = \frac{\tau U_\infty}{\ell}$$

$$\gamma = \text{specific heat ratio} = \frac{c_{p\infty}}{c_{v\infty}}$$

$$M_\infty = \text{Mach number} = \frac{U_\infty}{(\gamma p_\infty/\rho_\infty)^{1/2}}$$

$$Pr = \text{Prandtl number} = \frac{\mu_\infty c_{p\infty}}{\kappa_\infty}$$

$$Re_\mu = \text{first viscosity Reynolds number} = \frac{\rho_\infty U_\infty \ell}{\mu_\infty}$$

$$Re_\lambda = \text{second viscosity Reynolds number} = \frac{\rho_\infty U_\infty \ell}{\lambda_\infty}$$

have been previously encountered. Of these, only γ and Pr are properties solely of the fluid. Other forms of the equations introduce still other numbers, such as the Peclet number and, with buoyancy, the Grashof number. Note that we are not using the Eckert number, which normally is defined as $U_\infty^2/(c_{p\infty}\Delta T)$ where ΔT is typically the temperature difference across a boundary layer.

Two Reynolds numbers appear in the equations. Had Stokes' hypothesis ($\lambda = -2\mu/3$) been utilized to eliminate λ, only one Reynolds number would occur. In this circumstance, Re_μ would then be written as Re.

As discussed in Section 1.5, λ may be negative, which would result in Re_λ also being negative. (A negative Reynolds number can be avoided by using the bulk viscosity in place of λ.) A more interesting situation occurs when a gas, such as CO_2, is considered. In this case, Re_μ would exceed Re_λ by about three orders of magnitude. While both Reynolds numbers appear in the energy and momentum equations, the dependence on Re_λ disappears when the flow is incompressible and Φ_λ is zero. Thus, the effect of Re_λ (e.g., on the stability of a boundary layer) can be ascertained best in a high-speed boundary layer when M_∞ is supersonic or hypersonic and compressibility effects are important.

In addition to the equations, boundary conditions also introduce dimensionless parameters. For an impermeable wall, the skin friction at a wall location x is $\tau_w(x)$. In nondimensional form, this yields the local skin-friction coefficient

$$c_f = \frac{2\tau_w(x)}{(\rho U^2)_\infty} \tag{13.13}$$

For heat transfer at the wall, we have Fourier's equation in scalar form

$$q_w = -\kappa\left(\frac{\partial T}{\partial n}\right)_w \tag{13.14}$$

where the temperature derivative normal to the wall is evaluated in the fluid adjacent to the wall. The wall heat flux can also be written in terms of Newton's law of heat transfer

$$q_w = h_f(T_w - T_\infty) \tag{13.15}$$

where h_f is the film coefficient and $(T_w - T_\infty)$ is the temperature difference across a thermal boundary layer. Equating these two expressions yields the Nusselt number

$$Nu = \frac{h_f \ell}{\kappa} = \frac{\ell}{T_\infty - T_w}\left(\frac{\partial T}{\partial n}\right)_w \tag{13.16}$$

An alternative heat transfer coefficient is the Stanton number, defined by

$$St = \frac{h_f}{(\rho c_p U)_\infty} \tag{13.17}$$

where the two numbers are related by

$$Nu = St Re_\mu Pr \tag{13.18}$$

Discussion

Of the various dimensionless numbers, four are of primary importance in most gas flows. The first is the ratio of specific heats and is a molecular parameter. It has the approximate values given by

$$\begin{aligned} \gamma &= 5/3, \quad \text{monatomic gas} \\ &= 7/5, \quad \text{diatomic gas} \\ &= 9/7, \quad \text{triatomic gas} \end{aligned}$$

and $\gamma \leq 9/7$ for molecules with four or more atoms. These estimates assume all rotational modes are fully excited and all vibrational modes are unexcited. For example, when a diatomic molecule has some vibrational excitation, γ then falls between 7/5 and 9/7. In particular, for molecules consisting of a large number of atoms, γ approaches unity with increasing temperature.

The second property parameter is the Prandtl number, which only appears in the energy equation. Its variation is given by

$$Pr = \begin{cases} \ll 1, & \text{liquid metals} \\ O(1), & \text{most gases} \\ \gg 1, & \text{oils} \end{cases}$$

where $Pr \cong 0.71$ for air. It plays an important role in heat transfer; e.g., a large value for Pr (in which case κ is small) implies a relatively small rate of heat transfer. The Prandtl number is a function of the temperature, since μ, c_p, and κ are generally functions only of the temperature. Quite frequently, however, Pr has a negligible variation with T over a large temperature range. This insensitivity, for instance, holds for air. Consequently, a constant Prandtl number approximation is often warranted. The reason for $Pr \cong$ constant is that $c_p \cong$ constant and μ and κ have a similar dependence on the temperature.

The third parameter is the Mach number. Flow regimes are characterized by its value:

$$M = \begin{cases} <0.4, & \textit{incompressible} \\ <1, & \textit{subsonic} \\ \sim 1, & \textit{transonic} \\ >1, & \textit{supersonic} \\ \gg 1, & \textit{hypersonic} \end{cases}$$

Here, M is the maximum value of the Mach number in the flow field, since a vehicle in supersonic or hypersonic flight may well have one or more subsonic flow regions.

The most important parameter for a viscous flow, of course, is the Reynolds number, which here is based on a characteristic dimension of the body. In practice, it can range from below unity to above 10^9. For instance, for bacteria and aerosols, and in tribology, it is a small number. On the other hand, for a sizable vehicle in supersonic flight it is quite large.

13.4 BULK VISCOSITY

A relation was obtained in Section 1.5 between the viscous stress tensor and the rate of deformation tensor under the assumptions of linearity and isotropy. This relation, Equation (1.54), is used in both the momentum and energy equations, and is the reason for the appearance of μ and λ, or the bulk viscosity μ_b, in these equations. The λ, or μ_b, parameters always appear in conjunction with $\nabla \cdot \vec{w}$, and a large dilatation is required for the bulk viscosity terms to be of possible interest in a compressible flow.

The physical interpretation of the bulk viscosity in a gas flow is associated with the relaxation of internal, rotational and vibrational, modes of polyatomic molecules. It is known from kinetic theory, and confirmed by experiment (Prangsma et al., 1973), that the bulk viscosity is zero for a monatomic gas. At room temperature, diatomic gases, such as O_2, N_2, CO, and NO, are fully excited rotationally but possess negligible vibrational excitation. As a consequence, only rotation contributes to μ_b, and the μ_b/μ ratio is of unity order; e.g., this ratio is about 2/3 for air. The number of collisions required for rotational energy equilibration is about four or five, whereas vibrational energy relaxation typically requires thousands of collisions. Consequently, when the vibrational mode(s) is partly, or fully, excited, μ_b/μ is large compared to unity. For instance, at room temperature, CO_2 has a value of about 2.1×10^3 (Tisza, 1942) for the ratio. Moreover, the magnitude of this difference between rotational and vibrational relaxation is justified by kinetic theory (Monchik, et al., 1963).

Our interest in λ, or μ_b, is limited to continuum, chemically inert flows that are governed by the Navier–Stokes equations. The discussion, therefore, is not relevant to a dense gas (Hanley and Cohen, 1976) or to a liquid. In both cases, the physical interpretation for the bulk viscosity differs from that of a simple molecular collisional relaxation process. For example, μ_b is proportional to the square of the density for argon (Madigosky, 1967) at relatively large density values. Our discussion is based on Emanuel (1998), which can be consulted for additional details and references.

The practical question arises as to when the bulk viscosity terms in the Navier–Stokes equations are significant. For instance, Van Dyke (1962) showed that these terms are of third order for a compressible laminar boundary layer and thus quite inconsequential. In a hypersonic laminar boundary layer analysis for flow over a flat plate, Emanuel (1992) had difficulty in justifying the importance of these terms, even when the gas was CO_2. (A hypersonic Mach number was required, which would cause other phenomena, e.g., radiation and dissociation, to dominate effects associated with the bulk viscosity.) In general, Stokes' hypothesis of $\mu_b = 0$ appears to be reasonable for laminar and turbulent boundary, or free-shear, layers in air. Of course, many Navier–Stokes codes have been using Stokes' hypothesis *a priori*, with results that generally compare favorably with experiment.

One might anticipate that the bulk viscosity terms may be of some importance in a supersonic flow in which vibrational relaxation is present. This type of relaxation process, for example, occurs in air and other polyatomic gases, downstream of a relatively intense shock wave. In fact, vibrational relaxation properties are evaluated downstream of the incident, and reflected, normal shocks in shock tube experiments. The downstream flow is modeled with the Euler equations in conjunction with rate equations for the vibrational energy. Thus, in flows where the shear stress is negligible, the value of the bulk viscosity, along with other transport properties, is taken as zero, and any significant relaxation process is modeled with a rate equation.

So far, our discussion has yet to identify flow situations where the bulk viscosity terms play a significant role. Three cases are suggested. In the first, a weak ultrasonic signal is propagated one-dimensionally through a gas and the wave amplitude attenuation and the frequency dispersion are measured. Both the attenuation and dispersion depend, in part, on μ_b. In fact, this type of experiment is used to determine the bulk viscosity. Quite often, the product of frequency, ω, and the relaxation time, $\omega\tau$, is of order unity. In this circumstance, the corresponding theory yields a complex valued bulk viscosity (Bauer, 1965; Kneser, 1965). This result, however, is not appropriate for the Navier–Stokes equations, since the theory is based on an isentropic energy equation. Tisza (1942) has pointed out that a local thermodynamic equilibrium condition, $\omega\tau \ll 1$, is essential for experimental results to yield a value for μ_b appropriate for the Navier–Stokes equations. (This condition is always satisfied in a low-frequency limit.) A perturbation analysis of these equations yields a low-frequency limit formula (Emanuel, 1998) for μ_b. This analysis further demonstrates the nature of the entropy generation. Although the experiments by Prangsma et al. (1973) were performed at relatively low temperatures, where only rotational relaxation is present, the agreement with the low-frequency theory is quite good. For decades, this experimental technique was plagued with uncertainty. The work by Prangsma et al., however, demonstrates that these difficulties have been overcome. With their approach, the experimental uncertainty should be further reduced at higher temperatures, where vibrational relaxation dominates.

The low-frequency perturbation analysis is not limited to a laminar flow but should hold for the turbulent flows. This is especially true for air near room temperature, since $\omega\tau$ is generally small (Emanuel, 1998). The local equilibrium condition of Tisza, however, is less certain at appreciably higher temperatures, where vibrational relaxation should yield a relatively large μ_b/μ value. Nevertheless, when the Tisza condition does hold, the combination of the perturbation analysis, a positive entropy generation, and ultrasonic experiments yields a self-consistent formulation for a bulk viscosity that is appropriate to the Navier–Stokes equations.

Suppose $\omega\tau$ is not small compared to unity. We then have a complex-valued μ_b, local thermodynamic equilibrium cannot be assumed, and the conventional kinetic theory formulas for the other transport coefficients are in question. This is the case for the thermal conductivity of a polyatomic gas, which, in part, depends on the relaxation of its internal modes. Moreover, the excellent theory (Kneser, 1965; Bauer, 1965) utilized when $\omega\tau$ is of order unity does not reduce to the nonisentropic perturbation theory in the $\omega \to 0$ limit.

A second flow where the bulk viscosity is important is in shock wave structure. There are two serious flaws, however, in this circumstance. The linearity assumption used in the derivation of the stress-rate of strain relation is generally invalid because of the gradients inside a very thin shock wave. This appears to be the case even for a weak normal shock in argon (Garen et al., 1977). Secondly, the equilibrium shock jump conditions are unaltered by the value of any transport coefficient and thus yield no information about them.

The last example where the bulk viscosity may be important arises in the force and moment analysis presented in the last section of the next chapter. Aside from this one possible application, the fuss over the bulk viscosity appears to be a case of theoreticians looking for something to endlessly argue about.

Since its introduction by George Stokes, over one and a half centuries ago, bulk viscosity has been controversial. Indeed, Stokes himself appears to have changed his mind, perhaps more than once, about this transport property. Aside from love of argumentation, two reasons are suggested for the controversy. Unlike all other transport and thermodynamic properties, the foregoing ultrasonic technique is the only method, so far, that has been developed to deduce values for μ_b. As mentioned, the method had been plagued by a large experimental uncertainty. Moreover, the experimental data are sparse, especially above room temperature, where vibrational relaxation starts to occur in many polyatomic molecules. At least one additional technique (Emanuel and Argrow, 1994) would be of value.

In the equation just above Equation (1.57), the average normal stress, $\sigma_{ii}/3$, appears. In the hydrostatic situation this average equals $-p$. In a nonhydrostatic situation, however, $\sigma_{ii}/3$ does not represent a measurable pressure and should not be confused with the thermodynamic pressure. In this regard, note that $\sigma_{ii}/3$ does not appear in the Navier–Stokes equations, and Equations (1.54) and (1.55) are obtained without its introduction. As with σ_{ii}, one can define a variety of kinetic theory parameters that are not measurable in a continuum flow.

13.5 VISCOUS FLOW IN A HEATED DUCT

Introductory Discussion

Viscous flow in a duct with heating or cooling occurs inside the tubes of a heat exchanger, in a gas pipeline, in various experimental devices, etc. Among the latter is the circular test section of a plasma gun (Candler, 1997) and the duct of a molecular beam device (Rohrs et al., 1995). Our discussion focuses on the molecular beam device, where heating of a gas occurs in a circular duct. The influence coefficient approach is used, which is not limited to just heating or to a circular duct. This is a one-dimensional approximation that deals with average properties at any cross section of the duct. A constant cross-sectional area duct with just heat transfer is called Rayleigh flow; its counterpart with just skin friction is called Fanno flow. The basic theory for the influence coefficient method was pioneered by Shapiro (1953).

Molecular beam devices are used to study molecular energy exchange at very low pressures and temperatures. Between the plenum, where the gas to be examined is generated, and the vacuum chamber is a nozzle. Generally, the nozzle is just a small hole of diameter d in a metal (tungsten or

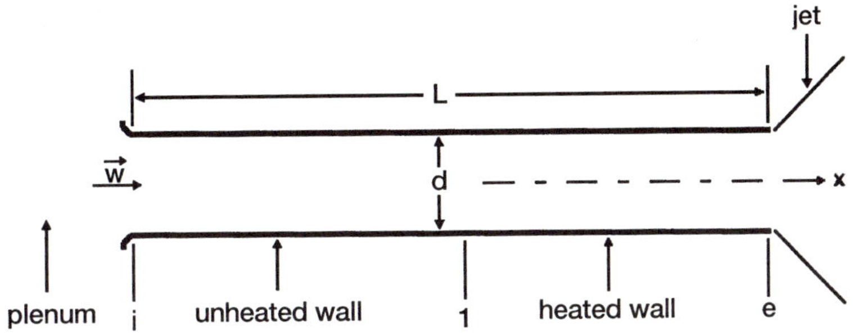

FIGURE 13.1 Schematic of a heated tube.

silicon carbide). Downstream of the hole, in a vacuum chamber, is the molecular beam, which actually is a low-pressure, supersonic jet or plume. Figure 13.1 is a sketch of a tube based on the one described in Rohrs et al. (1995). In their experiment, the gas is helium, which is seeded with a small molar fraction (probably below 1%) of a vapor consisting of large molecules. It is the energy exchange properties of the vapor molecules that are investigated in the downstream plume. Between stations 1 and e, the wall is resistively heated with an electric current to a high temperature. A typical plenum pressure is about 1 bar, while the downstream ambient pressure ranges from 0.1 to 0.5 torr. Thus, the pressures satisfy $p_i > p_e >> p_a$, and the flow should be choked at the exit of the tube.

An accurate representation for the viscosity and thermal conductivity of a gas is

$$\mu = c_\mu T^{\omega_\mu} \tag{13.19a}$$

$$\kappa = c_\kappa T^{\omega_\kappa} \tag{13.19b}$$

where μ is in Pa-s, κ is in J/m-K-s, and T is in degrees Kelvin. The helium constants are based on data in Svehla (1962):

$$c_\mu = 5.0385 \times 10^{-7}, \qquad \omega_\mu = 0.649 \tag{13.20a}$$

$$c_\kappa = 3.9261 \times 10^{-3}, \qquad \omega_\kappa = 0.649 \tag{13.20b}$$

With helium considered as a perfect gas, and with $\omega_\mu = \omega_\kappa$, the Prandtl number

$$Pr = \frac{\mu c_p}{\kappa} = \frac{\gamma R}{\gamma - 1}\frac{c_\mu}{c_\kappa} = \frac{2}{3} \tag{13.21}$$

is a constant.

In accord with the preceding discussion, a steady continuum flow of a perfect Newtonian gas is assumed. With a rounded entrance, entrance effects, such as a vena contracta, can be disregarded. It is anticipated that both frictional and heat transfer effects are important; hence, a combined Rayleigh/Fanno flow analysis is in order using the influence coefficient method. (See Problem 9.2 for Rayleigh/Fanno flow of a calorically imperfect gas.)

A crucial consideration is whether the flow in the tube is laminar or turbulent. The Reynolds number is likely to be relatively small, since d is only about 1 mm (Rohrs et al., 1995). In turn, the Reynolds number depends on the inlet Mach number M_i. To obtain this relation, we start with

$$Re = \frac{\rho d w}{\mu} = \frac{4\dot{m}}{\pi d \mu} \tag{13.22}$$

where $4\dot{m}/(\pi d)$ is a constant along the tube. Since T increases with x, except possibly near the exit where it may have a Rayleigh-flow type of maximum, the Reynolds number is a maximum at the entrance. With the assumption of isentropic flow between the plenum, denoted with a zero subscript, and station i, the mass flow rate is given by

$$\dot{m} = \frac{M_i}{X_i^{(\gamma+1)/[2(\gamma-1)]}}\left(\frac{\gamma}{RT_o}\right)^{1/2} p_o\left(\frac{\pi}{4}d^2\right) \tag{13.23}$$

where

$$X = 1 + \frac{\gamma - 1}{2}M^2 \tag{13.24}$$

The inlet viscosity is written as

$$\mu_i = c_\mu T_i^{\omega_\mu} = c_\mu T_o^{\omega_\mu} X_i^{-\omega_\mu} \tag{13.25}$$

This results in

$$Re_i = \frac{p_o d}{c_\mu T_o^{\omega_\mu}}\left(\frac{\gamma}{RT_o}\right)^{1/2} \frac{M_i}{X_i^{(\gamma+1)/[2(\gamma-1)] - \omega_\mu}} \tag{13.26a}$$

With

$$\gamma = \frac{5}{3}, \qquad p_o = 10^5 \text{ Pa}, \qquad T_o = 300 \text{ K}, \qquad d = 1.07 \times 10^{-3} \text{ m} \tag{13.27}$$

we obtain

$$Re_i = \frac{8584 M_i}{\left(1 + \frac{1}{3}M_i^2\right)^{1.351}} \tag{13.26b}$$

(Hereafter, all numerical values are based on Rohrs et al., 1995.) When $M_i = 0.3$, the inlet Reynolds number is about 2474. For M_i values below 0.3, we may assume laminar flow and take inlet conditions as stagnation, or plenum, conditions. Above a value of 0.3, the flow should be transitional or turbulent, and inlet and plenum conditions are isentropically related.

Since the tube has an L/d of 40 and the flow is subsonic, we anticipate that the gas temperature at the exit may not be too different from the exit wall temperature T_{we}. With $T_o = 300$ K, $T_{we} \cong 1500$ K and $M_e = 1$, we assume T_{oe} is about 1200 K, and a Rayleigh line analysis yields $M_i \cong 0.23$. (A larger value for T_{oe} would further reduce M_i.) With friction included, this inlet value would be even smaller. In the latter analysis, we therefore assume a laminar flow and ignore the expansion from the plenum to the inlet station. In this circumstance, Equation (13.23) becomes

$$\dot{m} = 2.224 \times 10^{-2} \frac{p_o}{T_o^{1/2}} d^2 M_i \tag{13.28a}$$

and with Equations (13.27)

$$\dot{m} = 1.470 \times 10^{-4} M_i \tag{13.28b}$$

For instance, if $M_i = 0.1$, then the mass flow rate is only 1.47×10^{-5} kg/s or 52.9 g/h.

FORMULATION

The flow is analyzed using the influence coefficient equations (Emanuel, 1986)

$$\frac{2}{M}\frac{dM}{dx} = -\frac{WX}{ZT_o}\frac{dT_o}{dx} - \frac{\gamma M^2 X}{Z}\left(\frac{4c_f}{d}\right) \tag{13.29}$$

$$\frac{1}{p}\frac{dp}{dx} = \frac{\gamma M^2 X}{ZT_o}\frac{dT_o}{dx} + \frac{\gamma M^2 (W - M^2)}{2Z}\left(\frac{4c_f}{d}\right) \tag{13.30}$$

$$\frac{1}{T}\frac{dT}{dx} = -\frac{(1-\gamma M^2)X}{ZT_o}\frac{dT_o}{dx} + \frac{\gamma(\gamma-1)M^4}{2Z}\left(\frac{4c_f}{d}\right) \tag{13.31}$$

where

$$X = 1 + \frac{\gamma - 1}{2}M^2, \qquad W = 1 + \gamma M^2, \qquad Z = M^2 - 1 \tag{13.32}$$

Only the M and T equations are coupled and require a simultaneous solution. It is convenient, however, to also include the pressure equation, since its solution then provides easy evaluation of other parameters, such as the density, or the stagnation pressure

$$p_o(x) = pX^{\gamma/(\gamma-1)} \tag{13.33}$$

which is given by an isentropic, [illegible] relation. Similarly

$$T_o = TX \tag{13.34}$$

and its differential form

$$\frac{dT_o}{T_o} = \frac{dT}{T} + \frac{(\gamma - 1)M}{X}dM \tag{13.35}$$

is consistent with Equations (13.29) and (13.31).

Suitable equations must be established for $4c_f/d$ and $(1/T_o)dT_o/dx$. We proceed to this task with the understanding that this is the least precise part of the formulation. The reason is that the relations we adopt were primarily developed for constant property, incompressible, fully developed pipe flow. Comparable, well-established, laminar and turbulent compressible flow relations do not appear to be available.

Since the flow is presumed laminar, we utilize

$$c_f = \frac{16}{Re} \tag{13.36}$$

for the skin-friction coefficient. (The incompressible turbulent skin-friction coefficient is adequate for a subsonic flow; Humble et al., 1951; Kennan and Neumann, 1946.) This equation becomes

$$\frac{4c_f}{d} = \frac{16\pi\mu}{\dot{m}} \tag{13.37}$$

and is used in Equations (13.29)–(13.31).

For the heat transfer, we write dq as

$$dq = dh_o = \frac{\gamma R}{\gamma - 1} dT_o \tag{13.38}$$

and Newton's formula

$$\dot{m}dq = \pi d\hat{h}(T_w - T_{ad})dx \tag{13.39}$$

where $\hat{h}$ is the convective heat transfer coefficient, with units of J/m²-K-s, and T_{ad} is the adiabatic wall temperature. With the approximation of a unity value for the recovery factor (Shapiro, 1953, p. 243), T_{ad} equals T_o. The Prandtl number, Nusselt number

$$Nu = \frac{\hat{h}d}{\kappa} \tag{13.40}$$

and Equation (13.34) are introduced into the above, with the result

$$\frac{1}{T_o}\frac{dT_o}{dx} = \frac{\pi\mu Nu}{\dot{m}PrX}\left(\frac{T_w}{T} - X\right) \tag{13.41}$$

With incompressible flow in a constant heat flux heated tube (Mills, 1995), the local Nusselt number is equal to 4.364.

Other approaches have been discussed; e.g., see Shapiro (1953). These include the use of Reynolds' analogy, a constant wall temperature, or a constant wall heat flux. These approximations are usually made in the interest of simplicity.

Equations (13.29)–(13.31) now become

$$\frac{dM}{dz} = -\frac{\pi L\mu}{2\dot{m}}\frac{M}{Z}\left[\frac{Nu}{Pr}W\left(\frac{T_w}{T} - X\right) + 16\gamma M^2 X\right] \tag{13.42a}$$

$$\frac{dp}{dz} = -\frac{\pi\gamma L\mu}{\dot{m}}\frac{pM^2}{Z}\left[\frac{Nu}{Pr}\left(\frac{T_w}{T} - X\right) + 8(W - M^2)\right] \tag{13.42b}$$

$$\frac{dT}{dz} = \frac{\pi L\mu}{\dot{m}}\frac{T}{Z}\left[\frac{Nu}{Pr}(\gamma M^2 - 1)\left(\frac{T_w}{T} - X\right) + 8\gamma(\gamma - 1)M^4\right] \tag{13.42c}$$

where $z = (x/L)$. At the exit, we have

$$z = 1, \qquad M_e = 1, \qquad Z_e = 0 \tag{13.43}$$

With Z equally to zero in the denominator, we expect that each of the square brackets is zero when the Mach number is unity. All three square brackets reduce to

$$\frac{T_{we}}{T_e} = \frac{\gamma + 1}{2} - \frac{8\gamma Pr}{Nu_e} \tag{13.44a}$$

and for helium, this becomes

$$\frac{T_{we}}{T_e} = \frac{4}{3} - \frac{80}{9Nu_e} \tag{13.44b}$$

Based on the device described in Rohrs et al. (1995), we assume for the wall's temperature distribution

$$T_w = T_{wi} + \frac{z}{z_1}(T_{w1} - T_{wi}), \qquad 0 \le z \le z_1 \tag{13.45a}$$

$$T_w = T_{we} - (T_{we} - T_{w1})\left(\frac{1 - z}{1 - z_1}\right)^2, \qquad z_1 \le z \le 1 \tag{13.45b}$$

where z_1 is the location where the electrical heating starts. For the unheated section, a linear variation is used, where T_{w1} is not expected to be much larger than T_{wi}. A quadratic variation is used for the heated section, with

$$T_w(1) = T_{we}, \qquad \left.\frac{dT_w}{dz}\right|_{z=1} = 0 \tag{13.46}$$

Thus, the wall temperature increases rapidly just downstream of z_1. The following data are used in the subsequent computations:

$$L = 4 \times 10^{-2}\ \text{m}, \qquad T_{wi} = 300\ \text{K}, \qquad z_1 = 0.425, \qquad T_{w1} = 350\ \text{K},$$

$$T_{we} = 1500\ \text{K}, \qquad M(0) = M_i, \qquad p(0) = p_o, \qquad T(0) = T_o$$

along with Equations (13.27). Equations (13.42) have the form

$$\frac{dy_i}{dz} = f(y_1, y_2, y_3, z) \tag{13.47}$$

where the y_i represent M, p, and T. A fourth-order Runga–Kutta scheme with a fixed step size and double precision is utilized.* Actually, the step size is reduced by a factor of 10 for Mach numbers

* I am indebted to Mr. D. S. Gathright for the computations and the associated figures.

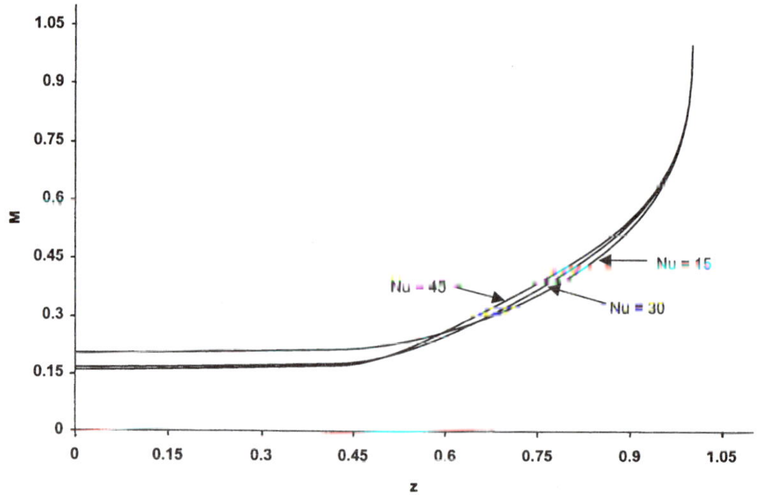

FIGURE 13.2 Mach number distribution for various Nusselt numbers.

larger than 0.97. The solution is an iterative one with a guessed value, to five significant digits, for M_i. Convergence is expected to occur when M_e is unity.

These calculations demonstrate that Equations (13.44a,b) are not satisfied, regardless of the constant value for the Nusselt number. In order for any of the square brackets to go to zero as $M \to 1$, the factor $(T_w/T) - X$ must have a sign change near the end of the duct. This, however,

Nu	M_i	Re_i
15	0.20223	1733
30	0.16483	1412
45	0.15803	1354

does not happen even with a Nusselt number as large as 90.

Figure 13.2 shows M vs. z for various Nusselt numbers. The initial Mach number and Reynolds number are given in the above table, while the mass flow rate is provided by Equation (13.28). The Mach number does not change significantly until after the gas reaches the heated section, and there is little variation between the three cases. The Mach number gradient, dM/dz, is infinite at the end of the duct, where $M = 1$ and the square bracketed terms in Equations (13.42) are finite.

Figures 13.3 and 13.4 show the pressure and temperature variation. Both are negatively infinite at the end of the duct and show little change until after the heated section is encountered. The temperature has a maximum value typical of a thermally choked Rayleigh flow. Neither the pressure nor temperature profiles change much for Nusselt numbers larger than 45. The peak temperature in the 15 and 45 Nusselt number cases is 949 and 1281 K, respectively, with correspondingly lower exit temperatures. (With a Nusselt number of 4.364, the exit temperature is only 360 K.) Thus, the exit temperature, regardless of Nusselt number, is well below the exit wall temperature of 1500 K. For example, it is only 1110 K when $Nu = 45$.

From the temperature profiles, it would appear that there is a reversal in the direction of heat transfer near the end of the duct. Figure 13.5 shows the stagnation temperature profiles, which steadily increase. Consequently, there is no reversal in the heat transfer direction. Finally, Figure 13.6 shows the stagnation pressure profiles, which behave as expected.

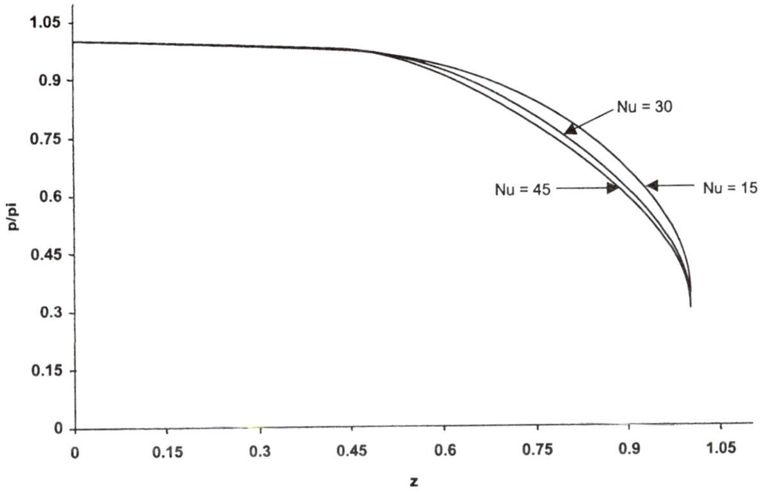

FIGURE 13.3 Pressure distribution for various Nusselt numbers.

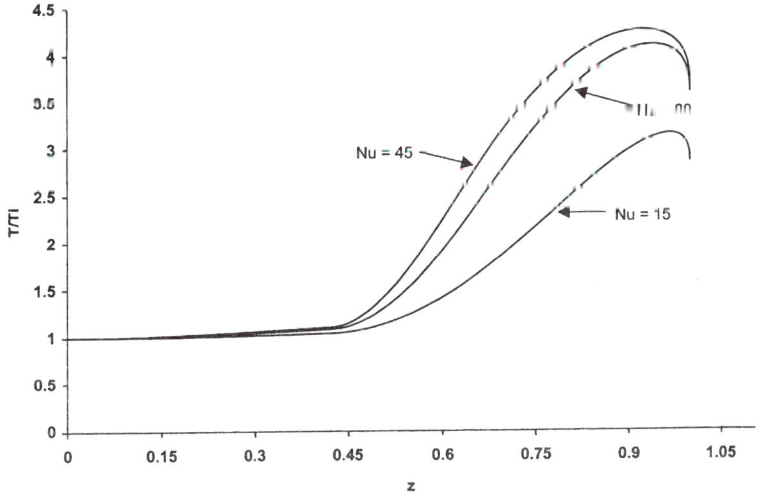

FIGURE 13.4 Temperature distribution for various Nusselt numbers.

It is somewhat surprising that the exit temperature should be considerably lower than T_{we}. This stems, in part, from the very small value, near the duct's exit, for the factor

$$(\gamma M^2 - 1)\left(\frac{T_w}{T} - X\right)$$

that appears in Equation (13.42c). Equally surprising is the inability to attain the expected indeterminacy condition, "0/0," at the exit of the duct. This condition regularly occurs in contoured nozzle flows when $M = 1$. Here, however, the exit of the duct is abrupt and the transition between subsonic and supersonic flows is not smooth. Rather, the sonic exit condition corresponds to a limit point.

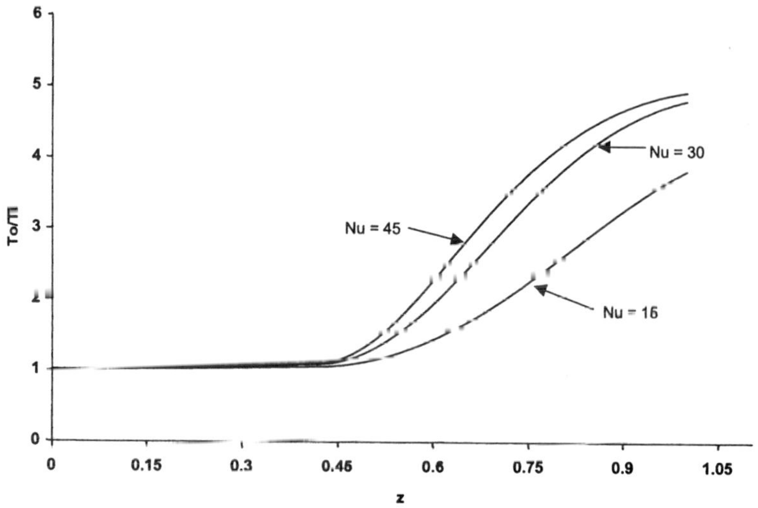

FIGURE 13.5 Stagnation temperature distribution for various Nusselt numbers.

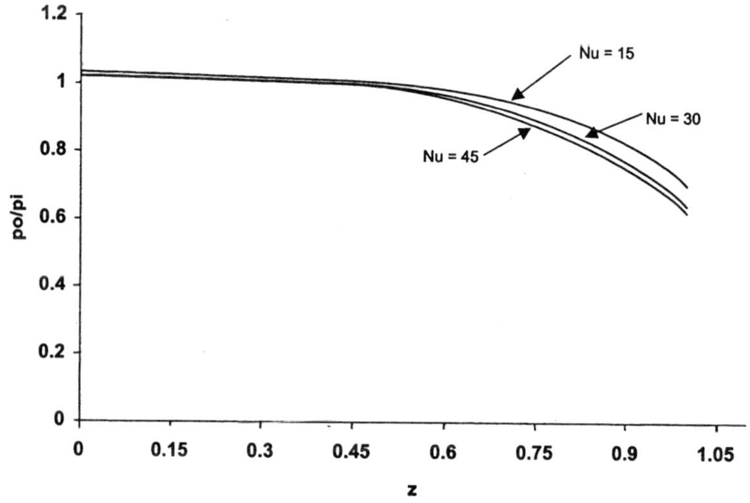

FIGURE 13.6 Stagnation pressure distribution for various Nusselt numbers.

REFERENCES

Bauer, H.J., "Phenomenological Theory of the Relaxation Phenomena in Gases," *Physical Acoustics,* Vol. II, Part A, Academic Press, 1965, pp. 47–131.

Candler, G.V., Laux, C.O., Gessman, R.J., and Kruger, C.H., "Numerical Simulation of a Nonequilibrium Nitrogen Plasma Experiment," AIAA 97–2365 (1997).

Emanuel, G., *Gasdynamics: Theory and Applications,* AIAA Education Series, New York, 1986, Chapter 8.

Emanuel, G., "Effect of Bulk Viscosity on a Hypersonic Boundary Layer," *Phys. Fluids A* 4, 491 (1992).

Emanuel, G., "Bulk Viscosity in the Navier–Stokes Equations," *Int. J. Eng. Sci.* 36, 1313 (1998).

Emanuel, G. and Argrow, B.M., "Linear Dependence of the Bulk Viscosity on Shock Wave Thickness," *Phys. Fluids* 6, 3203 (1994).

Garen, W., Synofzik, R., and Wortberg, G., "Experimental Investigation of the Structure of Weak Shock Waves in Noble Gases," *Prog. in Astro. and Aeron.,* Vol. 51, Part 1, AIAA, 519–528 (1977).

Hanley, H.J.M. and Cohen, E.G.D., "Analysis of the Transport Coefficients for Simple Dense Fluids: The Diffusion and Bulk Viscosity Coefficients," *Physica* 83A, 215 (1976).

Humble, L.V., Lowdermilk, W.H., and Desmon, L.G., "Measurements of Average Heat-Transfer and Friction Coefficients for Subsonic Flow of Air in Smooth Tubes at High Surface and Fluid Temperatures," NACA Lewis Flight Prop. Lab., Report 1020 (1951).

Kennan, J.H. and Neumann, E.P., "Measurements of Friction in a Pipe for Subsonic Flow of Air," *ASME Trans.* 68, pp. A-91–A-100 (1946).

Kneser, H.O., "Relaxation Processes in Gases," *Physical Acoustics,* Vol. II, Part A, Academic Press, New York, 1965, pp. 133–202.

Madigosky, W.M., "Density Dependence of the Bulk Viscosity in Argon," *J. Chem. Phys.* 46, 4441 (1967).

Mills, A.F., *Basic Heat and Mass Transfer,* Irwin, Chicago, 1995, pp. 235–238.

Monchik, L., Yun, K.S., and Mason, E.A., "Formal Kinetic Theory of Transport Phenomena in Polyatomic Gas Mixtures," *J. Chem. Phys.* 39, 654 (1963).

Prangsma, G.J., Alberga, A.H., and Beenakker, J.J.M., "Ultrasonic Determination of the Volume Viscosity of N_2, CO, CH_4, and CD_4 between 77 and 300 K," *Physica* 6, 278 (1973).

Rohrs, H.W., Wickham-Jones, C.T., Ellison, G.B., Berry, D., and Argrow, B.M., "Fourier Transform Infrared Absorption Spectroscopy of Jet-Cooled Radicals," *Rev. Sci. Instrum.* 66, 2430 (1995).

Shapiro, A.H., *The Dynamics and Thermodynamics of Compressible Fluid Flow,* Vol. I, The Ronald Press Co., New York, 1953.

Svehla, R.A., "Estimated Viscosities and Thermal Conductivities of Gases at High Temperatures," NASA TR R-132 (1962).

Tisza, L., "Supersonic Absorption and Stokes' Viscosity Relation," *Phys. Rev.* 61, 531 (1942).

Van Dyke, M., "Second-Order Compressible Boundary Layer Theory with Application to Blunt Bodies in Hypersonic Flow," *Hypersonic Flow Research,* Academic Press, New York, 1962, pp. 37–76.

PROBLEMS

13.1 Consider a three-dimensional, compressible flow in which the velocity and temperature depend only on the radius r and on t. Other quantities, such as pressure and density, may also depend on the angles θ and ϕ. Assume that μ, λ, c_p, and κ depend only on T. Use spherical coordinates, and write the equations for:

(a) rate-of-deformation tensor, $\overleftrightarrow{\varepsilon}$

(b) rotation tensor, $\overleftrightarrow{\omega}$

(c) velocity gradient tensor, $\nabla\vec{w}$

(d) continuity equation

(e) vorticity

(f) stress/strain relation

(g) momentum equation in scalar form

(h) dissipation function

(i) energy equation

Simplify your results as much as possible.

13.2 **(a)** Continue with the cylindrical coordinate example of Section 13.2 by writing the conservation equations in scalar form.

(b) Utilize the results of part (a) to obtain the governing equations for steady axisymmetric flow using the coordinates shown in the sketch. Insert the σ parameter in the equations so that they hold for two-dimensional as well as axisymmetric flow. Allow for swirl but assume the vorticity only has a nonzero component in the $\hat{e}_\theta$ direction. Determine v_θ and use this result to simplify the equations.

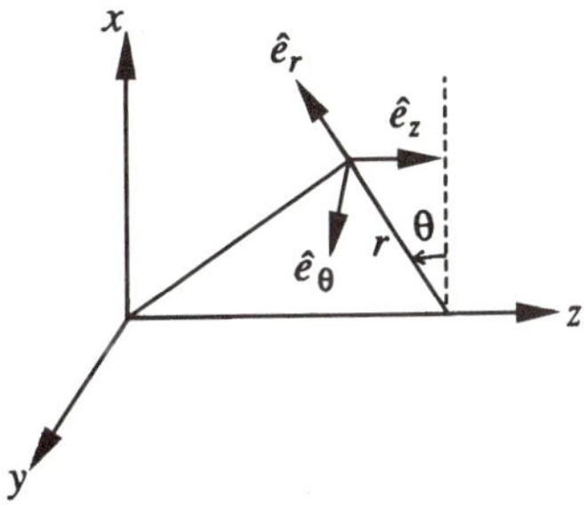

13.3 Continue with Problem 13.2(a) with the assumptions that the axisymmetric flow is without swirl, depends only on r and t, and $v_z = 0$.

(a) Determine the governing equations and $\vec{\omega}$.

(b) Assume ρ, μ, κ, and c_p are constants. Integrate the continuity and momentum equations, thereby determining v_r and p. Simplify the energy equations as much as possible.

13.4 Apply the results of Problem 13.3 to the incompressible flow between two concentric cylinders as shown in the sketch. The radius of the inner cylinder oscillates in accordance with

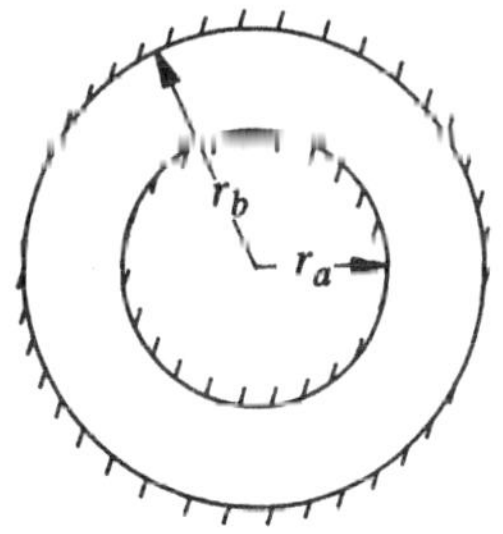

$$r_a = r_{ao}(1 + \varepsilon \sin \omega t)$$

where r_{ao}, ε, and ω are constants, and the flow has a time-averaged constant pressure p_{ao}. Assume ω is small compared to unity and determine $v_r(r, t)$, $p(r, t)$, and $r_b(t)$. The motion of the wall at r_b is such that the assumption of a cylindrically symmetric motion is preserved.

13.5 Derive the governing scalar equations for steady, two-dimensional or axisymmetric flow in natural coordinates using the nomenclature of the adjoining sketch. (The end product of this problem is the natural coordinate version of the compressible Navier–Stokes equations.) Assume no swirl and utilize for the scale factors (see Section 5.7):

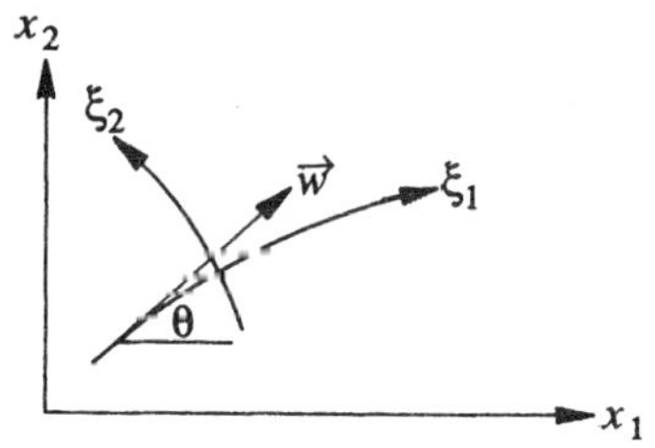

$$\frac{1}{h_2}\frac{\partial h_1}{\partial \xi_2} = -\frac{\partial \theta}{\partial \xi_1}, \qquad \frac{1}{h_1}\frac{\partial h_2}{\partial \xi_1} = \frac{\partial \theta}{\partial \xi_2}$$

$$h_3 = x_2^{\sigma}, \qquad \frac{1}{h_1}\frac{\partial h_3}{\partial \xi_1} = \sigma \sin\theta, \qquad \frac{1}{h_2}\frac{\partial h_3}{\partial \xi_2} = \sigma\cos\theta$$

$$\frac{\partial x_2}{\partial \xi_1} = h_1 \sin\theta, \qquad \frac{\partial x_2}{\partial \xi_2} = h_2\cos\theta$$

$$\frac{\partial}{\partial s} = \frac{1}{h_1}\frac{\partial}{\partial \xi_1}, \qquad \frac{\partial}{\partial n} = \frac{1}{h_2}\frac{\partial}{\partial \xi_2}$$

The independent variables should be s and n, while the dependent variables are ρ, h, p, w, θ, and x_2. First, develop formulas for ε_{ij}, $\frac{D\phi}{Dt}$ (ϕ is a scalar), $\frac{D\vec{w}}{Dt}$, $\nabla \cdot \vec{w}$, Φ, $\vec{q}$, and $\nabla \cdot \vec{q}$. Next, write the viscous terms as

$$\nabla \cdot \overleftrightarrow{\tau} = F_1\hat{e}_s + F_2\hat{e}_n$$

where, for $i = 1, 2$,

$$F_i = \mu f_i^{\mu} + \mu_b f_i^{\mu_b} + \frac{\partial \mu}{\partial s} f_i^{\mu_s} + \frac{\partial \mu_b}{\partial s} f_i^{\mu_{b,s}} + \frac{\partial \mu}{\partial n} f_i^{\mu_n} + \frac{\partial \mu_b}{\partial n} f_i^{\mu_{b,n}} + \frac{\sigma\mu}{x_2} f_i^{\sigma}$$

and μ_b is the bulk viscosity. (The F_i components here differ from the F_i^s in Appendix G.) Use the foregoing results to obtain the governing equations. Can the continuity equation be integrated?

13.6 Continue with Problem 13.5 and develop an equation for the vorticity $\vec{\omega}$. Utilize Equation (4.19) and derive the equation for the change of vorticity along a streamline

$$\frac{\partial \omega}{\partial s} = \frac{1}{\rho}\frac{\partial(w, \rho)}{\partial(s, n)} + \frac{\sigma \omega \sin\theta}{x_2} + \frac{V}{\rho w}$$

where thermal and caloric state equations are not assumed and V (for viscosity) is

$$V = \frac{\partial \theta}{\partial n}F_2 + \frac{\partial F_2}{\partial s} + \frac{\partial \theta}{\partial s}F_1 - \frac{\partial F_1}{\partial n}$$

13.7 Consider the structure of a steady, normal shock wave. Upstream and downstream conditions are denoted with subscripts 1 and 2, respectively, and the velocity is written as

$$\vec{w} = u(x)\hat{1}_x$$

(a) Use the theory in Chapter 3 to show that

$$\Delta s = s_2 - s_1 = \int_{-\infty}^{\infty} \dot{s}_{irr}\frac{dx}{u}$$

Hence, only irreversible processes contribute to the entropy change across the shock.

(b) Derive an equation for $s(x) - s_1$ that shows the dependence on κ, μ, and λ.

(c) Determine the condition for s to have a maximum value inside the shock and interpret the physical significance of this condition.

13.8 Continue with Problem 13.7. Do not assume constant properties or Stokes' hypothesis.

(a) Integrate each conservation equation once.

(b) Set the Prandtl number, defined here as $(2\mu + \lambda)c_p/\kappa$, equal to unity and integrate the energy equation.

(c) Assume a perfect gas and $\mu' = 2\mu + \lambda = C_{\mu'}T$, where $C_{\mu'}$ is a constant, and integrate the momentum equation. Introduce the ratio of flow speeds, u_2/u_1, to simplify your results. How would you choose the integration constant?

13.9 Use the results and assumptions of Problems 13.7 and 13.8. Determine an algebraic equation for the rate of entropy production in the shock with the form

$$\frac{C_{\mu'}\dot{s}_{irr}}{R^2\rho_1} = f\left(\frac{u}{u_1}; \frac{u_2}{u_1}, \gamma\right)$$

13.10 Consider a thermally perfect gas situated above an infinite horizontal flat plate. Let z be a coordinate normal to the plate and include gravity with the approximation that $\vec{g}$ is a constant. Further, assume the gas is quiescent and that Fourier's equation holds.

(a) Determine the governing equations.

(b) With

$$\kappa = C_\kappa T^\omega$$

where C_κ and ω are constants ($\omega > -1$), determine $T(z)$ and $p(z)$.

(c) The requirement for a stable atmosphere at altitude z_1 is that $(ds/dz)_1 > 0$. Determine a temperature lapse rate, $(dT/dz)_1$, condition so that the atmosphere is stable at this altitude.

13.11 Start with Equations (13.6) and consider the unsteady, one-dimensional motion of a perfect gas. Simplify the notation by introducing $\vec{w} = u(x, t)\hat{1}_x$ and $\mu' = 2\mu + \lambda$.

(a) Introduce Equations (13.10) with $\tau = (\ell / U_\infty)$ and obtain nondimensional equations that exhibit their dependence on the parameters

$$M_\infty^2 = \frac{\rho_\infty U_\infty^2}{\gamma p_\infty}, \qquad Re = \frac{\rho_\infty U_\infty \ell}{\mu'_\infty}, \qquad Pr = \frac{\gamma R}{\gamma - 1}\frac{\mu'_\infty}{\kappa_\infty}$$

where the infinity subscript denotes a reference condition.

(b) Show that the governing equations, obtained in part (a), are invariant under a Galilean transformation (see Section 1.2). As a consequence, if $u(x)$ is a stationary solution, such as for a standing shock wave, then $u(x, t) = u(x - Ut)$ is a solution of the unsteady equations, where U is the constant speed with which the disturbance propagates in the direction of increasing x.

(c) Let $u_o(x)$, $\rho_o(x), \ldots$, be a stationary solution. Derive the first-order perturbation equations for $u_1, \ldots, \mu'_1, \kappa_1$, where

$$u = u_o + u_1 = u_o(x - Ut) + u_1(x, t)$$
$$\rho = \rho_o + \rho_1 = \rho_o(x - Ut) + \rho_1(x, t)$$
$$\vdots$$

What is the condition for viscous dissipation to appear in the first-order equations? (Although similar to Problem 2.15, this problem differs from it in that u_o is not identically zero.)

13.12 Parabolic coordinates ξ_i are utilized:

$$x_1 = \frac{1}{2}(\xi_1^2 - \xi_2^2), \qquad x_2 = \xi_1 \xi_2 \sin \xi_3, \qquad x_3 = \xi_1 \xi_2 \cos \xi_3$$

These coordinates are orthogonal and the surfaces ξ_1 = constant, ξ_2 = constant are paraboloids of revolution about the x_1-axis, while ξ_3 is the azimuthal angle about x_1. Assume axisymmetric flow with swirl. Thus, the velocity components v_i are functions of t, ξ_1, and ξ_2.

(a) Use Appendix G to obtain h_i, $\nabla\phi$, $\nabla^2\phi$, $\nabla \cdot \vec{A}$, $\nabla \times \vec{A}$, $\partial \hat{e}_j / \partial \xi_i$, $D\phi/Dt$, $D\vec{w}/Dt$, ε_{ij}, Φ, and the F_i^s. Simplify results as much as possible.

(b) Obtain the governing viscous equations using parabolic coordinates.

(c) Obtain the vorticity.

13.13 Continue with Problem 13.12 but, in addition, assume a steady, inviscid flow without swirl of a perfect gas.

(a) Write down the Euler equations.

(b) Further assume the flow is irrotational and isoenergetic. Introduce a velocity potential ϕ and obtain explicit equations for v_1, v_2, p, and ρ in terms of ϕ.

(c) Derive the PDE for ϕ.

(d) Determine under what conditions the PDE for ϕ is elliptic, parabolic, or hyperbolic. For this evaluation, the Mach number needs to be introduced.

(e) Assume a uniform flow

$$\vec{w} = w_1 \hat{\imath}_1 = \text{constant}$$

and derive the corresponding ϕ in terms of parabolic coordinates. Show that this $\phi(\xi_1, \xi_2)$ satisfies the PDE obtained in part (c).

13.14 Derive the ε_{ij} equations contained in Appendix G.

13.15 Consider steady, two-dimensional or axisymmetric, compressible, viscous flow. Neglect body forces and assume a Newtonian/Fourier fluid, that w_3 is a constant when $\sigma = 0$, and that the flow may have swirl when $\sigma = 1$. Utilize the coordinate system developed in Section 5.6 and note that $\partial(\;)/\partial x_3$ is zero, except for several $\hat{\imath}_i$ derivatives.

(a) In preparation for the later analysis, develop equations for

$$\frac{\partial \hat{\imath}_i}{\partial x_j}, \quad \nabla \cdot \vec{w}, \quad \vec{\omega}, \ \nabla \vec{w}, \quad \overleftrightarrow{\varepsilon}, \quad \vec{F}^s = 2\nabla \cdot (\mu \overleftrightarrow{\varepsilon}), \quad \Phi$$

and $D(\;)/Dt$ of a scalar.

(b) Derive the conservation equations starting with those given in Chapter 2. Your final result for momentum should be in scalar form. Utilize $\vec{F}^s$, and simplify your equations as much as possible.

(c) Assume inviscid, adiabatic flow and compare your conservation equations with Equations (5.67). Explain any differences.

(d) For $\sigma = 1$ and $\mu = \mu(x_1)$, show that Equation (5.75) satisfies the $\hat{\imath}_3$ momentum equation.

14 Force and Moment Analysis

14.1 PRELIMINARY REMARKS

Forces and moments are important in engineering, especially in fluid dynamics and aerodynamics. In most textbooks, their evaluation is typically done in a piecemeal fashion, e.g., providing an analytical or empirical formula for the drag of a body with a particularly simple shape, such as a sphere or a circular cylinder. A more general approach is taken here in which a variety of topics is surveyed. For example, these include the apparent mass of an object and the force and moment components associated with supersonic flight. The range of topics encompasses incompressible to supersonic (or hypersonic) flows.

The more general methods to be discussed are associated with the momentum theorem. It is derived from Newton's second law in the next section. Section 14.3 evaluates the surface integral in the momentum theorem, while Section 14.4 derives the angular momentum theorem. The analysis of many of the subsequent topics is directly based on the momentum theorem and its angular momentum counterpart.

Sections 14.5 through 14.9 deal with incompressible flows. In sequence, we discuss hydrostatics, simple duct flow, acyclic motion, jet-plate interaction, and a syringe with a hypodermic needle.

A variety of approaches are presented that are applicable to a compressible flow. Some of these are inviscid, since pressure forces often dominate. In this circumstance, a correction for the skin friction can be added after the inviscid calculation is completed. Several useful methods are not discussed. For instance, the simplest approach, when applicable, is supersonic thin airfoil theory (Liepmann and Roshko, 1957, Section 4.17). In this technique, the lift and drag of a thin airfoil is decomposed into angle of attack, camber, and thickness contributions, which are additive. Leung and Emanuel (1995) provide a relatively simple method for obtaining the wave and viscous drag (as well as the heat transfer) for a cone and wedge, at zero incidence, in a hypersonic flow. The shock is attached and a laminar boundary layer is presumed. In a separate analysis (Bae and Emanuel, 1991), a procedure is established for the evaluation of the thrust, lift, pitching moment, and heat transfer associated with a special type of asymmetric nozzle that can be used with a scramjet engine. Other articles that deal with the application of Newtonian impact theory for a hypersonic flow are by Jaslow (1968) and by Pike (1974). A more comprehensive treatment of the forces and moments in a hypersonic flow is provided by Rasmussen (1994). The book by Nielsen (1988) can be consulted if the forces and moments on a missile are of interest.

The first topic, in Section 14.10, is shock expansion theory. This section also discusses wave interference. Section 14.11 is concerned with the forces that act on an aerosol particle in a nonuniform flow. The role of entropy production is described in Section 14.12, while the forces and moments on a supersonic vehicle is the topic of the last section.

As in other chapters, the emphasis is on concepts, limitations, and the way principles, or concepts, are formulated. Actual engineering design studies are best performed using specially developed computer codes or CFD software.

14.2 MOMENTUM THEOREM

As indicated in Figure 14.1, a solid, impermeable, motionless object, whose surface is denoted as S_w, is fully immersed in a gaseous or liquid fluid. The moving fluid exerts a force $\vec{F}_w$ and a moment $\vec{M}_w$ that acts on the S_w surface. The overall fluid plus solid body system is enclosed by the surface S_∞. As indicated in the figure, S_∞ may be a large surface, well removed from S_w. The subsequent

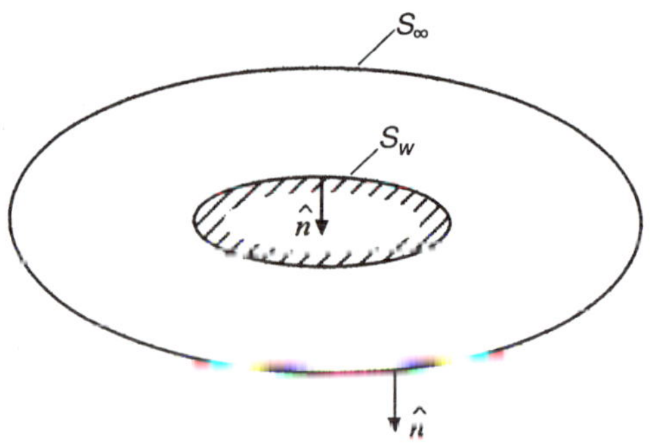

FIGURE 14.1 Schematic of a solid body immersed in a fluid.

derivation, however, does not require this. In fact, later sections discuss flows in which S_∞ is in close proximity to, or even adjacent to, S_w. Fluid crosses S_∞, both S_w and S_∞ are stationary, and the sum of the two surfaces constitutes the control surface (CS). Thus, the control volume (CV) represents the moving fluid between the two surfaces. The fluid motion may be steady or unsteady, incompressible or compressible, inviscid or viscous, and irrotational or rotational. For instance, shock waves and turbulent shear layers may be present. If the fluid is a gas, it need not be perfect; if it is a liquid, it need not be Newtonian.

From Chapter 2, we recall that Newton's second law, in an inertial frame, can be written as

$$\frac{D}{Dt}\int_V \rho\vec{w}dv = \int_V \rho\vec{F}_b dv + \oint_S \vec{\sigma} ds \tag{14.1}$$

where the right-hand side represents the vector sum of the applied forces that act on the system, which here is the fluid inside V. The first term on the right side represents any body force that acts throughout V. When this is due to gravity, we set $\vec{F}_b = \vec{g}$, where $\vec{g}$ is the acceleration due to gravity. The rightmost term represents the applied surface force at S_w and at S_∞ that acts on the fluid in V.

Specific Forms for the Momentum Theorem

At a given instant of time, the volume V and bounding surface S are identified as a fixed CV and CS, respectively. The transport theorem, with $\psi = \rho\vec{w}$, yields for the left side of Equation (14.1)

$$\frac{D}{Dt}\int_V \rho\vec{w}dv = \int_V \frac{\partial(\rho\vec{w})}{\partial t}dv + \oint_S \rho\vec{w}(\vec{w}\cdot\hat{n})ds \tag{14.2}$$

A simple, but effective, form for the momentum theorem can be written as

$$\sum\vec{F} = \int_{CV} \frac{\partial(\rho\vec{w})}{\partial t}dv + \oint_{CS} \rho\vec{w}(\vec{w}\cdot\hat{n})ds \tag{14.3}$$

where the left side represents the sum of all of the applied forces that act on the fluid inside the control volume.

In order to focus on the applied pressure and viscous forces that act only on the wall, Equations (14.1) and (14.2) are combined to yield

$$\int_{CV} \frac{\partial(\rho\vec{w})}{\partial t}dv + \oint_{CS} \rho\vec{w}\vec{w}\cdot\hat{n}ds = \int_{CV} \rho\vec{F}_b dv + \oint_{CS} \vec{\sigma} ds \tag{14.4}$$

The stress vector provides the pressure and shear forces that act on S_w; i.e.,

$$\vec{F}_w = -\oint_{S_w} \vec{\sigma} ds \tag{14.5}$$

which is a general equation for the force on a solid surface. The minus sign stems from our interest in the fluid force that acts on the solid body. As previously shown, the stress vector is given by

$$\vec{\sigma} = \hat{n} \cdot \overleftrightarrow{\tau} - p\hat{n} \tag{14.6}$$

where $\overleftrightarrow{\tau}$ is the viscous stress tensor. Consequently, we have

$$\oint_{CS} \vec{\sigma} ds = \oint_{S_\infty} (\hat{n} \cdot \overleftrightarrow{\tau} - p\hat{n}) ds - \vec{F}_w \tag{14.7}$$

Since $\vec{w} \cdot \hat{n} = 0$ on S_w, which is stationary, we obtain for an inviscid or viscous flow

$$\oint_{CS} \rho \vec{w}(\vec{w} \cdot \hat{n}) ds = \oint_{S_\infty} \rho \vec{w}(\vec{w} \cdot \hat{n}) ds \tag{14.8}$$

for the momentum flux term in Equation (14.4). This equation now reduces to

$$\vec{F}_w = \int_{CV} \left[\rho \vec{F}_b - \frac{\partial(\rho \vec{w})}{\partial t}\right] dv - \oint_{S_\infty} [p\hat{n} + \rho \vec{w}(\vec{w} \cdot \hat{n}) - \hat{n} \cdot \overleftrightarrow{\tau}] ds \tag{14.9}$$

which is a general form for the momentum theorem. The force $\vec{F}_w$ is therefore provided by a volumetric integral plus an integral over a stationary surface. When the interior of S_w does not contain a solid body, $\vec{F}_w = 0$. This form of the theorem will appear in both the incompressible and compressible flow sections.

The $\vec{F}_w$ force can be decomposed into various components by dotting it with a unit vector. For instance, let $\vec{w}_\infty = w_\infty \hat{1}_x$ be the freestream velocity. The drag is then given by

$$\vec{F}_w \cdot \hat{1}_x = \int_{CV} \left[\rho \vec{F}_b \cdot \hat{1}_x - \frac{\partial(\rho \vec{w} \cdot \hat{1}_x)}{\partial t}\right] dv - \oint_{S_\infty} \left[p\hat{n} \cdot \hat{1}_x + \rho(\vec{w} \cdot \hat{1}_x)(\vec{w} \cdot \hat{n}) - (\hat{n} \cdot \overleftrightarrow{\tau}) \cdot \hat{1}_x\right] ds$$

since $\hat{1}_x$ is a constant vector.

If $\vec{F}_b = \vec{g}$, the CV integral in Equation (14.9) can be written as

$$\int_{CV} \left[\rho \vec{F}_b - \frac{\partial(\rho \vec{w})}{\partial t}\right] dv = m\vec{g} - \frac{d}{dt}\int_{CV} \rho \vec{w} dv \tag{14.10}$$

where the mass of fluid in the CV is

$$m = \int_{CV} \rho dv \tag{14.11}$$

When the flow is steady and the body force $\vec{F}_b$ is negligible, we have

$$\vec{F}_w = -\oint_{S_\infty} [p\hat{n} + \rho\vec{w}(\vec{w}\cdot\hat{n}) - \hat{n}\cdot\overleftrightarrow{\tau}]ds \tag{14.12}$$

Detailed flow conditions within the *CV* are now unnecessary; only conditions on S_∞ are required. This is a particularly useful result because these conditions can be measured, or, in certain situations, estimated. Note that viscous effects are not necessarily confined to the $\hat{n}\cdot\overleftrightarrow{\tau}$ term. Moreover, the inviscid result is not necessarily obtained by just deleting this term. This is because the Euler and Navier–Stokes solutions differ for p, ρ, and $\vec{w}$ throughout the *CV* and S_∞. For instance, the pressure and flux integrals do not have the same value for an attached inviscid solution as compared to a detached viscous solution.

14.3 SURFACE INTEGRAL

The three surface terms on the right side of Equation (14.12) are individually discussed. Two cases are to be distinguished. In the first, S_∞ is well removed from S_w. As will become apparent in later sections, situations commonly arise in which part of S_∞ is adjacent to a solid surface. This case is distinguished by denoting this part of the surface as S'_w. The rest of S_∞ is not adjacent to a solid surface, and S'_w is viewed an open surface, i.e., a surface with one or more edges. For instance, S'_w may be adjacent to the interior surface of a duct. Along with inlet and exit surfaces, S'_w constitutes S_∞. Since there is no body internal to S_∞, $\vec{F}_w$ in Equation (14.9) is zero. Of course, there is a fluid force on the interior surface of the duct, as discussed in Section 14.6.

Pressure Force

When S_∞ is well removed from S_w, the pressure at S_∞ can be approximated as a constant, with the result

$$\oint_{S_\infty} p_\infty \hat{n}ds = p_\infty \oint_{S_\infty} \hat{n}ds = 0 \tag{14.13}$$

The rightmost equality stems from the divergence theorem. It is important to note that the integral is not zero if a shock wave intersects S_∞. When part of S_∞ is adjacent to a solid surface, this part of the integral supplies the pressure force on S'_w.

Momentum Flux Force

We next discuss the flux term in Equation (14.12). On a solid stationary surface, $\vec{w}\cdot\hat{n}$ is zero for both inviscid and viscous flows. On S_∞, this term can be evaluated using a surface, such as the one sketched in Figure 14.2. Surface 1 is far upstream of the body and is normal to the freestream velocity $\vec{w}_\infty$, as is surface 2. If the freestream flow is supersonic, surface 1 need not be far upstream.

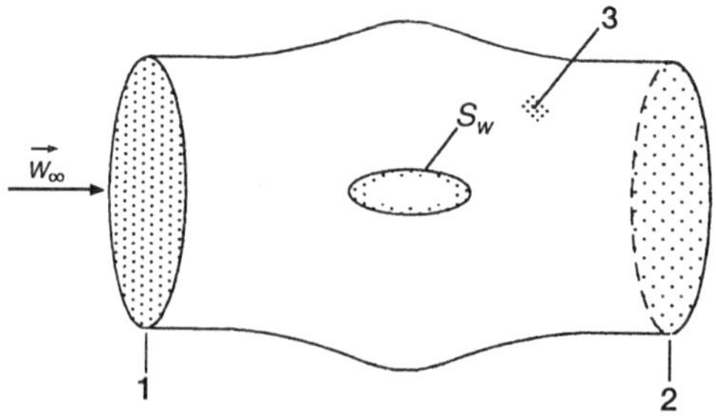

FIGURE 14.2 Solid body immersed in a large diameter streamtube.

Surface 3 is chosen as a streamtube. The convenience of using a streamtube for the lateral surface in a steady flow is illustrated in Problems 14.11 and 14.12. If the flow is unsteady, then a lateral, unsteady surface consisting of streaklines is appropriate. This surface consists of particle paths that have originated from a fixed curve that is located in surface 1. When the unsteadiness is largely confined to a region near the body and its wake, then a sufficiently distant steady streamtube can still be used. In any case, $\vec{w} \cdot \hat{n} = 0$ and

$$\int_{S_3} \rho\vec{w}(\vec{w} \cdot \hat{n})ds = 0 \tag{14.14}$$

Consequently, we obtain

$$\oint_{S_\infty} \rho\vec{w}(\vec{w} \cdot \hat{n})ds = \int_{S_1+S_2} \rho\vec{w}(\vec{w} \cdot \hat{n})ds \tag{14.15}$$

For instance, with a uniform freestream flow, the S_1 integral is given by

$$\int_{S_1} \rho\vec{w}(\vec{w} \cdot \hat{n})ds = -\rho_\infty A_1 w_\infty \vec{w}_\infty \tag{14.16}$$

where A_1 is the cross-sectional area at station 1, and the minus sign stems from

$$\hat{n}_1 = -\frac{\vec{w}_\infty}{w_\infty} \tag{14.17}$$

The S_2 integral is not as easily evaluated, since the wake shed by the body alters $\vec{w}$ on S_2. Nevertheless, Equation (14.15) states that the net flux of momentum through S_∞ is just the net flux through $S_1 + S_2$.

Viscous Force

Finally, the $\hat{n} \cdot \overleftrightarrow{\tau}$ term is examined. When S_∞ is well removed from S_w, the velocity gradient on S_∞ is orders of magnitude smaller than the gradient in a boundary-layer flow that is adjacent to a surface. Consequently, the viscous force on S_∞

$$\oint_{S_\infty} \hat{n} \cdot \overleftrightarrow{\tau} ds \cong 0 \tag{14.18}$$

is negligible. This result holds even in the wake region, providing the part of S_∞ that is crossed by the wake is sufficiently far downstream.

An expression is developed for $\hat{n} \cdot \overleftrightarrow{\tau}$ that will hold for a stationary solid surface, S'_w, or a part of S_∞ that is not well removed from S_w. From the Newtonian relation for the viscous stress tensor in Chapter 1

$$\overleftrightarrow{\tau} = 2\mu\overleftrightarrow{\varepsilon} + \left(\mu_b - \frac{2}{3}\mu\right)(\nabla \cdot \vec{w})\overleftrightarrow{I} \tag{14.19}$$

we have

$$\hat{n} \cdot \overleftrightarrow{\tau} = 2\mu\hat{n} \cdot \overleftrightarrow{\varepsilon} + \left(\mu_b - \frac{2}{3}\mu\right)(\nabla \cdot \vec{w})\hat{n} \tag{14.20}$$

where μ_b is the bulk viscosity. From Warsi (1993), we obtain (see Problem 14.1)

$$\hat{n} \cdot \overset{\leftrightarrow}{\varepsilon} = \frac{1}{2}\vec{\omega} \times \hat{n} + (\nabla \cdot \vec{w})\hat{n} + (\hat{n} \times \nabla) \times \vec{w} \tag{14.21}$$

where the vorticity is

$$\vec{\omega} = \nabla \times \vec{w} \tag{14.22}$$

Equation (14.20) now has the form

$$\hat{n} \cdot \overset{\leftrightarrow}{\tau} = \mu\vec{\omega} \times \hat{n} + \left(\mu_b + \frac{4}{3}\mu\right)(\nabla \cdot \vec{w})\hat{n} + 2\mu(\hat{n} \times \nabla) \times \vec{w} \tag{14.23}$$

This relation is not limited to a solid wall, but holds at any point, in any $\hat{n}$ direction, in a steady or unsteady, laminar, transitional, or turbulent flow field. It is not utilized until the final section; however, it is of central importance in that section. Although the velocity is zero on a stationary wall, this does not mean that any of the terms on the right side of Equation (14.23) are necessarily zero, since $\vec{w}$ appears in a derivative form. Nevertheless, we demonstrate that the two rightmost terms, in fact, are zero on a stationary wall in a steady flow.

Skin Friction

The above assertion is first established for the term containing $\nabla \cdot \vec{w}$. From continuity, we have

$$\nabla \cdot \vec{w} = -\frac{1}{\rho}\frac{D\rho}{Dt} = -\frac{1}{\rho}\left(\frac{\partial \rho}{\partial t} + \vec{w} \cdot \nabla\rho\right) \tag{14.24a}$$

and, on a solid wall in a steady flow, both $\partial\rho/\partial t$ and $\vec{w}$ are zero; hence,

$$\nabla \cdot \vec{w}\big|_{S'_w} = 0 \tag{14.24b}$$

The evaluation on S'_w of the rightmost term in Equation (14.23) starts with a form of Stokes' theorem

$$\oint_C d\vec{r} \times \vec{A} = \int_S (\hat{n} \times \nabla) \times \vec{A}\,ds \tag{14.25}$$

where $\vec{A}$ is an arbitrary vector and C is a simple closed curve, or curves, that bounds the open surface S. Let $\vec{A} = \vec{w}$ and C' be the bounding curve, or curves, of the wall surface S'_w, with the result

$$\oint_{C'} d\vec{r} \times \vec{w} = \int_{S'_w} (\hat{n} \times \nabla) \times \vec{w}\,ds \tag{14.26}$$

Since $\vec{w} = 0$ on S'_w, we have

$$\int_{S'_w} (\hat{n} \times \nabla) \times \vec{w}\,ds = 0 \tag{14.27}$$

and the integrand is zero because S'_w is arbitrary. This relation and Equation (14.24b) yield the elegant result

$$\int_{S'_w} \hat{n} \cdot \overleftrightarrow{\tau} ds = \int_{S'_w} \mu \vec{\omega} \times \hat{n} ds \quad \textbf{(14.28)}$$

for the viscous force, or skin friction, on a solid surface.

With a different version of Stokes' theorem, we can show that (Warsi, 1993)

$$(\vec{\omega} \cdot \hat{n})_{S'_w} = 0 \quad \textbf{(14.29)}$$

Hence, $\vec{\omega}$ is tangent to the wall, as is $\vec{\omega} \times \hat{n}$, where this cross product is aligned with the flow direction. Since $\vec{\omega}$ and $\hat{n}$ are perpendicular to each other, $|\vec{\omega} \times \hat{n}|$ has a maximum value of $|\vec{\omega}|$ on S'_w. These observations become clearer if a two-dimensional flow

$$\vec{w} = u\hat{\imath}_x + v\hat{\imath}_y, \qquad \hat{n} = \hat{\imath}_y \quad \textbf{(14.30)}$$

is considered, where $y = 0$ on a planar wall. We then have

$$\vec{\omega} = \left(\frac{\partial v}{\partial x} - \frac{\partial u}{\partial y}\right)\hat{\imath}_z \quad \textbf{(14.31a)}$$

$$\vec{\omega} \times \hat{n} = -\left(\frac{\partial v}{\partial x} - \frac{\partial u}{\partial y}\right)\hat{\imath}_x \quad \textbf{(14.31b)}$$

and with $\partial v/\partial x$ equal to zero on the wall,

$$\int_{S'_w} \hat{n} \cdot \overleftrightarrow{\tau} ds = \hat{\imath}_x \oint_{S'_w} \mu \frac{\partial u}{\partial y} ds \quad \textbf{(14.32)}$$

This is the expected result for the integrated skin friction in the x-direction.

An alternate approach would consider the magnitude of the skin friction

$$F_{sf} = \left| \int_{S'_w} \hat{n} \times \vec{F}_w ds \right| \quad \textbf{(14.33)}$$

that acts on the S'_w surface. From Equations (14.5) and (14.6), we have

$$F_{sf} = \int_{S'_w} |\hat{n} \times \vec{\sigma}| ds \quad \textbf{(14.34)}$$

where

$$\hat{n} \times \vec{\sigma} = \hat{n} \times (\hat{n} \cdot \overleftrightarrow{\tau}) \quad \textbf{(14.35)}$$

With the aid of Equation (14.23), this becomes

$$\hat{n}\times\vec{\sigma} = \mu\hat{n}\times(\vec{\omega}\cdot\hat{n}) = \mu[(\hat{n}\cdot\hat{n})\vec{\omega}-(\hat{n}\cdot\vec{\omega})\hat{n}] = \mu\vec{\omega} \tag{14.36}$$

We therefore obtain

$$F_{sf} = \int_{S'_w}\mu|\vec{\omega}|ds \tag{14.37}$$

which is in accord with Equation (14.28).

Comments

In the shock-free case, the surface integral in Equation (14.12) reduces to the momentum flux integral

$$\oint_{S_\infty}[p\hat{n}+\rho\vec{w}(\vec{\omega}\cdot\hat{n})-\hat{n}\cdot\overleftrightarrow{\tau}]ds = \int_{S_1+S_2}\rho\vec{w}(\vec{\omega}\cdot\hat{n})ds \tag{14.38}$$

for the type of flow sketched in Figure 14.2. As discussed in Problem 14.2, it is possible to simplify the CV integral in Equation (14.9) differently from that given by Equation (14.10). The result, however, is not an alternate form of the momentum theorem, since the $\vec{F}_w$ term cancels.

Suppose there is a shock wave inside the CV, as pictured in Figure 14.3. In this circumstance, is Equation (14.12) still valid? Denote by $S_{-\infty}$ the part of S_∞ that is upstream of the shock plus the surface S_-, which is just upstream of the shock. Similarly, denote as $S_{+\infty}$ the CS downstream of the shock; this surface encloses the bullet. Inside $S_{-\infty}$, there is a uniform freestream flow; hence,

$$\vec{F}_w = 0, \qquad \overleftrightarrow{\tau} = 0, \qquad \oint_{S_{-\infty}}p_\infty\hat{n}ds = 0 \tag{14.39a}$$

$$(\rho w^2)_\infty\oint_{S_{-\infty}}\hat{1}_x\cdot\hat{n}ds = (\rho w^2)_\infty\hat{1}_x\cdot\oint_{S_{-\infty}}\hat{n}ds = 0 \tag{14.39b}$$

and Equation (14.12) reduces to an identity when $S_{-\infty}$ is the CS. Next, we consider the CS consisting of $S_- + S_+$, which encloses the shock wave. For this surface,

$$\vec{F}_w = 0, \qquad \overleftrightarrow{\tau} = 0 \tag{14.40}$$

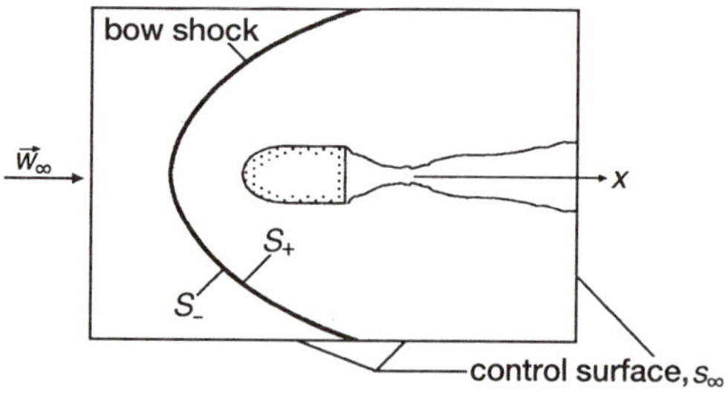

FIGURE 14.3 Control surfaces for a bullet in supersonic flight.

and the unit normal vector, $\hat{n}$, associated with S_- is equal and opposite to the one associated with S_+. Equation (14.12) reduces to

$$\int_{S_-} [p_\infty + \rho_\infty \vec{w}_\infty(\vec{w}_\infty \cdot \hat{n})]ds = \int_{S_+} [p_+\hat{n} + \rho_+\vec{w}_+(\vec{w}_+ \cdot \hat{n})]ds \quad \textbf{(14.41)}$$

This relation is equivalent to the momentum jump condition for a shock. Equation (14.12) holds for the force on the bullet if $S_{+\infty}$ is the control surface. Since S_∞ equals $S_{-\infty} + S_{+\infty}$, Equation (14.41) means that Equation (14.12) also holds for S_∞.

14.4 ANGULAR MOMENTUM

With the aid of Equation (2.14) and the transport theorem, we have

$$\int_{CV}\left[\frac{\partial(\rho\vec{r}\times\vec{w})}{\partial t} - \rho\vec{r}\times\vec{F}_b\right]dv = \oint_{CS} [\vec{r}\times\overleftrightarrow{\sigma} - \rho\vec{r}\times\vec{w}(\vec{w}\cdot\hat{n})]\,ds \quad \textbf{(14.42)}$$

As evident from Equation (14.5), the moment, or torque, exerted by the fluid on the body in Figures 14.1 and 14.2 is defined as

$$\vec{M}_w = -\oint_{S_w} \vec{r}\times\overleftrightarrow{\sigma}\,ds \quad \textbf{(14.43)}$$

With the aid of Equation (14.6), the *CS* integral becomes

$$\oint_{CV} [\vec{r}\times\overleftrightarrow{\sigma} - \rho\vec{r}\times\vec{w}(\vec{w}\cdot\hat{n})]ds$$

$$= -\vec{M}_w - \oint_{S_\infty} [p\vec{r}\times\hat{n} + \rho\vec{r}\times\vec{w}(\vec{w}\cdot\hat{n}) - \vec{r}\times(\hat{n}\cdot\overleftrightarrow{\tau})]ds \quad \textbf{(14.44)}$$

We thus obtain

$$\vec{M}_w = \oint_{CV}\left[\rho\vec{r}\times\vec{F}_b - \frac{\partial(\rho\vec{r}\times\vec{w})}{\partial t}\right]dv$$

$$-\oint_{S_\infty} [p\vec{r}\times\hat{n} + \rho\vec{r}\times\vec{w}(\vec{w}\cdot\hat{n}) - \vec{r}\times(\hat{n}\cdot\overleftrightarrow{\tau})]ds \quad \textbf{(14.45)}$$

which parallels Equation (14.9). As with this equation, $\vec{M}_w$ can be decomposed into various components, such as a pitching moment.

For a steady flow without a fluid body force, the above equation simplifies to

$$\vec{M}_w = -\oint_{S_\infty} [p\vec{r}\times\hat{n} + \rho\vec{r}\times\vec{w}(\vec{w}\cdot\hat{n}) - \vec{r}\times(\hat{n}\cdot\overleftrightarrow{\tau})]ds \quad \textbf{(14.46)}$$

which parallels Equation (14.12). As with Equation (14.12), only conditions on S_∞ are needed for the $\vec{M}_w$ evaluation. If the pressure is constant on S_∞, we obtain

$$\oint_{S_\infty} p_\infty\vec{r}\times\hat{n}ds = -p_\infty\oint_{S_\infty} \hat{n}\times\vec{r}\,ds = -p_\infty\int_{CV} \nabla\times\vec{r}\,dv = 0 \quad \textbf{(14.47)}$$

where a form of the divergence theorem

$$\oint_S \hat{n} \times \vec{A} ds = \int_V \nabla \times \vec{A} dv \tag{14.48}$$

is used, and the curl of $\vec{r}$ is identically zero. In this case, $\vec{M}_w$ is simply given by the angular momentum flux and viscous terms.

Since the value of $\vec{M}_w$ depends on the location of the origin for $\vec{r}$, we examine the change in $\vec{M}_w$ when a different origin is chosen. Suppose it is moved by a distance $\vec{r}_0$; i.e.,

$$\vec{r} = \vec{\bar{r}} + \vec{r}_0 \tag{14.49}$$

where $\vec{\bar{r}}$ is the position vector from the new origin. Since the right side of Equation (14.25) is linear in $\vec{r}$, we have

$$\vec{M}_w = \vec{\bar{M}}_w + \vec{M}_0 \tag{14.50}$$

where the new angular momentum, $\vec{\bar{M}}_w$, is given by Equation (14.46) with $\vec{r}$ replaced by $\vec{\bar{r}}$. Because $\vec{r}_0$ is a constant, $\vec{M}_0$ is given by

$$\vec{M}_0 = \vec{r}_0 \times \left\{ \int_{CV} \left[\rho \vec{F}_b - \frac{\partial(\rho \vec{w})}{\partial t} \right] dv - \oint_{S_\infty} [p\hat{n} + \rho \vec{w}(\vec{w} \cdot \hat{n}) - (\hat{n} \cdot \overleftrightarrow{\tau})] ds \right\} = \vec{r}_0 \times \vec{F}_w \tag{14.51}$$

We thus obtain

$$\vec{M}_w = \vec{\bar{M}}_w + \vec{r}_0 \times \vec{F}_w \tag{14.52}$$

The line of action for the force can be determined by evaluating $\vec{r}_0$ for the $\vec{\bar{M}}_w = 0$ condition [see part (d) of Problem 14.8].

14.5 HYDROSTATICS

This section discusses the hydrostatic force, where we have

$$\rho = \text{constant}, \qquad \vec{w} = 0, \qquad \vec{F}_b = \vec{g}, \qquad \frac{\partial}{\partial t} = 0 \tag{14.53}$$

and the fluid is not in motion. Equation (14.9) reduces to

$$\int_{CV} \rho \vec{g} dv = \oint_{S_\infty} p\hat{n} ds \tag{14.54}$$

where S_∞ is a closed surface without an interior solid body. The left side equals $m\vec{g}$, where m is the mass inside the *CV*. Consequently, we have

$$m\vec{g} = \oint_{S_\infty} p\hat{n} ds \tag{14.55}$$

and the weight of fluid is balanced by the pressure force on S_∞. From the divergence theorem

$$\oint_S \phi \hat{n}\, ds = \int_V \nabla \phi dv \tag{14.56}$$

we have

$$\oint_{S_\infty} p\hat{n} ds = \int_{CV} \nabla p dv \tag{14.57}$$

and Equation (14.54) becomes

$$\int_{CV} (\rho \vec{g} - \nabla p) dv = 0 \tag{14.58}$$

Since the *CV* is arbitrary, the well-known result is obtained as

$$\nabla p = \rho \vec{g} \tag{14.59}$$

for the hydrostatic pressure.

Example

As an illustration, we evaluate the force $\vec{F}_w$ exerted on one side of the curved surface sketched in Figure 14.4. The hydrostatic pressure is given by

$$p = p_{amb} + \rho g z \tag{14.60}$$

where the ambient pressure is at the surface. The surface integral in Equation (14.55) is written as

$$\oint_{S_\infty} p\hat{n}\, ds = \vec{F}_w - p_{ac} A_{ac} \hat{1}_z - \hat{1}_x \int_{A_{bc}} p ds \tag{14.61}$$

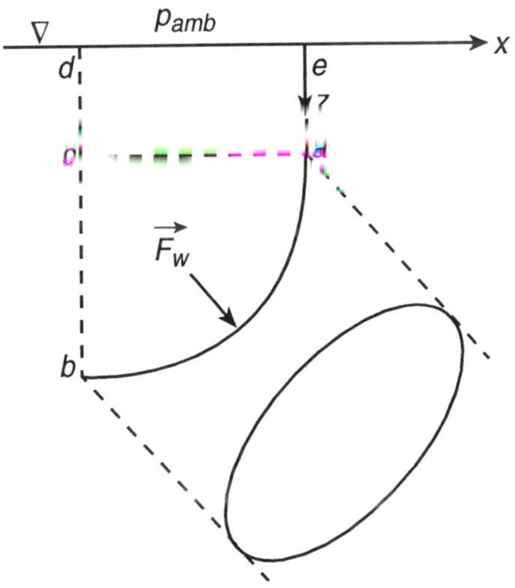

FIGURE 14.4 Force on a submerged curved surface.

where

$$p_{ac} = p_{amb} + \rho g z_a \tag{14.62}$$

and A_{ac} and A_{bc} are the projected areas of the curved surface. We thus obtain

$$\vec{F}_w = (mg + p_{ac} A_{ac})\hat{\imath}_z + \hat{\imath}_x \int_{A_{bc}} p ds \tag{14.63}$$

where m is the mass of fluid in the abc volume. As shown by fluid mechanic textbooks, further simplification is certainly possible.

14.6 FLOW IN A DUCT

A liquid is in steady, inviscid flow through a horizontal duct, as sketched in Figure 14.5. Because the duct is horizontal, gravity can be neglected. We are to determine the F_x and F_y force components acting on the duct in terms of inlet conditions, p_1 and w_1, and known geometrical parameters.

From continuity, we readily obtain

$$w_2 = \frac{A_1}{A_2} w_1 \tag{14.64}$$

Because the flow is steady and inviscid, Bernoulli's equation

$$p_1 + \frac{1}{2}\rho w_1^2 = p_2 + \frac{1}{2}\rho w_2^2 \tag{14.65a}$$

applies. This can be written as

$$p_2 = p_1 + \frac{1}{2}\rho w_1^2 \left[1 - \left(\frac{A_1}{A_2}\right)^2\right] \tag{14.65b}$$

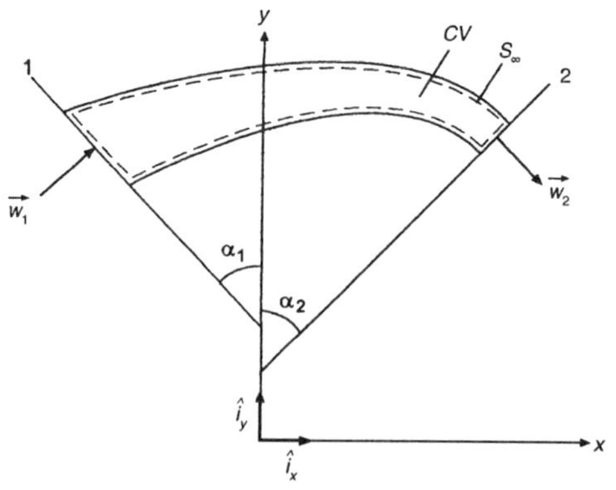

FIGURE 14.5 Incompressible flow inside a curved duct.

The dashed line in Figure 14.5 represents the *CS*. Since the *CV* does not contain a solid body, $\vec{F}_w$ in Equation (14.9) is zero. This equation then reduces to

$$\oint_{S_\infty} [p\hat{n} + \rho\vec{w}(\vec{w} \cdot \hat{n})]\, ds = 0 \tag{14.66}$$

On the inner surface of the duct, $\vec{w} \cdot \hat{n} = 0$, and the force on the wall of the duct is

$$\vec{F} = \int_{S'_w} p\hat{n}\, ds \tag{14.67}$$

with

$$\hat{n}_1 = -\frac{\vec{w}_1}{w_1}, \qquad \hat{n}_2 = \frac{\vec{w}_2}{w_2} \tag{14.68}$$

Equation (14.66) becomes

$$p_1A_1\hat{n}_1 + \vec{F} + p_2A_2\hat{n}_2 + \rho A_1 w_1^2\hat{n}_1 + \rho A_2 w_2^2\hat{n}_2 = 0 \tag{14.69a}$$

or

$$\vec{F} = -A_1(p_1 + \rho w_1^2)\hat{n}_1 - A_2(p_2 + \rho w_2^2)\hat{n}_2 \tag{14.69b}$$

From the figure, we have

$$\hat{n}_1 \cdot \hat{1}_x = -\cos\alpha_1, \qquad \hat{n}_1 \cdot \hat{1}_y = -\sin\alpha_1 \tag{14.70a}$$

$$\hat{n}_2 \cdot \hat{1}_x = \cos\alpha_2, \qquad \hat{n}_2 \cdot \hat{1}_y = -\sin\alpha_2 \tag{14.70b}$$

The two force components are given by

$$F_x = \vec{F} \cdot \hat{1}_x = A_1(p_1 + \rho w_1^2)\cos\alpha_1 - A_2(p_2 + \rho w_2^2)\cos\alpha_2 \tag{14.71a}$$

$$F_y = \vec{F} \cdot \hat{1}_y = A_1(p_1 + \rho w_1^2)\sin\alpha_1 + A_2(p_2 + \rho w_2^2)\sin\alpha_2 \tag{14.71b}$$

Since

$$p_2 + \rho w_2^2 = p_1 + \frac{1}{2}\rho w_1^2\left[1 + \left(\frac{A_1}{A_2}\right)^2\right] \tag{14.72}$$

we obtain the result

$$\frac{F_x}{p_1A_1} = \left(1 + \frac{\rho w_1^2}{p_1}\right)\cos\alpha_1 - \frac{A_2}{A_1}\left\{1 + \frac{\rho w_1^2}{2p_1}\left[1 + \left(\frac{A_1}{A_2}\right)^2\right]\right\}\cos\alpha_2 \tag{14.73a}$$

$$\frac{F_y}{p_1 A_1} = \left(1 + \frac{\rho w_1^2}{p_1}\right)\sin\alpha_1 + \frac{A_2}{A_1}\left\{1 + \frac{\rho w_1^2}{2p_1}\left[1 + \left(\frac{A_1}{A_2}\right)^2\right]\right\}\sin\alpha_2 \tag{14.73b}$$

The nondimensional force components are thus functions of α_1, α_2, A_2/A_1, and $\rho w_1^2/p_1$, as would be expected from dimensional analysis.

As a check, we consider a straight duct, where

$$\alpha_1 = \alpha_2 = 0 \tag{14.74}$$

In this case, we have

$$F_x = (pA + \rho w^2 A)_1 - (pA + \rho w^2 A)_2 \tag{14.75a}$$

$$F_y = 0 \tag{14.75b}$$

where $pA + \rho w^2 A$ is called the impulse function.

14.7 ACYCLIC MOTION

PRELIMINARY REMARKS

An ideal fluid is incompressible and inviscid. An acyclical motion refers to an ideal fluid for which there is a single-valued velocity potential. In particular, we are concerned with an infinite expanse of fluid, with no bounding surface, that contains a finite-sized, solid body. The body is in rectilinear motion with a velocity $\vec{w}_b(t)$, which is time-dependent, in a fluid that is otherwise quiescent. In contrast to the generality of the momentum theorem, the many restrictions in this subsection will limit our presentation to a cursory overview. An introductory discussion is contained in Panton (1984), while Karamcheti (1980) presents a more comprehensive, but still introductory, treatment. For an advanced treatment, the reader may consult the references contained in Sherwood and Stone (1997).

In view of the velocity potential requirement, the fluid motion is irrotational. Moreover, the theory is not applicable to viscous flows, since these are generally rotational. The single-valued restriction means, e.g., that a two-dimensional flow about an infinite cylinder is not permissible. As we know, this type of flow may possess an arbitrary amount of circulation about a path that encloses the cylinder; consequently, the potential function is not single-valued. In the three-dimensional case, however, with a finite-sized body, the circulation about an arbitrary closed path is zero, and a potential function is single-valued. We also ignore the hydrostatic pressure, which may result in a buoyancy force that can be treated separately.

Let m_b be the mass of a solid body. A first step is to write Newton's second law for the body as

$$m_b \frac{d\vec{w}_b}{dt} = \vec{F}_p + \vec{F}_b \tag{14.76}$$

where $\vec{F}_b$ is the external force that accelerates, or decelerates, the body, and $\vec{F}_p$ is the pressure force on the body induced by its own motion. When $\vec{w}_b$ is independent of time, $d\vec{w}_b/dt$ equals zero, and

$$\vec{F}_b = -\vec{F}_p \tag{14.77}$$

which is in accord with Newton's third law, which states that for every action (force) there is an equal and opposite reaction. From the point of view of the body, the flow is steady, and the pressure force in a steady, potential flow is zero. Hence, both $\vec{F}_b$ and $\vec{F}_p$ are zero. When $d\vec{w}_b/dt$ is nonzero, however, the flow is unsteady, and Bernoulli's equation, Equation (5.19), now contains an unsteady term. In this case, the pressure force $\vec{F}_p$ is nonzero, and $\vec{F}_b$ is similarly nonzero. Physically, $\vec{F}_b$ represents the force on the body that is required to accelerate it and the fluid that is adjacent to the body. As noted, this force is zero only when the fluid motion is steady.

Flow about a Sphere

The simplest case is that of a sphere of radius a (Karamcheti, 1980). In this case, we obtain the pressure force from an unsteady potential flow solution

$$\vec{F}_p = -\frac{1}{2}V_b\rho\frac{d\vec{w}_b}{dt} \tag{14.78}$$

where the volume of the sphere is

$$V_b = \frac{4}{3}\pi a^3 \tag{14.79}$$

Let

$$m' = \frac{1}{2}V_b\rho \tag{14.80}$$

and Equation (14.76) becomes

$$\vec{F}_b = (m_b + m')\frac{d\vec{w}_b}{dt} \tag{14.81}$$

If the magnitude of $d\vec{w}_b/dt$ is positive, then $\vec{F}_b$ is the force acting on the sphere in order for it to accelerate. In part, this force also overcomes the unsteady drag of the sphere. The component of $\vec{F}_b$ associated with m' is the force required to accelerate the fluid that is adjacent to the sphere. This component stems from the unsteady pressure force.

The m' mass is called the apparent mass; it is also referred to as the added or virtual mass. For a sphere, and only for a sphere, m' equals half of the volume of the sphere times the density of the surrounding fluid. Hence, for a sphere in water, the apparent mass force is important. On the other hand, for a dense solid body in air, the force is quite negligible (see Problem 14.6). For a body with an arbitrary shape, m' generalizes to a symmetric second-order tensor, called the induced-mass tensor.

14.8 JET-PLATE INTERACTION

We evaluate the force $\vec{F}_w$ of a jet that impinges on a plate; see Figure 14.6. Aside from incompressibility, the jet flow is assumed to be steady, inviscid, two-dimensional, and independent of gravity. The pressure outside the jet, in regions I, II, and III, is a constant, p_∞. Bernoulli's equation holds and, consequently, the flow speed, on the surface of the jet that is exposed to p_∞, is w_∞. The jet bifurcates along the stagnation streamline with part of the flow crossing station 1 and the remainder crossing station 2. At these stations, which are far removed from the plate, the flow has

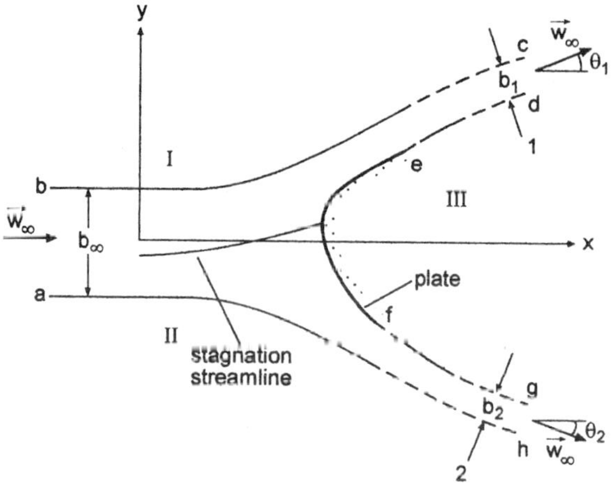

FIGURE 14.6 Schematic of a jet-plate interaction.

a uniform speed of w_∞. At these two asymptotic stations, the flow angles θ_i and the plate angles at points e and f are respectively equal only when the plate is of infinite lateral extent. The control surface is taken as $abc\ldots gha$.

By continuity, the b values are related by

$$b_\infty = b_1 + b_2 \tag{14.82}$$

In this analysis, it is simpler to use Equation (14.3) for the momentum theorem with the unsteady term deleted. Since the pressure in regions I, II, and III is p_∞, the only applied force is due to the plate; i.e.,

$$\sum \vec{F} = -\vec{F}_w = -F_{wx}\hat{\imath}_x - F_{wy}\hat{\imath}_y \tag{14.83}$$

where the minus signs stem from the definition of $\vec{F}_w$ as the fluid force exerted on the plate. Equation (14.3) reduces to

$$-F_{wx}\hat{\imath}_x - F_{wy}\hat{\imath}_y = \rho \sum_{\infty,1,2} \int \vec{w}(\vec{w}\cdot\hat{n})\,ds \tag{14.84}$$

The various factors inside the integral are given by

$$\vec{w}_\infty = w_\infty\hat{\imath}_x, \qquad (\vec{w}\cdot\hat{n})_\infty = -w_\infty, \qquad \int_\infty ds = b_\infty$$

$$\vec{w}_1 = w_\infty(\cos\theta_1\hat{\imath}_x + \sin\theta_1\hat{\imath}_y), \qquad (\vec{w}\cdot\hat{n})_1 = w_\infty, \qquad \int_1 ds = b_1$$

$$\vec{w}_2 = w_\infty(\cos\theta_2\hat{\imath}_x - \sin\theta_2\hat{\imath}_y), \qquad (\vec{w}\cdot\hat{n})_2 = w_\infty, \qquad \int_2 ds = b_2$$

with the result

$$-F_{wx} = -\rho w_\infty^2 b_\infty + \rho w_\infty^2 b_1\cos\theta_1 + \rho w_\infty^2 b_2\cos\theta_2$$

$$-F_{wy} = -\rho w_\infty^2 b_1\sin\theta_1 - \rho w_\infty^2 b_2\sin\theta_2$$

These relations can be written in terms of nondimensional force component coefficients

$$\bar{F}_{wx} = \frac{F_{wx}}{\frac{1}{2}\rho w_\infty^2 b_\infty} = 2\left(1 - \frac{b_1}{b_\infty}\cos\theta_1 - \frac{b_2}{b_\infty}\cos\theta_2\right) \tag{14.85a}$$

$$\bar{F}_{wy} = \frac{F_{wy}}{\frac{1}{2}\rho w_\infty^2 b_\infty} = 2\left(\frac{b_1}{b_\infty}\sin\theta_1 - \frac{b_2}{b_\infty}\sin\theta_2\right) \tag{14.85b}$$

per unit width.

If the flow field is symmetric about the x-axis, then

$$b_1 = b_2 = \frac{1}{2}b_\infty, \qquad \theta_1 = \theta_2 = \theta \tag{14.86a}$$

and

$$\bar{F}_{wx} = 2(1 - \cos\theta), \qquad \bar{F}_{wy} = 0 \tag{14.86b}$$

For an infinite flat plate, $\theta = \pi/2$ and $\bar{F}_{wx} = 2$. This force just equals the momentum flux, $w_\infty(\rho w_\infty)b_\infty$, of the jet.

More generally, the force coefficients depend on b_i and θ_i, which, in turn, depend on the shape and location, relative to the jet, of the plate. These parameters can be found from experiment or computationally determined. In the latter case, the hodograph transformation (Chow et al., 1995) can be used. As discussed in Chapter 7, the independent variables are w and the flow angle θ, while a stream function is the dependent variable. An explicit shape for the control surface boundary in the hodograph plane, however, requires a flat plate, in contrast to the curved one sketched in Figure 14.6. Nevertheless, this approach is useful in the analysis of flaps and thrust reversers (Chow et al., 1995). The stream function is governed by a second-order, linear PDE, which is readily solved with a centered finite difference scheme. To obtain the solution in the physical plane, the transformation equations are then numerically integrated. The force on the plate can be obtained by integrating the pressure on the plate, which is obtained with the aid of Bernoulli's equation. Hence, the momentum theorem provides a useful check on the computational analysis.

14.9 SYRINGE WITH A HYPODERMIC NEEDLE

The last incompressible example evaluates the force required by a plunger in a syringe; see Figure 14.7. The plunger, or piston, is moving into a liquid-filled cylinder at a constant speed w_p. (The analysis is readily modified for a plunger moving in the opposite direction.) The plunger forces the liquid through a slender needle with an inside diameter d. The flow inside the needle and cylinder is (quasi-) steady and laminar; i.e., the Reynolds numbers for the needle and cylinder are assumed to be below about 2300. The length ℓ is considered to be appreciably larger than the entrance length; i.e., entrance length effects are neglected. Frictional losses in the cylinder and at the entrance to the needle are neglected. Thus, only the friction loss inside the needle is evaluated. Many of these assumptions are unnecessary; however, they both clarify and simplify the analysis. Moreover, our results are realistic, since the wall shear in the needle is the dominant loss mechanism. Our goal is to evaluate the needle drag, or skin friction, F_d, and the magnitude of the force $\vec{F}_p$ required by the plunger.

From continuity, we readily obtain

$$\bar{w} = \bar{w}_2 = \bar{w}_3 \tag{14.87}$$

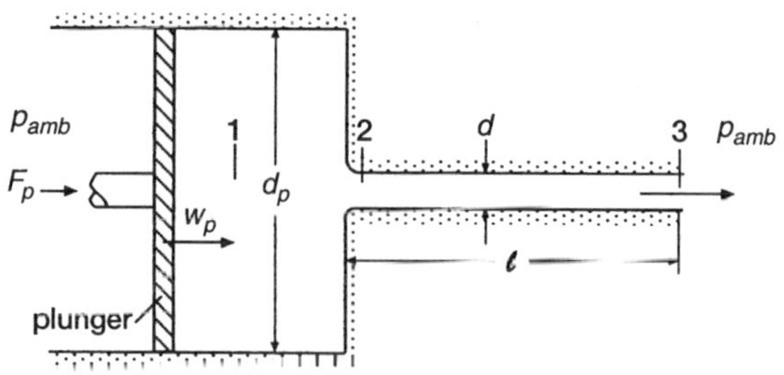

FIGURE 14.7 Sketch of a syringe with a hypodermic needle.

for the average flow speed

$$\bar{w} = \frac{1}{A}\int_A w\, dA \tag{14.88}$$

in the needle. A standard energy equation between stations 2 and 3 is

$$\frac{p_2}{\rho} + \frac{1}{2}\alpha_2 \bar{w}^2 = \frac{p_3}{\rho} + \frac{1}{2}\alpha_3 \bar{w}^2 + g h_L \tag{14.89}$$

where α is the kinetic energy correction factor

$$\alpha = \frac{1}{A\bar{w}^3}\int_A w^3 dA \tag{14.90}$$

and h_L is the head loss. The device is assumed to be horizontal; hence, any (very minor) gravitational force can be neglected. The velocity profile at the needle's entrance, station 2, is taken to be uniform, while at station 3 it is assumed to be a fully developed parabolic profile. Hence, the α_i in Equation (14.89) are

$$\alpha_2 = 1, \qquad \alpha_3 = 2$$

and the pressure drop in the needle is given by

$$\Delta p_d = p_2 - p_3 = \frac{1}{2}\rho \bar{w}^2 + \rho g h_L \tag{14.91}$$

For laminar pipe flow, the head loss can be written as

$$g h_L = \frac{64}{Re}\frac{\ell}{d}\frac{\bar{w}^2}{2} \tag{14.92}$$

where the Reynolds number has its usual definition

$$Re = \frac{\rho d \bar{w}}{\mu} \tag{14.93}$$

and $64/Re$ is the friction factor. With a fully developed flow between stations 2 and 3, the drag force is

$$F_d = \pi \ell d \tau_w = (\pi \ell d)\left(\frac{16}{Re}\right)\left(\frac{1}{2}\rho \bar{w}^2\right) \tag{14.94}$$

Bernoulli's equation holds between stations 1 and 2,

$$\frac{p_1}{\rho} + \frac{1}{2}w_p^2 = \frac{p_2}{\rho} + \frac{1}{2}\bar{w}^2 \tag{14.95a}$$

Since $\bar{w} \gg w_p$, we have

$$p_1 - p_2 = \frac{1}{2}\rho \bar{w}^2 \tag{14.95b}$$

The force on the plunger is

$$F_p = A_p(p_1 - p_{amb}) = \frac{\pi}{4}d_p^2(p_1 - p_2 + p_2 - p_{amb}) \tag{14.96a}$$

With $p_3 = p_{amb}$, this becomes

$$F_p = \frac{\pi}{4}d_p^2\left(\frac{1}{2}\rho \bar{w}^2 + \Delta p_d\right) \tag{14.96b}$$

With the aid of Equations (14.91) and (14.92), we obtain

$$F_p = \frac{\pi}{4}d_p^2\left(\frac{1}{2}\rho \bar{w}^2\right)\left(2 + \frac{64}{Re}\frac{\ell}{d}\right) \tag{14.96c}$$

Equations (14.94) and (14.96c) achieve the stated objective.

The ratio of the two forces provides the interesting result

$$\frac{F_d}{F_p} = \left(\frac{d}{d_p}\right)^2 \frac{\frac{64}{Re}\frac{\ell}{d}}{2 + \frac{64}{Re}\frac{\ell}{d}} \tag{14.97}$$

As suggested by Problem 14.7, $(64/Re)(\ell/d)$ typically exceeds unity, and the rightmost factor is then near unity. On the other hand, $(d/d_p)^2$ is exceedingly small, and consequently $F_p \gg F_d$. Although the drag force inside the needle is negligible compared to F_p, its inclusion in the analysis is crucial. It corresponds to the large pressure drop, Δp_d, across the needle. In turn, this results in a relatively high pressure inside the cylinder and, therefore, a sizable plunger force F_p. Alternatively, the plunger force is required to accelerate the fluid in the cylinder that passes through the needle. This is a sizable acceleration, since the flow speed goes from w_p to a much larger value of $\bar{w}$ in the needle.

The validity of a uniform flow at station 2 might be questioned. This approximation corresponds to $\alpha_2 = 1$, as compared, e.g., to a parabolic profile with $\alpha_2 = 2$. The chosen value for α_3 leads to the 2 that appears in the $2 + (64/Re)(\ell/d)$ term. We observe that the choice of a value for α_2 is not significant as long as $(64/Re)(\ell/d)$ is large compared to unity.

The pressure

$$p_2 = p_3 + \frac{1}{2}\rho\bar{w}^2\left(1 + \frac{64}{Re}\frac{\ell}{d}\right)$$

can be viewed as an estimate of the average pressure that acts on the contraction between the cylinder and needle. The ratio

$$\frac{p_2 - p_3}{p_1 - p_3} = \frac{1 + \frac{64}{Re}\frac{\ell}{d}}{2 + \frac{64}{Re}\frac{\ell}{d}}$$

thus describes how the pressure varies in the syringe; its value is typically close to 1/2. An estimate for the force on the contraction

$$\frac{\pi}{4}(d_p^2 - d^2)p_2$$

is quite unnecessary for the preceding analysis.

The discussion in this section can be contrasted with an inviscid analysis of steady flow through a sudden contraction. In this case, $\bar{w} \gg w_p$ should not be used, and a quite different result is obtained $(p_2 - p_3)/(p_1 - p_3)$ (see Problem 14.16).

14.10 SHOCK-EXPANSION THEORY

This theory assumes steady, inviscid, supersonic, two-dimensional flow. Although not essential, a perfect gas is usually assumed. In its simplest form, it utilizes a mixture of planar, oblique, attached shock waves and centered Prandtl–Meyer expansions. For instance, Figure 14.8(a) shows a quadrilateral airfoil in which the bow and tail shocks are attached and the Mach numbers M_1 and M_3 are supersonic. The theory actually requires these conditions. We have uniform flow in regions 1 through 4 with a constant pressure along the wall in each of these regions. The lift and drag coefficients are thus easily calculated. Note that this evaluation does not involve any tail shock or slipstream calculations.

The assumption of uniform flow along the four planar surfaces of Figure 14.8(a) is not always correct. For instance, if surface 1 is relatively short, as sketched in Figure 14.8(b), the wave reflected from the upper shock may impinge on surface 2, thereby altering the pressure along the downstream part of this surface. (The interaction of an expansion with a planar shock is discussed in detail in Chapter 11, while Problem 14.14 deals with the possible impingement of the reflected wave on surface 2.) The reflected wave, however, is generally quite weak (Eggers et al., 1955). Consequently, in its simplest form, the theory ignores the effect of reflected waves (see Section 6.3).

The presence of shocks guarantees a positive drag for any configuration. Physically, this is apparent for the airfoil sketched in Figure 14.8(a) from the relatively high pressure on surfaces 1 and 3 and the relatively low pressure on surfaces 2 and 4. On the other hand, the lift may be positive or negative. Although not apparent in the analysis, there is a downwash velocity component whenever the lift is positive. This holds for any freestream speed, including hypersonic, and is a direct requirement of Newton's third law.

As with supersonic thin airfoil theory, the method ignores contributions caused by a viscous boundary layer. These contributions are in the form of skin friction and a boundary-layer displacement

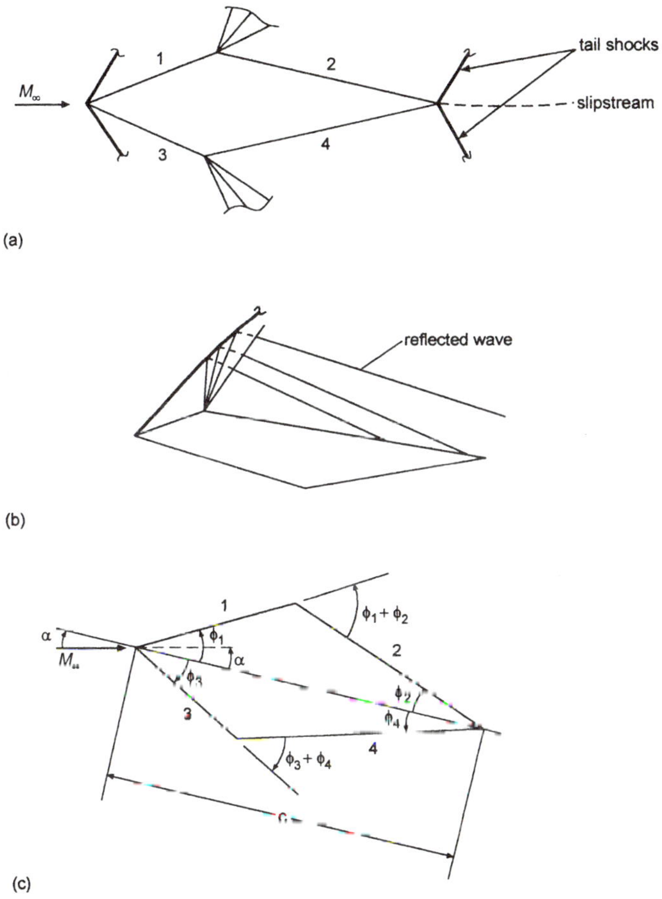

FIGURE 14.8 Shock-expansion sketches for a quadrilateral airfoil.

thickness that alters the location and strength of the shock and expansion waves. Generally, the effect of the boundary layer is quite small; e.g., the skin friction contribution to the drag is much less than that of the pressure. There are, however, two situations when the effect of the boundary layer is important and cannot be ignored. If boundary-layer separation occurs on one of the surfaces, the inviscid flow sketched in Figure 14.8(a) is inappropriate. As in subsonic flow, separation is most likely to occur at a relatively large angle of attack. Secondly, if the airfoil is slender (i.e., its maximum thickness to chord length is small), the strengths of the shock and expansion waves are also small. (A small angle of attack is again assumed.) On the other hand, the integrated skin friction contribution, especially if most of the boundary layer is turbulent, may become significant. These comments apply equally well to thin airfoil theory.

Lift and Drag Coefficients of a Quadrilateral Airfoil

Emanuel (1986), Appendix G, provides general lift and drag relations for the quadrilateral airfoil pictured in Figure 14.8(c) when the reflected waves are weak or do not intersect the airfoil. The relevant coefficients, per unit span, are defined as

$$c_\ell = \frac{2\ell}{(\rho w^2)_\infty c}, \qquad c_d = \frac{2d}{(\rho w^2)_\infty c} \tag{14.98}$$

where ℓ and d are the lift and drag, c is the chord length, and

$$(\rho w^2)_\infty = \gamma p_\infty M_\infty^2 \tag{14.99}$$

Since a constant value is presumed for the ratio of specific heats γ, a perfect gas is assumed. The overall lift and drag are written as

$$\ell = \sum_{i=1}^{4} \ell_i, \qquad d = \sum_{i=1}^{4} d_i \tag{14.100}$$

where, for instance, for surface 1 we have

$$\ell_1 = -p_1 L_1 \cos(\phi_1 - \alpha), \qquad d_1 = p_1 L_1 \sin(\phi_1 - \alpha), \qquad L_1 = \frac{\sin\phi_2}{\sin(\phi_1 + \phi_2)} c \tag{14.101}$$

Here, L_1 is the length of surface 1 and α is the angle of attack. We thus obtain

$$c_\ell = \frac{2}{\gamma M_\infty^2} \left\{ \frac{1}{\sin(\phi_1 + \phi_2)} \left[\sin\phi_2 \cos(\phi_1 - \alpha) \frac{p_1}{p_\infty} + \sin\phi_1 \cos(\phi_2 + \alpha) \frac{p_2}{p_\infty} \right] \right.$$
$$\left. + \frac{1}{\sin(\phi_3 + \phi_4)} \left[\sin\phi_4 \cos(\phi_3 + \alpha) \frac{p_3}{p_\infty} + \sin\phi_3 \cos(\phi_4 - \alpha) \frac{p_4}{p_\infty} \right] \right\} \tag{14.102}$$

$$c_d = \frac{2}{\gamma M_\infty^2} \left\{ \frac{1}{\sin(\phi_1 + \phi_2)} \left[\sin\phi_2 \sin(\phi_1 - \alpha) \frac{p_1}{p_\infty} - \sin\phi_1 \sin(\phi_2 + \alpha) \frac{p_2}{p_\infty} \right] \right.$$
$$\left. + \frac{1}{\sin(\phi_3 + \phi_4)} \left[\sin\phi_4 \sin(\phi_3 - \alpha) \frac{p_3}{p_\infty} - \sin\phi_3 \sin(\phi_4 - \alpha) \frac{p_4}{p_\infty} \right] \right\} \tag{14.103}$$

Evaluation of the two coefficients requires a knowledge of γ, M_∞, α, ϕ_i, and p_i/p_∞ for $i = 1, \ldots, 4$. The pressure ratios, of course, are determined by shock and Prandtl–Meyer formulas. When the chord is defined as passing through the leading and trailing edges of the airfoil, as in Figure 14.8(c), the lift coefficient may not be zero when $\alpha = 0$. In fact, it may be positive or negative. With $\alpha \geq 0$, there is an attached shock upstream of surface 1 when $\phi_1 > \alpha$; otherwise, there is a Prandtl–Meyer expansion. A similar statement holds for surface 3. Problem 14.8 develops the corresponding equation for the pitching moment.

Comments

Shock-expansion theory is applicable to airfoils with curved surfaces, as sketched in Figure 14.9. The lift and drag coefficients are obtained as integrals over the surface that involve the appropriate components of the pressure force. The bow shock must be attached; hence, the sharp leading edge, and the effect of reflected waves, should be considered for a relatively thick airfoil. Generally, shock-expansion theory provides results in good accord with more exact inviscid method-of-characteristic calculations. Further discussion can be found in the papers by Eggers et al. (1955) and Mahony (1955), and in the report by Waldman and Probstein (1957).

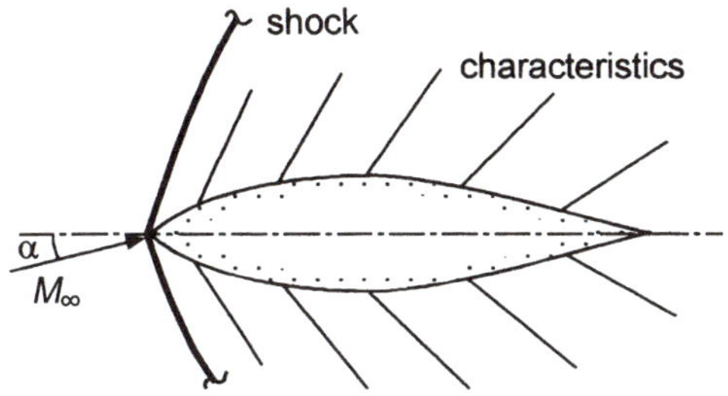

FIGURE 14.9 Shock-expansion schematic for a lens-shaped airfoil.

INTERFERENCE

We utilize simple shock-expansion theory to obtain the drag for a quadrilateral airfoil that is symmetric about the x-axis. In view of symmetry and a zero incidence, there is no lift, and only wave drag is present. As we shall see, the magnitude of the drag is strongly influenced by interference, or an interaction, between the expansion and the shock waves. The interaction stems from the shape of the wave-generating body and may occur at some distance from the surface. Nevertheless, this interaction is associated with the pressure force on the surface. It is not necessary, however, to analyze the complicated interference flow field, as is done in Chapter 11, when shock-expansion theory can directly provide the surface pressure force. In this case, the wave drag can be associated with the airfoil's thickness. Problem 14.17 investigates the simpler case of the wave drag of a symmetric wedge, where there is no interference. The results of this problem show a strong dependence on the freestream Mach number and a faster-than linear increase with the half-angle, which represents the wedge's thickness.

An introductory discussion of interference can be found in Liepmann and Roshko (1957, Section 4.19), which examines the interference that occurs with a Busemann biplane. A weak wave approximation is used for the internal flow between the two symmetric, triangular-shaped airfoils. When there is wave cancellation, i.e., favorable interference, the wave drag goes to zero in the weak wave approximation. The drag, however, is sensitive to the ratio of the gap between airfoils to the chord length.

The illustrative example considered here is for an external flow. It is more realistic than that for a Busemann biplane, in part because the weak wave approximation is not utilized. For simplicity, the presentation is limited to steady, supersonic, inviscid, two-dimensional flow of a perfect gas. These assumptions similarly apply to the analysis of the Busemann biplane. As before, we assume the tail shocks are attached and the Mach number, M_1, behind the attached bow shock is supersonic. As we shall see, the attached tail shock condition is important. The centered expansion waves interact with both the bow and tail shocks. The strength and nature (i.e., favorable or unfavorable interference) of the interaction depends on the relative strengths of the waves and their locations with respect to each other.

Figure 14.10 shows a sketch of the upper half of the symmetric airfoil. Its geometry is controlled by the thickness parameter

$$\tau = \frac{2h}{c} \tag{14.104}$$

and the half angle ϕ_1. The planform area, projected frontal area, and cross-sectional area are all fixed once τ and the chord length are specified. We thus examine the dependence of c_d on ϕ_1 with fixed values for M_∞ and τ. This is equivalent to moving the shoulder that separates regions 1 and 2 along a line parallel to the x-axis.

With the symmetry values

$$\alpha = 0, \qquad \phi_3 = \phi_1, \qquad \phi_4 = \phi_2, \qquad p_3 = p_1, \qquad p_4 = p_2 \tag{14.105}$$

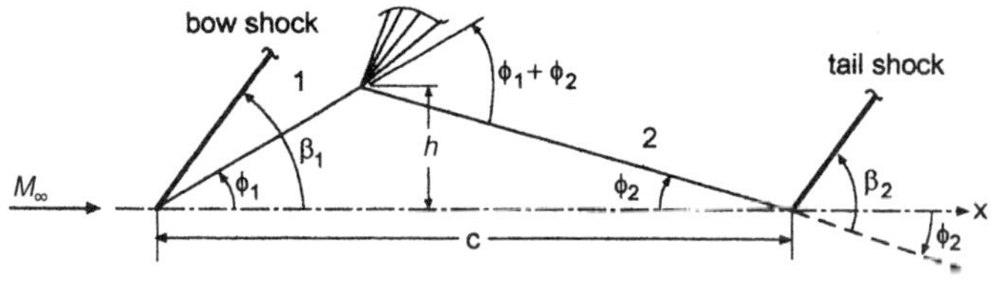

FIGURE 14.10 Symmetric quadrilateral airfoil at zero incidence; only the upper half is shown.

Equation (14.103) reduces to

$$c_d = \frac{4}{\gamma M_\infty^2}\frac{\sin\phi_1\sin\phi_2}{\sin(\phi_1+\phi_2)}\left(\frac{p_1}{p_\infty}-\frac{p_2}{p_\infty}\right) \tag{14.106a}$$

The oblique shock and isentropic relations

$$\frac{p_1}{p_\infty} = \frac{2}{\gamma+1}\left(\gamma M_\infty^2\sin^2\beta_1-\frac{\gamma-1}{2}\right) \tag{14.107a}$$

$$\frac{p_2}{p_\infty} = \frac{\frac{p_2}{p_{01}}}{\frac{p_1}{p_{01}}}\frac{p_1}{p_\infty} = \frac{p_1}{p_\infty}\left(\frac{1+\frac{\gamma-1}{2}M_1^2}{1+\frac{\gamma-1}{2}M_2^2}\right)^{\gamma/(\gamma-1)} \tag{14.107b}$$

then yield

$$c_d = \frac{8}{\gamma(\gamma+1)M_\infty^2}\frac{\sin\phi_1\sin\phi_2}{\sin(\phi_1+\phi_2)}\left(\gamma M_\infty^2\sin^2\beta_1-\frac{\gamma-1}{2}\right)\left[1-\left(\frac{1+\frac{\gamma-1}{2}M_1^2}{1+\frac{\gamma-1}{2}M_2^2}\right)^{\gamma/(\gamma-1)}\right] \tag{14.106b}$$

From Figure 14.10, we obtain

$$\tan\phi_2 = \frac{\tau\tan\phi_1}{2\tan\phi_1-\tau} \tag{14.108}$$

for the half angle ϕ_2. The maximum allowable ϕ_1 value occurs when $M_1 = 1$. For example, with $\gamma = 1.4$, $\phi_1(M_1 = 1)$ is 22.5° when $M_\infty = 2$ and is 38.5° when $M_\infty = 4$. At the other extreme, ϕ_1 is a minimum, ϕ_{1d}, when the tail shocks detach. With given values for γ, M_∞, and τ, the angle ϕ_1 is computationally increased from its ϕ_{1d} value but terminates well before the $M_1 = 1$ condition is reached.

With γ, M_∞, and ϕ_1 known, Appendix C is used to evaluate β_1, while M_1 is given by the oblique shock relation

$$M_1 = \frac{1}{\sin(\beta_1-\phi_1)}\left[\frac{1+\frac{\gamma-1}{2}M_\infty^2\sin^2\beta_1}{\gamma M_\infty^2\sin^2\beta_1-\frac{\gamma-1}{2}}\right]^{1/2} \tag{14.109}$$

With τ specified, the angle ϕ_2 is next determined by Equation (14.108), after which M_2 is found by iteratively solving

$$\nu(M_2) = \nu(M_1) + \phi_1 + \phi_2 \tag{14.110}$$

where ν is the Prandtl–Meyer function. At this point, c_d is provided by Equation (14.106b). The minimum value for ϕ_1 is found iteratively using the above, starting with Equation (14.108), in conjunction with the detachment equations for the tail shocks. These are

$$\sin\beta_{2d} = \left(\frac{\gamma+1}{4\gamma M_2^2}\left\{M_2^2 - \frac{4}{\gamma+1} + \left[M_2^4 + 8\frac{\gamma-1}{\gamma+1}M_2^2 + \frac{16}{\gamma+1}\right]^{1/2}\right\}\right)^{1/2} \tag{14.111a}$$

$$\tan\phi_{1d} = \frac{\tau(M_2^2\sin^2\beta_{2d} - 1)}{2(M_2^2\sin^2\beta_{2d} - 1) - \tau\left[1 + \left(\frac{\gamma+1}{2} - \sin^2\beta_{2d}\right)M_2^2\right]\tan\beta_{2d}} \tag{14.111a}$$

The last equation is obtained by eliminating $\tan\phi_{2d}$ from Equation (14.108) and the oblique shock relation

$$\tan\phi_{2d} = \frac{M_2^2\sin^2\beta_{2d} - 1}{\tan\beta_{2d}\left[1 + \left(\frac{\gamma+1}{2} - \sin^2\beta_{2d}\right)M_2^2\right]} \tag{14.112}$$

Figures 14.11(a–c) show results* for $\gamma = 1.4$, $M_\infty = 2, 3, 4$, and $\tau = 0.05, 0.1, 0.15, 0.2$. A comparison of the three panels indicates that c_d decreases with increasing M_∞ for given values of τ and ϕ_1. The actual drag, however, which is proportional to $M_\infty^2 c_d$, increases. The general trend that a thicker airfoil has a larger drag is readily evident.

For a given curve, the smallest ϕ_1 value shown is for tail shock detachment, while the largest ϕ_1 value is well below where M_1 equals unity. The angle ϕ_{1d} depends strongly on τ but very weakly on M_∞. Thus, when $\tau = 0.05$, ϕ_{1d} is approximately 1.47°, while $\phi_{1d} \cong 6.4°$ when $\tau = 0.2$ for M_∞ values in the 2 to 4 range. The airfoil is diamond shaped when $\phi_2 = \phi_1 = \phi_{dm}$, or

$$\tan\phi_{dm} = \tau \tag{14.113}$$

For the curves shown, the airfoil is diamond-shaped when

τ	ϕ_{dm}
0.05	2.862°
0.10	5.711°
0.15	8.531°
0.20	11.31°

Roughly, ϕ_{dm} is about twice ϕ_{1d}. The difference between ϕ_{dm} and ϕ_{1d} is small for a thin airfoil; e.g., $\phi_{dm} - \phi_{1d} \cong 1.4°$ when $\tau = 0.05$.

* I am indebted to Mr. T.-L. Ho, who performed the calculations and generated the figure.

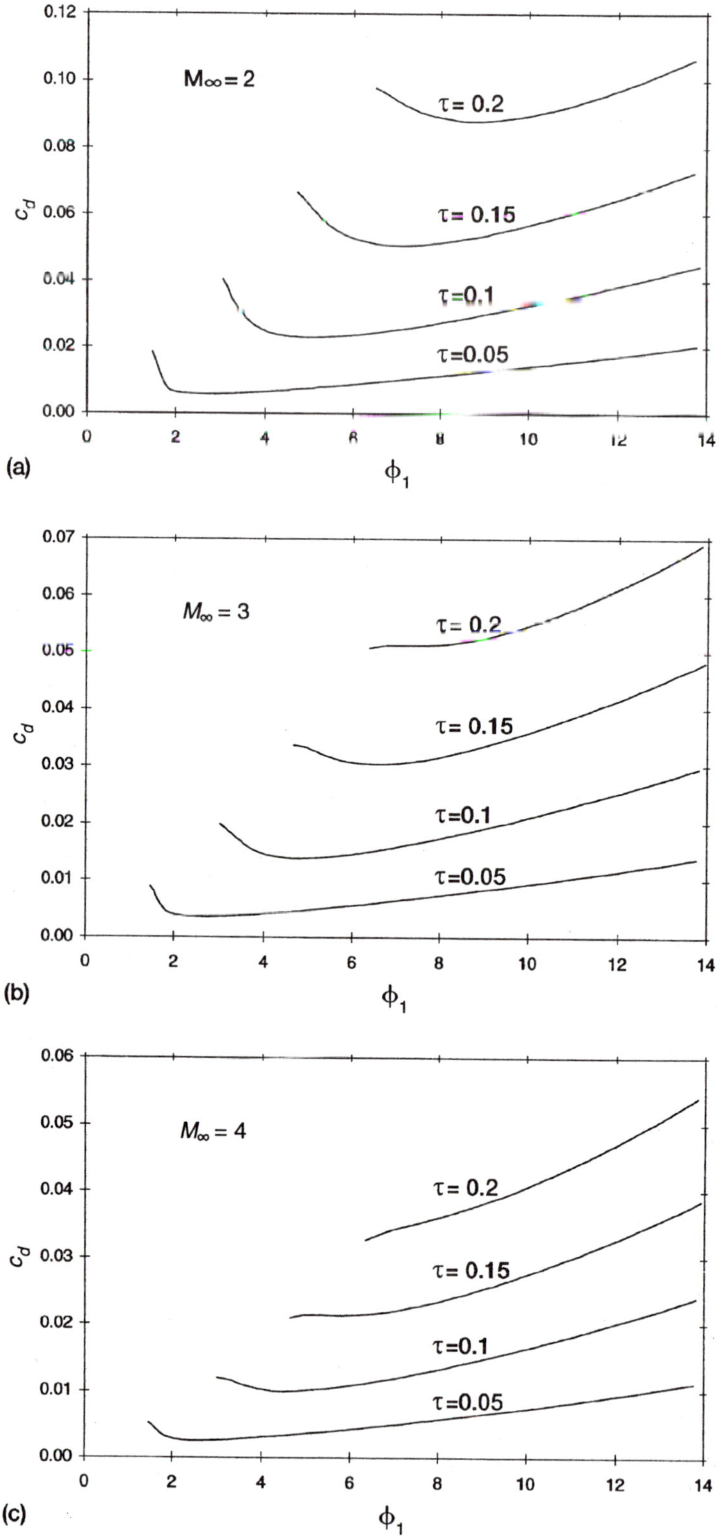

FIGURE 14.11 The drag coefficient vs. ϕ_1 when $\gamma = 1.4$; (a) $M_\infty = 2$, (b) $M_\infty = 3$, (c) $M_\infty = 4$.

The strength of the expansion and shock waves varies with ϕ_1. For instance, when ϕ_1 is large, e.g., well above ϕ_{dm}, the bow shocks are relatively strong, while the tail shocks are weak. Similarly, when ϕ_1 is near ϕ_{1d}, the tail shocks are relatively strong, and the bow shocks are weak. In these two extremes, the centered expansion is closer to the stronger shock and can preferentially weaken it. When $\phi_1 = \phi_{dm}$, the bow and tail shocks are roughly of comparable strength.

The strength of the expansion, measured by $(p_1 - p_2)/p_1$, only depends on M_1 and $\phi_1 + \phi_2$. This strength has a modest variation with ϕ_1, except when c_d shows a rapid increase near ϕ_{1d}. This increase is prominent at all Mach numbers when τ is small. In this circumstance, as ϕ_1 decreases toward ϕ_{1d}, ϕ_2 — and therefore $\phi_1 + \phi_2$ — increases rapidly. In turn, this results in a small value for p_2, relative to p_1, and a rapidly increasing value for c_d.

Overall, the effect of interference is significant. It is most pronounced when τ is small, where a change in ϕ_1 can result in a substantial change in c_d. For small τ, c_d has a minimum value slightly above ϕ_{1d}. This condition, however, occurs at a ϕ_1 value close to where the rapid rise in c_d takes place. A thin component, such as a fin or strut, may be sensitive to this effect.

Interference is still important at moderate or large τ values. For a moderate value of τ, the drag coefficient has a minimum value when ϕ_1 is between ϕ_{1d} and ϕ_{dm}. At a large τ value, c_d is a minimum at, or near, ϕ_{1d}. In this case, the airfoil has an arrowhead cross-sectional shape. Again, the solution is sensitive to ϕ_1. Should ϕ_1 be below ϕ_{1d}, the tail shocks are no longer attached and shock-expansion theory is invalid. In this circumstance, the tail shocks would intersect the airfoil, downstream of the expansions, and probably cause boundary-layer separation.

As is evident, the analysis does not consider the reflection of the expansion wave from the bow shock wave. The weak reflected wave may possibly impinge on surface 2 when $\phi_1 > \phi_{dm}$. As indicated in Chapter 11, if impingement does occur, it will be with an expansion wave that will increase the drag coefficient.

14.11 FORCES ON A PARTICLE

Preliminary Remarks

In this chapter, so far, only rather simple examples have been used, such as incompressible flow in a duct or supersonic flow over a quadrilateral airfoil. Flow fields of practical interest are generally much more complicated, and their analysis with a large computer code is often necessary. For instance, consider the still relatively simple case of uniform flow about a smooth sphere or circular cylinder. Suppose the Reynolds number is such that the boundary layer on the forward part of the sphere or cylinder is laminar and the early part of the wake is also laminar. The downstream wake, however, can be transitional and then become turbulent. The flow is thus inherently unsteady, even though all boundary conditions are steady. Only when the Reynolds number is relatively small is the flow steady and without a wake. At the other extreme in flow speed, a supersonic flow often contains a complicated shock wave system with regions of separated flow. Nevertheless, the study of simple flow fields is essential for developing solution techniques and understanding the concepts involved, but these flows may be of limited practical interest.

An approach analogous to what is discussed in this section is sometimes appropriate for flows encountered in practice. We undertake the inverse task of establishing the motion of a particle subject to known forces. This inverse approach is of practical relevance. A similar procedure is used in Chapter 12 when discussing internal ballistics. In both cases, the analysis is directly based on Newton's second law.

There are many situations when a gas flow contains liquid or solid particles. A common occurrence is with liquid droplets or ice crystals inside a cloud or when they fall to earth. Another example is a solid rocket propulsion engine, which frequently contains small amounts of a metal additive, usually aluminum or boron, in the otherwise rubber-like propellant. The highly exothermic oxidation of the metal additive increases the specific thrust of the engine. When the engine is firing, small micron-size aluminum or boron oxide particles form in the hot gas inside the plenum chamber. These particles may hit and erode the wall of the nozzle, especially at the throat where ablative inserts are sometimes used. Other applications include the atomized mist of a liquid fuel (e.g., diesel fuel) in a combustion chamber, centrifuge operation as in a cyclone separator, or in a seeded turbulent flow, where the motion of the small seed particles is monitored by laser beams.

For simplicity, the equation of motion is obtained for a single, spherical, aerosol-type particle. (Surface tension results in a spherical shape for a small liquid droplet.) The effect of the particle on the gas flow is ignored. In fact, the gas flow is viewed as if the particle is not present. These assumptions hold for small particles whose number density in the flow is not too large. The gas flow itself may be steady or unsteady and subsonic or supersonic. Actually, what is relevant is the velocity of the particle relative to the adjacent gas flow. This relative velocity is typically much smaller than the gas velocity. Our discussion is primarily based on Hinze (1959) and Hourng and Emanuel (1987), which can be consulted for additional references.

Drag Force

We consider a sphere, or particle, of diameter d_p with a velocity $\vec{w}_p$. At first, the particle is assumed to be in a flow with uniform upstream conditions, denoted by an infinity subscript. A coordinate system fixed to the sphere allows us to introduce a conventional drag coefficient

$$C_D = \frac{2|\vec{F}_d|}{\rho_\infty (w_\infty - w_p)^2 (\pi d_p^2/4)} \tag{14.114}$$

where ρ_∞ is the freestream fluid density, $\vec{w}_\infty - \vec{w}_p$ is the relative velocity, and $\pi d_p^2/4$ is the projected frontal area of the particle. In a uniform flow, the drag, $\vec{F}_d$, on the particle is aligned with the relative velocity. For a particle with a small diameter and with a small relative speed, the drag coefficient depends on a particle Reynolds number

$$Re_p = \frac{\rho_\infty d_p |\vec{w}_\infty - \vec{w}_p|}{\mu_\infty} \tag{14.115}$$

where the density and viscosity are evaluated in the uniform upstream flow. When Re_p is less than unity, we have Stokes flow in which the pressure and viscous forces dominate. In this type of flow, the drag coefficient is approximately

$$C_D = \frac{24}{Re_p} \tag{14.116}$$

and, since $Re_p \le 1$, it is quite large. In this regime, the flow is laminar and the fluid does not separate from the surface of the sphere, except at the downstream stagnation point. The particle thus induces a nonuniform pressure field in its vicinity in an otherwise globally uniform pressure field. In Stokes flow, 1/3 of the drag is due to the induced pressure field; the balance of the drag is viscous.

Of interest, however, is a flow where the particle Reynolds number may be below or above unity, up to about 10^2; consequently, Equation (14.116) requires considerable modification. At a Reynolds number near 20, separation occurs and there is a wake downstream of the sphere. Once a wake is present, the pressure force starts to become more prominent, since the downstream wake region, near the body, has a relatively low pressure. For instance, by the time the Reynolds number is 10^3, the forward part of the sphere has a laminar boundary layer and the form (or pressure) drag, associated with the separated flow, dominates the now quite small viscous drag.

In view of our interest in a Reynolds number that may exceed unity, the drag force in Equation (14.112) is written as

$$\vec{F}_d = \frac{1}{2}\rho\frac{\pi d_p^2}{4}(w-w_p)^2 C_D \frac{\vec{w}-\vec{w}_p}{|\vec{w}-\vec{w}_p|} = \frac{\pi}{8}\rho d_p^2[\operatorname{sgn}(w-w_p)](w-w_p)(\vec{w}-\vec{w}_p)C_D \tag{14.117}$$

where the sgn function is the sign of its argument. Thus, if $w - w_p$ is negative, its sgn is −1. Formulas for C_D are provided shortly. Typically, the flow of interest may not have a nice uniform upstream state. Consequently, ρ and $\vec{w}$ now refer to the fluid density and velocity at the center of the sphere as if the sphere were not present. Equation (14.117) is thus approximate.

If the relative velocity is small compared to the speed of sound, there is little error in replacing ρ_∞ with ρ. However, if this velocity is large, say supersonic, then the error can be significant. Equation (14.117) provides the force on a sphere when it is in a steady, uniform flow. When this is not the case, correction terms, discussed shortly, are required.

Newton's Second Law

Suppose the particle is subject to an imposed pressure gradient, distinct from the one induced by the particle when it is in a uniform flow. The imposed pressure on the surface of the sphere is generally not uniform or symmetric. From Equation (14.57), we see that the pressure gradient force can be written as

$$\vec{F}_p = -\frac{\pi d_p^3}{6}\nabla p \tag{14.118}$$

where the volume of the sphere is $\pi d_p^3/6$. Since the effect of the induced pressure field is accounted for in $\vec{F}_d$, $\vec{F}_p$ is only associated with the imposed pressure field.

There is an apparent mass force on the sphere, given by

$$\vec{F}_a = \frac{1}{2}\frac{\pi d_p^3}{6}\rho\left(\frac{D\vec{w}}{Dt} - \frac{D\vec{w}_p}{Dt}\right) \tag{14.119}$$

This relation is in accord with the earlier discussion, where the sphere was in a quiescent fluid; hence, the acceleration term of the fluid, $D\vec{w}/Dt$, was not present.

When the flow is unsteady, the flow pattern near the particle deviates from that of a steady flow. This gives rise to a history-dependent so-called Basset force

$$\vec{F}_B = \frac{3}{2}d^2(\pi\rho\mu)^{1/2}\int_0^t\left(\frac{D\vec{w}}{Dt} - \frac{D\vec{w}_p}{Dt}\right)\frac{dt'}{(t-t')^{1/2}} \tag{14.120}$$

If the flow is steady, or if the particle is not highly accelerated relative to the flow, this term can be neglected. The relative acceleration can become significant for a small, high-density particle in an unsteady gas flow.

Finally, if a particle is in a shear flow with a uniform pressure, there is a lift force that is normal to the relative velocity and is in the direction of the shear with an increasing flow speed. This direction is denoted with the unit vector $\hat{e}$. We thus have

$$\vec{F}_\ell = 0.514(\rho\mu)^{1/2}\pi d_p^2(w-w_p)\kappa^{1/2}\hat{e} \tag{14.121}$$

where κ is the magnitude of the linear velocity gradient, i.e., the shear, and 0.514 is a nondimensional constant. (For an updated discussion on this topic, see Legendre and Magnaudet, 1997.)

Newton's second law for the particle now has the form

$$\frac{\pi d_p^3}{6}\rho_p \frac{D\vec{w}_p}{Dt} = \vec{F}_d + \vec{F}_p + \vec{F}_a + \vec{F}_B + \vec{F}_\ell \quad \textbf{(14.122)}$$

The left side is referred to as the inertia force. If this were the only force present, $\vec{w}_p$ would be a constant, and the particle would move in a straight line in accord with Newton's first law. The first part of the apparent mass force can be incorporated with the pressure gradient force, while the second part can be included with the inertia force. A buoyancy force has not been included, since it was found to be inconsequential in the centrifuge study, described shortly.

The right side of Equation (14.122) represents the vector sum of the applied forces that act on the sphere. These forces should be physically independent of each other. To examine this point, let us compare them with the drag, $\vec{F}_d$. Remember that this force is associated with a motionless sphere in a steady, uniform flow. Clearly, the buoyancy force and unsteady Basset force differ from the drag. The apparent mass force accelerates the surrounding fluid and also differs from the drag. Finally, the pressure gradient and lift forces can be viewed as correction terms that account for the nonuniform flow conditions that may be imposed on the sphere.

Drag Coefficient

Over the years, a number of measurements of the drag coefficient of a sphere in a uniform flow have been performed (Morsi and Alexander, 1972). Here, we prefer the empirical correlation of Henderson (1976, 1977), which compares well with experiment and covers the continuum, slip, transition, and molecular flow regimes. It is valid up to a Reynolds number Re_p of about 200 when the wake of the sphere starts to transition from laminar to turbulent. The correlation depends on the ratio of specific heats γ, the particle Reynolds number, the particle-to-gas temperature ratio T_p/T, and a particle Mach number

$$M_p = \frac{|\vec{w} - \vec{w}_p|}{a} \quad \textbf{(14.123)}$$

where a is the speed of sound in the gas. For $M_p \leq 1$, the drag coefficient is

$$\begin{aligned} C_D = {} & 24\left(Re_p + \left(\frac{\gamma}{2}\right)^{1/2} M_p\left\{4.33 + \left(\frac{3.65 - 1.53T_p/T}{1 + 0.353T_p/T}\right)\exp\left[-0.247\left(\frac{2}{\gamma}\right)^{1/2}\frac{Re_p}{M_p}\right]\right\}\right)^{-1} \\ & + \exp\left(-\frac{0.5M_p}{Re_p^{1/2}}\right)\left[\frac{4.5 + 0.38(0.03Re_p + 0.48Re^{1/2}{}_p)}{1 + 0.03Re_p + 0.48Re_p^{1/2}} + 0.1M_p^2 + 0.2M_p^8\right] \\ & + 0.6\left(\frac{\gamma}{2}\right)^{1/2} M_p\left[1 - \exp\left(-\frac{M_p}{Re_p}\right)\right] \end{aligned} \quad \textbf{(14.124a)}$$

Note that when M_p is zero, we recover Equation (14.116) plus a positive correction term that only depends on Re_p. For $M_p \geq 1.75$, it is given by

$$C_D = \frac{0.9 + \frac{0.34}{M_p^2} + 1.86\left(\frac{M_p}{Re_p}\right)^{1/2}\left[2 + \frac{4}{\gamma M_p^2} + 1.058\frac{1}{M_p}\left(\frac{2T_p}{\gamma T}\right)^{1/2} - \frac{4}{\gamma^2 M_p^4}\right]}{1 + 1.86(M_p/Re_p)^{1/2}} \quad \textbf{(14.124b)}$$

In the M_p region between 1 and 1.75, the coefficient is linearly interpolated using the relation

$$C_D(M_p, Re_p) = C_D(1.0, Re_p) + (4/3)(M_p - 1)[C_D(1.75, Re_p) - C_D(1.0, Re_p)] \qquad \textbf{(14.124c)}$$

where $C_D(1.0, Re_p)$ is calculated using Equation (14.124a) with $M_p = 1.0$, and $C_D(1.75, Re_p)$ is calculated using Equation (14.124b) with $M_p = 1.75$. When M_p exceeds unity, the flow is supersonic relative to the particle and there is a detached bow shock wave upstream of the sphere. In this circumstance, part of the drag is wave drag. For instance, if we set

$$\gamma = 1.4, \qquad \frac{T_p}{T} = 1, \qquad M_p = 1.75, \qquad Re_p = 1$$

we obtain

$$C_D = 2.71$$

which is significantly smaller than its Stokes flow counterpart. As pointed out earlier, the $(w_\infty - w_p)^2$ normalization, however, would be significantly larger in the supersonic case.

Supersonic Vortex Centrifuge

Hourng (1986) and Hourng and Emanuel (1987) examined the feasibility of using a supersonic potential vortex as a centrifuge. This is a two-dimensional, homentropic flow where the streamlines are concentric circular arcs. On a streamline, fluid properties are constant; e.g., we have

$$wr = \text{constant} \qquad \textbf{(14.125)}$$

where w is the flow speed and r is the distance from the symmetry axis. There is a minimum radius r_m on which w has a finite maximum value and the Mach number is infinite. As r increases, w and M decrease, while the pressure increases. At a radius r^*, the Mach number equals unity; beyond this radius the flow is subsonic. Only the supersonic region is explored, since particle separation efficiency increases with Mach number.

If only the inertia force is present, a particle would travel in a straight line and soon leave the vortex flow. It would be collected in a constant pressure region, which bounds the outer edge of the jet. Thus, the other forces typically try to keep the particle inside the vortex flow. For instance, the viscous force does this by trying to minimize the relative velocity. Similarly, the pressure gradient force is radially inward, toward the symmetry axis.

A supersonic potential vortex is generated by a curved subsonic-supersonic nozzle. (Problem 14.9 deals with the analysis of a potential vortex.) The design of the nozzle is based on characteristic theory (Hourng, 1986). A schematic is shown as Figure 14.12, where $r_c > r_m$ and $r^* > r_a$. The nozzle/vortex combination can also be used as an aerodynamic window for the transmission of a high-power laser beam (Emanuel, 1986, Chapter 18).

Of particular interest is the particle trajectory of a micron-sized sphere. Parametric calculations (Hourng and Emanuel, 1987) were performed that included all of the above forces except for the Basset force, since the flow is steady. These calculations revealed that the effect on the trajectory of the temperature of the particle was inconsequential, and that M_p was quite subsonic. A comparative analysis of the forces on the right side of Equation (14.122) showed that the viscous drag force was clearly dominant for the large range of cases that were examined. (The calculation would have been greatly simplified had this been known beforehand.) As previously noted, the buoyancy force is negligible. Thus, the trajectory of a particle stems from a balance of the inertia and drag forces.

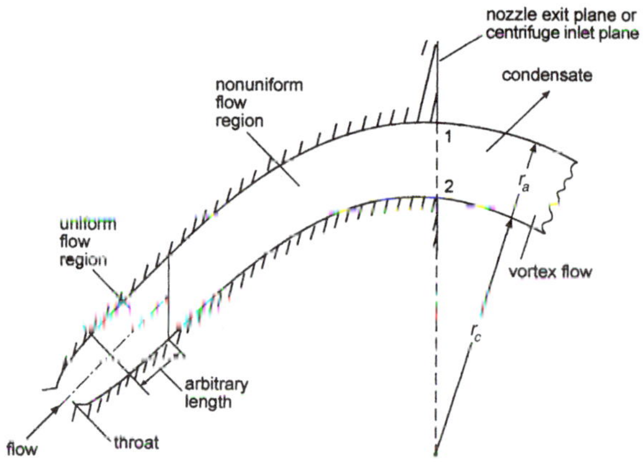

FIGURE 14.12 Asymmetric nozzle schematic in planar view.

When written in scalar form with unsteady terms and $\vec{F}_B$ deleted, Equation (14.122) provides the trajectory of a small, spherical particle, where Equation (14.124a) provides the drag coefficient. The equations are nondimensionalized and conveniently written with cylindrical polar coordinates. Two coupled, first-order PDEs are thereby obtained. This system of equations turns out to be hyperbolic, where the characteristic lines are the twice-repeated path lines of the particle. As a consequence, the equations reduce to three first-order ODEs, where one ODE is the equation for the path of the particle. These equations, with appropriate initial conditions, are readily solved using a fourth-order Runge–Kutta numerical scheme.

14.12 ENTROPY GENERATION

Preliminary Remarks

The concept of wave drag is associated with the entropy production caused by shock waves. For example, if the airfoil pictured in Figure 14.8(a) were in a subsonic, potential flow, it would have no drag and any lift would require circulation about the airfoil. The strength of the circulation is determined by the Kutta trailing-edge condition. By way of contrast, in a supersonic flow with an attached bow shock, there is no Kutta condition or a corresponding circulation, and the airfoil has drag.

As mentioned earlier in this chapter, the drag of a supersonic thin airfoil can be decomposed into angle of attack, camber, and thickness contributions. None of these geometric effects, however, is explicitly related to the wave drag. What then is the relationship between shock wave and boundary-layer entropy production and the drag? Another question that arises is whether or not the tail shocks sketched in Figure 14.8(a) contribute to the drag. Since they do not enter into the shock-expansion lift and drag computations, we might anticipate that the attached tail shocks do not contribute to the drag.

These questions are addressed by a result published by Oswatitsch (1980), which is an English translation of an earlier article. A generalization of this analysis is presented.

Assumptions and Restrictions

A steady, viscous flow of a gas about a finite-sized body is assumed, as sketched in Figure 14.1. Far upstream of the body there is a uniform flow and the surface S_∞ is (temporarily) well-removed from the surface S_w of the solid body. The $\vec{F}_b$ force is negligible, and the S_w surface is adiabatic. The wake downstream of the body may be turbulent, but with steady mean values. It is worth noting that the analysis, for some time, will not assume a perfect gas.

Aside from drag, the lift and a side force are also discussed. More importantly, the analysis accounts for all dissipative or entropy-producing processes, including boundary layers and the wave phenomenon associated with a base region, as occurs on bullets, waveriders, and missiles. The total drag for a supersonic vehicle can be subdivided into viscous, wave, form, and the drag due to lift. All these are accounted for, and the analysis also applies when the flow is subsonic and shock waves are not present.

The disturbance caused by a moving solid body decays with distance when measured from the body. The rate of decay is more rapid in a subsonic flow than in a supersonic flow, since the rate of decay of shock waves is relatively slow. (This is evident from the ground-level sonic boom of a high-flying, supersonic aircraft.) Nevertheless, because of the finite size of the body, shock waves ultimately do decay. In contrast to a semi-infinite cone or wedge, a finite-sized body in a supersonic flow always produces expansion waves. As discussed earlier, these waves interact with the shock waves and help to gradually weaken them until they become acoustic waves.

Our basic approach is to utilize the momentum theorem in conjunction with several integral relations that stem from Chapters 2 and 3. The only thermodynamic variables that directly appear in the mass, momentum, and energy equations are the density, pressure, and enthalpy. To introduce the entropy, $\tilde{s}$, the enthalpy is replaced with the second law equation

$$dh = Td\tilde{s} + \frac{dp}{\rho} \tag{14.126}$$

The same enthalpy replacement procedure is used when deriving Crocco's relation in Chapter 5.

Small Perturbation Analysis

We presume the S_∞ surface is sufficiently far removed from the body such that conditions on it are a small perturbation of freestream conditions. On S_∞, we thus write

$$p' = p - p_\infty, \qquad \rho' = \rho - \rho_\infty, \qquad \tilde{s}' = \tilde{s} - \tilde{s}_\infty, \qquad \ldots \tag{14.127a}$$

$$\vec{w}' = \vec{w} - \vec{w}_\infty \tag{14.127b}$$

where a prime denotes a small perturbation value. The S_w and S_∞ surfaces are not perturbed.

We can be somewhat more precise about the location of S_∞. For instance, consider the flow field about a supersonic bullet, as sketched in Figure 14.13. For axisymmetric flow, the principal features are a curved, detached bow shock, subsonic flow in a region adjacent to the nose of the

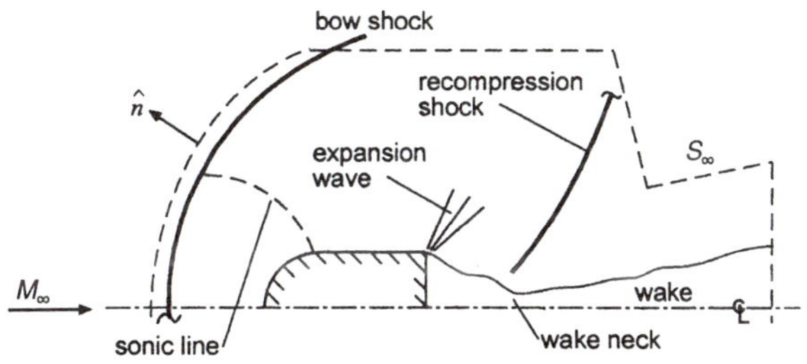

FIGURE 14.13 Flow field about a bullet in supersonic flight. Only the upper half of the axisymmetric flow is shown.

bullet, a relatively weak expansion that starts near the bullet's base, a recompression shock wave, and a turbulent wake that gradually spreads and decays downstream of its neck. The S_∞ surface is shown as a dashed line. It is chosen such that the viscous force term, $\hat{n} \cdot \overleftrightarrow{\tau}$, is negligible on it, and the relative magnitude of the perturbation variables $(\rho'/\rho_\infty,\ldots,u'/w_\infty)$ is small compared to unity. Note that perturbation values are zero on the part of S_∞ upstream of the bow shock. Also observe that S_∞ is extended in the downstream direction in order to allow the wake sufficient time to decay. This particular choice for S_∞ is not unique. For instance, if a larger S_∞ surface is used, the decrease in the magnitude of the perturbation is compensated for by the increase in the surface area of the integral. As we shall later see, a simpler choice for S_∞ is advisable.

It is convenient to introduce a Cartesian basis such that

$$\vec{w}_\infty = w_\infty \hat{1}_x \tag{14.128a}$$

$$\vec{w}' = u'\hat{1}_x + v'\hat{1}_y + w'\hat{1}_z \tag{14.128b}$$

where the z-coordinate velocity component w' should not be confused with $|\vec{w}'|$. The drag, lift, and side force then are

$$D = \vec{F}_w \cdot \hat{1}_x, \qquad L = \vec{F}_w \cdot \hat{1}_z, \qquad Y = \vec{F}_w \cdot \hat{1}_y \tag{14.129}$$

We also need

$$w^2 = (w_\infty + u')^2 + v'^2 + w'^2 = w_\infty^2 + 2w_\infty u' + \text{HOT}$$

where HOT stands for higher-order terms.

The momentum theorem now has the form

$$\vec{F}_w = -\oint_{S_\infty} [p\hat{n} + \rho\vec{w}(\vec{w} \cdot \hat{n})]ds \tag{14.130}$$

Flow conditions on S_∞ are assumed to be steady and inviscid. Consequently, continuity and the energy equation can be written in integral form [see Problem 14.15 for Equation (14.131b)]

$$\oint_{S_\infty} \rho(\vec{w} \cdot \hat{n})ds = 0, \qquad \oint_{S_\infty} \rho h_o(\vec{w} \cdot \hat{n})ds = 0 \tag{14.131a,b}$$

Note that the mass integral is not zero if the addition of engine fuel is considered and the energy integral is not zero if the wall of the body is not adiabatic or if hot engine exhaust gas is considered. The continuity relation becomes

$$\oint_{S_\infty} (\rho_\infty + \rho')(\vec{w}_\infty + \vec{w}') \cdot \hat{n}ds = 0 \tag{14.132}$$

or

$$\oint_{S_\infty} (\rho_\infty\vec{w}_\infty \cdot \hat{n} + \rho_\infty\vec{w}' \cdot \hat{n} + \rho'\vec{w}_\infty \cdot \hat{n} + \rho'\vec{w}' \cdot \hat{n})ds = 0 \tag{14.133}$$

where the higher-order term, $\rho'\vec{w}' \cdot \hat{n}$, can be neglected. The first term is evaluated as

$$\oint_{S_\infty} \rho_\infty \vec{w}_\infty \cdot \hat{n} ds = \rho_\infty \vec{w}_\infty \cdot \oint_{S_\infty} \hat{n} ds = 0 \tag{14.134}$$

We thus obtain the first-order perturbation result

$$\oint_{S_\infty} (\rho_\infty \vec{w}' \cdot \hat{n} + \rho' \vec{w}_\infty \cdot \hat{n}) ds = 0 \tag{14.135}$$

for continuity.

Similarly, the energy equation, which assumes an adiabatic wall, becomes

$$\oint_{S_\infty} (\rho_\infty + \rho')(h_{o\infty} + h'_o)(\vec{w}_\infty \cdot \hat{n} + \vec{w}' \cdot \hat{n}) ds = 0 \tag{14.136}$$

As before, higher-order terms are neglected and

$$\oint_{S_\infty} \rho_\infty h_{o\infty} \vec{w}_\infty \cdot \hat{n} ds = \rho_\infty h_{o\infty} \vec{w}_\infty \cdot \oint_{S} \hat{n} ds = 0 \tag{14.137}$$

Equation (14.136) reduces to

$$\oint_{S_\infty} (\rho_\infty h'_o \vec{w}_\infty \cdot \hat{n} + h_{o\infty}\{\rho_\infty \vec{w}' \cdot \hat{n} + \rho' \vec{w}_\infty \cdot \hat{n}\}) ds = 0 \tag{14.138}$$

where the integral of the term in braces is zero by virtue of Equation (14.135). With h_o defined by

$$h_o = h + \frac{1}{2} w^2$$

we have

$$h'_o = h' + w_\infty u' + \text{HOT} \tag{14.139}$$

Equation (14.126) can be written as

$$h' = T_\infty \tilde{s}' + \frac{1}{\rho_\infty} p' \tag{14.140}$$

By combining the above, Equation (14.138) yields

$$\oint_{S_\infty} (\rho_\infty T_\infty \tilde{s}' + p' + \rho_\infty w_\infty u')(\hat{1}_x \cdot \hat{n} ds) = 0 \tag{14.141}$$

where $\vec{w}_\infty \cdot \hat{n} = w_\infty \hat{1}_x \cdot \hat{n}$.

Next, Equation (14.130) is written as

$$\vec{F}_w = -\oint_{S_\infty} [(p_\infty + p')\hat{n} + (\rho_\infty + \rho')(\vec{w}_\infty + \vec{w}')(\vec{w}_\infty \cdot \hat{n} + \vec{w}' \cdot \hat{n})]ds \tag{14.142a}$$

which simplifies to

$$\vec{F}_w = -\oint_{S_\infty} (p'\hat{n} + \rho_\infty w_\infty \vec{w}'\hat{1}_x \cdot \hat{n} + \vec{w}_\infty\{\rho_\infty \vec{w}' \cdot \hat{n} + \rho' \vec{w}_\infty \cdot \hat{n}\})ds \tag{14.142b}$$

where, again, the term in braces integrates to zero. We thus have

$$\vec{F}_w = -\oint_{S_\infty} (p'\hat{n} + \rho_\infty w_\infty \vec{w}'\hat{1}_x \cdot \hat{n})ds \tag{14.142c}$$

for the net force exerted by the fluid on S_w.

The drag is now written as

$$D = -\oint_{S_\infty} (p' + \rho_\infty w_\infty u')\hat{1}_x \cdot \hat{n}ds \tag{14.143a}$$

With the aid of Equation (14.141), this simplifies to

$$D = \rho_\infty T_\infty \oint_{S_\infty} \tilde{s}'\hat{1}_x \cdot \hat{n}ds \tag{14.143b}$$

where $\hat{1}_x \cdot \hat{n}$ is the cosine of the included angle between these two unit vectors. Hence, the drag is related to the entropy production within S_∞, which, in turn, determines $\tilde{s}'$ on S_∞. This fundamental relation also can be written as

$$\begin{aligned} D &= \rho_\infty T_\infty \oint_{S_\infty} (\tilde{s}\hat{1}_x \cdot \hat{n} - \tilde{s}_\infty \hat{1}_x \cdot \hat{n})ds \\ &= \rho_\infty T_\infty \oint_{S_\infty} \tilde{s}\hat{1}_x \cdot \hat{n}ds - \rho_\infty T_\infty \tilde{s}_\infty \hat{1}_x \cdot \oint_{S_\infty} \hat{n}ds \\ &= \rho_\infty T_\infty \oint_{S_\infty} \tilde{s}\hat{1}_x \cdot \hat{n}ds \end{aligned} \tag{14.143c}$$

Observe that if the entropy is a constant on S_∞, the drag is then zero. In other words, a steady, inviscid, homentropic flow is drag-free. This is expected for what is, in effect, a potential flow. This relation clarifies the aerodynamic effect of shock waves. Regardless of location, any shock wave inside S_∞ contributes to the vehicle's drag. Consequently, the tail shocks in Figure 14.8(a), in fact, contribute to the wave drag. This point is also evident from the previous interference discussion.

Earlier, the location of S_∞ was discussed in a somewhat heuristic manner. Suppose a second surface S'_∞ is considered, where S'_∞ is larger than S_∞. The value for the drag should be the same for each surface; hence,

$$\hat{1}_x \cdot \oint_{S'_\infty} \tilde{s}\hat{n}ds = \hat{1}_x \cdot \oint_{S_\infty} \tilde{s}\hat{n}ds \tag{14.144}$$

With the aid of the divergence theorem, we observe that $\nabla\tilde{s}$ is zero in the region between the two surfaces. This means that there is negligible entropy production outside of S_∞. Ultimately, this requirement could be used to determine an approximate minimum size for S_∞.

We compute the lift and side forces as follows:

$$L = -\oint_{S_\infty} (p'\hat{1}_z \cdot \hat{n} + \rho_\infty w_\infty w'\hat{1}_x \cdot \hat{n})ds \tag{14.145}$$

$$Y = -\oint_{S_\infty} (p'\hat{1}_y \cdot \hat{n} + \rho_\infty w_\infty v'\hat{1}_x \cdot \hat{n})ds \tag{14.146}$$

In contrast to the drag analysis, Equation (14.141) does not simplify either relation. We note, however, that the sum of the force components equals

$$D + L + Y = -\oint_{S_\infty} [p'\vec{1} \cdot \hat{n} + \rho_\infty w_\infty (u' + v' + w')\hat{1}_x \cdot \hat{n}]ds \tag{14.147}$$

where

$$\vec{1} = \hat{1}_x + \hat{1}_y + \hat{1}_z \tag{14.148}$$

Comments

The form of the equations for D, L, and Y suggests that a rectangular parallelpiped might be used for S_∞ with the body at the origin; see Figure 14.14. The six planar surfaces are labeled S_{x-}, $S_{x+}, \ldots, S_{z+}$, and on these surfaces, we have:

$$\begin{array}{llll}
\hat{1}_x \cdot \hat{n} = -1, & \hat{1}_y \cdot \hat{n} = 0, & \hat{1}_z \cdot \hat{n} = 0, & S_{x-} \\
\hat{1}_x \cdot \hat{n} = 1, & \hat{1}_y \cdot \hat{n} = 0, & \hat{1}_z \cdot \hat{n} = 0, & S_{x+} \\
\hat{1}_x \cdot \hat{n} = 0, & \hat{1}_y \cdot \hat{n} = -1, & \hat{1}_z \cdot \hat{n} = 0, & S_{y-} \\
\hat{1}_x \cdot \hat{n} = 0, & \hat{1}_y \cdot \hat{n} = 1, & \hat{1}_z \cdot \hat{n} = 0, & S_{y+} \\
\hat{1}_x \cdot \hat{n} = 0, & \hat{1}_y \cdot \hat{n} = 0, & \hat{1}_z \cdot \hat{n} = -1, & S_{z-} \\
\hat{1}_x \cdot \hat{n} = 0, & \hat{1}_y \cdot \hat{n} = 0, & \hat{1}_z \cdot \hat{n} = 1, & S_{z+}
\end{array} \tag{14.149}$$

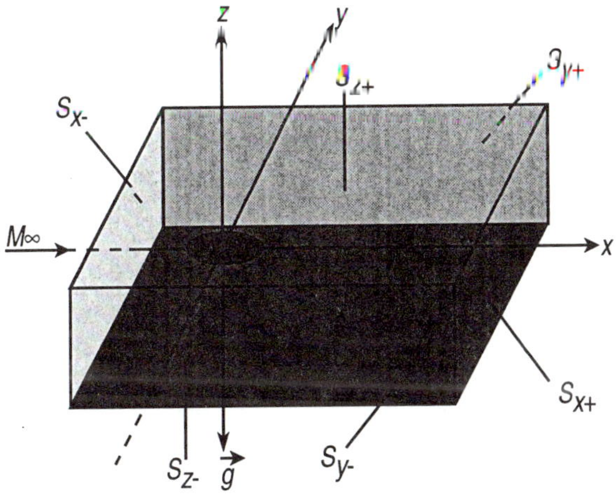

FIGURE 14.14 S_∞ as a parallelpiped surface.

Consequently, the force components simplify to

$$D = -\rho_\infty T_\infty \iint\limits_{S_{x-}} \tilde{s}' dydz + \rho_\infty T_\infty \iint\limits_{S_{x+}} \tilde{s}' dydz \tag{14.150a}$$

$$L = \iint\limits_{S_{z-}} p' dxdy - \iint\limits_{S_{z+}} p' dxdy + \rho_\infty w_\infty \iint\limits_{S_{x-}} w' dydz - \rho_\infty w_\infty \iint\limits_{S_{x+}} w' dydz \tag{14.150b}$$

$$Y = \iint\limits_{S_{y-}} p' dxdy - \iint\limits_{S_{y+}} p' dxdy + \rho_\infty w_\infty \iint\limits_{S_{x-}} v' dydz - \rho_\infty w_\infty \iint\limits_{S_{x+}} v' dydz \tag{14.150c}$$

These equations hold for all Mach numbers. For the lift, w' is conceptually similar to the downwash, or upwash, velocity component. From Equation (14.140), observe that the p' can be replaced with

$$p' = \rho_\infty h' - \rho_\infty T_\infty \tilde{s}' = \frac{\gamma R}{\gamma - 1} \rho_\infty T' - \rho_\infty T_\infty \tilde{s}' \tag{14.151}$$

where R is the gas constant.

Equations (14.150) further simplify if, at a supersonic Mach number, the disturbance only intersects the S_{x+} downstream surface. In this circumstance, we have

$$D = \rho_\infty T_\infty \iint\limits_{S_{x+}} \tilde{s}' dydz \tag{14.152a}$$

$$L = -\rho_\infty w_\infty \iint\limits_{S_{x+}} w' dydz \tag{14.152b}$$

$$Y = -\rho_\infty w_\infty \iint\limits_{S_{x+}} v' dydz \tag{14.152c}$$

and the body may still have roll and yaw angles, as well as an incidence angle. Observe that $\tilde{s}'$ is nonnegative on S_{x+}; hence, D is positive. For positive lift, w' is negative, as expected. These relations clarify the fundamental difference between drag and the other forces. Drag is associated with entropy production; the lift and side forces are associated with momentum transfer.

Despite its generality and elegance, Oswatitsch's (1980) method does not appear to have resulted in its application to practical flows. Undoubtedly, determining perturbation quantities on S_∞ is a formidable task. While Equations (14.152) are elegant in their simplicity, an obvious detriment is that S_{x+} must be sufficiently far downstream of the body for a perturbation solution to hold. This large separation distance, made larger by the presence of shock waves, is a serious difficulty for CFD codes. In the next section, however, formulas are derived for the force and moment components on a supersonic vehicle that avoids this difficulty.

One exception is the transonic analysis of Inger (1993), where the drag of a supercritical airfoil in inviscid flow is evaluated. Figure 14.15 is a sketch showing the principal features of the flow field in which the freestream Mach number has a high subsonic value. Note that the shock is normal to the wall at the wall and has zero strength where it intersects the sonic line. Because the shock is curved, the flow downstream of it is rotational.

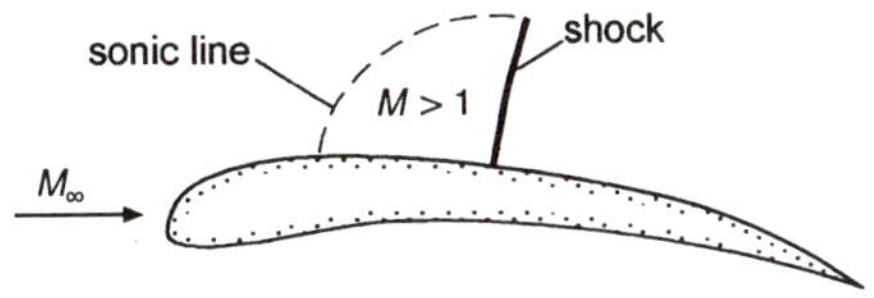

FIGURE 14.15 Sketch of a supercritical airfoil in transonic flow.

14.13 FORCES AND MOMENTS ON A SUPERSONIC VEHICLE

Introductory Remarks

There has been an effort over many decades to analyze wind tunnel wake data for the lift and drag and to decompose the drag into profile, induced, vortex, entropy, enthalpy, etc. components. More recent contributions (e.g., see Chatterjee and Janus, 1995; Cummings et al., 1996; Hunt et al., 1997; Nikfetrat et al., 1992; van Dam and Nikfetrat, 1992; and Wu et al., 1979) have increased our understanding of drag and our CFD capabilities. Nevertheless, Takahashi (1997) has discussed several difficulties inherent in this approach. Historically, drag has been analyzed by decomposing it into somewhat arbitrary components. This is often done by introducing a potential function, stream function, and the vorticity, and analyzing the flow in a wake or cross-flow plane downstream of the vehicle. This is generally preferable to a difficult integration of the pressure and skin friction over the vehicle's surface. In turn, a mean steady flow is required when there is turbulence, whose analysis still requires an ad hoc treatment. Moreover, these studies typically only deal with subsonic and/or transonic flows.

Our discussion is an outgrowth of the above references and the preceding section. Major differences are that a perturbation procedure is not utilized; all force components, not just the drag, and the moment of momentum are evaluated; the flow may be unsteady; and the viscous force is treated differently. The analysis is thus a generalization of previous work. By not using a perturbation procedure, the downstream surface can be relocated in close proximity to the vehicle, which may be an aircraft or a missile. As with Equations (14.152), a near-field solution is required that encompasses this plane, which we now rename the cross-flow plane, S_c; see Figure 14.16. In a supersonic flow, a solution is only required over that portion of S_c enclosed by the curve C_c. This curve is generated by the intersection of the bow shock and the cross-flow plane. The volumetric extent of an Euler or Navier–Stokes solution is thereby limited, and its generation is now practical. Although the discussion presumes a CFD solution, cross-flow plane experimental data may also be used. If the data in this plane are for a viscous flow, the total drag contains wave, lift, form, and viscous contributions, as in the preceding section. These contributions, of course, also include interference effects.

In order to focus the discussion, a supersonic vehicle in steady flight is assumed. The flow field downstream of the bow shock, however, may be unsteady, e.g., in the wake. Thus, the bow shock and cross-flow plane are steady and in fixed positions relative to the vehicle. For purposes of simplicity, the presence of engines is ignored. This is often done in preliminary aerodynamic studies. Engines can be modeled, e.g., by placing mass, momentum, and energy sources at their location (Cummings et al., 1996).

The drag is not subdivided into the components mentioned earlier, which, from a CFD viewpoint, would represent a misplaced emphasis. Instead, the focus is on the use of primitive variables, such as the pressure, density, and velocity, rather than on a stream function, potential function, or the vorticity. This is also convenient from a CFD viewpoint, which generally provides a near-field solution in terms of primitive variables. As a consequence, the cross-flow plane and near-field solution match in the choice of dependent and independent variables. As a practical matter, the two computations should have a compatible grid structure. We also note that the cross-flow plane approach is free of singularities, inconsistencies, and paradoxes to the extent that the near-field solution is also free of these difficulties.

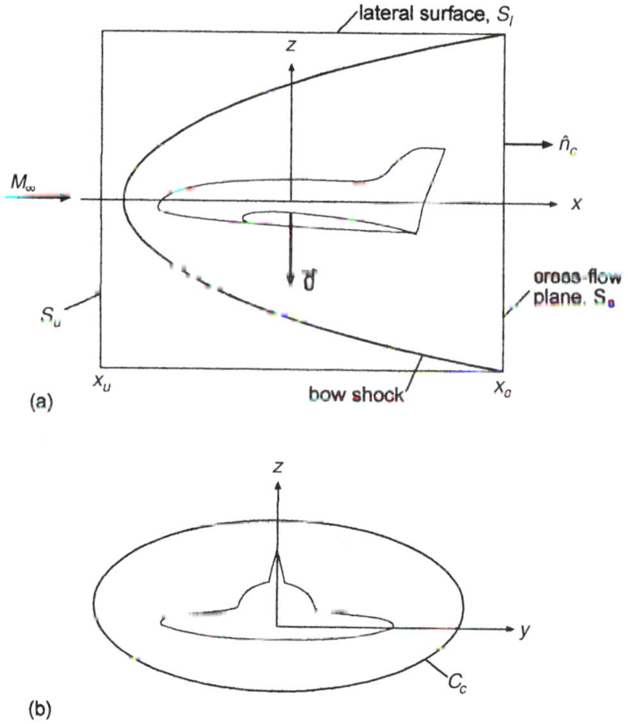

FIGURE 14.16 Supersonic aircraft with a control surface; (a) side view, (b) rear view.

Since the boundary layer and the wake will involve turbulent flow, either some type of Reynolds averaging or a Direct Numerical Simulation (DNS) is required for the near-field solution. The DNS approach (Moin and Mahesh, 1998) utilizes the unsteady Navier–Stokes equations, thereby avoiding the closure problem that is associated with any averaging procedure. As long as the solution in the cross-flow plane, and, if necessary, in the control volume, is physically correct, we are not concerned with how turbulence is modeled. This aspect is therefore not discussed.

A Cartesian coordinate system, sketched in Figure 14.16, is used in which the x-axis is parallel to $\vec{w}_\infty$, and $\vec{g}$ is opposite to the positive z-axis. The control surface consists, in part, of the S_u and S_c planes, where S_u is upstream of the bow shock. The lateral surface, S_ℓ, is a streamtube that passes through C_c. The S_u and S_ℓ surfaces could be replaced with one just upstream of the bow shock. The current approach, however, is analytically simpler. In particular, the shape of the bow shock, except for its intersection with the cross-flow plane, is not required. In this regard, shock waves may occur inside the control volume. These shocks may intersect with the bow shock, the vehicle, or the cross-flow plane.

Our goal is to reduce the equations for the components of the force and for the components of the moment of momentum to a practical computational form, akin to what is done when developing a panel method code.

Force Components

Because of the simplicity of S_∞, Equation (14.9) can be written as

$$\vec{F}_w = -\int_{CV} \frac{\partial(\rho\vec{w})}{\partial t} dv + \oint_{S_\infty} \left[-p\hat{n} - \rho\vec{w}(\vec{w}\cdot\hat{n}) + \hat{n}\cdot\overleftrightarrow{\tau}\right] ds \qquad \textbf{(14.153a)}$$

The unsteady integrand can be replaced with one that is free of unsteady terms. The derivation utilizes continuity to eliminate $\partial\rho/\partial t$ and a viscous form of Crocco's equation to eliminate $\partial\vec{\omega}/\partial t$. The complicated result, however, is not worth pursuing. With the Cartesian coordinate system in Figure 14.16, we obtain

$$\vec{F}_w = p_\infty A_c \hat{1}_x - p_\infty \int_{S_\ell} \hat{n}\,ds + (\rho w^2)_\infty A_c \hat{1}_x - \int_{A_c} \left[p\hat{1}_x + \rho\vec{w}(\vec{w}\cdot\hat{1}_x) - \hat{1}_x \cdot \overleftrightarrow{\tau} \right] ds - \int_{CV} \frac{\partial(\rho\vec{w})}{\partial t} dv \tag{14.153b}$$

where A_c is the area in the cross-flow plane enclosed by C_c. It is also the S_u area. Since

$$\oint_{S_\infty} \hat{n}\,ds = 0 \tag{14.154a}$$

implies

$$\int_{S_\ell} \hat{n}\,ds = 0 \tag{14.154b}$$

we have

$$\vec{F}_w = A_c(p + \rho w^2)_\infty \hat{1}_x - \int_{A_c} \left[p\hat{1}_x + \rho u\vec{w} - \hat{1}_x \cdot \overleftrightarrow{\tau} \right] ds - \int_{CV} \frac{\partial(\rho\vec{w})}{\partial t} dv \tag{14.153c}$$

where

$$\vec{w} = u\hat{1}_x + v\hat{1}_y + w\hat{1}_z \tag{14.155}$$

The first term on the right side is recognized as the impulse function for the freestream flow through A_c.

Equation (14.23), with $\hat{n} = \hat{1}_x$, is used for the evaluation of the $\hat{1}_x \cdot \overleftrightarrow{\tau}$ term. We thus obtain

$$\hat{1}_x \cdot \overleftrightarrow{\tau} = A_x\hat{1}_x + A_y\hat{1}_y + A_z\hat{1}_z \tag{14.156}$$

where (see Problem 14.10)

$$A_x = \left(\mu_b - \frac{2}{3}\mu\right)\nabla\cdot\vec{w} + 2\mu\frac{\partial u}{\partial x} \tag{14.157a}$$

$$A_y = \mu\left(\frac{\partial u}{\partial y} + \frac{\partial v}{\partial x}\right) \tag{14.157b}$$

$$A_z = \mu\left(\frac{\partial u}{\partial z} + \frac{\partial w}{\partial x}\right) \tag{14.157c}$$

Note that all of the terms in the area integral in Equation (14.153c) are evaluated at $x = x_c$ and over the A_c area, where $ds = dydz$. The numerical values for p, the components of $\vec{w}$, and the derivatives of these components, evaluated in the cross-flow plane, come from a numerical near-field solution of the Navier–Stokes or Euler equations. Similarly, the volumetric integral can be written as

$$\int_{CV} \frac{\partial(\rho\vec{w})}{\partial t} dv = \hat{1}_x \int_{CV} \frac{\partial(\rho u)}{\partial t} dv + \hat{1}_y \int_{CV} \frac{\partial(\rho v)}{\partial t} dv + \hat{1}_z \int_{CV} \frac{\partial(\rho w)}{\partial t} dv \quad \textbf{(14.158)}$$

These integrals extend from the bow shock to the S_c surface, exclusive of the vehicle itself. The right side of Equation (14.153c) is evaluated at a given instant of time. At that time, part of the flow in the cross-flow plane may be laminar, transitional, or turbulent. Remember that Equation (14.23) is a general viscous relation. If a steady, mean, viscous flow is utilized, then the volumetric integrals are zero, but a Reynolds stress term needs to be incorporated.

With Equations (14.129), the force components have the computationally convenient form

$$D = (p + \rho w^2)_\infty A_c - \int_{A_c} (p + \rho u^2 - A_x) dydz - \int_{CV} \frac{\partial(\rho u)}{\partial t} dv \quad \textbf{(14.159a)}$$

$$L = -\int_{A_c} (\rho uw - A_z) dydz - \int_{CV} \frac{\partial(\rho w)}{\partial t} dv \quad \textbf{(14.159b)}$$

$$Y = -\int_{A_c} (\rho uv - A_y) dydz - \int_{CV} \frac{\partial(\rho v)}{\partial t} dv \quad \textbf{(14.159c)}$$

In the drag formula,

$$(p + \rho w^2)_\infty = (\gamma + 1) p_\infty M_\infty^2 \quad \textbf{(14.160)}$$

and the right-hand side consists of a difference of impulse functions, a viscous term, and an unsteady term.

Discussion

If the flow is steady and incompressible, the drag simplifies to

$$D = (p + \rho w^2)_\infty A_c - \int_{A_c} (p + \rho u^2) dydz - 2\int_{A_c} \mu \frac{\partial u}{\partial x} dydz$$

Thus, D is given by an impulse function difference and a viscous term, which would not be present in a Euler calculation. More generally, there is a wake momentum deficit, which appears as a reduced value for $p + \rho u^2$. When the cross-flow plane is far downstream of the vehicle, the deficit appears as a reduced value for ρu^2. In any case, a sharply reduced value for $p + \rho u^2$ means a large drag. It is important to note that the shear layers in the wake, which stem from separated boundary layers, are a contributor to the momentum deficit. Viscous effects, therefore, are not solely confined to the A_x term.

The bulk viscosity only appears in A_x. As in shock wave structure studies, μ_b is important here and the correct value for it should be used. This is not provided by Stokes' hypothesis, i.e., $\mu_b = 0$. For air at room temperature, μ_b is approximately equal to $2\mu/3$, and A_x is dominated by the $2\mu \times \partial u/\partial x$ term. At significantly higher temperatures, such as about 800 K, μ_b for air should start to greatly exceed μ (Emanuel, 1998), and the divergence term becomes important. Temperatures in excess of 800 K occur in hypersonic boundary and shear layer flows. As discussed in Chapter 13, there is a rapid increase with temperature in μ_b relative to μ when the vibrational modes of O_2 and N_2 become active. The importance of μ_b for the drag, however, does not necessarily extend to a near-field Navier–Stokes solution, even at elevated temperatures. Generally, the terms in the compressible Navier–Stokes equations that contain μ_b have a negligible impact on the solution (Van Dyke, 1962).

In the lift equation, the ρuw term represents the downwash when w is negative. Downwash, of course, contributes a positive lift force. There is a coupling between drag and the lift through the $p + \rho u^2$ and ρuw terms. This coupling is referred to as lift-induced drag. A positive lift can be generated by the unsteady term, as caused, e.g., by an oscillating wing. For this, asymmetry inside the control volume in the z direction is required. This type of lift is also coupled to the drag, since $\partial(\rho w)/\partial t$ and $\partial(\rho u)/\partial t$ are not independent of each other. When the flow field and vehicle are symmetric about the x, z plane, the side force is zero. This stems from the symmetry properties about the x,z plane of $\rho, u, v, \ldots$, which causes each term in Y to integrate to zero.

The integrands in Equations (14.159) can be used for diagnostic purposes. For example, suppose the profile of the $(p + \rho u^2 - A_x)$ integrand in the cross-flow plane exhibits a steep minimum, which would correspond to a large drag contribution. Streamlines from this region of the cross-flow plane can be extended in the upstream direction to where they encounter that portion of the vehicle responsible for the drag contribution. A redesign might reduce the drag. This approach requires modification if the drag increase is caused by shock waves internal to the control volume. Nevertheless, a systematic analysis of integrand profiles in the cross-flow plane and the $\partial(\rho\vec{w})/\partial t$ profiles in the control volume should assist in the analysis and possible improvement of aerodynamic performance.

Moment Components

With

$$\oint_{\Gamma_\infty} p_\infty \vec{r} \times \hat{n}\, ds = 0, \qquad (\vec{w} \cdot \hat{n})_{S_\ell} = 0, \qquad (\overleftrightarrow{\tau})_{S_u} = (\overleftrightarrow{\tau})_{S_\ell} = 0 \tag{14.161}$$

Equation (14.45) becomes

$$\vec{M}_w = -\int_{A_c} (p - p_\infty)\vec{r} \times \hat{1}_x\, ds + (\rho w^2)_\infty \int_{A_u} \vec{r} \times \hat{1}_x\, ds - \int_{A_c} \rho u \vec{r} \times \vec{w}\, ds + \int_{A_c} \vec{r} \times \left(\hat{1}_x \cdot \overleftrightarrow{\tau}\right) ds - \int_{CV} \frac{\partial(\rho \vec{r} \times \vec{w})}{\partial t} dv \tag{14.162}$$

where A_u has the same area as A_c but is located in the S_u plane. We also utilize

$$\vec{r} = x\hat{1}_x + y\hat{1}_y + z\hat{1}_z, \qquad x = x_c, \qquad \hat{n} = \hat{1}_x, \qquad ds = dydz \tag{14.163}$$

to assist in obtaining

$$\vec{r} \times \hat{1}_x = z\hat{1}_y - y\hat{1}_z \tag{14.164a}$$

$$\vec{r} \times \vec{w} = (yw - zv)\hat{1}_x + (zu - x_c w)\hat{1}_y + (x_c v - yu)\hat{1}_z \tag{14.164b}$$

$$\vec{r} \times \left(\hat{1}_x \cdot \overleftrightarrow{\tau}\right) = (yA_z - zA_y)\hat{1}_x + (zA_x - x_c A_z)\hat{1}_y + (x_c A_y - yA_x)\hat{1}_z \tag{14.164c}$$

$$\int_{A_c} \vec{r}\, dydz = (x_c\hat{1}_x + y_{ce}\hat{1}_y + z_{ce}\hat{1}_z)A_c \tag{14.164d}$$

where the centroid of the A_c surface is

$$y_{ce} = \frac{1}{A_c}\int_{A_c} y dydz, \qquad z_{ce} = \frac{1}{A_c}\int_{A_c} z dydz \tag{14.164e}$$

With the foregoing relations, the moment of momentum can be written as

$$\vec{M} = M_x\hat{1}_x + M_y\hat{1}_y + M_z\hat{1}_z \tag{14.165}$$

where

$$M_x = \int_{A_c} [(zv - yw)\rho u + yA_z - zA_y]dydz + \int_{CV}\left[z\frac{\partial(\rho v)}{\partial t} - y\frac{\partial(\rho w)}{\partial t}\right]dv \tag{14.166a}$$

$$\begin{aligned} M_y = {} & (p + \rho w^2)_\infty A_c z_{ce} + \int_{A_c} [-pz + (x_c w - zu)\rho u + zA_x - x_c A_z]dydz \\ & + \int_{CV}\left[x_c\frac{\partial(\rho w)}{\partial t} - z\frac{\partial(\rho u)}{\partial t}\right]dv \end{aligned} \tag{14.166b}$$

$$\begin{aligned} M_z = {} & -(p + \rho w^2)_\infty A_c y_{ce} + \int_{A_c} [py + (yu - x_c v)\rho u + x_c A_y - yA_x]dydz \\ & + \int_{CV}\left[y\frac{\partial(\rho u)}{\partial t} - x_c\frac{\partial(\rho v)}{\partial t}\right]dv \end{aligned} \tag{14.166c}$$

The pitching moment is M_y, and its downwash contribution is provided by the $(\rho uw)x_c$ term. For positive lift, on average w is negative, and the downwash moment contribution is also negative when x_c is positive. Note the constant contribution to M_y and M_z that is proportional to $(\gamma + 1)$ $p_\infty M_\infty^2 A_c$. If the bounding curve C_c in the cross-flow plane is symmetric, e.g., with respect to the z-axis, then y_{ce} is zero.

Comments

The computation of the force and moment components has been reduced to several integrations in a cross-flow plane and a control volume. The same information can be obtained by performing a variety of integrations over the surface of a vehicle, after a CFD solution has been obtained. There may be some inherent advantage, however, to using our approach. It is global, in which the data can be conveniently stored for later retrieval and processing. Additional information, such as obtaining stability derivatives, may be possible.

A possible application would be to use this approach as a link between a near-field CFD solution and an optimization code, such as a multidisciplinary design optimization or multicriteria optimization code. Geometric and weight, etc., constraints can be imposed. The optimization, e.g., may be for maximum range or minimum drag. An iterative procedure between the CFD, cross-flow plane, and optimization routines could be used to alter the configuration, thereby generating a vehicle with improved performance.

REFERENCES

Bae, Y.-Y. and Emanuel, G., "Performance of an Aerospace Plane Propulsion Nozzle," *J. Aircraft* 28, 113 (1991).

Chatterjee, A. and Janus, J.M., "On the Use of a Wake Integral Method for Computational Drag Analysis," AIAA 95-0535, (1995).

Chow, W.L., Ke, Z.P., and Lu, J.Q., "The Interaction Between a Jet and a Flat Plate — An Inviscid Analysis," *J. Fluids Eng.* 117, 623 (1995).

Cummings, R.M., Giles, M.B., and Shrinivas, G.N., "Analysis of the Elements of Drag in Three-Dimensional Viscous and Inviscid Flows," AIAA 96-2482 (1996).

Eggers, Jr., A.J., Savin, R.C., and Syvertson, C.A., "The Generalized Shock Expansion Method and Its Application to Bodies Traveling at High Supersonic Air Speeds," *J. Aeronaut. Sci.* 22, 231–238, 248 (1955).

Emanuel, G., *Gasdynamics: Theory and Applications,* AIAA Educational Series, Washington, D.C., 1986.

Emanuel, G., "Bulk Viscosity in the Navier–Stokes Equations," *Int. J. of Eng. Sci.* 36, 1313 (1998).

Henderson, C.B., "Drag Coefficients of Spheres in Continuum and Rarefield Flows," *AIAA J.* 14, 707 (1976).

Henderson, C.B., "Reply by Author to M. J. Walsh," *AIAA J.* 15, 895 (1977).

Hinze, J.O., *Turbulence,* McGraw-Hill Book Co., New York, 1959, Section 5-7

Hourng, L.-W., "Study of a Supersonic Vortex Centrifuge," Ph.D. Dissertation, University of Oklahoma, Norman, OK, 1986.

Hourng, L.W. and Emanuel, G., "Particle Motion in a Supersonic Vortex Flow," *J. Aerosol Sci.* 18, 369 (1987).

Hunt, D.L., Cummings, R.M., and Giles, M.B., "Determination of Drag from Three-Dimensional Viscous and Inviscid Flowfield Computations," AIAA 97-2257 (1997).

Inger, G.R., "Application of Oswatitsch's Theorem to Supercritical Airfoil Drag Calculation," *J. Aircraft* 30, 415 (1993).

Jaslow, H., "Aerodynamic Relationships Inherent in Newtonian Impact Theory," *AIAA J.* 6, 608 (1968).

Karamcheti, K., *Principles of Ideal-Fluid Aerodynamics,* R. E. Krieger Pub. Co., Malabar, FL, 1980, Chapter 10.

Legendre, D. and Magnaudet, J., "A Note on the Lift Force on a Spherical Bubble or Drop in a Low-Reynolds-Number Shear Flow," *Phys. Fluids* 9, 3572 (1997).

Leung, K.K. and Emanuel, G., "Hypersonic Inviscid Flow over a Wedge and Cone," *J. Aircraft* 32, 385 (1995).

Liepmann, H.W. and Roshko, A., *Elements of Gasdynamics,* John Wiley, New York, 1957.

Mahony, J.J., "A Critique of Shock-Expansion Theory," *J. Aeronaut. Sci.* 22, 673–680, 720 (1955).

Moin, P. and Mahesh, K., "Direct Numerical Simulation: A Tool in Turbulence Research," *Annu. Rev. Fluid Mech.* 30, 539 (1998).

Morsi, S. A. and Alexander, A.J., "An Investigation of Particle Trajectories in Two-Phase Flow Systems," *J. Fluid Mech.* 55, 193 (1972).

Nielsen, J.N., *Missile Aerodynamics,* Nielsen Engineering & Research, Inc., Mountain View, CA, 1988.

Nikfetrat, K., van Dam, C.P., Vijgen, P. M. H. W., and Chang, I. C., "Prediction of Drag at Subsonic and Transonic Speeds Using Euler Methods," AIAA 92-0169 (1992).

Oswatitsch, K., "The Drag as Integral of the Entropy Flow," in *Contributions to the Development of Gasdynamics,* Friedr. Vieweg & Sohn, Braunschweig, 1980, pp. 2–5.

Panton, R.L., *Incompressible Flow,* John Wiley, New York, 1984, Section 19.12.

Pike, J., "Moments and Forces on General Convex Bodies in Hypersonic Flow," *AIAA J.* 12, 1241 (1974).

Rasmussen, M., *Hypersonic Flow,* John Wiley, New York, 1994.

Sherwood, J.D. and Stone, H.A., "Added Mass of a Disc Accelerating Within a Pipe," *Phys. Fluids* 9, 3141 (1997).

Takahashi, T.T., "On the Decomposition of Drag Components from Wake Flow Measurements," AIAA 97-0717 (1997).

van Dam, C.P. and Nikfetrat, K., "Accurate Prediction of Drag Using Euler Methods," *J. Aircraft* 29, 516 (1992).
Van Dyke, M., "Second-Order Compressible Boundary Layer Theory with Application to Blunt Bodies in Hypersonic Flow," *Hypersonic Flow Research,* edited by F. R. Riddell, Academic Press, New York, 1962, pp. 37–76.
Waldman, G.D. and Probstein, R.F., "An Analytic Extension of the Shock–Expansion Method," Wright Air Development Center TN 57-214 (May 1957) or AD-130751.
Warsi, Z.U.A., *Fluid Dynamics,* CRC Press, Boca Raton, FL, 1993.
Wu, J.C., Hackett, J.E., and Lilley, D.E., "A Generalized Wake–Integral Approach for Drag Determination in Three-Dimensional Flows," AIAA 79-0279 (1979).

PROBLEMS

14.1 Use an orthonormal basis to verify Equation (14.21).

14.2 Evaluate the *CV* integral in Equation (14.9) using the divergence theorem and the differential form of the momentum equation. What happens when this relation is substituted into Equation (14.9)? Can you demonstrate that this result is an identity?

14.3 Assume steady, inviscid flow, without an $\vec{F}_b$ force, and a uniform freestream velocity. Also assume that S_∞ does not contain a solid body but that there may be a solid body adjacent to part of S_∞. Derive the relations

$$\oint_{S_\infty} [(p-p_\infty)\hat{n} + \rho(\vec{w}-\vec{w}_\infty)(\vec{w}\cdot\hat{n})]ds = 0$$

$$\oint_{S_\infty} [(p-p_\infty)\vec{r}\times\hat{n} + \rho\vec{r}\times\vec{w}(\vec{w}\cdot\hat{n})]ds = 0$$

Show why $\vec{r}\times\vec{w}$, in the second equation, cannot be written as $\vec{r}\times(\vec{w}-\vec{w}_\infty)$.

14.4 There is steady, compressible flow inside a duct with a straight axis. The gas need not be perfect, the cross-sectional area $A(x)$ may vary smoothly with x and need not be circular, the flow may be viscous with heat transfer, and shock waves may be present.

(a) Denote flow conditions in a cross-sectional plane with a tilde; e.g., $\tilde{p}(x,y,z)$ and $\tilde{w}(x,y,z) = \tilde{u}\hat{1}_x + \tilde{v}_y\hat{1}_y + \tilde{v}_z\hat{1}_z$, where y and z are Cartesian coordinates in the cross-sectional plane. How are averaged (one-dimensional) values for $p(x)$, $\rho(x)$, and $w(x)$ to be obtained?

(b) Assume average values are known, via the part (a) procedure, at the inlet and exit planes of the duct. Inside the duct, the flow may be viscous, with shock waves, etc. Use the momentum theorem to show that the force on the inside wall of the duct is $F_w = F_2 - F_1$, where F is the impulse function and the 1 and 2 subscripts denote average conditions at the inlet and exit planes, respectively. This result is especially useful for Fanno and Rayleigh flows.

14.5 The drag coefficient C_D for a sphere in a steady, incompressible, laminar flow is about 0.4 for the Reynolds number range

$$10^3 \le Re = \frac{2a\rho w}{\mu} \le 2\times 10^5$$

where a is the radius of the sphere. For potential flow about a sphere, the pressure coefficient on the surface of the sphere is

$$C_p = \frac{2(p-p_\infty)}{\rho w_\infty^2} = 1 - \frac{9}{4}\sin^2\theta$$

where θ is measured from the positive x-axis; i.e., θ equals π at the upstream stagnation point. Laminar flow measurements are in rough accord with this C_p over the forward part of the sphere. Downstream of separation, however, C_p varies with θ from –0.6 to about –0.35. (This is a relatively small variation.) Obtain a crude, single-number estimate for a downstream C_p value by modeling the pressure on the forward-facing part of the sphere as if it were in potential flow and assuming a constant pressure p_{aft} on the aft part. Utilize the above 0.4 drag coefficient value.

14.6 A large, spherical, hot air balloon rises vertically from the ground. Assume the diameter d of the balloon and its internal hot air temperature T_{ha} are constants. For simplicity, also assume the surrounding atmosphere has a constant sea-level pressure p_∞ and temperature T_∞, where $T_{ha} > T_\infty$ and $(p_{ha} = p_\infty)$. Let m_b be the mass of the balloon (including its passengers and gondola, but excluding the hot air) and m_{ha} be the hot air mass; both masses are assumed constant.

(a) Write Newton's second law for the balloon. Include gravity, buoyancy, and the apparent mass. Simplify your result and determine, as functions of time, the altitude z_b, vertical speed w_b, and the magnitude of the acceleration a_b of the balloon.

(b) Determine $a_b(t)$ with and without the apparent mass contribution, for the following data:

$$d = 15 \text{ m}, \quad m_b = 200 \text{ kg}, \quad T_{ha} = 127\ ^\circ\text{C}, \quad p_\infty = 1.01 \times 10^5 \text{ Pa},$$

$$\rho_\infty = 1.23 \text{ kg/m}^3, \quad \mu_\infty = 1.78 \times 10^{-5} \text{ kg/m-s}$$

(c) Evaluate the Reynolds number for the balloon, including the effect of apparent mass on its flow speed, as a function of time. After one second of flight, is the flow about the balloon laminar or turbulent? (Use the data in part b.) Redo Newton's second law of part (a), but now include a drag coefficient term. This term has viscous and pressure (form drag) contributions. Is the pressure contribution redundant with that from the apparent mass? Explain.

14.7 A syringe with a hypodermic needle contains water ($\rho = 10^3$ kg/m^3, $\mu = 1.13 \times 10^{-3}$ Pa-s). The diameters are $d = 0.3$ mm and $d_p = 8$ mm, while the length of the needle is 5 cm. For a volumetric flow rate of 4×10^{-7} m^3/s, determine w_p, w, and the Reynolds numbers for the needle and the cylinder, in order to verify that the flow is laminar. Check that the entrance length, ℓ_e, given by $\ell_e = 0.06\, Red$, is smaller than ℓ. Determine Δp_d, F_d, and F_p.

14.8 Consider a quadrilateral airfoil whose lift and drag coefficients are given by Equations (14.102) and (14.103). Introduce a Cartesian coordinate system whose origin is at the leading edge of the airfoil and where the x-axis is aligned with the freestream velocity.

(a) Develop an equation for the force $\vec{F}$, per unit span, on the airfoil in terms of c_ℓ and c_d.

(b) For visualization purposes only, assume $\phi_1 > \alpha > 0$. Develop an equation for the pitching moment $\vec{M}_w$, per unit span, about the leading edge. Use the convention that a nose up moment is positive, and remember that the unit normal vector $\hat{n}$ is into the airfoil.

(c) Write this result as a pitching moment coefficient

$$c_{m_{LE}} = \frac{2M_w}{(\rho w^2)_\infty c^2}$$

The $c_{m_{LE}}$ equation should be similar in appearance to Equations (14.102) and (14.103).

(d) Determine the vector $\vec{r}_0$ that is perpendicular to the line along which $\vec{F}$ acts. Your answer should be in terms of c_ℓ, c_d, and $c_{m_{LE}}$.

14.9 Use planar cylindrical coordinates, r and θ, to obtain algebraic equations for a perfect gas potential vortex. The equations should be for ρ, p, T, u, and θ, where v and u are the r and θ components of the velocity, respectively. Assume M_2 and r_c are known; see Figure 14.12. Your answer should be in terms of stagnation conditions, γ, r_m/r, and u_m, where r_m is the radius where u has its maximum value u_m. Determine equations for r_m and u_m.

14.10 **(a)** Start with Equation (14.23) and derive Equations (14.156) and (14.157).
(b) Derive Equation (14.164d).

14.11 Consider steady, incompressible, two-dimensional flow past a symmetric (with respect to the x-axis) cylinder. The pressure at stations 1 and 2 is the same. Use the momentum theorem to derive formulas for the drag, d, assuming (a) the lateral surface is a streamtube and (b) the lateral surface is parallel to the freestream velocity.

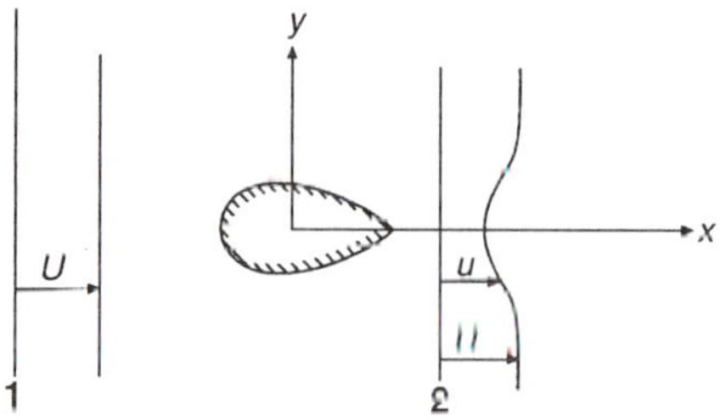

14.12 Continue with Problem 14.11 when the wake velocity profile has the simple shape shown in the sketch. The parameter c is a function of y_w/L. Write the drag as a drag coefficient, per unit depth,

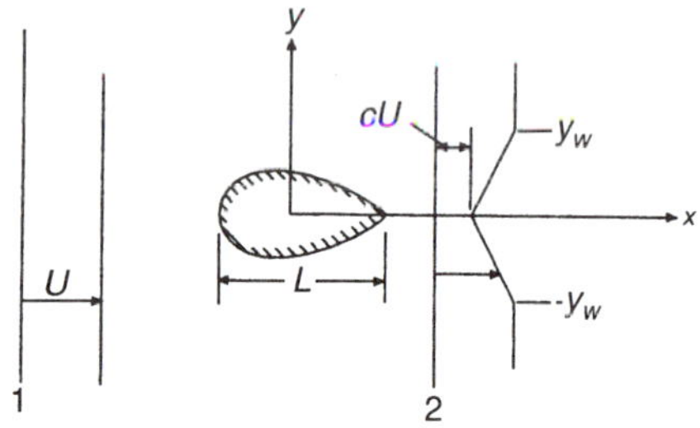

$$c_d = \frac{d}{\frac{1}{2}\rho U^2 L}$$

and determine c_d as a function of c and y_w/L.

14.13 A pitched baseball ($m = 0.1453$ kg, $d = 7.378 \times 10^{-2}$ m) initially travels at 40.23 m/s (90 mph) in air ($\rho = 1.177$ kg/m^3, $\mu = 1.846 \times 10^{-5}$ Pa-s).
(a) Determine the initial value for the Reynolds number.
(b) Determine the time, t_{ff}, it takes the baseball to travel 18.3 m (60 ft) in free flight; i.e., no external forces act on the moving baseball.
(c) Neglect the effects of rotation and of gravity and assume laminar flow with a drag coefficient of about 0.45, which is approximately correct for a smooth sphere at the Reynolds number of part (a). Determine the flight time, t_{lam}.
(d) Assume turbulent flow with a drag coefficient of about 0.15, and determine the time of flight, t_{turb}. How significant is the difference between t_{lam} and t_{turb}?

14.14 The sketch shows the leading and trailing edges of the expansion and the right-running C_- characteristic that starts where the LE and shock intersect. Additional notation is provided by Figure 14.10. If point e is downstream of point c, then the reflected wave does not alter the pressure distribution on the surface of the quadrilateral.

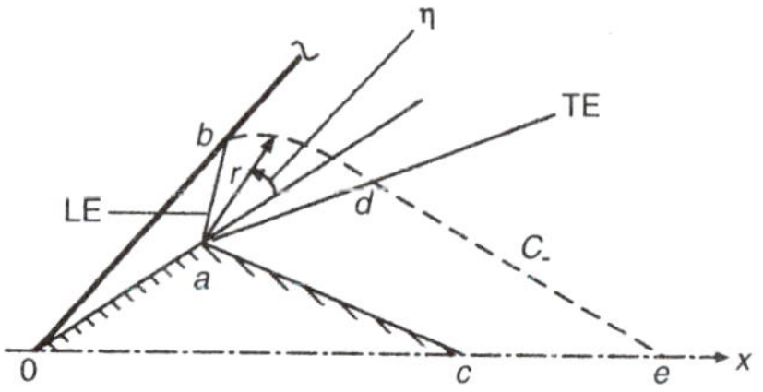

(a) Establish gas dynamic equations, in the proper sequence, for determining M_1 and M_2. Assume use of a chart for the shock wave angle β_1 and a normal shock table.
(b) Establish equations for x_a/c, y_a/c, x_b/c, and y_b/c in terms of known parameters, such as τ, ϕ_1,....
(c) The equation for the coordinates, r, η, of a point on the C_- characteristic inside the expansion is

$$\left(\frac{r}{r_b}\right)^2 = \left\{\frac{\left(\frac{\gamma+1}{2X_1}\right)^{1/2}}{\cos\left[\left(\frac{\gamma-1}{\gamma+1}\right)^{1/2}\left(\nu_1 + \frac{\pi}{2} - \eta\right)\right]}\right\}^{(\gamma+1)/(\gamma-1)} \frac{\left[\frac{\gamma-1}{2}\frac{M_1^2-1}{X_1}\right]^{1/2}}{\sin\left[\left(\frac{\gamma-1}{\gamma+1}\right)^{1/2}\left(\nu_1 + \frac{\pi}{2} - \eta\right)\right]}$$

where $\nu_1 = \nu(M_1)$ is the Prandtl–Meyer function. Derive this equation. Use this result to obtain an equation for r_d/c. Determine equations for x_d/c and y_d/c.

(d) Establish an equation for x_e/c.

(e) With

$$\gamma = 1.4, \qquad M_\infty = 3, \qquad \phi_1 = 12°, \qquad \tau = 0.15$$

determine values for:

$$x_a/c, \quad y_a/c, \quad x_b/c, \quad y_b/c, \quad x_d/c, \quad y_d/c, \quad x_e/c$$

Repeat the above calculation with $M_\infty = 2$.

14.15 Show the equivalence of Equation (14.131b) and $Dh_o/Dt = 0$.

14.16 Consider steady, inviscid, incompressible flow through a sudden contraction (see Figure 14.7). Assume the inlet and exit areas, A_1 and A_2, are known, along with ρ, w_1, and $p_2(= p_{amb})$. The cross-sectional areas need not be circular. Determine the normalized magnitude of the force $F_c/(p_2A_2)$ on the contraction and the pressure ratio $(p_c - p_2)/(p_1 - p_2)$, where p_c is an estimate of the average pressure on the contraction and does not equal the ambient pressure.

14.17 The wave drag is to be evaluated for a symmetric wedge of half-angle θ when the two shocks are attached and planar. Ignore viscous effects and the base drag.

(a) Derive an equation for the drag coefficient that only depends on γ, M_∞, θ, and β.

(b) For the values

$$\gamma = 1.4, \quad M_\infty = 2, 4, 6, \quad \theta = 2°(2°)22°$$

determine β and c_d in tabular form and plot c_d vs. θ for the three freestream Mach numbers.

PART IV

Exact Solutions for a Viscous Flow

Outline of Part IV

The study of viscous flow is largely a product of the 20th century. Of course, selected results were known at the turn of the century, like the Navier–Stokes equations, Hagen–Poiseuille flow, and Rayleigh flow; however, these were isolated results. Nothing in the way of a systematic, consistent theory existed. Equally important, what was known often had no bearing on any technology. The one relevant technology was hydraulics, which was primarily empirical. Starting with Prandtl and his discovery of the boundary layer, viscous flow has become a major component of fluid dynamics.

The three short chapters in Part IV are limited to large Reynolds number laminar flows. Low Reynolds number flows are interesting but are limited in application to tribology and aerosol aerodynamics.

As we have seen, the governing equations are coupled and nonlinear. Consequently, no general solution is possible. Exact solutions exist only for special cases and these frequently require an incompressible flow with constant fluid properties. However, the primary simplification is usually a reduction in the number of independent variables. Chapters 15 through 17, therefore, consider flows that are independent of at least two of the x, y, z, and t variables.

The term "exact solution" is used rather loosely. For instance, if the partial differential equations can be reduced to ordinary differential equations, we consider the problem as solved even though the ordinary differential equations may require numerical integration. A second caveat is that the solution may be exact only in a limited region of the flow, such as in the immediate vicinity of a stagnation point. Finally, the exact designation does not necessarily extend to the solution of the energy equation.

It is not possible, or useful, to attempt to survey all known exact solutions. Instead, we concentrate on three that are the most physically interesting examples. Chapter 15 considers Rayleigh flow, also known as Stokes' first problem. It is the only unsteady one among the examples. Since the fluid is incompressible, the energy equation is decoupled from the others. Nevertheless, we shall solve the energy equation, thereby determining the temperature field.

The second exact solution is for Couette flow, Chapter 16. Strictly speaking, this flow is incompressible, since the density is constant along a streamline. In our approach, however, the density varies from streamline to streamline, thereby resulting in a variable property flow. Here too, we solve the energy equation. A different approach would be followed if the density were a constant everywhere.

The last example, in Chapter 17, is for incompressible two-dimensional or axisymmetric stagnation point flow. For this flow, the solution for the pressure and velocity is limited to the immediate vicinity of the stagnation point. Additional assumptions are required for the solution of the energy equation for the temperature field.

15 Rayleigh Flow

15.1 PRELIMINARY REMARKS

Initially, there is a quiescent incompressible fluid in the half space above an infinite, smooth flat plate. At time $t = 0$, the plate impulsively moves parallel to itself at a constant speed U, as sketched in Figure 15.1. Because of viscosity, the fluid adjacent to the plate also moves with speed U. With increasing distance above the plate, the speed gradually decreases to zero. We choose the x-axis to be in the direction of motion and the y-axis to be perpendicular to the plate.

An infinity subscript is used to denote constant conditions as $y \to \infty$. Before any motion, the plate and fluid have a uniform temperature, T_∞. After $t = 0$, we assume an adiabatic wall; other wall temperature assumptions are possible, such as a constant value T_∞.

Because of its conceptual simplicity, a number of Rayleigh flow generalizations have been investigated. For example, the case of incompressible turbulent flow has been studied (Crow, 1968). A second category is provided by bodies that impulsively move in a laminar incompressible fluid in a direction perpendicular to their cross section (Batchelor, 1954; Wu and Wu, 1964), such as a circular or rectangular cylinder that impulsively moves along its axis. In each of these cases, the flow inside and outside the cylinder has been determined. Another example would be a wedge consisting of two semi-finite planar walls that move parallel to the intersection line where the walls meet. An additional example is provided by Problem 15.4. A third category allows the fluid to be compressible (Van Dyke, 1952) (see Problem 15.1). Viscous dissipation then heats the fluid, which induces a velocity component normal to the wall that results in a shock wave motion that is directed away from the wall.

The above extensions are all mathematically quite complicated. We, thus, focus on the classical problem outlined in the opening two paragraphs. From this description, the following simplifications are self-evident:

$$\frac{\partial}{\partial x} = \frac{\partial}{\partial z} = 0, \qquad w = 0 \text{ (the } z \text{ component of } \vec{w})$$

Thus, y and t are the only independent variables. The continuity equation is

$$\frac{\partial \rho}{\partial t} + \frac{\partial \rho v}{\partial y} = 0$$

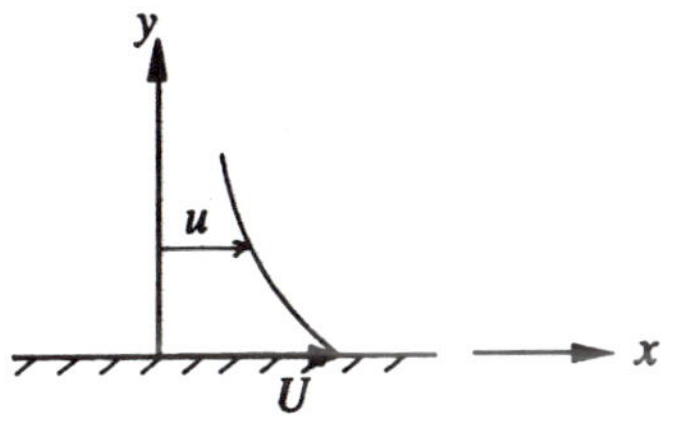

FIGURE 15.1 Rayleigh flow schematic.

and if the flow were compressible, v would not be zero. However, for an incompressible flow, $v = 0$ and the x-momentum and energy equations reduce to

$$\frac{\partial u}{\partial t} = \frac{\partial}{\partial y}\left(\frac{\mu}{\rho}\frac{\partial u}{\partial y}\right) \tag{15.1}$$

$$\frac{\partial}{\partial t}(h - p/\rho) = \frac{\partial}{\partial y}\left(\frac{\kappa}{\rho}\frac{\partial T}{\partial y}\right) + \frac{\mu}{\rho}\left(\frac{\partial u}{\partial y}\right)^2 \tag{15.2}$$

All variables are dimensional, and the rightmost term in the energy equation is the viscous dissipation function divided by ρ.

We are to obtain a solution when

$$t \geq 0, \qquad y \geq 0$$

subject to the initial and boundary conditions

$$\begin{aligned} u(y,0) &= 0, & u(0,t) &= U, & u(\infty,t) &= 0 \\ T(y,0) &= T_\infty, & \frac{\partial T}{\partial y}(0,t) &= 0, & T(\infty,t) &= T_\infty \end{aligned} \tag{15.3}$$

The above equations and associated conditions are sufficient only for determining u and T. To eliminate the enthalpy, we set

$$h = c_p T, \qquad c_p = \text{constant}$$

Since the pressure does not appear in Equation (15.1), we can set it equal to p_∞ and replace Equation (15.2) with

$$\frac{\partial T}{\partial t} = \frac{\partial}{\partial y}\left(\alpha\frac{\partial T}{\partial y}\right) + \frac{\nu}{c_p}\left(\frac{\partial u}{\partial y}\right)^2 \tag{15.4}$$

where

$$\alpha = \text{thermal diffusivity} = \frac{\kappa}{c_p \rho} \tag{15.5a}$$

$$\nu = \text{kinematic viscosity} = \frac{\mu}{\rho} \tag{15.5b}$$

To effect a simple solution, we further assume ν and α are constants. This uncouples the momentum equation from the energy equation and further simplifies both equations. In the next section, the solution for u is first obtained, after which the temperature solution is found. The section concludes by discussing vorticity and viscous dissipation.

15.2 SOLUTION

SOLUTION OF THE MOMENTUM EQUATION

The momentum equation, which is linear,

$$\frac{\partial u}{\partial t} = \nu \frac{\partial^2 u}{\partial y^2} \tag{15.6}$$

is transformed into an ordinary differential equation (ODE) by the substitution

$$\eta = \frac{y}{2(\nu t)^{1/2}}, \qquad \frac{u}{U} = f(\eta) \tag{15.7}$$

where η is a nondimensional similarity variable. The resulting equation is

$$f'' + 2\eta f' = 0 \tag{15.8a}$$

where a prime denotes differentiation with respect to η. The boundary conditions are

$$f(0) = 1, \qquad f(\infty) = 0 \tag{15.8b}$$

which are consistent with those for $u(y, t)$. By means of a similarity substitution, we reduced the PDE for u and its boundary and initial conditions into an ODE problem. As we will see, this method of solution is frequently encountered in viscous flows. The similarity solution of Equation (15.8) is

$$\frac{u}{U} = f(\eta) = 1 - \text{erf}\,\eta \tag{15.9}$$

where the error function is defined by

$$\text{erf}\,\eta = \frac{2}{\pi^{1/2}} \int_0^\eta e^{-z^2} dz \tag{15.10}$$

and z is a dummy integration variable.

As η increases, the error function approaches unity, as is evident in Figure 15.2. There is thus a viscous layer whose thickness is of order $\eta = 2$, since erf(2) = 0.995. We denote the thickness of the layer as δ, in which case

$$\frac{\delta}{2(\nu t)^{1/2}} = 2$$

or

$$\delta = 4(\nu t)^{1/2} \tag{15.11}$$

This thickness increases as $t^{1/2}$ as more of the fluid is brought under the influence of the moving wall.

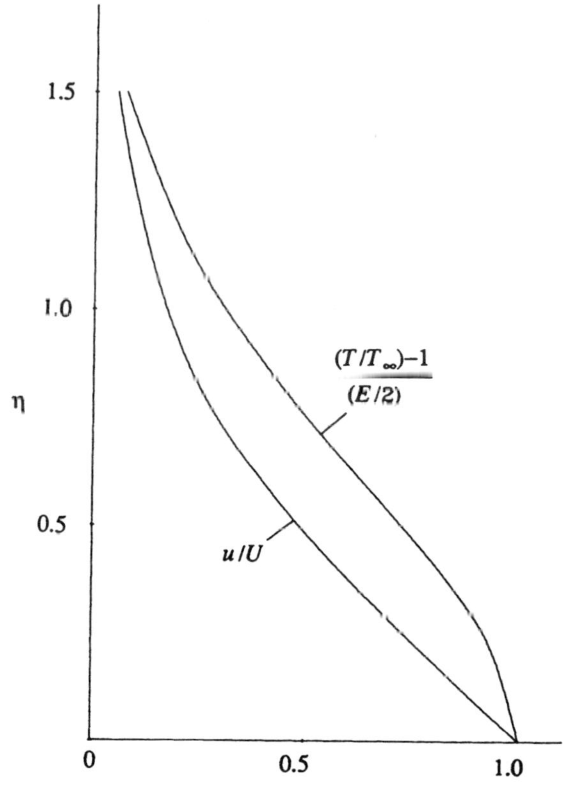

FIGURE 15.2 Flow speed and temperature profiles.

The skin friction is given by

$$\tau_w = -\mu\left(\frac{\partial u}{\partial y}\right)_w = -\frac{\mu U}{2(\nu t)^{1/2}}\left(\frac{df}{d\eta}\right)_w = \frac{\mu U}{(\pi \nu t)^{1/2}}$$

while the skin-friction coefficient is

$$c_f = \frac{2\tau_w}{\rho U^2} = \frac{2}{U}\left(\frac{\nu}{\pi t}\right)^{1/2} \tag{15.12}$$

Thus, both τ_w and c_f are infinite at $t = 0$ and decrease as $t^{-1/2}$. If we utilizc Equation (15.11) to eliminate t, we have

$$c_f = \frac{2}{U}\left(\frac{\nu}{\pi}\right)^{1/2}\frac{4\nu^{1/2}}{\delta} = \frac{8/\pi^{1/2}}{(U\delta/\nu)}$$

Hence, c_f varies inversely with the Reynolds number $U\delta/\nu$.

Solution of the Energy Equation

We attempt to reduce Equation (15.4) to an ordinary differential equation by introducing the similarity variable η. We, therefore, use

$$T^* = \frac{T}{T_\infty}$$

$$\frac{\partial}{\partial t} = \frac{\partial \eta}{\partial t}\frac{\partial}{\partial \eta} = -\frac{1}{2}\frac{y}{2v^{1/2}t^{3/2}}\frac{\partial}{\partial \eta} = -\frac{\eta}{2t}\frac{\partial}{\partial \eta}$$

$$\frac{\partial}{\partial y} = \frac{\partial \eta}{\partial y}\frac{\partial}{\partial \eta} = \frac{1}{2(vt)^{1/2}}\frac{\partial}{\partial \eta}$$

$$\frac{\partial^2}{\partial y^2} = \frac{1}{2(vt)^{1/2}}\frac{\partial}{\partial y}\frac{\partial}{\partial \eta} = \frac{1}{4vt}\frac{\partial^2}{\partial \eta^2}$$

to obtain

$$-\frac{\eta}{2t}T_\infty\frac{\partial T^*}{\partial \eta} = \alpha\frac{T_\infty}{4vt}\frac{\partial^2 T^*}{\partial \eta^2} + \frac{v}{c_p}\frac{U^2}{4vt}\left(\frac{df}{d\eta}\right)^2$$

However, we have

$$\frac{df}{d\eta} = -\frac{2}{\pi^{1/2}}e^{-\eta^2}$$

so that

$$-\eta\frac{dT^*}{d\eta} = \frac{\alpha}{2v}\frac{d^2T^*}{d\eta^2} + \frac{U^2}{2c_pT_\infty}\frac{4}{\pi}e^{-2\eta^2}$$

which simplifies to

$$\frac{d^2T^*}{d\eta^2} + 2Pr\,\eta\frac{dT^*}{d\eta} = -\frac{4}{\pi}PrE\,e^{-2\eta^2} \qquad (15.13)$$

where the Eckert number E is U^2/c_pT_∞. This Eckert number is similar to the square of a Mach number and must be small compared to unity for the incompressible assumption to be valid. In this equation, the second derivative term represents conduction, the first derivative term represents convection, and the inhomogeneous term on the right side represents viscous dissipation. This term is negligible when E is near zero. As η increases, it rapidly decreases in magnitude to zero from its wall value of $-4PrE/\pi$. The boundary conditions are

$$\frac{dT^*}{d\eta}(0) = 0, \qquad T^*(\infty) = 1 \qquad (15.14)$$

Thus, the similarity substitution also reduces the energy equation to an ODE problem.

To simplify the subsequent analysis, we assume the Prandtl number is unity, which does not alter the physics of the problem. (Problem 15.3 considers a nonunity Prandtl number.) With this alteration, the left side of the energy equation is identical in form to the momentum equation; hence, we have two linearly independent solutions for the homogeneous part of Equation (15.13):

$$T_1^* = \operatorname{erf}\eta, \qquad T_2^* = 1$$

The method of variation of parameters can be used for the particular solution of the inhomogeneous equation that will satisfy Equations (15.14). We thus have

$$T^* = c + \frac{2E}{\pi^{1/2}}\left(\int_0^\eta e^{-z^2}\operatorname{erf} z\, dz - \operatorname{erf}\eta\int_0^\eta e^{-z^2}dz\right)$$

where c is a constant and again z is a dummy integration variable. Since

$$\int_0^\eta e^{-z^2}dz = \frac{\pi^{1/2}}{2}\operatorname{erf}\eta$$

and

$$\int_0^\eta e^{-z^2}\operatorname{erf} z\, dz = \frac{\pi^{1/2}}{4}(\operatorname{erf}\eta)^2$$

we obtain

$$T^* = c - \frac{E}{2}(\operatorname{erf}\eta)^2$$

The error function satisfies

$$\operatorname{erf}\eta = \begin{cases} 0, & \eta = 0 \\ 1, & \eta \to \infty \end{cases}$$

Consequently, when $\eta \to \infty$, we have

$$c = 1 + \frac{E}{2}$$

with the final result for the temperature

$$\frac{T - T_\infty}{T_\infty} = \frac{E}{2}[1 - (\operatorname{erf}\eta)^2] \tag{15.15}$$

The derivative of temperature with respect to η yields

$$\frac{1}{T_\infty}\frac{dT}{d\eta} = -\frac{2E}{\pi^{1/2}}e^{-\eta^2}\operatorname{erf}\eta$$

and satisfies the adiabatic wall condition. Hence, the Nusselt and Stanton numbers are zero.

As with u, the decrease in temperature with increasing η is rapid (see Figure 15.2); e.g., for $\eta = 2$, we have

$$\frac{T - T_\infty}{T_\infty} = 9.34 \times 10^{-3}(E/2)$$

Hence, both the velocity and thermal layers on the plate have thickness on the order of $\eta = 2$. Since the velocity layer varies as erf η, while the thermal layer varies as $(\text{erf } \eta)^2$, the thermal layer is slightly thicker than the velocity layer, as is evident in Figure 15.2.

Vorticity and Entropy Production

Because of the incompressible assumption, the similarity solution is valid, provided the Eckert number E is small compared to unity. Inasmuch as the Prandtl number is unity, E is the only free parameter in the problem. We thus evaluate the behavior of the vorticity and entropy production in terms of E, with E being small. For the vorticity, we have

$$\vec{\omega} = \nabla \times \vec{w} = -\frac{\partial u}{\partial y}\hat{1}_z = \frac{U}{(\pi \nu t)^{1/2}} e^{-\eta^2}\hat{1}_z = \left(\frac{c_p T_\infty E}{\pi \nu t}\right)^{1/2} e^{-\eta^2}\hat{1}_z \tag{15.16}$$

For the entropy production, we need

$$\nabla T = \frac{\partial T}{\partial y}\hat{1}_y = -\frac{U^2 e^{-\eta^2}\text{erf } \eta}{c_p(\pi \nu t)^{1/2}}\hat{1}_y$$

and

$$\Phi = \mu\left(\frac{\partial u}{\partial y}\right)^2 = \frac{\rho U^2}{\pi t}e^{-2\eta^2}$$

Equation (3.17b) then becomes

$$\rho T \dot{s}_{irr} = \frac{\rho U^2}{\pi t}e^{-2\eta^2} + \frac{\kappa U^4 e^{-2\eta^2}(\text{erf } \eta)^2}{c_p^2 \pi \nu t T}$$

which results in

$$\frac{\dot{s}_{irr}}{c_p} = \frac{E\, e^{-2\eta^2}}{\pi t\left\{1 + \frac{E}{2}\left[1 - (\text{erf } \eta)^2\right]\right\}}\left(1 + \frac{E(\text{erf } \eta)^2}{1 + \frac{E}{2}[1 - (\text{erf } \eta)^2]}\right) \tag{15.17}$$

where $Pr = 1$ and Equation (15.15) are utilized.

Since

$$1 - \text{erf } \eta \sim \frac{1}{\pi^{1/2}\eta}e^{-\eta^2}, \qquad \eta \to \infty$$

the quantities u/U, T/T_∞, ω, and $\dot{s}_{irr}$ all decay as $e^{-\eta^2}$ or $e^{-2\eta^2}$ as η decreases. This rapid exponential decay is typical of both steady and unsteady high Reynolds number boundary layers. Within the viscous layer, the ordering with respect to a small value for E is seen to be

$$\frac{u}{U} = O(1)$$

$$\frac{T - T_\infty}{T_\infty} = O(E)$$

$$\omega = O(E^{1/2})$$

$$\dot{s}_{irr} = O(E)$$

Actually, the part $\dot{s}_{irr}$ due to viscous dissipation is $O(E)$, while the heat conduction part is smaller, since it is of $O(E^2)$. (It is worth noting that both contributions to $\dot{s}_{irr}$ are positive, in accord with the second law.) Hence, the entropy production is primarily due to viscous dissipation in an incompressible Rayleigh flow.

REFERENCES

Batchelor, G.K., "The Skin Friction on Infinite Cylinders Moving Parallel to Their Length," *Q. J. Mech. Appl. Math.* VII (2), 179 (1954).

Carslaw, H.S. and Jaeger, J.C., *Conduction of Heat in Solids,* 2nd ed., Oxford at the Clarendon Press, 1959.

Crow, S., "Turbulent Rayleigh Shear Flow," *J. Fluid Mech.* 32, 113 (1968).

Van Dyke, M., "Impulsive Motion of an Infinite Plate in a Viscous Compressible Fluid," *ZAMP* III, 343 (1952).

Wu, J.C. and Wu, T. Yao-Tsu, "Generalized Rayleigh's Problem in Viscous Flows," Douglas Aircraft Report SM-45868 (March 1964).

PROBLEMS

15.1 Derive the governing equations for compressible Rayleigh flow of a perfect gas and establish appropriate initial and boundary conditions.

15.2 Verify Equation (13.9) for the Rayleigh flow solution of Section 15.2.

15.3 Assume a constant Prandtl number, where $0 < Pr \leq 2$ and $T^*(0, t) = 1$ for $t \geq 0$.

(a) Determine

$$T^* = T^*(\eta; Pr, E)$$

where your solution will involve a quadrature defined by

$$I(\eta; a, b) = \int_0^\eta e^{-a^2 z^2} \operatorname{erf}(bz)\, dz$$

and where a and b are constants.

(b) Determine the heat transfer to the wall q_w.

15.4 Consider incompressible Rayleigh flow in which the lower wall impulsively moves with a constant speed U_a. There is an upper wall, at a distance ℓ from the lower wall, that impulsively moves in the same direction and starts at the same time as the lower wall but with a constant speed U_b.

(a) Determine the solution for $u(y, t)$ when $0 \leq y \leq \ell$ and $t \geq 0$. (Hint: use Carslaw and Jaeger, 1959).

(b) Determine the vorticity $\omega(y, t)$ and use this result to determine the skin-friction coefficients

$$c_{fa} = \frac{2\tau_{wa}}{\rho U_a^2}, \qquad c_{fb} = \frac{2\tau_{wb}}{\rho U_b^2}$$

To simplify the c_{fa} and c_{fb} results, introduce the theta function

$$\theta_3(\beta; T) = 1 + 2\sum_{n=1}^{\infty} e^{-n^2\pi^2 T}\cos(2\pi n\beta)$$

where T is a variable, not the temperature.

16 Couette Flow

16.1 PRELIMINARY REMARKS

Consider two infinite planar, parallel smooth walls separated by a distance ℓ, as sketched in Figure 16.1. The upper wall moves with a constant speed U_∞ relative to the lower wall. As shown, an x,y coordinate system is used, where x is in the direction of motion of the upper wall. The walls are at temperatures T_w and T_∞. We emphasize the similarity with a boundary layer by utilizing a wall and freestream notation that is common in boundary-layer analysis. Couette flow is sometimes viewed as a simple model for a laminar boundary layer.

In contrast to Rayleigh flow, we assume a gas with the properties

$$c_p = \text{constant}, \qquad \frac{\mu}{\mu_\infty} = \frac{T}{T_\infty}, \qquad \frac{\kappa}{\kappa_\infty} = \frac{T}{T_\infty} \tag{16.1}$$

Hence, the Prandtl number is a constant, although not necessarily unity. Although more realistic, the case where μ and κ are proportional to T^ω, with ω constant, is appreciably more complicated and is not considered.

For this steady flow, we have the simplifications

$$\frac{\partial}{\partial t} = \frac{\partial}{\partial x} = \frac{\partial}{\partial z} = 0, \quad w = 0 \text{ (the } z \text{ component of } \vec{w})$$

From continuity

$$\nabla \cdot (\rho \vec{w}) = \frac{d(\rho v)}{dy} = 0$$

we conclude that $v = 0$, since the density is not constant with y. Thus, the flow speed u and density are functions only of y, and the continuity equation is identically satisfied.

The only nonzero components of the rate-of-deformation tensor are

$$e_{xy} = e_{yx} = \frac{1}{2}\frac{du}{dy}$$

and, furthermore,

$$\frac{D(\;)}{Dt} = 0, \qquad \nabla \cdot \vec{w} = 0, \qquad \Phi = \mu\left(\frac{du}{dy}\right)^2$$

Consequently, the second viscosity coefficient does not enter into the formulation. [This would not be the case if the problem were generalized to include blowing and suction on opposite walls (Gonzalez amd Emanuel, 1993).] Because

$$\frac{D\rho}{Dt} = 0$$

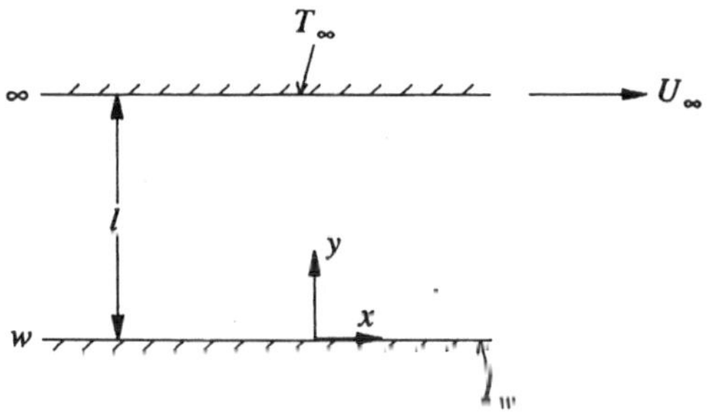

FIGURE 16.1 Couette flow schematic.

the flow is, by definition, incompressible, although the density varies from streamline to streamline; we refer to this example as a variable property flow, since μ and κ also vary with y.

In the next section, we obtain a solution for $u(y)$ and $T(y)$ subject to the conditions

$$T(0) = T_w, \qquad T(\ell) = T_\infty, \qquad u(0) = 0, \qquad u(\ell) = U_\infty$$

Consequently, there is heat transfer at both walls given by

$$q_w = -\left(\kappa \frac{dT}{dy}\right)_w, \qquad q_\infty = -\left(\kappa \frac{dT}{dy}\right)_\infty$$

where a negative value for q means the heat transfer is in the negative y direction. At this time, q_w and q_∞ may have either sign; however, the net heat transfer at the walls must balance the heat generated in the gas by viscous dissipation. In addition, we derive the skin-friction coefficient, Stanton number, and Reynolds analogy. (Still other Couette flow properties are found in Problems 3.3, and 16.2 through 16.5.) The final section deals with an adiabatic lower wall, where the recovery temperature and the recovery factor for this wall are derived.

16.2 SOLUTION

The F_i^s are given by

$$F_x^s = \frac{d}{dy}\left(\mu \frac{du}{dy}\right), \qquad F_y^s = F_z^s = 0$$

and the momentum equation in the x direction reduces to

$$\frac{d}{dy}\left(\mu \frac{du}{dy}\right) = 0$$

Since the only nonzero shear stress is

$$\tau_{xy} = \tau_{yx} = \mu \frac{du}{dy}$$

we obtain

$$\mu \frac{du}{dy} = \tau_w = \left(\mu \frac{du}{dy}\right)_w \tag{16.2}$$

Hence, the shear stress is a constant throughout the flow. This relation will be integrated later, after a relation for the temperature is found. Of course, if μ is assumed constant, then u varies linearly with y:

$$\frac{u}{U_\infty} = \frac{y}{\ell}$$

For the y momentum equation, we have

$$\frac{\partial p}{\partial y} = 0$$

and the pressure is constant and can be set equal to p_∞.

Equation (13.6c) for energy reduces to

$$\Phi = \nabla \cdot \vec{q}$$

or

$$\mu\left(\frac{du}{dy}\right)^2 = \frac{dq}{dy} = -\frac{d}{dy}\left(\kappa\frac{dT}{dy}\right) \tag{16.3}$$

Since the leftmost term is positive, we see that q, which may be positive or negative, increases with y. Since $\mu\, du/dy$ is a constant, we can write

$$\mu\left(\frac{du}{dy}\right)^2 = \mu\left(\frac{du}{dy}\right)^2 + u\frac{d}{dy}\left(\mu\frac{du}{dy}\right) = \frac{d}{dy}\left(\mu u\frac{du}{dy}\right)$$

In combination with Equation (16.3), we obtain

$$\frac{d}{dy}\left(\kappa\frac{dT}{dy}\right) + \frac{d}{dy}\left(\mu u\frac{du}{dy}\right) = 0$$

which integrates to

$$\kappa\frac{dT}{dy} + \frac{\mu}{2}\frac{du^2}{dy} = \kappa_w\left(\frac{dT}{dy}\right)_w = -q_w \tag{16.4}$$

This relation can be written as

$$q = u\tau + q_w = u\tau_w + q_w$$

which exhibits the balance between heat generated by viscous dissipation, $u\tau$, and lateral conductive heat transfer. Since $u\tau$ is nonnegative, this relation further verifies that q increases with y, reaching a value of $U_\infty\tau_w + q_w$ at the upper wall.

To aid in the integration of Equation (16.4), we utilize Equations (16.1), multiply Equation (16.4) by c_p, and introduce the Prandtl number, to obtain

$$\frac{d(c_pT)}{dy} + \frac{d}{dy}\left(\frac{1}{2}Pru^2\right) = -\frac{Prq_w}{\mu}$$

This relation is integrated from the lower wall:

$$c_p(T - T_w) + \frac{1}{2}Pru^2 = -Prq_w\int_0^y \frac{dy'}{\mu(y')}$$

where a prime indicates a dummy integration variable. However, from Equation (16.2), we have

$$\int_0^y \frac{dy'}{\mu} = \int_0^u \frac{du}{\tau_w} = \frac{u}{\tau_w}$$

which result in

$$c_p(T - T_w) + \frac{1}{2}Pru^2 = -\frac{Prq_w}{\tau_w}u \tag{16.5}$$

Thus, T varies quadratically with u. There is a corresponding quadratic relation associated with a laminar boundary layer that is due to Crocco.

Nondimensionalization and Reynolds' Analogy

At this time it is convenient to introduce nondimensional parameters and variables. For the variables, we use

$$Y = \frac{y}{\ell}, \qquad V = \frac{u}{U_\infty}, \qquad \theta = \frac{T - T_w}{T_\infty - T_w}$$

Note that the θ transformation cannot be used if $T_\infty = T_w$. In this case it would be replaced with $\theta = (T/T_\infty)$. At the two walls, we have

$$\begin{aligned} V(0) &= 0, \qquad \theta(0) = 0 \\ V(1) &= 1, \qquad \theta(1) = 1 \end{aligned} \tag{16.6}$$

With the assumption of a perfect gas, we utilize γ, Pr, and

$$\begin{aligned} M_\infty &= \frac{U_\infty}{(\gamma R T_\infty)^{1/2}} \\ Re_\infty &= \frac{\rho_\infty U_\infty \ell}{\mu_\infty} \\ c_f &= \frac{2\tau_w}{\rho_\infty U_\infty^2} \\ St &= \frac{q_w}{\rho_\infty c_p U_\infty \Delta T} \end{aligned} \tag{16.7}$$

These are the principal nondimensional parameters for Couette flow and for a compressible boundary layer. Except for the Stanton number definitions, they were previously encountered in Chapter 13.

In the Stanton number, there are three possible choices for ΔT, which can be written as

$$\begin{aligned} \Delta T_\infty &= T_w - T_\infty \\ \Delta T_r &= T_w - T_r \\ \Delta T_o &= T_w - T_{o\infty} \end{aligned} \tag{16.8}$$

The first of these is the simplest; it coincides with the film coefficient definition, Equation (13.15). However, when $\Delta T_\infty = 0$, as can happen, this Stanton number is infinite, unless $q_w = 0$. Thus, the first of the above ΔT_∞ choices is rejected.

In ΔT_r the recovery temperature T_r is used, which will be defined in the next section. We thus base the Stanton number definition on ΔT_o, where T_w and $T_{o\infty}$ are the stagnation temperatures at the respective walls. In particular, $T_{o\infty}$ is the stagnation temperature of the fluid that is adjacent to the upper wall; i.e.,

$$T_{o\infty} = T_\infty\left(1 + \frac{\gamma - 1}{2}M_\infty^2\right) \tag{16.9}$$

There is one dominant reason for this choice; it coincides with the Stanton number definition used for a compressible boundary layer. In addition, this Stanton number is well defined when $\Delta T_\infty = 0$.

It is also convenient to define the parameters

$$a = \frac{\gamma - 1}{2} Pr M_\infty^2, \qquad b = \frac{T_w}{T_\infty}$$

With these definitions, the q_w/τ_w ratio that appears in Equation (16.5) can be written as

$$\frac{q_w}{\tau_w} = -\frac{[(1-b)Pr + a]U_\infty}{a}\frac{St}{c_f} \tag{16.10}$$

and Equation (16.5) becomes

$$(1-b)\theta + aV^2 = [(1-b)Pr + a]\frac{2St}{c_f}V \tag{16.11}$$

Equations (16.6) now yield, at the upper wall,

$$\frac{c_f}{2St} = \frac{(1-b)Pr + a}{1 - b + a} \tag{16.12}$$

where the ratio on the left side is called Reynolds' analogy; it represents a skin friction to heat transfer ratio. We have shifted from considering τ_w and q_w as the key unknown parameters to c_f and St. In the incompressible limit, $M_\infty = 0$, and when $Pr = 1$, or when $T_\infty = T_w$, this ratio is unity. Since $a \geq 0$, the ratio may be less than unity. Observe that the Reynolds number does not appear in this equation; it will appear in the analysis shortly.

Skin-Friction Coefficient and Stanton Number

Equation (16.11) provides one equation for c_f and St. For a second equation, we need V as a function of Y, or vice versa. With

$$\mu = \mu_\infty \frac{T}{T_\infty} = \mu_\infty[b + (1-b)\theta]$$

and a similar relation for κ, Equation (16.4) is converted to

$$\frac{\mu_\infty}{Pr}\left(\frac{\gamma R}{\gamma-1}\right)[b+(1-b)\theta]\frac{T_\infty}{\ell}\frac{d(T/T_\infty)}{dY}+\mu_\infty[b+(1-b)\theta]\frac{U_\infty^2}{\ell}V\frac{dV}{dY}$$
$$=\frac{\gamma R T_\infty}{\gamma-1}\left(\frac{\rho_\infty U_\infty}{Pr}\right)[(1-b)Pr+a]St$$

which simplifies to

$$(1-b)\frac{d\theta}{dY}+2aV\frac{dV}{dY}=\frac{(1-b)Pr+a}{b+(1-b)\theta}Re_\infty St$$

We eliminate θ with the aid of Equation (16.11), to obtain

$$\frac{dV}{dY}=\frac{(1-b)Pr+a}{1-b+a}\frac{Re_\infty St}{b+(1-b+a)V-aV^2}$$

This equation is easily integrated, from the lower wall, with the result

$$Y=\frac{1-b+a}{(1-b)Pr+a}\frac{V}{Re_\infty St}\left[b+\frac{1}{2}(1-b+a)V-\frac{a}{3}V^2\right]$$

With the aid of Equations (16.6), we now have

$$Re_\infty St=\frac{(1-b+a)\left(1+b+\frac{a}{3}\right)}{2[(1-b)Pr+a]} \tag{16.13}$$

and the $Y = Y(V)$ equation simplifies to

$$Y=\frac{2}{1+b+\frac{a}{3}}V\left[b+\frac{1}{2}(1-b+a)V-\frac{a}{3}V^2\right] \tag{16.14}$$

Similarly, Equations (16.11) and (16.12) become

$$\theta=\frac{1}{1-b}V(1-b+a-aV) \tag{16.15}$$

$$c_f=\frac{1+b+\frac{a}{3}}{Re_\infty} \tag{16.16}$$

Equations (16.14) and (16.15) implicitly provide the velocity and temperature profiles, while Equations (16.13) and (16.16) provide the Stanton number and skin-friction coefficient. These later equations provide c_f and St in terms of γ, Pr, M_∞, Re_∞, and T_w/T_∞, or in terms of a, b, Pr, and Re_∞. All these parameters are nonnegative, with $\gamma \geq 1$. Observe that both c_f and St decrease with the Reynolds number as $1/Re_\infty$. (For a laminar boundary layer, the dependence for both parameters is $1/Re^{1/2}$. However, for a boundary layer, the Reynolds number is based on the length along the wall. Consequently, the comparison is not really appropriate.) Both c_f and St increase with a or with $(\gamma - 1)M_\infty^2$. The skin-friction coefficient also increases with b; thus, a cold wall minimizes the skin-friction coefficient. The variation of St with b is more complicated, as discussed in Problem 16.2.

16.3 ADIABATIC WALL

We now replace $T(0) = T_w$ with the $(dT/dy) = 0$ condition at the lower wall. From Equation (16.13) with $St = 0$, we obtain

$$1 + a = b \tag{16.17}$$

as the condition for an adiabatic lower wall. The temperature of the gas, T_r, that is adjacent to the adiabatic wall is obtained from this relation by setting $T_w = T_r$, with the result

$$\frac{T_r}{T_\infty} = 1 + \frac{\gamma - 1}{2} PrM_\infty^2 = 1 + a \tag{16.18}$$

As noted in conjunction with Equations (16.8), T_r is referred to as the recovery temperature (of the gas) or the adiabatic wall temperature. When $Pr > 1$, T_r exceeds the stagnation temperature of the gas, $T_{o\infty}$, at the upper wall. When $Pr < 1$, T_r is less than $T_{o\infty}$; T_r equals $T_{o\infty}$ when Pr is unity. For many gases, Pr is near 0.7. Recall that the temperature is a balance between conduction and the heat generated by viscous dissipation. Since $Pr = c_p\mu/\kappa$, the balance favors conduction when $Pr < 1$, thereby resulting in a reduced gas temperature at the lower wall.

We next obtain the adiabatic wall temperature distribution from Equation (16.15) as

$$\frac{T}{T_\infty} = 1 + a(1 - V^2) \tag{16.19}$$

We utilize Equations (16.1) and write Equation (16.2) as

$$\frac{\mu_\infty U_\infty}{\ell} \frac{T}{T_\infty} \frac{dV}{dY} = \tau_r \tag{16.20}$$

where the r subscript, hereafter, denotes an adiabatic lower wall. In combination with Equation (16.19), we integrate this equation, to obtain

$$Re_\infty c_{fr} Y = 2(1 + a)V - \frac{2}{3}aV^3 \tag{16.21}$$

This relation is evaluated at the upper wall, which yields the skin-friction coefficient for the lower wall

$$c_{fr} = \frac{2(1 - 4a/3)}{Re_\infty} \tag{16.22}$$

(Since Equation (16.16) is not based on an adiabatic wall condition, Equation (16.22) is not obtained by combining Equations (16.16) and (16.17).) An adiabatic wall condition is possible only when

$$c_{fr} \geq 0$$

or

$$0 \leq a \leq \frac{3}{4}$$

If a exceeds 3/4, the heat produced by viscous dissipation is too large to permit the lower wall to remain adiabatic in a steady flow.

The adiabatic wall parameters, τ_r and T_r, can be viewed as reference quantities and used when the lower wall is not adiabatic. For instance, the parameter r

$$r = \frac{T_r - T_\infty}{T_{o\infty} - T_\infty} \tag{16.23}$$

is called the recovery factor. With Equation (16.18) we have

$$r = \frac{T_\infty + T_\infty Pr\left(\frac{\gamma-1}{2}\right)M_\infty^2 - T_\infty}{T_\infty + T_\infty \frac{\gamma-1}{2}M_\infty^2 - T_\infty} = Pr \tag{16.24}$$

for Couette flow.

Starting with Equations (16.10) and (16.18), we obtain

$$q_w = \frac{c_p \tau_w}{U_\infty Pr}(T_w - T_r) \tag{16.25}$$

for a wall with heat transfer, where the $(c_p \tau_w)/(U_\infty Pr)$ coefficient is positive. The intuitive presumption that the lower wall heats the gas if $T_w > T_{o\infty}$, or if $T_w > T_\infty$, is therefore incorrect. No heat transfer occurs when $T_w = T_r$. Of course, if $Pr = 1$, then $T_r = T_{o\infty}$ and q_w can be written as

$$q_w = \frac{c_p \tau_w}{U_\infty}(T_w - T_{o\infty}) = \frac{\tau_w}{U_\infty}(h_{ow} - h_{o\infty}) \tag{16.26}$$

and the lower wall heats the gas if $T_w > T_{o\infty}$.

REFERENCE

Gonzalez, H. and Emanuel, G., "Effect of Bulk Viscosity on Couette Flow," *Phys. Fluids A* 5, 1267 (1993).

PROBLEMS

16.1 Derive:

(a) Equation (16.11), starting from Equation (16.5).

(b) Equation (16.21) and (16.22), starting from Equation (16.20).

16.2 **(a)** Use Equation (16.13) to determine the behavior of $Re_\infty St$ in terms of a, b, and Pr.

(b) Determine an equation for b^*, which is the value when $Re_\infty St$ is an extremum (a maximum, minimum, or horizontal inflection point) with respect to b.

(c) Consider three separate cases: $a = 0$, $Pr = 1$, and $Pr > 1$. What type of extremum do we have for $a = 0$ and $Pr = 1$? What can you say for $Pr > 1$?

(d) Draw a sketch of $Re_\infty St$ vs. b when $a = 3$, $Pr = 0.5$. Label all significant points, such as the zero values of $Re_\infty St$.

16.3 **(a)** In the diabatic wall case, can the temperature have a maximum or minimum value in the $0 < Y < 1$ range?

(b) If the answer is yes, can the temperature have either a maximum or a minimum value?

16.4 For a diabatic lower wall, tabulate Y and θ vs. V for $V = 0(0.2)1$. For the parameters, utilize

$$\gamma = 1.4, \quad Pr = 0.7, \quad M_\infty = 0.1, 1, 10, \quad b = 0, 1/2, 1, 2$$

Also, tabulate $Re_\infty c_f$, $Re_\infty St$, and $c_f/(2St)$.

16.5 Show that the viscous dissipation Φ and entropy production $\dot{s}_{irr}$ have minimum values at

$$Y^* = \frac{1}{a}\frac{1-b+a}{1+b+\frac{a}{3}}\left[b + \frac{(1-b+a)^2}{6a}\right]$$

where $0 < Y^* < 1$ when b is in the range $1 - a < b < 1 + a$. Can you explain these results?

16.6 Consider an extension of planar Couette flow in which a perfect gas is confined to a region between two concentric circular cylinders, as shown in the sketch. The inner cylinder is fixed, while the outer cylinder has a constant speed U_∞.

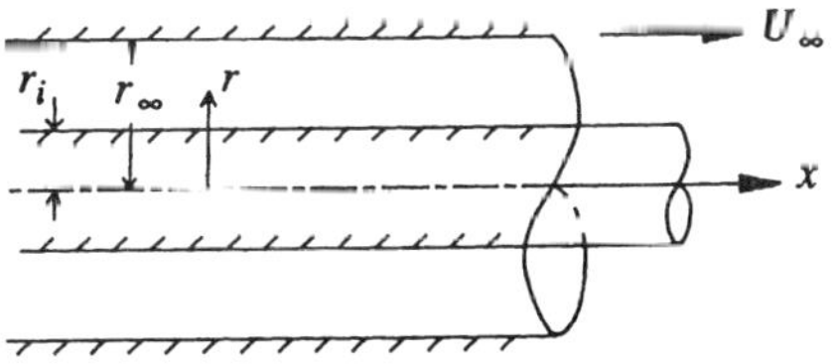

(a) With the aid of Section 5.6, write the governing equations for this problem.

(b) With the following variables and parameters

$$R = \frac{r - r_i}{r_\infty - r_i}, \qquad V = \frac{u}{U_\infty}, \qquad \theta = \frac{T - T_i}{T_\infty - T_i}$$

$$b = \frac{T_i}{T_\infty}, \qquad c = \frac{r_i}{r_\infty}$$

derive Equation (16.12).

(c) Use Equations (16.1) to derive the cylinder equivalent of Equation (16.14).

17 Stagnation Point Flow

17.1 PRELIMINARY REMARKS

We will obtain the solution for a two-dimensional or axisymmetric (without swirl) stagnation point in a steady, incompressible flow. This flow is also known as Hiemenz flow, since Hiemenz first investigated it in 1911. In conjunction with the incompressible assumption, we also assume that c_p, μ, and κ are constants. Furthermore, we later assume that the kinematic viscosity, ν is small; i.e., the flow is at a large Reynolds number. As shown in Figure 17.1, there is uniform flow far upstream of the stagnation point. The location of the inviscid stagnation point is the origin of the coordinate system, where x is a radial coordinate in the axisymmetric case.

The incompressible assumption is often reasonable even in a compressible flow. To illustrate this, let us consider a blunt body in hypersonic flow in air. A sketch of the flow field is shown in Figure 17.2. Point d denotes the location just downstream of the normal part of the bow shock wave. The streamline through this point wets the body. Between point d and the stagnation point, viscous and heat transfer effects are negligible, except in a thin boundary layer on the body. Hence, the flow external to the boundary layer between these two points is isentropic. In the $M_\infty \to \infty$ limit, we obtain for the stagnation point density ρ_o

$$\frac{\rho_o}{\rho_d} = \left(1 + \frac{\gamma - 1}{2} M_d^2\right)^{1/(\gamma-1)} = \left[\frac{(\gamma+1)^2}{4\gamma}\right]^{1/(\gamma-1)} = 1.073$$

where the Mach number at point d, M_d, equals $[(\gamma - 1)/(2\gamma)]^{1/2}$. Consequently, when $\gamma = 7/5$ there is only a 7% change in the density along the streamline from point d to the inviscid stagnation point.

In contrast to Rayleigh and Couette flows, stagnation point flow is of considerable practical importance. It serves as the starting solution for the boundary layer downstream of a stagnation point. It is also of interest in its own right, since a knowledge of conditions at a stagnation point is often essential. This is especially true for heat transfer, since the peak heat transfer frequently occurs at the stagnation point. An exception to this occurs when shock wave interference is present. A second exception occurs when there is a shoulder with a small radius of curvature. This latter configuration is discussed in Section 21.6.

In the next section, we formulate the problem and provide the inviscid solution. The following two sections, respectively, provide the viscous solution for the velocity components and for the temperature. The solution for the components requires a numerical solution of an ordinary differential equation.

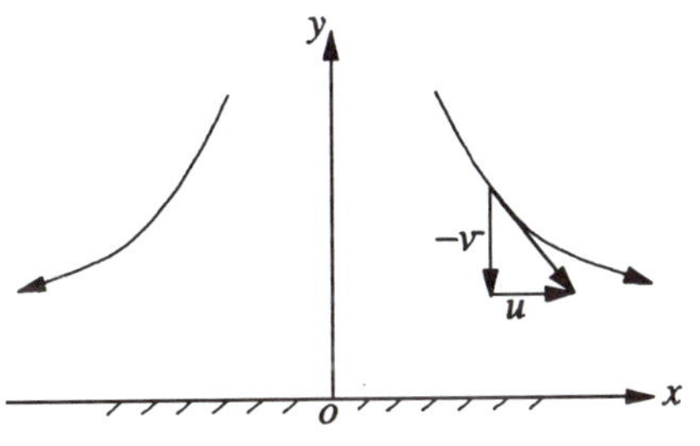

FIGURE 17.1 Schematic for a stagnation point flow; x is a radial coordinate when the flow is axisymmetric.

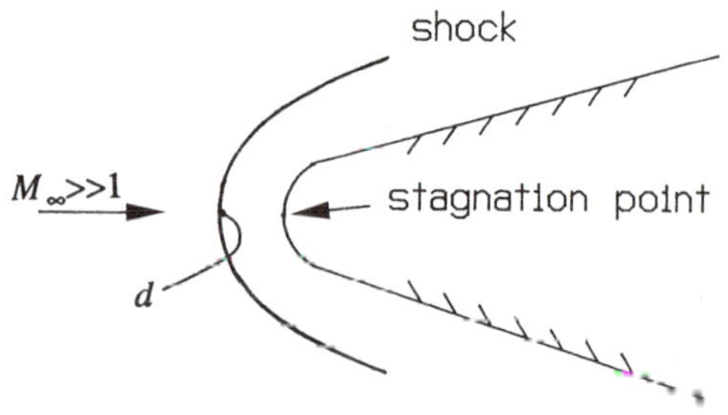

FIGURE 17.2 Blunt body in a hypersonic flow.

(Tabulated results are provided in Section 21.8.) On the other hand, the solution for the temperature is in terms of a quadrature.

17.2 FORMULATION

With the foregoing assumptions, including no dependence on the azimuthal angle, the governing equations for incompressible, axisymmetric flow have the form

$$\frac{\partial(xu)}{\partial x} + \frac{\partial(xv)}{\partial y} = 0 \tag{17.1a}$$

$$u\frac{\partial u}{\partial x} + v\frac{\partial u}{\partial y} = -\frac{1}{\rho}\frac{\partial p}{\partial x} + \nu\left(\nabla_a^2 u - \frac{u}{x^2}\right) \tag{17.1b}$$

$$u\frac{\partial v}{\partial x} + v\frac{\partial v}{\partial y} = -\frac{1}{\rho}\frac{\partial p}{\partial y} + \nu\nabla_a^2 v \tag{17.1c}$$

$$\rho c_p\left(u\frac{\partial T}{\partial x} + v\frac{\partial T}{\partial y}\right) = u\frac{\partial p}{\partial x} + v\frac{\partial p}{\partial y} + \kappa\nabla_a^2 T + 2\mu\left[\left(\frac{\partial u}{\partial x}\right)^2 + \frac{u^2}{x^2} + \left(\frac{\partial v}{\partial y}\right)^2\right] + \mu\left(\frac{\partial u}{\partial y} + \frac{\partial v}{\partial x}\right)^2 \tag{17.1d}$$

where the axisymmetric form for the Laplacian operator is

$$\nabla_a^2 = \frac{\partial^2}{\partial x^2} + \frac{\partial^2}{\partial y^2} + \frac{1}{x}\frac{\partial}{\partial x}$$

By introducing σ (see Equation (5.65)), these equations can be written as

$$\frac{\partial(x^\sigma u)}{\partial x} + \frac{\partial(x^\sigma v)}{\partial y} = 0 \tag{17.2a}$$

$$u\frac{\partial u}{\partial x} + v\frac{\partial u}{\partial y} = -\frac{1}{\rho}\frac{\partial p}{\partial x} + \nu\left[\nabla^2 u + \sigma\frac{\partial}{\partial x}\left(\frac{u}{x}\right)\right] \tag{17.2b}$$

$$u\frac{\partial v}{\partial x}+v\frac{\partial v}{\partial y} = -\frac{1}{\rho}\frac{\partial p}{\partial y}+\nu\left(\nabla^2 v+\frac{\sigma}{x}\frac{\partial v}{\partial x}\right) \tag{17.2c}$$

$$c_p\left(u\frac{\partial T}{\partial x}+v\frac{\partial T}{\partial y}\right) = \frac{1}{\rho}\left(u\frac{\partial p}{\partial x}+v\frac{\partial p}{\partial y}\right)+\frac{\kappa}{\rho}\left(\nabla^2 T+\frac{\sigma}{x}\frac{\partial T}{\partial x}\right)$$
$$+2\nu\left[\left(\frac{\partial u}{\partial x}\right)^2+\sigma\left(\frac{u}{x}\right)^2+\left(\frac{\partial v}{\partial y}\right)^2\right]+\nu\left(\frac{\partial u}{\partial y}+\frac{\partial v}{\partial x}\right)^2 \tag{17.3}$$

where the two-dimensional Laplacian operator in Equations (17.2) and (17.3) is

$$\nabla^2 = \frac{\partial^2}{\partial x^2}+\frac{\partial^2}{\partial y^2}$$

Equations (17.2) and (17.3) hold for both two-dimensional and axisymmetric flow. Because ρ and μ are constants, the continuity and momentum equations are uncoupled from the energy equation.

Stream Function Formulation

The continuity equation is satisfied by introducing a stream function defined by

$$u = \frac{1}{x^\sigma}\frac{\partial \psi}{\partial y} = \frac{1}{x^\sigma}\psi_y, \qquad v = -\frac{1}{x^\sigma}\psi_x \tag{17.4}$$

We introduce ψ into the momentum equations and eliminate p by cross differentiation to obtain, after some algebra,

$$\left(\psi_y\frac{\partial}{\partial x}-\psi_x\frac{\partial}{\partial y}\right)\nabla^2\psi+\frac{\sigma}{x}\left(-3\psi_y\psi_{xx}-2\psi_y\psi_{yy}+\psi_x\psi_{xy}+\frac{3}{x}\psi_x\psi_y\right)$$
$$= \nu x^\sigma\nabla^4\psi+\sigma\nu\frac{\partial}{\partial x}\left[-2\nabla^2\psi+3\left(\frac{\psi_x}{x}\right)\right] \tag{17.5}$$

where

$$\nabla^4 = \frac{\partial^4}{\partial x^4}+2\frac{\partial^4}{\partial x^2\partial y^2}+\frac{\partial^4}{\partial y^4}$$

and

$$-3\psi_y\psi_{xx}-2\psi_y\psi_{yy}+\psi_x\psi_{xy}+\frac{3}{x}\psi_x\psi_y = \psi_y\left[-2\nabla^2\psi-\psi_y\frac{\partial}{\partial x}\left(\frac{\psi_x}{\psi_y}\right)+\frac{3}{x}\psi_x\right]$$

Consequently, the three partial differential equations are reduced to a single fourth-order equation for the stream function.

After a stream function solution is found that is consistent with the boundary conditions, the velocity components u and v are determined by Equations (17.4). The pressure is then found by integrating both scalar momentum equations, written as

$$\begin{aligned}\frac{1}{\rho}p_x &= -uu_x - vu_y + \nu\left[\nabla^2 u + \sigma\left(\frac{u_x}{x} - \frac{u}{x^2}\right)\right] \\ \frac{1}{\rho}p_y &= -uv_x - vv_y + \nu\left(\nabla^2 v + \frac{\sigma}{x}v_x\right)\end{aligned} \tag{17.6}$$

Remember that Bernoulli's equation, which is much simpler, cannot be used when a flow is viscous. A consistent solution can be obtained from Equations (17.6), because we have used

$$p_{xy} = p_{yx}$$

in deriving Equation (17.5). In other words, dp is an exact differential.

Once u, v, and p are known, the skin friction and pressure force can be computed. Recall that the drag (associated with the pressure force) is zero in a two-dimensional potential flow but is not zero in a viscous flow. In the last section, the temperature field is found by solving the energy equation. After this, the heat transfer to the wall is determined.

Potential Flow Solution

Before embarking on a viscous flow analysis, it is necessary to examine the two-dimensional, potential (or inviscid) flow solution for the flow around a circular cylinder. With $\sigma = \nu = 0$, Equation (17.5) simplifies to

$$\left(\psi_y \frac{\partial}{\partial x} - \psi_x \frac{\partial}{\partial y}\right)\nabla^2 \psi = 0$$

or

$$\nabla^2 \psi_i = 0 \tag{17.7}$$

where the i subscript denotes an inviscid flow. For flow around a cylinder of radius R and without circulation, we have

$$\psi_i = U_\infty\left(\tilde{y} - \frac{R^2\tilde{y}}{\tilde{x}^2 + \tilde{y}^2}\right)$$

where the $\tilde{x}, \tilde{y}$ coordinates are defined in Figure 17.3. The first term on the right side represents a uniform flow which prevails far upstream; the second term is a doublet at the origin. Here we merely quote results, since this flow is a familiar one. We transform to the body-oriented x,y

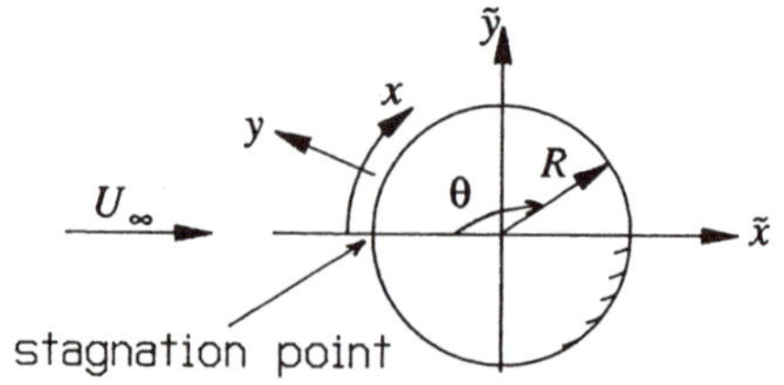

FIGURE 17.3 Inviscid flow about a circular cylinder.

coordinates of Figure 17.3 by means of the substitution

$$y = -\frac{\tilde{x}}{\cos\theta} - R, \qquad x = R\theta = -R\tan^{-1}(\tilde{y}/\tilde{x})$$

or its inverse

$$\tilde{x} = -(y+R)\cos\frac{x}{R}, \qquad \tilde{y} = (y+R)\sin\frac{x}{R}$$

to obtain

$$\psi_i = U_\infty \frac{y^2 + 2yR}{y+R}\sin\frac{x}{R} \tag{17.8}$$

Since we are only interested in the flow in the vicinity of the stagnation point, we expand this relation for large R to obtain

$$\psi_i \cong 2\frac{U_\infty}{R}xy = axy \tag{17.9}$$

where a has the value of $2U_\infty/R$. This is a simple but approximate potential flow solution in the vicinity of a two-dimensional stagnation point when the upstream flow is uniform. Equation (17.9) actually does not satisfy the upstream uniform flow condition, whereas that given by Equation (17.8) exactly satisfies this condition.

The sketch in Figure 17.1 is somewhat deceptive. If the wall were truly planar and infinite, there would not be a single stagnation point; in fact, we would have $\vec{w} = 0$ everywhere. For a single stagnation point to occur at a fixed location, the wall should have a finite radius of curvature R at the stagnation point; hence, there is a unique value for a.

The generalization of Equation (17.9) to include axially symmetric flow about a sphere of radius R yields

$$\psi_i = \frac{a}{1+\sigma}x^{1+\sigma}y \tag{17.10}$$

where a is still $2U_\infty/R$. We thus obtain for the inviscid velocity components

$$u_i = \frac{1}{x^\sigma}\frac{\partial \psi_i}{\partial y} = \frac{a}{1+\sigma}x$$
$$v_i = -\frac{1}{x^\sigma}\frac{\partial \psi_i}{\partial x} = -ay \tag{17.11}$$

where u_i is nonnegative and v_i is nonpositive, in accordance with Figure 17.1. Observe that v_i is zero on the wall, whereas u_i is not, as expected for an inviscid flow. From Bernoulli's equation

$$\frac{1}{2}(u_i^2 + v_i^2) + \frac{p_i}{\rho} = \frac{p_o}{\rho}$$

where p_o is the inviscid pressure at the stagnation point, we have

$$p_i = p_o - \frac{1}{2}\rho\left[\left(\frac{ax}{1+\sigma}\right)^2 + (ay)^2\right] \tag{17.12}$$

for the pressure.

17.3 VELOCITY SOLUTION

We are now in a position to attempt a solution for the viscous flow. Let us surmise that the inviscid streamlines are approximately correct. The main effect of viscosity is to deflect the streamlines slightly outward away from the wall. Hence, we try a solution, based on Equation (17.10), of the form

$$\psi = \frac{a}{1+\sigma}x^{1+\sigma}F(y) \tag{17.13}$$

for the viscous flow. This is substituted into Equation (17.5), which yields the ordinary differential equation

$$F'''' + \frac{a}{\nu}FF''' + (\sigma-1)\frac{a}{\nu}F'F'' = 0 \tag{17.14}$$

where a prime denotes differentiation with respect to y. We also have for the velocity components

$$u = \frac{1}{x^{\sigma}}\psi_y = \frac{a}{1+\sigma}xF'$$

$$v = -\frac{1}{x^{\sigma}}\psi_x = -aF$$

On the body, where $y = 0$, the no-slip condition results in

$$F_w = F(0) = 0, \qquad F'_w = F'(0) = 0$$

and a nonzero wall shear

$$\tau_w = \mu\left(\frac{\partial u}{\partial y} + \frac{\partial v}{\partial x}\right)_w = \frac{a\mu}{1+\sigma}xF'' \tag{17.15}$$

where $F''_w \neq 0$. Away from the stagnation point, we require that

$$\psi \sim \psi_i \quad \text{as} \quad y \to \infty$$

In accordance with Equation (17.10), this is equivalent to

$$F(y) \to \infty \quad \text{as} \quad y \to \infty$$

or

$$F'(\infty) = 1$$

Since F varies as y for a large y, we also have

$$F^{(n)}(\infty) = 0, \qquad n = 2, 3, \ldots \tag{17.16}$$

for the higher-order derivatives, a result which will be needed shortly. While the viscous solution exactly satisfies the governing equations and the no-slip condition, it only satisfies the approximate freestream condition provided by Equation (17.10). In this regard the solution is a local one, valid only in the vicinity of the stagnation point.

We can integrate Equation (17.14) once by first writing it as

$$dF''' + \frac{a}{\nu}[FdF'' + (\sigma - 1)F'dF'] = 0$$

since a $1/dy$ factor cancels. The middle term is integrated by parts:

$$\int FdF'' = FF'' - \int F'dF' = FF'' - \frac{1}{2}F'^2$$

$$u = F, \quad dv = dF''$$

$$du = F'dy, \quad v = F'' = \frac{dF'}{dy}$$

where udv represents the first integrand. Hence, we obtain

$$F''' + \frac{a}{\nu}\left(FF'' - \frac{1}{2}F'^2 + \frac{\sigma - 1}{2}F'^2\right) = c$$

where c is an integration constant. Since σ is either zero or one, the coefficient F'^2 of becomes

$$-\frac{1}{2} + \frac{\sigma - 1}{2} = \frac{\sigma - 2}{2} = -\frac{1}{1 + \sigma}$$

and we have

$$F''' + \frac{a}{\nu}\left(FF'' + \frac{1}{1 + \sigma}F'^2\right) = c$$

We utilize Equation (17.16) and evaluate the left side as $y \to \infty$, to obtain

$$c = -\frac{1}{1 + \sigma}\frac{a}{\nu}$$

The equation

$$F''' + \frac{a}{\nu}\left[FF'' + \frac{1}{1 + \sigma}(1 - F'^2)\right] = 0 \tag{17.17}$$

and its boundary conditions

$$F_w = F'_w = 0, \qquad F'(\infty) = 1$$

determine F. The solution depends on the a/ν and σ parameters.

Derivation of the Falkner–Skan Equation

In the foregoing, neither F nor y is nondimensional. It is, therefore, desirable to introduce new dimensionless variables, η and f, as follows:

$$\eta = \left(\frac{a}{\nu}\right)^{1/2} y$$

$$f = \left(\frac{a}{\nu}\right)^{1/2} F$$

$$F' = \frac{dF}{dy} = \frac{dF}{df}\frac{df}{d\eta}\frac{d\eta}{dy} = \left(\frac{\nu}{a}\right)^{1/2}\left(\frac{a}{\nu}\right)^{1/2} f' = f' \tag{17.18}$$

$$F'' = \frac{dF'}{dy} = \frac{dF'}{df'}\frac{df'}{d\eta}\frac{d\eta}{dy} = \left(\frac{a}{\nu}\right)^{1/2} f''$$

$$F''' = \frac{dF''}{dy} = \frac{dF''}{df''}\frac{df''}{d\eta}\frac{d\eta}{dy} = \frac{a}{\nu} f'''$$

where a prime now denotes differentiation with respect to η, and η is proportional to y, while f is similarly proportional to F. We, thus, obtain in place of Equation (17.17)

$$\frac{a}{\nu} f''' + \frac{a}{\nu}\left[\left(\frac{\nu}{a}\right)^{1/2} f \left(\frac{a}{\nu}\right)^{1/2} f'' + \frac{1}{1+\sigma}(1 - f'^2)\right] = 0$$

which simplifies to

$$f''' + ff'' + \frac{1}{1+\sigma}(1 - f'^2) = 0 \tag{17.19a}$$

with the boundary conditions

$$f_w = f'_w = 0, \qquad f'(\infty) = 1 \tag{17.19b}$$

Equations (17.19) now depend on only one parameter, σ. The equations represent a two-point boundary value problem where the solution can be numerically obtained. For this integration a value must be guessed for f''_w. The integration proceeds from $\eta = 0$ and must be performed repetitively until the condition $f'(\infty) = 1$ is satisfied. This equation with the definition

$$\beta = \frac{1}{1+\sigma} \tag{17.20}$$

is known as the Falkner–Skan equation. It is further discussed in Chapter 21, where tabulated results are provided.

Discussion

The corresponding stream function and velocity components are

$$\psi = \frac{(a\nu)^{1/2}}{1+\sigma} x^{1+\sigma} f(\eta)$$
$$u = \frac{1}{x^\sigma}\frac{\partial \psi}{\partial y} = \frac{1}{x^\sigma}\frac{\partial \psi}{\partial \eta}\frac{d\eta}{dy} = \frac{a}{1+\sigma} x f'$$
$$v = -\frac{1}{x^\sigma}\psi_x = -(a\nu)^{1/2} f \tag{17.21}$$

Observe that u can be written as

$$\frac{u}{u_i} = f'$$

It is customary to use the 99% value for u/u_i to evaluate a velocity thickness δ of the viscous layer. The δ thickness is arbitrary, since the 99% value is arbitrary. The resulting thickness, however, is not sensitive to this value, providing it is not chosen as 100%. With a 99% value for u/u_i, η is

$$\eta = \left(\frac{a}{\nu}\right)^{1/2}\delta = \begin{cases} 2.40, & \sigma = 0 \\ 2.75, & \sigma = 1 \end{cases}$$

Since δ is independent of x, we have the important result that the viscous boundary layer has a constant finite thickness in the stagnation point region. By introducing $a = 2U_\infty/R$, we obtain

$$\frac{\delta}{R} = \frac{2.4 + 0.35\sigma}{(2Re_\infty)^{1/2}} \tag{17.22a}$$

where

$$Re_\infty = U_\infty R/\nu$$

Equation (17.22a) holds even if the Reynolds number is not large, e.g., the flow of oil about a circular cylinder. If Re_∞ is large, however, then δ/R is small relative to unity. We are, in effect, dealing with a thin laminar boundary layer. Hereafter, we shall presume the Reynolds number to be large or ν to be small and, consequently, v is much smaller than u.

It is convenient in boundary-layer theory to normalize δ by x, with the result

$$\frac{\delta}{x} = \frac{2.4 + 0.35\sigma}{(Re_x)^{1/2}}\left(\frac{R}{2x}\right)^{1/2} \tag{17.22b}$$

which is, in terms of a local Reynolds number,

$$Re_x = U_\infty x/\nu$$

By utilizing Equations (17.6) and (17.21), we obtain for the η pressure gradient, which is equivalent to the y pressure gradient,

$$\frac{1}{\rho}p_\eta = -a\nu(f'' + ff')$$

With the aid of Equation (17.19a), we have for the x pressure gradient

$$\frac{1}{\rho}p_x - \frac{a^2}{1+\sigma}x\left(f''' + ff'' - \frac{1}{1+\sigma}f'^2\right) = -\frac{a^2}{(1+\sigma)^2}x$$

Since ν is small, we have

$$\frac{1}{\rho}p_\eta \cong 0$$

To a first approximation, the pressure does not vary laterally across the viscous layer. We can readily integrate these equations, to obtain

$$p = p_o - \frac{1}{2}\rho\left(\frac{ax}{1+\sigma}\right)^2 \qquad \textbf{(17.23)}$$

which matches the inviscid pressure, Equation (17.12), at the wall.

The wall shear stress is given by Equation (17.15) as

$$\tau_w = \frac{a}{1+\sigma}(a\nu)^{1/2}\rho x f''_w$$

Hence, the skin-friction coefficient is

$$c_f = \frac{2\tau_w}{\rho u_i^2} = \frac{2}{\rho}\frac{1}{\left(\frac{ax}{1+\sigma}\right)^2}\frac{a}{1+\sigma}(a\nu)^{1/2}\rho x f''_w = \frac{1+\sigma}{(Re_x)^{1/2}}\left(\frac{2R}{x}\right)^{1/2} f''_w \qquad \textbf{(17.24)}$$

where the rightmost term has a form consistent with Equation (17.22b). It is customary to use u_i, the inviscid speed external to the boundary layer, rather than U_∞ for the normalization in c_f. Also note that c_f is infinite at the stagnation point even though τ_w is zero. This is a result of the normalization with u_i.

17.4 TEMPERATURE SOLUTION

We now consider Equation (17.3) for the temperature. We introduce η, f, and Equation (17.23) for the pressure, to obtain

$$c_p\left[\frac{a}{1+\sigma}xf'\frac{\partial T}{\partial x} - (a\nu)^{1/2}f\left(\frac{a}{\nu}\right)^{1/2}\frac{\partial T}{\partial \eta}\right] = -\frac{a}{1+\sigma}xf'\frac{a}{(1+\sigma)^2}x + \frac{\kappa}{\rho}\left[\frac{\partial^2 T}{\partial x^2} + \left(\frac{a}{\nu}\right)\frac{\partial^2 T}{\partial \eta^2} + \frac{\sigma}{x}\frac{\partial T}{\partial x}\right]$$

$$+ 2\nu\left[\left(\frac{a}{1+\sigma}\right)^2 f'^2 + \frac{\sigma}{x^2}\left(\frac{a}{1+\sigma}\right)^2 x^2 f'^2 + a^2 f'^2\right]$$

$$+ \nu\left[\frac{a}{1+\sigma}\left(\frac{a}{\nu}\right)^{1/2} xf''\right]^2$$

which rearranges to

$$c_p\left(\frac{x}{1+\sigma}f'\frac{\partial T}{\partial x} - f\frac{\partial T}{\partial \eta}\right) = \frac{c_p}{Pr}\frac{\partial^2 T}{\partial \eta^2} + \left(\frac{ax}{1+\sigma}\right)^2\left[f''^2 - \frac{1}{(1+\sigma)a}f'\right]$$
$$+ \frac{\nu}{a}\left[\frac{c_p}{Pr}\left(\frac{\partial^2 T}{\partial x^2} + \frac{\sigma}{x}\frac{\partial T}{\partial x}\right) + \frac{4+2\sigma}{1+\sigma}a^2 f'^2\right] \quad \textbf{(17.25)}$$

This equation consists of four types of terms: (i) convective terms on the left side, (ii) an $x^2 f'$ work term, (iii) the $1/Pr$ heat conduction terms, and (iv) two viscous dissipation terms. As a first approximation that is in accordance with the pressure, all terms proportional to ν are ignored.

Observe that T can be a function only of η providing the terms proportional to $(ax)^2$ are neglected. This customary approximation further limits the region of validity of the solution to small x values. We thus assume a similarity solution of the form, $T = T(\eta)$, which eliminates the $\partial T/\partial x$ terms, to obtain in place of Equation (17.25)

$$\frac{d^2 T}{d\eta^2} + Prf\frac{dT}{d\eta} = 0 \quad \textbf{(17.26)}$$

Only convection, given by the rightmost term, and conduction are represented in this equation. In the vicinity of a stagnation point viscous dissipation, the work term and all gradients in the x direction are small. The equation is a linear one that will be solved after considering boundary conditions.

This equation requires that the boundary conditions at the wall and external to the thermal layer satisfy the similarity assumption. These conditions cannot depend on x. If the wall is made of a poorly conducting material, then T_w = constant or $q_w = 0$ are appropriate. On the other hand, if the wall is made of a highly conductive material, such as oxygen-free copper, then a constant wall temperature with x may be restricted to a narrow region in the immediate vicinity of the stagnation point. In any case, we assume a constant wall temperature. Consequently, the region of validity of the solution for the temperature may be less than that for the velocity components. It is worth noting in this context that f is an exact solution of the continuity and momentum equations. (Earlier, when dealing with the momentum equations, p_η was set equal to zero. However, this is a matter of convenience, not of necessity, as shown by Problem 17.3.) As a result of approximating the boundary conditions and neglecting numerous terms, the temperature solution is an approximate one.

For the external inviscid flow, we assume

$$T = T_\infty = \text{constant}, \qquad \eta \to \infty$$

This is justified by the incompressible, constant property assumptions, and the small flow speed in the vicinity of the stagnation point. It is thus convenient to introduce

$$\theta = \frac{T - T_w}{T_\infty - T_w} \quad \textbf{(17.27)}$$

and observe that θ is a function only of η, since both T_∞ and T_w are constant. (If $T_w = T_\infty$, then T/T_w would be used for θ.) As a result, Equation (17.26) becomes

$$\frac{d^2\theta}{d\eta^2} + Prf(\eta)\frac{d\theta}{d\eta} = 0 \quad \textbf{(17.28a)}$$

with the boundary conditions

$$\theta_w = 0, \qquad \theta(\infty) = 1 \tag{17.28b}$$

[If an adiabatic wall is assumed, then $(d\theta/d\eta)_w = 0$ and $\theta(\eta) = 1$ is the solution.] A first integral is readily obtained as

$$\frac{d\left(\frac{d\theta}{d\eta}\right)}{\frac{d\theta}{d\eta}} = -Prfd\eta$$

$$\frac{d\theta}{d\eta} = c_1 \exp\left(-Pr\int_0^{\eta} f(\eta')\,d\eta'\right)$$

where c_1 is an integration constant. A second integration results in

$$\theta = c_2 + c_1 \int_0^{\eta} \exp\left(-Pr\int_0^{\eta''} f\,d\eta'\right) d\eta''$$

where η' and η'' are dummy integration variables. Evaluation with the boundary conditions finally yields

$$\theta = c_1 \int_0^{\eta} \exp\left(-Pr\int_0^{\eta''} f\,d\eta'\right) d\eta'' \tag{17.29a}$$

where the two constants are

$$c_1 = \left[\int_0^{\infty} \exp\left(-Pr\int_0^{\eta''} f\,d\eta'\right) d\eta''\right]^{-1}$$
$$c_2 = 0 \tag{17.29b}$$

Observe that θ and c_1 depend not only on Pr but also, through f, on the dimensionality parameter σ.

The wall heat flux is given by

$$q_w = -\kappa\left(\frac{\partial T}{\partial y}\right)_w = -\kappa(T_\infty - T_w)\left(\frac{a}{\nu}\right)^{1/2}\left(\frac{d\theta}{d\eta}\right)_w$$

However, $d\theta/d\eta$ at the wall is c_1, which results in

$$q_w = \frac{2^{1/2} c_p \rho U_\infty}{Pr(Re_\infty)^{1/2}} c_1 (T_w - T_\infty)$$

In terms of a film coefficient, we have

$$q_w = h_f(T_w - T_\infty)$$

which yields

$$h_f = \frac{2^{1/2} c_p \rho U_\infty}{Pr(Re_\infty)^{1/2}} c_1$$

The Stanton and Nusselt numbers are then given by

$$St = \frac{h_f}{\rho c_p U_\infty} = \frac{2^{1/2} c_1}{Pr(Re_\infty)^{1/2}} \tag{17.30a}$$

$$Nu = StPrRe_\infty = c_1(2Re_\infty)^{1/2} \tag{17.30b}$$

Problem 21.15 shows that $c_1 = G_w'$ with the additional Chapter 21 assumption of $Pr = 1$. The quantity G_w' is a nondimensional stagnation temperature gradient at the wall and is given later in Table 21.7. It depends on $\beta(= 1/2$ for $\sigma = 1$ and 1 for $\sigma = 0)$ and a temperature parameter, g_w, which here equals T_w/T_∞.

REFERENCE

Hiemenz, K., "Dic Grenzschicht an einem in den gleichförmigen Flüessigkeitsstrom eingetauchten geraden Kreiszylinder," *Dingl. Polytechn. J.* 326, 321 (1911).

PROBLEMS

17.1 Derive Equation (17.5) for steady, incompressible, two-dimensional or axisymmetric flow. In terms of ψ, ν, and σ, determine equations p_x, p_y, and ω Simplify your results as much as possible.

17.2 Derive Equation (17.14), starting with Equations (17.15) and (17.13).

17.3 For stagnation point flow, derive an exact result for the pressure with the form

$$\frac{2(p_0 - p)}{\rho U_\infty^2} = fcn\left(f, f', \frac{x}{(1+\sigma)R}, Re_\infty\right)$$

17.4 Show that Equation (17.10) with $\sigma = 1$ satisfies Equation (17.5).

17.5 The simplest relation for $f(\eta)$ that satisfies the f_w, f_w', and f_w'' the wall condition is

$$f = \frac{1}{2} f_w'' \eta^2$$

where f_w'' is 1.2326 for a two-dimensional flow and 0.9277 for an axisymmetric flow.

(a) Use this approximation to obtain the dependence of $Re_\infty^{1/2}$ St on Pr for both a two-dimensional and an axisymmetric flow.

(b) Explain why the above f can be used even though the $f'(\infty) = 1$ condition is not satisfied.

17.6 Consider two-dimensional stagnation point flow in which the freestream speed has an angle α with respect to the wall.

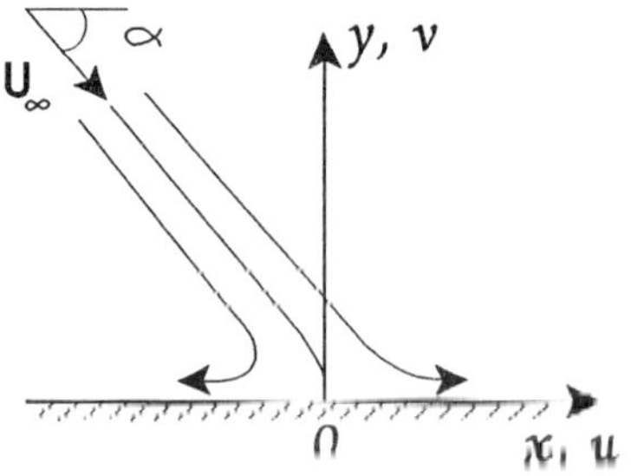

(a) Show that

$$u = a(x + 2y \cot\alpha), \qquad v = -ay$$

satisfies the freestream orientation angle α and Equations (17.2). Determine $p(x, y)$. Note that u, v, and p correspond to inviscid values even though they satisfy the Navier–Stokes equations.

(b) Derive the counterpart to Equations (17.9).

(c) Utilize a generalization ψ_i; i.e., use

$$\psi = axF(y) + 2aG(y)\cot\alpha$$

along with Equations (17.5) to determine equations for F and $\bar{G}$.

(d) Determine wall and freestream boundary conditions for F and $\bar{G}$. In addition, establish the condition that the wall attachment point is located at $x = 0$.

(e) Integrate the F and $\bar{G}$ equations once. Perform a second integration of the G equation by setting $G = \bar{G}'$, where $(\ )' = d(\ )/dy$.

(f) Transform the $G(y)$ and $F(y)$ equations to $g(\eta)$ and $f(\eta)$ as is done by Equations (17.18) for F. Is $g(\eta)$ nondimensional?

PART V

Laminar Boundary-Layer Theory for Steady Two-Dimensional or Axisymmetric Flow

Outline of Part V

The vast majority of flows, both gas and liquid, have a large Reynolds number because μ is very small for most common fluids. As a consequence, viscosity plays no role in the bulk of the flow field, which can then be determined by potential flow theory for an irrotational flow or the Euler equations if the flow is rotational. The Euler equations are simply obtained by setting $Re \to \infty$ in the dimensionless momentum and energy equations. This limit does not affect the continuity equation but does decrease the differential order by one of the momentum and energy equations. Consequently, the no-slip and temperature wall conditions can no longer be satisfied. The resolution of this dilemma was theoretically discovered by Prandtl in 1904. His discovery is called boundary-layer theory; it is the principal topic of Part V. It is worth noting that direct experimental verification of laminar, incompressible boundary-layer flow over a flat plate first occurred in 1924 (Van der Hegge-Zijnen, 1924), two decades after its conception. During the next 20 years, however, a number of experimentalists have reconfirmed the theory.

Overall, the subject matter of Part V represents the results of about 90 years of intensive research by fluid dynamicists. In fact, for the first four decades the research effort focused almost exclusively on incompressible flows. During much of this time span, laminar boundary-layer theory was the preeminent research topic in fluid dynamics, despite the fact that viscous flows often do not fall into this category. However, there are several sound reasons, which are still relevant, for this emphasis.

First, the idea of a thin viscous wall region, for both laminar and turbulent layers, provided a conceptual and analytical framework for simplifying the otherwise intractable governing (Navier–Stokes) equations. The resulting laminar boundary-layer equations are still nonlinear but, nevertheless, are amenable to analysis. Consequently, some exact and a number of approximate methods of solution were developed for these equations. Second, the concept led to a wide variety of experiments verifying the laminar theory. An approach thereby developed in which theory and experiment went hand-in-hand and were successfully compared with each other. Third, the concept and techniques of the theory became the cornerstone for research in other areas. For instance, the use of integral methods for turbulent boundary layers was first initiated in the laminar theory (Pohlhausen, 1921). The subject matter of Part V is the basis of much of fluid dynamics as well as other diverse subjects.

In addition to the assumptions first discussed in Part I, we now assume a steady, laminar two-dimensional or axisymmetric flow of a perfect gas or an incompressible fluid. Even these limitations are insufficient. Thus, Part V does not consider wall suction or blowing or interaction phenomena, which would occur with a shock wave or in hypersonic flow. Of course, three-dimensional boundary layers and shock wave interaction are important, but these subjects are not elementary and fall outside the scope of Part V.

Although our mode of presentation is not a historical one, the subject matter of Chapter 18 represents the beginning of the theory as originally conceived by Prandtl in the first decade of the 20th century. This chapter, therefore, considers incompressible flow over a semi-infinite flat plate. Chapter 19 briefly discusses boundary layers from the viewpoint of matched asymptotic expansions. This sets the stage for a more formal presentation, in Chapter 20, of the first-order incompressible theory. Chapter 21 continues with an extensive account of the corresponding first-order compressible theory. A variety of flows are discussed in Chapter 22 for which the Chapter 21 theory is applied. The presentation concludes in Chapter 23 with a discussion of second-order, compressible boundary-layer theory.

A number of books on viscous flow are listed in the references below. Of these, the one by Schlichting is the best known. Its first edition, which appeared in the early 1950s, was an important contribution to the subsequent development of the subject, since it was an authoritative monograph on the subject. The book is in its seventh edition and is still used both as a text and perhaps more often as a primary reference, since it contained a comprehensive treatment of the subject at the time of its publication. This edition should be the last one, as H. Schlichting died in 1982.

REFERENCES

Lagerstrom, P.A., "Laminar Flow Theory," *Theory of Laminar Flows,* edited by F.K. Moore, High Speed Aerodynamics and Jet Propulsion, Vol. IV, Princeton University Press, 1964, pp. 20–285.
Pohlhausen, K., "Zur Näherungsweisen Integration der Differentialgleichung der Laminaren Reibungsschicht," *Z. Angew. Math. Mech.* 1, 252 (1921).
Rosenhead, L., ed., *Laminar Boundary Layers,* Oxford University Press, New York, 1963.
Schetz, J.A., *Boundary Layer Analysis,* Prentice-Hall, Englewood Cliffs, NJ, 1993.
Schlichting, H., *Boundary-Layer Theory,* 7th ed., McGraw-Hill Book Co., New York, 1979.
Sherman, F.S., *Viscous Flow,* McGraw-Hill, New York, 1990.
Van der Hegge-Zijnen, B.G., "Measurements of the Velocity Distribution in the Boundary Layer along a Plane Surface," Thesis, Delft, 1924.
White, F.M., *Viscous Fluid Flow,* 2nd ed., McGraw-Hill, New York, 1991.

18 Incompressible Flow over a Flat Plate

18.1 PRELIMINARY REMARKS

In this chapter, we analyze the most elementary of all boundary-layer flows, in which the fluid is incompressible, has constant properties, and has a constant freestream speed U_∞. It flows over a semi-infinite planar wall of zero thickness that is aligned with the freestream velocity. Consequently, the flow is two-dimensional, as shown in Figure 18.1. The leading edge of the plate is at the origin of the coordinate system and is infinitely sharp, and only the flow over the upper surface needs to be considered. In the region away from the wall, the flow is essentially inviscid. However, because of the no-slip wall boundary condition, there is a viscous layer adjacent to the wall, which is the subject of our discussion.

In the next section, a derivation of the flat plate boundary-layer equations is given that is reminiscent of Prandtl's original derivation (Prandtl, 1928). The last section provides a similarity solution of the boundary-layer equations.

18.2 DERIVATION OF THE BOUNDARY-LAYER EQUATIONS

Because the flow is incompressible and the fluid properties are constant, the energy equation is decoupled from the continuity and momentum equations and need not be considered (see Problem 18.3). We begin with the appropriate dimensional governing equations that appear at the start of Section 13.2. After the foregoing assumptions are introduced, we have

$$\frac{\partial u}{\partial x} + \frac{\partial v}{\partial y} = 0$$

$$\rho\left(u\frac{\partial u}{\partial x} + v\frac{\partial u}{\partial y}\right) = -\frac{\partial p}{\partial x} + \mu\left(2\frac{\partial^2 u}{\partial x^2} + \frac{\partial^2 u}{\partial y^2} + \frac{\partial^2 v}{\partial x \partial y}\right)$$

$$\rho\left(u\frac{\partial v}{\partial x} + v\frac{\partial v}{\partial y}\right) = -\frac{\partial p}{\partial y} + \mu\left(\frac{\partial^2 v}{\partial x^2} + \frac{\partial^2 u}{\partial x \partial y} + 2\frac{\partial^2 v}{\partial y^2}\right)$$

The viscous terms are simplified by utilizing continuity, as follows:

$$2\frac{\partial^2 u}{\partial x^2} + \frac{\partial^2 u}{\partial y^2} + \frac{\partial^2 v}{\partial x \partial y} = \frac{\partial^2 u}{\partial x^2} + \frac{\partial^2 u}{\partial y^2} + \frac{\partial}{\partial x}\left(\frac{\partial u}{\partial x} + \frac{\partial v}{\partial y}\right) = \frac{\partial^2 u}{\partial x^2} + \frac{\partial^2 u}{\partial y^2}$$

$$\frac{\partial^2 v}{\partial x^2} + \frac{\partial^2 u}{\partial x \partial y} + 2\frac{\partial^2 v}{\partial y^2} = \frac{\partial^2 v}{\partial x^2} + \frac{\partial^2 v}{\partial y^2} + \frac{\partial}{\partial y}\left(\frac{\partial u}{\partial x} + \frac{\partial v}{\partial y}\right) = \frac{\partial^2 v}{\partial x^2} + \frac{\partial^2 v}{\partial y^2}$$

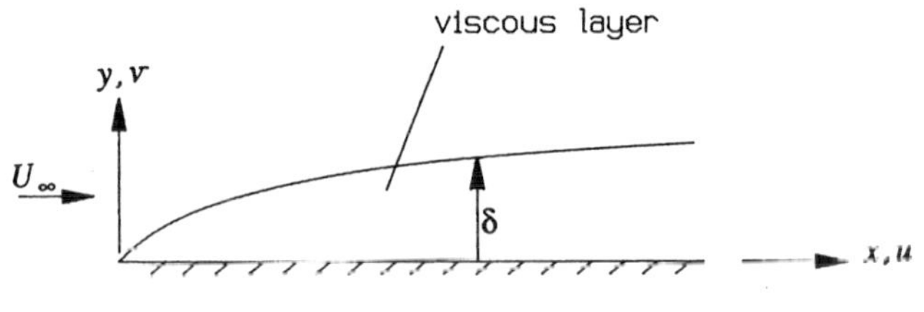

FIGURE 18.1 Flow over a flat plate.

After dividing by ρ, the momentum equations become

$$u\frac{\partial u}{\partial x}+v\frac{\partial u}{\partial y} = -\frac{\partial}{\partial x}\left(\frac{p}{\rho}\right)+\nu\left(\frac{\partial^2 u}{\partial x^2}+\frac{\partial^2 u}{\partial y^2}\right)$$

$$u\frac{\partial v}{\partial x}+v\frac{\partial v}{\partial y} = -\frac{\partial}{\partial y}\left(\frac{p}{\rho}\right)+\nu\left(\frac{\partial^2 v}{\partial x^2}+\frac{\partial^2 v}{\partial y^2}\right)$$

Along with continuity, these are the Navier–Stokes equations for the problem at hand. These equations are subject to wall and freestream conditions given by

$$u(x,0) = 0, \quad v(x,0) = 0, \quad x \ge 0$$
$$u(x,\infty) = U_\infty, \quad v(x,\infty) = 0$$

We nondimensionalize the variables, as was done in Equations (13.10), to obtain

$$\frac{\partial u^*}{\partial x^*}+\frac{\partial v^*}{\partial y^*} = 0 \tag{18.1}$$

$$u^*\frac{\partial u^*}{\partial x^*}+v^*\frac{\partial u^*}{\partial y^*} = -\frac{\partial p^*}{\partial x^*}+\frac{1}{Re}\nabla^2 u^* \tag{18.2}$$

$$u^*\frac{\partial v^*}{\partial x^*}+v^*\frac{\partial v^*}{\partial y^*} = -\frac{\partial p^*}{\partial y^*}+\frac{1}{Re}\nabla^2 v^* \tag{18.3}$$

$$u^*(x^*,0) = 0, \quad v^*(x^*,0) = 0, \quad x^* \ge 0$$
$$u^*(x^*,\infty) = 1, \quad v^*(x^*,\infty) = 0 \tag{18.4}$$

where

$$p^* = \frac{p}{\rho U_\infty^2}, \qquad Re = \frac{U_\infty \ell}{\nu}, \qquad \nabla^2 = \frac{\partial^2}{\partial x^{*2}}+\frac{\partial^2}{\partial y^{*2}}$$

Thus, only one parameter, a Reynolds number, appears in the nondimensional problem. This parameter, however, is an artificial one, since the distance ℓ is arbitrary. In fact, we could have normalized x and y in such a way that a Reynolds number would not appear in the equations.

Typical values for the constants in a Reynolds number for air would be

$$U_\infty \sim 1\,\text{m/s}, \quad \ell \sim 1.5\ \text{m}, \quad \nu = 1.5 \times 10^{-5}\ \text{m}^2/\text{s}$$

Hence, $1/Re$ is about 10^{-5}. Thus, the $\nabla^2 u^*$ and $\nabla^2 v^*$ viscous terms are very small, except in a thin layer adjacent to the wall where they are significant. Away from the wall, the Euler equations accurately hold. For this flow, the solution to the Euler equations is simple:

$$u^* = 1, \qquad v^* = 0, \qquad p^* = \frac{p}{\rho U_\infty^2} = \text{constant} \tag{18.5}$$

As shown in Figure 18.1, we consider the thin viscous layer to have a thickness of δ, in which case we expect $\ell \gg \delta$. With the foregoing discussion in mind, we can estimate the magnitude of the individual terms in the governing equations. (A more formal derivation is provided in Chapter 20.) We set

$$u^* = O(1), \qquad x^* = O(1), \qquad y^* = O(\delta)$$

and obtain from Equation (18.1)

$$\frac{O(1)}{O(1)} + \frac{v^*}{O(\delta)} = 0$$

so that

$$v^* = O(\delta)$$

As expected, the velocity component perpendicular to the wall is small.

We next examine Equation (18.3), with the result

$$O(1)\frac{O(\delta)}{O(1)} + O(\delta)\frac{O(\delta)}{O(\delta)} = -\frac{\partial p^*}{\partial y^*} + \frac{1}{Re}\left[\frac{O(\delta)}{O(1)} + \frac{O(\delta)}{O(\delta^2)}\right]$$

which simplifies to

$$O(\delta) = -\frac{\partial p^*}{\partial y^*} + \frac{1}{Re}O\left(\frac{1}{\delta}\right)$$

Since we expect the transverse pressure gradient to be small, in accordance with Equations (18.5), we obtain

$$Re \cong O\left(\frac{1}{\delta^2}\right), \qquad \frac{\partial p^*}{\partial y^*} = O(\delta)$$

We now examine Equation (18.2), which yields

$$O(1)\frac{O(1)}{O(1)} + O(\delta)\frac{O(1)}{O(\delta)} = -\frac{\partial p^*}{\partial x^*} + O(\delta^2)\left[\frac{O(1)}{O(1)} + \frac{O(1)}{O(\delta^2)}\right]$$

or

$$O(1) = \frac{\partial p^*}{\partial x} + O(1)$$

In other words, $\partial p^*/\partial x^*$ is of $O(1)$. Observe that the $\partial^2 u^*/\partial x^{*2}$ term is much smaller than the $\partial^2 u^*/\partial y^{*2}$ term.

With the result that $Re \gg 1$, Equations (18.1) to (18.3) simplify to

$$\frac{\partial u^*}{\partial x^*} + \frac{\partial v^*}{\partial y^*} = 0$$

$$u^*\frac{\partial u^*}{\partial x^*} + v^*\frac{\partial u^*}{\partial y^*} = -\frac{\partial p^*}{\partial x^*} + \frac{1}{Re}\frac{\partial^2 u^*}{\partial y^{*2}}$$

$$\frac{\partial p^*}{\partial y^*} = 0$$

Thus, the pressure depends only on x^* and is obtained from a solution of the Euler equations. For a flat plate, however, the pressure is constant. The flat plate boundary-layer equations finally become

$$\frac{\partial u^*}{\partial x^*} + \frac{\partial v^*}{\partial y^*} = 0 \quad \textbf{(18.6)}$$

$$u^*\frac{\partial u^*}{\partial x^*} + v^*\frac{\partial u^*}{\partial y^*} = \frac{1}{Re}\frac{\partial^2 u^*}{\partial y^{*2}} \quad \textbf{(18.7)}$$

These are two equations for u^* and v^* subject to boundary conditions [Equations (18.4)].

Observe that no simplification occurred in continuity; all of it transpired in the momentum equations. Furthermore, no further simplification can be made, since all retained terms are of $O(1)$ in the viscous region. In this region, the left side of Equation (18.7) represents the convective momentum of the fluid, which is retarded by the viscous stress term on the right side. Notice that the number of second-order derivative terms in the x-momentum equation has decreased from two to one. Consequently, the equation has changed type, going from elliptic to parabolic. This is a general result; i.e., the elliptic Navier–Stokes equations are simplified to parabolic boundary-layer equations. In this regard, numerical methods that are suitable for the Navier–Stokes equations may not be suitable for the boundary-layer equations, and vice versa.

18.3 SIMILARITY SOLUTION

The length ℓ that appears in the Reynolds number represents an arbitrary distance measured from the tip to a fixed point on the plate. Since the foregoing ordering requires $Re \gg 1$, the boundary-layer equations are not valid in the vicinity of the plate's tip. In particular, the assumption that v is small compared to u does not hold near the tip. In this region the full governing equations are required. We thus imagine we are at a station well downstream of the tip.

We actually do not know the location of this station; it might be at an x^* of 20 or 2×10^6. In either case, the solution would appear to be the same. In other words, we expect u^* to be a function only of $y^*/\delta(x^*)$. This motivates us to try the following transformation:

$$\xi = x^*, \qquad \eta = c_1\frac{y^*}{x^{*m}}, \qquad \psi = c_2 x^{*n} f(\eta) \tag{18.8}$$

where c_1, c_2, m, and n are constants that are to be determined. (Notice that this transformation also holds for the stagnation point flow of Chapter 17 with $m = 0$ and $n = 1 + \sigma$.) Although the stream function ψ is constant on streamlines, it is introduced in order to satisfy the continuity equation. Thus, only the momentum equation will require further consideration. As will be apparent shortly, it is convenient to replace ψ with a different stream function. Despite its name, the new function, f, is not constant on streamlines.

The subsequent analysis is expedited by introducing the following derivative transformation:

$$\begin{aligned}
\frac{\partial}{\partial x^*} &= \frac{\partial \xi}{\partial x^*}\frac{\partial}{\partial \xi} + \frac{\partial \eta}{\partial x^*}\frac{\partial}{\partial \eta} = \frac{\partial}{\partial \xi} - \frac{c_1 m y^*}{x^{*m+1}}\frac{\partial}{\partial \eta} = \frac{\partial}{\partial \xi} - \frac{m\eta}{\xi}\frac{\partial}{\partial \eta}\\
\frac{\partial}{\partial y^*} &= \frac{\partial \xi}{\partial y^*}\frac{\partial}{\partial \xi} + \frac{\partial \eta}{\partial y^*}\frac{\partial}{\partial \eta} = \frac{c_1}{\xi^m}\frac{\partial}{\partial \eta}\\
\frac{\partial^2}{\partial y^{*2}} &= \frac{c_1^2}{\xi^{2m}}\frac{\partial^2}{\partial \eta^2}
\end{aligned} \tag{18.9}$$

The velocity components now become

$$u^* = \frac{\partial \psi}{\partial y^*} = \frac{c_1}{\xi^m}\frac{\partial \psi}{\partial \eta} = c_1 c_2 \xi^{n-m} f'$$

$$v^* = -\frac{\partial \psi}{\partial x^*} = -\frac{\partial \psi}{\partial \xi} + \frac{m\eta}{\xi}\frac{\partial \psi}{\partial \eta} = c_2 \xi^{n-1}(-nf + m\eta f')$$

where f' is $df/d\eta$. We also need the derivatives

$$\frac{\partial u^*}{\partial x^*}, \quad \frac{\partial u^*}{\partial y^*}, \quad \frac{\partial^2 u^*}{\partial y^{*2}}$$

which are determined as follows:

$$\begin{aligned}
\frac{\partial u^*}{\partial x^*} &= c_1 c_2 \xi^{n-m-1}[(n-m)f' - m\eta f'']\\
\frac{\partial u^*}{\partial y^*} &= c_1^2 c_2 \xi^{n-2m} f''\\
\frac{\partial^2 u^*}{\partial y^{*2}} &= c_1^3 c_2 \xi^{n-3m} f'''
\end{aligned}$$

After insertion of the above into Equation (18.7), we have

$$c_1^2c_2^2\xi^{2n-2m-2}[(n-m)f'^2 - mnf'f''] + c_1^2c_2^2\xi^{2n-2m-1}(-nff'' + m\eta f'f'') = \frac{c_1^3c_2}{Re}\xi^{n-3m}f'''$$

which simplifies to

$$\frac{c_1}{c_2Re}\xi^{1-m-n}f''' + nff'' + (m-n)f'^2 = 0$$

For a similarity solution this equation cannot depend on ξ. We thereby require

$$m + n = 1$$

In the formulation, there are more free constants than are needed. Hence, without loss of generality, we can set

$$c_1 = c_2nRe$$
$$m - n = 0$$

thereby obtaining the equation

$$f''' + ff'' = 0 \tag{18.10}$$

which was first deduced by Blasius (1908). This equation is identical to Equation (17.19a) when β is set equal to zero. Of course, flow over a flat plate is not the same as stagnation point flow; nevertheless, the similarity in the physical processes results in analogous mathematics.

The above analysis results in

$$m = n = \frac{1}{2}, \qquad c_1 = \frac{1}{2}c_2Re$$

where c_2 is still arbitrary. We now require ψ to be independent of the arbitrary length ℓ. Since the only other length scale in the problem is ν/U_∞, we replace ℓ with this quantity, which is equivalent to setting $Re = 1$. (A different Reynolds number, large compared to unity, is introduced in the next equation.) We thereby obtain

$$\psi = \left(\frac{2U_\infty x}{\nu}\right)^{1/2} f(\eta) = (2\,Re_x)^{1/2} f \tag{18.11}$$

where we have set

$$c_2 = 2^{1/2}$$
$$Re_x = \frac{U_\infty x}{\nu}$$

thus obtaining for c_1 and η

$$c_1 = 2^{-1/2}$$

$$\eta = \frac{1}{2^{1/2}} \frac{y/\ell}{(x/\ell)^{1/2}} = \frac{y}{(2x\ell)^{1/2}} = \left(\frac{U_\infty}{2\nu x}\right)^{1/2} y \tag{18.12}$$

If we view η as $y/\delta(x)$, we have a first rough estimate, given by $\eta = 1$, of the boundary-layer thickness

$$\delta \cong \left(\frac{2\nu x}{U_\infty}\right)^{1/2}$$

or, in nondimensional form,

$$\frac{\delta}{x} \cong \left(\frac{2}{Re_x}\right)^{1/2} \tag{18.13}$$

For the velocity components, with $\ell = \nu/U_\infty$, we have

$$\frac{u}{U_\infty} = \frac{U_\infty \ell}{2^{1/2}\nu} 2^{1/2} f' = f' \tag{18.14}$$

and

$$\frac{v}{U_\infty} = 2^{1/2}\left(\frac{\ell}{x}\right)^{1/2}\left(\frac{1}{2}\right)(-f + \eta f') = \frac{\eta f' - f}{(2Re_x)^{1/2}} \tag{18.15}$$

Since both u and v are zero on the wall, where y and η are zero, we obtain

$$f(0) = f'(0) = 0 \tag{18.16a}$$

Away from the wall y and, thus, η become infinite, resulting in

$$f'(\infty) = 1 \tag{18.16b}$$

Discussion

A solution of the Blasius equation, subject to Equations (18.16), is numerically obtained. Tables of f, f', and f'' vs. η can be found in many viscous flow textbooks. From such tables we obtain the velocity boundary-layer thickness δ, defined as the y value where $f' = 0.99$. This value occurs at the viscous edge of the boundary layer, where $\eta_{ev} = 3.472$ or

$$\frac{\delta}{x} = \frac{4.910}{Re_x^{1/2}} \tag{18.17a}$$

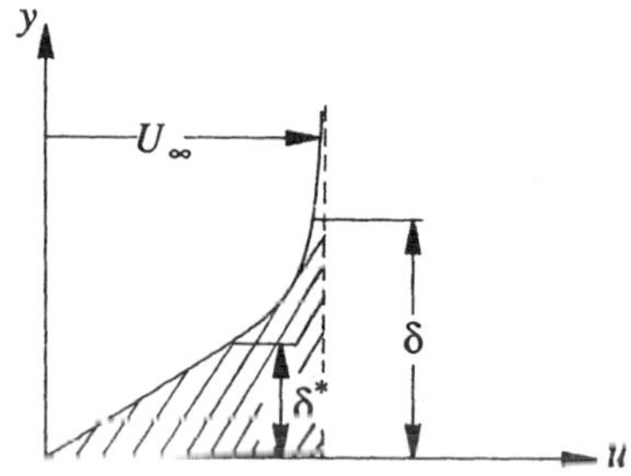

FIGURE 18.2 Velocity profile schematically showing δ and δ^*.

An alternate form for this relation is

$$Re_x^{1/2}\frac{\delta}{x} = 4.910 \quad \textbf{(18.17b)}$$

where this relation replaces the approximate result, Equation (18.13).

An equally important boundary-layer thickness, the displacement thickness δ^*, is defined with reference to Figure 18.2. The mass flux across any x-station can be written as

$$\int_0^\infty \rho u \, dy = \int_0^\infty \rho U_\infty \, dy - \rho U_\infty \delta^*$$

where the rightmost term represents the shaded area in the figure. This relation becomes

$$\delta^*(x) = \int_0^\infty \left(1 - \frac{u}{U_\infty}\right) dy \quad \textbf{(18.18)}$$

where x is held fixed inside the integral. However, with

$$\frac{u}{U_\infty} = f'$$

and

$$dy = \left(\frac{2\nu x}{U_\infty}\right)^{1/2} d\eta$$

we obtain

$$\delta^* = \int_0^\infty (1 - f')\left(\frac{2\nu x}{U_\infty}\right)^{1/2} d\eta = \left(\frac{2\nu x}{U_\infty}\right)^{1/2} \{\eta|_0^\infty - [f(\infty) - f(0)]\}$$

or

$$\frac{\delta^*}{x} = \left(\frac{2}{Re_x}\right)^{1/2} (\eta - f)_{\eta \to \infty}$$

It can be shown that

$$f = \eta - C_v, \quad \eta \rightarrow \infty \tag{18.19}$$

where the velocity parameter, C_v, is a constant. (When an incompressible flow has a pressure gradient, as in Chapter 20, C_v depends on the pressure gradient parameter. For a compressible flow, as in Chapter 21, it also depends on a stagnation temperature ratio.) Here it equals

$$C_v = 1.217$$

so that

$$\frac{\delta^*}{x} = \frac{1.721}{Re_x^{1/2}} \tag{18.20}$$

Thus, the displacement thickness is about 1/3 of the velocity thickness.

Finally, the wall shear stress is

$$\tau_w = \mu\left(\frac{\partial u}{\partial y}\right)_w = \mu\left(U_\infty \frac{df'}{d\eta}\frac{\partial \eta}{\partial y}\right)_w = \mu U_\infty f''_w \left(\frac{U_\infty}{2\nu x}\right)^{1/2}$$

which results in a local skin-friction coefficient

$$c_f = \frac{2\tau_w}{\rho U_\infty^2} = \left(\frac{2}{Re_x}\right)^{1/2} f''_w = \frac{0.6641}{Re_x^{1/2}} \tag{18.21}$$

where the wall shear is $f''_w = f''(0) = 0.4696$.

REFERENCES

Blasius, H., "Grenzschichten in Flüssigkeiten mit kleiner Reibung," *Z. Math., Phys.* 56, 1 (1908).

Prandtl, L., "Über Flüssigkeitsbewegung bei sehr kleiner Reibung," III Intern. Math. Kongr. Heidelberg, 1904. Reprinted as NACA Tech. Memo. No. 452 (1928).

PROBLEMS

18.1 Consider air at 300 K moving over a flat plate at 50 m/s. Determine δ^* and c_f at a point on the plate 1 m from the leading edge. Determine v/U_∞ at this station at the 99% edge of the boundary layer.

18.2 **(a)** Determine the vorticity $\vec{\omega}$ and dissipation Φ using boundary-layer variables for the boundary-layer flow of this chapter.

(b) Simplify your result so that only the dominant term appears in $\vec{\omega}$ and in Φ.

18.3 **(a)** Derive the energy equation in similarity form using $T(\eta)$ as the dependent variable. Assume constant properties, unity Prandtl number, and $c_p T$ for h.

(b) Transform to $\theta(\eta)$, where

$$\theta = \frac{h_o - h_{ow}}{h_{o\infty} - h_{ow}}, \qquad h_o = h + \frac{1}{2}u^2$$

and establish the boundary conditions for θ.

(c) Determine the solution for θ for a diabatic wall. Rewrite this solution for $(T - T_w)/(T_\infty - T_w)$, where this ratio is a function of f' and the Eckert number $U_\infty^2/[c_p(T_\infty - T_w)]$. Note that this number can be large in an incompressible flow when $T_\infty - T_w$ is small.

18.4 **(a)** Use results from the preceding two problems to obtain the boundary-layer approximation for the rate of entropy production in the form

$$\frac{x\dot{s}_{irr}}{c_p U_\infty} = F\left(\eta;\, E, \frac{T_\infty}{T_w}\right)$$

(b) Evaluate $x\dot{s}_{irr}/(c_p U_\infty)$ at the wall and at the outer edge of the boundary layer.

(c) Rederive the part (a) result when $T_w = T_\infty$.

19 Large Reynolds Number Flow

19.1 PRELIMINARY REMARKS

In the subsequent discussion we assume a fluid where c_p and Pr are of order unity, but where μ is very small compared to unity. Thus, μ, λ, and κ all have very small values of a comparable magnitude. In the limit of these transport properties becoming zero, the governing equations become the Euler equations. In our discussion, the energy equation is to be included in both sets of equations. Since the Euler equations are first-order, partial differential equations (PDEs), the no-slip and wall temperature conditions cannot be satisfied. The decrease in order of the momentum and energy equations means the problem is singular for both equations.

In the next section, we discuss matched asymptotic expansions for a viscous flow. This method provides a mathematical formalism for treating singular perturbation problems. The last section provides the conservation equations in an orthogonal, body-oriented coordinate system. This form for the equations is an essential first step in developing boundary-layer theory. Thus, the function of this chapter is to provide background for the discussion in later chapters.

Before embarking on a discussion of matched asymptotic expansions for a boundary layer, it is instructive to first consider several elementary perturbation examples. In order to contrast a regular perturbation problem with a singular one, the first example briefly focuses on the former. We then proceed to discuss several slightly different singular perturbation problems.

Illustrative Example 1

A perturbation problem requires a small positive parameter, ε, where the placement of the parameter in the equation is all-important. For instance, suppose we have the differential equation

$$\frac{d^2y}{dx^2} - \varepsilon^2 y = 0 \tag{19.1}$$

subject to the boundary conditions

$$y(0) = 1, \qquad y(1) = 0 \tag{19.2}$$

This is a two-point boundary value problem. Observe that as $\varepsilon \to 0$ the resultant ordinary differential equation (ODE) remains second-order and can still satisfy both boundary conditions. This property classifies it as a regular perturbation problem.

The exact solution of the above equations is readily shown to equal

$$y(x) = \frac{\exp[\varepsilon(1-x)] - \exp[\varepsilon(x-1)]}{\exp\varepsilon - \exp(-\varepsilon)} \tag{19.3}$$

If this solution is expanded for a small value of ε, we obtain

$$
\begin{aligned}
y &= \frac{\left[1 + \varepsilon(1-x) + \frac{1}{2}\varepsilon^2(1-x)^2 + \cdots\right] - \left[1 - \varepsilon(1-x) + \frac{1}{2}\varepsilon^2(1-x)^2 - \cdots\right]}{\left(1 + \varepsilon^2 + \frac{1}{2}\varepsilon^2 + \cdots\right) - \left(1 - \varepsilon + \frac{1}{2}\varepsilon^2 - \cdots\right)} \\
&= (1-x)\frac{1 + \frac{1}{6}\varepsilon^2(1-x)^2 + \cdots}{1 + \frac{1}{6}\varepsilon^2 + \cdots} \\
&= 1 - x + \frac{1}{6}\varepsilon^2 x(1-x)(x-2) + O(\varepsilon^4)
\end{aligned}
\tag{19.4}
$$

Thus, the solution has a straightforward expansion in powers of ε^2. This results also can be obtained by assuming an expansion of the form

$$
y(x, \varepsilon) = y_1(x) + \varepsilon^2 y_2(x) + \varepsilon^4 y_3(x) + \cdots \tag{19.5}
$$

This series is then substituted into Equations (19.1) and (19.2). Equating like powers of ε^2 yields the following sequence of problems:

$$
\begin{aligned}
&\frac{d^2y_1}{dx^2} = 0, && y_1(0) = 1, && y_1(1) = 0 \\
&\frac{d^2y_2}{dx^2} = y_1(x), && y_2(0) = 0, && y_2(1) = 0 \\
&\frac{d^2y_3}{dx^2} = y_2(x), && y_3(0) = 0, && y_3(1) = 0 \\
&\quad\vdots && \quad\vdots && \quad\vdots
\end{aligned}
$$

[Had the series expansion, Equation (19.5), included odd powers of ε, we would find the associated $y_i(x)$ to be identically equal to zero.] The solution of these problems, when substituted into Equation (19.5), results in Equation (19.4). Thus, y_1 equals $1 - x$, etc.

Illustrative Example 2

Suppose we retain Equations (19.2) but move the ε^2 to the derivative term in Equation (19.1); i.e.,

$$
\varepsilon^2\frac{d^2y}{dx^2} - y = 0 \tag{19.6}
$$

We again have a two-point boundary value problem, but the highest-order derivative is now multiplied by a small parameter. The exact solution is given by

$$
y = \frac{\exp[(1-x)/\varepsilon] - \exp[(x-1)/\varepsilon]}{\exp(1/\varepsilon) - \exp(-1/\varepsilon)} \tag{19.7}
$$

For $0 < \varepsilon \ll 1$, the solution is sketched in Figure 19.1, and y undergoes a rapid decrease when x is near the origin. In contrast to Equation (19.1), Equation (19.6) becomes $y = 0$ when $\varepsilon \to 0$, and

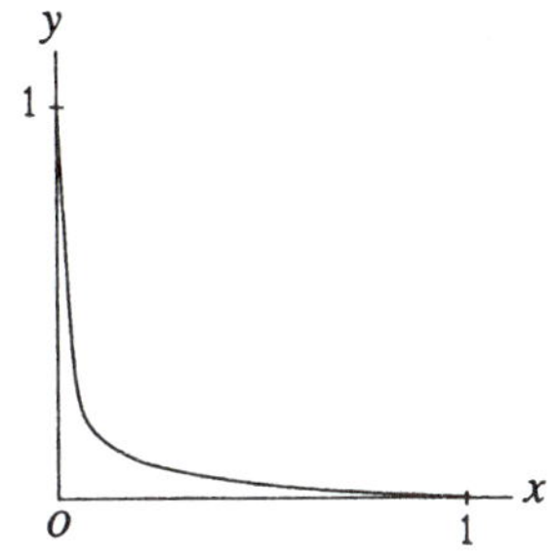

FIGURE 19.1 Solution of the singular perturbation example.

the $y(0) = 1$ boundary condition cannot be satisfied. Thus, a single expansion of the form of Equation (19.5) cannot be used, and we have a singular perturbation problem.

The method of matched asymptotic (or inner and outer) expansions is used for this type of problem. Instead of a single expansion, such as Equation (19.5), two expansions are required. One of these, called the outer expansion, will hold where the variation in $y(x)$ is small or moderate. The inner expansion will hold where the variation in $y(x)$ is extreme; this is the region near the origin in Figure 19.1. These expansions are referred to as solutions. Moreover, we typically obtain only the first term in each of the expansions; these are nevertheless also called solutions. Because each of the expansions is an asymptotic expansion that holds in the $\varepsilon \to 0$ limit, the leading, or first-order, terms are the dominant ones and generally provide an excellent approximation to the exact solution, when such a solution is known. This dominance increases, and the approximation improves, as ε becomes smaller.

The term "order" has two distinct meanings. The order of an ODE or PDE refers to the number of derivatives in the term with the largest number. In addition, an approximation is referred to as first-order, second-order, etc., depending on the number of terms retained. The definition to be used is clear from the context.

Let x_o, y_o replace x, y after the $\varepsilon \to 0$ limit is applied to Equations (19.2) and (19.6). Thus, x_o, y_o are the outer region variables. As indicated earlier, we have

$$y_o(x_o) = 0$$

for the outer solution, which only satisfies the $y_o(1) = 0$ boundary condition.

The structure of the inner layer that is adjacent to $x = 0$ is obtained by expanding this region with the transformation

$$\bar{x} = x/\varepsilon^n, \qquad \bar{y} = y$$

$$\frac{d}{dx} = \frac{1}{\varepsilon^n}\frac{d}{d\bar{x}}, \qquad \frac{d^2}{dx^2} = \frac{1}{\varepsilon^{2n}}\frac{d^2}{d\bar{x}^2} \tag{19.8}$$

where n is to be determined, and $\bar{x}, \bar{y}$ are called the inner variables. By dividing x with ε^n, where $n > 0$, we enlarge the region near the origin so that $\bar{x}$ is of $O(1)$ in this region. Since $\bar{y}$ has not been written in a form such as Equation (19.5), we are only obtaining the first (dominant) term of the inner expansion.

With this transformation, Equation (19.6) becomes

$$\varepsilon^{2(1-n)}\frac{d^2\bar{y}}{d\bar{x}^2} - \bar{y} = 0$$

It is necessary to set $n = 1$ in order to retain the highest-order derivative term. The boundary conditions now have the form

$$\bar{y}(0) = 1, \quad \bar{y}(1/\varepsilon) = 0$$

We next take the $\varepsilon \to 0$ limit, which normally simplifies the equation, although in this case no change occurs. The $x = 1$ boundary condition, however, becomes $\bar{y}(\infty) = 0$. It is easy to see that the differential equation

$$\frac{d^2\bar{y}}{d\bar{x}^2} - \bar{y} = 0$$

has the solution

$$\bar{y} = \exp(-\bar{x})$$

that satisfies both $\bar{y}$ boundary conditions. In terms of the original variables, we have

$$y = \exp(-x/\varepsilon) \tag{19.9}$$

for the inner layer. Hence, there is an inner solution and an outer one, given by $y = 0$. The two solutions asymptotically approximate the exact solution. In this case, a uniformly valid composite solution is just Equation (19.9). This composite solution can also be obtained by expanding the exact solution for small ε, with the result

$$y \cong \frac{\exp[(1-x)/\varepsilon]}{\exp(1/\varepsilon)} = \exp(-x/\varepsilon)$$

We observe that the essence of this approach is to rescale the dependent and independent variables so that the structure of the inner layer is obtained. Generally, the original variables are satisfactory for the region outside this layer. This outer solution is often relatively simple because of the reduced order of the governing equations. In this example, for instance, it is given by $y = 0$.

Illustrative Example 3

In the next example, Equation (19.6) is modified to

$$\varepsilon^2 \frac{d^2y}{dx^2} + \frac{dy}{dx} - y = 0 \tag{19.10}$$

and Equations (19.2) are retained as the boundary conditions. The exact solution is found as

$$y = \frac{\exp[-m_-(1-x)] - \exp[-m_+(1-x)]}{\exp(-m_-) - \exp(-m_+)}$$

where

$$m_\pm = \frac{1}{2\varepsilon^2}\left[-1 \pm (1+4\varepsilon^2)^{1/2}\right]$$

For the first term of the outer expansion, we set

$$x_o = x, \quad y_o = y$$

and take the $\varepsilon \to 0$ limit to obtain

$$\frac{dy_o}{dx_o} - y_o = 0$$

The single boundary condition for this equation is $y_o(1) = 0$, which yields

$$y_o(x_o) = 0$$

as was the case in the preceding example.

The earlier transformation, Equations (19.8), is applied to Equation (19.10), with the result

$$\varepsilon^{2-2n}\frac{d^2\bar{y}}{d\bar{x}^2} + \varepsilon^{-n}\frac{d\bar{y}}{d\bar{x}} - \bar{y} = 0$$

To avoid an infinity when $\varepsilon \to 0$, we multiply by ε^n:

$$\varepsilon^{2-n}\frac{d^2\bar{y}}{d\bar{x}^2} + \frac{d\bar{y}}{d\bar{x}} - \varepsilon^n\bar{y} = 0$$

To retain the highest-order derivative term, we set $n = 2$ to obtain

$$\frac{d^2\bar{y}}{d\bar{x}^2} + \frac{d\bar{y}}{d\bar{x}} - \varepsilon^2\bar{y} = 0$$

We now take the $\varepsilon \to 0$ limit, which yields the first-order, inner-layer equation

$$\frac{d^2\bar{y}}{d\bar{x}^2} + \frac{d\bar{y}}{d\bar{x}} = 0 \tag{19.11}$$

Thus, the addition of the *dy*/*dx* term to Equation (19.6) alters the value of *n* as well as the form of the equation for the inner layer. This equation satisfies the boundary conditions

$$\bar{y}(0) = 1$$
$$\bar{y}(\infty) = y_o(x_o), \quad x_o = 0 \tag{19.12}$$

The $\bar{y}(\infty)$ condition is expressed somewhat differently from that used in the previous example in order to introduce the concept of matching two asymptotic expansions. Equation (19.12) is a matching condition and can be expressed as follows: The outer limit of the inner expansion, $\bar{y}(\infty)$, equals the inner limit of the outer expansion, $y_o(x_o)$. Observe that the $x = 1$ point has become $\bar{x} = (1/\varepsilon^2) \to \infty$, while the $x_o = 0$ point remains $\bar{x} = 0$ when $\varepsilon \to 0$. Equation (19.12) provides what otherwise would be a missing boundary condition for Equation (19.11) and, moreover, ensures that, at least to first order, the two expansions overlap. For the specific case under consideration, Equation (19.12) becomes $\bar{y}(\infty) = 0$.

Equation (19.11), along with its boundary conditions, yields

$$\bar{y} = e^{-\bar{x}}$$

or in terms of the original variables

$$y = e^{-x/\varepsilon^2}$$

Although this result closely resembles the one in the preceding example, the dependence on ε is different; i.e., the inner layer here is much thinner. Since this result also approximates the exact solution in the outer region, it is a uniformly valid approximate solution. For instance, it differs from the $y(1) = 0$ condition by an exponentially small amount, $\exp(-1/\varepsilon^2)$.

Illustrative Example 4

As a final example, we address a question of uniqueness that is raised in Chapter 23. A generic differential equation is utilized as the starting point:

$$(a+bx)^2\frac{d^2y}{dx^2} + a_1(a+bx)\frac{dy}{dx} + a_2y = 0 \tag{19.13}$$

where a_1, a_2, a, and b are constants, and Equations (19.2) still provide the boundary conditions. Incidentally, this ODE is referred to as Legendre's linear equation (Murphy, 1960).

Suppose a term, such as $(1+\varepsilon x)^{-n}$, is a coefficient of the dy/dx or y terms, where n is a positive integer. The question to be examined is whether or not it makes any difference if the asymptotic expansions are first performed with $(1+\varepsilon x)^{-n}$ removed by cross-multiplication or by the expansion

$$\frac{1}{(1+\varepsilon x)^n} = 1 - n\varepsilon x + O(\varepsilon^2)$$

Any difference between the two approaches, however, does not appear until second-order terms are considered. Hence, the inner and outer expansions must be developed, at least, to second order.

As an illustrative example, unfortunately, it is easy to generate cases of overwhelming comlexity when starting from Equation (19.13). For example, suppose the constants are chosen as

$$a = \varepsilon^{1/2}, \quad a_1 = 0, \quad a_2 = 1, \quad b = 2\varepsilon^{3/2}$$

thus yielding

$$\varepsilon\frac{d^2y}{dx^2} + \frac{y}{(1+2\varepsilon x)^2} = 0 \tag{19.14}$$

which certainly is simple in appearance. We seek a real solution to this linear equation that satisfies Equations (19.2). Toward this end, the substitution (Rainville, 1943)

$$z = \frac{1}{2\varepsilon^{3/2}}\ln(1+2\varepsilon x)$$

yields an equation

$$\frac{d^2y}{dz^2} - 2\varepsilon^{3/2}\frac{dy}{dz} + y = 0$$

with the constant coefficients. As usual, the solution is obtained by setting $y = e^{mz}$. Hence, the exact solution to Equations (19.2) and (19.14) is obtained as

$$y = (1 + 2\varepsilon x)\frac{\sin(\chi_1 - \chi)}{\sin\chi_1} \tag{19.15}$$

where

$$\chi(x) = \frac{(1-\varepsilon^3)^{1/2}}{2\varepsilon^{3/2}}\ln(1 + 2\varepsilon x)$$

$$\chi_1 = \chi(1) = \frac{(1-\varepsilon^2)^{1/2}}{2\varepsilon^{3/2}}\ln(1 + 2\varepsilon)$$

For a small positive ε, we obtain the expansions

$$\frac{(1-\varepsilon^3)^{1/2}}{2\varepsilon^{3/2}} = \frac{1}{2}\varepsilon^{-3/2} - \frac{1}{4}\varepsilon^{3/2} + \cdots$$

$$\ln(1 + 2\varepsilon x) = 2\varepsilon x - 2\varepsilon^2 x^2 + \cdots$$

$$\ln(1 + 2\varepsilon) = 2\varepsilon - 2\varepsilon^2 + \cdots$$

$$\chi = \frac{x}{\varepsilon^{1/2}} - \varepsilon^{1/2}x^2 + \cdots$$

$$\chi_1 = \frac{1}{\varepsilon^{1/2}} - \varepsilon^{1/2} + \cdots$$

Thus, as $\varepsilon \to 0$, $\sin\chi_1$ oscillates with increasing rapidity between -1 and $+1$. The $\sin(\chi_1 - \chi)$ factor similarly oscillates when $0 < x \le 1$. The two oscillations are not in phase, and, consequently, y oscillates wildly between $-\infty$ and $+\infty$ as $\varepsilon \to 0$ for all x in the range, $0 < x \le 1$. (In the $\varepsilon \to 0$ limit, Equation (19.15) is a space-filling curve.)

A case much more amenable to analysis stems from the choice

$$a = \varepsilon, \quad a_1 = 1, \quad a_2 = 0, \quad b = \varepsilon^2$$

which yields

$$\varepsilon\frac{d^2y}{dx^2} + \frac{1}{1 + \varepsilon x}\frac{dy}{dx} = 0 \tag{19.16}$$

The balance of this section focuses on this equation. To obtain its exact solution, we write it as

$$\varepsilon\frac{dy'}{y'} + \frac{1}{\varepsilon}\frac{d(1 + \varepsilon x)}{1 + \varepsilon x} = 0$$

where the notation $y' = (dy/dx)$ is introduced. A first integration results in

$$y' = \frac{c_1}{(1+\varepsilon x)^{1/\varepsilon^2}}$$

where c_1 is an integration constant. A second integration and application of the boundary conditions yields the exact solution

$$y = \frac{1}{1-(1+\varepsilon)^{(1-\varepsilon^2)/\varepsilon^2}}\left[1 - \frac{(1+\varepsilon)^{(1-\varepsilon^2)/\varepsilon^2}}{(1+\varepsilon x)^{(1-\varepsilon^2)/\varepsilon^2}}\right] \tag{19.17}$$

To expand this for small ε, we write

$$A = (1+\varepsilon x)^{(1-\varepsilon^2)/\varepsilon^2}$$

and

$$\ln A = \frac{1-\varepsilon^2}{\varepsilon^2}\ln(1+\varepsilon x) = \frac{1-\varepsilon^2}{\varepsilon^2}\left(\varepsilon x - \frac{1}{2}\varepsilon^2 x^2 + \cdots\right) = \frac{x}{\varepsilon} - \frac{1}{2}x^2 + O(\varepsilon)$$

We thus have

$$(1+\varepsilon x)^{(1-\varepsilon^2)/\varepsilon^2} = e^{-x^2/2}e^{x/\varepsilon}[1+O(\varepsilon)]$$
$$(1+\varepsilon)^{(1-\varepsilon^2)/\varepsilon^2} = e^{-1/2}e^{1/\varepsilon}[1+O(\varepsilon)]$$

and Equation (19.17) can be simplified to

$$y = \frac{1}{1-e^{-1/2}e^{1/\varepsilon}}\left(1 - \frac{e^{-1/2}e^{1/\varepsilon}}{e^{-x^2/2}e^{x/\varepsilon}}\right) \cong \frac{1-e^{-(1-x^2)/2}e^{(1-x)/\varepsilon}}{1-e^{-1/2}e^{1/\varepsilon}}$$

Note that this approximation still satisfies Equations (19.2). For x in the range $0 \le x < 1$ and with $\varepsilon \to 0$, the $e^{(1/\varepsilon)}$ and $e^{(1-x)/\varepsilon}$ factors become infinite. Hence, to all orders in ε^n, $n \ge 0$, we have

$$y \cong e^{x^2/2}e^{-x/\varepsilon} \tag{19.18}$$

This result demonstrates the presence of a single boundary layer, which is located at $x = 0$. It satisfies the conditions

$$y(0) = 1$$
$$y(1) \to 0, \quad \varepsilon \to 0$$

We now write Equation (19.16) two ways, i.e., as

$$\varepsilon(1+\varepsilon x)y'' + y' = 0 \tag{19.19a}$$

denoted hereafter as case I, and as

$$\varepsilon y'' + (1 - \varepsilon x + \varepsilon^2 x^2 - \cdots)y' = 0 \tag{19.19b}$$

denoted as case II, but with the $O(\varepsilon^3)$ terms deleted. We obtain, to second order, the outer and matching inner expansions for both cases. For this, it is convenient to introduce the type of notation later utilized in Chapter 23.

We thus write the outer expansion as

$$y(x, \varepsilon) \sim Y_1(x) + \varepsilon^m Y_2(x) + \cdots, \qquad m \geq 0 \tag{19.20}$$

where the tilde means asymptotic to …, when $\varepsilon \to 0$. For case I, we have

$$\varepsilon Y_1'' + \varepsilon^2 x Y_1'' + \varepsilon^{1+m} Y_2'' + \varepsilon^{2+m} x Y_2'' + Y_1' + \varepsilon^m Y_2' = 0 \tag{19.21}$$

where still higher-order terms on the left side are not shown. Examination of this relation yields, with $\varepsilon \to 0$,

$$Y_1' = 0, \qquad Y_1(1) = 0 \tag{19.22a}$$

or $Y_1(x) = 0$. The Y_2 equation becomes

$$\varepsilon Y_2'' + \varepsilon^2 x Y_2'' + Y_2' = 0$$

with m being arbitrary. By setting $\varepsilon \to 0$, we have

$$Y_2' = 0, \qquad Y_2(1) = 0 \tag{19.22b}$$

or $Y_2(x) = 0$. The fact that m is arbitrary indicates that it can be set equal to zero, and the Y_2 term is irrelevant. Hence, the case I outer expansion is

$$y(x, \varepsilon) \sim 0 \tag{19.23}$$

to all orders in ε^m.

The substitution of Equation (19.20) into Equation (19.19b) yields

$$\varepsilon Y_1'' + \varepsilon^{1+m} Y_2'' + (1 - \varepsilon x + \varepsilon^2 x^2)(Y_1' + \varepsilon^m Y_2') = 0$$

We readily obtain Equations (19.22) and, therefore, Equation (19.23) is the outer expansion for both cases.

For the inner expansion, the independent variable is stretched:

$$\bar{x} = \frac{x}{\varepsilon^n}, \qquad n > 0$$

in order to analyze the region where a boundary layer occurs. Equation (19.16) becomes

$$\varepsilon^{1-n} \frac{d^2 y}{d\bar{x}^2} + \frac{1}{1 + \varepsilon^{1+n}\bar{x}} \frac{dy}{d\bar{x}} = 0$$

The inner expansion is written as

$$y(\bar{x}, \varepsilon) \sim y_1(\bar{x}) + \varepsilon^m y_2(\bar{x}) + \cdots, \qquad m \geq 0 \tag{19.24}$$

which yields

$$\varepsilon^{1-n}\left(\frac{d^2 y_1}{d\bar{x}^2} + \varepsilon^m \frac{d^2 y_2}{d\bar{x}^2} + \cdots\right) + \frac{1}{1 + \varepsilon^{1+n}\bar{x}}\left(\frac{dy_1}{d\bar{x}} + \varepsilon^m \frac{dy_2}{d\bar{x}} + \cdots\right) = 0 \tag{19.25}$$

Note that the m exponent is not the same as that used in Equation (19.20).

The dominant case I terms in Equation (19.25) are

$$(\varepsilon^{1-n} + \varepsilon^2\bar{x})(y_1'' + \varepsilon^m y_2'') + y_1' + \varepsilon^m y_2' = 0 \tag{19.26}$$

where $y' = (dy/d\bar{x})$. To preserve the structure of the boundary layer, the coefficient of y_1'' must be of order unity. Hence, we set $n = 1$ and the equation simplifies to

$$(1 + \varepsilon^2\bar{x})(y_1'' + \varepsilon^m y_2'') + y_1' + \varepsilon^m y_2' = 0$$

which yields, when $\varepsilon \to 0$,

$$y_1'' + y_1' = 0, \quad y_1(0) = 1, \quad y_1(\infty) = 0$$

In a more formal treatment, the $y_1(\infty)$ boundary condition would be left unspecified. It would be determined later by matching with the outer expansion. In view of Equation (19.23), the matching step is trivial and here is bypassed.

The solution for y_1 is

$$y_1 = e^{-\bar{x}} \tag{19.27}$$

Similarly, the y_2 equation is

$$\varepsilon^2\bar{x}y_1'' + \varepsilon^m y_2'' + \varepsilon^m y_2' = 0, \quad y_2(0) = 0, \quad y_2(\infty) = 0$$

An inhomogeneous term is required; otherwise, $y_2(\bar{x}) = 0$. To retain the y_2'' term, set $m = 2$ so that

$$y_1'' + y_2' = -\bar{x}y_1'' \tag{19.28a}$$

$$= -\bar{x}e^{-\bar{x}} \tag{19.28b}$$

Two linearly independent solutions of the reduced (homogeneous) equation are 1 and $e^{-\bar{x}}$. The variation-of-parameter method then yields

$$y_2(\bar{x}) = c_1 + c_2 e^{-\bar{x}} + \frac{1}{2}(\bar{x}^2 + 2\bar{x} + 2)e^{-\bar{x}}$$

The above boundary conditions result in

$$y_2(0) = c_1 + c_2 + 1 = 0$$
$$y_2(\infty) = c_1 = 0$$

thereby yielding

$$y_2 = \bar{x}\left(1 + \frac{1}{2}\bar{x}\right)e^{-\bar{x}} \tag{19.29}$$

Equations (19.24), (19.27), and (19.29) result in

$$y \sim e^{-\bar{x}} + \varepsilon^2\bar{x}\left(1 + \frac{1}{2}\bar{x}\right)e^{-\bar{x}} + \cdots$$

for the second-order inner expansion, which also represents a uniformly valid expansion. In terms of x, this becomes

$$y \sim e^{-x/\varepsilon} + x\left(\varepsilon + \frac{1}{2}x\right)e^{-x/\varepsilon} + \cdots \tag{19.30}$$

which superficially appears to differ from Equation (19.18). However, when x is of $O(1)$, both expansions yield $y \sim 0$, while when x is small both yield

$$y \sim \left(1 + \frac{1}{2}x^2\right)e^{-x/\varepsilon}$$

when $\varepsilon \to 0$.

For case II, Equation (19.25) becomes

$$\varepsilon^{1-n}y_1'' + \varepsilon^{1-n+m}y_2'' + y_1' + \varepsilon^m y_2' - \varepsilon^{1+n}\bar{x}y_1' - \varepsilon^{1+n+m}\bar{x}y_2' = 0$$

We again require $n = 1$, which also results in Equation (19.27) for y_1. For y_2, we have

$$\varepsilon^m y_2'' + \varepsilon^m y_2' - \varepsilon^2\bar{x}y_1' - \varepsilon^{2m}\bar{x}y_2' = 0$$

and again $m = 2$. We thus have

$$y_2'' + y_2' = \bar{x}y_1'$$

for the second-order variable.

In contrast to Equation (19.28), the inhomogeneous term $\bar{x}y_1'$ differs from $-\bar{x}y_1''$. Thus, cases I and II yield different equations for the second-order inner variable. This represents the nonuniqueness referred to earlier. This nonuniqueness stems from alternate approaches when expanding coefficients in order to obtain second- or higher-order terms. As is evident in this example, it does not occur when deriving the first-order equations. The nonuniqueness problem, fortunately, appears to be of little, if any, consequence. This is evident when the inhomogeneous terms are evaluated; they both yield $-\bar{x}e^{-\bar{x}}$. Hence, for both cases, Equation (19.30) is the uniformly valid asymptotic expansion. To the author's knowledge, this conclusion has not been generally established.

19.2 MATCHED ASYMPTOTIC EXPANSIONS

The basic ideas of the preceding examples are applied to the conservation equations. The small parameters are μ, λ, and κ, which usually are of comparable magnitude. However, the first step is to introduce a body oriented coordinate system for the equations. In such a system one coordinate, x_3, is zero on the body and increases outward into the flow. Without loss of generality, the coordinate system may be orthogonal, although this is not essential.

An outer solution is then obtained by setting $\mu = \lambda = \kappa = 0$ and only using a velocity tangency condition on the surface of the body. The solution is provided by the Euler equations and corresponds to an inviscid, adiabatic flow. If the inviscid flow is irrotational, then a potential flow solution can be used.

Quite often a full-blown solution to the Euler equations is not essential. This is because the inviscid, adiabatic flow may satisfy one or more additional assumptions. For instance, these might encompass the irrotational, homenergetic, or perfect gas assumption. A Bernoulli equation is often appropriate, as will be exemplified in the next chapter.

There are several difficulties that may arise with the Euler solution. It need not be unique; e.g., consider potential flow about a circular cylinder where the circulation is a free parameter. However, the most serious difficulty is that the Euler equations cannot assess when or where boundary-layer separation is to occur. Consequently, a physically wrong Euler solution can be obtained. This difficulty is not easily overcome, since the boundary-layer equations require an *a priori* inviscid solution. Finally, at a sufficiently high Reynolds number ($\cong$ 100) the flow may be unsteady, even though freestream and boundary conditions are both steady. An example of this is the Kármán vortex street behind a circular cylinder.

One remedy to these difficulties is to utilize the unsteady governing (Navier–Stokes) equations directly, thereby bypassing the need for solutions of the Euler and boundary-layer equations. In recent years this has become the central focus of computational fluid dynamics. This is the most suitable approach for treating boundary-layer separation. Numerical solutions to the governing equations, however, come at a steep price. A large amount of computational time and storage is required, and often the solution is hard to interpret, validate, or understand in physical terms.

After obtaining the appropriate inviscid solution, we transform the body-oriented conservation equations in order to magnify the viscous and heat-conducting layer adjacent to the body. This is done by introducing inner variables (Lagerstrom, 1964)

$$\bar{x}_1 = x_1, \quad \bar{x}_2 = x_2, \quad \bar{x}_3 = \frac{x_3}{\mu_r^{1/2}} \tag{19.31}$$

where μ_r is a constant reference viscosity value that also typifies the magnitudes of λ and κ. The velocity component v_3 is also stretched:

$$\bar{v}_3 = \frac{v_3}{\mu_r^{1/2}} \tag{19.32}$$

that is perpendicular to the body. This stretching is essential if continuity is to be satisfied. If this is not done, we would have

$$\frac{\partial u^*}{\partial x^*} = 0$$

instead of

$$\frac{\partial u^*}{\partial x^*} + \frac{\partial v^*}{\partial y^*} = 0$$

for flow over a flat plate. We also set

$$\bar{\mu} = \frac{\mu}{\mu_r}, \qquad \bar{\lambda} = \frac{\lambda}{\mu_r}, \qquad \bar{\kappa} = \frac{\kappa}{c_{pr}\mu_r}$$

where c_{pr} is a reference specific heat and $\bar{\kappa}$ is equivalent to the inverse of the Prandtl number.

After the barred inner variables are introduced, the $\mu_r \to 0$ limit is taken. This is equivalent to letting the Reynolds number become infinite for the flow in the viscous layer. While this limit process simplifies the equations, it does so with the structure of the viscous layer correctly preserved. In contrast to the first $\mu \to 0$ limit, which yields the Euler equations, we now obtain the first-order boundary-layer equations. After these equations are found, it is usually convenient to return to the original, unbarred variables. These simplified viscous equations will now explicitly contain μ.

The boundary conditions for the boundary-layer equations are the no-slip and temperature, or temperature gradient, wall conditions. By moving away from the wall in a normal direction, the $\bar{x}_3 \to \infty$ boundary conditions are obtained. These are provided by evaluating the solution of the Euler equations *at the wall*. This last step is a matching condition that is analogous to Equations (19.12).

Discussion

The basis of the foregoing procedure is the application of matched asymptotic expansions to the governing equations. It was discovered and first used for laminar viscous flows, but subsequently it has been utilized for analyzing a wide variety of problems. It is important to realize that the two separate expansions (or solutions) are not being "patched" together, where the dependent variables, and possibly their first derivatives, are made continuous at some intervening point. Instead, we have two expansions that interlock with each other through the boundary conditions, both at the wall and far from the wall.

In boundary-layer theory, the small parameter that orders both expansions is $Re^{-1/2}$. The leading term of the outer expansion is provided by the Euler equations and holds throughout the flow field except in the viscous region adjacent to the wall. In this region the leading term of the inner expansion is provided by the boundary-layer equations. The two expansions interlock with each other through their boundary conditions and the pressure gradient. Their regions of validity overlap, thereby making it possible to also obtain a uniformly valid composite expansion. However, a composite expansion is not germane to our purposes, which is ultimately to determine the skin friction, various boundary-layer thickness, and the heat transfer at the wall.

In some problems there may be more than two expansions. For example, suppose we have a large Reynolds number flow of a highly conducting metal, such as liquid sodium. In this circumstance, the Prandtl number is nearly zero and would represent a second small parameter. Inside the viscous layer, adjacent to the wall, there is a very thin thermal sublayer. Since no highest-order derivative is lost when the $Pr \to 0$ limit is taken, the analysis of the sublayer represents a regular perturbation problem.

For an incompressible flow with constant properties, only the convective terms are nonlinear in the continuity and momentum equations. (In a compressible flow, other terms are also nonlinear.) The Euler and boundary-layer equations, both of which contain the convective terms, are thus nonlinear. First-order boundary-layer theory is therefore nonlinear, whereas the second- and higher-order theory is linear. The awkwardness of retaining nonlinear terms in only the first-order theory is actually an important virtue. It ensures that significant nonlinear effects are incorporated from the start; consequently, the first-order theory provides an excellent first approximation. The second-order theory, which incorporates additional second-order terms, generally provides only a small correction. Therefore, in the next three chapters we concentrate on the first-order theory, while Chapter 23 discusses the second-order theory.

19.3 GOVERNING EQUATIONS IN BODY-ORIENTED COORDINATES

As shown in Figure 19.2, we consider an internal or external flow over a smooth two-dimensional or axisymmetric body. The sign convention used for the longitudinal curvature k is indicated in the figure. If the center of the radius of curvature is inside the wall, then $k > 0$; otherwise, k is negative. Figure 19.3 shows the body-oriented coordinates s, n, where s is defined only on the surface of the body, while the n coordinate yields straight lines that are normal to the surface. These coordinates are *not* natural coordinates; they are valid only in the immediate vicinity of the body. If the flow is two-dimensional, then r and r_w become y and y_w, respectively.

Appendix H summarizes the conservation equations based on these coordinates. Item A provides the assumptions that apply throughout the table. In item B, note that k, r_w, and θ are not independent of each other but are related by

$$\theta = \sin^{-1} r_w', \qquad k = -\frac{r_w''}{(1 - r_w'^2)^{1/2}}$$

However, the radius r refers to a point in the flow field and is a function of both s and n. Observe that a prime denotes differentiation with respect to s, and that θ is positive in the counterclockwise direction. Hence, $d\theta$, as sketched in Figure 19.3, is negative. For convenience, the scale factors h_i

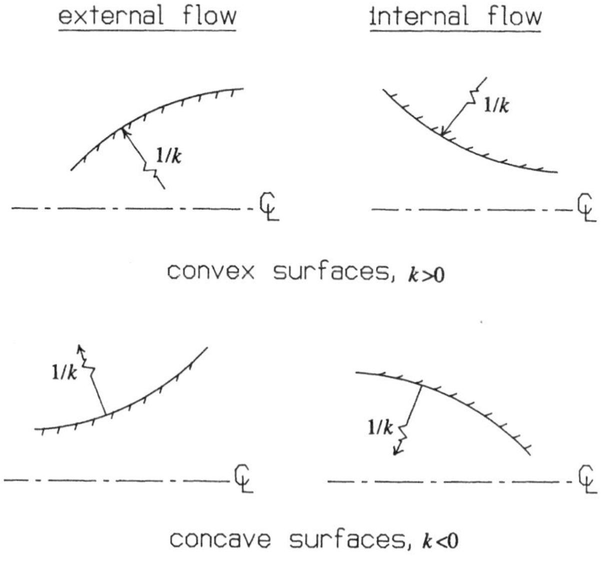

FIGURE 19.2 Curvature schematic of a convex or concave surface (Back, 1973).

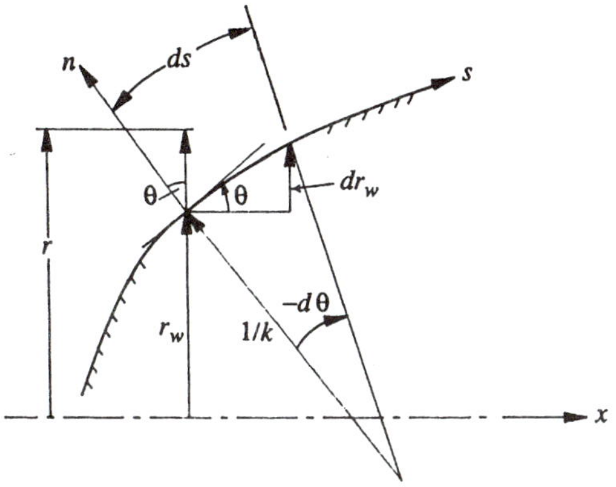

FIGURE 19.3 Nomenclature for a body-oriented coordinate system (Van Dyke, 1963).

and their derivatives, along with the derivatives of the basis vectors, are provided. (See Appendix G for the formula for the derivative of the basis vectors.)

The dynamic quantities of interest are listed in item *C*. For notational simplicity, the scale factors h_n and h_ϕ are replaced by unity and r^σ, respectively. Because s and n are not natural coordinates, the velocity has a nonzero component in the ∂_n direction. Items *D*, *E*, and *F* summarize the conservation equations along with the shear stress and heat transfer at the wall, the Euler equations, and the incompressible conservation equations, respectively.

Much of the material in Appendix H is based on Appendix G. On the other hand, the analysis in the remaining chapters is based on Appendix H.

REFERENCES

Back, L.H., "Transonic Laminar Boundary Layers with Surface Curvature," *Int. J. Heat Mass Transfer* 16, 1745 (1973).

Lagerstrom, P.A., "Laminar Flow Theory," *Theory of Laminar Flows,* edited by F.K. Moore, High Speed Aerodynamics and Jet Propulsion, Vol. IV, Princeton University Press, 1964, pp. 20–285.

Murphy, G.M., *Ordinary Differential Equations and Their Solutions,* D. Van Nostrand Co., New York, 1960, p. 87.

Rainville, E.D., *Intermediate Course in Differential Equations,* John Wiley, New York, 1943, p. 13.

Van Dyke, M., "Higher Approximations in Boundary-Layer Theory. Part I. General Analysis," *J. Fluid Mech.* 14, 161 (1963).

PROBLEMS

19.1 Derive the equations for $\partial \hat{e}_j / \partial \xi_i$ and for the ε_{ij} given in Appendix H.

19.2 Derive the equations for $D\vec{w}/Dt$, $\nabla \cdot \vec{w}$, F_i^s, Φ, and $\nabla^2 T$ as shown in Appendix H.

19.3 Consider the equation

$$\varepsilon^2 \frac{d^3 y}{dx^3} - \frac{dy}{dx} = -1$$

with the boundary conditions

$$y(0) = 1, \quad \frac{dy}{dx}(0) = -1, \quad y \sim x \text{ as } x \to \infty$$

(a) Determine the exact solution. Where is the inner region when $\varepsilon \to 0$?

(b) Determine the first approximation to the inner and outer expansions, where each approximation has an undetermined integration constant.

(c) Write the outer expansion in terms of inner variables and evaluate at $\bar{x} = 0$. Next, evaluate the inner expansion using inner variables as $\bar{x} \to \infty$. Determine the unknown constants by comparing the two expansions. Can you write a uniformly valid result that compares favorably with the exact solution in the $\varepsilon \to 0$ limit?

19.4 A problem in heat convection results in the following energy equation:

$$c\frac{d^2\theta}{ds^2} + \frac{d\theta}{ds} - \beta\theta = 0$$

$$\theta(0) = \theta_r, \quad \theta(1) = \theta_R$$

where

$$\theta = \frac{T - T_r}{T_r}, \quad s = 1 - \frac{x}{L}, \quad \varepsilon = \frac{1}{RePr}$$

and β is a nondimensional constant. Assume ε is small and that a boundary layer occurs along the wall which starts at $s = 0$.

(a) Determine the leading term of the outer expansion

$$\theta = \theta_o(s) + O(\varepsilon)$$

(b) For the inner expansion, introduce

$$\theta = \bar{\theta}_o(\eta) + O(\varepsilon), \quad \eta = \frac{s}{\varepsilon^m}$$

and determine and solve the equation for θ_o. In doing so, evaluate any unknown integration constants by matching and write the solution in terms of s.

(c) To determine the overlap, $(\theta_o)_{inner}$, write θ_o in terms of inner variables and set $\varepsilon = 0$. A composite is then formed from

$$\theta_{c,o}(s) = \theta_o(s) + \bar{\theta}_o(s) - [\theta_o(s)]_{inner}$$

Does this composite satisfy both boundary conditions?

20 Incompressible Boundary-Layer Theory

20.1 PRELIMINARY REMARKS

The first step of the procedure outlined in Section 19.2 has been accomplished, i.e., writing the conservation equations in body-oriented coordinates. The Euler equations are readily obtained from Appendix H as

$$\frac{\partial}{\partial s}(r^\sigma u) + \frac{\partial}{\partial n}(h_s r^\sigma v) = 0$$

$$h_s \frac{Du}{Dt} + k\,uv + \frac{\partial}{\partial s}\left(\frac{p}{\rho}\right) = 0$$

$$\frac{Dv}{Dt} - \frac{ku^2}{h_s} + \frac{\partial}{\partial n}\left(\frac{p}{\rho}\right) = 0$$

where

$$\frac{D}{Dt} = \frac{u}{h_s}\frac{\partial}{\partial s} + v\frac{\partial}{\partial n}$$

Chapter 17 shows there is no need to include the energy equation, which is decoupled from the others. Furthermore, there is no need to assume either a thermal or caloric state equation; i.e., the fluid may be a liquid or an imperfect gas. (Problem 20.2 derives a consistent form for the energy equation.) As Chapter 17 further indicates, a single, third-order PDE is obtained by introducing a stream function, to satisfy continuity, after which the pressure is eliminated from the momentum equations by cross-differentiation. The solution of the Euler equations, when evaluated at the wall, will be denoted with a subscript e. In addition, the e means the edge of the thin boundary layer. Uniform freestream conditions are denoted with an infinity subscript.

Suppose a solution to the Euler equations is known. With uniform freestream conditions, the solution is irrotational and thus satisfies Bernoulli's equation

$$p + \frac{1}{2}\rho(u^2 + v^2) = p_\infty + \frac{1}{2}\rho U_\infty^2$$

On the wall, we have

$$p_e = p_\infty + \frac{1}{2}\rho U_\infty^2 - \frac{1}{2}\rho u_e^2$$

since $v_e = 0$. The pressure gradient along the wall is thus given by

$$\left(\frac{\partial p}{\partial s}\right)_e = -\rho\left(u\frac{\partial u}{\partial s}\right)_e \tag{20.1}$$

where u_e stems from a solution of the Euler equations. As we shall see, the only information needed from the Euler equations is $u_e(s)$.

In the next two sections, we derive and solve the incompressible first-order, boundary-layer equations.

20.2 PRIMITIVE VARIABLE FORMULATION

Normally, μ equals $\mu(T)$ whether the fluid is a gas or a liquid. However, in a constant density flow the changes in temperature are generally quite small and the assumption of a constant μ value is appropriate. The situation here is different from the Couette flow of Chapter 16, where the density is constant along streamlines but can vary substantially across streamlines.

A primitive variable formulation is utilized where these are the original dependent variables in the conservation equations. Thus, p, u, and v are primitive variables, whereas a stream function is not. We continue to follow the procedure in Section 19.2 by setting

$$\mu = \mu_r = \text{constant}, \qquad n = \mu^{1/2}\bar{n}, \qquad v = \mu^{1/2}\bar{v} \tag{20.2}$$

We thus obtain from Appendix H

$$h_s = 1 + \mu^{1/2}k\bar{n}$$

$$r = r_w + \mu^{1/2}\bar{n}\cos\theta$$

$$\frac{\partial}{\partial n} = \mu^{-1/2}\frac{\partial}{\partial \bar{n}}$$

$$\frac{D}{Dt} = \frac{u}{h_s}\frac{\partial}{\partial s} + \bar{v}\frac{\partial}{\partial \bar{n}}$$

$$F_s^s = \frac{2}{h_s}\left[\frac{\mu}{h_s}\frac{\partial}{\partial s}\left(\frac{\partial u}{\partial s} + \mu^{1/2}k\bar{v}\right) + \frac{\mu^{1/2}}{2}\frac{\partial}{\partial \bar{n}}\left(\frac{h_s}{\mu^{1/2}}\frac{\partial u}{\partial \bar{n}} - ku + \mu^{1/2}\frac{\partial \bar{v}}{\partial s}\right) + \cdots\right]$$

$$F_n^s = \frac{\mu}{h_s^2}\frac{\partial}{\partial s}\left(\frac{h_s}{\mu^{1/2}}\frac{\partial u}{\partial \bar{n}} - ku + \mu^{1/2}\frac{\partial \bar{v}}{\partial s}\right) + 2\mu^{1/2}\frac{\partial^2 \bar{v}}{\partial \bar{n}^2} + \cdots$$

Next, we retain only the leading μ term in each relation:

$$h_s = 1 + O(\mu^{1/2})$$

$$F_s^s = \frac{\partial^2 u}{\partial \bar{n}^2} + O(\mu^{1/2})$$

$$F_n^s = \mu^{1/2}\left(\frac{\partial^2 u}{\partial s\partial \bar{n}} + 2\frac{\partial^2 \bar{v}}{\partial \bar{n}^2} + \frac{\sigma}{r_w}\frac{dr_w}{ds}\frac{\partial u}{\partial \bar{n}}\right) + O(\mu)$$

where the $\partial(\ )/\partial n$ relation and the substantial derivative are unaltered, except for setting $h_s = 1$.

The equations for continuity and momentum now become

$$\frac{\partial}{\partial s}(r_w^\sigma u) + \frac{\partial}{\partial \bar{n}}(r_w^\sigma \bar{v}) + \sigma O(\mu^{1/2}) = 0$$

$$\rho\left(u\frac{\partial u}{\partial s} + \bar{v}\frac{\partial u}{\partial \bar{n}}\right) = -\frac{\partial p}{\partial s} + \frac{\partial^2 u}{\partial \bar{n}^2} + O(\mu^{1/2})$$

$$\rho\mu\left(u\frac{\partial \bar{v}}{\partial s} + \bar{v}\frac{\partial \bar{v}}{\partial \bar{n}}\right) - \rho\mu^{1/2} k u^2 = -\frac{\partial p}{\partial \bar{n}} + \mu\left(\frac{\partial^2 u}{\partial s \partial \bar{n}} + 2\frac{\partial^2 \bar{v}}{\partial \bar{n}^2} + \frac{\sigma}{r_w}\frac{dr_w}{ds}\frac{\partial u}{\partial \bar{n}}\right) + O(\mu^{3/2})$$

We set $\mu = 0$, to obtain

$$\frac{\partial}{\partial s}(r_w^\sigma u) + \frac{\partial}{\partial \bar{n}}(r_w^\sigma \bar{v}) = 0 \qquad \textbf{(20.3a)}$$

$$\rho\left(u\frac{\partial u}{\partial s} + \bar{v}\frac{\partial u}{\partial \bar{n}}\right) = -\frac{dp}{ds} + \frac{\partial^2 u}{\partial \bar{n}^2} \qquad \textbf{(20.3b)}$$

$$\frac{\partial p}{\partial \bar{n}} = 0 \qquad \textbf{(20.3c)}$$

In view of the above equations, p can depend only on s, as already indicated in Equation (20.3b). The pressure is thus constant across the boundary layer, and dp/ds is given by Equation (20.1). This identification is reasonable, since Equations (20.3a,b) can determine only two variables, u and $\bar{v}$. In other words, the inviscid flow imposes its wall pressure gradient on the boundary layer.

At this point it is useful to return to the original n and v variables, with the result

$$\frac{\partial}{\partial s}(r_w^\sigma u) + \frac{\partial}{\partial n}(r_w^\sigma v) = 0 \qquad \textbf{(20.4a)}$$

$$\rho\left(u\frac{\partial u}{\partial s} + v\frac{\partial u}{\partial n}\right) = -\frac{dp_e}{ds} + \mu\frac{\partial^2 u}{\partial n^2} \qquad \textbf{(20.4b)}$$

These are the boundary-layer equations in primitive variable form. As noted in Section 19.2, they depend on μ (or a Reynolds number), in contrast to Equations (20.3), which do not explicitly depend on μ. They represent a generalization of Equations (18.6) and (18.7) by allowing for a pressure gradient, through dp_e/ds, and an axisymmetric flow, through r_w^σ. While they are simpler than their Navier–Stokes predecessors, they are still two coupled, nonlinear PDEs. Their solution is far from trivial, although, of course, they can be numerically solved with the aid of a computer.

We follow tradition, however, and introduce new dependent and independent variables in the next section. Our approach and nomenclature will be reminiscent of that used in Chapter 18. We thus obtain a single PDE whose form expedites the subsequent development of a similarity solution. This similarity solution is not a general one; it holds under rather restrictive conditions. Nevertheless, its importance cannot be overemphasized. For many decades, before computational fluid dynamics, similarity solutions were the cornerstone of boundary-layer theory.

20.3 SOLUTION OF THE BOUNDARY-LAYER EQUATIONS

To solve Equations (20.4), we introduce a stream function

$$u = \frac{1}{r_w^\sigma}\frac{\partial \psi}{\partial n}, \qquad v = -\frac{1}{r_w^\sigma}\frac{\partial \psi}{\partial s} \tag{20.5}$$

which satisfies continuity. We also introduce new independent variables

$$\xi(s) = \rho\mu\int_o^s u_e r_w^{2\sigma}\, ds, \qquad \eta(s,n) = \frac{\rho u_e r_w^\sigma n}{(2\xi)^{1/2}} \tag{20.6}$$

where ξ is a dimensional scaled wall length, with units of [(kg-m$^{\sigma-1}$)/s]^2, and η is a nondimensional normal coordinate. Here ξ is measured from the start of the boundary layer, which is either a sharp leading edge or a stagnation point. A typical example would be the boundary layer on an airfoil. In this case, ξ would be measured from the forward stagnation point on the airfoil. To determine ξ, an inviscid solution is required that provides u_e along the surface starting at the stagnation point and covering the surface to the point of interest. An inviscid solution of a global character is thus needed. Observe that ξ is proportional to μ and, thus, ξ typically has a small magnitude. Although $\xi^{-1/2}$ appears in the η equation, φ and η are independent variables. Simply view the ξ in the η equation as a function of s.

For a flat plate, with $\sigma = 0$ and $u_e = U_\infty$, Equations (20.6) reduce to

$$\xi = \rho\mu U_\infty s$$

$$\eta = \frac{\rho U_\infty n}{(2\rho\mu U_\infty s)^{1/2}} = \left(\frac{U_\infty}{2\nu s}\right)^{1/2} n = \left(\frac{U_\infty s}{2\nu}\right)^{1/2}\frac{n}{s}$$

Thus, η is proportional to the square root of a Reynolds number and is a generalization of the η in Equation (18.12).

To introduce ξ and η into the momentum equation will require a considerable analytical effort. This effort is expedited by first determining a number of partial derivative relations as follows:

$$\frac{\partial \eta}{\partial n} = \frac{\rho u_e r_w^\sigma}{(2\xi)^{1/2}}$$

$$\frac{\partial \eta}{\partial s} = \rho n\left[\frac{1}{(2\xi)^{1/2}}\frac{d(u_e r_w^\sigma)}{ds} - \frac{u_e r_w^\sigma n}{(2\xi)^{3/2}}\frac{d\xi}{ds}\right] = \eta\frac{d\ell n(u_e r_w^\sigma)}{ds} - \frac{\eta}{2\xi}\frac{d\xi}{ds}$$

$$\frac{\partial \xi}{\partial s} = \frac{d\xi}{ds} = \rho\mu u_e r_w^{2\sigma}$$

$$\frac{\partial}{\partial s} = \frac{\partial \xi}{\partial s}\frac{\partial}{\partial \xi} + \frac{\partial \eta}{\partial s}\frac{\partial}{\partial \eta} = \frac{d\xi}{ds}\left(\frac{\partial}{\partial \xi} - \frac{\eta}{2\xi}\frac{\partial}{\partial \eta}\right) + \frac{d\ell n(u_e r_w^\sigma)}{ds}\eta\frac{\partial}{\partial \eta}$$

$$\frac{\partial}{\partial n} = \frac{\partial \eta}{\partial n}\frac{\partial}{\partial \eta} = \frac{\rho u_e r_w^\sigma}{(2\xi)^{1/2}}\frac{\partial}{\partial \eta}$$

$$\frac{\partial^2}{\partial n^2} = \frac{(\rho u_e r_w^\sigma)^2}{2\xi}\frac{\partial^2}{\partial \eta^2} \tag{20.7}$$

Various functions of s, such as u_e and r_w, appear on the right sides. As we will see, there is no need to convert these quantities to ξ; such a conversion, in fact, is not algebraically feasible.

We next use Equations (20.5) and (20.7) to convert u and v to derivatives of the stream function. We thus have

$$u = \frac{\rho u_e}{(2\xi)^{1/2}} \psi_\eta \tag{20.8a}$$

$$v = -\frac{1}{r_w^\sigma}\left[\frac{d\xi}{ds}\left(\psi_\xi - \frac{\eta}{2\xi}\psi_\eta\right) + \frac{d\ell n(u_e r_w^\sigma)}{ds}\eta\psi_\eta\right] \tag{20.8b}$$

where

$$\psi_\xi = \frac{\partial \psi}{\partial \xi}, \qquad \psi_\eta = \frac{\partial \psi}{\partial \eta}$$

At this point, it is convenient and conventional to introduce a new nondimensional stream function $f(\xi, \eta)$ that depends on both ξ and η. This is done by means of

$$\frac{u}{u_e} = \frac{\partial f}{\partial \eta} = f_\eta$$

which is a generalization of Equation (18.14). We determine v in terms of f by first relating f to ψ as follows:

$$u = u_e \frac{\partial f}{\partial \eta} = \frac{\rho u_e}{(2\xi)^{1/2}} \frac{\partial \psi}{\partial \eta}$$

The rightmost equation can be integrated, to yield

$$\psi = \frac{(2\xi)^{1/2}}{\rho} f \tag{20.9}$$

where the function of integration is set equal to zero. With the aid of Equation (20.8b), we thereby obtain

$$v = -\frac{1}{\rho r_w^\sigma (2\xi)^{1/2}}\left[\frac{d\xi}{ds}(f + 2\xi f_\xi - \eta f_\eta) + 2\xi\eta\frac{d\ell n(u_e r_w^\sigma)}{ds} f_\eta\right] \tag{20.10}$$

for v, which represents a generalization of Equation (18.15).

With the above results, we now determine the various terms in Equation (20.4b) as follows:

$$u = u_e f_\eta$$

$$\frac{\partial u}{\partial s} = \frac{du_e}{ds} f_\eta + u_e \frac{d\xi}{ds}\left(f_{\xi\eta} - \frac{\eta}{2\xi} f_{\eta\eta}\right) + u_e \frac{d\ell n(u_e r_w^\sigma)}{ds} \eta f_{\eta\eta}$$

$$\frac{\partial u}{\partial n} = \frac{\rho u_e^2 r_w^\sigma}{(2\xi)^{1/2}} f_{\eta\eta}$$

$$\mu \frac{\partial^2 u}{\partial n^2} = \frac{\rho^2 \mu u_e^3 r_w^{2\sigma}}{2\xi} f_{\eta\eta\eta} = \frac{\rho u_e^2}{2\xi}\frac{d\xi}{ds} f_{\eta\eta\eta}$$

$$u\frac{\partial u}{\partial s} + v\frac{\partial u}{\partial n} = u_e f_\eta \left[\frac{du_e}{ds} f_\eta + u_e \frac{d\xi}{ds}\left(f_{\xi\eta} - \frac{\eta}{2\xi} f_{\eta\eta}\right) + u_e \frac{d\ell n\,(u_e r_w^\sigma)}{ds} \eta f_{\eta\eta}\right]$$

$$-\frac{\rho u_e^2 r_w^\sigma}{(2\xi)^{1/2}} f_{\eta\eta} \frac{1}{\rho r_w^\sigma (2\xi)^{1/2}}\left[\frac{d\xi}{ds}(f + 2\xi f_\xi - \eta f_\eta) + 2\xi\eta \frac{d\ell n(u_e r_w^\sigma)}{ds} f_\eta\right]$$

$$= u_e^2\left[\frac{1}{u_e}\frac{du_e}{ds} f_\eta^2 - \frac{1}{2\xi}\frac{d\xi}{ds} f f_{\eta\eta} + \frac{d\xi}{ds}(f_\eta f_{\xi\eta} - f_\xi f_{\eta\eta})\right]$$

With these relations, the momentum equation becomes

$$\rho u_e^2\left[\frac{1}{u_e}\frac{du_e}{ds} f_\eta^2 - \frac{1}{2\xi}\frac{d\xi}{ds} f f_{\eta\eta} + \frac{d\xi}{ds}(f_\eta f_{\xi\eta} - f_\xi f_{\eta\eta})\right] = \rho u_e \frac{du_e}{ds} + \frac{\rho u_e^2}{2\xi}\frac{d\xi}{ds} f_{\eta\eta\eta}$$

which simplifies to

$$f_{\eta\eta\eta} + f f_{\eta\eta} + \beta(1 - f_\eta^2) = 2\xi\,(f_\eta f_{\xi\eta} - f_\xi f_{\eta\eta}) \tag{20.11}$$

where

$$\beta(\xi) = \frac{2\xi}{u_e}\frac{\frac{du_e}{ds}}{\frac{d\xi}{ds}} = 2\frac{d\ell n u_e}{d\ell n \xi} = -\frac{2\left(\frac{dp}{ds}\right)_e}{\rho u_e^3 r_w^{2\sigma}}\int_0^s u_e r_w^{2\sigma}\, ds \tag{20.12}$$

Equation (20.11) is the conventional laminar boundary-layer equation for an incompressible flow. Observe that it is free of parameters such as σ, ρ, μ, r_w, k, θ, and u_e. Some of these parameters, however, affect β through ξ. The parameter β is called the pressure gradient parameter since it represents the effect of the pressure gradient on the boundary layer. It is fully determined by a solution of the Euler equations when evaluated at the wall and is of major importance in the theory. Observe that in a favorable pressure gradient, where $(dp/ds)_e$ is negative, β is positive. In this circumstance, the pressure gradient is accelerating the fluid in the boundary layer, which will not separate from a smooth wall. On the other hand, in an unfavorable pressure gradient, β is negative. The pressure gradient now retards the flow and the layer may separate from the wall. For a flow in which p_e is constant along streamlines, such as in a uniform or potential vortex flow, β is zero.

The boundary conditions on f are

$$f(\xi,0) = f_\eta(\xi,0) = 0, \qquad f_\eta(\xi,\infty) = 1$$

The first two conditions ensure that u and ψ are zero on the wall. (This justifies setting the function of integration equal to zero in Equation (20.9).) For v to be zero on the wall, we should have the above wall conditions along with

$$f_{\xi}(\xi,0) = 0$$

However, this relation is unnecessary since it is redundant with $f(\xi,0) = 0$.

An initial condition for f at some ξ value is also required. Typically, this is the stagnation point solution of Chapter 17. This solution is also an exact solution of the boundary-layer equations with β given by Equation (17.20). Thus, the airfoil boundary layer mentioned earlier would utilize $\beta = 1$ and the $f = f(\eta)$ solution from Chapter 17 at the stagnation point where ξ and s are zero. The special algorithms developed for the numerical integration of the boundary-layer equations are a subject in themselves that is discussed at the end of Chapter 21. Instead, we will concentrate on the possibility of a similarity solution.

Similarity Solution

By a similarity solution we mean a solution in which f depends only on η. Such a solution requires

$$f_{\xi} = 0, \qquad \frac{d\beta}{d\xi} = 0$$

which results in the equation (Falkner and Skan, 1931)

$$f''' + ff'' + \beta(1 - f'^2) = 0 \tag{20.13}$$

where $f' = df/d\eta$ and β is a constant. The boundary conditions are

$$f(0) = f'(0) = 0, \qquad f'(\infty) = 1 \tag{20.14}$$

Of course, an ordinary differential equation is much easier to solve, analytically or numerically, than a partial differential equation.

We previously encountered three special cases of the Falkner–Skan equation, namely, $\beta = 0$, 1/2, 1. These are outlined in Table 20.1, along with several other cases which stem from

$$\beta = \frac{d\ell n u_e^2}{d\ell n \xi} = \text{constant}$$

This yields an external flow speed given by

$$u_e(s) = Ks^{\beta/(2-\beta)} \tag{20.15}$$

where K is an arbitrary constant.

Accurate results from numerically solving the Falkner–Skan equation are shown in Table 20.2 (Bae and Emanuel, 1989). The skin friction and skin-friction coefficient are given by

$$\tau_w = \mu\left(\frac{\partial u}{\partial n}\right)_w = \frac{\rho\mu u_e^2 r_w^{\sigma}}{(2\xi)^{1/2}} f_w'' \tag{20.16a}$$

$$c_f = \frac{2\tau_w}{\rho u_e^2} = \mu r_w^{\sigma}\left(\frac{2}{\xi}\right)^{1/2} f_w'' \tag{20.16b}$$

TABLE 20.1
Special Cases of the Falkner–Skan Equation (White, 1991)

β	Type of Flow
–0.1988	boundary-layer separation
0	flat plate
0.5	axisymmetric stagnation point
1	two-dimensional stagnation point
$-2 \leq \beta \leq 0$	flow around an expansion corner of angle $\pi\beta/2$:
$0 \leq \beta \leq 2$	flow against a wedge of half-angle $\pi\beta/2$:
4	doublet flow near a plane wall
5	doublet flow near a 90° corner
∞	flow toward a point sink:

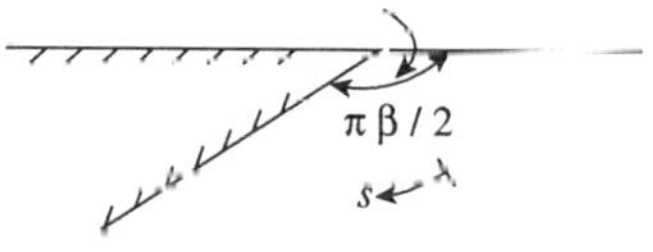

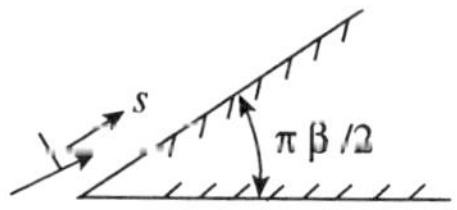

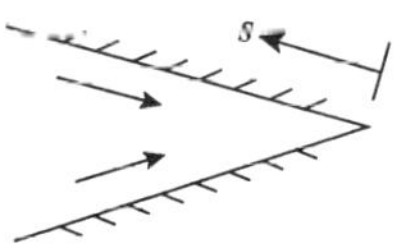

TABLE 20.2
Boundary-Layer Parameters vs. β for Incompressible Laminar Flow (Bae and Emanuel, 1989)

β	f_w''	C_v	η_{ev}
–0.1988	0	2.358	4.788
0	0.4696	1.217	3.472
0.5	0.9277	0.8047	2.750
1	1.233	0.6479	2.379
2	1.687	0.4975	1.948
5	2.616	0.3334	1.382
10	3.675	0.2408	1.022
15	4.492	0.1980	0.8474
20	5.181	0.1721	0.7398

where ξ is given by Equation (20.6), and the second column in the table lists f_w'. The third column in the table provides C_v, which is defined by Equation (18.19). The final column lists η_{ev}, which represents the boundary-layer edge of the velocity profile where $f'(\eta_{ev}) = 0.99$. The manner of utilization of the C_v and η_{ev} parameters is discussed in the next chapter, which treats the more general case of a compressible boundary layer.

The solution of the Falkner–Skan equation, subject to Equations (20.14), is unique when $\beta \geq 0$. For $\beta < 0$ the solution is not unique (Libby and Liu, 1968). An additional condition that f'' decreases as rapidly as possible when η increases is then arbitrarily imposed when $\beta_{sp} \leq \beta < 0$, where β_{sp} denotes the value of β where f_w'' first becomes zero. This value is referred to as the separation value, since the boundary layer separates from the wall when the skin friction first becomes zero.

Vorticity

We conclude this chapter with a brief discussion of boundary-layer vorticity. Appendices G and H are used to obtain

$$\vec{\omega} = \nabla \times \vec{w} = \frac{1}{1+kn}\left[\frac{\partial v}{\partial s} - (1+kn)\frac{\partial u}{\partial n} - ku\right]\hat{e}_\phi$$

for the vorticity in a two-dimensional or axisymmetric flow. A vortex filament is thus a circle centered about the symmetry axis in an axisymmetric flow. With the scaling provided by Equations (20.2), we have

$$\vec{\omega} = -\frac{\partial u}{\partial n}\hat{e}_\phi \tag{20.17}$$

to first order. Hence, the magnitude of the vorticity at the wall is

$$\omega_w = -\frac{\tau_w}{\mu}$$

It is of interest to examine how the vorticity changes for a fluid particle inside a boundary layer. For this, we determine $D\omega/Dt$. This relation is obtained by first multiplying Equation (20.4b) with $\partial(\;)/\partial n$ while assuming that μ is constant. We thereby obtain

$$\rho\left(\frac{\partial u}{\partial n}\frac{\partial u}{\partial s} + u\frac{\partial^2 u}{\partial s\,\partial n} + \frac{\partial v}{\partial n}\frac{\partial u}{\partial n} + v\frac{\partial^2 u}{\partial n^2}\right) = \mu\frac{\partial^3 u}{\partial n^3}$$

Observe that the left side can be written as

$$\rho\left(u\frac{\partial}{\partial s} + v\frac{\partial}{\partial n}\right)\frac{\partial u}{\partial n} + \rho\frac{\partial u}{\partial n}\left(\frac{\partial u}{\partial s} + \frac{\partial v}{\partial n}\right) = -\rho\frac{D\omega}{Dt} - \rho\omega\left(\frac{\partial u}{\partial s} + \frac{\partial v}{\partial n}\right)$$

where the first-order relation, Equation (20.17), is used for ω. To eliminate the divergence-like terms, Equation (20.4a) is written as

$$\frac{\partial u}{\partial s} + \frac{\partial v}{\partial n} = -\frac{\sigma}{r_w}\frac{dr_w}{ds}u$$

By combining these results, we have

$$\frac{D\omega}{Dt} = \nu\frac{\partial^2 \omega}{\partial n^2} + \frac{\sigma}{r_w}\frac{dr_w}{ds}u\omega \tag{20.18}$$

Hence, there is lateral diffusion of vorticity away from the wall with ν as the effective diffusion coefficient. In the rightmost term the factor, $r_w^{-1}(dr_w/ds)u$, stems from the continuity equation for an axisymmetric flow. This term accounts for any lateral stretching, or contraction, of the circular vortex filaments.

Part (a) of Problem 4.5 provides a general kinematic equation for $D\vec{\omega}/Dt$ that should be consistent with the one above. In an incompressible fluid with μ being constant, the equation in the problem reduces to

$$\frac{D\vec{\omega}}{Dt} = \nu\nabla^2\vec{\omega} + \vec{\omega}\cdot(\nabla\vec{w}) \tag{20.19}$$

where the rightmost term represents vortex stretching, which can occur in a three dimensional or axisymmetric flow. In a two-dimensional or axisymmetric flow, with

$$\vec{w} = u\hat{e}_s + v\hat{e}_n$$

$$\vec{\omega} = \omega\hat{e}_\phi$$

$$\nabla = \hat{e}_s\frac{\partial}{\partial s} + \hat{e}_n\frac{\partial}{\partial n} + \frac{\hat{e}_\phi}{r_w^\sigma}\frac{\partial}{\partial\phi}$$

this term becomes

$$\vec{\omega}\cdot(\nabla\vec{w}) = \sigma(u\sin\theta + v\cos\theta)\frac{\omega}{r_w}\hat{e}_\phi$$

where various vector relations in part B of Appendix H are utilized. In the boundary layer large Reynolds number limit, we obtain

$$\nabla^2\vec{\omega} = \frac{\partial^2\omega}{\partial n^2}\hat{e}_\phi$$

and

$$\vec{\omega}\cdot(\nabla\vec{w}) = \frac{\sigma}{r_w}\frac{dr_w}{ds}u\omega\hat{e}_\phi$$

since $\sin\theta = dr_w/ds$. In this circumstance, Equation (20.19) reduces to Equation (20.18) as anticipated.

REFERENCES

Bae, Y.-Y. and Emanuel, G., "Boundary-Layer Tables for Similar Compressible Flow," *AIAA J.* 27, 1163 (1989).

Falkner, V.M. and Skan, S.W., "Some Approximate Solutions of the Boundary Layer Equations," *Philos. Mag.* 12, 865 (1931).

Libby, P.A. and Liu, T.M., "Some Similar Laminar Flows Obtained by Quasilinearization," *AIAA J.* 6, 1541 (1968).

White, F.M., *Viscous Fluid Flow,* 2nd ed., McGraw-Hill, New York, 1991.

PROBLEMS

20.1 Equation (17.8) provides the stream function for inviscid, incompressible flow about a circular cylinder of radius R. Use this result to obtain $\beta(s)$, where the arc length s is measured along the cylinder's surface from the forward stagnation point. Sketch β vs. s/R for the upper half of the cylinder. Interpret your results for β at $s/R = 0$, $\pi/2$, and π.

20.2 **(a)** Assume κ is a small constant, comparable to μ in magnitude, and develop a form of the energy equation that is consistent with Equation (20.4).

(b) Assume $h = c_p T$, with c_p being constant, and introduce Equations (20.5) and (20.6). Determine the energy equation that is equivalent to Equation (20.11) using

$$\theta = \frac{T - T_w}{T_e - T_w}$$

where T_e and T_w may be functions of s.

(c) What are the conditions for θ to have a similarity solution?

20.3 Consider incompressible flow about a symmetric wedge (see Equation (20.15) and Table 20.1) with $0 \le \beta \le 2$ and with $\beta = -0.1988$ included for convenience.

(a) Determine an equation for the integrated skin friction from the tip of the wedge to a point s on the upper surface, where s is nondimensional and less than unity.

(b) Evaluate the nondimensional integrated skin friction

$$\frac{1}{(\rho\mu K^3)^{1/2}} \int_0^s \tau_w \, ds$$

when $\beta = -0.1988$, 0, 0.5, 1, and 2. Explain your $\beta = -0.1988$ and 2 results.

20.4 An incompressible, axisymmetric flow has

$$u_e = u_{e1} \ell n(s + s_1), \qquad r_w = r_1 s^{1/2}$$

where u_{e1}, s_1, and r_1 are positive constants and s is nondimensional.

(a) Determine equations for $\xi(s)$, $\eta(s, n)$ and $\beta(s)$. Simplify your results.

(b) With $s = s_1 = 1.04633$, determine $\beta, \xi/(\rho\mu u_{e1} r_1^2)$, and $(\rho u_{e1}/\mu)^{1/2} c_f$.

20.5 For the Falkner–Skan equation

$$f''' + ff'' + \beta(1 - f'^2) = 0$$

determine the two leading, nonzero terms of the Taylor series expansion

$$f = a_0 + a_1\eta + a_2\eta^2 + \cdots \tag{1}$$

that holds near the wall but does not hold for large η. Equation (1) should satisfy

$$f_w = 0, \qquad f'_w = 0, \qquad f''_w \neq 0$$

and the coefficients of the two leading terms should only depend on β and f''_w.

20.6 **(a)** For an incompressible, viscous fluid, use the results of Problem 20.5 to determine an approximate solution, valid near a stagnation point, for w^2. Retain only the leading term, in η, for u and the leading term for v. Your final result should be in terms of

$$\frac{w}{U_\infty} = W, \qquad \frac{x}{R} = X, \qquad \frac{y}{R} = Y, \qquad Re_\infty = \frac{RU_\infty}{\nu}$$

(b) Evaluate the total pressure

$$p_t = p + \frac{1}{2}\rho w^2$$

in the vicinity of the stagnation point. Use nondimensional variables, including

$$c_{p_t} = \frac{2(p_t - p_0)}{\rho U_\infty^2}$$

where p_0 is the pressure at the stagnation point.

21 Compressible Boundary-Layer Theory

21.1 PRELIMINARY REMARKS

As in the preceding chapter, a steady, large Reynolds number laminar flow about a two-dimensional or axisymmetric body is considered. From the outset, we assume the fluid is a perfect gas and, on the streamline or stream surface that wets the body, the inviscid flow is isentropic and has a constant stagnation enthalpy.

First-order boundary-layer theory is again the topic of this chapter. For this theory, only the inviscid solution for the stream surface that wets the body is required. Second-order theory would account for the possibility of gradients of stagnation enthalpy and entropy that are transverse to the body. The higher-order theory also accounts for other effects, such as body curvature.

As a consequence of the high Reynolds number ordering of effects, the first-order theory is applicable under a wide range of conditions, including inviscid flows that are rotational and may not be homentropic or isoenergetic. For example, a flow generated by a curved shock wave, which is rotational and not homentropic, is encompassed within the first-order theory. In other words, the effect on the skin friction or heat transfer of a rotational flow, external to the boundary layer, is a second-order correction. As evident from Chapter 20, the dominant effect considered by the first-order theory is the pressure gradient along the wall. To explicitly account for second-order effects, we must resort to the theory in Chapter 23 or a numerical solution of the Navier–Stokes equations.

Most boundary-layer flows are transitional or turbulent, except near their origin where they are laminar. Since the 1950s, however, high-speed flight at an elevated altitude has become a major concern of engineers. Under these conditions the boundary layer remains laminar much longer (Sternberg, 1952). Larger portions of the structure are thus covered by a laminar boundary layer because of surface cooling, boundary-layer acceleration, and a relatively low Reynolds number associated with the low density of the upper atmosphere. For instance, all three factors are present during atmospheric reentry of a missile or spacecraft.

The subsequent theory will utilize the $Re \to \infty$ limit. In a real flow, when a Reynolds number based on wall length becomes sufficiently large, the boundary layer becomes transitional. This effect is ignored since transition does not start until the Reynolds number is of $O(10^6)$ or higher.

The first-order boundary-layer equations for a compressible flow are derived in the next section. A transformation is performed in Section 21.3 that resembles the one in Section 20.3. To simplify the presentation, we assume from the start a similarity form for the solution. A series of sections then follows that focuses on the similarity equations and the concept of local similarity. In the last of these sections a critique of this approach is discussed. Overall, the importance of these concepts justifies the emphasis given to the similarity form of the boundary-layer equations. Analytical results are more readily obtained with this approach than with any other. As a consequence, Sections 21.7 to 21.10 provide a simple and quick method for obtaining values for the skin friction, heat transfer, and a variety of boundary-layer thicknesses. Such values are particularly useful when rough estimates or dominant trends are desired. The last two sections discuss (numerical) solutions to the nonsimilar equations.

Because of the number of complex equations in this chapter, Appendix I summarizes this material, except for that in the final two sections, which discuss nonsimilar flows. Chapter 22 discusses a variety of applications of the similarity theory.

Wall Pressure Gradient

We conclude our remarks by obtaining the pressure gradient along the wall. As before, stagnation conditions are denoted by a zero subscript, whereas an e subscript denotes the outer edge of the boundary layer. In first-order theory, this later condition is provided by a solution of the Euler equations when evaluated at the surface of the body. Thus, although this solution may not be homenergetic or homentropic, the inviscid stream surface or streamline that wets the body has constant values for the stagnation enthalpy and the entropy. Hence, for this stream surface, we have

$$h_{oe} = \frac{\gamma R}{\gamma - 1} T_e + \frac{1}{2} u_e^2 = \text{constant} \tag{21.1}$$

along with the isentropic relation

$$\frac{T_e}{T_{oe}} = \left(\frac{p_e}{p_{oe}}\right)^{(\gamma-1)/\gamma} \tag{21.2}$$

and the thermal state equation

$$p_e = \rho_e R T_e \tag{21.3}$$

Equation (21.2) is differentiated, with the result

$$\frac{1}{T_{oe}} \frac{dT_e}{ds} = \frac{\gamma-1}{\gamma} \frac{1}{p_{oe}} \left(\frac{p_e}{p_{oe}}\right)^{[(\gamma-1)/\gamma]-1} \frac{dp_e}{ds} = \frac{\gamma-1}{\gamma} \frac{1}{p_{oe}} \frac{p_{oe}}{p_e} \frac{T_e}{T_{oe}} \frac{dp_e}{ds}$$

or

$$\frac{dT_e}{ds} = \frac{\gamma-1}{\gamma} \frac{T_e}{p_e} \frac{dp_e}{ds} = \frac{\gamma-1}{\gamma R} \frac{1}{\rho_e} \frac{dp_e}{ds}$$

This is combined with the differential form of Equation (21.1) to yield

$$\frac{dp_e}{ds} = -\rho_e u_e \frac{du_e}{ds} \tag{21.4}$$

for the pressure gradient along the surface in the streamwise direction.

21.2 BOUNDARY-LAYER EQUATIONS

As in Section 20.2, we utilize Appendix H, to obtain

$$h_s = 1$$

$$h_\phi = [r_w(s)]^\sigma$$

$$\frac{D}{Dt} = u\frac{\partial}{\partial s} + v\frac{\partial}{\partial n}$$

$$F_s^s = \frac{\partial}{\partial n}\left(\mu \frac{\partial u}{\partial n}\right)$$

$$\Phi = \mu\left(\frac{\partial u}{\partial n}\right)^2$$

$$\kappa \nabla^2 T + \frac{\partial \kappa}{\partial n}\frac{\partial T}{\partial n} + \frac{\partial \kappa}{\partial s}\frac{\partial T}{\partial s} = \frac{\partial}{\partial n}\left(\kappa \frac{\partial T}{\partial n}\right)$$

at a large Reynolds number, where higher-order terms have been omitted. Hence, the first-order boundary-layer equations in primitive variable form are

$$\frac{\partial}{\partial s}(r_w^\sigma \rho u) + \frac{\partial}{\partial n}(r_w^\sigma \rho v) = 0 \tag{21.5}$$

$$\rho \frac{Du}{Dt} = -\frac{dp_e}{ds} + \frac{\partial}{\partial n}\left(\mu \frac{\partial u}{\partial n}\right) \tag{21.6}$$

$$\rho \frac{Dh}{Dt} = u\frac{dp_e}{ds} + \frac{\partial}{\partial n}\left(\kappa \frac{\partial T}{\partial n}\right) + \mu\left(\frac{\partial u}{\partial n}\right)^2 \tag{21.7}$$

Observe that the terms containing the second viscosity coefficient, λ, are of higher order and do not contribute to either the momentum or energy equations. Furthermore, use has been made of the result that $\partial p/\partial n = 0$. Finally, note that the energy equation still has terms for convection, pressure work, transverse heat conduction, and viscous dissipation.

Before introducing new independent variables, we transform Equation (21.7) into a simpler form. For this, we introduce $h = c_p T$ and the stagnation enthalpy

$$h_o = h + \frac{1}{2}u^2$$

which does not include the higher order $v^2/2$ term and, unlike h_{oe}, h_o is not constant. The energy equation thus becomes

$$\rho \frac{Dh_o}{Dt} - \rho u \frac{Du}{Dt} = u\frac{dp_e}{ds} + \frac{\partial}{\partial n}\left[\frac{\kappa}{c_p}\left(\frac{\partial h_o}{\partial n} - u\frac{\partial u}{\partial n}\right)\right] + \mu\left(\frac{\partial u}{\partial n}\right)^2$$

We now introduce the Prandtl number and eliminate the Du/Dt term by means of the momentum equation, with the result

$$\rho \frac{Dh_o}{Dt} = \frac{\partial}{\partial n}\left(\frac{\mu}{Pr}\frac{\partial h_o}{\partial n}\right) - \frac{\partial}{\partial n}\left(\frac{\mu}{Pr}u\frac{\partial u}{\partial n}\right) + \frac{\partial}{\partial n}\left(\mu u\frac{\partial u}{\partial n}\right)$$

where

$$u\frac{\partial}{\partial n}\left(\mu \frac{\partial u}{\partial n}\right) + \mu\left(\frac{\partial u}{\partial n}\right)^2 = \frac{\partial}{\partial n}\left(\mu u \frac{\partial u}{\partial n}\right)$$

We thus obtain for the energy equation

$$\rho \frac{Dh_o}{Dt} = \frac{\partial}{\partial n}\left\{\mu\left[\frac{1}{Pr}\frac{\partial h_o}{\partial n} + \left(1 - \frac{1}{Pr}\right)u\frac{\partial u}{\partial n}\right]\right\} \tag{21.8}$$

Transformation of the Independent Variables

New independent variables are utilized by means of the transformation

$$\xi(s) = \int_0^s (\rho\mu u)_e r_w^{2\sigma}\, ds \tag{21.9}$$

$$\eta(s, n) = \frac{r_w^\sigma \rho_e u_e}{(2\xi)^{1/2}} \int_0^n \frac{\rho}{\rho_e}\, dn \tag{21.10}$$

This transformation was developed over a number of years by Dorodnitsyn, Mangler, Howarth, Stewartson, Levy, Lees, and possibly others. Its attribution varies from author to author. It is designed to reduce the axisymmetric case to the two-dimensional case and to remove compressibility effects. As we shall see, the final version of the equations is independent of σ and the density, and one of the equations resembles Equation (20.3). Aside from the foregoing important reasons, the transformation resembles Equations (20.6) and results in a generalization of the Falkner–Skan equation. As in Section 20.3, ξ is dimensional, whereas η is nondimensional.

In the subsequent analysis we will need the following derivatives:

$$\frac{d\xi}{ds} = (\rho\mu u)_e r_w^{2\sigma}$$

$$\frac{\partial\eta}{\partial n} = \frac{r_w^\sigma \rho u_e}{(2\xi)^{1/2}}$$

$$\frac{\partial}{\partial n} = \frac{\partial\xi}{\partial n}\frac{\partial}{\partial\xi} + \frac{\partial\eta}{\partial n}\frac{\partial}{\partial\eta} = \frac{r_w^\sigma \rho u_e}{(2\xi)^{1/2}}\frac{\partial}{\partial\eta}$$

$$\frac{\partial}{\partial s} = \frac{\partial\xi}{\partial s}\frac{\partial}{\partial\xi} + \frac{\partial\eta}{\partial s}\frac{\partial}{\partial\eta}$$

Observe that ρ in $\partial\eta/\partial n$ has no e subscript and $\partial\eta/\partial s$ is not listed. At this time, the $\partial\eta/\partial s$ derivative cannot be evaluated, since ρ/ρ_e in the integrand of Equation (21.10) is an unknown function of s and n. As evident by the second of Equations (20.7), this difficulty does not occur for an incompressible flow. (Problem 21.9 provides $\partial\eta/\partial s$ when the flow is compressible.) Fortunately, $\partial\eta/\partial s$ cancels in the subsequent analysis; its evaluation is therefore unnecessary.

Transformation of the Equations and Boundary Conditions under a Similarity Assumption

The introduction of φ, η, and h_o have paved the way for assuming a similarity solution of the form

$$\frac{u}{u_e} = \frac{df}{d\eta} = f'(\eta) \tag{21.11}$$

$$\frac{h_o}{h_{oe}} = g(\eta) \tag{21.12}$$

where f and g depend only on η and h_{oe} is a constant. (The nonsimilar equations are given in Section 21.6.) A compressible flow stream function is also introduced as

$$\frac{\partial \psi}{\partial s} = -r_w^\sigma \rho v, \qquad \frac{\partial \psi}{\partial n} = r_w^\sigma \rho u \tag{21.13}$$

that satisfies the continuity equation. This stream function is a provisional variable; it will not appear in the final formulation.

We must relate ψ to f in order to obtain a relation for v. In turn, v is needed in order to transform the substantial derivative to the new variables. The ψ, f relation stems from

$$u = u_e f' = \frac{1}{r_w^\sigma \rho}\frac{\partial \psi}{\partial n} = \frac{u_e}{(2\xi)^{1/2}}\frac{\partial \psi}{\partial \eta}$$

or

$$\frac{\partial \psi}{\partial \eta} = (2\xi)^{1/2}\frac{df}{d\eta}$$

This integrates to

$$\psi = (2\xi)^{1/2} f \tag{21.14}$$

where the function of integration is set equal to zero, thereby allowing $\psi = 0$ on the body or along an axis of symmetry. Thus, v is given by

$$v = -\frac{1}{r_w^\sigma \rho}\frac{\partial \psi}{\partial s} = -\frac{1}{r_w^\sigma \rho}\left[\frac{d\xi}{ds}\frac{\partial \psi}{\partial \xi} + \frac{\partial \eta}{\partial s}\frac{\partial \psi}{\partial \eta}\right]$$

or

$$v = -\frac{1}{r_w^\sigma \rho}\left[\frac{d\xi/ds}{(2\xi)^{1/2}} f + (2\xi)^{1/2}\frac{\partial \eta}{\partial s} f'\right] \tag{21.15}$$

The substantial derivative now becomes

$$\begin{aligned}\frac{D}{Dt} &= u_e f'\left(\frac{\partial \eta}{\partial s}\frac{\partial}{\partial \eta} + \frac{d\xi}{ds}\frac{\partial}{\partial \xi}\right) - \frac{1}{r_w^\sigma \rho}\left[\frac{d\xi/ds}{(2\xi)^{1/2}} f + (2\xi)^{1/2}\frac{\partial \eta}{\partial s} f'\right]\frac{r_w^\sigma \rho u_e}{(2\xi)^{1/2}}\frac{\partial}{\partial \eta} \\ &= u_e \frac{d\xi}{ds}\left(f'\frac{\partial}{\partial \xi} - \frac{f}{2\xi}\frac{\partial}{\partial \eta}\right)\end{aligned} \tag{21.16}$$

Since v originally only appears in $D(\ )/Dt$, we see that $\partial \eta/\partial s$ is unnecessary.

With the foregoing, Equation (21.6) becomes

$$\rho u_e \frac{d\xi}{ds}\left(\frac{du_e}{d\xi} f'^2 - \frac{u_e}{2\xi} f f''\right) = \rho_e u_e \frac{d\xi}{ds}\frac{du_e}{d\xi} + \frac{r_w^\sigma \rho u_e}{(2\xi)^{1/2}}\frac{\partial}{\partial \eta}\left[\frac{\mu r_w^\sigma \rho u_e^2}{(2\xi)^{1/2}} f''\right]$$

which simplifies to

$$\frac{du_e}{d\xi}f'^2 - \frac{u_e}{2\xi}ff'' = \frac{\rho_e}{\rho}\frac{du_e}{d\xi} + \frac{u_e}{2\xi\rho_e\mu_e}\frac{\partial}{\partial\eta}(\rho\mu f'')$$

or

$$\left(\frac{\rho\mu}{\rho_e\mu_e}f''\right)' + ff'' + \frac{2\xi}{u_e}\frac{du_e}{d\xi}\left(\frac{\rho_e}{\rho} - f'^2\right) = 0$$

To eliminate the ρ_e/ρ factor, we use

$$\frac{\rho_e}{\rho} = \frac{p_e T}{pT_e} = \frac{T}{T_e} = \frac{h}{h_e} = \frac{h_o - \frac{1}{2}u^2}{h_{oe} - \frac{1}{2}u_e^2} = \frac{h_{oe}g - \frac{1}{2}u_e^2 f'^2}{h_{oe} - \frac{1}{2}u_e^2} \quad \textbf{(21.17)}$$

to obtain

$$\left(\frac{\rho\mu}{\rho_e\mu_e}f''\right)' + ff'' + \beta(g - f'^2) = 0 \quad \textbf{(21.18)}$$

The pressure gradient parameter is given by

$$\beta = \frac{2\xi}{u_e}\frac{du_e}{d\xi}\frac{T_{oe}}{T_e} = \frac{T_{oe}}{T_e}\frac{d\ell n u_e^2}{d\ell n \xi} \quad \textbf{(21.19)}$$

since

$$\frac{h_{oe}}{h_{oe} - \frac{1}{2}u_e^2} = \frac{T_{oe}}{T_e}$$

Aside from the T_{oe}/T_e factor, this definition is compatible with Equation (20.12). In order for the similarity assumption to hold, we must have

$$\beta = \text{constant} \quad \textbf{(21.20)}$$

and

$$\frac{\rho\mu}{\rho_e\mu_e} = C(\eta) \quad \textbf{(21.21)}$$

When C is assumed to be a constant, it is referred to as the Chapman–Rubesin parameter (Chapman and Rubesin, 1949).

Equation (21.21) yields $C = 1$ if

$$\mu \sim T$$

since $\rho \sim T^{-1}$. We thereby obtain

$$f''' + ff'' + \beta(g - f'^2) = 0 \tag{21.22}$$

in place of Equation (21.18). The transformed version of Equation (21.8), for the energy, is

$$\rho u_e \frac{d\xi}{ds}\left(0 - \frac{f}{2\xi}h_{oe}g'\right) = \frac{r_w^\sigma \rho u_e}{(2\xi)^{1/2}}\frac{\partial}{\partial \eta}\left\{\mu\left[\frac{h_{oe}}{Pr}\frac{r_w^\sigma \rho u_e}{(2\xi)^{1/2}}g' + \left(1 - \frac{1}{Pr}\right)u_e f' u_e \frac{r_w^\sigma \rho u_e}{(2\xi)^{1/2}}f''\right]\right\}$$

which simplifies to

$$\frac{\partial}{\partial \eta}\left\{\frac{\rho\mu}{\rho_e\mu_e}\left[\frac{1}{Pr}g' + \left(1 - \frac{1}{Pr}\right)\frac{u_e^2}{h_{oe}}f'f''\right]\right\} + fg' = 0 \tag{21.23}$$

We previously required the Chapman–Rubesin parameter to be unity; hence, the energy equation becomes

$$\left(\frac{1}{Pr}g'\right)' + fg' + \left[\left(1 - \frac{1}{Pr}\right)\frac{u_e^2}{h_{oe}}f'f''\right]' = 0 \tag{21.24}$$

This highly transformed version of the boundary-layer energy equation still includes the effects of convection, work, conduction, and viscous dissipation. None of these processes has been discarded. Although a thermally and calorically perfect gas and $\mu \sim T$ are assumed, the Prandtl number need not be constant. It becomes a constant only if $\kappa \sim T$. If Pr is not constant, it must be a function only of η. In this case, a similarity solution is possible only if u_e is a constant.

Subsonic flows where u_e is constant are limited to those above a flat plate at zero incidence, uniform flows parallel to a cylinder, and to cylindrically symmetric flows, such as a vortex flow, in which the streamlines are circles. The number of supersonic flows is somewhat larger. In addition to the foregoing cases, these include the flow behind a planar oblique shock wave, the flow along a planar wall downstream of a Prandtl–Meyer expansion or compression, and the flow along a cone at zero incidence with an attached shock wave. These supersonic flows need hold only locally. For instance, a hypersonic vehicle may have several different regions where u_e is constant. An additional case occurs in the hypersonic limit on a slender body where u_e is essentially a constant.

Alternatively, if $u_e(\xi)$ is not constant, then similarity requires that $Pr = 1$. Hereafter, we will concentrate on this situation for which the above equation reduces to

$$g'' + fg' = 0 \tag{21.25}$$

It is important to note that setting $Pr = 1$ does not mean we are ignoring any process, such as viscous dissipation. Instead, various processes combine to partially cancel each other. One of these processes is the viscous dissipation, which constitutes part of the $u\partial u/\partial n$ term in Equation (21.8). For air, $Pr \cong 0.7$, and viscous dissipation is thus somewhat underestimated by using $Pr = 1$.

It is often convenient to replace g with

$$G = \frac{g - g_w}{1 - g_w} = \frac{h_o - h_{ow}}{h_{oe} - h_{ow}} = \frac{T_o - T_w}{T_{oe} - T_w} \tag{21.26}$$

where

$$h_{ow} = c_p T_w$$

From Equation (21.12), the g_w parameter is

$$g_w = \frac{h_{ow}}{h_{oe}} = \frac{T_w}{T_{oe}} \tag{21.27}$$

The final form is

$$f''' + ff'' + \beta[g_w + (1 - g_w)G - f'^2] = 0 \tag{21.28}$$

$$G'' + fG' = 0 \tag{21.29}$$

for the similar, compressible, boundary-layer equations. The boundary conditions for these equations are

$$f(0) = f_w = 0, \qquad f'(0) = f'_w = 0, \qquad G(0) = G_w = 0,$$
$$f'(\infty) = 1, \qquad G(\infty) = 1 \tag{21.30}$$

The introduction of G has simplified the temperature boundary condition at the wall. Moreover, G and f' now have the same wall and infinity conditions. (We utilize this fact in the next section.) Equations (21.28)–(21.30) provide the basis for much of the discussion in the rest of this chapter. All variables and the two constant parameters are nondimensional. Observe that there is no direct dependence on γ, M_e, σ, ρ/ρ_e, or a Reynolds number.

21.3 SOLUTION OF THE SIMILARITY EQUATIONS

Solutions of Equations (21.28) through (21.30) are shown in Figures 21.1 and 21.2 for a few β and g_w values.* It is conventional to refer to these curves as showing the velocity, f', and stagnation enthalpy, G, profiles in the η, or transformed, plane. (Comparable solutions in the physical plane are presented in Section 21.7.) When $\beta = 0$ the solution no longer depends on g_w; hence, the $\beta = 0$ curves hold for all g_w values in both figures. Moreover, the $\beta = 0$ curves are identical, since $G = f'$ satisfies Equations (21.29) and (21.30). In other words, the $G(\eta)$ and $f'(\eta)$ profiles are the same when $\beta = 0$.

As discussed in Bae and Emanuel (1989), f' exceeds unity when $\beta > 0$ and $g_w > 1$. This results in a maximum value for f', as is evident for the $g_w = 3$ curve in Figure 21.1. This phenomenon is referred to as velocity overshoot. It occurs when there is a hot wall, i.e., $g_w > 1$, with an accelerating boundary layer, i.e., $\beta > 0$. In this circumstance, the density near the wall is less than ρ_e thereby allowing u to exceed u_e.

Figure 21.3 shows additional velocity profiles when $\beta = 1.5$. We observe that the $g_w = 1.5$ curve has a small, but discernible overshoot. The amount of velocity overshoot increases steadily with g_w when $g_w > 1$. It is worth noting that velocity overshoot can occur even when the edge Mach number M_e goes to zero. In short, this phenomenon is not associated with viscous dissipation.

* I am indebted to Dr. Y.-Y. Bae for many of the figures and tables in this chapter. The computer code that generated these figures and tables is briefly described in Bae and Emanuel (1989).

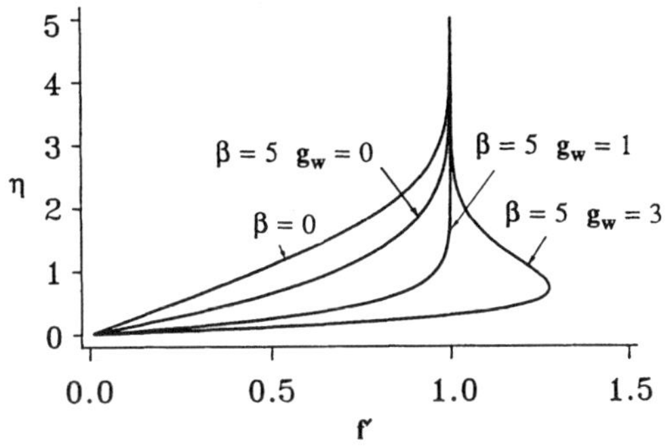

FIGURE 21.1 Velocity profiles in the transformed plane.

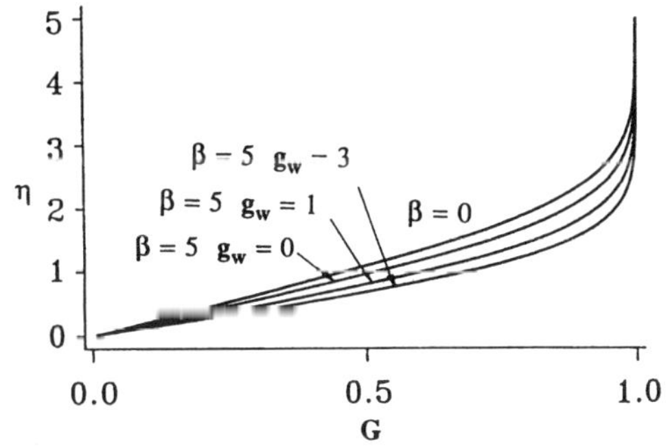

FIGURE 21.2 Total enthalpy profiles in the transformed plane.

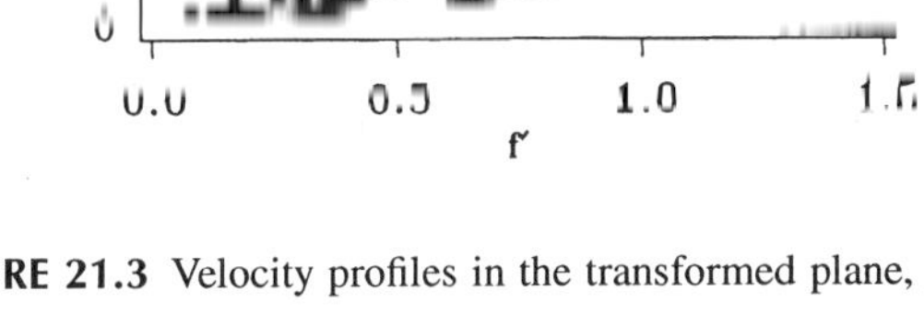

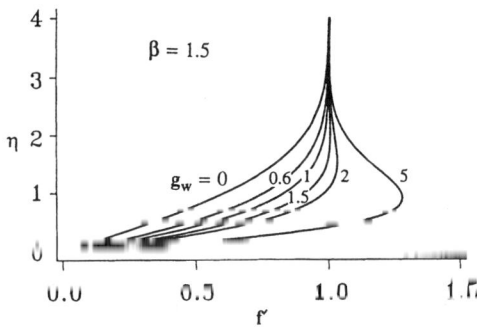

FIGURE 21.3 Velocity profiles in the transformed plane, $\beta = 1.5$.

Observe from Figure 21.2 that G does not exhibit any overshoot. This is a general result for G, and consequently G increases monotonically through the boundary layer. However, this does not mean that the temperature has no overshoot. In fact, the temperature may exhibit a very pronounced overshoot. With Equations (21.17) and (21.26), the temperature is given by

$$\frac{T}{T_e} = \frac{g_w + (1 - g_w)G - [u_e^2/(2h_{oe})]f'^2}{1 - [u_e^2/(2h_{oe})]} \tag{21.31}$$

where

$$\frac{u_e^2}{2h_{oe}} = \frac{\gamma-1}{2}\frac{M_e^2}{1+\frac{\gamma+1}{2}M_e^2}$$

If the derivative of the temperature is set equal to zero:

$$\frac{1}{T_e}\frac{\partial T}{\partial \eta} = \left(1+\frac{\gamma-1}{2}M_e^2\right)\left[(1-g_w)G' - \frac{(\gamma-1)M_e^2}{1+\frac{\gamma-1}{2}M_e^2}f'f''\right] = 0$$

we obtain

$$G' = \frac{(\gamma-1)M_e^2}{\left(1+\frac{\gamma-1}{2}M_e^2\right)(1-g_w)}f'f''$$

as the condition for determining the η value where T is a maximum. As a consequence of this condition, one can show that velocity and temperature overshoot cannot simultaneously occur (see Problem 21.17).

A simple estimate for the magnitude of the overshoot can be obtained by choosing $\beta = 0$. With this value, Equations (21.28)–(21.30) are independent of g_w. Consequently, G and f' satisfy the same differential equation as well as the same boundary conditions; hence, $G = f'$ and $G' = f''$. [In this case, Equation (21.31) shows that T varies quadratically with u, as mentioned in Section 16.2.] The above condition then simplifies to

$$G = \frac{1+\frac{\gamma-1}{2}M_e^2}{(\gamma-1)M_e^2}(1-g_w)$$

A maximum temperature occurs within the boundary layer when $\beta = 0$, providing the right side is between zero and one. For instance, let us assume a relatively cold wall with $g_w = 0.2$, $\gamma = 1.4$, and the values of 2 and 7 for M_e. With the aid of the $\beta = 0$ curve in Figure 21.2, we obtain at the maximum temperature location

$$\eta \cong 2.4, \qquad G = f' = 0.900 \qquad \text{when} \qquad M_e = 2$$

and

$$\eta \cong 0.9, \qquad G = f' = 0.441 \qquad \text{when} \qquad M_e = 7$$

The corresponding maximum temperatures are

$$\frac{T}{T_e} \cong 1.01 \qquad \text{when} \qquad M_e = 2$$

and

$$\frac{T}{T_e} \cong 4.06 \qquad \text{when} \qquad M_e = 7$$

These values are to be compared with the wall temperature ratio given by

$$\frac{T_w}{T_e} = \frac{T_w}{T_{oe}}\frac{T_{oe}}{T_e} = g_w\left(1+\frac{\gamma-1}{2}M_e^2\right) = \begin{cases} 0.36, & M_e = 2 \\ 2.16, & M_e = 7 \end{cases}$$

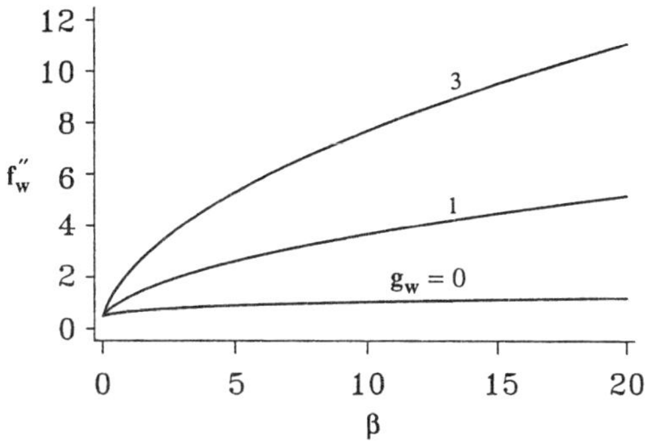

FIGURE 21.4 The parameter f''_w vs. β.

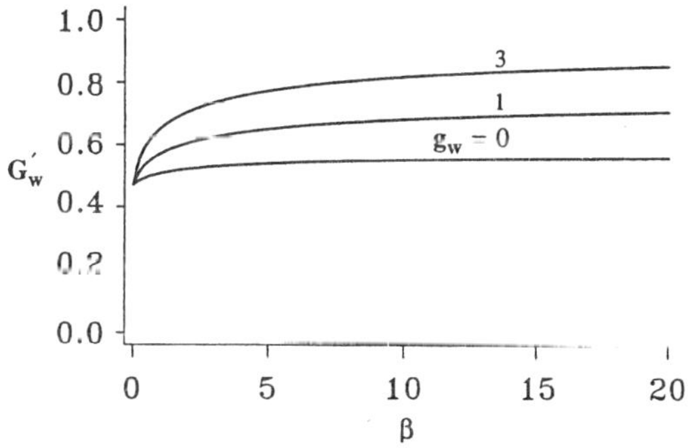

FIGURE 21.5 The parameter G'_w vs. β.

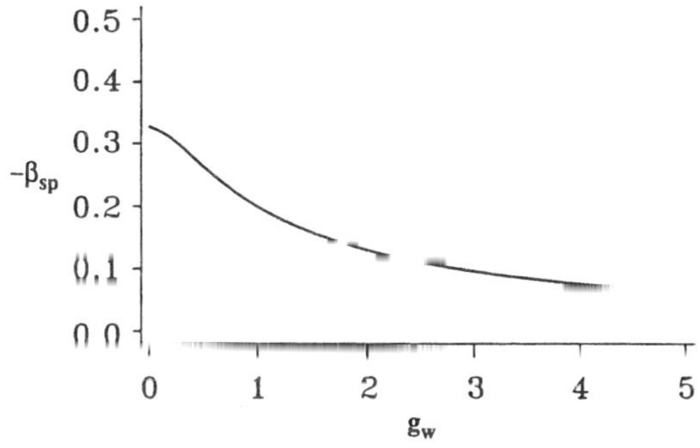

FIGURE 21.6 The separation parameter β_{sp} vs. g_w.

Hence, a sizable temperature overshoot occurs when M_e is in the high supersonic or hypersonic range. In contrast to velocity overshoot, temperature overshoot is caused by viscous dissipation, in which the shearing motion inside a high-speed boundary layer generates an appreciable amount of heat.

Figures 21.4 and 21.5 show f''_w and G'_w vs. β for several values of g_w. The trends exhibited in these figures will be more meaningful later after the skin-friction coefficient and Stanton number are introduced. The separation value of β, denoted as β_{sp}, is shown as $-\beta_{sp}$ vs. g_w in Figure 21.6. Recall that this is the β value for which f''_w first becomes zero. When $g_w = 1$, β_{sp} equals –0.1998, in accordance with Table 20.2. As g_w decreases, the density of the fluid increases and the boundary layer is less prone to separate. The density increase results in an increase in the tangential component of the momentum, which helps keep the boundary layer from a change in direction

21.4 SOLUTION OF THE ENERGY EQUATION

Observe that G can be eliminated from Equations (21.28) and (21.29) with the result that f is determined by a fifth-order equation. In view of the two boundary conditions for G, this step is never taken since these boundary conditions cannot be transferred to f.

A more useful observation is that the energy equation can be integrated if $f(\eta)$ is known. The first integration yields

$$G'(\eta) = G'_w \exp\left(-\int_0^\eta f\, d\eta'\right)$$

A second integration results in the quadrature solution

$$G(\eta) = G_w + G'_w \int_0^\eta \exp\left(-\int_0^{\eta''} f\, d\eta'\right) d\eta'' \tag{21.32a}$$

where η' and η'' are dummy integration variables. Since G_w is zero, we obtain

$$G(\eta) = G'_w \int_0^\eta \exp\left(-\int_0^{\eta''} f\, d\eta'\right) d\eta'' \tag{21.32b}$$

We evaluate $G(\eta)$ at infinity, with the result for the heat transfer parameter

$$G'_w = \left[\int_0^\infty \exp\left(-\int_0^{\eta''} f\, d\eta'\right) d\eta''\right]^{-1} \tag{21.32c}$$

Equations (21.32) constitute a formal solution of the energy equation.

With this solution, the heat flux at the wall is given by

$$\begin{aligned} q_w &= -\kappa_w \left(\frac{\partial T}{\partial n}\right)_w = -\frac{(\kappa r^\sigma \rho)_w u_e T_{oe}}{(2\xi)^{1/2}} g'_w \\ &= -\frac{(\kappa r^\sigma \rho)_w u_e T_{oe}}{(2\xi)^{1/2}} (1 - g_w) G'_w \end{aligned} \tag{21.33}$$

where g'_w is the gradient of the gas temperature adjacent to the wall:

$$g'_w = \frac{1}{T_{oe}} \left(\frac{\partial T}{d\eta}\right)_w = (1 - g_w) G'_w \tag{21.34}$$

Later analysis, in Section 21.8, will show that G_w' is always positive. As a consequence, when $g_w < 1$, q_w is negative and the heat transfer is from the gas to the wall. When $g_w = 1$, the wall is adiabatic and when $g_w > 1$, q_w is positive. Thus, the direction of the heat transfer depends on T_w/T_{oe}, not on T_w/T_e, although the difference disappears in a low-speed flow. This conclusion requires a unity value for the Prandtl number and therefore a unity value for the recovery factor.

The solution provided by Equations (21.32) is deceptive in that it depends on f, whereas f, which is given by Equation (21.28), depends on G. However, there are two circumstances in which Equation (21.28) becomes decoupled from Equation (21.29), and Equations (21.32) are then particularly useful. This decoupling occurs when $\beta = 0$ or $g_w = 1$. Both cases are discussed in later sections.

21.5 THE β AND g_w PARAMETERS

We observe that the momentum equation is uncoupled from the energy equation when $g_w = 1$ or $\beta = 0$. In this circumstance, we have the Falkner–Skan equation and Equations (21.32). In general, however, a solution depends on specific values for β and g_w. We therefore discuss these parameters in this section.

The g_w Parameter

The temperature ratio g_w, which equals T_w/T_{oe}, can range from near zero to a value in excess of unity. These two extremes are referred to as the cold and hot wall cases, respectively. In terms of a wall to boundary layer edge temperature ratio, we have

$$\frac{T_w}{T_e} = g_w\left(1 + \frac{\gamma - 1}{2} M_e^2\right)$$

In a high-speed flow, heat transfer from the gas to the wall is the general rule. For instance, with $\gamma = 1.4$ and $M_e = 4$, T_e can exceed 10^3 K when air goes through a relatively strong upstream shock wave. In order for T_w/T_e to equal 0.5, e.g., intensive wall cooling with $g_w = 0.119$ is required if wall ablation is to be avoided. The hot wall case can occur with the flow of a relatively cold gas, e.g., as occurs in a cryogenic wind tunnel. As discussed in Section 21.9, an adiabatic wall has $g_w = 1$.

The β Parameter

For an adverse pressure gradient, β is negative and a lower limit, β_{sp}, occurs when f_w'' first becomes zero. At this value, the wall shear is zero and the boundary layer separates from the wall. (Aside from this condition, separation also occurs when the wall terminates or has an abrupt slope change. In general, boundary layers are unable to negotiate sharp convex or concave wall turns.) In addition, our discussion does not encompass solutions with reverse flow (Libby and Liu, 1968; Rogers, 1969), for which $\beta < \beta_{sp}$. The boundary-layer equations, in their present form, are not appropriate for a separated flow.

With a favorable pressure gradient, values for β that exceed 10 or more are possible. These values are present in regions where the inviscid flow is highly accelerated. Two situations where this can occur is in the throat region of a supersonic nozzle and on the shoulder of a blunt body in a supersonic flow. This latter case is discussed in the next section.

We now follow Emanuel (1984) in order to develop a computationally suitable form for β, Equation (21.19), where ξ is given by Equation (21.9). In its current form, β depends on an integral and its analytical or numerical evaluation is quite awkward.

We begin by observing that the inviscid streamline or stream surface along the wall is isentropic, so that

$$T_e = \frac{T_{oe}}{1+\frac{\gamma-1}{2}M_e^2}$$

$$p_e = \frac{p_{oe}}{\left(1+\frac{\gamma-1}{2}M_e^2\right)^{\gamma/(\gamma-1)}}$$

$$u_e = (\gamma R T_e)^{1/2} M_e = (\gamma R T_{oe})^{1/2}\frac{M_e}{\left(1+\frac{\gamma-1}{2}M_e^2\right)^{1/2}} \tag{21.35}$$

We also use $C = 1$, to obtain

$$\rho_e\mu_e = \rho_w\mu_w = \frac{\rho_w}{\rho_e}\rho_e\mu_w = \frac{T_e}{T_w}\frac{p_e}{RT_e}\mu_w = \frac{1}{R}\left(\frac{\mu}{T}\right)_w p_e \tag{21.36}$$

An isothermal wall is assumed, which enables us to obtain for ξ and $d\ell n\xi$

$$\xi = \frac{1}{R}\left(\frac{\mu}{T}\right)_w (\gamma R T_{oe})^{1/2} p_{oe}\int_0^s r_w^{2\sigma}F(M_e)\,ds \tag{21.37}$$

$$d\ell n\xi = \frac{r_w^{2\sigma}F ds}{\int_0^s r_w^{2\sigma}F\,ds} \tag{21.38}$$

where

$$F = M_e\left(1+\frac{\gamma-1}{2}M_e^2\right)^{-(3\gamma-1)/[2(\gamma-1)]} \tag{21.39}$$

Hence, $d\ell n u_e^2$ and β can be written as

$$d\ell n u_e^2 = 2\frac{dM_e}{M_e} - \frac{(\gamma-1)M_e^2}{1+\frac{\gamma-1}{2}M_e^2}\frac{dM_e}{M_e} = \frac{2dM_e}{M_e\left(1+\frac{\gamma-1}{2}M_e^2\right)}$$

$$\beta = \left(1+\frac{\gamma-1}{2}M_e^2\right)\frac{2\frac{dM_e}{ds}}{M_e\left(1+\frac{\gamma-1}{2}M_e^2\right)}\frac{1}{\frac{d\ell n\xi}{ds}}$$

$$= \frac{2}{M_e}\frac{dM_e}{ds}\frac{\int_0^s r_w^{2\sigma}F ds}{r_w^{2\sigma}F(M_e)} \tag{21.40}$$

In the integrand, F is a function of the wall arc length s, since M_e depends on s. We occasionally encounter the $F(M_e)$ function. With $\gamma = 1.4$, it has a maximum value of 0.495 when $M = \gamma^{-(1/2)}$ (=0.845) and is zero when M_e is zero or infinite.

Let us now consider β as a function of the arc length along the wall. A differential equation for β is obtained by first solving for the integral

$$2\int_0^s r_w^{2\sigma} F\,ds = \frac{M_e}{\frac{dM_e}{ds}} r_w^{2\sigma} F\beta$$

We differentiate with respect to s, with the result

$$2r_w^{2\sigma}F = r_w^{2\sigma}F\beta - \frac{M_e}{(dM_e/ds)^2}\frac{d^2M_e}{ds^2}r_w^{2\sigma}F\beta + 2\sigma\frac{M_e}{dM_e/ds}r_w\frac{dr_w}{ds}F\beta$$
$$+\frac{M_e}{dM_e/ds}r_w^{2\sigma}\frac{dF}{ds}\beta + \frac{M_e}{dM_e/ds}r_w^{2\sigma}F\frac{d\beta}{ds}$$

Since

$$\frac{dF}{ds} = \frac{1-\gamma M_e^2}{M_e\left(1+\frac{\gamma-1}{2}M_e^2\right)}F\frac{dM_e}{ds} \tag{21.41}$$

this simplifies to

$$\frac{1}{\beta}\frac{d\beta}{ds} = \frac{2-\beta}{\beta}\frac{1}{M_e}\frac{dM_e}{ds} + \frac{1}{(dM_e/ds)}\frac{d^2M_e}{ds^2} - \frac{2\sigma}{r_w}\frac{dr_w}{ds} + \frac{\gamma M_e^2-1}{1+\frac{\gamma-1}{2}M_e^2}\frac{1}{M_e}\frac{dM_e}{ds}$$

or, in its final form,

$$\frac{1}{\beta}\frac{d\beta}{ds} = \frac{1}{dM_e/ds}\frac{d^2M_e}{ds^2} + 2\left(\frac{1}{\beta} - \frac{1-\frac{\gamma+1}{4}M_e^2}{1+\frac{\gamma-1}{2}M_e^2}\right)\frac{1}{M_e}\frac{dM_e}{ds} - \frac{2\sigma}{r_w}\frac{dr_w}{ds} \tag{21.42}$$

This equation determines $\beta(s)$ once we have

$$r_w = r_w(s), \qquad M_e = M_e(s)$$

and an initial value

$$\beta_o = \beta(s_o)$$

For a two-dimensional body, the wall shape r_w is not required. The numerical integration of Equation (21.42) can be done simultaneously with the numerical evaluation of the boundary-layer equations. This differential form for β is much more convenient than its integral version, as will be evident when we utilize this relation in the next chapter.

The similarity assumption means that

$$\frac{d\beta}{ds} = 0 \tag{21.43}$$

in which case the inviscid flow along the wall should satisfy

$$\frac{d^2M_e}{ds^2} + 2\left(\frac{1}{\beta_o} - \frac{1-\frac{\gamma+1}{4}M_e^2}{1+\frac{\gamma-1}{2}M_e^2}\right)\frac{1}{M_e}\left(\frac{dM_e}{ds}\right)^2 - \frac{2\sigma}{r_w}\frac{dr_w}{ds}\frac{dM_e}{ds} = 0 \tag{21.44}$$

where β_o is a constant. Clearly, this relation holds when M_e is constant.

Aside from the constant M_e result, there is no general solution to this equation (numerical solutions are discussed in the next chapter); however, it can be solved in special cases. For instance, if the flow is incompressible, so that $M_e^2 \ll 1$, then M_e can be replaced by u_e. Equation (21.44) thus simplifies to

$$\frac{d^2u_e}{ds^2} + \frac{2(1-\beta_o)}{\beta_o}\frac{1}{u_e}\left(\frac{du_e}{ds}\right)^2 - \frac{2\sigma}{r_w}\frac{dr_w}{ds}\frac{du_e}{ds} = 0 \tag{21.45}$$

If we now assume

$$u_e = ks^m, \qquad r_w^2 = cs^n$$

where k, c, m, and n are constants, Equation (21.45) is satisfied when

$$\beta_o = \frac{2m}{1+m+\sigma n}$$

(Problem 21.10 contains an alternate solution.) When $\sigma = 0$, we obtain the Falkner–Skan similarity relation, Equation (20.15). It is not evident that the power law relation for $u_e(s)$ is consistent with the wall shape. Consistency, in fact, is not expected.

The similarity condition is certainly restrictive, and one might expect that similarity solutions are of little value. This is not the case because of the concept of local similarity, which we discuss next.

21.6 LOCAL SIMILARITY

Nonsimilar Boundary-Layer Equations

The transformed boundary-layer equations, without the similarity assumption, are

$$\left(\frac{\rho\mu}{\rho_e\mu_e}f_{\eta\eta}\right)_\eta + ff_{\eta\eta} + \beta(\xi)(g - f_\eta^2) = 2\xi(f_\eta f_{\xi\eta} - f_\xi f_{\eta\eta}) \tag{21.46a}$$

$$\left\{\frac{\rho\mu}{\rho_e\mu_e}\left[\frac{1}{Pr}g_\eta + \left(1-\frac{1}{Pr}\right)\frac{u_e^2}{h_{oe}}f_\eta f_{\eta\eta}\right]\right\}_\eta + fg_\eta = 2\xi(f_\eta g_\xi - f_\xi g_\eta) \tag{21.46b}$$

These equations are parabolic and can be numerically integrated by starting with a known η solution for f and g at a fixed ξ value. A stagnation point solution, as given in Chapter 17, can be used, where $\xi = 0$ at the stagnation point and the nonsimilar terms on the right sides initially are zero. For the integration, we must also supply relations for $\beta(\xi)$, $u_e(\xi)$, Pr, and $\rho\mu/(\rho\mu)_e$, in addition to the boundary and initial conditions.

Local Similarity Discussion

A primary requirement for local similarity is that the right sides of the above equations be approximately zero. (See Dewey and Gross, 1967, for a cogent discussion of local similarity as well as additional references.) This occurs when either ξ is near zero, as at a stagnation point, or when

$$f_\eta f_{\xi\eta} - f_\xi f_{\eta\eta} \cong 0$$
$$f_\eta g_\xi - f_\xi g_\eta \cong 0$$

Of course, these relations are satisfied whenever we have

$$f_\xi = 0, \qquad g_\xi = 0 \tag{21.47}$$

In addition, the coefficients of all f and g terms on the left side of Equations (21.46) can only depend on η. While not essential for similarity, this requirement is most easily met by assuming that the Prandtl number and Chapman–Rubesin parameter are both unity. Suffice it to say that for air under a wide range of flight conditions, these approximations are satisfactory for first estimates of the skin friction and heat transfer. Section 21.10 evaluates, in some detail, the $Pr = 1$ and $C = 1$ assumptions.

Of course, when the flow has constant values for Pr and C (where Pr and C need not be unity), u_e, and the boundary conditions, then similarity is not an assumption. In this case, β is zero and Equations (21.46) admit an exact similar solution for which Equations (21.47) then hold.

Consequently, the local similarity assumption only needs to be addressed when the inviscid flow is accelerating or decelerating. The decelerating case corresponds to $\beta < 0$ and is discussed later in the chapter. The rest of our discussion thus presumes a favorable pressure gradient, which corresponds to an accelerating boundary-layer flow.

In this circumstance, the local similarity assumption means that the boundary layer, as it changes along the wall in the streamwise direction, is able to readjust rapidly to changes in local flow conditions. It knows about changes in these conditions only through g_w and β. Because typical wall materials have relatively small thermal conductivities, g_w varies slowly with ξ. Generally, a smooth variation in g_w is accurately handled by the local similarity assumption.

Hypersonic Flow about a Blunted Circular Cylinder

By way of contrast, a rapid change of β is readily accomplished and is not an infrequent circumstance. Our present discussion is based on Marvin and Sinclair (1967), which does not assume $Pr = C = 1$ in the numerical part of their analysis. They experimentally and computationally examined hypersonic flow about a blunted circular cylinder with a rounded shoulder, as shown in Figure 21.7. Four model configurations were tested:

$$\frac{r_s}{R} = 0.5, 0.25, 0.15, 0.05$$

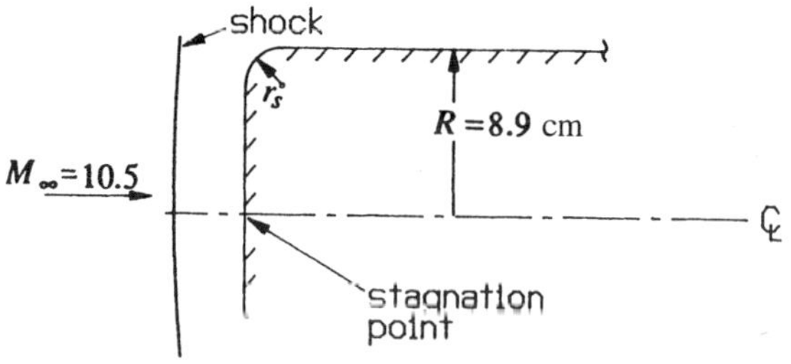

FIGURE 21.7 Blunted circular cylinder with a rounded shoulder in a hypersonic flow.

At the axisymmetric stagnation point, β is 0.5. As the flow moves radially outward, β, at first, increases slowly. However, as the shoulder is approached, the acceleration becomes more pronounced and β starts to rapidly increase. Several peaks occur near the shoulder, after which β usually decreases. It does not decrease in the $(r_s/R) = 0.5$ case, where the acceleration at the shoulder is quite gradual. On the other hand, the acceleration is rapid when $(r_s/R) = 0.05$, in which case β peaks at a value of about 12. The heating rate distribution is relatively constant on the flat face of the cylinder. However, when $(r_s/R) = 0.05$, it increases on the shoulder, reaching a value 50% larger than the stagnation point value.

The principal measurements are the heat transfer rate and pressure along the surface. This pressure is used to deduce u_e from which β is obtained. The computations involved two different locally similar formulations with different choices for the independent variables and a nonsimilar formulation. Except for the following item 5, our summary of this work is restricted to the transformation provided by Equations (21.9) and (21.10). (See Section 21.11 for additional comments.)

1. In terms of heat transfer along the body, both similar and nonsimilar solutions agreed to within 15% of the measurements, even at the highest acceleration rates. Agreement was better than 15% at the lower acceleration rates.
2. At the lower acceleration rates, the right sides of the nonsimilar equations are indeed negligible, as expected.
3. As the acceleration rate increases, the terms on the right sides increase; nevertheless, the similar and nonsimilar heat transfer rates depart quite slowly from each other. This favorable similarity result is partly due to the fact that regions of rapid acceleration can never persist very long.
4. Several boundary-layer thicknesses (later denoted as δ^*, θ, and ϕ) are compared with their nonsimilar counterparts. When $(r_s/R) = 0.05$, the displacement thickness δ^* becomes negative just upstream of the shoulder. Consequently, when the nonsimilar displacement thickness is zero, the relative error for the corresponding similar value is infinite. Aside from this peculiarity, the agreement between similar and nonsimilar thicknesses is quite good.
5. The two different local similarity transformations yielded essentially identical results when β was constant or nearly so. Thus, the local similarity solution for a flat plate, for instance, is insensitive to the transformation. On the other hand, in the region of the shoulder, the heat transfer and skin friction predictions were markedly different. Recall that the local similarity transformation used in Chapter 20 is especially effective for an incompressible, two-dimensional boundary layer. Because Equations (21.9) and (21.10) suppress compressibility and dimensionality effects, local similarity remains effective with ξ, η coordinates. Other transformations that do not suppress these effects are appreciably less successful in yielding a local similarity solution for a compressible flow, or an axisymmetric flow, when β is variable.

This discussion indicates the value and frequent suitability of the local similarity assumption when using ξ, η coordinates. With this assumption, g_w and β are given their local values that can continuously change with ξ. At each wall point the boundary-layer equations are solved as if the flow is a similar one. This concept is the basis of the next three sections.

21.7 BOUNDARY-LAYER PARAMETERS

This section considers a variety of topics. For instance, formulas are developed for various boundary-layer thicknesses, the skin friction, and the heat transfer rate. In addition, the transformation back to the physical plane is provided. While we assume a similar boundary layer with $Pr = C = 1$, none of the basic definitions, such as those for S, $\bar{x}$, or the boundary-layer thicknesses, require a similar boundary layer or a specific value for Pr or C.

We begin by noting that f and G depend on η and parametrically on g_w and β. Those quantities of interest that depend on γ and M_e will do so only through a single flow speed parameter S, defined by

$$S = \frac{\frac{\gamma-1}{2}M_e^2}{1+\frac{\gamma-1}{2}M_e^2} \tag{21.48}$$

The definition of S appears to coincide with τ, Equation (7.18b), used in hodograph theory. However, τ is an independent variable, whereas S is a parameter that is a known function of s.

A dimensional body surface length $\bar{x}$ is introduced (Back, 1970)

$$\bar{x}(s) = \frac{\xi}{(\rho\mu u)_e r_w^{2\sigma}} = \frac{\int_0^s (\rho\mu u)_e r_w^{2\sigma}\, ds}{(\rho\mu u)_e r_w^{2\sigma}} \tag{21.49}$$

where $\bar{x} = s$ whenever boundary-layer edge conditions and r_w are constant. Shortly, it will be convenient to define a Reynolds number based on $\bar{x}$. As will be the case for the Reynolds number, $\bar{x}$ depends only on the inviscid flow and r_w^σ. We will also need the density ratio

$$\frac{\rho_e}{\rho} = \frac{g_w + (1-g_w)G - Sf'^2}{1-S} \tag{21.50}$$

which stems directly from Equation (21.17) or (21.31). In contrast to f and G, ρ_e/ρ depends on S

PHYSICAL PLANE COORDINATES

To invert Equation (21.10), we differentiate both sides, keeping s fixed, to obtain

$$\frac{\partial \eta}{\partial n} = \frac{r_w^\sigma(\rho u)_e}{(2\xi)^{1/2}}\frac{\rho}{\rho_e}$$

We eliminate ξ with Equation (21.49) and rewrite this equation as

$$\frac{(\rho u)_e r_w^\sigma}{[(\rho\mu u)_e r_w^\sigma \bar{x}]^{1/2}}dn = 2^{1/2}\frac{\rho_e}{\rho}\frac{\partial \eta}{\partial n}dn$$

We introduce the Reynolds number

$$Re_{\bar{x}} = \frac{(\rho u)_e \bar{x}}{\mu_e} \tag{21.51}$$

and simplify to obtain, at a fixed s or ξ,

$$Re_{\bar{x}}^{1/2} \frac{dn}{\bar{x}} = 2^{1/2} \frac{\rho_e}{\rho} \frac{\partial \eta}{\partial n} dn = 2^{1/2} \frac{\rho_e}{\rho} d\eta \tag{21.52}$$

After integration, we have

$$Re_{\bar{x}}^{1/2} \frac{n}{\bar{x}} = 2^{1/2} \int_0^{\eta} \frac{\rho_e}{\rho} d\eta$$

or

$$\Upsilon n = \int_0^{\eta} \frac{\rho_e}{\rho} d\eta \tag{21.53}$$

where the convenient parameter, Υ, with dimensions of inverse length,

$$\Upsilon = \left(\frac{Re_{\bar{x}}}{2}\right)^{1/2} \frac{1}{\bar{x}} = \frac{(\rho u)_e r_w^{\sigma}}{(2\xi)^{1/2}}$$

is introduced. (The Υ parameter is just the coefficient of the integral in the definition of η, Equation (21.10).) The density ratio in Equation (21.53) is provided by Equation (21.50); consequently, n has a strong S dependence. (See Problem 21.9(a) for the evaluation of the right side of Equation (21.53).) Hereafter, we regard $\bar{x}$ and n, or Υn, as the coordinates for the physical plane.

Velocity Boundary-Layer Thickness δ

A velocity boundary-layer thickness δ is determined by setting $n = \delta$ when $f'(\eta) = 0.99$, which occurs at an η value denoted as η_{ev}. We thus have

$$\Upsilon \delta = \int_0^{\eta_{ev}} \frac{\rho_e}{\rho} d\eta \tag{21.54}$$

The 0.99 value is the conventional, but arbitrary, definition for η_{ev}; it applies when there is no velocity overshoot. When overshoot is present, let η_m be the η value where f' is a maximum. We now define η_{ev} as the smaller of the two η_{ev} values given by (Bae and Emanuel, 1989)

$$f'(\eta_{ev}) = 0.9 + 0.1 f'(\eta_m), \qquad \eta_{ev} > \eta_m \tag{21.55a}$$

$$f'(\eta_{ev}) = 1.01, \qquad \eta_{ev} > \eta_m \tag{21.55b}$$

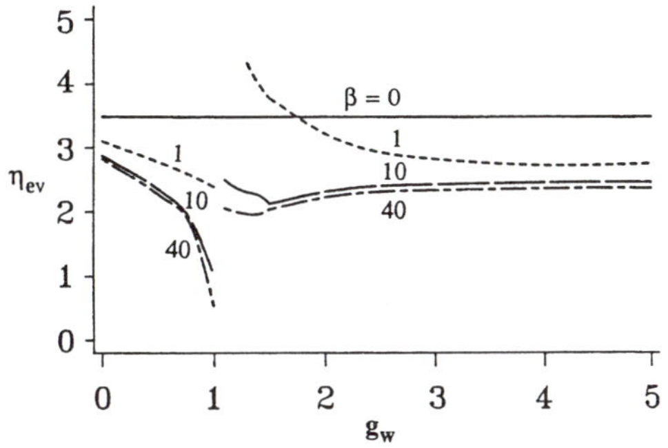

FIGURE 21.8 The parameter η_{ev} vs. g_w for several β values.

where both η_{ev} values are above the location η_m where f' is a maximum. For a large overshoot, as shown in Figure 21.3, Equation (21.55b) is used. Occasionally, the amount of overshoot is small, e.g., the $g_w = 1.5$ curve in Figure 21.3, with $f'(\eta_m)$ falling in between 1 and 1.01. In this circumstance, Equation (21.55a) is used. The rather arbitrary definition of η_{ev} is self-evident when there is no overshoot or when there is a significant amount of overshoot. However, Equation (21.55a), which is used when the overshoot is small, causes several discontinuities. These are evident in Figure 21.8 which shows η_{ev} vs. g_w for several β values. There is a discontinuity in η_{ev}, near $g_w = 1$, when η_{ev} shifts from its no overshoot definition to Equation (21.55a). At a still larger g_w value, η_{ev} has a discontinuous slope change when its definition shifts from Equation (21.55a) to (21.55b). The reason for the definition provided by Equation (21.55a) is discussed in the next section.

Thermal Boundary-Layer Thickness δ_T

A thermal boundary-layer thickness δ_t is determined by setting $n = \delta_t$ when $G(\eta) = 0.99$, which occurs when η equals η_{et}. As evident from Equation (21.54), we have

$$\Upsilon \delta_t = \int_0^{\eta_{et}} \frac{\rho_e}{\rho} d\eta \tag{21.56}$$

The δ and δ_t thicknesses are equal only when $\beta = 0$ (see Tables 21.4 and 21.5). When the Prandtl number and the Chapman–Rubesin constant are both equal to unity, and with $\beta > 0$, later results will indicate that $\delta_t > \delta$. In fact, for large β, δ_t may be several times larger than δ. Since $G(\eta_{et}) = 0.99$, δ_t is a stagnation enthalpy thickness and *not* a temperature thickness. Figure 21.9 shows η_{et} vs. g_w for the same β values used in Figure 21.8. In contrast to Figure 21.8, the curves are quite smooth. The reason for this is that G never exhibits overshoot and only one definition for η_{et} suffices. Also evident in both figures is that η_{ev} and η_{et} approach limiting values as $\beta \to \infty$.

Velocity and Total Enthalpy Profiles in the Physical Plane

The edge values, η_{ev} and η_{et}, are transformed quantities. For profiles in the physical plane, we should use n or, nondimensionally, Υn. Figures 21.10 and 21.11 show the velocity and stagnation enthalpy profiles for the same β and g_w values used in Figures 21.1 and 21.2.

As evident from Equation (21.53), Υn also depends on S and g_w in addition to β. (There is no g_w dependence when $\beta = 0$.) We have used $S = 0$, 0.5, and 0.9, which corresponds to $M_e = 0$, 2.236, and 6.708, respectively, when $\gamma = 1.4$. The curves in Figure 21.10 terminate abruptly when $\eta = \eta_{ev}$. Similarly, the curves in Figure 21.11 terminate when $\eta = \eta_{et}$. However, this figure uses

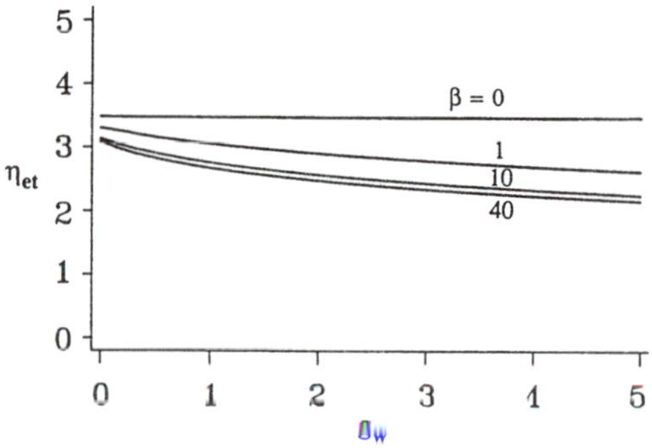

FIGURE 21.9 The parameter η_{et} vs. g_w for several β values.

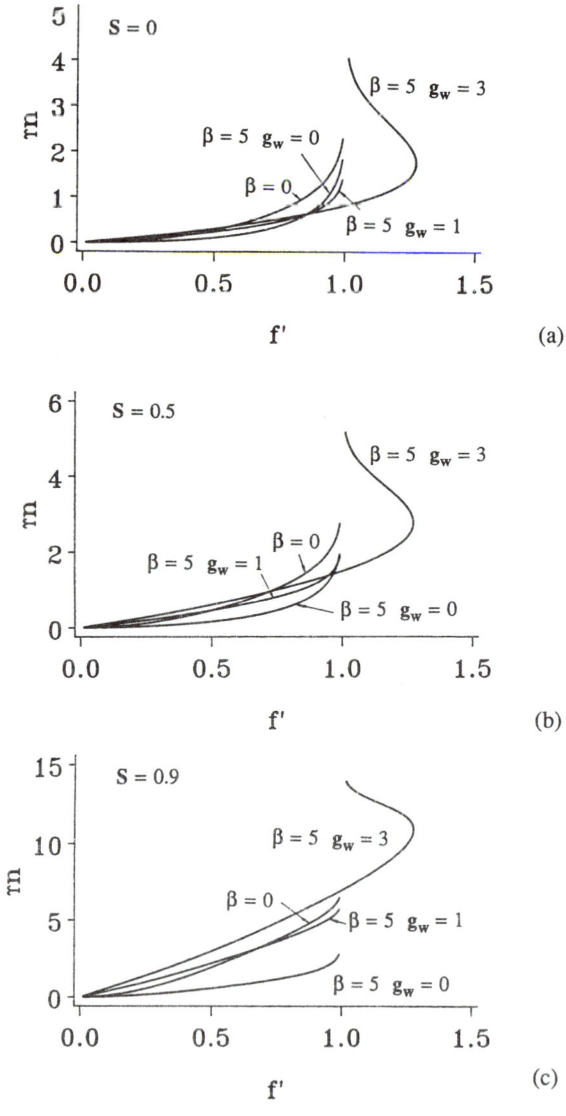

FIGURE 21.10 Velocity profiles in the physical plane; (a) $S = 0$; (b) $S = 0.5$; (c) $S = 0.9$.

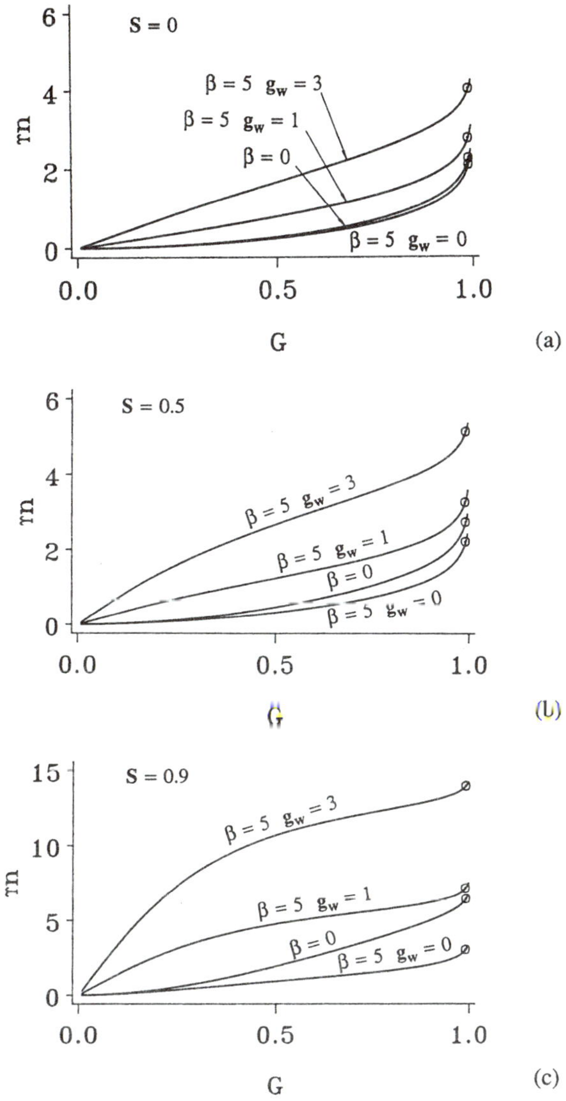

FIGURE 21.11 Total enthalpy profiles in the physical plane, where $G(\eta_{et}) = 0.995$ and the circles indicate $G(\eta_{et}) = 0.99$; (a) $S = 0$; (b) $S = 0.5$; (c) $S = 0.9$.

$G(\eta_{et}) = 0.995$ at termination; the circles near the terminal points of each curve indicate the usual $G(\eta_{et}) = 0.99$ value. Increasing the 0.99 value to 0.995 simply extends the curve a modest amount. The edge value of Υn is thus not sensitive to a small change in the 0.99 boundary-layer edge value for either f' or G.

Observe in both figures that a hot wall $(g_w > 1)$ has a substantially thicker boundary layer than does a cold wall and rapidly increases in thickness with S. Thus, a high-speed boundary layer on a hot wall is quite thick. This trend and the profile shapes are different from those in the transformed plane as shown in Figures 21.1 and 21.2. One of the reasons for this is that Equations (21.9) and (21.10) have removed, in the transformed plane, the sensitivity to compressibility effects. Note that a very cold wall (i.e., $g_w = 0$) has only a slight dependence on S.

Other Boundary-Layer Thicknesses

The thicknesses δ and δ_t do not fully characterize the boundary layer. We therefore introduce thicknesses for displacement, momentum defect, and stagnation enthalpy defect, as follows (Back, 1970):

$$\delta^* = \int_0^\infty \left(1 - \frac{\rho u}{(\rho u)_e}\right) dn \tag{21.57}$$

$$\theta = \int_0^\infty \frac{\rho u}{(\rho u)_e}\left(1 - \frac{u}{u_e}\right) dn \tag{21.58}$$

$$\phi = \int_0^\infty \frac{\rho u}{(\rho u)_e}\left(1 - \frac{h_o - h_{ow}}{h_{oe} - h_{ow}}\right) dn \tag{21.59}$$

Along with δ and δ_t, these are the most frequently encountered boundary-layer thicknesses. Their use is required in Section 21.12. The displacement thickness represents the change that would occur in the body size if the mass flux in the boundary layer were set equal to $(\rho u)_e$. The momentum defect thickness θ represents the loss of momentum flux ρu^2, relative to the adjacent inviscid flow, that occurs in the boundary layer. This loss is caused by skin friction. An analogous statement holds for ϕ. Observe that the stagnation enthalpy factor in Equation (21.59) can be written as

$$1 - \frac{h_o - h_{ow}}{h_{oe} - h_{ow}} = 1 - G = \frac{T_{oe} - T_o}{T_{oe} - T_w}$$

At a given wall location, only T_o varies with n. Thus, ϕ accounts for the variation of the stagnation temperature T_o as it changes from T_w to T_{oe}. For instance, if $T_w \ll T_{oe}$ and the boundary layer is highly conductive, i.e., $Pr \ll 1$, then T_o will be close to T_w for a significant portion of the layer. There is then a reduction of the stagnation temperature, due to wall cooling, and a large stagnation enthalpy defect relative to the freestream. In other words, G is near zero, and ϕ is relatively large.

The thicknesses given by Equations (21.57) to (21.59) can be put in nondimensional form with the aid of Equations (21.50) and (21.52). We thus obtain

$$\Upsilon\delta^* = \int_0^\infty \left[\frac{g_w + (1 - g_w)G - Sf'^2}{1 - S} - f'\right] d\eta \tag{21.60}$$

$$\Upsilon\theta = \int_0^\infty f'(1 - f')\, d\eta \tag{21.61}$$

$$\Upsilon\phi = \int_0^\infty f'(1 - G)\, d\eta \tag{21.62}$$

In contrast to δ, δ_t, and δ^*, the θ and ϕ thicknesses do not depend on S; nevertheless, they indirectly depend on M_e through Υ and β. If the boundary layer is nonsimilar, then f' is $\partial f/\partial\eta$ and the integrals are performed with ξ, or s, held fixed.

The Skin-Friction Coefficient

The skin friction is given by

$$\tau_w = \mu_w \left(\frac{\partial u}{\partial n}\right)_w = \frac{(\mu r^\sigma \rho)_w u_e^2}{(2\xi)^{1/2}} f_w''$$

We utilize the $C = 1$ assumption

$$(\rho\mu)_w = (\rho\mu)_e \tag{21.63}$$

and introduce $\bar{x}$, to obtain

$$\tau_w = \frac{r_w^\sigma(\rho\mu u^2)_e}{r_w^\sigma[2(\rho\mu u^2)_e \bar{x}]^{1/2}} f_w'' = \left[\frac{\mu_e}{2(\rho u^2)_e \bar{x}}\right]^{1/2} (\rho u^2)_e f'' \tag{21.64}$$

Aside from f_w'', τ_w depends only on the inviscid flow and r_w^σ. This will not be the case, e.g., if $\mu \sim T^\omega$ with $\omega \neq 1$. The local skin-friction coefficient is given by

$$c_f = \frac{2\tau_w}{(\rho u^2)_e} = \left(\frac{2}{Re_{\bar{x}}}\right)^{1/2} f_w'' \tag{21.65}$$

Hence, c_f is proportional to f_w'', and examination of Figure 21.4 shows that it increases rapidly with g_w and β when g_w exceeds unity. It also increases as $\bar{x}^{-1/2}$, as $\bar{x}$ decreases toward zero.

An average skin-friction coefficient is defined as

$$\bar{c}_f = \frac{1}{s}\int_0^s c_f\, ds = \frac{1}{s}\int_0^{\bar{x}} \left(\frac{2}{Re_{\bar{x}}}\right)^{1/2} f_w''(\bar{x}) \frac{ds}{d\bar{x}} d\bar{x}$$

When $\beta = 0$, i.e., a flat plate or a uniform flow parallel to a cylinder, this yields

$$\bar{c}_f = \frac{1}{s}\left(\frac{2\mu_e}{\rho_e u_e}\right)^{1/2} f_w'' \int_0^s \frac{ds}{s^{1/2}} = 2c_f(s)$$

We thus have the result that the average skin-friction coefficient is twice the local value.

The shear stress inside the boundary layer is simply given by

$$\tau = \mu \frac{\partial u}{\partial n}$$

where Equation (21.64) provides the wall value. As you might expect, τ and the vorticity ω are related, as demonstrated in part (d) of Problem 21.8. Moreover, this problem shows that under certain conditions, τ and ω have a zero value and a change in sign *inside* the layer. These zero values occur when there is velocity overshoot. In this circumstance, neither τ nor ω goes to zero monotonically as η increases to infinity. This is in contrast to an incompressible boundary layer in which τ and ω monotonically decrease to zero as η increases.

The Stanton Number

In the boundary layer, the stagnation enthalpy is

$$h_o = h_{ow} + (h_{oe} - h_{ow})G$$

Thus, the temperature is given by

$$T = \frac{1}{c_p}\left(h_o - \frac{1}{2}u_e^2 f'^2\right) = \frac{1}{c_p}\left[h_{ow} + (h_{oe} - h_{ow})G - \frac{1}{2}u_e^2 f'^2\right]$$

where h_{ow} and u_e may be functions of s, and h_{oe} is a constant. Hence, we have

$$\frac{\partial T}{\partial n} = \frac{1}{c_p}\left[(h_{oe} - h_{ow})\frac{\partial G}{\partial n} - u_e^2 f'\frac{\partial f'}{\partial n}\right]$$

and at the wall this becomes

$$\left(\frac{\partial T}{\partial n}\right)_w = \frac{h_{oe} - h_{ow}}{c_p}\left(\frac{\partial G}{\partial n}\right)_w = \frac{h_{oe} - h_{ow}}{c_p}\frac{r_w^\sigma \rho_w u_e}{(2\xi)^{1/2}}G_w'$$

The heat transfer at the wall is

$$q_w = -\kappa_w\left(\frac{\partial T}{\partial n}\right)_w = \frac{\kappa_w}{c_p}(h_{ow} - h_{oe})\frac{r_w^\sigma \rho_w u_e}{(2\xi)^{1/2}}G_w'$$

We now introduce $Pr = 1$, Equation (21.63), and $\bar{x}$ to obtain for the heat flux

$$q_w = \left[\frac{\mu_e}{2(\rho u)_e \bar{x}}\right]^{1/2}(h_{ow} - h_{oe})(\rho u)_e G_w'$$

The Stanton number is thus given by

$$St = \frac{q_w}{(h_{ow} - h_{oe})(\rho u)_e} = \frac{G_w'}{(2Re_{\bar{x}})^{1/2}} \qquad \textbf{(21.66)}$$

where

$$h_{ow} - h_{oe} = c_p(T_w - T_{oe})$$

The Stanton number is therefore proportional to G_w' and Figure 21.5 shows that it is relatively insensitive to changes in β and g_w, in contrast to c_f. Since G_w' is positive, so is the Stanton number. The direction of the heat transfer, however, is provided by $1 - g_w$, as discussed in Section 21.4.

An average Stanton number is defined as

$$\overline{St} = \frac{1}{s}\int_0^s St\, ds = \frac{1}{s}\int_0^s \frac{G_w'(\bar{x})}{(2Re_{\bar{x}})^{1/2}}\frac{ds}{d\bar{x}}d\bar{x}$$

which similarly reduces to $\overline{St} = 2St(s)$ when $\beta = 0$.

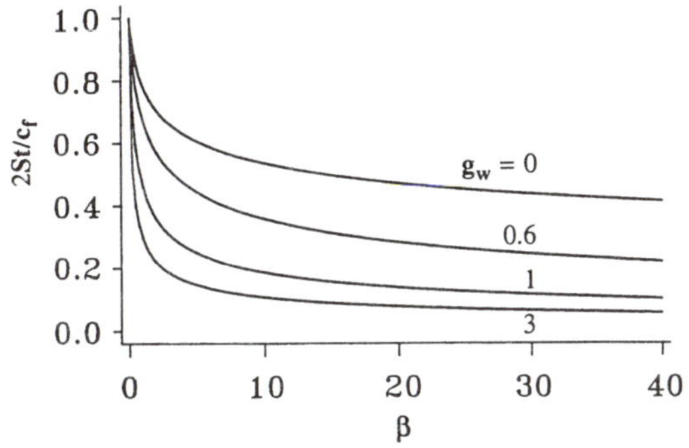

FIGURE 21.12 Reynolds' analogy with $Pr = 1$.

Our choice for a ΔT in the Stanton number coincides with ΔT_o, Equations (16.8), now written as $T_w - T_{oe}$. Since $Pr = 1$, the stagnation enthalpy difference in Equation (21.66) coincides with that in Equation (16.26). As a consequence of $Pr = 1$, the recovery temperature, Equation (16.18), does not explicitly appear because it equals T_{oe}.

Reynolds' Analogy

With c_f and the Stanton number, Reynolds' analogy takes the form

$$\frac{2St}{c_f} = \frac{G'_w}{f''_w} \tag{21.67}$$

This result, which is independent of S, is shown in Figure 21.12. Observe the steep decrease in heat transfer relative to the skin friction as β or g_w increases. The decrease in heat transfer is especially pronounced when $\beta \le 2$. Reynolds' analogy is used to provide a heat transfer estimate after c_f is evaluated. As evident from the figure, the estimate should not assume that the right side of Equation (21.67) is constant if the flow is compressible with a changing value of β and/or g_w. A Prandtl number correction to this equation is provided in Section 21.10.

21.8 COMPREHENSIVE TABLES

There are many publications dealing with one or more aspects of similar boundary-layer theory, including the occasional presentation of tabulated results. These come in two forms: (a) parameters directly involved in the solution of the differential equations, such as f''_w and G'_w and (b) derived parameters, such as a momentum thickness. Some of the derived parameters depend on S as well as β and g_w. A comprehensive table of derived results is, therefore, three dimensional and, in fact, does not exist. One of the more extensive tabulations is due to Back (1970). However, his three-dimensional table is scanty and lacks results for negative β and for $g_w > 1$. Furthermore, the spacing on g_w and S is nonuniform and inadequate for interpolation or extrapolation. These remarks are not criticisms, since a comprehensive three-dimensional table is a prohibitive undertaking, and, even if it existed, would be difficult and awkward to use.

Bae and Emanuel (1989) have shown that a three-dimensional table is unnecessary; comprehensive and accurate results only require a series of two-dimensional tables. This section presents these tables and the analysis behind them. The full paper that this reference is based on can be consulted for a discussion of the rather elaborate algorithm developed for numerically solving the similar boundary-layer equations.

Integral Relations

The subsequent derivation utilizes the following integrals, all of which are exact:

$$\int_0^\eta d\eta = \eta \tag{21.68a}$$

$$\int_0^\eta f'd\eta = \int_0^f df = f \tag{21.68b}$$

$$\int_0^\eta f''d\eta = \int_0^{f'} df' = f' \tag{21.68c}$$

$$\int_0^\eta f'''d\eta = \int_{f''_w}^{f''} df'' = f'' - f''_w \tag{21.68d}$$

$$\int_0^\eta ff''d\eta = \int_0^{f'} f df' = ff' - \int_0^\eta f'^2 d\eta \tag{21.68e}$$

$$\int_0^\eta Gf'd\eta = fG + G' - G'_w \tag{21.68f}$$

$$\int_0^\eta f'^2 d\eta = \frac{1}{1+\beta}(f'' + ff' + \beta g_w \eta - f''_w) + \frac{\beta(1-g_w)}{1+\beta}\int_0^\eta G d\eta \tag{21.68g}$$

$$\int_0^\eta f'^3 d\eta = \frac{2}{1+2\beta} \times \left[f'f'' + \frac{1}{2}ff'^2 + \beta g_w f + \beta(1-g_w)(fG + G' - G'_w) - \int_0^\eta f''^2 d\eta\right] \tag{21.68h}$$

For these integrals, we set $f_w = f'_w = 0$; thus, transpiration cooling and wall suction are not included. The first few integrals are self-evident. Equations (21.68e,f) stem from an integration by parts. In the case of Equation (21.68f), this is followed by using Equations (21.29) to replace the fG' integrand with $-G''$. To obtain Equation (21.68g), multiply Equation (21.28) by $d\eta$ and integrate from zero to η. The desired result requires the use of Equations (21.68d,e). The derivation of Equation (21.68h) is the subject of Problem 21.4.

Relations for ϕ, θ, and δ^*

We begin by replacing the upper integration limits in Equations (21.60) to (21.62) with η. After the integrals are analytically evaluated, we proceed to take the $\eta \to \infty$ limit. The simplest derivation is for ϕ, which we write as

$$\Upsilon\phi = \left(f - \int_0^\eta Gf'd\eta\right)_{\eta\to\infty}$$

where Equation (21.68b) is used. With the aid of Equation (21.68f), we have

$$\Upsilon\phi = (f - Gf - G' + G'_w)_{\eta\to\infty}$$

Since

$$G(\infty) = 1, \qquad G'(\infty) = 0$$

this reduces to the elegant result

$$\Upsilon\phi = G'_w \tag{21.69}$$

Recall from Equation (18.19) that

$$C_v(\beta, g_w) = \eta - f, \qquad \eta\to\infty \tag{21.70}$$

where now the velocity parameter, C_v, depends on both β and g_w. Consequently, we have

$$f'(\infty) = 1, \qquad f''(\infty) = 0 \tag{21.71}$$

In view of the G integral in Equation (21.68g), it is convenient to introduce a second parameter

$$C_t(\beta, g_w) = \eta - \int_0^\eta G d\eta, \qquad \eta\to\infty \tag{21.72}$$

where C_t is a thermal layer constant whose role is analogous to C_v. In fact, Tables 21.2 and 21.3 show that $C_v(0, g_w) = C_t(0, g_w) = 1.2168$.

With Equations (21.68b) and (21.68g), Equation (21.61) becomes

$$\Upsilon\theta = \frac{1}{1+\beta}\left[(1+\beta)f - f'' - ff' - \beta g_w\eta + f''_w - \beta(1-g_w)\int_0^\eta G d\eta\right]_{\eta\to\infty}$$

Equations (21.70) to (21.72) reduce this relation to

$$\Upsilon\theta = \frac{1}{1+\beta}\{f''_w - \beta[C_v - (1-g_w)C_t]\} \tag{21.73}$$

where the η terms on the right side cancel.

In a similar manner, we have for the displacement thickness

$$\Upsilon\delta^* = \frac{1}{(1-S)(1+\beta)}\left\{Sf''_w + [1+(1-S)\beta]g_w\eta - (1-S)(1+\beta)f \right.$$
$$\left. - S(f'' + ff') + (1-g_w)[1+(1-S)\beta]\int_0^\eta G d\eta\right\}_{\eta\to\infty}$$

When the right side is evaluated as before, we have

$$\Upsilon\delta^* = \frac{1}{(1-S)(1+\beta)}\{Sf_w'' + [1+(1-S)\beta][C_v - (1-g_w)C_t]\} \tag{21.74}$$

where, again, the η terms cancel.

Relations for δ and δ_t

Similar relations can be obtained for δ and δ_t. With the aid of Equations (21.50) and (21.68g), the integrals in Equations (21.54) and (21.56) can be written as

$$\int_0^{\eta^*}\frac{\rho_e}{\rho}d\eta = \frac{g_w}{1-S}\eta^* - \frac{S}{(1-S)(1+\beta)}(f'' + ff' + \beta g_w\eta^* - f_w'')$$
$$+\frac{(1-g_w)[1+\beta(1-S)]}{(1-S)(1+\beta)}\int_0^{\eta^*}Gd\eta$$

where η^* is either η_{ev} or η_{et}. Although η^* is finite, we assume it is sufficiently large so that Equations (21.70) to (21.72) apply at η^*; the accuracy of this approximation is evaluated shortly. We thus obtain

$$\int_0^{\eta^*}\frac{\rho_e}{\rho}d\eta = \eta^* + \frac{1}{(1-S)(1+\beta)}\{S(f_w'' + C_v) - (1-g_w)[1+\beta(1-S)]C_t\}$$

Equations (21.54) and (21.56) now become

$$\Upsilon\delta = \eta_{ev}(\beta, g_w) + \frac{1}{(1-S)(1+\beta)}\{S(f_w'' + C_v) - (1-g_w)[1+\beta(1-S)]C_t\} \tag{21.75}$$

$$\Upsilon\delta_t = \eta_{et}(\beta, g_w) + \frac{1}{(1-S)(1+\beta)}\{S(f_w'' + C_v) - (1-g_w)[1+\beta(1-S)]C_t\} \tag{21.76}$$

where $G(\eta_{et}) = 0.99$ and η_{ev} is given by $f'(\eta_{ev}) = 0.99$ or by Equations (21.55). Thus, Equation (21.75) also holds for a boundary layer with velocity overshoot. As indicated, both η_{ev} and η_{et} depend on β and g_w.

We noted in the previous section the rapid thickening of the boundary layer with increasing M_e when g_w is not too small. The dominant Mach number behavior in the equations for δ^*, δ, and δ_t is provided by the $(1-S)^{-1}$ factor, which equals

$$\frac{1}{1-S} = 1 + \frac{\gamma-1}{2}M_e^2$$

It is this factor that is largely responsible for the boundary-layer thickening that is evident in Figures 21.10 and 21.11.

Boundary-Layer Tables

Equations (21.69) and (21.73) to (21.76) provide the final result for the various boundary-layer thicknesses. The dependence of these thicknesses on $(2/Re_{\bar{x}})^{1/2}\bar{x}$ is contained in Υ. While all five thicknesses depend on β and g_w, only δ, δ_t, and δ^* depend on S.

Observe that f''_w, G'_w, C_v, C_t, η_{ev}, and η_{et} are functions only of β and g_w. Thus, a two-dimensional table, with β and g_w as the entrees, is sufficient for each of these six parameters. With a set of tables, the five boundary-layer thicknesses can be determined, since S appears explicitly in the equations for $\Upsilon\psi$, where ψ now represents δ, δ_t, and δ^*. Additional tables for c_f and St are not required, since these parameters are given by Equations (21.65) and (21.66).

The full paper associated with Bae and Emanuel (1989) provides comprehensive tables for the above six parameters. These, along with a seventh table, are reproduced here as Tables 21.1 through 21.7. The seventh table, which is Table 21.1, shows β_{sp} and $0.5\beta_{sp}$ vs. g_w. This table is necessary since the other six tables provide results for these two negative β values. These tables are indeed comprehensive, covering the range

$$\beta_{sp} \le \beta \le 100, \qquad 0 \le g_w \le 5$$

Table 21.4 shows η_{ev}, whose value is affected by the definition in Equations (21.55). To the left of the solid lines shown in the table there is no velocity overshoot and the conventional η_{ev} definition applies. To the right of all solid lines there is appreciable overshoot and Equation (21.55b) is used. In the middle region, Equation (21.55a) holds.

Accuracy of the Tables

For two of the thickness parameters, δ and δ_t, we have replaced $\eta \to \infty$ with $\eta = \eta^*$. Since values for δ and δ_t are based on the parameters in Tables 21.2 to 21.6, we evaluate the relative errors (Bae and Emanuel, 1989)

$$E_v = \frac{|\delta_{ex} - \delta|}{\delta_{ex}} \times 10^2, \qquad E_t = \frac{|\delta_{tex} - \delta_t|}{\delta_{tex}} \times 10^2$$

for each of the 231 cases in Tables 21.2 to 21.6 at S values of 0, 0.5, and 0.9. Exact values for the two thicknesses, δ_{ex} and δ_{tex}, are directly determined by numerically evaluating Equations (21.54)

TABLE 21.1
β_{sp} vs. g_w

g_w	β_{sp}	$0.5\beta_{sp}$	β_{sp}[a]
0.0	–0.32650	–0.16325	–0.326
0.2	–0.30865	–0.15433	–0.3088
0.4	–0.27783	–0.13892	—
0.6	–0.24757	–0.12379	–0.246
0.8	–0.22115	–0.11058	—
1.0	–0.19884	–0.09942	–0.1988
1.5	–0.15735	–0.07867	—
2.0	–0.12950	–0.06475	–0.1295
3.0	–0.09521	–0.04760	—
4.0	–0.07511	–0.03756	—
5.0	–0.06199	–0.03099	—

[a] See Cohen and Reshotko (1956a).

TABLE 21.2
$C_v(\beta, g_w)$

β	g_w 0.0	0.2	0.4	0.6	0.8	1.0	1.5	2.0	3.0	4.0	5.0
SP	3.4554	2.9267	2.6691	2.5199	2.4246	2.3580	2.2597	2.2051	2.1473	2.1177	2.0989
0.5SP	1.3383	1.3814	1.4079	1.4615	1.5217	1.4408	1.4499	1.4541	1.4599	1.4594	1.4601
0.00	1.2168	1.2168	1.2168	1.2168	1.2168	1.2168	1.2168	1.2168	1.2168	1.2168	1.2168
0.25	1.1145	1.0767	1.0411	1.0075	0.9756	0.9453	0.8752	0.8119	0.7008	0.6051	0.5207
0.50	1.0529	0.9947	0.9416	0.8926	0.8473	0.8047	0.7085	0.6243	0.4807	0.3605	0.2566
0.75	1.0107	0.9391	0.8749	0.8165	0.7629	0.7135	0.6029	0.5073	0.3467	0.2138	0.0999
1.00	0.9793	0.8979	0.8261	0.7612	0.7023	0.6479	0.5279	0.4250	0.2535	0.1126	−0.0076
1.25	0.9549	0.8660	0.7883	0.7187	0.6557	0.5979	0.4711	0.3629	0.1837	0.0373	−0.0874
1.50	0.9350	0.8401	0.7577	0.6843	0.6183	0.5580	0.4260	0.3140	0.1290	−0.0216	−0.1496
1.75	0.9185	0.8186	0.7323	0.6561	0.5875	0.5250	0.3891	0.2740	0.0846	−0.0693	−0.1999
2.00	0.9044	0.8002	0.7108	0.6321	0.5615	0.4975	0.3582	0.2406	0.0476	−0.1089	−0.2415
3.00	0.8642	0.7476	0.6493	0.5638	0.4877	0.4190	0.2710	0.1470	−0.0553	−0.2187	−0.3567
4.00	0.8382	0.7133	0.6094	0.5197	0.4403	0.3689	0.2158	0.0881	−0.1196	−0.2868	−0.4279
5.00	0.8196	0.6887	0.5808	0.4881	0.4066	0.3334	0.1769	0.0467	−0.1644	−0.3342	−0.4772
10.00	0.7708	0.6233	0.5050	0.4052	0.3182	0.2408	0.0765	−0.0591	−0.2780	−0.4532	−0.6006
15.00	0.7483	0.5924	0.4694	0.3665	0.2772	0.1980	0.0307	−0.1070	−0.3287	−0.5060	−0.6550
20.00	0.7347	0.5734	0.4475	0.3428	0.2523	0.1721	0.0032	−0.1357	−0.3589	−0.5372	−0.6871
30.00	0.7186	0.5503	0.4211	0.3142	0.2223	0.1411	−0.0297	−0.1698	−0.3946	−0.5740	−0.7248
40.00	0.7090	0.5363	0.4050	0.2970	0.2042	0.1224	−0.0494	−0.1901	−0.4157	−0.5958	−0.7470
50.00	0.7024	0.5266	0.3939	0.2851	0.1918	0.1096	−0.0628	−0.2039	−0.4301	−0.6105	−0.7621
100.00	0.6862	0.5019	0.3659	0.2552	0.1607	0.0777	−0.0961	−0.2380	−0.4654	−0.6466	−0.7987

TABLE 21.3

$C_t(\beta, g_w)$

β	g_w 0.0	0.2	0.4	0.6	0.8	1.0	1.5	2.0	3.0	4.0	5.0
SP	2.1374	1.9006	.7[illegible]30	1.7328	1.6951	1.6690	1.6310	1.6101	1.5881	1.5767	1.5697
0.5SP	1.2570	1.2727	.[illegible]28	1.3026	1.3252	1.2952	1.3002	1.3023	1.3050	1.3051	1.3056
0.00	1.2168	1.2168	.[illegible]68	1.2168	1.2168	1.2168	1.2168	1.2168	1.2168	1.2168	1.2168
0.25	1.1829	1.1696	.[illegible]72	1.1456	1.1348	1.1246	1.1014	1.0810	1.0463	1.0175	0.9929
0.50	1.1623	1.1418	.[illegible]35	1.1070	1.0918	1.0779	1.0473	1.0214	0.9791	0.9454	0.9175
0.75	1.1479	1.1227	[illegible]08	1.0813	1.0637	1.0478	1.0135	0.9849	0.9393	0.9036	0.8745
1.00	1.1370	1.1085	[illegible]40	1.0625	1.0434	1.0262	0.9895	0.9594	0.9119	0.8753	0.8456
1.25	1.1285	1.0973	[illegible]08	1.0479	1.0276	1.0095	0.9713	0.9402	0.8916	0.8544	0.8244
1.50	1.1215	1.0881	1.0[illegible]02	1.0361	1.0150	0.9962	0.9569	0.9250	0.8756	0.8381	0.8079
1.75	1.1156	1.0805	1.[illegible]12	1.0263	1.0045	0.9852	0.9450	0.9126	0.8627	0.8249	0.7947
2.00	1.1106	1.0739	1.[illegible]36	1.0180	0.9956	0.9759	0.9350	0.9023	0.8519	0.8140	0.7837
3.00	1.0959	1.0547	1.[illegible]15	0.9938	0.9701	0.9493	0.9066	0.8729	0.8217	0.7835	0.7533
4.00	1.0861	1.0419	1.[illegible]69	0.9780	0.9534	0.9320	0.8884	0.8542	0.8027	0.7645	0.7344
5.00	1.0791	1.0325	0.[illegible]62	0.9665	0.9413	0.9195	0.8754	0.8410	0.7893	0.7512	0.7212
10.00	1.0599	1.0071	0.[illegible]74	0.9356	0.9092	0.8865	0.8413	0.8065	0.7548	0.7171	0.6876
15.00	1.0508	0.9947	0.[illegible]35	0.9209	0.8939	0.8710	0.8254	0.7906	0.7391	0.7017	0.6725
20.00	1.0452	0.9870	0.[illegible]48	0.9118	0.8846	0.8615	0.8158	0.7810	0.7297	0.6925	0.6635
30.00	1.0384	0.9775	0.[illegible]43	0.9007	0.8732	0.8500	0.8042	0.7695	0.7185	0.6816	0.6528
40.00	1.0344	0.9717	0.[illegible]78	0.8940	0.8664	0.8431	0.7973	0.7626	0.7118	0.6751	0.6465
50.00	1.0316	0.9677	0.[illegible]34	0.8893	0.8616	0.8383	0.7925	0.7579	0.7072	0.6706	0.6422
100.00	1.0241	0.9569	0.[illegible]117	0.8772	0.8494	0.8261	0.7803	0.7460	0.6958	0.6596	0.6315

TABLE 21.4
$\eta_{ev}(\beta, g_w)$

β	g_w 0.0	0.2	0.4	0.6	0.8	1.0	1.5	2.0	3.0	4.0	5.0
SP	5.9631	5.4208	5.1427	4.9755	4.8659	4.7879	4.6702	4.6036	4.5321	4.4940	4.4714
0.5SP	3.6642	3.7172	3.7465	3.8117	3.8828	3.7770	3.7828	3.7846	3.7874	3.7845	3.7840
0.00	3.4717	3.4717	3.4717	3.4717	3.4717	3.4717	3.4717	3.4717	3.4717	3.4717	3.4717
0.25	3.3039	3.2483	3.1934	3.1388	3.0842	3.0290	2.8872	2.7348	4.6741	3.8609	3.4840
0.50	3.2041	3.1160	3.0280	2.9386	2.8471	2.7501	2.4763	4.2142	3.3704	3.0558	2.8765
0.75	3.1390	3.0287	2.9175	2.8020	2.6784	2.5432	4.3225	3.5213	3.0066	2.7821	2.7663
1.00	3.0930	2.9670	2.8388	2.7023	2.5517	2.3794	3.7633	3.2089	2.8125	2.7155	2.7455
1.25	3.0591	2.9216	2.7801	2.6268	2.4522	2.2448	3.4573	3.0199	2.6841	2.6993	2.7213
1.50	3.0331	2.8867	2.7350	2.5678	2.3718	2.1311	3.2582	2.8882	2.6259	2.6809	2.6982
1.75	3.0125	2.8591	2.6994	2.5207	2.3057	2.0333	3.1149	2.7889	2.6142	2.6630	2.6771
2.00	2.9959	2.8369	2.6707	2.4826	2.2504	1.9479	3.0050	2.7101	2.6017	2.6462	2.6583
3.00	2.9523	2.7791	2.5970	2.3838	2.0993	1.6898	2.7293	2.5035	2.5560	2.5922	2.6002
4.00	2.9270	2.7460	2.5561	2.3302	2.0115	1.5129	2.5723	2.3961	2.5212	2.5544	2.5607
5.00	2.9102	2.7241	2.5298	2.2971	1.9563	1.3819	2.4671	2.3744	2.4949	2.5265	2.5321
10.00	2.8691	2.6706	2.4684	2.2263	1.8494	1.0217	2.2109	2.3094	2.4237	2.4529	2.4573
15.00	2.8508	2.6464	2.4414	2.1978	1.8164	0.8474	2.0996	2.2780	2.3906	2.4191	2.4231
20.00	2.8399	2.6317	2.4251	2.1809	1.7990	0.7398	2.0698	2.2588	2.3706	2.3988	2.4026
30.00	2.8270	2.6139	2.4054	2.1606	1.7792	0.6090	2.0483	2.2362	2.3469	2.3745	2.3781
40.00	2.8193	2.6031	2.3935	2.1484	1.7674	0.5296	2.0354	2.2226	2.3327	2.3600	2.3632
50.00	2.8141	2.5956	2.3853	2.1400	1.7594	0.4748	2.0265	2.2132	2.3229	2.3499	2.3527
100.00	2.7982	2.5744	2.3632	2.1179	1.7388	0.3374	2.0034	2.1884	2.2969	2.3229	2.3244

TABLE 21.5
$\eta_{et}(\beta, g_w)$

	g_w										
β	0.0	0.2	0.4	0.6	0.8	1.0	1.5	2.0	3.0	4.0	5.0
SP	5.3324	4.8710	4.6533	4.5293	4.4510	4.3965	4.3167	4.2725	4.2259	4.2014	4.1871
0.5SP	3.5622	3.5957	3.6167	3.6586	3.7061	3.6436	3.6513	3.6550	3.6600	3.6599	3.6607
0.00	3.4717	3.4717	3.4717	3.4717	3.4717	3.4717	3.4717	3.4717	3.4717	3.4717	3.4717
0.25	3.3959	3.3672	3.3403	3.3151	3.2914	3.2689	3.2174	3.1715	3.0925	3.0257	2.9680
0.50	3.3501	3.3061	3.2663	3.2299	3.1963	3.1653	3.0965	3.0371	2.9384	2.8581	2.7905
0.75	3.3183	3.2643	3.2165	3.1736	3.1345	3.0988	3.0206	2.9544	2.8464	2.7601	2.6883
1.00	3.2947	3.2334	3.1800	3.1325	3.0898	3.0510	2.9670	2.8967	2.7832	2.6935	2.6193
1.25	3.2762	3.2093	3.1516	3.1009	3.0555	3.0145	2.9264	2.8533	2.7362	2.6442	2.5687
1.50	3.2611	3.1897	3.1287	3.0754	3.0280	2.9854	2.8943	2.8191	2.6994	2.6060	2.5294
1.75	3.2485	3.1733	3.1096	3.0543	3.0054	2.9615	2.8680	2.7913	2.6697	2.5751	2.4979
2.00	3.2379	3.1594	3.0935	3.0365	2.9863	2.9413	2.8460	2.7681	2.6450	2.5496	2.4718
3.00	3.2069	3.1192	3.0469	2.9853	2.9316	2.8839	2.7839	2.7029	2.5762	2.4789	2.4001
4.00	3.1867	3.0928	3.0165	2.9522	2.8964	2.8472	2.7445	2.6618	2.5334	2.4352	2.3559
5.00	3.1722	3.0737	2.9947	2.9284	2.8713	2.8210	2.7166	2.6330	2.5035	2.4048	2.3253
10.00	3.1338	3.0327	2.9364	2.8655	2.8052	2.7526	2.6446	2.5590	2.4276	2.3283	2.2487
15.00	3.1159	2.9984	2.9089	2.8360	2.7744	2.7270	2.6116	2.5254	2.3936	2.2943	2.2148
20.00	3.1050	2.9835	2.8920	2.8180	2.7557	2.70[illegible]8	2.5915	2.5050	2.3732	2.2740	2.1947
30.00	3.0921	2.9653	2.8714	2.7962	2.7331	2.67[illegible]7	2.5679	2.4812	2.3491	2.2501	2.1710
40.00	3.0844	2.9[illegible]42	2.8590	2.7830	2.7195	2.6649	2.5537	2.4668	2.3347	2.2358	2.1568
50.00	3.0791	2.9465	2.8504	2.7740	2.7102	2.6534	2.5438	2.4569	2.3248	2.2259	2.1471
100.00	3.0605	2.9220	2.8247	2.7465	2.6828	2.62[illegible]1	2.5156	2.4292	2.2987	2.2003	2.1218

TABLE 21.6
$f''_w(\beta, g_w)$

β	g_w 0.0	0.2	0.4	0.6	0.8	1.0	1.5	2.0	3.0	4.0	5.0
SP	0.0001	0.0001	0.0001	0.0001	0.0000	0.0002	0.0000	0.0000	0.0000	0.0002	0.0000
0.5SP	0.4063	0.3743	0.3525	0.3178	0.2812	0.3203	0.3089	0.3024	0.2942	0.2918	0.2896
0.00	0.4696	0.4696	0.4696	0.4696	0.4696	0.4696	0.4696	0.4696	0.4696	0.4696	0.4696
0.25	0.5344	0.5757	0.6161	0.6555	0.6941	0.7319	0.8283	0.9121	1.0805	1.2399	1.3924
0.50	0.5811	0.6550	0.7262	0.7952	0.8623	0.9277	1.0849	1.2348	1.5177	1.7835	2.0366
0.75	0.6181	0.7198	0.8173	0.9112	1.0021	1.0904	1.3019	1.5026	1.8799	2.2332	2.5686
1.00	0.6489	0.7755	0.8963	1.0122	1.1241	1.2326	1.4916	1.7367	2.1963	2.6259	3.0332
1.25	0.6754	0.8249	0.9668	1.1027	1.2336	1.3603	1.6622	1.9473	2.4810	2.9792	3.4511
1.50	0.6987	0.8695	1.0310	1.1854	1.3338	1.4772	1.8185	2.1403	2.7420	3.3031	3.8342
1.75	0.7196	0.9104	1.0903	1.2618	1.4266	1.5857	1.9636	2.3196	2.9845	3.6039	4.1900
2.00	0.7386	0.9483	1.1456	1.3334	1.5135	1.6872	2.0996	2.4877	3.2118	3.8859	4.5236
3.00	0.8013	1.0790	1.3382	1.5836	1.8182	2.0439	2.5781	3.0793	4.0121	4.8790	5.6980
4.00	0.8502	1.1874	1.5003	1.7954	2.0769	2.3473	2.9857	3.5836	4.5946	5.7257	6.6992
5.00	0.8907	1.2816	1.6427	1.9824	2.3056	2.6158	3.3469	4.0307	5.2998	6.4765	7.5869
10.00	1.0308	1.6422	2.1980	2.7162	3.2066	3.6752	4.7751	5.7995	7.5948	9.4477	11.0994
15.00	1.1231	1.9114	2.6208	3.2789	3.8996	4.4915	5.8774	7.1656	9.5449	11.7428	13.8125
20.00	1.1935	2.1349	2.9757	3.7528	4.4843	5.1807	6.8089	8.3203	11.1091	13.6833	16.1063
30.00	1.2997	2.5045	3.5688	4.5472	5.4655	6.3382	8.3744	10.2614	13.7387	16.9457	19.9628
40.00	1.3805	2.8124	4.0670	5.2164	6.2930	7.3148	9.6959	11.9002	15.9592	19.7006	23.2193
50.00	1.4463	3.0815	4.5051	5.8057	7.0221	8.1755	10.8611	13.3453	17.9173	22.1299	26.0911
100.00	1.6701	4.1255	6.2186	8.1158	9.8829	11.5545	15.4370	19.0214	25.6091	31.6733	37.3727

TABLE 21.7
$G'_w(\beta, g_w)$

β	g_w 0.0	0.2	0.4	0.6	0.8	1.0	1.5	2.0	3.0	4.0	5.0
SP	0.2479	0.2826	0.3013	0.3128	0.3204	0.3259	0.3342	0.3389	0.3441	0.3468	0.3485
0.5SP	0.4530	0.4466	0.4425	0.4348	0.4263	0.4371	0.4354	0.4345	0.4334	0.4333	0.4331
0.00	0.4696	0.4696	0.4696	0.4696	0.4696	0.4696	0.4696	0.4696	0.4696	0.4696	0.4696
0.25	0.4846	0.4909	0.4970	0.5027	0.5082	0.5135	0.5258	0.5371	0.5573	0.5750	0.5909
0.50	0.4942	0.5045	0.5140	0.5228	0.5311	0.5390	0.5569	0.5729	0.6007	0.6247	0.6457
0.75	0.5012	0.5143	0.5262	0.5371	0.5473	0.5568	0.5783	0.5972	0.6298	0.6574	0.6816
1.00	0.5067	0.5219	0.5357	0.5482	0.5597	0.5705	0.5945	0.6156	0.6515	0.6817	0.7081
1.25	0.5111	0.5281	0.5433	0.5571	0.5698	0.5815	0.6075	0.6302	0.6687	0.7009	0.7289
1.50	0.5148	0.5334	0.5498	0.5646	0.5781	0.5906	0.6183	0.6423	0.6828	0.7167	0.7460
1.75	0.5179	0.5378	0.5553	0.5710	0.5853	0.5934	0.6275	0.6526	0.6948	0.7300	0.7604
2.00	0.5206	0.5417	0.5601	0.5766	0.5915	0.6052	0.6354	0.6615	0.7052	0.7414	0.7728
3.00	0.5289	0.5536	0.5748	0.5935	0.6104	0.6258	0.6594	0.6882	0.7361	0.7756	0.8096
4.00	0.5346	0.5619	0.5851	0.6054	0.6236	0.6401	0.6760	0.7066	0.7573	0.7990	0.8347
5.00	0.5389	0.5682	0.5929	0.6144	0.6335	0.6509	0.6885	0.7204	0.7731	0.8164	0.8534
10.00	0.5510	0.5866	0.6157	0.6406	0.6625	0.6822	0.7245	0.7600	0.8182	0.8657	0.9062
15.00	0.5572	0.5963	0.6277	0.6543	0.6776	0.6985	0.7431	0.7804	0.8413	0.8908	0.9330
20.00	0.5611	0.6027	0.6356	0.6633	0.6875	0.7091	0.7551	0.7935	0.8561	0.9069	0.9501
30.00	0.5660	0.6108	0.6456	0.6748	0.7001	0.7226	0.7704	0.8101	0.8748	0.9271	0.9716
40.00	0.5691	0.6160	0.6521	0.6821	0.7080	0.7311	0.7801	0.8206	0.8865	0.9399	0.9851
50.00	0.5712	0.6197	0.6567	0.6873	0.7137	0.7372	0.7869	0.8280	0.8948	0.9488	0.9947
100.00	0.5768	0.6297	0.6689	0.7012	0.7288	0.7533	0.8049	0.8475	0.9165	0.9722	1.0194

and (21.56). All (3 × 231) = 693 E_v values are below 0.9%, with the overwhelming majority considerably below this value. The largest values for E_t occur when $S = 0.9$ and $\beta = \beta_{sp}$. These are 1.40% and 1.11%, respectively, when $g_w = 0$ and 0.2. All other E_t values are below 0.9%, usually considerably so.

An important reason for the uniformly small E_v values is Equation (21.55a). Other relations were tried, including smooth interpolation formulas, but these resulted in substantially larger E_v values. With Equation (21.55a), the maximum E_v value in the middle region, between the solid lines, of Table 21.4 is only 0.285%. As a consequence of Equation (21.55a), however, η_{ev} does not have a smooth variation when $g_w > 1$, as shown in Figure 21.8. Only η_{ev} is subject to this type of behavior. The parameters in the other tables all have a smooth variation.

Another assessment of the accuracy is obtainable from the β_{sp} values for f''_w in Table 21.6. An occasional 1 or 2 appears in the fourth decimal place. (The fifth decimal place is used for rounding.) Finally, results have been compared with those previously published in Marvin and Sinclair (1967), Back (1970), Cohen and Reshotko (1956a), Pade et al. (1985), Back (1976), and Narayana and Ramamoorthy (1972). When results overlap, agreement is excellent. For instance, Table 21.1 lists the separation values of Cohen and Reshotko (1956a) in the last column.

Negative Thicknesses

From Equation (21.74), the displacement thickness δ^* is negative when

$$(1 - g_w)C_t - C_v > \frac{Sf''_w}{1 + (1 - S)\beta}$$

This thickness is negative when the boundary-layer mass flux exceeds $(\rho u)_e$, which occurs when the wall is highly cooled and β is large. In this case, the density in the boundary layer exceeds that in the freestream; nevertheless, the above inequality also holds when $g_w = 0.2$, $\beta = 1.25$, and S is 0.38 or less. Generally, negative δ^* conditions occur in the throat region of a highly cooled rocket nozzle or on the shoulder of a cooled body, as occurred in the experiment described in Section 21.6 when $r_s/R = 0.05$. When $g_w \geq 1$, the above inequality shows that δ^* cannot be negative. Similarly, the momentum defect thickness θ is negative when

$$C_v - (1 - g_w)C_t > \frac{f''_w}{\beta}$$

This inequality holds when there is sufficient velocity overshoot, for instance, when $g_w = 1.5$ and $\beta = 10$. It is easy to see that δ, δ_t, and ϕ are never negative.

Incompressible Limit

The incompressible limit is provided by $M_e = 0$ or $S = 0$. For Equations (20.13) and (21.28) to be in accord, $g(\eta)$ and g_w must equal unity. Because $S = 0$, $T = T_o$ holds throughout the boundary layer. Hence, the incompressible boundary layer of a perfect gas is isothermal and the Stanton number is zero. If heat transfer is to be considered, we must forgo the assumption of a thermally perfect gas. In this regard, observe that the incompressible analysis of Chapter 20 is not restricted to a perfect gas, and the associated boundary layer need not be isothermal nor the Stanton number equal to zero. Nevertheless, Equations (20.16b) and (21.65) for the skin friction coincide, and, for example, Equation (21.74) correctly simplifies to $\Upsilon\delta^* = C_v$ when $S = 0$ and $g_w = 1$.

21.9 ADIABATIC WALL

The formulation in the preceding sections assumes a known wall temperature. Alternatively, a nonzero wall heat flux could be prescribed, in which case G'_w is known. We obtain G_w from Equation (21.32a), evaluated at $\eta \to \infty$:

$$G_w = 1 - G'_w \int_0^\infty \exp\left(-\int_0^{\eta'} f d\eta'\right) d\eta''$$

which then determines the gas temperature at the wall, T_w. (This is not the recovery temperature defined in Section 16.3, which requires $q_w = 0$.) A solution is obtained by solving the heat conduction equation for the wall using G'_w and T_w as boundary conditions. For a self-consistent solution, an iterative procedure is required, involving both the gas flow and the conduction process within the wall.

Since the heat flux q_w is usually unknown, this procedure is generally not pursued. The one major exception is the adiabatic wall case. This assumption is often warranted because of the small thermal conductivity of the gas and of the wall material. For instance, composites, ceramics, and many other materials have relatively small thermal conductivities.

From Equations (21.33), we have $q_w = 0$ when $g_w = 1$. This is compatible with the unique solution

$$g(\eta) = 1 \qquad (21.77)$$

to Equation (21.25) and its boundary conditions

$$g'_w = 0, \qquad g(\infty) = 1$$

From Equation (21.12), we see that the adiabatic wall boundary layer of a perfect gas has a constant stagnation enthalpy, which equals h_{oe}. The temperature and density profiles across the layer are provided by Equation (21.50) with $g_w = 1$ as

$$\frac{T}{T_e} = \frac{\rho_e}{\rho} = \frac{1 - Sf'^2}{1 - S} \qquad (21.78)$$

Since $g_w = 1$, the velocity profile has no overshoot and the gas temperature varies monotonically from its wall-adjacent value of T_{oe} to its lower T_e value at the outer edge of the boundary layer. This profile is sketched in Figure 21.13, where a tilde denotes an adiabatic wall. Observe that the temperature has a maximum value and a zero gradient at the wall; hence, $q_w = 0$. This profile is in contrast to a solution with wall heat transfer, where the temperature may exhibit a pronounced overshoot in a high-speed flow. As with temperature overshoot, the increase in temperature as the adiabatic wall is approached is due to viscous dissipation. The wall's temperature itself is unknown. In Section 16.3 the gas temperature at the wall was referred to as the recovery temperature, T_r. As pointed out in that section, the identification, $T_r = T_{oe}$, requires $Pr = 1$.

With $g_w = 1$, Equation (21.28) reduces to the Falkner–Skan equation of Chapter 20. Thus, adiabatic wall values for the nonthermal parameters

$$C_v,\ \eta_{ev},\ f''_w,\ c_f,\ \Upsilon\delta,\ \Upsilon\delta^*,\ \Upsilon\theta$$

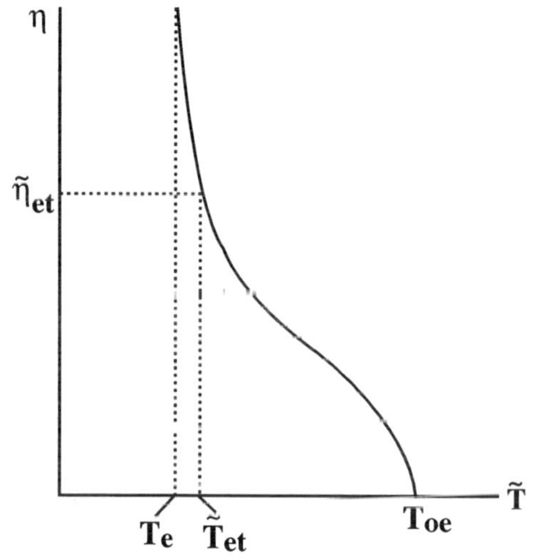

FIGURE 21.13 Schematic of the temperature profile for an adiabatic wall.

are based on the $g_w = 1$, column in Tables 21.2, 21.4, and 21.6. Values for thermal parameters that involve the energy equation are discussed shortly.

Despite referring to $g(\eta) = 1$ as the unique adiabatic wall solution, there is a second possibility that requires discussion. This is the $g_w = 1$ solution of Equations (21.28) to (21.30). The solution to the first of these equations yields the Falkner–Skan result of the above paragraph. With the diabatic wall condition, $G_w = 0$, a positive value is obtained for G_w', as shown in Table 21.7 and in Back (1970). Evidently, the $G_w = 0$ boundary condition does *not* correspond to an adiabatic wall. In this regard, Reynolds' analogy, Equation (21.67), yields a positive value for the Stanton number, which is incorrect for an adiabatic wall. Moreover, the right side of Equations (21.26), which defines G, is indeterminate, since both g_w and $g(\eta)$ equal unity when the wall is adiabatic. As shown in Problem 21.13, this indeterminacy can be resolved by setting $g_w = 1 - \varepsilon$, where ε is small compared to unity. In the process of obtaining the part (c) result of this problem, we also obtain $g(\eta) = 1$ as the adiabatic wall solution. The foregoing indeterminacy is a consequence of defining g and G in terms of h_o, which is constant in an adiabatic wall boundary layer. More simply, the energy equation reduces to $h_o = h_{oe}$ instead of Equation (21.29).

Adiabatic Wall Parameters

The remaining thermal parameters $\tilde{\eta}_{et}$, $\tilde{C}_t$, $\Upsilon\tilde{\delta}_t$, $\Upsilon\tilde{\phi}$, and $\tilde{S}t$ require revision. We discard $\tilde{C}_t$, since it is no longer needed, set $\tilde{S}t = 0$, and redefine the other three parameters in terms of a temperature profile. Recall that the profile is sketched in Figure 21.13, which shows the thermal boundary-layer edge at $\tilde{\eta}_{et}$, where the temperature is $\tilde{T}_{et}$. We utilize for this edge temperature the relation

$$\frac{T_{oe} - \tilde{T}_{et}}{T_{oe} - T_e} = \frac{\frac{T_{oe}}{T_e} - \frac{\tilde{T}_{et}}{T_e}}{\frac{T_{oe}}{T_e} - 1} = \frac{1 + \frac{\gamma-1}{2}M_e^2 - \frac{1 - S[f'(\eta_{et})]^2}{1-S}}{\frac{\gamma-1}{2}M_e^2} = [f'(\tilde{\eta}_{et})]^2$$

Since the argument of f' is $\tilde{\eta}_{et}$, an additional parameter table is avoided by the artifice of setting (Bae and Emanuel, 1989)

$$\tilde{\eta}_{et} = \eta_{ev}(\beta, 1) \tag{21.79}$$

This yields for the $\tilde{T}_{et}$ edge temperature

$$\frac{T_{oe}-\tilde{T}_{et}}{T_{oe}-T_e} = 0.99^2 = 0.9801 \tag{21.80}$$

and $T_{oe}-\tilde{T}_{et}$ is 98% of the overall temperature change $T_{oe}-T_e$.

Equations (21.56) and (21.78) now yield

$$\Upsilon\tilde{\delta}_t = \int_0^{\tilde{\eta}_{et}} \frac{\rho_e}{\rho}\,d\eta = \int_0^{\tilde{\eta}_{et}} \frac{1-Sf'^2}{1-S}\,d\eta$$

Equation (21.68g), with $g_w = 1$, becomes

$$\int_0^{\eta} f'^2 d\eta = \frac{1}{1+\beta} f''(ff' - \beta\eta - f''_w)$$

Combining the foregoing results in

$$\Upsilon\tilde{\delta}_t = \eta_{ev}(\beta,1) + \frac{S}{(1-S)(1+\beta)}(f''_w + C_v) \tag{21.81}$$

which is in accordance with Equation (21.76) when $g_w = 1$, and η_{et} is replaced with $\eta_{ev}(\beta,1)$.

For consistency with Equation (21.80), the G factor in ϕ, Equation (21.62), is replaced with

$$\frac{T_{oe}-\tilde{T}}{T_{oe}-T_e} = f'^2$$

Equation (21.62) now becomes

$$\Upsilon\tilde{\phi} = \int_0^{\infty} f'(1-f'^2)\,d\eta \tag{21.82}$$

The f'^3 part of the integrand can be replaced by Equation (21.68h), which introduces a numerically unevaluated integral whose integrand contains f''^2 (see Problem 21.4). Equation (21.82) is therefore not utilized for an adiabatic wall. In summary, we see that η_{et}, $\Upsilon\delta_t$, and $\Upsilon\phi$ are replaced with $\tilde{\eta}_{et}$, $\Upsilon\tilde{\delta}_t$, and $\Upsilon\tilde{\phi}$, respectively, where the last item has not been evaluated.

21.10 CRITIQUE OF THE PRANDTL NUMBER AND CHAPMAN–RUBESIN PARAMETER ASSUMPTIONS

The $Pr = 1$ and $C = 1$ approximations are useful for obtaining first estimates and establishing trends. Accurate values for the skin friction and heat transfer, however, require more precise values. For air, e.g., the Prandtl number is often given as 0.71 or 0.72. Similarly, the viscosity can be modeled as

$$\mu \sim T^{\omega}$$

although, for air, Sutherland's formula is often used. At moderate temperatures, $\omega = 0.7$ is a frequent choice for air.

Discussion of the Chapman–Rubesin Parameter

With the above relation and a perfect gas, we have

$$C(\eta) = \frac{\rho\mu}{\rho_e\mu_e} = \frac{T_e}{T}\left(\frac{T}{T_e}\right)^{\omega} = \left(\frac{T_e}{T}\right)^{1-\omega} \tag{21.83}$$

By definition, $C_e = 1$, whereas at the wall, we obtain

$$C_w = \left(\frac{1-S}{g_w}\right)^{1-\omega}$$

These relations hold for adiabatic (with $g_w = 1$) and diabatic walls. With $0 \le \omega \le 1$ and an adiabatic wall, C_w lies between zero and one, whereas for a diabatic wall, C_w exceeds unity when $g_w \le 1 - S$.

The value of C_w is of importance in view of its possible effect on the skin friction and heat transfer. With $C_e = 1$, the $C(\eta) = 1$ assumption can be expected to be accurate whenever $C_w = 1$. [In between its wall and edge values, $C(\eta)$ may differ from unity.] Aside from the $\omega = 1$ case, C_w is unity for a diabatic wall when $g_w = 1 - S$ or $T_w = T_e$. For an adiabatic wall and $\omega = 1$, we readily see that $C_w = 1$ when $M_e = 0$.

A simple derivation sheds further light on $C(\eta)$ when $Pr = 1$ and the wall is adiabatic. In this circumstance, Equation (21.77) is an exact solution of Equation (21.46b) and its boundary conditions. As noted in the preceding section, h_o is now a constant throughout the boundary layer and we obtain

$$C(\eta) = \left(\frac{1-S}{1-Sf'^2}\right)^{1-\omega}$$

which yields the above C_w wall value when $\eta = 0$. With $(1-S)^{1-\omega}$ as an estimate for an average $C(\eta)$ value, we observe that $C = 1$ is accurate for a low-speed flow but becomes progressively inaccurate as S approaches unity (Wortmann and Mills, 1971). For instance, with $\omega = 0.7$, C_w equals 0.81 and 0.50 when $S = 0.5$ and 0.9, respectively. (Problems 21.11 and 21.16 provide alternate approaches for assessing the $C = 1$ assumption.)

Prandtl Number Discussion

The value of the Prandtl number has little effect on the skin friction, since Pr does not directly appear in Equation (21.46a). Its only effect on the skin friction is through g and $C(\eta)$. This weak dependence of c_f on Pr is evident in a number of studies (Van Driest, 1952; Back and Witte, 1966).

On the other hand, the Prandtl number has a significant effect on the heat transfer. This is generally accommodated by using Colburn's analogy (Colburn, 1933), which modifies Equation (21.67) to

$$\frac{2St}{c_f} = Pr^{-2/3}\left(\frac{G'_w}{f''_w}\right) \tag{21.84a}$$

In situations when Reynolds' analogy is not appropriate, we would use

$$St = Pr^{-2/3} St(Pr = 1) \tag{21.84b}$$

where the rightmost Stanton number assumes $Pr = 1$. Problem 17.5 also predicts the $Pr^{-2/3}$ dependence for the Stanton number for a stagnation point flow. With $Pr = 0.71$, e.g., we have $Pr^{-2/3} = 1.256$ and the Stanton number needs to be increased by this factor, relative to its $Pr = 1$ value. Prandtl number corrections to the thermal boundary-layer thicknesses are usually not made. However, see Problem 21.3(f) for one correction formula.

Flat Plate Comparison

Van Driest (1952) performed extensive calculations for air over a flat plate using $Pr = 0.75$ and 1, Equation (21.83) with $\omega = 0.76$, and also Sutherland's equation, for $M = M_e$, with values ranging from zero to 20. His skin-friction coefficient and Stanton number results are shown in Figures 21.14 and 21.15 as the downward sloping curves. The uppermost of these curves is for $(T_w/T_e) = 0.25$ while the lower one is for $(T_w/T_e) = 1.0$. Both curves use Sutherland's equation and $Pr = 0.75$. Much later,

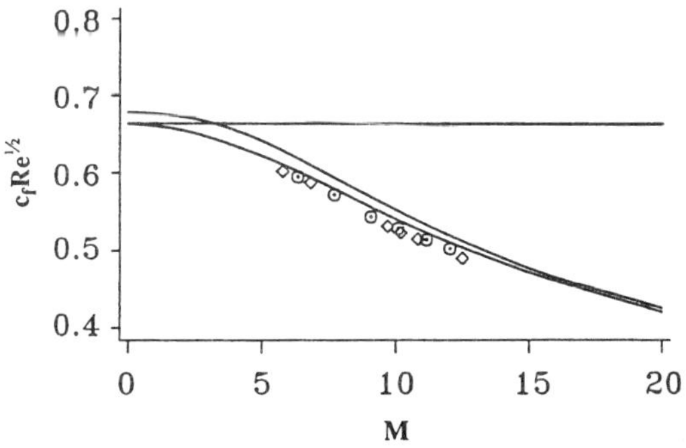

FIGURE 21.14 Local skin-friction coefficient vs. boundary-layer edge Mach number for flow over a flat plate (Bae, 1989)

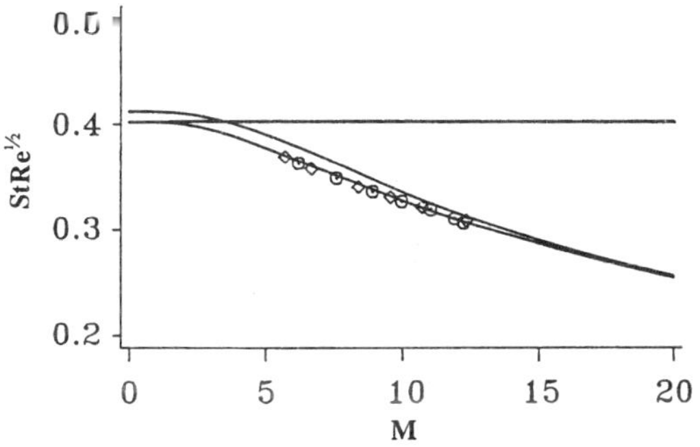

FIGURE 21.15 Local Stanton number vs. boundary-layer edge Mach number for flow over a flat plate (Bae, 1989).

Cook (1977) independently performed almost identical calculations; the circles and diamonds are his results when $(T_w/T_e) = 0.33$ and 1.0, respectively. The agreement between the separate calculations is excellent.

The horizontal line in Figure 21.14 stems from Equation (21.65) with

$$\beta = 0, \qquad f_w'' = 0.4696$$

so that

$$Re_{\bar{x}}^{1/2} c_f = 0.6641$$

The overprediction error in c_f, relative to the bottommost curve, at $M = 6$ is 8.8%. The horizontal line in Figure 21.15 represents Equation (21.84a), with $G_w = 0.4696$ and $Pr = 0.75$. Here the error at $M = 6$, relative to the bottommost curve, is 8.0%. It is important to note that the errors associated with both $Pr = 1$ and $C = 1$ are being simultaneously assessed. For both the skin friction and heat transfer, the errors at $M_e = 6$ are modest; they are even smaller at lower Mach numbers.

Comprehensive Comparison

The foregoing discussion is rather sketchy; e.g., only a flat plate flow is considered. Haridas (1995) provides a more comprehensive treatment. Equations (21.46) are used in similarity form

$$(Cf'')' + ff'' + \beta(g - f'^2) = 0 \tag{21.85a}$$

$$\left\{C\left[\frac{g'}{Pr} + 2\left(1 - \frac{1}{Pr}\right)Sf'f''\right]\right\}' + fg' = 0 \tag{21.85b}$$

with the boundary conditions

$$f(0) = f'(0) = 0, \qquad g(0) = g_w, \qquad f'(\infty) = g(\infty) = 1 \tag{21.85c}$$

The temperature and Chapman–Rubesin parameter are given by

$$\frac{T}{T_e} = \frac{g - Sf'^2}{1 - S}, \qquad C = \left(\frac{T}{T_e}\right)^{\omega - 1} \tag{21.86a,b}$$

and the computational parameter space is

$$\begin{aligned} \gamma &= 1.4 \\ M_e &= 0, 1, 3, 6 \\ g_w &= 0.2, 0.4, 0.6, 0.8 \\ \beta &= 0, 0.5, 1, 1.5, 2, 3, 4, 5 \\ Pr &= \omega = 0.7, 1 \end{aligned}$$

The subsequent discussion, including figures and the table, is entirely based on Haridas' study. Only the more significant findings, however, are presented here. For example, since the dependence on g_w is smooth, only results for g_w equal to 0.2 and 0.8 are given.

With the definitions

$$k_o = f, \quad k_1 = f', \quad k_2 = f'', \quad g_o = g, \quad g_1 = g' \tag{21.87}$$

the above relations are written as five first-order ODEs:

$$\frac{dk_o}{d\eta} = k_1, \quad \frac{dk_1}{d\eta} = k_2, \quad \frac{dg_o}{d\eta} = g_1 \tag{21.88a}$$

$$\frac{dk_2}{d\eta} = \frac{1}{g_o - Sk_1^2}\left[(1-\omega)(g_1 - 2Sk_1k_2)k_2 - \frac{(g_o - Sk_1^2)^{2-\omega}}{(1-S)^{1-\omega}}(k_ok_2 - \beta k_1^2 + \beta g_o)\right] \tag{21.88b}$$

$$\frac{dg_1}{d\eta} = \frac{1}{g_o - Sk_1^2}\left\{(1-\omega)(g_1 - 2Sk_1k_2)[g_1 + 2S(Pr-1)k_1k_2] - \frac{Pr}{(1-S)^{1-\omega}}(g_o - Sk_1^2)^{2-\omega}k_og_1\right\} - 2S(Pr-1)\left(k_2^2 + k_1\frac{dk_2}{d\eta}\right) \tag{21.88c}$$

These are subject to the boundary conditions

$$k_o(0) = 0, \quad k_1(0) = 0, \quad g_o(0) = g_w \tag{21.89}$$

$$k_1(\infty) = 1, \quad g_o(\infty) = 1 \tag{21.90a}$$

Equations (21.90a) are replaced with

$$k_2(0) = f_w'', \quad g_1(0) = g_w'' \tag{21.91}$$

These two conditions are iterated on, until

$$k_1(\eta^*) = 1, \quad g_o(\eta^*) = 1 \tag{21.90b}$$

are satisfied, where $\eta^* = 7$ is found to be sufficiently large enough to easily cover all cases within the parameter space. A fourth-order Runge–Kutta scheme is employed in conjunction with a modified Newton–Raphson method for the $k_2(0)$ and $g_1(0)$ iteration. All calculations are performed in double precision. A fixed $\Delta\eta$ step size is utilized, which yields six-digit agreement for all dependent variables with 10^{-3} and 10^{-2} $\Delta\eta$ step sizes. For the cases with $Pr = 1$ and $C = 1$, there is complete agreement with the earlier work of Bae and Emanuel (1989).

Figures 21.16 and 21.17, respectively, show the skin-friction coefficient and Stanton number vs. the pressure gradient parameter for g_w values of 0.2 and 0.8. The solid curves are for $M_e = 0$, 1, 3, and 6 when $Pr = \omega = 0.7$, while the dashed curves are for $Pr = \omega = 1$ and are independent of M_e. The curves consist of straight line segments that connect the above β values. (There is insufficient data to warrant spline-fit curves.)

The $Pr = \omega = 0.7$, $M_e = 0,1$ skin-friction curves are in accord with the $Pr = \omega = 1$ curve when $g_w = 0.8$. Since the spread with M_e in the 0.7 curves is modest, the $Pr = \omega = 1$ curves provide

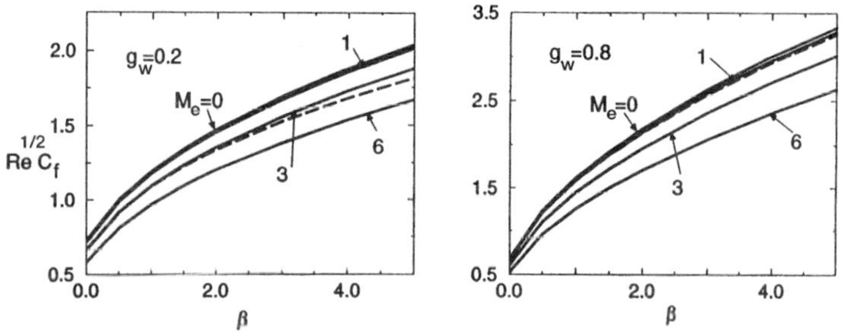

FIGURE 21.16 Skin-friction coefficient when $Pr = 1$, $\omega = 1$ (dashed curve), and when $Pr = 0.7$, $\omega = 0.7$ (solid curves) for $g_w = 0.2$ and 0.8.

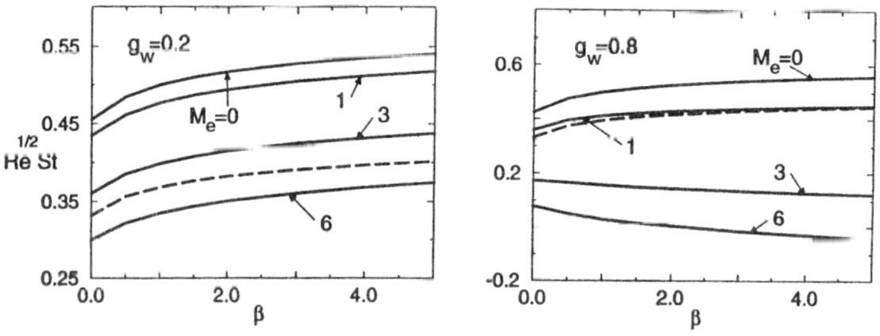

FIGURE 21.17 Stanton number when $Pr = 1$, $\omega = 1$ (dashed curve), and when $Pr = 0.7$, $\omega = 0.7$ (solid curves), for $g_w = 0.2$ and 0.8.

useful results at low M_e values and the correct trends as β increases. In part, this favorable comparison is due to the negligible dependence of the skin friction on the Prandtl number, which does not appear in the momentum equation. On the other hand, the $Pr = \omega = 1$ vs. the 0.7 Stanton number comparison is relatively poor at large g_w and M_e values. As will be shown, the low Mach number–Stanton number comparison improves significantly with the use of Colburn's analogy. This adjustment, however, cannot remove the disparity in trends with β as M_e increases when $g_w = 0.8$. Note that the wall is adiabatic when $M_e = 6, g_w = 0.8, Pr = \omega = 0.7$, and $\beta \cong 2$, while the corresponding $Pr = \omega = 1$ solution is not adiabatic.

Colburn's analogy is most simply given as Equation (21.84b) and does not require Reynolds' analogy for its application. It is nevertheless convenient to show a comparison in terms of a ratio of the Reynolds' analogy parameters:

$$E_{col} = \frac{Pr^{2/3}\left(2\frac{St}{c_f}\right)_{Pr\neq 1}}{\left(2\frac{St}{c_f}\right)_{Pr=1}} \tag{21.92}$$

where $E_{col} = 1$ when Colburn's analogy is exact. The two $2St/c_f$ factors are evaluated with the same β, g_w, and M_e values, and should also use the same value for ω. This latter step is unnecessary, however, since a comparison of cases with $Pr = 1$ and 0.7, both with $\omega = 0.7$, with cases where $\omega = 1$ and $Pr = 1$ shows a negligible ω effect (Haridas, 1995). Hence, Table 21.8 compares $Pr = \omega = 0.7$ cases with those for $Pr = \omega = 1$. Colburn's analogy at small M_e and β values is excellent. When $M_e = 1$, a g_w dependence is evident with best results when g_w is small. By the time $M_e = 3$, the analogy is marginally useful only when g_w is small.

TABLE 21.8
Colburn Analogy Error Ratio when $Pr = \omega = 0.7$ and $Pr = \omega = 1$

M_e	β	g_w	E_{col}
0	0	0.2	0.9857
		0.8	0.9916
	5	0.2	0.9468
		0.8	0.9604
1	0	0.2	0.9540
		0.8	0.8567
	5	0.2	0.9165
		0.8	0.7855
3	0	0.2	0.8611
		0.8	0.4562
	5	0.2	0.8311
		0.8	0.2379

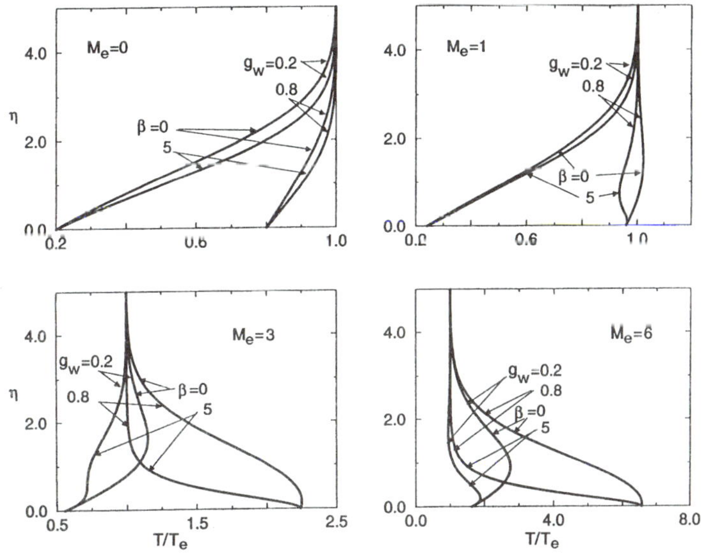

FIGURE 21.18 Temperature profiles when $Pr = 0.7$, $\omega = 0.7$, $g_w = 0.2$ and 0.8, $\beta = 0$ and 5, and for $M_e = 0$, 1, 3, 6.

Figure 21.18 shows temperature profiles for four M_e values when $g_w = 0.2$ and 0.8, $\beta = 0$ and 5, and $Pr = \omega = 0.7$. The $M_e = 0$ curves have the expected monotonic behavior. Panels for the larger M_e values show some temperature profiles with overshoot and some with undershoot. Overshoot, of course, is caused by viscous dissipation, which is most intense near the wall. Undershoot can occur when β is large and $M_e > 0$; i.e., there is a strong favorable pressure gradient that tends to cool the flow in the streamwise direction in the outer portion of the boundary layer. (These heating and cooling mechanisms should also occur in compressible transitional and turbulent boundary layers.) This streamwise cooling is evident from the fuller profiles for f' when the flow is accelerating (Haridas, 1995). By differentiating Equation (21.86a) with respect to η, we observe that the temperature extremums occur when

$$g' = 2Sf'f'' \qquad (21.93)$$

Moreover, *both* extremums can occur in a single profile, as is evident in the $M_e = 1$, $\beta = 5$, $g_w = 0.8$ and $M_e = 6$, $\beta = 5$, $g_w = 0.2$ profiles. In these double-reversal profiles, the heat transfer is into the wall, but there is a middle region, inside the boundary layer, where the transfer is in the opposite direction. In the latter instance, the heat transfer is into the cooler, outer portion of the boundary layer.

Double-reversal profiles are unexpected, since the corresponding g and f' profiles are sometimes monotonic, even with a double-reversal temperature profile. In the literature, profiles are occasionally shown for g, which represents the stagnation enthalpy, but neither f' nor g exhibits a profile with two extremums. The $Pr = \omega = 1$ temperature profiles are similar to those in Figure 21.18. The 0.7 curves are slightly smoother, however, since they correspond to a relatively large thermal conductivity. For example, the $M_e = 3, \beta = 5, g_w = 2$ curve has a double-reversal profile when $Pr = \omega = 1$ but does not when $Pr = \omega = 0.7$. Since the density is inverse to the temperature, a double-extremum profile can also occur for this parameter.

Laminar boundary-layer studies often invoke the Crocco–Busemann quadratic velocity relation for the temperature (White, 1991). This relation is exact when $Pr = \omega = 1$ and $\beta = 0$ but is inappropriate when there is a double reversal. This type of profile may not have been recognized previously because it does not occur in the commonly studied cases of an adiabatic wall or for flat plate or stagnation point flows. A second reason is that the first temperature reversal is close to the wall, as in the $M_e = 1$ and 3, $\beta = 5$, $g_w = 0.8$ profiles, or the undershoot is quite modest, as in the $M_e = 6, \beta = 5, g_w = 0.2$ profile. When the first reversal is close to the wall, it may be quite difficult to resolve with a nonsimilar boundary-layer code.

Viscous and thermal edges of the boundary layer are typically defined by

$$f'(\eta_{ev}) = 0.99, \qquad g(\eta_{et}) = 0.99 + 10^{-2} g_w$$

Figure 21.19 shows the Mach number when $\eta = \eta_{ev}$ for $M_e = 3$ and 6, $g_w = 0.2$ and 0.8, and $Pr = \omega = 0.7$. The edge value, $M_{\eta_{ev}}$, is close to 3 when $M_e = 3$ but is well below 6 when $M_e = 6$, except when β is large and g_w is small. A similar result is obtained for T/T_e and ρ/ρ_e, and with the thermal edge thickness η_{et} (Haridas, 1995). That is, when M_e is large, η_{ev} and η_{et} may significantly underpredict the layer's thickness. This result also holds for $Pr = \omega = 1$. The underprediction when $Pr = \omega = 0.7$ is, in part, caused by a small overshoot in the f' profile. Overshoot in this profile, however, does not occur when $Pr = \omega = 1$. In this case, the profiles go to unity at different rates. For instance, when $Pr = \omega = 1$, $M_e = 6$, $\beta = 0$, and $g_w = 0.8$, we have $T/T_e = 1.01$ at $\eta = 4.30$,

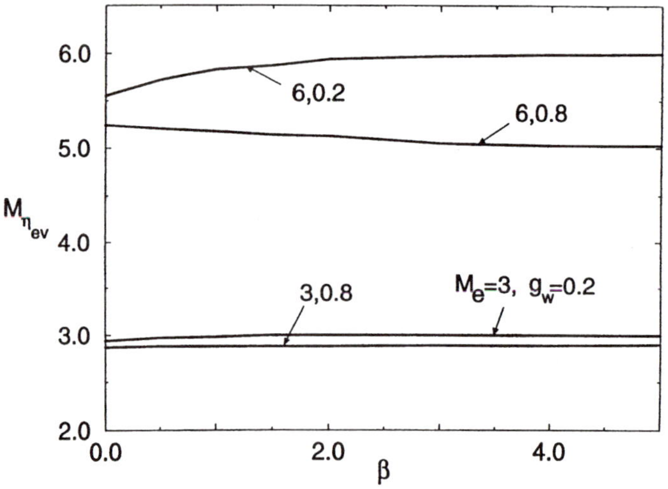

FIGURE 21.19 Mach number when $\eta = \eta_{ev}$ for $Pr = 0.7$, $\omega = 0.7$, $M_e = 3$ and 6, and $g_w = 0.2$ and 0.8.

but $f' = 0.99$ at $\eta = 3.47$ and $g = 0.99$ at $\eta = 2.77$. An underprediction by η_{ev} and η_{et} results in a similar underprediction for the viscous and thermal thicknesses, δ and δ_t.

Summarizing Discussion

In summary, we observe that the use of $Pr = 1$ does not cause a significant error in the skin friction, while the Colburn analogy yields accurate results for the Stanton number, especially at low Mach numbers, and at moderate supersonic Mach numbers for a cold wall. The use of $C = 1$ is also accurate at low Mach numbers but results in a slowly increasing error for an adiabatic wall as the edge Mach number increases. Nevertheless, both approximations, with the Colburn analogy, provide satisfactory results for a flat plate when $M_e \leq 6$. A similar statement holds for other wall configurations; e.g., Back and Witte (1966) show this to be the case for stagnation point flow. Additional discussion of these approximations can be found in Schlichting (1979), Dewey and Gross (1967), and Back (1970).

At Mach number values larger than six, the question of accuracy of the $Pr = 1$ and $C = 1$ approximations is less relevant, since Equations (21.46) may need to be replaced with a formulation that incorporates real-gas effects (Nagamatsu and Kim, 1986) and chemical reactions (Blottner, 1970). For air at a large M_e value, vibrational excitation, molecular dissociation, and the formation of nitrogen oxide should be considered. At somewhat higher Mach numbers, radiative heat transfer becomes important. The results shown in Figures 21.14 and 21.15 are theoretical; they should not be relied on at Mach numbers of 6 or more.

In conclusion, Tables 21.1 to 21.7, in conjunction with the equations of the last few sections and the Colburn analogy, provide an efficient means for obtaining estimates for various boundary-layer parameters when the external Mach number is subsonic or has a modest supersonic value. Of primary importance are the skin friction, heat transfer, and displacement thickness. Perhaps the most difficult aspect of these estimates is obtaining values for β. We illustrate the overall approach with several examples in the next chapter.

21.11 NONSIMILAR BOUNDARY LAYERS — I

A boundary layer may be nonsimilar because of rapid changes in the inviscid flow or because of wall changes, such as its shape or temperature distribution. These lead to changes in β and g_w. In addition, suction, transpiration cooling, or ablation may occur, in which case the boundary layer is generally nonsimilar and f_w and f'_w are not zero.

A variety of techniques has been developed for solving the nonsimilar, boundary-layer equations. These techniques can be subdivided into incompressible or compressible techniques, since incompressible techniques often are not suitable for the compressible equations. Approaches are further subdivided according to the choice of dependent and independent variables. Thus, one might solve Equations (21.5) to (21.7), which use primitive variables, or Equations (21.46) in which all dependent and independent variables are transformed. The choice for the transformed variables is by no means unique. In early studies, Crocco variables (Van Driest, 1952; Cook, 1977; Ruger, 1967) are sometimes used.

The discussion in this section focuses on numerical methods for directly solving the boundary-layer equations as PDEs. An alternative approach is to first reduce the equations to ODEs, which are then solved numerically. This topic is the subject of the next section.

Incompressible Flow

Smith and Clutter (1963) use a numerical technique to solve the transformed boundary-layer equations. A number of flows are solved, including that approaching a separation point and around a cylinder and a sphere. For the separation point, the Howarth flow speed is utilized:

$$u_e = 1 - \frac{1}{8}x$$

which results in an adverse pressure gradient. For some of the cases, such as a circular cylinder, experimental wall measurements provide u_e. A number of subsequent papers compare their results with those in Smith and Clutter (1963), which are viewed as providing exact numerical solutions.

Our discussion will focus on an approach for incompressible boundary layers given in Sparrow et al. (1970). (Also see the comments on this approach in Coxon and Parks, 1971; Sparrow, 1971; and Rogers, 1974.) After the method is presented, we briefly mention its extension to a compressible flow. With a unity Chapman–Rubesin constant, the incompressible version of Equation (21.46a) is

$$f_{\eta\eta\eta} + f f_{\eta\eta} + \beta(\xi)(1 - f_\eta^2) = 2\xi(f_\eta f_{\xi\eta} - f_\xi f_{\eta\eta}) \tag{21.94a}$$

with the boundary conditions

$$f(\xi, 0) = f_\eta(\xi, 0) = 0, \qquad f_\eta(\xi, \infty) = 1 \tag{21.94b}$$

We now introduce

$$g(\xi, \eta) = f_\xi \tag{21.95}$$

where g should not be confused with the g defined earlier. Equation (21.94a) becomes

$$f_{\eta\eta\eta} + f f_{\eta\eta} + \beta(1 - f_\eta^2) - 2\xi(f_\eta g_\eta - f_{\eta\eta} g) = 0 \tag{21.96}$$

where only derivatives with respect to η explicitly appear. This relation and Equations (21.94b) are differentiated with respect to ξ, with the result

$$g_{\eta\eta\eta} + f g_{\eta\eta} - 2(1 + \beta) f_\eta g_\eta + 3 f_{\eta\eta} g + \frac{d\beta}{d\xi}(1 - f_\eta^2) = 2\xi \frac{\partial}{\partial\xi}(f_\eta g_\eta - f_{\eta\eta} g) \tag{21.97a}$$

$$g(\xi, 0) = g_\eta(\xi, 0) = 0, \qquad g_\eta(\xi, \infty) = 0 \tag{21.97b}$$

The only nonsimilar terms are on the right side of Equations (21.94a) and (21.97a). For a nonsimilar flow we might use, as a first approximation, Equation (21.94a), with the right side set equal to zero. This results in the Falkner–Skan equation. A better approximation (hopefully) would use Equations (21.96) and (21.97) with only the right side of Equation (21.97a) set equal to zero. This assertion, however, is without mathematical justification. Intuitively, we might anticipate that the right side of Equation (21.97a) is relatively small compared to the left side in a nonsimilar flow, whereas the right side of Equation (21.94a) may not be as small compared to its left side. A better justification is that the solution of Equations (21.96) and (21.97) is in better accord with that of the Navier–Stokes equations than is the solution of Equations (21.94).

With the right side of Equation (21.97a) set equal to zero, Equations (21.96) and (21.97a) represent two coupled, ordinary differential equations of the sixth order for f and g. At a given wall location, values are required for the parameters ξ, β, and $d\beta/d\xi$ that appear in these equations. The equations are repetitively integrated numerically from the wall outward with assumed values for $f_{\eta\eta}(\xi, 0)$ and $g_{\eta\eta}(\xi, 0)$. The assumed values are correct when the external boundary conditions

$$f_\eta(\xi, \infty) = 1, \qquad g_\eta(\xi, \infty) = 0$$

are simultaneously satisfied.

This procedure can be extended by defining a third dependent variable

$$h = \frac{\partial g}{\partial \xi} = \frac{\partial^2 f}{\partial \xi^2}$$

and deriving an equation for h by differentiating Equation (21.97a) with respect to ξ, with the right side of this equation now retained. The only nonsimilar terms are placed on the right side of the h equation and they are set equal to zero. Thus, there are three coupled, ninth-order ODEs for which three infinity conditions must be simultaneously satisfied. The procedure appears to have been convergent when it was applied to a series of four flows (Sparrow et al., 1970). Local similarity thus becomes the first approximation in a systematic, but otherwise heuristic procedure.

Nath (1976) provides a modified version of the above approach that is computationally faster but more complex from the programming viewpoint. Good accuracy in the solution for the above four flows usually required the f, g, h ninth-order system of ODEs with three infinity conditions to be satisfied by iteration. For a compressible boundary layer, these numbers would be doubled. To the author's knowledge, neither the method nor its modifications appears to have been adapted to a compressible flow. In view of its complexity, this is not unexpected. It appears that the direct numerical solution of the primitive variable or transformed boundary-layer equations is both simpler and more efficient. This would certainly be the case, for instance, if nonequilibrium reactions need to be considered.

Compressible Flow

One of the earliest analyses of the compressible, nonsimilar boundary-layer equations is by Chapman and Rubesin (1949). Flow over a flat plate is considered when its wall temperature distribution is a prescribed polynomial. Since $\beta = 0$, the Blasius function $f(\eta)$ is a solution of Equation (21.46a). The constant, but nonunity, Chapman–Rubesin parameter is absorbed in normalizing the stream function. In their analysis, the energy equation is linear and the first step in its solution is the use of separation of variables.

Another early analysis of the nonsimilar compressible equations is provided by Bush (1961). The technique is unwieldy and apparently has not been further utilized, although Problem 21.3 is based on this approach. We previously discussed the experiment of Marvin and Sinclair (1967) in Section 21.6. They utilize the ξ,η independent variables that are favored in this chapter. Their numerical approach, which does not presume $Pr = 1$ or $C = 1$, is straightforward and effective. In addition to the foregoing references other nonsimilar studies are provided by Kramer and Lieberstein (1959), Smith and Clutter (1965), Sullivan (1970), Hsu and Liakopoulos (1982), Cebeci et al. (1983), Vasantha and Nath (1985), and [illegible] ([illegible]). These references, and the references contained therein, provide a diverse selection of numerical techniques as well as a selection of flow fields. This later point is evident by simply scanning their titles.

21.12 NONSIMILAR BOUNDARY LAYERS — II

General Discussion

Even in the elementary case of steady, incompressible flow over a flat plate, the boundary-layer equations are nonlinear PDEs. Blasius (1908) was able to reduce the problem to a nonlinear ODE for which approximate analytical and hand-generated numerical solutions could be obtained. A few years later, Hiemenz (1911) solved the problem of stagnation-point flow. There was, nevertheless, a pressing need for solving the incompressible boundary-layer equations for a surface with an arbitrary pressure gradient. Since Equations (21.5) and (21.6) are nonlinear, the method of solution would most likely be approximate. A key step was taken by von Kármán (1921), namely, to transform the streamwise momentum equation into an ODE by analytically integrating it across the boundary layer. Thus, the momentum equation is satisfied only on an averaged basis. This idea is the genesis of all integral methods. As later references will indicate, such methods are still popular since they provide an elegant and relatively simple way of obtaining nonsimilar solutions. It is worth noting that Kármán's integral equation holds for incompressible and compressible laminar and turbulent boundary layers. Shortly, we shall derive this equation.

In the journal article immediately following Kármán's paper, Pohlhausen (1921a) utilizes Kármán's equation to develop the first practical method for evaluating the skin friction for an incompressible, two-dimensional, laminar boundary layer. (Remember, this is a decade before the Falkner–Skan, 1931, publication.) In order to perform various integrals, such as the ones that yield the displacement and momentum thicknesses, the velocity profile, $u(s,n)/u_e(s)$, is written as an explicit, but approximate, function of $\eta = n/\delta(s)$. The e subscript here denotes the outer edge of the boundary layer, at $n = \delta$. A fourth-degree polynomial is utilized:

$$\frac{u}{u_e} = a\overline{\eta} + b\overline{\eta}^2 + c\overline{\eta}^3 + d\overline{\eta}^4 \tag{21.98}$$

which automatically satisfies $u(s,0) = 0$. The four coefficients are determined by three imposed conditions at $\overline{\eta} = 1$:

$$\frac{u}{u_e} = 1, \qquad \frac{\partial u}{\partial n} = 0, \qquad \frac{\partial^2 u}{\partial n^2} = 0 \tag{21.99}$$

and a wall compatibility condition

$$\left(\frac{\partial \tau}{\partial n}\right)_w = \frac{dp_e}{ds} \tag{21.100}$$

This relation is obtained by evaluating Equation (21.6) at the wall and by replacing $\mu(\partial u/\partial n)$ with the shear stress. (This replacement allows the momentum equation, on a time-averaged basis, to also hold for transitional and turbulent boundary layers.) It is through the compatibility relation that the pressure gradient enters into the formulation. Further details on Pohlhausen's approach are given at the end of the next subsection and can be found in many books, e.g., Schlichting (sixth edition), White (1991), and Walz (1969).

Integral methods for a steady, laminar boundary layer are provided by Walz (1969), Millikan (1932), Rott and Crabtree (1952), Libby et al. (1952), Tani (1954), Cohen and Reshotko (1956b), Pallone (1961), Libby and Fox (1965), and Lida and Fujimoto (1986), where Walz's monograph is primarily devoted to this topic. Millikan's (1932) early analysis is for incompressible, axisymmetric flow over a dirigible. A polynomial is used for the laminar portion of the boundary layer, while

$$\frac{u}{u_e} = \left(\frac{n}{\delta}\right)^{1/7}$$

is used for the turbulent portion. Rott and Crabtree (1952) survey the different approaches available at that time and investigate laminar, incompressible flow over a yawed, infinite circular cylinder. Libby et al. (1952) survey compressible integral methods and observe that Pohlhausen's approach provides reasonable results when the pressure gradient is favorable but is inadequate for an adverse gradient. This method is particularly poor for predicting the location of separation. In this paper, as well as in the others, adequacy is evaluated by comparing results with more exact boundary-layer solutions and with experimental results. A one-parameter formulation with a sixth-degree velocity profile is recommended by Libby et al. (1952).

The reason for the unsatisfactory adverse pressure gradient result is the use of Equation (21.100) (Tani, 1954), which only holds at the wall. Because separation largely depends on the overall velocity profile, Tani replaces the compatibility condition with a new integral relation, referred to as an energy integral. This relation is obtained by multiplying the longitudinal momentum equation by u and integrating with respect to n from zero to δ. Tani (1954) primarily deals with incompressible flow; the energy-integral relation should not be confused with an integrated form of the energy equation. Tani's method yields substantially better agreement for predicting separation. Cohen and Reshotko (1956b) treat the compressible boundary layer for a perfect gas by first transforming the equations to an incompressible form with a transformation somewhat different from that used in Section 21.2. Pallone (1961) uses a strip method for a compressible boundary layer in which the strips are in the flow direction. With a single strip, the method reduces to a standard integral method. The nonsimilar heat transfer and skin friction are evaluated (Pallone, 1961) along an impermeable wall that is located downstream of an injection-cooled surface. Libby and Fox (1965) develop a general integral moment method for a compressible, laminar boundary layer. As will be the case in the last subsection, they first transform the equations to ξ, η variables. Each of Equations (21.46) is multiplied by $\bar{\eta}^m$ and integrated from zero to $\eta = \eta_e(\xi)$, where η_e corresponds to $n = \delta$. Excellent results are obtained using m equal to zero and unity for both the transformed momentum and energy equations. A fourth-degree polynomial with one free parameter is used for u/u_e, while a fifth-degree polynomial with two free parameters is used for the stagnation enthalpy ratio, g. Lida and Fujimoto (1986) describe a numerically efficient method for a two-dimensional, incompressible laminar boundary layer. The unique feature of their approach is to use the energy-integral equation as a second ODE that is numerically integrated with the skin friction as an unknown. Three flows (flat plate, stagnation point, and the Howarth retarded flow) are utilized to develop approximate correlation formulas for various coefficients, such as δ^*/θ, in the two ODEs. As a consequence, a specific functional form for the velocity profile, such as Equation (21.98), is not required.

Truckenbrodt (1952), Rasmussen (1975), and Dey and Narasimha (1990) utilize the integral method for transitional and/or turbulent boundary layers. Lees and Reeves (1964) and Ko and Kubota (1969) apply the integral method to the interaction of a supersonic, laminar boundary layer with an impinging shock wave that not only can cause the boundary layer to separate but also to become transitional.

Kármán's Momentum Integral Equation

As previously indicated, we assume a compressible (or incompressible), steady boundary layer that may be laminar, transitional, or turbulent, or all three if the Reynolds number is sufficiently large. (Averaged equations are used if the boundary layer is transitional or turbulent.) At the edge of the boundary layer, where $n = \delta$, we prescribe

$$\rho = \rho_e, \qquad u = u_e, \qquad v = v_e, \qquad \tau = 0 \tag{21.101}$$

Our presentation first utilizes continuity to evaluate v_e, which is not zero. We shall need to evaluate the flux of mass across the boundary layer in order that v_e be given in terms of δ and δ^*. The momentum equation is then integrated across the boundary layer and, along with v_e and the momentum flux across the boundary layer, yields Kármán's equation. This subsection concludes with a brief description of how this equation is utilized to provide the skin friction.

Equation (21.5) is multiplied by dn and integrated from zero to δ, to yield

$$\int_0^\delta \frac{\partial}{\partial s}(r_w^\sigma \rho u)dn + \int_0^\delta \frac{\partial}{\partial n}(r_w^\sigma \rho v)dn = 0$$

or

$$r_w^\sigma(\rho v)_e = -\int_0^\delta \frac{\partial}{\partial s}(r_w^\sigma \rho u)dn \qquad \textbf{(21.102a)}$$

Leibniz's rule, see Section 1.6, can be used to obtain

$$\frac{d}{ds}\int_0^\delta r_w^\sigma \rho u\, dn = \int_0^\delta \frac{\partial}{\partial s}(r_w^\sigma \rho u)\, dn + r_w^\sigma(\rho u)_e \frac{d\delta}{ds}$$

On the other hand, the left side of the above equation can be written as

$$\frac{d}{ds}\int_0^\delta r_w^\sigma \rho u\, dn = \frac{d}{ds}\left(r_w^\sigma \int_0^\delta \rho u\, dn\right) = \sigma \frac{dr_w}{ds}\int_0^\delta \rho u\, dn + r_w^\sigma \frac{d}{ds}\int_0^\delta \rho u\, dn$$

Equation (21.102a) thus becomes

$$(\rho v)_e = -\frac{\sigma}{r_w}\frac{dr_w}{ds}\int_0^\delta \rho u\, dn - \frac{d}{ds}\int_0^\delta \rho u\, dn + (\rho u)_e \frac{d\delta}{ds} \qquad \textbf{(21.102b)}$$

The integrals are next eliminated in favor of boundary-layer thicknesses.

Equation (21.57) is approximated as

$$\delta^* = \int_0^\delta \left[1 - \frac{\rho u}{(\rho u)_e}\right] dn = \delta - \frac{1}{(\rho u)_e}\int_0^\delta \rho u\, dn$$

or

$$\int_0^\delta \rho u\, dn = (\rho u)_e(\delta - \delta^*) \qquad \textbf{(21.103a)}$$

This integral represents the mass flux, per unit depth, across the boundary layer when $\sigma = 0$. The mass flux for an axisymmetric flow is provided by the same integral multiplied by $2\pi r_w$. The s derivative of this equation produces

$$\frac{d}{ds}\int_0^\delta \rho u\, dn = \frac{d(\rho u)_e}{ds}(\delta - \delta^*) + (\rho u)_e\left(\frac{d\delta}{ds} - \frac{d\delta^*}{ds}\right) \qquad \textbf{(21.103b)}$$

These equations are substituted into Equation (21.102b), with the final result

$$(\rho v)_e = -\left[\frac{d(\rho u)_e}{ds} + \frac{\sigma}{r_w}\frac{dr_w}{ds}(\rho u)_e\right](\delta - \delta^*) + (\rho u)_e \frac{d\delta^*}{ds} \tag{21.104}$$

Equation (21.6) is now written as

$$\rho\left(u\frac{\partial u}{\partial s} + v\frac{\partial u}{\partial n}\right) = -\frac{dp_e}{ds} + \frac{\partial \tau}{\partial n}$$

We again multiply by dn and integrate, with the result

$$\int_0^\delta \rho u \frac{\partial u}{\partial s} dn + \int_0^\delta \rho v \frac{\partial u}{\partial s} dn = -\delta \frac{dp_e}{ds} - \tau_w \tag{21.105a}$$

where the rightmost of Equations (21.101) is used. The second integral on the left side can be integrated by parts:

$$\int_0^\delta \rho v \frac{\partial u}{\partial s} dn = (\rho u v)_e - \int_0^\delta u \frac{\partial(\rho v)}{\partial n} dn$$

However, continuity can be written as

$$\frac{\partial}{\partial n}(\rho v) = -\frac{1}{r_w^\sigma}\frac{\partial}{\partial s}(r_w^\sigma \rho u) = -\frac{\partial(\rho u)}{\partial s} - \frac{\sigma}{r_w}\frac{dr_w}{ds}(\rho u)$$

so that the above integral becomes

$$\int_0^\delta \rho v \frac{\partial u}{\partial n} dn = (\rho u v)_e + \frac{\sigma}{r_w}\frac{dr_w}{ds}\int_0^\delta \rho u^2 dn + \int_0^\delta u \frac{\partial(\rho u)}{\partial s} dn$$

This relation, along with Equation (21.4), is substituted into Equation (21.105a), with the result

$$\int_0^\delta \frac{\partial(\rho u^2)}{\partial s} dn + (\rho u v)_e + \frac{\sigma}{r_w}\frac{dr_w}{ds}\int_0^\delta \rho u^2 dn = (\rho u)_e \frac{du_e}{ds}\delta - \tau_w \tag{21.105b}$$

where two of the integrals combine to yield the leftmost one. Leibniz's rule provides

$$\frac{d}{ds}\int_0^\delta \rho u^2 dn = \int_0^\delta \frac{\partial(\rho u^2)}{\partial s} dn + (\rho u^2)_e \frac{d\delta}{ds}$$

so that Equation (21.105b) becomes

$$\frac{d}{ds}\int_0^\delta \rho u^2 dn + \frac{\sigma}{r_w}\frac{dr_w}{ds}\int_0^\delta \rho u^2 dn - (\rho u^2)_e \frac{d\delta}{ds} - (\rho u)_e \frac{du_e}{ds}\delta + (\rho u v)_e = -\tau_w \tag{21.105c}$$

and only a ρu^2 integral remains to be evaluated.

The momentum thickness, Equation (21.58), is approximate with

$$\theta = \int_0^\delta \frac{\rho u^2}{(\rho u)_e}\left(1 - \frac{u}{u_e}\right)dn = \frac{1}{(\rho u)_e}\int_0^\delta \rho u\, dn - \frac{1}{(\rho u^2)_e}\int_0^\delta \rho u^2 dn$$

We insert Equation (21.103a) and solve for the ρu^2 integral, to obtain

$$\int_0^\delta \rho u^2 dn = (\rho u^2)_e(\delta - \delta^* - \theta) \tag{21.106a}$$

where the integral represents the momentum flux. The s derivative of it yields

$$\frac{d}{ds}\int_0^\delta \rho u^2 dn = \frac{d(\rho u^2)_e}{ds}(\delta - \delta^* - \theta) + (\rho u^2)_e\left(\frac{d\delta}{ds} + \frac{d\delta^*}{ds} + \frac{d\theta}{ds}\right) \tag{21.106b}$$

Equations (21.104) and (21.106) are substituted into Equation (21.105c), with the result

$$\frac{d\theta}{ds} + \frac{d\ln\left[r_w^\sigma(\rho u^2)_e\right]}{ds}\theta + \frac{d\ln u_e}{ds}\delta^* = \frac{\tau_w}{(\rho u^2)_e} \tag{21.107a}$$

after some algebraic simplification. This is one form of Kármán's equation. If $\sigma = 0$ and the flow is incompressible, it reduces to the more familiar relation

$$\frac{d\theta}{ds} + (2\theta + \delta^*)\frac{1}{u_e}\frac{du_e}{ds} = \frac{\tau_w}{\rho u_e^2} \tag{21.107b}$$

When the external inviscid flow is isentropic at the wall and the gas is perfect, Equation (21.107a) becomes

$$\frac{d\theta}{ds} + \left[\frac{\sigma}{r_w}\frac{dr_w}{ds} + \frac{2 - M_e^2}{M_e\left(1 + \frac{\gamma-1}{2}M_e^2\right)}\frac{dM_e}{ds}\right]\theta + \frac{1}{M_e\left(1 + \frac{\gamma-1}{2}M_e^2\right)}\frac{dM_e}{ds}\delta^*$$

$$= M_e\frac{\left(1 + \frac{\gamma-1}{2}M_e^2\right)^{\gamma/(\gamma-1)}}{\gamma M_e^2}\frac{\tau_w}{p_{oe}} \tag{21.107c}$$

where use is made of Equations (21.35). Problem 21.25 obtains the energy equation counterpart to Equation (21.107a).

We briefly outline the Pohlhausen method of solution when the flow is incompressible and two-dimensional. Equations (21.98) and (21.99) yield

$$\frac{u}{u_e} = (2\bar{\eta} - 2\bar{\eta}^3 + \bar{\eta}^4) + \frac{\Lambda}{6}(\bar{\eta} - 3\bar{\eta}^2 + 3\bar{\eta}^3 - \bar{\eta}^4) \tag{21.108}$$

where the parameter Λ is

$$\Lambda = \frac{\delta^2}{\nu}\frac{du_e}{ds} \tag{21.109}$$

With this profile, the thicknesses and shear stress become

$$\begin{aligned} \delta^* &= \left(\frac{3}{10} - \frac{\Lambda}{120}\right)\delta \\ \theta &= \frac{1}{63}\left(\frac{37}{2} - \frac{\Lambda}{15} - \frac{\Lambda^2}{144}\right)\delta \\ \tau_w &= \mu\left(2 + \frac{\Lambda}{6}\right)\frac{u_e}{\delta} \end{aligned} \tag{21.110}$$

Thus, Equation (21.107b) becomes a messy first-order ODE for $\delta(s)$. Once solved, however, the skin friction is readily obtained as a function of s. Schlichting (sixth edition) organizes Equations (21.107b), (21.109), and (21.110) in a manner that expedites their solution. Although we have only discussed the simplest of cases, other integral approaches are conceptually similar in that a system of one or more ODEs is established where the solution can be obtained with a computer using, e.g., a Runge–Kutta integration procedure.

Integral Form of the Transformed Equations

As Marvin and Sinclair (1967) and Vasantha and Nash (1985) demonstrate, it is advantageous to start with the transformed nonsimilar equations. We thus write Equations (21.46) and their boundary conditions as follows:

$$(Cf_{\eta\eta})_\eta + ff_{\eta\eta} + \beta[g_w + (1 - g_w)G - f_\eta^2] = 2\xi(f_\eta f_{\xi\eta} - f_\xi f_{\eta\eta}) \tag{21.111a}$$

$$\frac{1}{Pr}\left\{C\left[G_\eta + \frac{2S(Pr-1)}{1-g_w}f_\eta f_{\eta\eta}\right]\right\}_\eta + fG_\eta = 2\xi(f_\eta G_\xi - f_\xi G_\eta) \tag{21.111b}$$

$$\begin{aligned} f(\xi, 0) = f_\eta(\xi, 0) = 0, \qquad & f_\eta(\xi, \infty) = 1, \\ G(\xi, 0) = 0, \qquad & G(\xi, \infty) = 1 \end{aligned} \tag{21.111c}$$

In the above, C is given by the relation

$$C(\xi, \eta) = \left[\frac{1 - S}{g_w + (1 - g_w)G - Sf_\eta^2}\right]^{1-\omega} \tag{21.111d}$$

which does not assume a similar boundary layer. We treat ω and Pr as constants, while β, g_w, and S are known functions of ξ. (It is possible to have the Prandtl number depend on η; however, the subsequent presentation is then more involved. An adiabatic wall can also be considered.) In writing Equations (21.111), we have assumed a perfect gas, $\mu \sim T^\omega$, and an inviscid flow that is isentropic and isoenergetic along the wall.

Equations (21.111a,b) are multiplied by $d\eta$ and integrated from the wall, where $\eta = 0$, to a location η. Subsequently, the η upper limit of the integrals will become infinite, thus allowing boundary-layer thicknesses such as δ^* and θ to be introduced. The arbitrarily defined thickness δ, however, does not appear in the formulation. In the interest of brevity, equations involving higher moments, as is done in Libby and Fox (1965), are not obtained. In addition to Equations (21.111c), we also impose the following self-evident boundary-layer edge conditions:

$$f_{\eta\eta}(\xi, \infty) = 0, \qquad G_{\xi}(\xi, \infty) = 0, \qquad G_{\eta}(\xi, \infty) = 0 \tag{21.112}$$

Based on Equations (21.70) and (21.72), the additional edge conditions are introduced:

$$f(\xi, \eta) = \eta - C_v(\beta, g_w) \tag{21.113a}$$

$$\int_0^{\eta} G\, d\eta = \eta - C_t(\beta, g_w) \tag{21.113b}$$

where $\eta \to \infty$, and C_v and C_t depends on ξ through β and g_w. These two relations are approximate; they stem from the earlier similarity analysis.

Each term is now systematically evaluated in Equation (21.111a), starting with the leftmost one and then moving rightward. Hence, we have

$$\int_0^{\eta} \frac{\partial}{\partial \eta}(Cf_{\eta\eta})d\eta = \int_{(Cf_{\eta\eta})_w}^{Cf_{\eta\eta}} d(Cf_{\eta\eta}) = Cf_{\eta\eta} - C_w f''_w \tag{21.114a}$$

where

$$f''_w = f_{\eta\eta w}, \qquad C_w = \left(\frac{1-S}{g_w}\right)^{1-\omega}$$

Later, when $\eta \to \infty$, we set

$$C(\xi, \infty) = 1$$

and use Equations (21.112) to evaluate $f_{\eta\eta}$ as zero. In view of Equation (21.68e), the second term in Equation (21.111a) becomes

$$\int_0^{\eta} ff_{\eta\eta} d\eta = ff_{\eta} - \int_0^{\eta} f_{\eta}^2 d\eta \tag{21.114b}$$

Equations (21.111c) and (21.113a) will result in ff_{η} equaling $\eta - C_v$ when $\eta \to \infty$, with all η containing terms canceling in Equation (21.111a). The next term on the left side of Equation (21.111a) becomes

$$\beta\int_0^{\eta} [g_w + (1-g_w)G - f_{\eta}^2]d\eta = \beta g_w \eta + \beta(1-g_w)\int_0^{\eta} Gd\eta - \beta\int_0^{\eta} f_{\eta}^2 d\eta \tag{21.114c}$$

where the G integral will be replaced with the aid of Equation (21.113b). For the rightmost term in Equation (21.111a), an integration by parts yields

$$\int_0^\eta f_\xi f_{\eta\eta} d\eta = f_\xi f_\eta - \int_0^\eta f_\eta f_{\xi\eta} d\eta$$

We thus obtain for the right side

$$\int_0^\eta (f_\eta f_{\xi\eta} - f_\xi f_{\eta\eta}) d\eta = 2\int_0^\eta f_\eta f_{\xi\eta} d\eta - f_\xi f_\eta$$

Since ξ and η are independent variables, this becomes

$$\int_0^\eta (f_\eta f_{\xi\eta} - f_\xi f_{\eta\eta}) d\eta = \frac{\partial}{\partial \xi}\int_0^\eta f_\eta^2 d\eta - f_\xi f_\eta$$

We set $\eta \to \infty$ and use Equations (21.111c) and (21.113a), to obtain

$$2\xi \int_0^\infty (f_\eta f_{\xi\eta} - f_\xi f_{\eta\eta}) d\eta = 2\xi \frac{\partial}{\partial \xi}\int_0^\infty f_\eta^2 d\eta + 2\xi \frac{dC_v}{d\xi}$$

where

$$\frac{dC_v}{d\xi} = \frac{\partial C_v}{\partial \beta}\frac{d\beta}{d\xi} + \frac{\partial C_v}{\partial g_w}\frac{dg_w}{d\xi}$$

although this C_v derivative will not appear in the final result.

Equations (21.114) are combined with $\eta \to \infty$, to yield, for Equation (21.111a)

$$2\xi\left(\frac{d}{d\xi}\int_0^\infty f_\eta^2 d\eta + \frac{dC_v}{d\xi}\right) + (1+\beta)\left(\int_0^\eta f_\eta^2 d\eta - \eta\right)_{\eta\to\infty} + C_v + \beta(1-g_w)C_t + C_w f_w'' = 0 \tag{21.115a}$$

Equation (21.61) for the momentum thickness can be written as

$$\Upsilon\theta = \left(\int_0^\eta (f_\eta - f_\eta^2) d\eta\right)_{\eta\to\infty} = \left(f - \int_0^\eta f_\eta^2 d\eta\right)_{\eta\to\infty} = \left(\eta - \int_0^\eta f_\eta^2 d\eta\right)_{\eta\to\infty} - C_v$$

or as

$$\left(\int_0^\eta f_\eta^2 d\eta - \eta\right)_{\eta\to\infty} = -\Upsilon\theta - C_v \tag{21.116a}$$

where Υ, θ, and C_v are functions only of ξ. The ξ derivative of this relation is

$$\frac{d}{d\xi}\int_0^\infty f_\eta^2 d\eta = -\frac{d}{d\xi}(\Upsilon\theta) - \frac{dC_v}{d\xi} \qquad \textbf{(21.116b)}$$

With Equations (21.116), Equation (21.115a) becomes

$$2\xi\frac{d}{d\xi}(\Upsilon\theta) + (1+\beta)\Upsilon\theta + \beta[C_v - (1-g_w)C_t] = C_w f_w'' \qquad \textbf{(21.115b)}$$

Recall that ξ is given in terms of the wall arc length s by Equation (21.9), and that Υ is

$$\Upsilon = \frac{r_w^\sigma(\rho u)_e}{(2\xi)^{1/2}} = \left(\frac{Re_{\bar{x}}}{2}\right)^{1/2}\frac{1}{\bar{x}} \qquad \textbf{(21.117a)}$$

Hence, we have

$$\frac{d\xi}{ds} = r_w^{2\sigma}(\rho\mu u)_e \qquad \textbf{(21.117b)}$$

$$\frac{d\Upsilon}{d\xi} = \Upsilon\left\{\frac{d\ln[r_w^\sigma(\rho u)_e]}{d\xi} - \frac{1}{2\xi}\right\} \qquad \textbf{(21.117c)}$$

In accordance with Kármán's equation, we replace f_w'' with the skin-friction coefficient, Equation (21.65). We thus obtain the final version for the momentum equation

$$\frac{d\theta}{d\xi} + \left\{\frac{\beta}{2\xi} + \frac{d\ln[r_w^\sigma(\rho u)_e]}{d\xi}\right\}\theta = \frac{\beta[C_v - (1-g_w)C_t]}{2\xi\Upsilon} + \frac{C_w}{2}\frac{\bar{x}c_f}{\xi} \qquad \textbf{(21.115c)}$$

where

$$\frac{\bar{x}}{\xi} = \frac{1}{(\rho\mu u)_e r_w^{2\sigma}}$$

Equations (21.35) can be used to replace ρ_e, u_e, S, ... with M_e-containing terms, as was done earlier for Kármán's equation, and Equation (21.42) can be used for β. For a flat plate, one can show that the above equation reduces to

$$\frac{d\theta}{ds} = \frac{C_w}{2}c_f$$

The foregoing type of analysis is repeated for Equation (21.111b). We thus obtain

$$\frac{1}{Pr}\int_0^\eta \frac{\partial}{\partial\eta}(CG_\eta)\,d\eta = \frac{1}{Pr}(CG_\eta - C_w G'_w)$$

$$\frac{1}{Pr}\int_0^\infty \frac{\partial}{\partial\eta}(CG_\eta)\,d\eta = -\frac{C_w G'_w}{Pr}$$

$$\frac{2S(Pr-1)}{Pr(1-g_w)}\int_0^\eta \frac{\partial}{\partial\eta}(f_n f_{\eta\eta})\,d\eta = \frac{2S(Pr-1)}{Pr(1-g_w)} f_n f_{\eta\eta}$$

or

$$\frac{2S(Pr-1)}{Pr(1-g_w)}\int_0^\infty \frac{\partial}{\partial\eta}(f_n f_{\eta\eta})\,d\eta = 0$$

This last result is the reason the Prandtl number does not have a stronger effect on the heat transfer than it does, and why a simple $Pr^{-2/3}$ factor is sufficient to correct Reynolds' analogy. The next term in Equation (21.111b) is evaluated as

$$\int_0^\eta fG_\eta\,d\eta = fG - \int_0^\eta Gf_\eta\,d\eta$$

To evaluate the rightmost integral, we introduce the stagnation enthalpy defect thickness in the form

$$\Upsilon\phi = \left(\int_0^\eta (f_\eta - Gf_\eta)d\eta\right)_{\eta\to\infty}$$

which becomes

$$\left(\int_0^\eta Gf_\eta\,d\eta - \eta\right)_{\eta\to\infty} = -\Upsilon\phi - C_v$$

We thus obtain

$$\int_0^\infty fG_\eta\,d\eta = \Upsilon\phi$$

The first term on the right side of Equation (21.111b) becomes

$$\int_0^\eta f_\eta G_\xi\,d\eta = fG_\xi - \int_0^\eta fG_{\xi\eta}\,d\eta$$

thereby yielding

$$\int_0^\eta (f_\eta G_\xi - f_\xi G_\eta)\,d\eta = fG_\xi - \int_0^\eta (fG_{\xi\eta} + f_\xi G_\eta)\,d\eta = fG_\xi - \frac{\partial}{\partial\xi}\int_0^\eta fG_\eta\,d\eta$$

In view of Equations (21.112) and (21.113a), the product fG_ξ is indeterminate as $\eta \to \infty$; we take its limiting value to be zero. Hence, we obtain

$$2\xi \int_0^\infty (f_\eta G_\xi - f_\xi G_\eta)\, d\eta = -2\xi \frac{d}{d\xi}(\Upsilon \phi)$$

By combining the relevant terms, we have

$$2\xi \frac{d}{d\xi}(\Upsilon \phi) + \Upsilon \phi = \frac{C_w G_w'}{Pr}$$

With the aid of Equations (21.66) and (21.117), this becomes

$$\frac{d\phi}{d\xi} + \frac{d\ln[r_w^\sigma(\rho u)_e]}{d\xi}\phi = \frac{C_w}{Pr}\frac{\bar{x}St}{\xi} \tag{21.118}$$

Equations (21.115c) and (21.118) represent two ODEs whose dependent variables are θ and ϕ. In conjunction with Equation (21.117b), these relations yield $\theta(s)$ and $\phi(s)$, provided c_f and St can be determined in terms of ξ or s. One way of doing this is to introduce velocity and stagnation enthalpy profiles, as is done in Libby and Fox (1965). Alternatively, correlation formulas can be introduced, as in Lida and Fujimoto (1986). The appearance of C_v and C_t in Equation (21.115c) can be viewed as being comparable to correlation formulas. Another possibility, suggested in the next paragraph, would combine the approaches of these two references.

In order to have a closed system of equations, two additional relations are required for the skin-friction coefficient and for the Stanton number. These equations can be obtained by taking one moment of Equations (21.111a,b). Thus, each equation can be multiplied by $\eta d\eta$ and then integrated from zero to infinity. Alternatively, Equations (21.111a,b) can be multiplied by $f'd\eta$ and then integrated. This approach represents a generalization of that used for the incompressible energy-integral equation. In either case, profiles such as Equation (21.108) are not required and the arbitrary boundary-layer thickness δ does not appear in the formulation. Of course, several new parameters, analogous to C_v and C_t, will require evaluation.

REFERENCES

Abramowitz, M. and Stegun, I.A. (eds.), *Handbook of Mathematical Functions,* NBS Applied Mathematics Series, 55, 1964.

Back, L.H., "Acceleration and Cooling Effects in Laminar Boundary Layers — Subsonic, Transonic and Supersonic Speeds," *AIAA J.* 8, 794 (1970).

Back, L.H., "Compressible Laminar Boundary Layers with Large Acceleration and Cooling," *AIAA J.* 14, 968 (1976).

Back, L.H. and Witte, A.B., "Prediction of Heat Transfer from Laminar Boundary Layer, with Emphasis on Large Free-Stream Velocity Gradients and Highly Cooled Walls," *J. Heat Transfer,* Ser. C 88, 249 (1966).

Bae, Y.-Y., "Performance of an Aero-Space Plane Propulsion Nozzle," Ph.D. Dissertation, University of Oklahoma (1989).

Bae, Y.-Y. and Emanuel, G., "Boundary Layer Tables for Similar Compressible Flow," *AIAA J.* 27, 1163 (1989).

Bansal, J.L., "On a Class of Non-Similar Solutions of Compressible Boundary Layer Equations," *Appl. Sci. Res.* 36, 117 (1980).

Blasius, H., "Grenzschichten in Flüssigkeiten mit kleiner Reibung," *Z. Math. Phys.* 56, 1 (1908).

Blottner, F.G., "Finite Difference Methods of Solutions of the Boundary-Layer Equations," *AIAA J.* 8 193 (1970).

Bush, W.B., "A Method of Obtaining an Approximate Solution of the Laminar Boundary-Layer Equations," *J. Aerosp. Sci.* 28, 350 (1961).

Cebeci, T., Stewarbson, K., and Brown, S.N., "Nonsimilar Boundary Layers on the Leeside of Cones at Incidence," *Comp. Fluids* 11, 175 (1983).

Chapman, D.R. and Rubesin, M.W., "Temperature and Velocity Profiles in the Compressible Laminar Boundary Layer with Arbitrary Distribution of Surface Temperature," *J. Aeronaut. Sci.* 16, 547 (1949).

Cohen, C.B. and Reshotko, E., "Similar Solutions for the Compressible Laminar Boundary Layer with Heat Transfer and Pressure Gradient," NACA TR 1293 (1956a).

Cohen, C.B. and Reshotko, E., "The Compressible Laminar Boundary Layer with Heat Transfer and Arbitrary Pressure Gradient," NACA TR 1294 (1956b).

Colburn, A.P., "A Method of Correlating Forced Convection Heat Transfer Data and a Comparison with Fluid Friction," *Trans. Am. Inst. Chem. Eng.* 29, 174 (1933).

Cook, W.J., "Correlation of Laminar Boundary-Layer Quantities for Hypersonic Flows," *AIAA J.* 14, 131 (1977).

Coxon, M. and Parks, E.K., "Comment on 'Local Nonsimilarity Boundary-Layer Solutions,'" *AIAA J.* 9, 1664 (1971).

Dewey, C.F., Jr. and Gross, J.F., "Exact Similar Solutions of the Laminar Boundary-Layer Equations," in *Advances in Heat Transfer,* edited by J.P. Hartnett and T.F. Irvine, Jr., Vol. 4, Academic Press, New York, 1967, pp. 317–446.

Dey, J. and Narasimha, R., "Integral Method for the Calculation of Incompressible Two-Dimensional Transitional Boundary Layers," *J. Aircraft* 27, 859 (1990).

Emanuel, G., "Supersonic Compressive Ramp without Laminar Boundary Layer Separation," *AIAA J.* 22, 29 (1984).

Falkner, V.M. and Skan, S.W., "Some Approximate Solutions of the Boundary Layer Equations," *Philos. Mag.* 12, 865 (1931).

Haridas, A.K., "Morphology of Compressible Laminar Boundary Layers," M.S. Thesis, University of Oklahoma, Norman, OK, 1995.

Hiemenz, K., "Die Grenzschicht an einem in den Gleichförmigen Flüessigkeitsstrom Eingetauchten geraden Kreiszylinder," *Dingl. Polytechn. J.* 326, 321 (1911).

Hsu, C.-C. and Liakopoulos, A., "Nonsimilar Solution of Compressible Laminar Boundary Layer Flows by a Semi-Discretization Method," in *Finite Element Flow Analysis,* edited by T. Kawai, North-Holland Publ. Co., New York, 1982, pp. 395–401.

Ko, D.R.S. and Kubota, T., "Supersonic Laminar Boundary Layer along a Two-Dimensional Adiabatic Curved Ramp," *AIAA J.* 7, 298 (1969).

Kramer, R.F. and Lieberstein, H.M., "Numerical Solution of the Boundary-Layer Equations without Similarity Assumptions," *J. Aerosp. Sci.* 26, 508 (1959).

Lees, L. and Reeves, B.L., "Supersonic Separated and Reattaching Laminar Flows: I. General Theory and Application to Adiabatic Boundary Layer-Shock Wave Interactions," *AIAA J.* 2, 1907 (1964).

Libby, P.A. and [illegible], H., "A [illegible] Method for Compressible Laminar Boundary Layers and Some Applications," *Int. J. Heat Mass Transfer* 8, [illegible] ([illegible]).

Libby, P.A. and Liu, T.M., "Some Similar Laminar Flows Obtained by Quasilinearization," *AIAA J.* 6, 1541 (1968).

Libby, P.A., Morduchow, M., and Bloom, M., "Critical Study of Integral Methods in Compressible Laminar Boundary Layers," NACA TN 2655 (1952).

Lida, S. and Fujimoto, A., "A Fast Approximate Solution of the Laminar Boundary-Layer Equations," *J. Fluids Eng.* 108, 200 (1986).

Marvin, J.G. and Sinclair, A.R., "Convection Heating in Regions of Large Favorable Pressure Gradient," *AIAA J.* 5, 1940 (1967).

Millikan, C.B., "The Boundary Layer and Skin Friction for a Figure of Revolution," *Trans. Am. Soc. Mech. Eng.* 54, 29 (1932).

Nagamatsu, H.T. and Kim, S.C., "Compressible Laminar Boundary with Real Gas Effects for Flight Conditions to $M_e = 8$ and $T_o = 2500$ K," AIAA-86-0305 (1986).

Narayana, C.L. and Ramamoorthy, P., "Compressible Boundary-Layer Equations Solved by the Method of Parametric Differentiation," *AIAA J.* 10, 1085 (1972).

Nath, G., "An Approximate Method for the Solution of a Class of Nonsimilar Laminar Boundary Layer Equations," *J. Fluids Eng.* 98, 292 (1976).
Pade, O., Postan, A., Anshelovitz, D., and Wolfshtein, M., "The Influence of Acceleration on Laminar Similar Boundary Layers," *AIAA J.* 23, 1469 (1985).
Pallone, A., "Nonsimilar Solutions of the Compressible-Laminar Boundary-Layer Equations with Applications to the Upstream-Transpiration Cooling Problem," *J. Aerosp. Sci.* 28, 449, 492 (1961).
Pohlhausen, K., "Zur Näherungsweisen Integration der Differentialgleichung der Laminaren Reibungsschicht," *Z. Angew. Math. Mech.* 1, 252 (1921a).
Pohlhausen, K., "Der Wärmeaustauch zwishen festern Körpenund Flüssigkeiten mit kleiner Reibung und kleiner Wärmeleitung," *Z. Angew. Math. Mech.* 1, 115 (1921b).
Rasmussen, M.L., "On Compressible Turbulent Boundary Layers in the Presence of Favorable Pressure Gradients," ASME 75 WA/HT-53 (1975).
Rogers, D.F., "Reverse Flow Solutions for Compressible Laminar Boundary-Layer Equations," *Phys. Fluids* 12, 517 (1969).
Rogers, D.F., "Further Comments on 'Local Nonsimilarity Boundary-Layer Solutions,'" *AIAA J.* 12, 1007 (1974).
Rott, N. and Crabtree, L.F., "Simplified Laminar Boundary-Layer Calculations for Bodies of Revolution and for Yawed Wings," *J. Aeronaut. Sci.* 19, 553 (1952).
Ruger, C.J., "Approximate Analytic Solutions for Nonsimilar Boundary Layers," *AIAA J.* 5, 923 (1967).
Schlichting, H., *Boundary-Layer Theory,* 7th ed., McGraw-Hill Book Co., New York, 1979.
Smith, A.M.O. and Clutter, D.W., "Solution of the Incompressible Laminar Boundary-Layer Equations," *AIAA J.* 1, 2062 (1963).
Smith, A.M.O. and Clutter, D.W., "Machine Calculation of Compressible Laminar Boundary Layers," *AIAA J.* 3, 639 (1965).
Spanier, J. and Oldham, K.B., *An Atlas of Functions,* Hemisphere Publ. Co., New York, 1987.
Sparrow, E.M., "Reply by Authors to Coxon and E. K. Parks," *AIAA J.* 9, 1664 (1971).
Sparrow, E.M., Quack, H., and Boerner, C.J., "Local Nonsimilarity Boundary-Layer Solutions," *AIAA J.* 8, 1936 (1970).
Sternberg, J., "A Free-Flight Investigation of the Possibility of High Reynolds Number Supersonic Laminar Boundary Layers," *J. Aeronaut. Sci.* 19, 721 (1952).
Sullivan, P.A., "Interaction of a Laminar Hypersonic Boundary Layer and a Corner Expansion Wave," *AIAA J.* 8, 765 (1970).
Tani, I., "On the Approximate Solution of the Laminar Boundary-Layer Equations," *J. Aeronaut. Sci.* 21, 487, 504 (1954).
Truckenbrodt, E., "Ein Quadraturverfahren zur Berechnung der Laminaren und Turbulenten Reibungsschicht bei ebener und Rotationssymmetrischer Strömung," *Ing. Arch.* 20, 211 (1952).
Van Driest, E.R., "Investigation of Laminar Boundary Layer in Compressible Fluids Using the Crocco Method," NACA TN 2597 (1952).
Vasantha, R. and Nath, G., "Unsteady Nonsimilar Compressible Laminar Two-Dimensional and Axisymmetric Boundary-Layer Flows," *Acta Mech.* 57, 215 (1985).
von Kármán, T., "Über Laminare und Turbulente Reibung," *ZAMM* 1, 233 (1921).
Walz, A., *Boundary Layers of Flow and Temperature,* MIT Press, Cambridge, MA, 1969.
White, F.M., *Viscous Fluid Flow,* 2nd ed., McGraw-Hill, New York, 1991.
Wortmann, A. and Mills, A.F., "Highly Accelerated Compressible Laminar Boundary Layers Flows with Mass Transfer," *J. Heat Transfer, Ser.* C 93, 281 (1971).

PROBLEMS

21.1 Start with Problem 13.5 and use the procedure in Section 20.2 to derive the first-order boundary-layer equations for a steady, compressible flow using natural coordinates. Do not retain second-order terms. Transform back to the original variables after the boundary-layer simplification is completed.

21.2 **(a)** Continue with Problem 21.1 and determine the solution for the streamline angle $\theta(s, n)$.

(b) Assume a perfect gas and $Pr = 1$. Use the momentum and energy equations to derive a simple equation for the stagnation enthalpy h_o.

(c) What are the boundary conditions for h_o and the flow speed w?

21.3 Consider the similarity case when $C = 1$ and $\beta = 0$. An approximate solution to the momentum equation is (Bush, 1961)

$$f' \cong \text{erf}(a\eta)$$

where a is a constant.

(a) Determine the most appropriate value for a.

(b) Utilize the repeated integral of the error function, $i^n \text{erfc } z$, and determine $f(\eta)$ and

$$I = \int_0^{\eta} f(\eta')\, d\eta'$$

(c) Start with Equation (21.46b) and the part (a) assumptions and derive an ODE for $G(\eta)$. Assume Pr is a nonunity constant.

(d) Use the method of variation of parameters to obtain a solution for G that satisfies $G_w = 0$ and $G(\infty) = 1$.

(e) Assume incompressible, adiabatic flow and show that $(\delta/\delta_t) = 1.04 \times Pr^{1/3}$ (Pohlhausen, 1921b).

21.4 Derive Equation (21.68h). Use this result to analytically evaluate the right side of Equation (21.82)

$$\int_0^{\infty} (f' - f'^3)\, d\eta$$

where your result will involve a quadrature whose integrand is f''^2.

21.5 Consider a very cold wall ($g_w \to 0$) at a two-dimensional or axisymmetric stagnation point.

(a) Assume $Pr = 0.71$ (Colburn's analogy) and tabulate $\delta_t/\delta, \delta^*/\delta, \theta/\delta, \phi/\delta, Re_{\bar{x}}^{1/2} c_f$, and $Re_{\bar{x}}^{1/2} St$.

(b) Assume the vehicle has a nose radius R and is flying at a Mach number M_∞, where $M_\infty > 1$. Determine equations for δ, τ_w, and q_w in terms of freestream conditions.

(c) Assume the vehicle of part (b) is flying at 20 km altitude with

$$\gamma = 1.4, \qquad M_\infty = 6, \qquad R = 10 \text{ cm}, \qquad g_w = 0$$

Determine δ, τ_w, and q_w at the axisymmetric stagnation point. (Use standard tables for atmospheric air properties at 20 km.)

21.6 Consider the flow of air at $M_\infty = 6$ over a highly cooled ($g_w \to 0$) semi-infinite flat plate.

(a) Assume $Pr = 0.71$ (Colburn's analogy) and tabulate δ_t/δ, δ^*/δ, θ/δ, ϕ/δ, $Re_{\bar{x}}^{1/2} c_f$, and $Re_{\bar{x}}^{1/2} St$.

(b) As in Problem 21.5, assume the plate is at an altitude of 20 km. Determine δ, τ_w, and q_w at a point 2 m from the leading edge.

21.7 Repeat Problem 21.6 but now assume the wall is adiabatic. Also, evaluate the temperature of the gas adjacent to the wall at a point 2 m from the leading edge.

21.8 **(a)** Show that the vorticity in the boundary layer is $2\varepsilon_{sn} - 2(\partial u/\partial n)$.

(b) Use the result of part (a) and Equations (20.2) to determine, to first order, the vorticity $\omega(\xi, \eta)$ in the boundary layer.

(c) Under what condition can ω have a zero value inside the boundary layer? What is the value of ω at the wall?

(d) Derive a relation for $\mu_e \omega / \tau$ that only depends on η and the g_w and S parameters.

21.9 (a) Start with Equation (21.53) and derive an algebraic equation for Υn, except for an integral of G.

(b) Check this result by taking its partial derivative with respect to n.

(c) Show that the s derivative of Equation (21.10) is

$$\frac{\partial \eta}{\partial s} = \frac{d\Upsilon}{ds}\frac{\rho}{\rho_e}n$$

Observe that this result cannot be obtained by directly differentiating Equation (21.10) with respect to s, since both ρ and ρ_e are functions of s. In particular, ρ depends on s through η.

(d) Derive a general algebraic relation between n and η for large η and compare your result with Equations (21.75) and (21.76).

(e) Derive the condition on M_e such that $\eta = \Upsilon n$ when η is large.

21.10 Use Equation (21.45) to determine $r_w = r_w(s)$ and β_o if $(u_e = ke^{ms})$, where k and m are constants.

21.11 Assume a similar boundary layer with $\beta = 0$, $Pr = 1$, $C = C(\eta)$, and Equation (21.83).

(a) Show that Equation (21.46a) becomes

$$(1-S)^{1-\omega}\left\{\frac{f''}{[g_w + (1-g_w)f' - Sf'^2]^{1-\omega}}\right\}' + ff'' = 0$$

(b) With the usual boundary conditions for f, numerically solve this equation for $\omega = 0.7$, $S = 0, 0.5, 0.9$, and $g_w = 0.2(0.2)\,1, 1.5, 2(1)5$.

(c) Tabulate C_w, f''_w, and G'_w and compare with the 0.4696 value of Tables 21.6 and 21.7.

21.12 Consider the $S \to 1$ limit.

(a) For an adiabatic and diabatic wall, determine the behavior of $\Upsilon\delta^*$, $\Upsilon\delta$, $\Upsilon\delta_t$, and p_{oe}/p in this limit.

(b) With $\omega < 1$, discuss what happens to the equation given in part (a) of Problem 21.11 when $(S \to 1)$. Introduce a small parameter ε and rescale η and f such that all boundary conditions on f can still be satisfied by the rescaled equation from part (a) of Problem 21.11. The rescaled equation and boundary conditions would provide a uniformly valid solution for a flat plate in the $S \to 1$ limit.

21.13 Consider the near adiabatic wall case

$$g_w = 1 - \varepsilon$$

where $|\varepsilon| \ll 1$ and ε may be positive or negative.

(a) Set

$$f = f_0(\eta) + \varepsilon f_1(\eta) + \cdots$$
$$G = G_0(\eta) + \varepsilon G_1(\eta) + \cdots$$

and show that f_0 is a solution to the Falkner–Skan equation.

(b) Determine the governing equations and boundary conditions for f_1 and G_0.

(c) Show that the heat transfer is given by

$$q_w = -\frac{(\kappa r^{\sigma}\rho)_w u_e T_{oe}}{(2\xi)^{1/2}}\varepsilon G'_{0w}$$

21.14 (a) Show that the local heat transfer to the wall can be written as

$$q_w = \frac{\kappa_e(T_w - T_{oe})}{Pr^{2/3}}\left(\frac{\gamma}{4RT_{oe}}\right)^{1/4}\left[\frac{p_{oe}M_e r_w^{2\sigma}F(M_e)}{\mu_e\int_0^s r_w^{2\sigma}F ds}\right]^{1/2}\frac{G'_w}{\left(1+\frac{\gamma-1}{2}M_e^2\right)^{(\gamma+1)/[4(\gamma-1)]}}$$

(b) Assume a flat plate with $Pr = 1$. Suppose T_w and κ_w are zero. Explain why q_w is finite and not zero.

21.15 Use the stagnation point theory in Chapter 17 to show that the constant c_1 in Equation (17.29b) is given by G'_w when the Prandtl number is unity.

21.16 Repeat Problem 21.11 but with Pr equal to a nonunity constant. In this case, Equations (21.46) do not reduce to the equation shown in Problem 21.11.

21.17 For an adiabatic wall, show that it is not possible to simultaneously have a boundary layer with both velocity and temperature overshoot.

21.18 (a) Determine formulas for the pitot pressure p_t, with the form

$$\frac{p_t}{p_e} = P(\eta, \gamma, S, g_w)$$

where p_t is the stagnation pressure when $M \le 1$ and is the Rayleigh pitot pressure when $M \ge 1$. This latter case assumes a normal shock upstream of the pitot probe and an inviscid flow between the shock and the probe's entrance.

(b) Consider airflow over a flat plate with $M_e = 1, 5$ and $g_w = 0.2, 1$. Plot and tabulate p_t/p_e vs. η and vs. n/s for these four cases when $Re_s = (\rho u/\mu)_\infty s = 10^5$. For the connection between n and η, use part (a) of Problem 21.9. For $f'(\eta)$ you may want to use the error function form given in Problem 21.3.

21.19 Consider Equations (21.28) and (21.29) for an adiabatic wall in the limit of $\beta \to \infty$.

(a) Introduce $\varepsilon = (1/\beta)$ and determine the first approximation $f_o(\eta)$ for the outer region. Use Table 21.2 to numerically estimate the constant of integration.

(b) Determine the first approximation $f_i(\bar{\eta})$ for the inner region, where $\bar{\eta}$ is the new indpendent variable.

(c) Determine a uniformly valid composite solution for u/u_e and T/T_e in terms of η.

21.20 Reconsider part (a) of Problem 21.19 with $g_w \ne 1$. Derive the first-order result

$$G_o = \frac{\text{erf}[(\eta - C_v)/2^{1/2}] + \text{erf}(C_v/2^{1/2})}{1 + \text{erf}(C_v/2^{1/2})}$$

where $C_v = C_v(\infty, g_w)$. Show that this result is spurious.

21.21 Use results of Problem 21.19 to obtain f''_w and $\Upsilon\delta^*$ when $g_w = 1$ and $\beta \to \infty$.

21.22 A compressible, two-dimensional boundary layer has $M_e = M_{e1}e^{as}$, where a is a constant and s is nondimensional. Assume the hypersonic limit

$$\frac{\gamma-1}{2}M_e^2 \gg 1$$

and use this to simplify the analysis.

(a) Use Equation (21.42) to determine an ODE for β. Integrate this equation subject to the initial condition, $\beta(0) = \beta_1$.

(b) Assume a constant wall temperature, and determine an approximate relation for $\xi(s)$. Write the constant coefficient as ξ_c.

(c) Determine how δ^* depends on s. Leave β, f''_w, C_v, and C_t in the equation, although the last three parameters depend on β and, therefore, on s. Other parameters, however, should be written in terms of s.

21.23 **(a)** Determine the time rate of change of T/T_e for a fluid particle in a steady laminar boundary layer. Your result should be in terms of $(\gamma RT_{oe})^{1/2}\beta^{-1}(dM/ds)$, g_w, S, G, f, and the η derivatives of G and f.

(b) Determine the condition for T/T_e to increase or decrease following a fluid particle.

21.24 Parts (a) and (b) of this problem obtain the energy equation counterpart to Equation (21.107a).

(a) Multiply Equation (21.5) by dn and integrate from zero to δ_t. Use the e subscript to denote conditions at $n = \delta_t$. Derive the $(\rho v)_e$ counterpart to Equation (21.104).

(b) Do not assume Pr is a constant, but assume h_{oe} is a constant while T_w can depend on s. Derive the energy equation counterpart to Equation (21.107a), starting with Equation (21.8) and using ϕ as the dependent variable.

(c) Obtain a general solution to the ϕ equation of part (b) that contains a quadrature. Evaluate the constant of integration by setting $s = 0$ at stagnation point.

(d) Utilize material from Chapter 17 and determine an algebraic equation for ϕ_o, where a zero subscript denotes conditions at a two-dimensional or axisymmetric stagnation point. Use Equation (21.66) and the ϕ_o result to obtain an equation for sSt.

22 Supersonic Boundary-Layer Examples

22.1 PRELIMINARY REMARKS

As an illustration of the local similarity theory in Chapter 21, we discuss several supersonic boundary-layer examples. The first one treats a thin airfoil at zero incidence in a uniform flow. A number of boundary-layer parameters are evaluated at the trailing edge of the airfoil. The second example utilizes a design approach for a compressive ramp that can be used to avoid laminar boundary-layer separation. A third example determines the shape of a wall with a zero value for the displacement thickness. A discussion of the performance of a scramjet propulsion nozzle constitutes the last example. Although this chapter focuses on compressible boundary layers, several problems deal with an incompressible boundary layer.

22.2 THIN AIRFOIL THEORY

A schematic of a lens-shaped airfoil is shown in Figure 22.1. The upper surface is given by

$$y = 4kx\left(1 - \frac{x}{c}\right) \tag{22.1}$$

where $k \ll 1$ and is nondimensional, c is the chord length, and at mid-chord, $y = kc$. (As long as the flow downstream of an attached forward shock is supersonic, the airfoil need not be symmetric about the x-axis. In our discussion, however, it is occasionally convenient to assume symmetry.) We begin by determining u_e, ξ, and β as functions of s on the upper surface using small perturbation theory for the flow field. Subsequently, we determine at the trailing edge various boundary-layer thicknesses and the skin-friction coefficient.

Solution for a Perturbation Potential

From small perturbation theory for supersonic inviscid flow, the perturbation potential satisfies the wave equation (see Liepmann and Roshko, 1957, Section 8.7)

$$\lambda^2 \phi_{xx} - \phi_{yy} = 0 \tag{22.2}$$

where

$$u = U_\infty + u' = U_\infty + \phi_x$$

$$v = \phi_y \tag{22.3}$$

and

$$\lambda = (M_\infty^2 - 1)^{1/2}$$

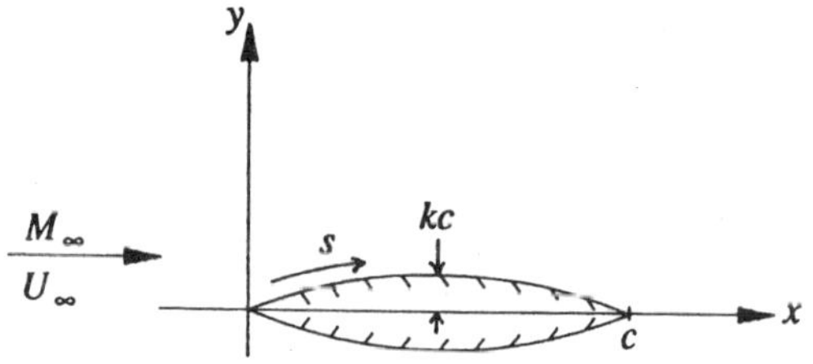

FIGURE 22.1 Lens-shaped airfoil in a supersonic flow.

The general solution to Equation (22.2) is

$$\phi = \phi(x - \lambda y) \tag{22.4}$$

which must satisfy the tangency condition, evaluated on the x-axis,

$$\frac{v(x,0)}{U_\infty} = \frac{dy}{dx} = 4k\left(1 - 2\frac{x}{c}\right) \tag{22.5}$$

From Equations (22.3) and (22.4), we have

$$v(x,0) = \frac{\partial\phi(x-\lambda y)}{\partial(x-\lambda y)}\frac{\partial(x-\lambda y)}{\partial y}\bigg|_{y=0} = -\lambda\frac{\partial\phi}{\partial x}(x,0) \tag{22.6}$$

We eliminate v from the above, to obtain

$$\frac{\partial\phi}{\partial x} = -\frac{4kU_\infty}{\lambda}\left(1 - 2\frac{x}{c}\right)$$

In view of Equation (22.4), ϕ is given by

$$\phi(x,y) = -\frac{4kU_\infty}{\lambda}\left[(x-\lambda y) - \frac{1}{c}(x-\lambda y)^2\right] \tag{22.7}$$

where the constant of integration is set equal to zero. For this perturbation solution to be valid, the coefficient $4k/\lambda$ must be small compared to unity. Consequently, $M_\infty^2 - 1$ cannot be small; i.e., the flow cannot be transonic.

Evaluation of the Pressure Gradient Parameter

From Equations (22.3) and (22.7), we obtain for the x and y velocity components

$$u = U_\infty\left\{1 - \frac{4k}{\lambda}\left[1 - \frac{2}{c}(x-\lambda y)\right]\right\}$$

$$v = 4kU_\infty\left[1 - \frac{2}{c}(x-\lambda y)\right] \tag{22.8}$$

The component of the velocity that is tangent to the surface of the airfoil is

$$u_e = (u^2 + v^2)^{1/2}$$

where the y in u and v is given by Equation (22.1). We thereby obtain

$$u_e(x) = U_\infty\left[1 - \frac{4k}{\lambda}\left(1 - \frac{2}{c}x\right)\right] \tag{22.9}$$

to $O(k)$. This result is consistent with Equation (22.3), with $y = 0$, since v^2 is of $O(k^2)$ and does not contribute to u_e. We readily determine

$$u_e(0) = U_\infty\left(1 - \frac{4k}{\lambda}\right), \qquad u_e(c/2) = U_\infty, \qquad u_e(c) = U_\infty\left(1 + \frac{4k}{\lambda}\right) \tag{22.10}$$

and the flow accelerates along the airfoil. In turn, this means a favorable pressure gradient and a positive value for β. Since the arc length s along the surface (see Figure 22.1) is to $O(k)$

$$s = \int_0^x \left[1 + \left(\frac{dy}{dx}\right)^2\right]^{1/2} dx = \int_0^x \left[1 + 16k^2\left(1 - \frac{2x}{c}\right)^2\right]^{1/2} dx \cong \int_0^x dx = x$$

we have

$$u_e(s) = U_\infty\left[1 - \frac{4k}{\lambda}\left(1 - 2\frac{s}{c}\right)\right] \tag{22.11}$$

To evaluate ξ, we will need ρ_e and T_e as functions of s. Because the inviscid flow is homentropic, it is convenient to introduce the Mach number M_e in order to utilize standard isentropic relations. With

$$T_e = \frac{T_{oe}}{1 + \frac{\gamma - 1}{2}M_e^2} \tag{22.12}$$

we have

$$u_e = (\gamma R T_e)^{1/2} M_e = M_e\left(\frac{\gamma R T_{oe}}{1 + \frac{\gamma-1}{2}M_e^2}\right)^{1/2}$$

We now write

$$M_e = M_\infty + M' \tag{22.13}$$

along with Equation (22.11), to obtain

$$M' = -\frac{M_\infty}{1 - S_\infty}\frac{4k}{\lambda}\left(1 - 2\frac{s}{c}\right) \tag{22.14}$$

where

$$U_\infty^2 = \gamma R T_\infty M_\infty^2$$

$$S_\infty = \frac{(\gamma-1)M_\infty^2/2}{1+\frac{\gamma-1}{2}M_\infty^2}$$

Equations (22.13) and (22.14) then result in

$$\left(1+\frac{\gamma-1}{2}M_e^2\right)^m = \left(1+\frac{\gamma-1}{2}M_\infty^2\right)^m\left[1-m(\gamma-1)M_\infty^2\frac{4k}{\lambda}\left(1-2\frac{s}{c}\right)\right]$$

to $O(k)$, where m is a constant. With m equal to -1 and $-(\gamma-1)^{-1}$, respectively, we obtain for T_e and ρ_e

$$T_e = T_{oe}\left(1+\frac{\gamma-1}{2}M_e^2\right)^{-1}$$

$$= T_\infty\left[1+(\gamma-1)M_\infty^2\frac{4k}{\lambda}\left(1-2\frac{s}{c}\right)\right] \tag{22.15}$$

$$\rho_e = \rho_{oe}\left(1+\frac{\gamma-1}{2}M_e^2\right)^{-1/(\gamma-1)}$$

$$= \rho_\infty\left[1+M_\infty^2\frac{4k}{\lambda}\left(1-2\frac{s}{c}\right)\right] \tag{22.16}$$

With $\mu \sim T$ and Equations (22.11), (22.15), and (22.16), we have

$$(\rho\mu u)_e = \rho_\infty\left[1+M_\infty^2\frac{4k}{\lambda}\left(1-2\frac{s}{c}\right)\right]\mu_\infty\left[1+(\gamma-1)M_\infty^2\frac{4k}{\lambda}\left(1-2\frac{s}{c}\right)\right]U_\infty\left[1-\frac{4k}{\lambda}\left(1-2\frac{s}{c}\right)\right]$$

$$= (\rho\mu U)_\infty\left[1+(\gamma M_\infty^2-1)\frac{4k}{\lambda}\left(1-2\frac{s}{c}\right)\right]$$

to $O(k)$. Equation (21.9), with $\sigma = 0$, thus becomes

$$\xi = (\rho\mu U)_\infty\int_0^s\left[1+(\gamma M_\infty^2-1)\frac{4k}{\lambda}\left(1-2\frac{s}{c}\right)\right]ds$$

$$= (\rho\mu U)_\infty\left\{\left[1+(\gamma M_\infty^2-1)\frac{4k}{\lambda}\right]s-(\gamma M_\infty^2-1)\frac{4k}{\lambda}\frac{s^2}{c}\right\} \tag{22.17}$$

where

$$\xi(0) = 0, \qquad \xi(c) = (\rho\mu U)_\infty c$$

The pressure gradient parameter can be written as

$$\beta = \frac{2\xi}{u_e}\frac{\frac{du_e}{ds}}{\frac{d\xi}{ds}}\frac{T_{oe}}{T_e} \tag{22.18}$$

From Equation (22.11), we have

$$\frac{du_e}{ds} = \frac{8U_\infty k}{c\lambda}$$

and, consequently, the other factors on the right side of Equation (22.18) need to be determined only to $O(1)$. Hence, with

$$\xi \cong (\rho\mu U)_\infty s$$

$$\frac{d\xi}{ds} = (\rho\mu U)_\infty$$

$$u_e \cong U_\infty$$

$$\frac{T_{oe}}{T_e} \cong \frac{T_{oe}}{T_\infty} = 1 + \frac{\gamma-1}{2}M_\infty^2$$

Equation (22.18) becomes

$$\beta = 16\frac{ks}{\lambda c}\frac{T_{oe}}{T_\infty} = 16k\frac{1+\frac{\gamma-1}{2}M_\infty^2}{(M_\infty^2-1)^{1/2}}\frac{s}{c} \tag{22.19}$$

Thus, the pressure gradient parameter increases linearly along the airfoil.

Tabular Results

Let us now determine, at the trailing edge of an adiabatic wall, several boundary-layer thicknesses of Section 21.7, normalized by the chord length, and c_f. We use local similarity theory along with the prescribed numerical values

$$\gamma = 1.4, \quad M_\infty = 1.988, \quad Re_\infty = \frac{\rho_\infty U_\infty c}{\mu_\infty} = 10^5, \quad k = 0.03, \quad g_w = 1$$

With this k value, a symmetric airfoil has a maximum thickness of 6% of the chord length.

We readily determine the following parameters:

$$\lambda = (M_\infty^2 - 1)^{1/2} = 1.718$$

$$\beta(c) = \frac{16 \times 0.03 \times (1 + 0.2 \times 1.988^2)}{1.718} = 0.500$$

$$\frac{u_e(c)}{U_\infty} = 1 + \frac{4 \times 0.03}{1.718} = 1.07$$

$$\frac{\mu_e(c)}{\mu_\infty} = \frac{T_e(c)}{T_\infty} = 1 - \frac{0.4 \times 1.988^2 \times 4 \times 0.03}{1.718} = 0.8896$$

$$\frac{\rho_e(c)}{\rho_\infty} = 1 - \frac{1.988^2 \times 4 \times 0.03}{1.718} = 0.7239$$

$$S_\infty = \frac{0.7904}{1.7904} = 0.4415$$

$$M'(c) = \frac{1.988 \times 4 \times 0.03}{0.5585 \times 1.718} = 0.2486$$

$$M_e(c) = 1.988 + 0.2486 = 2.236$$

$$S(c) = \frac{\frac{\gamma-1}{2} M_e^2(c)}{1 + \frac{\gamma-1}{2} M_e^2(c)} = 0.500$$

Note the difference between S_∞ and $S(c)$. We will also need ξ, $\bar{x}$, $Re_{\bar{x}}$, and Υ, all evaluated at the trailing edge. These are given by

$$\xi = (\rho\mu U)_\infty c = \mu_\infty^2 Re_\infty$$

$$\bar{x} = \frac{\xi}{(\rho\mu u)_e r_w^{2\sigma}} = \frac{(\rho\mu U)_\infty}{(\rho\mu u)_e} c = \frac{c}{0.7239 \times 0.8896 \times 1.07} = 1.45c$$

$$Re_{\bar{x}} = \frac{\rho_e u_e \bar{x}}{\mu_e} = \frac{0.7239 \times 1.07 \times 1.45}{0.8896} Re_\infty = 1.264 Re_\infty$$

$$\Upsilon = \left(\frac{Re_{\bar{x}}}{2}\right)^{1/2} \frac{1}{\bar{x}} = \frac{173.4}{c}$$

Let ψ represent any of the adiabatic wall boundary-layer thicknesses at the trailing edge of the airfoil, where the tilde notation of Section 21.9 is not used. We then have

$$\frac{\psi}{c} = \frac{\Upsilon\psi}{173.4}$$

With $\beta(c) = 0.5$, $g_w = 1$, and $S(c) = 0.5$, we use Tables 21.2, 21.4, and 21.6 to obtain the results that are summarized in Table 22.1.

Discussion

Since $\delta_t (= \tilde{\delta}_t)$ is determined by Equation (21.81), it equals δ. As expected, δ^* and θ are less than δ. The result that θ is substantially less than δ is typical of a high-speed boundary layer. The most important result is δ^*/c. Recall that y/c is only 0.03 at the half-chord location. Although δ^*/c is

TABLE 22.1
Boundary-Layer Values at the Trailing Edge of an Adiabatic Airfoil when $\beta = 0.5$, $g_w = 1$, and $S = 0.5$

δ/c	2.25×10^{-2}
δ_t/c	2.25×10^{-2}
δ^*/c	1.13×10^{-2}
θ/c	2.02×10^{-3}
f''_w	0.9277
c_f	3.69×10^{-3}

small, it is still a substantial fraction of the maximum airfoil thickness and would contribute to the form drag of the airfoil.

As a consequence of the monotonic increase in δ^* along the airfoil, an inviscid flow based on the airfoil's thickness plus the displacement thickness will have a reduced pressure gradient and, thus, a reduced β value. Examination of Equation (21.74), with fixed g_w and S values, shows that a corrected calculation will have a larger δ^*/c value than the one given in Table 22.1. A more accurate computation would use second-order boundary-layer theory, thereby accounting for the effects of both longitudinal curvature and displacement thickness. A second possibility would be to use a nonsimilar approach while simultaneously accounting for the displacement thickness. This is done by using either shock-expansion theory or small perturbation theory to establish an equation for β, with respect to s, that simultaneously accounts for the effects of airfoil thickness and of the boundary-layer displacement. No iteration would be involved, since the solution for the inviscid flow is coupled to that of the boundary layer in a self-consistent fashion. Finally, the full viscous governing equations can be used. With a refined mesh in the viscous region, this approach should provide the most accurate result. Various boundary-layer thicknesses can be obtained numerically from this solution at arbitrary locations along the airfoil.

22.3 COMPRESSIVE RAMP

GENERAL DISCUSSION

Figure 22.2 is a sketch of a curved ramp in a supersonic flow with a planar upstream wall. The compressive turn may be abrupt or gradual. If the corner is contoured, it generally consists of a circular arc that connects two planar walls. A turn angle of only 5 to 10° is usually sufficient to cause a laminar boundary layer to separate from the wall. Depending on the turn angle, length of the downstream wall, etc., the separated layer may or may not reattach. A lambda shock system forms if reattachment occurs. Further discussion of supersonic laminar boundary-layer flow over a compressive ramp can be found in Lewis et al. (1968).

Upstream of the ramp in Figure 22.2 there is a plate, with a sharp leading edge, of length s_1 that is aligned with the freestream. The ramp itself starts at the origin with a zero slope, and the x coordinate is aligned with the freestream. A compression wave is generated by the ramp that gradually becomes an oblique shock wave. The leading edge of the compression is a Mach line labeled OC. The shock starts on this Mach line at a point removed from the origin (Johannesen, 1952), for example, point C. Above this point the shock gradually increases in strength as compressive wave Mach lines merge with it. However, for the inviscid flow streamlines that pass through OC, the flow is homentropic.

Because the upstream plate is aligned with the flow, its leading edge causes, at most, a weak shock wave that we will ignore. However, it does have a laminar boundary layer with a β value

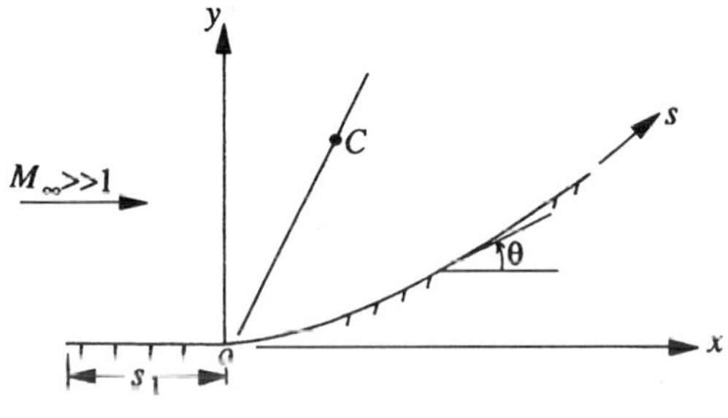

FIGURE 22.2 Schematic of a compressive ramp in a supersonic flow.

of zero. If point C is interior to the layer, there will be shock wave–boundary layer interaction. In this situation, the boundary layer experiences a relatively sudden adverse pressure gradient and may separate. Because of the interaction, the point of separation is upstream of the origin and the separated flow results in an oblique shock wave. (Normally, the boundary layer is thin and point C is well outside it. In a low density flow, however, the boundary layer can be thick and shock formation may start inside it.)

On the other hand, point C may fall outside the boundary layer. This condition can be met by either reducing s_1 or increasing, at the origin, the initial radius of curvature of the ramp. If the boundary layer with point C external to it does not separate, then the above interaction is avoided. This avoidance condition can occur if $-\beta$ is not too large along the ramp.

Examination of the compressible boundary-layer equations shows that the separation condition, $f''_w = 0$, occurs at a value, denoted as β_{sp}, that depends only on g_w. This result presumes that the Prandtl and Chapman–Rubesin numbers are unity, while γ is arbitrary. For an adiabatic wall, β_{sp} is –0.1988 (see Table 21.1). The more negative β_{sp}, the more intense the adverse pressure gradient can be without separation. Thus, the boundary layer is most likely to stay attached to a cold wall. With a cold wall, g is small, relative to unity, throughout much of the boundary layer. As is evident from Equation (21.22), this reduces the importance of the pressure gradient term. As a consequence, a cold wall can accommodate a larger adverse pressure gradient without separation.

There is a variety of applications where compressive ramps are required in a supersonic flow. Two examples are engine inlets and wind tunnel diffusers. For instance, on the supersonic B-1 aircraft, the engine inlet is approximately a two-dimensional diffuser. Of course, any engine inlet operates as a diffuser, especially when the flight speed is supersonic.

For efficient engine operation it is desirable that the boundary layer does not separate near the inlet. (This can lead to "buzz" and excessive inlet flow spillage.) In addition, the shock system should be as weak as possible. A weak system minimizes the mass-averaged loss of stagnation pressure. Recall that this loss is quite large across a normal shock at high Mach numbers. For instance, with $\gamma = 7/5$ the stagnation pressure ratio across a normal shock is

$$\frac{p_{o2}}{p_{o1}} = 6.17 \times 10^{-2}, \qquad M_1 = 5$$

$$= 1.54 \times 10^{-2}, \qquad M_1 = 7$$

One way of avoiding such a severe loss is to use an oblique shock system in preference to a single normal shock. The average stagnation pressure loss can be further reduced by having the shocks first form at some distance from the wall. As previously noted, this helps avoid shock wave–boundary layer interaction and boundary-layer separation. When the shock system forms away from the wall, some of the flow is compressed isentropically. In addition, for the flow that passes through the weak part of the shock wave, just above point C in Figure 22.2, the compression is nearly isentropic. The diffuser, or inlet, efficiency improves to the extent that more of the flow is compressed isentropically, or nearly so.

Emanuel (1984) presents a technique for the design of a compressive ramp that avoids boundary-layer separation and shock wave interaction. The approach can be used in a variety of situations. However, for purposes of brevity, our discussion is limited to the theory for a two-dimensional compressive ramp in a uniform supersonic flow. Since the loss of stagnation pressure can be considerable when the inlet Mach number is large, the approach to be outlined may prove advantageous even if the ramp only provides a modest Mach number reduction. The remainder of the inlet or diffuser would then be of a conventional design.

Our purpose in this section is to design, without laminar boundary-layer separation, a supersonic compressive ramp. Moreover, the design is to be an optimum one in that the compression is to be as rapid as possible, thus minimizing the length of the ramp wall.

Conceptually, the solution is simple. We presume a known value for g_w and select from Figure 21.6 or Table 21.1 a β value, denoted as β_o, that just exceeds β_{sp}. The theory in Section 21.5 is then used to determine $M_e(s)$, which, in turn, determines the inviscid wall contour. In the balance of this section, we present the consequences of this procedure.

Design Procedure

Since β_o is a constant, a global similarity solution is obtained. (As we will see, this is not precisely correct at the start of the ramp.) With $\sigma = 0$, we obtain from Equation (21.44)

$$\frac{d^2M_e}{ds^2} + 2\left(\frac{1}{\beta_o} - \frac{1 - \frac{\gamma+1}{4}M_e^2}{1 + \frac{\gamma-1}{2}M_e^2}\right)\frac{1}{M_e}\left(\frac{dM_e}{ds}\right)^2 = 0 \qquad (22.20)$$

which holds for a subsonic or a supersonic flow. (It also holds for an expansive as well as a compressive wall turn.) The solution of this equation will parametrically depend on γ and β_o and two initial conditions. These are imposed at the start of the curved ramp:

$$M_e(s_1) = M_1, \quad \frac{dM_e}{ds}(s_1) = M_1' \qquad (22.21)$$

where $M_1 = M_\infty$.

While a flat plate can represent the upstream wall for a wind tunnel diffuser, it is also essential from a theoretical viewpoint. By its presence, an infinite value for M_1' is avoided. Furthermore, boundary layer theory is not valid at the plate's leading edge, which must precede the start of the ramp.

From the theory of characteristics (Emanuel, 1986, Section 16.1), we know that derivatives can be discontinuous across Mach lines. Most frequently this occurs on the leading and trailing edges of expansion or compression waves. This is the case on the leading edge Mach line, OC, where $(dM_e/ds) = 0$ upstream of this line and dM_e/ds is negative, with a finite value, immediately downstream of it. This result is true even though the wall slope is continuous at the origin. As a consequence, β changes discontinuously at the start of the ramp from zero to β_o, where β_o is slightly negative. Such a change results in nonsimilar effects, which are restricted to a region just downstream of the origin and are expected to rapidly decay (Smith and Clutter, 1963).

The initial conditions, Equations (22.21), are evaluated just downstream of the origin, where $M_1 = M_\infty$. For M_1', we use Equation (21.40):

$$\beta_o = \frac{2}{M_1}M_1' \frac{\int_0^{s_1} r_w^{2\sigma} F ds}{r_w^{2\sigma} F}$$

which simplifies to

$$M_1' = \frac{M_\infty \beta_o}{2s_1} \tag{22.22}$$

since

$$\frac{\int_0^{s_1} r_w^{2\sigma} F\, ds}{r_w^{2\sigma} F} - s_1$$

Thus, $|M_1'|$ decreases as s_1 increases. Physically, this is because the boundary layer at the origin is thicker and more readily separates as s_1 increases. (On average, a thin boundary layer is more energetic than a thick one.) In other words, a thin boundary layer can accommodate a relatively large $|M_1'|$ value, other factors being the same.

We now proceed to integrate Equation (22.20) by setting

$$\phi = \frac{dM_e}{ds}$$

and

$$\frac{d^2 M_e}{ds^2} = \frac{d\phi}{ds} = \frac{d\phi}{dM_e}\frac{dM_e}{ds} = \phi\frac{d\phi}{dM_e}$$

to obtain

$$\frac{1}{\phi}\frac{d\phi}{dM_e} + \frac{2}{M_e}\left(\frac{1}{\beta_o} - \frac{1 - \frac{\gamma+1}{4}M_e^2}{1 + \frac{\gamma-1}{2}M_e^2}\right) = 0$$

Upon integration, we obtain

$$\int_{M_1'}^{\phi} \frac{d\phi}{\phi} = -\frac{2}{\beta_o}\int_{M_\infty}^{M_e} \frac{dM}{M} + 2\int_{M_\infty}^{M_e} \frac{1 - \frac{\gamma+1}{4}M^2}{1 + \frac{\gamma-1}{2}M^2}\frac{dM}{M}$$

The rightmost integral becomes

$$2\int \frac{1 - \frac{\gamma+1}{4}M^2}{1 + \frac{\gamma-1}{2}M^2}\frac{dM}{M} = \ell n[MF(M)] + \text{constant}$$

where F is given by Equation (21.39). After simplification, we have

$$\frac{dM_e}{ds} = \kappa M_e^{1-2/\beta_o} F(M_e) \tag{22.23}$$

where Equation (22.22) is used and the constant κ is

$$\kappa = \frac{\beta_o}{2s_1}\frac{M_\infty^{2/\beta_o}}{F(M_\infty)} \tag{22.24}$$

A second integration would result in a quadrature solution that is not needed.

The coordinates of the ramp are parametrically represented by

$$x_w = x(s), \qquad y_w = y(s) \tag{22.25}$$

where s is arc length along the ramp and now is measured from the origin. Let θ be the wall slope, as shown in Figure 22.2, so that

$$dx_w = \cos\theta\, ds, \qquad dy_w = \sin\theta\, ds \tag{22.26}$$

Eliminate ds between Equations (22.23) and (22.26) to obtain, after integration,

$$x_w = \frac{1}{\kappa}\int_{M_\infty}^{M_e} \cos\theta \frac{M^{(2/\beta_o)-1}}{F(M)}\, dM$$

$$y_w = \frac{1}{\kappa}\int_{M_\infty}^{M_e} \sin\theta \frac{M^{(2/\beta_o)-1}}{F(M)}\, dM$$

These results actually hold for subsonic as well as supersonic flow and for a contoured ramp that produces an expansion or a compression (Emanuel, 1984). In the case of an expansion, β_o is a positive constant. Furthermore, the theory can be extended to ramps with an axisymmetric configuration (Emanuel, 1984).

The wall geometry must be consistent with the inviscid flow along the ramp. However, we know from gas dynamics that a homentropic, simple wave, supersonic, two-dimensional compression or expansion is governed by

$$\theta = \nu(M_\infty) - \nu(M_e)$$

where ν is the Prandtl–Meyer function, Equation (8.29). Thus, x_w is given by

$$x_w = \frac{1}{\kappa}\int_{M_\infty}^{M_e} \cos[\nu(M_\infty) - \nu(M)] \frac{M^{(2/\beta_o)-1}}{F(M)}\, dM \tag{22.27}$$

with a similar relation for y_w. A solution of these equations is provided by a straightforward numerical integration. The result is the wall shape with M_e as an intermediate parameter. Of course, the ramp must terminate at an angle θ where M_e is still supersonic. It is also worth noting that x_w and y_w are conveniently normalized by s_1. We thereby obtain

$$X_w = \frac{x_w}{s_1} = \frac{2F(M_\infty)}{\beta_o M_\infty^{2/\beta_o}}\int_{M_\infty}^{M_e} \cos[\nu(M_\infty) - \nu(M)]\frac{M^{(2/\beta_o)-1}}{F(M)}dM$$

with a similar result for $Y_w = (y_w/s_1)$. Thus, x_w and y_w linearly scale with s_1.

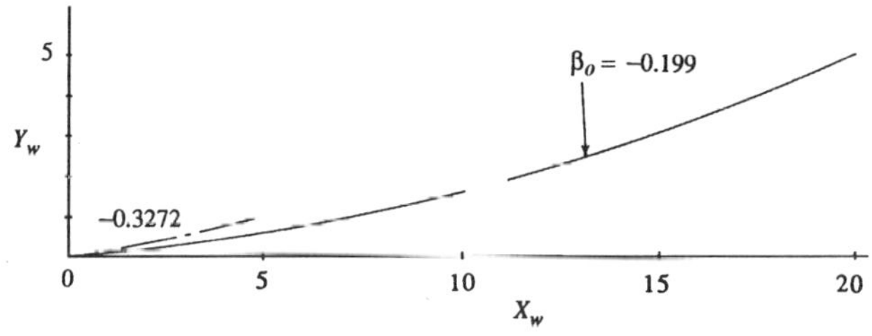

FIGURE 22.3 Optimum wall shape when β_o is −0.1988 or −0.3272. At the start of the ramp M_∞ is 2, while at the end M_e is 1.4 (Emanuel, 1984).

Figure 22.3 shows the nondimensional wall shape for two cases, given by

$$\gamma = 1.4, \qquad M_\infty = 2, \qquad \beta_o = \begin{cases} -0.3272, & g_w = 0 \\ -0.1988, & g_w = 1 \end{cases}$$

where the $g_w = 1$ case is for an adiabatic wall. The two curves terminate when the wall Mach number has decreased to 1.4, where $X_w \cong 20$ and $Y_w \cong 4.5$ for the $g_w = 1$ case.

The ramp size is dramatically reduced by wall cooling. This sensitivity to β_o is primarily due to the M^{2/β_o} factor in the integrand of Equation (22.27). The result is in accordance with the experimental observation that a boundary layer on a cold wall is less prone to separate. Since a laminar boundary layer results in a wall that may be of excessive length, it would be of interest to reexamine this procedure for a turbulent boundary layer. In this circumstance the ramp size should be significantly reduced.

22.4 ZERO DISPLACEMENT THICKNESS WALL SHAPE

Normally, a cooled wall can have a negative displacement thickness over a limited region, where the wall is cold and the pressure gradient parameter is sufficiently positive. At the start and end of the region, δ^* would pass through zero. In line with the discussion near the end of Section 21.8, we now show that it is possible to shape a cold wall so that δ^* is zero along its entire length. In contrast to a compressive ramp, such a wall is of limited utility. It may be of interest, however, in second-order boundary-layer studies. (For possible use in boundary-layer transition studies, Problem 22.6 considers the shape a wall would have whose momentum defect thickness is zero.)

Design Procedure

Equation (21.74), with $\delta^* = 0$, yields

$$Sf_w'' + [1 + (1 - S)\beta][C_v - (1 - g_w)C_t] = 0 \tag{22.28}$$

It is easy to show that this equation is satisfied only if

$$0 < \beta, \qquad 0 \le g_w < 1$$

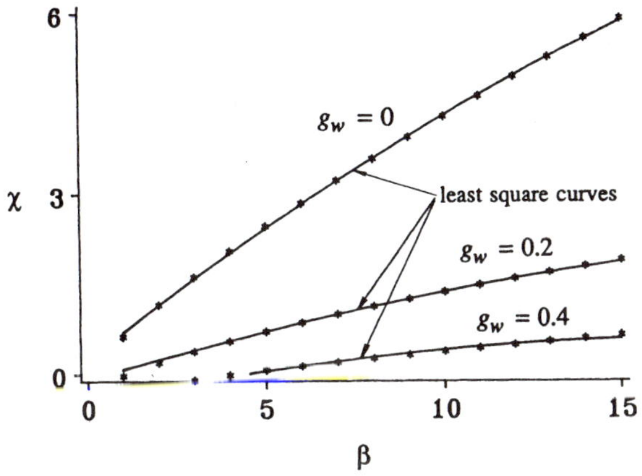

FIGURE 22.4 χ vs. β for $g_w = 0, 0.2, 0.4$ and $\gamma_w = 1.4$.

Equation (21.48) is used to replace S with M_e. Upon solving for $(\gamma - 1)M_e^2/2$, we obtain

$$\chi(\beta; g_w) = \frac{\gamma - 1}{2} M_e^2 = \frac{1+\beta}{\dfrac{f''_w}{(1-g_w)C_t - C_v} - 1} \tag{22.29}$$

In this relation, g_w is a prescribed constant parameter, whereas β must vary along the wall if M_e is to change. Figure 22.4 shows χ vs. β for three values of g_w and $\gamma = 1.4$. The asterisks are exact values, whereas the curves represent the quadratic

$$\chi = a + b\beta + c\beta^2 \tag{22.30}$$

The a, b, and c coefficients depend on g_w and are determined by a least-square fit to the points located at $\beta = 1, 2, \ldots, 15$.* Thus, for $g_w = 0.2$, the least-square fit yields

$$\chi = -0.008981 + 0.17857\beta - 0.0031197\beta^2$$

Note that as g_w increases, larger β values are required for χ to be positive. In view of its definition in terms of M_e, Equation (22.29), only nonnegative χ values are of interest.

For simplicity, we set $\sigma = 0$ in Equation (21.44), with the result

$$\frac{1}{\beta}\frac{d\beta}{ds} = \frac{1}{\dfrac{dM_e}{ds}}\frac{d^2M_e}{ds^2} + 2\left[\frac{1}{\beta} - \frac{1 - \dfrac{\gamma+1}{4}M_e^2}{1 + \dfrac{\gamma-1}{4}M_e^{22}}\right]\frac{1}{M_e}\frac{dM_e}{ds} \tag{22.31}$$

for a two-dimensional flow. Equation (22.29) is now written as

$$M_e = \left(\frac{2}{\gamma - 1}\right)^{1/2} \chi^{1/2}$$

---

* I am indebted to Dr. H.-K. Park for the calculations in this section and for the associated figures.

and its first two s derivatives are

$$\frac{dM_e}{ds} = \frac{1}{2}\left(\frac{2}{\gamma-1}\right)^{1/2}\chi^{-1/2}\chi_\beta\frac{d\beta}{ds}$$

$$\frac{d^2M_e}{ds^2} = \frac{1}{4}\left(\frac{2}{\gamma-1}\right)^{1/2}\chi^{-3/2}\left[(2\chi\chi_{\beta\beta}-\chi_\beta^2)\left(\frac{d\beta}{ds}\right)^2 + 2\chi\chi_\beta\frac{d^2\beta}{ds^2}\right]$$

where

$$\chi_\beta = \frac{d\chi}{d\beta}, \qquad \chi_{\beta\beta} = \frac{d^2x}{d\beta^2}$$

These relations are inserted into Equation (22.31), to obtain

$$\frac{d^2\beta}{ds^2} = J\left(\frac{d\beta}{ds}\right)^2 \tag{22.32}$$

where

$$J(\beta;\, g_w) = \frac{1}{\beta\chi_\beta}(\chi_\beta - \beta\chi_{\beta\beta}) + \frac{\chi_\beta}{\chi}\left(\frac{\frac{3}{2}-\frac{1}{\gamma-1}\chi}{1+\chi} - \frac{1}{\beta}\right) \tag{22.33a}$$

With Equation (22.30), J simplifies to

$$J = \frac{b}{(b+2c\beta)\beta} + \frac{b+2c\beta}{\chi}\left(\frac{\frac{3}{2}-\frac{1}{\gamma-1}\chi}{1+\chi} - \frac{1}{\beta}\right) \tag{22.33b}$$

and this approximation is used in all subsequent calculations. Figure 22.5 shows J vs. β for the same γ and g_w values used in Figure 22.4.

The shape of the wall is determined by a numerical integration. For this, it is convenient to introduce

$$B = \frac{d\beta}{ds}$$

and Equation (22.32) becomes

$$\frac{dB}{ds} = \frac{dB}{d\beta}\frac{d\beta}{ds} = B\frac{dB}{d\beta} = JB^2$$

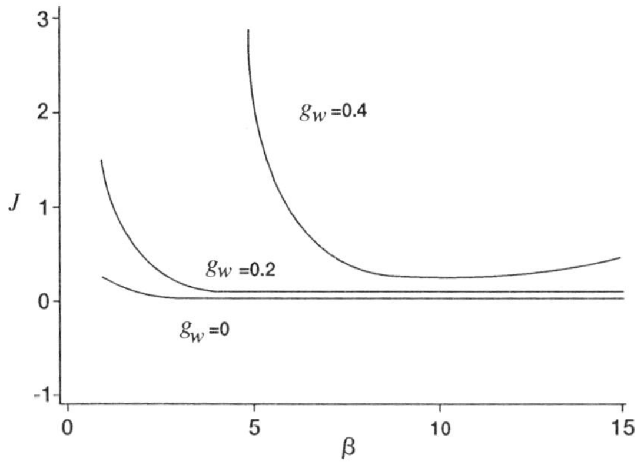

FIGURE 22.5 J vs. β for $g_w = 0$, 0.2, 0.4 and $\gamma = 1.4$.

or

$$\frac{dB}{d\beta} = JB \tag{22.34}$$

where J is a function only of β for a given g_w value. A Cartesian coordinate system is used whose origin is at the start of the zero displacement thickness boundary layer. This point is denoted with a zero subscript. Moreover, we assume that $M_{e0} \geq 1$ in order to take advantage of the analytical simplicity inherent in a supersonic flow. Downstream of this point the boundary-layer flow accelerates; thus, β and M_e increase with s. The resulting curved wall therefore generates a noncentered Prandtl–Meyer expansion.

The equations for the wall shape are

$$\frac{dx}{ds} = \cos\theta, \qquad \frac{dy}{ds} = -\sin\theta$$

which become

$$\frac{dx}{d\beta} = \frac{\cos\theta}{B} \tag{22.35a}$$

$$\frac{dy}{d\beta} = -\frac{\sin\theta}{B} \tag{22.35b}$$

where θ is given by the Prandtl–Meyer relation

$$\theta = \nu(M_e) - \nu(M_{e0})$$

Equations (22.34) and (22.35) are simultaneously integrated, starting with the initial conditions

$$B(\beta_o) = \left(\frac{d\beta}{ds}\right)_0, \qquad x(\beta_0) = 0, \qquad y(\beta_0) = 0 \tag{22.36}$$

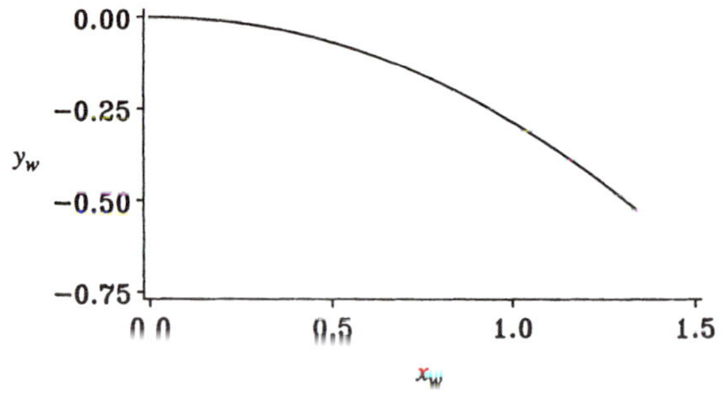

FIGURE 22.6 Wall shape when $\gamma = 1.4$, $M_{e0} = 1.5$, $(dM_e/ds)_0 = 1$, and $g_w = 0.2$.

Thus, B and the independent variable β are introduced for computational reasons; the resulting system of equations is easily solved numerically. With $\gamma = 1.4$ and given values for g_w and M_{e0}, Figure 22.4 can be used to determine β_0. The quantity $(d\beta/ds)_0$ is a free parameter that must be positive. A large value for this parameter results in a more rapidly turning wall. The computation terminates when $\beta = 15$, which is the upper limit used in the least-square fit that yielded Equation (22.30).

Figure 22.6 shows the wall shape when $g_w = 0.2$. At the start of the wall, $\theta = 0°$, $M_{e0} = 1.5$, $(d\beta/ds)_0 = 1.5$, and $\beta_0 = 3.202$, while at the end, $\theta = 39.2°$, $M_e = 3.072$, and $\beta = 15$. (If the upstream wall is a flat plate with $\beta = 0$, there will be a nonsimilar region during which the flow adjusts to its zero displacement thickness condition.) The wall has a similar appearance for other values of M_{e0} and $(dM_e/ds)_0$. In spite of the g_w dependence shown in Figures 22.4 and 22.5, the wall shape itself is barely influenced by g_w. The wall shape for g_w equal to 0 or 0.4 nearly overlays the one shown in Figure 22.6. An important difference, however, is that as g_w increases the wall length rapidly decreases. For example, with the same conditions used for Figure 22.6, except with $g_w = 0.4$, we have $M_e = 1.74$, $\theta = 7.07°$, $x_w = 0.2184$, and $y_w = -0.00130$ at the terminating point, where β again equals 15. Thus, a zero displacement thickness wall no longer exists well before g_w equals unity.

22.5 PERFORMANCE OF A SCRAMJET PROPULSION NOZZLE

An asymmetric thrust nozzle is generally used with a scramjet engine. This type of air-breathing engine is supposed to operate in the upper atmosphere when the flight Mach number is in excess of four. The engine's efficiency rapidly decreases below this Mach number and the engine is incapable of operation at subsonic Mach numbers. The combustion process is supersonic and the Mach number at the inlet to the nozzle is also supersonic. It should be noted that, at the time of writing, a scramjet engine that can actually *accelerate* a vehicle at hypersonic flight speeds has yet to be demonstrated. Remember that the drag (wave, viscous, form, …) of a vehicle flying in air at a high Mach number is formidable. On the other hand, in contrast to a rocket engine, there are limitations on how much thrust is available from the engine. Simply producing thrust on a test stand is not the issue; the thrust must exceed the drag in order to attain hypersonic cruise conditions. This limitation is discussed shortly.

A convenient model for an asymmetric nozzle can be based on minimum length nozzle (MLN) theory. This section utilizes the figures and analysis of Bae and Emanuel (1991), which in turn stems from Bae's (1989) dissertation. There is, unfortunately, a mixup in most of the figures and their captions in the journal version. For this reason and for the considerable amount of material in the dissertation not discussed in the abbreviated journal article, the dissertation is recommended if the reader is interested in pursuing this topic.

MLN theory minimizes the length of the supersonic portion of a nozzle subject to the constraints of inviscid flow and a uniform flow in the exit plane. This directly implies a shock-free flow inside the nozzle. A basic reference for the theory is Emanuel (1986, Chapter 17) with additional material

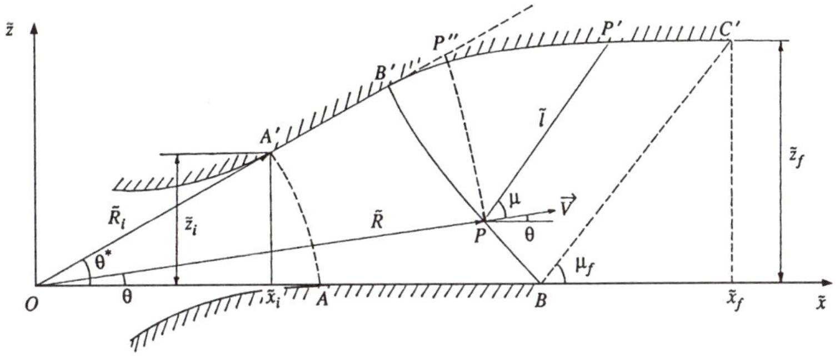

FIGURE 22.7 Schematic and part of the upstream combustor for a curved inlet MLN with $M_i > 1$.

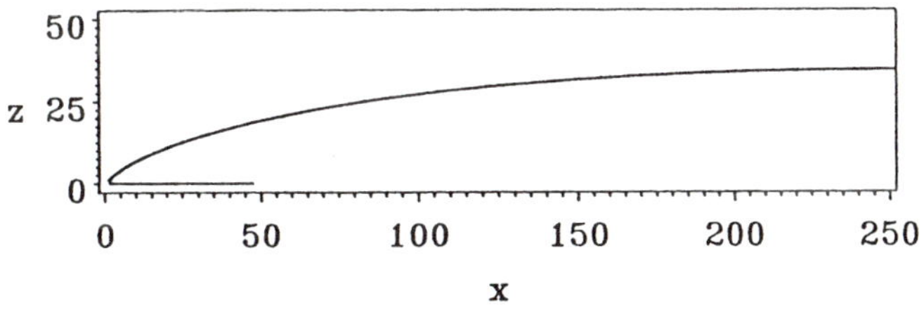

FIGURE 22.8 Actual shape of the nozzle when $M_i = 2$, $M_f = 6$, and $\gamma = 1.4$.

in Argrow and Emanuel (1988), Aldo and Argrow (1995), Argrow and Emanuel (1991), and Ho and Emanuel (2000).

As elsewhere in this chapter, a perfect gas with unity values for Pr and C is assumed in Bae and Emanuel (1991). A hypersonic vehicle of the type under discussion generally has a diffuser and a nozzle that can be approximated as two dimensional. Because of the high operational altitude, ρ_∞ is small, and the nozzle inlet height, $\tilde{z}_i$ in Figure 22.7, is also small. A characteristic Reynolds number Re_o is based on $\tilde{z}_i$ and on stagnation conditions at the nozzle's inlet; hence, the flow speed is $(2h_o)^{1/2}$. In Bae and Emanuel, $\tilde{z}_i$ is only 4 cm and Re_o is 1.3×10^5. For this type of Reynolds number definition, a laminar flow in the nozzle might be anticipated.

An arc of a circle, AA' in Figure 22.7, is used for the inlet, where the flow in region $AA'B'BA$ is a cylindrical , inviscid source flow. Region $BB'C'B$ is a simple wave region, where downstream of BC', which is a Mach line, the flow is uniform and parallel to the $\tilde{x}$-axis. On the circular inlet the Mach number, M_i, is sonic or supersonic. When sonic, the wall has a shoulder at A', otherwise, the contour is smooth, as shown in the figure. The lower wall of the nozzle is AB, which is planar and which ends at B. The upper wall usually has a planar section, $A'B'$, which has an angle θ^* relative to the lower wall. The upper wall then has a contoured section, $B'C'$, where the slope at C' is parallel to AB. For a scramjet engine, the exit Mach number is typically rather large and, therefore, the exit plane Mach angle μ_f is small. For instance, for the baseline case in Bae and Emanuel, the full nozzle, to scale, is shown in Figure 22.8. The inlet height of 4 cm is not discernible, nor is the $A'B'$ wall segment, and the upper wall is too long to be practical. Thus, the quite modest effect on performance of severe upper wall truncation is established in this paper.

There are other possible MLN configurations; this one, however, is unique in that an analytical solution exists. This solution provides the flow field as well as the wall shape. MLN theory provides the correct value for the inlet wall angle θ^*, as shown in Figure 22.9 for various M_i and $\gamma = 1.4$. For example, if $M_i = 2$ and M_f is between 2 and about 4, θ^* is provided by the dashed curve, and points A' and B' coincide. For M_f in excess of 4, θ^* is given by the nearly horizontal line and

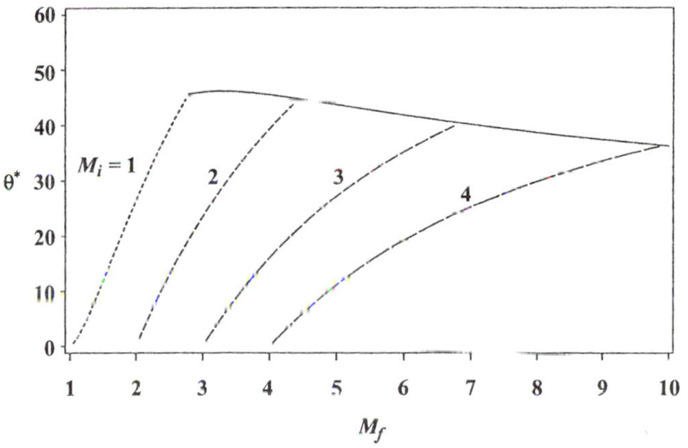

FIGURE 22.9 Initial expansion angle θ^* vs. the exit Mach number M_f for various inlet Mach numbers and $\gamma = 1.4$.

there is a planar wall section at the start of the nozzle's upper wall. For scramjet nozzles, M_f is expected to be sufficiently large such that the upper wall has a planar section.

The detailed inviscid theory in Bae and Emanuel (1991), or Bae (1989), enables explicit algebraic relations for β to be established. There is a separate β relation for each of the three wall segments. For instance, along wall a (i.e., wall AB) the pressure gradient parameter is

$$\beta = \frac{2X^2(\xi_i + KJ)}{K(M^2 - 1)} \tag{22.37}$$

where the constant K depends on γ, M_i, p_o, ..., ξ_i represents the length of the boundary layer that is upstream of A, and

$$X = 1 + \frac{\gamma - 1}{2} M^2 \tag{22.38a}$$

$$J = \ln\left[\frac{M_i}{M}\left(\frac{X}{X_i}\right)^{1/2}\right] + \frac{\gamma + 1}{2(\gamma - 1)}\left(\frac{1}{X_i} - \frac{1}{X}\right) \tag{22.38b}$$

At point A, we have

$$\beta_A = \frac{2X_i^2 \xi_i}{K(M_i^2 - 1)} \tag{22.39}$$

and ξ_i is taken as zero if M_i is unity. Although the AB wall is a flat plate, the Mach number and pressure along it are associated with a cylindrical source flow. The pressure gradient parameter varies with M and parametrically depends on γ, M_i, K, and ξ_i. Figure 22.10 shows how β varies along the three wall segments, where walls b and c are, respectively, $A'B'$ and $B'C'$. The curves are for conditions specified in the figure caption (i.e., a nominal case) and the three curves, per panel, are for different ξ_i values, where the upper wall ξ_i' value equals ξ_i. As the length of the upstream boundary layer increases, so does ξ_i. This effect is especially significant for the two planar wall segments. Remember that β is large when the pressure is rapidly decreasing along a wall. This effect is most pronounced

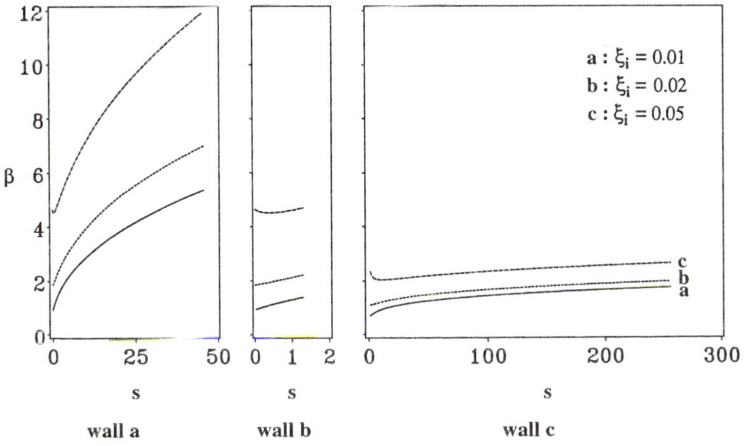

FIGURE 22.10 Distribution of β along the nozzle walls when $M_i = 2$, $M_f = 6$, and $\gamma = 1.4$ for three ξ_i values, where $\xi_i' = \xi_i$.

along wall *a*. Note the discontinuity β in at point B', where the wall slope is continuous but its curvature is not. For a short distance downstream of B', the boundary layer is nonsimilar.

Viscous, and when appropriate inviscid, nondimensional parametric results are provided for the thrust, lift, heat transfer, pitching moment, and a variety of boundary-layer thicknesses. In addition to global results, wall distributions of the thrust, heat transfer, etc. are given. The analysis demonstrates that the nozzle produces a considerable lift force whose magnitude may actually exceed that of the thrust and a significant pitching moment. The lift result is real but deceptive. Upstream of the nozzle, there will be a negative lift force of the same magnitude. Thus, the nozzle does not supplement the lift of the rest of the vehicle.

The thrust rapidly decreases as M_i increases. This stems from the assumption of a constant stagnation pressure, p_o, at the inlet of the nozzle. Thus, as M_i increases, the static pressure force on the nozzle's wall decreases. Consequently, the thrust is a maximum when M_i is sonic. The heat transfer is most intense on the upstream edges of walls *a* and *b*. There is little loss in thrust when the upper wall is significantly truncated.

All results in Bae and Emanuel are nondimensional. For example, forces are normalized per unit depth of nozzle, with $p_o\tilde{R}_i\theta^*$, where $\tilde{R}_i$ is the radius of arc AA'. As a consequence, the thrust per unit nozzle exit area, in a plane normal to the jet, is considered. This parameter can be viewed as the thrust throughput and is of major importance from a vehicle system viewpoint. If the thrust is below flight requirements, the only options are to reduce the overall drag, increase the thrust per unit area, increase the area, or a composite of all three. Increasing the nozzle exit area in turn means increasing the vehicle's projected frontal area with a corresponding increase in drag, especially wave drag. In Section 14.10, e.g., we saw how sensitive the wave drag of a wedge is to changes in its thickness. Of course, the depth can be increased, but this simply increases the thrust and drag proportionately.

Alternatively, the thrust per unit area can be increased by, e.g., decreasing M_i to near unity. Another possibility would be to increase p_o. The freestream stagnation pressure is given by

$$p_{o\infty} = p_\infty\left(1 + \frac{\gamma - 1}{2}M_\infty^2\right)^{\gamma/(\gamma-1)} \tag{22.40}$$

For Mach 5 flight at 30.5 km (100 kft), $p_{o\infty}$ is 5.66×10^5 Pa, which is not very large. This value then decreases across the bow shock and across the diffuser's inlet shock system. Inside the combustor, there is a further decrease. From a one-dimensional viewpoint, the influence coefficient method of Section 13.5 yields

$$\frac{dp_o}{p_o} = -\frac{\gamma M^2}{2}\left(\frac{dT_o}{T_o} + 4c_f\frac{dx}{d}\right) \tag{22.41}$$

where, of course, both dT_o/T_o and $4c_f dx/d$ are positive, with dT_o/T_o substantially so because of the highly exothermic combustor reactions. This relation is independent of any changes in the combustor's cross-sectional area. Of importance is the magnification of $(dT_o/T_o + 4c_f dx/d)$ by the $\gamma M^2/2$ coefficient, since the Mach number is supersonic. Thus, the shock, combustion, and frictional losses represent a major hurdle for scramjet engines. Of course, $p_{o\infty}$ can be increased by flying at a lower altitude, thereby increasing p_∞. This, however, also increases the drag.

The thrust throughput for a scramjet engine can be estimated, under relatively favorable assumptions, for a variety of flight conditions. When this is done, it is found that this parameter is about two orders of magnitude smaller than that for a rocket engine, going as far back as the V-2 engine. It will therefore be difficult to develop scramjet engines that can accelerate an air-breathing vehicle at hypersonic speeds.

REFERENCES

Aldo, A.C. and Argrow, B.M., "Dense Gas Flow in Minimum Length Nozzles," *J. Fluids Eng.* 117, 270 (1995).

Argrow, B.M. and Emanuel, G., "Comparison of Minimum Length Nozzles," *J. Fluids Eng.* 110, 283 (1988).

Argrow, B.M. and Emanuel, G., "Computational Analysis of the Transonic Flow Field of Two-Dimensional Minimum Length Nozzles," *J. Fluids Eng.* 113, 479 (1991).

Bae, Y.-Y., "Performance of an Aerospace Plane Propulsion Nozzle," Ph.D. Dissertation, University of Oklahoma, 1989.

Bae, Y.-Y. and Emanuel, G., "Performance of an Aerospace Plane Propulsion Nozzle," *J. Aircraft* 28, 113 (1991).

Emanuel, G., "Supersonic Compressive Ramp without Laminar Boundary-Layer Separation," *AIAA J.* 22, 29 (1984).

Emanuel, G., *Gasdynamics: Theory and Applications,* AIAA Education Series, Washington, D.C., 1986.

Ho, T.-L. and Emanuel, G., "Design of a Nozzle Contraction for Uniform Sonic Throat Flow," *AIAA J.* 38, 720 (2000).

Johannesen, N.H., "Experiments on Two-Dimensional Supersonic Flow in Corners and over Concave Surfaces," *Philos. Mag.* 43, 568 (1952).

Lewis, J.E., Kubota, T., and Lees, L., "Experimental Investigation of Supersonic Laminar, Two-Dimensional Boundary-Layer Separation in a Compression Corner with and without Cooling," *AIAA J.* 6, 7 (1968).

Liepmann, H.W. and Roshko, A., *Elements of Gasdynamics,* John Wiley, New York, 1957.

Smith, A.M.O. and Clutter, D.W., "Solution of the Incompressible Laminar Boundary-Layer Equations," *AIAA J.* 1, 2062 (1963).

PROBLEMS

22.1 One measure of the mass flow rate inside a boundary layer is

$$\dot{m}_\delta = \frac{1}{r_w^\sigma}\int_0^\delta \rho u r^\sigma dn \cong \int_0^\delta \rho u \; dn$$

where the last equality requires $r_w \gg \delta$ when $\sigma = 1$.

(a) Derive the approximate result

$$\dot{m}_\delta = \frac{(2\xi)^{1/2}}{r_w^\sigma}(\eta_{ev} - C_v)$$

Roughly speaking, how does $\dot{m}_\delta$ vary with s?

(b) For the example in Section 22.2, what is $\dot{m}_\delta$ at the end of the airfoil when normalized by $\rho_\infty U_\infty$ times the airfoil's projected frontal area? Utilize the boundary layer on both airfoil surfaces.

22.2 Consider a compressive ramp with an adiabatic wall. Conditions correspond to those in Section 22.3; i.e.,

$$\gamma = 1.4, \qquad M_\infty = 2, \qquad M_f = 1.4$$

where the ramp terminates when $M_e = M_f$. With $s_1 = 2 \times 10^{-2}$ m and $Re_\infty = 10^5$, where Re_∞ is based on s_1, estimate δ_f^* at the end of the ramp.

22.3 Consider a flat plate of length c at an angle of attack, α, in an incompressible flow with a freestream speed U_∞. The exact potential flow solution that satisfies the Kutta condition yields

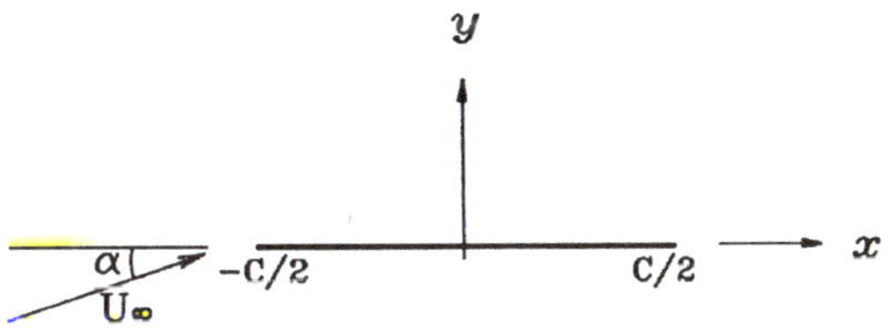

$$\frac{u(x,0)}{U_\infty} = \cos\alpha \pm \sin\alpha\left(\frac{1-X}{1+X}\right)^{1/2}, \qquad X \equiv \frac{2x}{c}$$

for the flow speed on the upper (+ sign) and lower (– sign) surfaces.

(a) Determine the location X_o of the stagnation point on the lower surface as a function of α.

(b) Determine $\beta(s)$ for the boundary layer downstream of the lower surface stagnation point.

(c) Introduce the assumption that α is small and obtain $\beta(X)$. Interpret β at the stagnation point, at the wall location where β has a minimum value, and at the trailing edge.

(d) Use the results of part (c) to evaluate [illegible] $Re_x^{1/2}$ [illegible] at X [illegible] 0.9571 when α is 5°.

22.4 Continue with Problem 20.1 for incompressible flow about a circular cylinder. Consider only the upstream half of the cylinder, of unit depth, where $-\pi/2 \le \theta \le \pi/2$, and ignore any question of boundary-layer separation.

(a) Derive separate equations for the pressure drag, C_{Dp}, and viscous drag, C_{Dv}, where C_D is the force in the freestream direction divided by $\rho U_\infty^2 A_p/2$ and A_p is the projected frontal area.

(b) Write $C_D = C_{Dp} + C_{Dv}$ and estimate the value for C_D when the fluid is air, U_∞ = 50 m/s, $d = 2$ cm, $p_\infty = 10^5$ N/m^2, and $\rho = 1.2$ kg/m^3.

22.5 A slender wedge is in an $M_\infty = 4$ air wind tunnel flow whose plenum condition is $p_o = 4$ atm and $T_o = 1050$ K. Consider only the first centimeter of the upper surface of the wedge whose length along its upstream edge is b. Evaluate the skin-friction drag, pressure drag, and heat transfer for this surface when its inclination angle, relative to the freestream velocity, is 0 or 20°. Assume the surface is kept at 300 K and use Colburn's analogy.

22.6 **(a)** As in Section 22.4, establish the differential equations for the shape of a wall whose momentum defect thickness is zero. Assume $\sigma = 0$ and the initial Mach number M_{eo} is sonic or supersonic.

(b) Numerically determine the wall shape when

$$\gamma = 1.4, \qquad g_w = 2, \qquad M_{e0} = 1.3, \qquad \left(\frac{dM_e}{ds}\right)_0 = 0.5$$

22.7 The displacement thickness has a maximum negative value when $g_w = 0$ and $\beta \to \infty$.

(a) In this limit, show that

$$\Upsilon\delta^* = C_v - C_t$$

for any S value, and that

$$C_v - C_t \cong -0.3465$$

(b) With $\mu = C_\mu T$, show that

$$Re_{\bar{x}} = \frac{1}{C_\mu}\left(\frac{\gamma}{RT_{oe}}\right)^{1/2} \frac{p_{oe}}{T_{oe}} \frac{M_e}{\left(1 + \frac{\gamma-1}{2}M_e^2\right)^{(3-\gamma)/[2(\gamma-1)]}} \bar{x}$$

where

$$\bar{x} = \frac{1}{r_w^{2\sigma}F(M_e)}\int_0^s r_w^{2\sigma}F\,ds$$

(c) We apply the above to the throat of a small rocket nozzle, where, for simplicity, a two-dimensional flow is assumed. We thus have the following conditions:

$$\sigma = 0, \qquad M_e = 1, \qquad p_{oe} = 2\times 10^6 \text{ Pa}, \qquad T_{oe} = 2861\text{K},$$

$$W = 21.87 \text{ kg/kmol}, \qquad \gamma = 1.229, \qquad s = 0.2 \text{ m}, \qquad C_\mu = 5\times 10^{-8}\,\frac{\text{Pa-s}}{\text{K}}$$

Estimate $\bar{x}$ and determine $Re_{\bar{x}}$, Υ, and δ^*, where δ^* assumes $g_w = 0$ and $\beta \to \infty$.

22.8 Consider the steady, homentropic flow of a perfect gas in a two-dimensional or axisymmetric nozzle. As shown in the sketch, the wall contour is a hyperbola, with an asymptotic angle θ, whose equation is

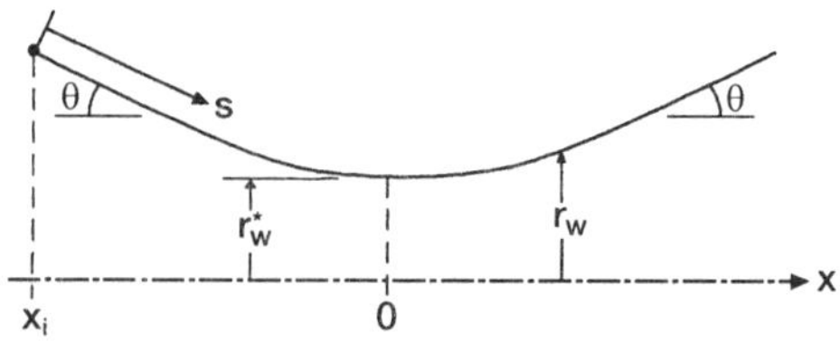

$$r_w^2 - ax^2 = r_w^{*2}, \qquad a = \tan^2\theta$$

(a) Introduce $X = (x/r_w^*)$, and $R = (r/r_w^*)$, and determine $S = (s/r_w^*)$ as a function of X, where s is measured from the inlet at x_i.

(b) Assume flow conditions are uniform in any plane that is perpendicular to the X-axis. Assume subsonic flow for negative X and supersonic flow for positive X. Develop a relation between M, which is equivalent to M_e, the inviscid Mach number at any nozzle cross section, and X. Use this relation to obtain equations for dM/dX and d^2M/dX^2.

(c) Develop expressions for dM/ds, d^2M/dS^2, dX/dS, d^2X/dS^2, and $d(\ln r)/dS$.

(d) Utilize the foregoing results and start with Equation (21.42) to obtain an equation for $d\beta/dS$ that depends on M, X, and various constants.

(e) With M_i small and positive and a favorable pressure gradient, β is positive everywhere. Evaluate the derivative $(dM/dX)^*$. Consider the $\sigma = 0$ case, and derive the condition such that $d\beta/dS$ is finite at the throat. Next, determine an appropriate value for β_i. Results for all parts of this problem should be simplified as much as possible.

22.9 Consider the flow of perfect gas in a conical duct with a half-angle of α; see the sketch below. The inlet and exit planes are spherical caps with radii of r^* and r_f, respectively. Treat the inviscid flow as a spherical source flow, where $M^* = 1$ and $M_f > 1$.

(a) Obtain the inviscid solution for p, ρ, T, r/r^*, and w in terms of γ, R, p_o, T_o, and M.

(b) Determine the pressure force F_p exerted by the fluid on the wall in the z direction by an integration over the wall. Your final answer should involve only γ, M_f, p_o, r^*, and α, and should involve an integral of the form

$$I = \int_1^{M_f} f(\gamma, M)\, dM \qquad (1)$$

(c) Determine F_p using the momentum theorem of Chapter 14, where your result should parallel the part (b) answer as much as possible. Establish that the two answers for F_p are consistent. Determine F_p when $\gamma = 1.4$, $p_o = 30$ atm, $\alpha = 15°$, $r^* = 8$ cm, and the conventional (one-dimensional) area ratio is 15.08.

(d) Consider the laminar boundary layer on the inside of the cone that starts as $s = 0$. Write the radial coordinate measured from the symmetry axis as $\hat{r}$. Assume the viscosity is given by $C_\mu T$, where C_μ is a constant. Derive an equation for $\xi(M)$, where $M = M_e$, that depends on constants and an integral of the form of Equation (1) above. To evaluate the integral, hereafter set $\gamma = (9/7)$. With this γ value, obtain $\xi(M)$. Obtain an equation for $\beta(M)$ and evaluate $\beta^* = \beta(1)$.

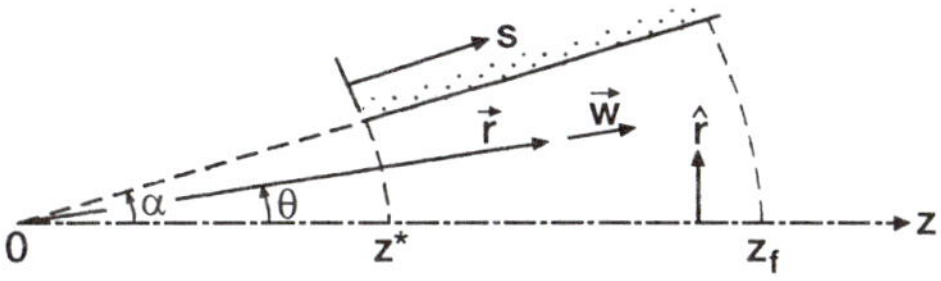

(e) Assume CO_2 gas with the following data:

$$\gamma = \frac{9}{7}, \quad p_o = 30 \text{ atm}, \quad T_o = 10^3 \text{ K}, \quad \alpha = 15°, \quad r^* = 8 \text{ cm},$$

$$\frac{A_f}{A^*} = 15.08, \quad C_\mu = 4.983 \times 10^{-8}\ \frac{\text{Pa-s}}{\text{K}}, \quad T_w = 400 \text{ K}$$

Determine the skin-friction coefficient at the exit of the nozzle.

23 Second-Order Boundary-Layer Theory

23.1 PRELIMINARY REMARKS

In view of its remarkable success, it is natural to ask if boundary-layer theory can be extended or generalized. For instance, suppose there is a curved shock wave, Figure 5.2, that generates vorticity. This vorticity, which is largely external to the boundary layer, can be expected to have some effect on the wall's heat transfer and skin friction. As a second example, suppose a body possesses both longitudinal and transverse curvature. Aside from its effect on the pressure gradient parameter, the surface curvature of the body will also alter the skin friction and heat transfer.

The above are referred to as second-order effects; there are others that will be discussed shortly. They usually occur in combination, although the analysis, to some extent, will treat them separately. Analytical and physical insights are gained into the mechanisms involved in a laminar viscous flow by examining these effects.

Later we will write an asymptotic expansion for u as

$$u(s,n) = u_1(s,\bar{n}) + \varepsilon u_2(s,\bar{n}) + \cdots \tag{23.1}$$

where ε is a small parameter,

$$\varepsilon = \left(\frac{\mu_\infty}{\rho_\infty U_\infty \ell}\right)^{1/2} = \frac{1}{Re^{1/2}} \tag{23.2}$$

ℓ is a characteristic length, and u_1 and u_2 are, respectively, the first- and second- order terms in the expansion. For example, the length ℓ could be the nose radius of a blunt body or arc length along the wall for a pointed body. As in Chapter 19, $n = \varepsilon\bar{n}$ and s, $\bar{n}$, u_1, and u_2 are of $O(1)$ in the boundary layer. The quantity u_2 will be decomposed in Section 23.5 in a way that accounts for all second-order effects. Third-order terms, which would be proportional to Re^{21}, are not considered. The analysis requires equations similar to Equation (23.1) for the other dependent boundary layer variables. For the same variables a separate set of asymptotic expansions will be written for the external inviscid flow. The small parameter in both sets of expansions is ε.

The balance of this section is devoted to additional discussion, starting with a catalog of second-order effects. It concludes by providing the governing equations in a suitable, body-oriented, nondimensional form. The next four sections then provide a detailed description of the general theory with emphasis on the derivation of the appropriate equations and their associated boundary conditions. This will be done in a computationally suitable format whenever possible. In contrast to Chapter 21, unity values for the Chapman–Rubesin parameter and Prandtl number are not assumed nor is a similarity form assumed. The Prandtl number, however, is assumed to be constant. Since the gas is perfect, μ and κ then have the same temperature dependence.

The last three sections are largely devoted to a treatment of compressible flow over a flat plate or along a circular cylinder. In order to utilize the theory in Chapter 21, unity values are here assumed for the Prandtl number and Chapman–Rubesin parameter. Section 23.8 concludes with a brief discussion of general trends and is not limited to flow over a flat plate or along a cylinder.

Second-Order Effects

Our presentation often closely follows the approach and nomenclature developed by Van Dyke (1962, 1963, 1969). For a compressible flow he classifies second-order phenomena as seven independent effects:

A. Curvature
   1. Longitudinal
   2. Transverse
B. Interaction with the external flow
   3. Displacement
   4. External entropy gradient
   5. External stagnation enthalpy gradient
C. Noncontinuum surface effects
   6. Velocity slip
   7. Temperature jump

The first two items refer to the geometry of the wall or body, while item 2 is not present in a two-dimensional flow. A semi-infinite flat plate has zero values for both curvatures. Realistic shapes, however, are subject to both effects.

The displacement effect, item 3, is caused by the first-order boundary layer slightly displacing the inviscid flow, by an amount δ^* from the surface of the body.

A primary cause for an external entropy gradient is a curved shock wave, downstream of which the flow is rotational and not homentropic. Stagnation enthalpy gradients, item 5, are typically produced by nonuniform heat addition or combustion, and the inviscid downstream flow is not homenergetic. These gradients are particularly evident in combustion processes where the fuel and oxidizer are not premixed; examples are jet engines, rocket engines, and chemical lasers with supersonic mixing. Nonuniform heat addition without combustion, for example, occurs when a nonuniform laser beam is used to excite molecules in a flowing gas; a prime example is laser isotope separation processes. Although external vorticity is often present, since it is produced by both nonuniform heat addition or combustion and a curved shock, it is not in the above list. By Crocco's equation for a steady flow with no body forces, Equation (5.30), the vorticity is determined by the entropy and stagnation enthalpy gradients. Consequently, it is not an independent effect. In an unsteady flow, external vorticity is an independent effect and should be added to the above list.

Effects associated with the second viscosity coefficient, or the bulk viscosity, are third-order (Van Dyke, 1962). For some gases, such as N_2O and CO_2 (Emanuel, 1990), the bulk viscosity μ_b is three to four orders of magnitude larger than μ. In the momentum and energy equations, μ_b is multiplied by the dilatation, $\nabla \cdot \vec{w}$. In a high-speed boundary layer, where $\nabla \cdot \vec{w}$ can be significant, it may be necessary to retain the bulk viscosity terms when dealing with the above gases (Emanuel, 1992). In effect, these terms should be promoted to second or even first order. Despite its name, in a CO_2 hypersonic boundary layer μ_b produces an increase in the heat transfer but no change in the skin friction (Emanuel, 1992). In this chapter, however, μ_b/μ, or λ/μ, is presumed to be of order unity. This assumption is appropriate for all monatomic gases, air, and some other polyatomic gases (see Section 13.4).

Velocity Slip and Temperature Jump

As the Reynolds number decreases, noncontinuum effects may appear in the form of velocity slip and a temperature jump at the wall, items 6 and 7 in the above list. These effects depend on the value of the Knudsen number,

$$Kn = \frac{\lambda}{\ell}$$

where λ is the molecular mean-free path. We evaluate λ in the freestream and obtain, from kinetic theory (Chapman and Cowling, 1960),

$$\lambda = \frac{1}{0.499}\left(\frac{\pi\gamma}{8}\right)^{1/2}\left(\frac{\mu}{\rho a}\right)_\infty$$

where a is the speed of sound. With this relation a Knudsen number can be written as

$$Kn = 1.26\lambda^{1/2}\frac{M_\infty}{Re}$$

where Re is based on ℓ, which here is a wall length. For example, if $\gamma = 1.4$, $M_\infty = 5$, and $Re = 10^4$, we have $Kn = 7.5 \times 10^{-4}$. Continuum and slip flow, respectively, occur when the Knudsen number is in the ranges

$$0 \le Kn < 10^{-4}, \text{ continuum flow}$$
$$10^{-4} \le Kn < 10^{-2}, \text{ slip-flow}$$

Hence, the above supersonic boundary layer is squarely in the slip-flow regime at this wall location. As the Reynolds number increases, however, the boundary layer gradually changes from one with velocity slip and temperature jump to one without these effects. An incompressible boundary layer has $Kn = 0$ and thus does not exhibit velocity slip or temperature jump, regardless of the Reynolds number.

Once again let us consider the boundary layer on a flat plate. A boundary-layer thickness is physically more appropriate for the length ℓ than is distance along the plate. Since the momentum thickness provides the simplest result, we set ℓ equal to θ. From Equation (21.73), we have

$$\Upsilon\theta = f''_w = 0.4696$$

where $\bar{x} = s$ and

$$\Upsilon = \left(\frac{Re_s}{2}\right)^{1/2}\frac{1}{s}$$

Although $\ell = \theta$, the Reynolds number is still based on the dimensional wall length s. The corresponding Knudsen number, which differs from the preceding one, is

$$\tilde{K}n = \frac{1.89\gamma^{1/2}M}{Re_r^{1/2}}$$

Hence, when ℓ equals θ, an $Re_s^{1/2}$ dependence is obtained, indicating a second-order effect.

Discussion

Second-order boundary-layer theory was developed with some difficulty in the 1950s and early 1960s. The initial controversies were resolved by the systematic application of the method of matched asymptotic expansions (Van Dyke, 1963) to the incompressible Navier–Stokes equations. Third-order theory has received little attention, in part because of the considerable complexity of the second-order theory. It has been further argued that a meaningful third-order theory should be based on the (still uncertain) Burnett equations (Van Dyke, 1962) rather than on the Navier-Stokes equations.

Despite its promising beginning and the importance of the phenomena involved, second-order theory has not had a major impact on fluid dynamics. One reason is that the first-order theory often provides adequate results for engineering purposes. The complexity of second-order theory is another factor. This is evident in the dissertation of Werle (1968), where a variety of incompressible, second-order solutions is obtained. Werle discusses the lack of validity of the second-order theory in the vicinity of a separation point, the difficulty in solving the equations, and the presence of singularities and eigensolutions (discussed in Section 23.8). A final comment would note the widespread development of Navier–Stokes computer codes that simply bypass boundary-layer theory in its entirety.

In spite of these comments, we examine second-order theory because of the understanding it provides of viscous phenomena. We are also interested in the analytical methodology involved in applying matched asymptotic expansions to a significant, albeit complicated, problem. Some of its complexity is mitigated by retaining, whenever possible, the approach, notation, and most of the assumptions of the preceding chapters. We thus assume a Newtonian/Fourier gas that is thermally and calorically perfect. The laminar flow is steady, compressible, and two-dimensional or axisymmetric, and the bounding wall is smooth and impermeable. As with the first-order theory, in certain circumstances the second-order inner equations may possess a similarity solution (Werle, 1968; Afzal, 1976). We will see how such a solution naturally arises for the example considered in the last few sections. Finally, more cross-referencing of equations will be done than in the previous chapters. This sometimes results in a wordy text but is essential for technical clarity.

Validation

Comparisons with numerical solutions of the governing (Navier–Stokes) equations are warranted, since the theory represents an approximate solution of these equations. For instance, Schlichting (1979) contains a comparison of the first- and second-order theories with numerical solutions of the Navier–Stokes equations. Incompressible, two-dimensional flow is considered in the vicinity of the stagnation point of a parabola, at zero incidence, and with a uniform freestream. The error in the skin-friction coefficient at the stagnation point for the first-order theory, relative to a Navier–Stokes solution, equals 2% when $Re = 1.5 \times 10^5$ where this Reynolds number is based on the freestream speed and the radius of curvature of the parabola at its apex. This remarkably small error vanishes as the Reynolds number becomes infinite. The second-order solution, which encompasses the first-order result but also accounts for longitudinal curvature and the displacement thickness, however, is accurate to 2% down to a Reynolds number of only 10^2. Hence, an important contribution of second-order theory is to provide accurate results at Reynolds numbers lower than those attainable with the first-order theory.

There are exceedingly few comparisons with either compressible Navier–Stokes solutions or with experiments. In the latter case, the second-order effects may be small and several often simultaneously occur. (We return to this topic at the end of the chapter.) Moreover, depending on what is being measured, these effects tend to cancel each other. Thus, the experimental signal-to-noise ratio is a considerable problem. Another difficulty is that one measured quantity may agree with theory but a different quantity may not. For instance, Fannelop and Flügge-Lotz (1966) observe poor agreement for the heat transfer but good agreement with respect to pressure for measurements made in a hypersonic flow across a circular cylinder.

To put this in focus, we quote an early assessment by Van Dyke (1969):

> Convincing quantitative experimental confirmation (or refutation) of the validity of higher order boundary-layer theory has not yet been achieved. Measurements are meager at low speeds, where the theory is nearly unassailable. At high speeds, where we have seen that kinetic theory casts some doubt upon even the second approximation, experimental data are more numerous but often in disagreement. Nevertheless, most experiments seem to show at least qualitative accord with the predictions of second-order theory.

A more current assessment apparently has not been made, and consequently the status of the theory with regard to its limitations is still unsatisfactory.

Normalization

The conservation, or Navier–Stokes, equations of Appendix H are utilized, since they are written in the requisite body-oriented coordinate system. The normalization, i.e., nondimensionalization, provided by Equations (13.10) is used, but with the constants λ_∞ and $Pr\kappa_\infty/c_p$ replaced with μ_∞. Hence, only one Reynolds number appears, and s^*, n^*, r^*, and $1/k^*$ are all normalized by a characteristic length, ℓ. In the normalization, all infinity-subscripted quantities are dimensional constants that are associated with the freestream flow. Inviscid quantities along the wall are still denoted with an e subscript and are functions only of s. If the freestream is nonuniform, the infinity-subscripted quantities represent a single fixed point in this flow. The gradients that may be present in a nonuniform freestream are invariably very mild compared to those in a boundary layer. Hence, a nonuniform freestream is always considered inviscid; i.e., it is independent of a Reynolds number.

In addition to Equations (13.10), it is convenient, at this time, to introduce

$$\omega = \frac{U_\infty}{\ell}\omega^*, \qquad h = h_\infty h^* = c_{p\infty}T_\infty h^*$$
$$h_o = h_\infty h_o^* = h_\infty\left(1 + \frac{\gamma-1}{2}M_\infty^2\right)h_o^*, \qquad \tilde{s} = \tilde{s}_\infty + c_{p\infty}\tilde{s}^* \tag{23.3}$$

where ω is the scalar vorticity, and a tilde is used to distinguish the entropy from the s coordinate. The particular normalization that we are using is not unique and does not necessarily lead to the simplest form for some of the resulting equations. It has several advantages, however, such as highlighting the important role of the freestream Mach number and systematically using a fixed freestream state for the normalization. As in Section 13.3, the asterisk notation is hereafter omitted for all nondimensional quantities.

Governing Equations

With κ replaced with $\mu c_p/Pr$, we obtain for the governing equations

$$\frac{\partial}{\partial s}(r^\sigma \rho u) + \frac{\partial}{\partial n}(h_s r^\sigma \rho v) = 0 \tag{23.4a}$$

$$\rho\left(\frac{Du}{Dt} + \frac{kuv}{h_s}\right) = -\frac{1}{\gamma M_\infty^2 h_s}\frac{\partial p}{\partial s} + \frac{1}{Re}\left[F_s^s + \frac{1}{h_s}\frac{\partial}{\partial s}(\lambda\nabla\cdot\vec{w})\right] \tag{23.4b}$$

$$\rho\left(\frac{Dv}{Dt} - \frac{ku^2}{h_s}\right) = -\frac{1}{\gamma M_\infty^2}\frac{\partial p}{\partial n} + \frac{1}{Re}\left[F_n^s + \frac{\partial}{\partial n}(\lambda\nabla\cdot\vec{w})\right] \tag{23.4c}$$

$$\rho\frac{DT}{Dt} = \frac{\gamma-1}{\gamma}\frac{Dp}{Dt} + \frac{1}{PrRe}\left(\mu\nabla^2 T + \frac{\partial\mu}{\partial n}\frac{\partial T}{\partial n} + \frac{1}{h_s^2}\frac{\partial\mu}{\partial s}\frac{\partial T}{\partial s}\right)$$
$$+\frac{(\gamma-1)M_\infty^2}{Re}\Phi \tag{23.4d}$$

In the above, all variables and parameters are nondimensional, k, r_w, and θ are functions only of the arc length along the wall, the $PrRe$ product is the Peclet number, and

$$\vec{w} = u\hat{e}_s + v\hat{e}_n \tag{23.5a}$$

$$\frac{D}{Dt} = \frac{u}{h_s}\frac{\partial}{\partial s} + v\frac{\partial}{\partial n} \tag{23.5b}$$

$$h_s = 1 + kn \tag{23.5c}$$

$$r = r_w + n\cos\theta \tag{23.5d}$$

$$M_\infty = \frac{U_\infty}{(\gamma p_\infty/\rho_\infty)^{1/2}} \tag{23.5e}$$

$$Pr = \frac{\mu c_p}{\kappa} = \left(\frac{\mu c_p}{\kappa}\right)_\infty \tag{23.5f}$$

To Equations (23.4), we append

$$p = \rho T \tag{23.6}$$

for a perfect gas and consider μ as a known function of temperature.

Van Dyke (1962) has shown in the hypersonic limit of $M_\infty \to \infty$ that the viscous flow field depends on

$$\varepsilon = \frac{[(\gamma - 1)M_\infty^2]^{\omega/2}}{Re^{1/2}}$$

rather than on the parameter defined by Equation (23.2). The ω exponent stems from writing $\mu(T)$ as T^ω. Thus, the flow depends on the above combination instead of M_∞ and Re separately. However, we will not be concerned with this similitude in the subsequent analysis, since we consider M_∞ to be of $O(1)$, as has been the case in the preceding chapters.

23.2 INNER EQUATIONS

BOUNDARY-LAYER EQUATIONS

A small parameter ε is defined by Equation (23.2), and the boundary-layer scaling

$$n = \varepsilon\bar{n}, \qquad v = \varepsilon\bar{v}$$

is introduced into Equations (23.4). Only terms of $O(1)$ and $O(\varepsilon)$ are retained; hence, the following terms of $O(\varepsilon^2)$ are discarded:

$$\frac{1}{Reh_s}\frac{\partial}{\partial s}(\lambda\nabla\cdot\vec{w}), \qquad \rho\frac{Dv}{Dt}, \qquad \frac{1}{Re}\left[F_n^s + \frac{\partial}{\partial n}(\lambda\nabla\cdot\vec{w})\right], \qquad \frac{1}{PrReh_s^2}\frac{\partial\mu}{\partial s}\frac{\partial T}{\partial s}$$

In addition, we utilize

$$\frac{1}{h_s} = 1 - \varepsilon k \bar{n}$$

$$\frac{1}{r} = \frac{1}{r_w} - \frac{\varepsilon \bar{n}\cos\theta}{r_w^2}$$

$$r^\sigma = r_w^\sigma + \sigma\varepsilon\bar{n}\cos\theta$$

$$F_s^s = \frac{1}{\varepsilon^2}\frac{\partial}{\partial \bar{n}}\left(\mu\frac{\partial u}{\partial \bar{n}}\right) + \frac{1}{\varepsilon}\left[\left(2k + \frac{\sigma\cos\theta}{r_w}\right)\mu\frac{\partial u}{\partial \bar{n}} - k\frac{\partial}{\partial \bar{n}}(\mu u)\right]$$

$$\nabla^2 T = \frac{1}{\varepsilon^2}\frac{\partial^2 T}{\partial \bar{n}^2} + \frac{r_w^\sigma}{\varepsilon}\left(k + \frac{\sigma\cos\theta}{r_w}\right)\frac{\partial T}{\partial \bar{n}}$$

$$\Phi = \frac{\mu}{\varepsilon^2}\left(\frac{\partial u}{\partial \bar{n}}\right)^2 - \frac{2k\mu u}{\varepsilon}\frac{\partial u}{\partial \bar{n}}$$

in which only relevant terms are retained. Equations (23.4) thus become

$$\frac{\partial}{\partial s}\left[r_w^\sigma \rho u\left(1 + \frac{\sigma\varepsilon\bar{n}\cos\theta}{r_w}\right)\right] + \frac{\partial}{\partial \bar{n}}\left\{r_w^\sigma \rho\bar{v}\left[1 + \varepsilon\left(k + \frac{\sigma\cos\theta}{r_w}\right)\bar{n}\right]\right\} = 0 \tag{23.7a}$$

$$\rho\frac{du}{dt} = -\frac{1}{\gamma M_\infty^2}\frac{\partial p}{\partial s} + \frac{\partial}{\partial \bar{n}}\left(\mu\frac{\partial u}{\partial \bar{n}}\right) + \varepsilon\left[k\rho u\left(\bar{n}\frac{\partial u}{\partial s} - \bar{v}\right) + \frac{k\bar{n}}{\gamma M_\infty^2}\frac{\partial p}{\partial s}\right.$$
$$\left. + \left(2k + \frac{\sigma\cos\theta}{r_w}\right)\mu\frac{\partial u}{\partial \bar{n}} - k\frac{\partial}{\partial \bar{n}}(\mu u)\right] \tag{23.7b}$$

$$\frac{\partial p}{\partial \bar{n}} = \gamma M_\infty^2 \varepsilon k \rho u^2 \tag{23.7c}$$

$$\rho\frac{dT}{dt} = \frac{\gamma-1}{\gamma}\frac{dp}{dt} + \frac{1}{Pr}\frac{\partial}{\partial \bar{n}}\left(\mu\frac{\partial T}{\partial \bar{n}}\right) + (\gamma - 1)M_\infty^2\mu\left(\frac{\partial u}{\partial \bar{n}}\right)^2 + \varepsilon\left[k\bar{n}u\left(\rho\frac{\partial T}{\partial s} - \frac{\gamma-1}{\gamma}\frac{\partial p}{\partial s}\right)\right.$$
$$\left. + \frac{r_w^\sigma}{Pr}\left(k + \frac{\sigma\cos\theta}{r_w}\right)\mu\frac{\partial T}{\partial \bar{n}} - 2(\gamma-1)M_\infty^2 k\mu u\frac{\partial u}{\partial \bar{n}}\right] \tag{23.7d}$$

where the inner region substantial derivative is

$$\frac{d}{dt} = u\frac{\partial}{\partial s} + \bar{v}\frac{\partial}{\partial \bar{n}}$$

A nondimensional version of the first-order boundary-layer equations in Chapter 21 can be obtained from Equations (21.7) by setting $\varepsilon = 0$.

Discussion

Equations (23.7) are combined first- and second-order boundary-layer equations. The terms containing k and σ/r_w provide for the effects of longitudinal and transverse curvature, respectively. The method of treatment of the other second-order effects is discussed in Section 23.5. A number of authors (Back, 1973; Fannellop and Flügge-Lotz, 1966; Van Tassell and Taulbee, 1971; Kleinstreuer and Eghlima, 1985) have directly utilized Equations (23.7) for numerically computing boundary-layer solutions. As with the first-order equations, these are still parabolic and are about as difficult to numerically solve as are the first-order equations.

Observe that Equation (23.7c) means there may be a transverse pressure gradient across the boundary layer. For a convex surface (see Figure 19.2), $k > 0$ and the pressure increases with $\bar{n}$, whereas it decreases for a concave surface. In the convex case, an increasing pressure is required to balance the centrifugal force, which tends to separate the flow from the wall. Observe that the transverse pressure gradient is particularly significant when the product, kM_ω^2, is large. This occurs at supersonic or hypersonic Mach numbers and when the normalized radius of curvature of the wall is small.

As pointed out by Kleinstreuer and Eghlima (1985), the second-order equations are not unique. (The first-order equations, however, are unique.) To illustrate this point, suppose Equation (23.4b) is multiplied by h_s, which is given by Equation (23.5c). In this case, the terms of $O(\varepsilon)$ in Equation (23.7b) would be different; e.g., the term containing $k\bar{n}(\partial p/\partial s)$ would not be present. The equation corresponding to Equation (23.7b), while different in appearance, is still correct to second order and would yield a theoretical formulation equivalent to Equation (23.7b).

Inner Expansion

For some time, it has been known (Van Dyke, 1975) that the first correction to first-order boundary-layer theory only involves terms containing integer powers of ε. In other situations, terms proportional to a $\ln\varepsilon$, for example, may appear (Van Dyke, 1975). It is through the matching conditions, discussed in Section 23.4, that the need for such terms is usually noticed. Hence, we assume integer powers of ε for the inner asymptotic expansions (Van Dyke, 1962)

$$u(s, n;\varepsilon) \sim u_1(s, \bar{n}) + \varepsilon u_2(s, \bar{n}) + \cdots \tag{23.8a}$$

$$\bar{v}(s, n;\varepsilon) \sim v_1(s, \bar{n}) + \varepsilon v_2(s, \bar{n}) + \cdots \tag{23.8b}$$

$$p(s, n;\varepsilon) \sim p_1(s, \bar{n}) + \varepsilon p_2(s, \bar{n}) + \cdots \tag{23.8c}$$

$$\rho(s, n;\varepsilon) \sim \rho_1(s, \bar{n}) + \varepsilon \rho_2(s, \bar{n}) + \cdots \tag{23.8d}$$

$$T(s, n;\varepsilon) \sim t_1(s, \bar{n}) + \varepsilon t_2(s, \bar{n}) + \cdots \tag{23.8e}$$

and

$$\mu(s, n;\varepsilon) = \mu(T) = \mu(t_1 + \varepsilon t_2 + \cdots) \sim \mu(t_1) + \frac{d\mu}{dt_1}\varepsilon t_2 + \cdots = \mu_1 + \varepsilon\mu_1' t_2 + \cdots \tag{23.9}$$

These expansions can be substituted into either Equations (23.4) or (23.7). Since the later equations are simpler, they are used. For the first-order equations, we thus obtain

$$\frac{\partial}{\partial s}(r_w^\sigma \rho_1 u_1) + \frac{\partial}{\partial \bar{n}}(r_w^\sigma \rho_1 v_1) = 0 \tag{23.10a}$$

$$\rho_1 \frac{du_1}{d_1 t} = -\frac{1}{\gamma M_\infty^2} \frac{\partial p_1}{\partial s} + \frac{\partial}{\partial \bar{n}}\left(\mu_1 \frac{\partial u_1}{\partial \bar{n}}\right) \tag{23.10b}$$

$$\frac{\partial p_1}{\partial \bar{n}} = 0 \tag{23.10c}$$

$$\rho \frac{dt_1}{d_1 t} = \frac{\gamma - 1}{\gamma} \frac{dp_1}{d_1 t} + \frac{1}{Pr} \frac{\partial}{\partial \bar{n}}\left(\mu_1 \frac{\partial t_1}{\partial \bar{n}}\right) + (\gamma - 1) M_\infty^2 \mu_1 \left(\frac{\partial u_1}{\partial \bar{n}}\right)^2 \tag{23.10d}$$

and, from Equation (23.6),

$$p_1 = \rho_1 t_1 \tag{23.10e}$$

These equations represent the conventional, nonsimilar, boundary-layer equations for a perfect gas. The notation

$$\frac{d}{d_i t} = u_i \frac{\partial}{\partial s} + v_i \frac{\partial}{\partial \bar{n}}, \qquad i = 1, 2 \tag{23.11a}$$

is introduced for the expansion

$$\frac{d}{dt} \sim \frac{d}{d_1 t} + \varepsilon \frac{d}{d_2 t} + \cdots \tag{23.11b}$$

of the inner region substantial derivative. Note that Equations (23.10c) and (23.10e) provide

$$\frac{1}{t_1} \frac{\partial t_1}{\partial \bar{n}} = -\frac{1}{\rho_1} \frac{\partial \rho_1}{\partial \bar{n}} \tag{23.12a}$$

and

$$\frac{\partial p_1}{\partial s} = \frac{dp_1}{ds} = t_1 \frac{\partial \rho_1}{\partial s} + \rho_1 \frac{\partial t_1}{\partial s} \tag{23.12b}$$

For the second-order equations, we have

$$\frac{\partial}{\partial s}\left\{ r_w^\sigma \left[u_1 \rho_2 + \rho_1 u_2 + \frac{\sigma}{r_w} \bar{n} \rho_1 u_1 \cos\theta \right]\right\} + \frac{\partial}{\partial \bar{n}}\left\{ r_w^\sigma \left[v_1 \rho_2 + \rho_1 v_2 + \left(k + \frac{\sigma \cos\theta}{r_w}\right) \bar{n} \rho_1 v_1 \right]\right\} = 0 \tag{23.13a}$$

$$\rho_1 \left(\frac{du_1}{d_2 t} + \frac{du_2}{d_1 t}\right) + \frac{du_1}{d_1 t} \rho_2 + \frac{1}{\gamma M_\infty^2} \frac{\partial p_2}{\partial s} - \frac{\partial}{\partial \bar{n}}\left(\mu_1' \frac{\partial u_1}{\partial \bar{n}} t_2 + \mu_1 \frac{\partial u_2}{\partial \bar{n}}\right)$$
$$= k \rho_1 u_1 \left(\bar{n} \frac{\partial u_1}{\partial s} - v_1\right) + \frac{k \bar{n}}{\gamma M_\infty^2} \frac{\partial p_1}{\partial s} + \left(2k + \frac{\sigma \cos\theta}{r_w}\right) \mu_1 \frac{\partial u_1}{\partial \bar{n}} - k \frac{\partial}{\partial \bar{n}} (\mu_1 u_1) \tag{23.13b}$$

$$\frac{\partial p_2}{\partial \bar{n}} = \gamma M_\infty^2 k \rho_1 u_1^2 \tag{23.13c}$$

$$\rho_1\left(\frac{dt_1}{d_2t}+\frac{dt_2}{d_1t}\right)+\frac{dt_1}{d_1t}\rho_2-\frac{\gamma-1}{\gamma}\left(\frac{dp_1}{d_2t}+\frac{dp_2}{d_1t}\right)-\frac{1}{Pr}\frac{\partial}{\partial \bar{n}}\left(\mu_1'\frac{\partial t_1}{\partial \bar{n}}t_2+\mu_1\frac{\partial t_2}{\partial \bar{n}}\right)$$
$$+(\gamma-1)M_\infty^2\left[\mu_1'\left(\frac{\partial u_1}{\partial \bar{n}}\right)^2 t_2+2\mu_1\frac{\partial u_1}{\partial \bar{n}}\frac{\partial u_2}{\partial \bar{n}}\right] = k\bar{n}u_1\left(\rho_1\frac{\partial t_1}{\partial s}-\frac{\gamma-1}{\gamma}\frac{\partial p_1}{\partial s}\right)$$
$$+\frac{1}{Pr}\left(k+\frac{\sigma\cos\theta}{r_w}\right)\mu_1\frac{\partial t_1}{\partial \bar{n}}-2(\gamma-1)M_\infty^2 k\mu_1 u_1\frac{\partial u_1}{\partial \bar{n}} \tag{23.13d}$$

$$p_2 = t_1\rho_2+\rho_1 t_2 \tag{23.13e}$$

The inhomogeneous terms can be simplified and p_2 can be eliminated by introducing Equations (23.10), (23.13c), and (23.13e). For instance, Equation (23.10a) can be written as

$$\frac{\partial}{\partial \bar{n}}(\rho_1 v_1) = -\frac{\partial(\rho_1 u_1)}{\partial s}-\frac{\sigma\sin\theta}{r_w}\rho_1 u_1$$

resulting in Equation (23.13a) becoming

$$\frac{\partial}{\partial s}[r_w^\sigma(u_1\rho_2+\rho_1 u_2)]+\frac{\partial}{\partial \bar{n}}[r_w^\sigma(v_1\rho_2+\rho_1 v_2)] =$$
$$k\bar{n}\frac{\partial(\rho_1 u_1)}{\partial s}+\frac{\sigma\bar{n}\sin\theta\cos\theta}{r_w}\rho_1 u_1 - r_w^\sigma\left(k+\frac{\sigma\cos\theta}{r_w}\right)\rho_1 v_1 \tag{23.14a}$$

Similarly, Equations (23.13b) and (23.13d) become

$$\rho_1\left(\frac{du_1}{d_2t}+\frac{du_2}{d_1t}\right)+\frac{du_1}{d_1t}\rho_2-\frac{\partial}{\partial \bar{n}}\left(\mu_1'\frac{\partial u_1}{\partial \bar{n}}t_2+\mu_1\frac{\partial u_2}{\partial \bar{n}}\right)+\frac{1}{\gamma M_\infty^2}\left[\frac{\partial(t_1\rho_2)}{\partial s}+\frac{\partial(\rho_1 t_2)}{\partial s}\right]$$
$$=\left(k+\frac{\sigma\cos\theta}{r_w}\right)\mu_1\frac{\partial u_1}{\partial \bar{n}}-k\rho_1 v_1\frac{\partial}{\partial \bar{n}}(\bar{n}u_1)+k\left[\bar{n}\frac{\partial}{\partial \bar{n}}\left(\mu_1\frac{\partial u_1}{\partial \bar{n}}\right)-\mu_1' u_1\frac{\partial t_1}{\partial \bar{n}}\right] \tag{23.14b}$$

and

$$\rho_1\left(\frac{dt_1}{d_2t}+\frac{dt_2}{d_1t}\right)+\frac{dt_1}{d_1t}\rho_2-\frac{1}{Pr}\frac{\partial}{\partial \bar{n}}\left(\mu_1'\frac{\partial t_1}{\partial \bar{n}}t_2+\mu_1\frac{\partial t_2}{\partial \bar{n}}\right)$$
$$-\frac{\gamma-1}{\gamma}\left\{\frac{dp_1}{ds}u_2+u_1\left[\frac{\partial(t_1\rho_2)}{\partial s}+\frac{\partial(\rho_1 t_2)}{\partial s}\right]\right\}-(\gamma-1)M_\infty^2\left[\mu_1'\left(\frac{\partial u_1}{\partial \bar{n}}\right)^2 t_2+2\mu_1\frac{\partial u_1}{\partial \bar{n}}\frac{\partial u_2}{\partial \bar{n}}\right]=$$
$$-k\bar{n}\rho_1 v_1\frac{\partial t_1}{\partial \bar{n}}+(\gamma-1)M_\infty^2 k\left[\left(\bar{n}\frac{\partial u_1}{\partial \bar{n}}-2u_1\right)\mu_1\frac{\partial u_1}{\partial \bar{n}}+\rho_1 u_1^2 v_1\right]$$
$$+\frac{1}{Pr}\left[\left(k+\frac{\sigma\cos\theta}{r_w}\right)\mu_1\frac{\partial t_1}{\partial \bar{n}}+k\bar{n}\frac{\partial}{\partial \bar{n}}\left(\mu_1\frac{\partial t_1}{\partial \bar{n}}\right)\right] \tag{23.14c}$$

In addition, Equations (23.10c), (23.10e), (23.13c), and (23.13e) yield

$$\frac{\partial}{\partial \bar{n}}(t_1\rho_2 + \rho_1 t_2) = \gamma M_\infty^2 k \rho_1 u_1^2 \tag{23.14d}$$

[In the interest of symmetry, Equation (23.12a) has not been introduced.] Equations (23.14) are four coupled, linear PDEs for u_2, v_2, ρ_2, and t_2. The pressure p_2 is given by Equation (23.13e). Each PDE possesses inhomogeneous terms, written on the right side, that depend on the solution of the first-order inner equations. The coefficients of the second-order terms on the left side also depend on this solution.

We conclude by providing relations for the first- and second-order scalar vorticities. We start with (see Problem 21.8)

$$\omega = -\frac{1}{1+kn}\left[(1+kn)\frac{\partial u}{\partial n} + ku - \frac{\partial v}{\partial s}\right] \tag{23.15a}$$

and scale ω as

$$\bar{\omega} = \varepsilon\omega$$

with the result

$$\bar{\omega} = -(1 - \varepsilon k\bar{n} + \cdots)\left[(1+\varepsilon k\bar{n})\frac{\partial u}{\partial \bar{n}} + \varepsilon ku - \varepsilon^2\frac{\partial \bar{v}}{\partial s}\right]$$

With the aid of Equations (23.8a,b), this becomes

$$\begin{aligned}\bar{\omega} &= (-1 + \varepsilon k\bar{n} - \cdots)\left[(1+\varepsilon k\bar{n})\left(\frac{\partial u_1}{\partial \bar{n}} + \varepsilon\frac{\partial u_2}{\partial \bar{n}} + \cdots\right) + \varepsilon k u_1 + \cdots + -\varepsilon^2\frac{\partial v_1}{\partial s} - \cdots\right] \\ &= -\frac{\partial u_1}{\partial \bar{n}} + \varepsilon\left(ku_1 + \frac{\partial u_2}{\partial \bar{n}}\right) + \cdots\end{aligned} \tag{23.15b}$$

We next write

$$\bar{\omega}(s, n; \varepsilon) \sim \omega_1(s, \bar{n}) + \varepsilon\omega_2(s, \bar{n}) + \cdots \tag{23.16a}$$

and compare with Equation (23.15b), to obtain

$$\omega_1 = -\frac{\partial u_1}{\partial \bar{n}} \tag{23.16b}$$

$$\omega_2 = -ku_1 - \frac{\partial u_2}{\partial \bar{n}} \tag{23.16c}$$

Note that ω_2 depends on the longitudinal curvature of the wall.

23.3 OUTER EQUATIONS

These equations hold in the region external to the boundary layer, where the flow is considered inviscid and adiabatic. In this region, s, n, r, θ, k, and h_s are unaltered, while for the dependent variables, we write

$$u(s, n;\varepsilon) \sim U_1(s, n) + \varepsilon U_2(s, n) + \cdots \tag{23.17a}$$

$$v(s, n;\varepsilon) \sim V_1(s, n) + \varepsilon V_2(s, n) + \cdots \tag{23.17b}$$

$$p(s, n;\varepsilon) \sim P_1(s, n) + \varepsilon P_2(s, n) + \cdots \tag{23.17c}$$

$$\rho(s, n;\varepsilon) \sim R_1(s, n) + \varepsilon R_2(s, n) + \cdots \tag{23.17d}$$

$$T(s, n;\varepsilon) \sim T_1(s, n) + \varepsilon T_2(s, n) + \cdots \tag{23.17e}$$

where ε is still defined by Equation (23.2). Observe that none of the dependent or independent variables are stretched.

First-Order Equations

The above expansions are substituted into Equations (23.4) and (23.6), with the following first-order result:

$$\frac{\partial}{\partial s}(r^\sigma R_1 U_1) + \frac{\partial}{\partial n}(h_s r^\sigma R_1 V_1) = 0 \tag{23.18a}$$

$$R_1\left(\frac{DU_1}{D_1 t} + \frac{kU_1V_1}{h_s}\right) + \frac{1}{\gamma M_\infty^2}\frac{1}{h_s}\frac{\partial P_1}{\partial s} = 0 \tag{23.18b}$$

$$R_1\left(\frac{DV_1}{D_1 t} - \frac{kU_1^2}{h_s}\right) + \frac{1}{\gamma M_\infty^2}\frac{\partial P_1}{\partial n} = 0 \tag{23.18c}$$

$$R_1\frac{DT_1}{D_1 t} - \frac{\gamma-1}{\gamma}\frac{DP_1}{D_1 t} = 0 \tag{23.18d}$$

$$P_1 = R_1 T_1 \tag{23.18e}$$

In the above, h_s and r are given by Equations (23.5c,d), respectively. The first- and second-order substantial derivatives, for the outer region, are

$$\frac{D}{D_i t} = \frac{U_i}{h_s}\frac{\partial}{\partial s} + V_i\frac{\partial}{\partial n}, \qquad i = 1, 2 \tag{23.19a}$$

where

$$\frac{D}{Dt} \sim \frac{D}{D_1 t} + \varepsilon\frac{D}{D_2 t} + \cdots \tag{23.19b}$$

Although their appearance may be unfamiliar, Equations (23.18) are just the Euler equations for a perfect gas.

Integrals of the First-Order Equations

We expand a nondimensional stream function, ψ, as

$$\psi(s, n;\varepsilon) \sim \Psi_1(s, n) + \varepsilon\Psi_2(s, n) + \cdots \tag{23.20}$$

Because Equations (23.4a) and (23.18a) have the same form, we have

$$\frac{\partial\psi}{\partial n} = r^\sigma \rho u, \qquad \frac{\partial\psi}{\partial s} = -h_s r^\sigma \rho v \tag{23.21a}$$

$$\frac{\partial\Psi_1}{\partial n} = r^\sigma R_1 U_1, \qquad \frac{\partial\Psi_1}{\partial s} = -h_s r^\sigma R_1 V_1 \tag{23.21b}$$

Since the first-order stream function is constant along first-order streamlines, we can write

$$\frac{D\Psi_1}{D_1 t} = \frac{U_1}{h_s}\frac{\partial\Psi_1}{\partial s} + V_1\frac{\partial\Psi_1}{\partial n} = 0 \tag{23.21c}$$

This PDE can be solved using the characteristic theory of Appendix E. Of course, the velocity components, $U_1(s, n)$ and $V_1(s, n)$, must first be known.

The nondimensional (see Equations (23.3)) entropy and stagnation enthalpy are also expanded as

$$\bar{s}(s, n;\varepsilon) \sim S_1(s, n) + \varepsilon S_2(s, n) + \cdots \tag{23.22a}$$

$$h_o(s, n;\varepsilon) \sim H_1(s, n) + \varepsilon H_2(s, n) + \cdots \tag{23.22b}$$

where $\bar{s}$ and h_o are given by

$$\bar{s} = \frac{1}{\gamma}\ln\frac{p}{\rho^\gamma} \tag{23.23a}$$

$$h_o = \frac{1}{X_\infty}\left[T + \frac{\gamma-1}{2}M_\infty^2(u^2+v^2)\right] \tag{23.23b}$$

with

$$X_\infty = 1 + \frac{\gamma-1}{2}M_\infty^2 \tag{23.23c}$$

By expanding the right side of Equations (23.23a,b), we obtain (see Problem 23.1)

$$S_1 = \frac{1}{\gamma}\ln\frac{P_1}{R_1^\gamma} \tag{23.24a}$$

$$S_2 = \frac{1}{\gamma}\frac{P_2}{P_1} - \frac{R_2}{R_1} \tag{23.24b}$$

$$H_1 = \frac{1}{X_\infty}\left[T_1 + \frac{\gamma - 1}{2} M_\infty^2 (U_1^2 + V_1^2)\right] \tag{23.25a}$$

$$H_2 = \frac{1}{X_\infty}[T_2 + (\gamma - 1)M_\infty^2 (U_1 U_2 + V_1 V_2)] \tag{23.25b}$$

Equations (23.18) are just the Euler equations for an isentropic and isoenergetic flow. Consequently, both S_1 and H_1 are constants along the first-order streamlines; i.e.,

$$S_1 = S_1(\Psi_1), \qquad H_1 = H_1(\Psi_1) \tag{23.26a}$$

or

$$\frac{DS_1}{D_1 t} = 0, \qquad \frac{DH_1}{D_1 t} = 0 \tag{23.26b}$$

The functions $S_1(\Psi_1)$ and $H_1(\Psi_1)$ are determined in the upstream flow, or in the case of S_1, just downstream of a shock wave, if one is present. Equations (23.26a), in conjunction with Equations (23.24a) and (23.25a), thus constitute two integrals of the first-order outer equations (see Problem 23.1). Because of Equations (23.26a), we can write

$$\frac{\partial S_1}{\partial n} = \frac{\partial \Psi_1}{\partial n}\frac{dS_1}{d\Psi_1} = r^\sigma R_1 U_1 S_1' \tag{23.27a}$$

and

$$\frac{\partial S_1}{\partial s} = \frac{\partial \Psi_1}{\partial s}\frac{dS_1}{d\Psi_1} = -h_s r^\sigma R_1 V_1 S_1' \tag{23.27b}$$

where $S_1' = dS/d\Psi_1$. There are similar equations for H_1. The S_1 and H_1 derivatives, when evaluated at the wall, $S_1'(0) = S_{1w}'$ and $H_1'(0) = H_{1w}'$, will play an important role in the subsequent analysis.

If the flow is rotational, the first-order vorticity is related to S_1' and H_1' by Crocco's equation. We first write the vorticity expansion

$$\omega(s, n; \varepsilon) \sim \Omega_1(s, n) + \varepsilon\Omega_2(s, n) + \cdots \tag{23.28}$$

and with the aid of Equation (5.30), we obtain (see Problem 23.1)

$$(\gamma - 1)M_\infty^2 \Omega_1 = r^\sigma R_1 (T_1 S_1' - X_\infty H_1') \tag{23.29}$$

Because R_1 and T_1 are not solely functions of Ψ_1, Ω_1, is not constant along streamlines. Knowledge of two of the Ω_1, S_1', H_1' functions is sufficient to determine the remaining one.

Normal Derivatives Evaluated at the Wall

In addition to quantities such as $U_1(s, 0)$, we will later need the normal derivatives of U_1, V_1, P_1, R_1, and T_1 when evaluated at the wall. This evaluation requires the boundary condition established in Section 23.4,

$$V_1(s,0) = 0 \tag{23.30}$$

which yields

$$\left.\frac{\partial V_1}{\partial s}\right|_w = 0$$

From Equations (23.18a) and (23.18c), we readily obtain

$$\left.\frac{\partial P_1}{\partial n}\right|_w = \gamma M_\infty^2 k(R_1 U_1^2)_w \tag{23.31a}$$

$$\left.\frac{\partial V_1}{\partial n}\right|_w = -\frac{1}{r_w^\sigma R_1(s,0)}\frac{d}{ds}(r_w^\sigma R_1 U_1)_w \tag{23.31b}$$

Note that r and h_s, respectively, become r_w and unity at the wall.

From Equation (23.15a), we have, for the first-order vorticity,

$$\Omega_1 = -\frac{\partial U_1}{\partial n} - \frac{k}{h_s}U_1 + \frac{1}{h_s}\frac{\partial V_1}{\partial s}$$

We eliminate Ω_1 from this relation and Equation (23.29) and set $n = 0$, to obtain

$$\left.\frac{\partial U_1}{\partial n}\right|_w = -\left\{kU_1 + \frac{r_w^\sigma R_1}{(\gamma-1)M_\infty^2}[T_1 S_1' - X_\infty H_1']\right\}_w \tag{23.31c}$$

We next differentiate Equation (23.25a) with respect to n, utilize the form of Equation (23.27b) for $\partial H_1/\partial n$, and set $n = 0$, with the result

$$\left.\frac{\partial T_1}{\partial n}\right|_w = [(\gamma-1)M_\infty^2 kU_1^2 + r_w^\sigma P_1 U_1 S_1']_w \tag{23.31d}$$

Finally, Equations (23.18e), (23.31a), and (23.31d) are used to obtain

$$\left.\frac{\partial R_1}{\partial n}\right|_w = \left[\frac{R_1 U_1}{T_1}(M_\infty^2 kU_1 - r_w^\sigma P_1 S_1')\right]_w \tag{23.31e}$$

Equations (23.31) provide the desired derivatives that are normal to the wall.

Second-Order Equations

In contrast to the second-order inner equations, the second-order outer equations are relatively easily obtained because the Re^{-1} terms in Equations (23.4) are of third order with respect to Equations (23.17). The second-order equations, therefore, are still inviscid; in fact, they are the small perturbation equations for an inviscid flow, with ε as the small perturbation parameter. We thus have

$$\frac{\partial}{\partial s}[r^{\sigma}(U_1 R_2 + R_1 U_2)] + \frac{\partial}{\partial n}[h_s r^{\sigma}(V_1 R_2 + R_1 V_2)] = 0 \tag{23.32a}$$

$$R_1\left[\frac{DU_1}{D_2 t} + \frac{DU_2}{D_1 t} + \frac{k}{h_s}(U_1 V_2 + V_2 U_2)\right] + R_2\left(\frac{DU_1}{D_1 t} + \frac{k}{h_s}U_1 V_1\right) + \frac{1}{\gamma M_\infty^2 h_s}\frac{\partial P_2}{\partial s} = 0 \tag{23.32b}$$

$$R_1\left(\frac{DV_1}{D_2 t} + \frac{DV_2}{D_1 t} - \frac{2k}{h_s}U_1 U_2\right) + R_2\left(\frac{DV_1}{D_1 t} - \frac{k}{h_s}U_1^2\right) + \frac{1}{\gamma M_\infty^2}\frac{\partial P_2}{\partial n} = 0 \tag{23.32c}$$

$$R_1\left(\frac{DT_1}{D_2 t} + \frac{DT_2}{D_1 t}\right) + R_2\frac{DT_1}{D_1 t} - \frac{\gamma - 1}{\gamma}\left(\frac{DP_1}{D_2 t} + \frac{DP_2}{D_1 t}\right) = 0 \tag{23.32d}$$

$$P_2 = R_1 T_2 + T_1 R_2 \tag{23.32e}$$

Alternate forms for these equations are possible. For instance, P_1 and P_2 can be eliminated with the use of Equations (23.18e) and (23.32e). Moreover, the factors multiplying R_2 in Equations (23.32b to d) can be replaced by utilizing the first-order equations. Equations (23.32a to d) are linear equations, but, in contrast to their boundary-layer counterparts, they are homogeneous regardless of the form used.

Integrals of the Second-Order Equations

Equations (23.4), with $(1/Re) = 0$, admit as integrals

$$\tilde{s}(s, n;\varepsilon) = \tilde{s}(\psi)$$
$$h_o(s, n;\varepsilon) = h_o(\psi)$$

Consequently, ψ, $\tilde{s}$, and h_o are equal to the sum of the $O(1)$ and $O(\varepsilon)$ terms on the right side of Equations (23.20) and (23.22). Hence, we can write

$$\frac{D\tilde{s}}{Dt} = \frac{DS_1}{D_1 t} + \varepsilon\frac{DS_2}{D_1 t} + \varepsilon\frac{DS_1}{D_2 t} = 0$$

with a similar equation for h_o. In view of Equation (23.26b), we have

$$\frac{DS_2}{D_1 t} = -\frac{DS_1}{D_2 t} \tag{23.33}$$

with a similar relation involving H_1 and H_2. Thus, the second-order entropy is not constant along the first-order streamlines, and the first-order entropy is not constant along the second-order streamlines. With the aid of Equations (23.27), the right side becomes

$$\frac{DS_1}{D_2t} = \frac{U_2}{h_s}\frac{\partial S_1}{\partial s} + V_2\frac{\partial S_1}{\partial n} = r^\sigma R_1(U_1V_2 - V_1U_2)S_1'$$

The second-order entropy S_2 (and H_2) is a function only of Ψ_1 and Ψ_2. We first evaluate (see Problem 23.2)

$$\frac{D\Psi_2}{D_1t} = \frac{U_1}{h_s}\frac{\partial \Psi_2}{\partial s} + V_1\frac{\partial \Psi_2}{\partial n} = r^\sigma R_1(V_1U_2 - U_1V_2)$$

Except for the inhomogeneous term, this is the same PDE as Equation (23.21c). To determine $\Psi_2(s,n)$, the PDE is solved by means of the characteristic theory in Appendix E. The left side of Equation (23.33) now becomes

$$\frac{DS_2}{D_1t} = \frac{\partial S_2}{\partial \Psi_1}\frac{D\Psi_1}{D_1t} + \frac{\partial S_2}{\partial \Psi_2}\frac{D\Psi_2}{D_1t} = r^\sigma R_1(V_1U_2 - U_1V_2)\frac{\partial S_2}{\partial \Psi_2}$$

and Equation (23.33) can be written as

$$\frac{\partial S_2}{\partial \Psi_2} = \frac{\partial S_1}{\partial \Psi_1} = S_1'$$

Integration with Ψ_1 held fixed yields the integral

$$S_2 = \Psi_2 S_1'(\Psi_1) + \tilde{S}(\Psi_1)$$

The function of integration, $\tilde{S}$, equals zero by evaluating S_2 at the location, e.g., just downstream of a shock wave, where $(\tilde{s} = S_1)$. (The corresponding stagnation enthalpy equation is $H_2 = \Psi_2 H_1'$.)

Final Form for the Second-Order Equations

With the aid of Equation (23.24b), we obtain

$$\frac{P_2}{P_1} \quad \gamma\frac{R_2}{R_1} \quad \gamma\Psi_2 S_1'$$

As demonstrated in Problem 23.2, we finally have

$$P_2 = -\frac{\gamma}{\gamma - 1}R_1[(T_1S_1' - X_\infty H_1')\Psi_2 + (\gamma - 1)M_\infty^2(U_1U_2 + V_1V_2)] \tag{23.34a}$$

$$R_2 = -\frac{R_1}{(\gamma - 1)T_1}[(\gamma T_1S_1' - X_\infty H_1')\Psi_2 + (\gamma - 1)M_\infty^2(U_1U_2 + V_1V_2)] \tag{23.34b}$$

$$T_2 = X_\infty H'\Psi_2 - (\gamma - 1)M_\infty^2(U_1U_2 + V_1V_2) \tag{23.34c}$$

These relations assume a perfect gas but otherwise are quite general in the sense that a specific flow configuration, such as flow over a cylinder, has not been assumed.

In view of Equations (23.34), the unknown variables are reduced to U_2, V_2, and Ψ_2. (We are, of course, assuming that a solution of the first-order equations is available.) Problem 23.4 shows that this group can be further reduced to just Ψ_2. One can show that the equation for Ψ_2 is a second-order, linear PDE without any inhomogeneous terms. Consequently, a single PDE provides the general equation for the second-order outer region flow. Boundary conditions for the U_2, V_2,... variables are discussed in the next section.

23.4 BOUNDARY AND MATCHING CONDITIONS

Since the wall is impermeable, we have Equation (23.30) and

$$v_1(s,0) = 0, \qquad v_2(s,0) = 0 \tag{23.35a,b}$$

The first-order outer variables also satisfy the possibly nonuniform freestream conditions. Consequently, $U_2, \ldots, T_2$ go to zero in the freestream.

VELOCITY SLIP AND TEMPERATURE JUMP

In dimensional form, these conditions can be written as (Street, 1960)

$$u = \left\{\frac{\mu}{p}\left[a_1(RT)^{1/2}\frac{\partial u}{\partial n} + \frac{3}{4}R\frac{\partial T}{\partial n}\right]\right\}_w$$

$$T = T_w(s) + c_1\left[\frac{\mu}{p}(RT)^{1/2}\frac{\partial T}{\partial n}\right]_w$$

where a_1 and c_1 are coefficients that, respectively, depend on the velocity and thermal accommodation coefficients. Hence, a_1 and c_1 are treated as $O(1)$ constants; e.g., if the accommodation coefficients are unity, then $a_1 = 1.25$ and $c_1 = 2.35$ (Street, 1960). (The a_1 and c_1 coefficients become infinite if the accommodation coefficients are zero. Realistic values for the accommodation coefficients, however, are near unity, not zero.) With our nondimensionalization, these relations become

$$u(s,0) = \varepsilon^2\gamma^{1/2}M_\infty\left[\frac{\mu}{p}\left(a_1T^{1/2}\frac{\partial u}{\partial n} + \frac{3}{4\gamma^{1/2}M_\infty}\frac{\partial T}{\partial s}\right)\right]_w \tag{23.36a}$$

$$T(s,0) = T_w + c_1\varepsilon^2\gamma^{1/2}M_\infty\left(\frac{\mu}{p}T^{1/2}\frac{\partial T}{\partial n}\right)_w \tag{23.36b}$$

We next introduce $\bar{n}$ and Equations (23.8) and (23.9) with the following first- and second-order inner results:

$$u_1(s,0) = 0 \tag{23.37a}$$

$$u_2(s,0) = a_1 \gamma^{1/2} M_\infty \left(\frac{\mu_1 t_1^{1/2}}{p_1} \frac{\partial u_1}{\partial \tilde{n}} \right)_w \tag{23.37b}$$

$$t_1(s,0) = T_w(s) \tag{23.38a}$$

$$t_2(s,0) = c_1 \gamma^{1/2} M_\infty \left(\frac{\mu_1 t_1^{1/2}}{p_1} \frac{\partial t_1}{\partial \tilde{n}} \right)_w \tag{23.38b}$$

The $\partial T / \partial s$ term in Equation (23.36a) is of third order and does not contribute to Equations (23.37). For an adiabatic wall, Equations (23.38) are replaced with

$$\left. \frac{\partial t_1}{\partial \tilde{n}} \right|_w = 0, \qquad \left. \frac{\partial t_2}{\partial \tilde{n}} \right|_w = 0 \tag{23.38c}$$

We thus obtain the usual wall conditions for the first-order inner equations. As noted in Section 23.1, there is no velocity slip or temperature jump in an incompressible flow. The second-order importance of these slip-type conditions increases with $\gamma^{1/2} M_\infty$ but decreases if the wall is highly cooled, since t_1 and μ_1 both become small for a gas.

Restricted Matching Principle

The expansions, Equations (23.8), hold in the viscous layer, whereas Equations (23.17) hold in the external flow. Thus, inner variables are generally unable to satisfy upstream (i.e., freestream) or any other condition imposed outside the viscous layer. On the other hand, the outer variables are unable to satisfy Equations (23.37) and (23.38) at the wall. Matching provides the missing boundary conditions for both expansions. This is possible because the region of validity of each set of expansions overlaps at the outer edge of the boundary layer. In our case, matching, which hinges on the expansions overlapping, is justified by obtaining physically and mathematically appropriate boundary conditions. In situations where matching appears to be impossible, either terms such as those containing a logarithm, are missing, or an additional buffer layer may be required.

Matching is usually based on the restricted matching principle of Lagerstrom (Van Dyke, 1962, 1963; Davis and Flügge-Lotz, 1964; Fannelop and Flügge-Lotz, 1965), which states:

> m-term inner expansion of the p-term outer expansion = p-term outer expansion of the m-term inner expansion

In an application, we would set $p = m$ for the mth-order boundary-layer approximation. For the pth-order outer flow approximation, we set $m = p - 1$. As we shall see, this outer flow procedure only yields results for V_2. We next derive matching conditions for u, p, ρ, and T; later, the condition for v is separately discussed.

Matching Conditions for u, p, ρ, and T

As pointed out by Davis and Flügge-Lotz (1964), the restricted matching principle is sometimes ambiguous. Instead, we utilize the equivalent, but straightforward, procedure of these authors. Let us begin by matching the two expansions for u at the outer edge of the boundary layer, where both expansions are presumed valid. For the inner one, we have

$$u \sim u_1(s, \bar{n}) + \varepsilon u_2(s, \bar{n}) + \cdots, \qquad \bar{n} \to \infty \tag{23.39}$$

where $\bar{n} \to \infty$ represents the outer edge of the inner layer. A Taylor series expansion about $n = 0$ is used for the outer expansion; i.e.,

$$u \sim U_1(s,0) + n\frac{\partial U_1}{\partial n}(s,0) + \frac{n^2}{2}\frac{\partial^2 U_1}{\partial n^2}(s,0) + \cdots + \varepsilon U_2(s,0) + \varepsilon n\frac{\partial U_2}{\partial n}(s,0) + \cdots \tag{23.40}$$

where each term in Equation (23.17a) is separately expanded. We shift the outer expansion to inner variables

$$u \sim U_1(s,0) + \varepsilon\bar{n}\frac{\partial U_1}{\partial n}(s,0) + \varepsilon^2\frac{\bar{n}^2}{2}\frac{\partial^2 U_1}{\partial n^2}(s,0) + \cdots + \varepsilon U_2(s,0) + \varepsilon^2\bar{n}\frac{\partial U_2}{\partial n}(s,0) + \cdots$$

in order to compare the two expansions. (Note that derivatives, such as $\partial U_1/\partial n$, are not transformed, since they only depend on s.) Upon comparison, the $O(1)$ and $O(\varepsilon)$ terms yield

$$u_1(s,\infty) = U_1(s,0) \tag{23.41a}$$

$$u_2(s,\bar{n}) \sim \bar{n}\frac{\partial U_1}{\partial n}(s,0) + U_2(s,0), \quad \bar{n} \to \infty \tag{23.42a}$$

The same two relations are obtained from the restricted matching principle with $m = p = 1$ and $m = p = 2$, respectively. In the process of obtaining Equation (23.42a), a Taylor series expansion is used, which requires that u be analytic with respect to n. Hereafter, analyticity is assumed whenever a Taylor series expansion is utilized.

If matching is to be done in terms of outer variables, i.e., n instead of $\bar{n}$, then Equation (23.40) is unchanged but Equation (23.39) has to be written in terms of n and expanded for small ε. Several difficulties are encountered, such as an indeterminate term in the inner expansion and an infinite series of $O(1)$ terms in the outer expansion. Consequently, this approach is not pursued.

Equation (23.41a), at a fixed s value, has the form, $u_1(s,n) = U_{1w} = \text{constant}$, with $\bar{n} \to \infty$. This result is in accordance with first-order boundary-layer theory. On the other hand, the u_2 condition involves an $\bar{n}$ term that dominates as $\bar{n} \to \infty$. Nevertheless, the $U_2(s,0)$ term *must* be retained; it yields a displacement matching condition, as discussed in the next section. Note that both the first- and second-order outer solutions, evaluated at the wall, are required for u_2. These observations will similarly apply to the other dependent variables.

The procedure used to obtain Equations (23.41a) and (23.42a) is repeated for p, ρ, and T, with the $O(1)$ result

$$p_1(s,\infty) = P_1(s,0) \tag{23.41b}$$

$$\rho_1(s,\infty) = R_1(s,0) \tag{23.41c}$$

$$t_1(s,\infty) = T_1(s,0) \tag{23.41d}$$

and the $O(\varepsilon)$ result

$$p_2(s,\bar{n}) \sim \bar{n}\left.\frac{\partial P_1}{\partial n}\right|_w + P_2(s,0), \quad \bar{n} \to \infty \tag{23.42b}$$

$$\rho_2(s, \bar{n}) \sim \bar{n} \left. \frac{\partial R_1}{\partial n} \right|_w + R_2(s, 0), \qquad \bar{n} \to \infty \tag{23.42c}$$

$$t_2(s, \bar{n}) \sim \bar{n} \left. \frac{\partial T_1}{\partial n} \right|_w + T_2(s, 0), \qquad \bar{n} \to \infty \tag{23.42d}$$

In view of Equations (23.10e) and (23.18e), Equation (23.41b) provides an inconsequential redundancy for the pressure.

Equations (23.41) represent the conventional boundary conditions for the first-order boundary-layer equations in which $U_1(s, 0), \ldots, T_1(s, 0)$ are known functions that stem from a solution of the first-order (Euler) outer equations. Furthermore, Equations (23.10c), (23.18b), and (23.41b) provide the usual pressure gradient condition

$$\frac{dp_1}{ds} = -\gamma M_\infty^2 \left(R_1 U_1 \frac{\partial U_1}{\partial s} \right)_w \tag{23.43}$$

which can be used to replace the first-order pressure gradient that appears on the left side of Equation (23.14c). Equations (23.35a) and (23.41) provide sufficient boundary conditions for the $\bar{n}$ variation of Equations (23.10). Initial conditions at some s value, e.g., a stagnation point, may also be required. Initial conditions, however, are not necessary for a similarity solution.

Equation (23.41b), e.g., can be written as

$$p_1(s, \bar{n}) \sim P_1(s, 0), \qquad \bar{n} \to \infty$$

Hence, Equations (23.41) and (23.42) are asymptotic boundary conditions that hold when $\bar{n} \to \infty$. Equations (23.42) can be written in a more explicit form by combining them with Equations (23.31). (It is for this purpose that Equations (23.31) were originally obtained.) We thus have

$$u_2(s, \bar{n}) \sim -\bar{n} \left\{ kU_1 + \frac{r_w^\sigma R_1}{(\gamma - 1) M_\infty^2} [T_1 S_1' - X_\infty H_1'] \right\}_w + U_2(s, 0) \tag{23.44a}$$

$$p_2(s, \bar{n}) \sim \gamma M_\infty^2 k \bar{n} (R_1 U_1^2)_w + P_2(s, 0) \tag{23.44b}$$

$$\rho_2(s, \bar{n}) \sim \bar{n} \left[\frac{R_1 U_1}{T_1} (M_\infty^2 k U_1 - r_w^\sigma P_1 S_1') \right]_w + R_2(s, 0) \tag{23.44c}$$

$$t_2(s, \bar{n}) \sim \bar{n} \; [(\gamma - 1) M z_\infty^2 k U_1^2 + r_w^\sigma U_1 P_1 S'_1]_w + T_2(s, 0) \tag{23.44d}$$

where $\bar{n} \to \infty$. These boundary conditions consist of a term, which is multiplied by $\bar{n}$, that depends on the first-order outer solution. In particular, the entropy S'_{1w} and stagnation enthalpy H'_{1w} gradients are encountered, where the derivatives are with respect to the first-order stream function Ψ_1. In addition, the second-order outer solution, evaluated at the wall, is required.

The leading terms in the expansions, Equations (23.16a) and (23.28), for the scalar vorticity can be matched using a procedure analogous to the one to be used for v. This will not be done, because the vorticity is an auxiliary quantity for which boundary conditions are unnecessary.

Matching Condition for v

Matching requires v, not $\bar{v}$; hence, Equation (23.8b) is written as

$$v \sim \varepsilon v_1(s,\bar{n}) + \varepsilon^2 v_2(s,\bar{n}) + \cdots, \qquad \bar{n} \to \infty$$

As before, this expansion is compared with its outer expansion, when written in terms of $\bar{n}$:

$$v \sim V_1(s,0) + \varepsilon\bar{n}\frac{\partial V_1}{\partial n}(s,0) + \varepsilon v_2(s,0) + \cdots$$

We thus obtain Equation (23.30) and

$$v_1(s,\bar{n}) \sim \bar{n}\left.\frac{\partial V_1}{\partial n}\right|_w + V_2(s,0), \qquad \bar{n} \to \infty \tag{23.45}$$

Since an explicit $\bar{n} \to \infty$ condition for v_1 is not required, this relation will be used to derive a wall boundary condition for V_2. This can also be obtained using the restricted matching principle with $m = 1$ and $p = 2$.

An alternate form for Equation (23.45) is obtained by differentiating, with respect to $\bar{n}$, the lead term in both v expansions and letting $\varepsilon \to 0$:

$$\frac{\partial v}{\partial \bar{n}} \sim \varepsilon\frac{\partial V_1}{\partial n}(s,\varepsilon\bar{n}) \sim \varepsilon\frac{\partial V_1}{\partial n}(s,0)$$

$$\frac{\partial v}{\partial \bar{n}} \sim \varepsilon\frac{\partial v_1}{\partial n}(s,\varepsilon\bar{n}) \sim \varepsilon\frac{\partial v_1}{\partial n}\left(s,\frac{n}{\varepsilon}\right) \sim \varepsilon\frac{\partial v_1}{\partial \bar{n}}(s,\infty)$$

Consequently, we have

$$\frac{\partial V_1}{\partial n}(s,0) = \frac{\partial v_1}{\partial \bar{n}}(s,\infty)$$

and the equation for $V_2(s,0)$ becomes

$$V_2(s,0) \sim v_1(s,\bar{n}) - \bar{n}\frac{\partial v_1}{\partial \bar{n}}, \qquad \bar{n} \to \infty \tag{23.46}$$

where the right side depends only on the first-order boundary-layer solution. In Section 23.6 (see also Problem 23.6), we show that the right side has the form of a constant divided by $s^{1/2}$ for a flat plate. For this to be the case, the v_1 term must be linear in $\bar{n}$ in such a way that the $\bar{n}$-containing terms cancel. Equation (23.46) provides for the effect of the displacement thickness, which manifests itself here as a displacement speed $V_2(s,0)$ and is equivalent to a distribution of sources or sinks along the wall. Sources normally occur, but sinks occur when δ^* is negative, as previously discussed in Section 21.8.

Alternatively, outer and inner expansions for the stream function can be written as

$$\psi \sim \Psi_1 + \varepsilon \Psi_2 + \cdots$$
$$\psi \sim \varepsilon \psi_1 + \varepsilon^2 \psi_2 + \cdots$$

which yields

$$\Psi_2(s, 0) \sim \psi_1(s, \bar{n}) - \bar{n}\frac{\partial \psi_1}{\partial \bar{n}}, \qquad \bar{n} \to \infty$$

Thus, the second-order outer stream function, $\Psi_2(s,0)$, is *not* zero along the wall but varies in a manner that provides for the displacement effect caused by the presence of the first-order boundary layer.

An additional form for $V_2(s,0)$ is obtained by combining Equations (23.31b) and (23.45), with the result

$$V_2(s, 0) \sim \frac{\bar{n}}{r_w^\sigma R_1(s,0)} \frac{d}{ds}(r_w^\sigma R_1 U_1)_w + v_1(s, \bar{n}), \quad \bar{n} \to \infty \tag{23.47}$$

If the arc length derivative is zero, as it would be for uniform flow over a flat plate, this term is then indeterminate, since $\bar{n} \to \infty$. In this circumstance, Equation (23.46) should be used for $V_2(s, 0)$.

As discussed in Section 19.2, the first-order outer equations do not have a unique solution, since they only satisfy a velocity tangency condition at the wall. An analogous situation holds for the second-order outer equations, for which only Equation (23.46) or (23.47) holds at the wall. Thus, improving the accuracy of the approximation by going to second order does not alter this fundamental nonuniqueness associated with the Euler equations.

23.5 DECOMPOSITION OF THE SECOND-ORDER BOUNDARY-LAYER EQUATIONS

SUMMARY OF FIRST- AND SECOND-ORDER RESULTS

We succinctly summarize the theory before breaking new ground, starting with the first-order outer equations given by Equations (23.10). At the wall, these satisfy Equation (23.30), while away from the wall they satisfy the imposed freestream conditions, which may be nonuniform. Equations (23.24a) and (23.25a) provide two first-order integrals of the motion.

Equations (23.10) are the first-order inner equations, in which dp_1/ds is given by Equation (23.43). An unspecified set of initial conditions may be required, while the wall boundary conditions are provided by Equations (23.35a), (23.37a), and (23.38a), or an adiabatic wall condition. Equations (23.41) apply as the outer edge of the boundary layer is approached; there is no outer edge condition for v_1.

Equations (23.32) represent the second-order outer problem. The variables $U_2, \ldots, T_2$ go to zero as the freestream is approached. There is only one wall condition, given by either Equation (23.46) or (23.47).

Finally, Equations (23.13) or (23.14) govern the second-order inner problem. The associated inner variables go to zero at the upstream location of the first-order initial conditions, except when a second-order similarity solution is found. Equations (23.35b), (23.37b), and (23.38b) apply at the wall, while Equations (23.44) hold when $\bar{n} \to \infty$.

Decomposition

As previously noted, the second-order inner equations, Equations (23.14), and their boundary conditions are linear, thus allowing for superposition of solutions (Rott and Leonard, 1959). To take advantage of this fact, the left side of the first three of these equations is written in operator form:

$$A(u_2, v_2, \rho_2) \equiv \frac{\partial}{\partial s}[r_w^\sigma(u_1\rho_2 + \rho_1 u_2)] + \frac{\partial}{\partial \bar{n}}[r_w^\sigma(v_1\rho_2 + \rho_1 v_2)] \quad \mathbf{(23.48a)}$$

$$B(u_2, v_2, \rho_2, t_2) \equiv \rho_1\left(\frac{du_1}{d_2t} + \frac{du_2}{d_1t}\right) + \frac{du_1}{d_1t}\rho_2 - \frac{\partial}{\partial \bar{n}}\left(\mu_1\frac{\partial u_2}{\partial \bar{n}} + \mu_1'\frac{\partial u_1}{\partial \bar{n}}t_2\right) + \frac{1}{\gamma M_\infty^2}\frac{\partial}{\partial s}(t_1\rho_2 + \rho_1 t_2) \quad \mathbf{(23.48b)}$$

$$C(u_2, v_2, \rho_2, t_2) \equiv \rho_1\left(\frac{dt_1}{d_2t} + \frac{dt_2}{d_1t}\right) + \frac{dt_1}{d_1t}\rho_2 - \frac{\gamma - 1}{\gamma}u_1\frac{\partial}{\partial s}(t_1\rho_2 + \rho_1 t_2) - \frac{1}{Pr}\frac{\partial}{\partial \bar{n}}\left(\mu_1\frac{\partial t_2}{\partial \bar{n}} + \mu_1'\frac{\partial t_1}{\partial \bar{n}}t_2\right) + (\gamma - 1)M_\infty^2\left[\left(R_1U_1\frac{\partial U_1}{\partial s}\right)_w u_2 - \mu_1'\left(\frac{\partial u_1}{\partial \bar{n}}\right)^2 t_2 - 2\mu_1\frac{\partial u_1}{\partial \bar{n}}\frac{\partial u_2}{\partial \bar{n}}\right] \quad \mathbf{(23.48c)}$$

where Equation (23.11a) defines $d(\ )/d_it$. The A, B, and C are linear differential operators whose unknown functions are shown as their arguments. Observe that only first-order derivatives of second-order variables are present in A. Second-order derivatives, $\partial^2 u_2/\partial\bar{n}^2$ and $\partial^2 t_2/\partial\bar{n}^2$, are, respectively, present in B and C. Hence, the system is fifth-order with respect to its $\bar{n}$ derivatives. The variable v_2 enters both B and C through the $d(\ )/d_2t$ derivative.

Equation (23.14d) is not written in operator form, since it can be integrated (Van Dyke, 1962). We start with

$$p_2(s, \bar{n}) = f(s) + \gamma M_\infty^2 k\int_0^{\bar{n}}\rho_1 u_1^2 d\bar{n}$$

where f is a function of integration. The integral is written as

$$\begin{aligned}\int_0^{\bar{n}}\rho_1 u_1^2 d\bar{n} &= \int_0^{\bar{n}}[\rho_1 u_1^2 - (R_1U_1^2)_w]\,d\bar{n} + (R_1U_1^2)_w\bar{n} \\ &= \bar{n}(R_1U_1^2)_w - \int_0^\infty[(R_1U_1^2)_w - \rho_1 u_1^2]d\bar{n} \\ &\quad + \int_{\bar{n}}^\infty[(R_1U_1^2)_w - \rho_1 u_1^2]d\bar{n}\end{aligned}$$

with the result

$$p_2 = f + \gamma M_\infty^2 k\bar{n}(R_1U_1^2)_w + \gamma M_\infty^2 k\int_{\bar{n}}^{\infty}[(R_1U_1^2)_w - \rho_1 u_1^2]\,d\bar{n}$$
$$-\gamma M_\infty^2 k\int_0^{\infty}[(R_1U_1^2)_w - \rho_1 u_1^2]\,d\bar{n}$$

The function of integration is evaluated by utilizing Equation (23.44b), with $\bar{n}\to\infty$:

$$\gamma M_\infty^2 k\bar{n}(R_1U_1^2)_w + P_2(s,0) = f + \gamma M_\infty^2 k\bar{n}(R_1U_1^2)_w$$
$$-\gamma M_\infty^2 k\int_0^{\infty}[(R_1U_1^2)_w - \rho_1 u_1^2]\,d\bar{n}$$

to obtain

$$f - \gamma M_\infty^2 k\int_0^{\infty}[(R_1U_1^2)_w - \rho_1 u_1^2]\,d\bar{n} = P_2(s,0) = (T_1R_2 + R_1T_2)_w$$

This yields the elegant result (Van Dyke, 1963)

$$p_2(s,\bar{n}) = t_1(s,\bar{n})\rho_2(s,\bar{n}) + \rho_1(s,\bar{n})t_2(s,\bar{n}) \tag{23.49a}$$

$$= \gamma M_\infty^2 kn(R_1U_1^2)_w + (T_1R_2 + R_1T_2)_w + \gamma M_\infty^2 k\int_{\bar{n}}^{\infty}[(R_1U_1^2)_w - \rho_1 u_1^2]\,d\bar{n} \tag{23.49b}$$

Thus, an explicit solution for p_2 is available that satisfies its boundary condition. When $\bar{n} = 0$, Problem 23.5 shows how this equation can be written in terms of the boundary-layer thicknesses introduced in Section 21.7.

Equation (23.49a) is a linear relation that, in principle, can be used to eliminate either ρ_2 or t_2 from the *A*, *B*, and *C* operators. This will be done in Section 23.8, where t_2 will be replaced. The reason for this choice is that t_2 primarily appears only in *C*, whereas ρ_2 appears in all three operators. Moreover, the quantity $(t_1\rho_2 + \rho_1 t_2)$ can be eliminated with the use of Equation (23.49b). On the other hand, the second-order inner boundary conditions become somewhat more complicated, since in their initial form they are in terms of t_2.

Following Van Dyke (1962), we decompose $u_2(s,\bar{n})$ into seven terms, one for each of the effects listed in Section 23.1. We thus write

$$u_2 = u^{(\ell)} + \sigma u^{(t)} + a_1u^{(s)} + c_1u^{(T)} + S'_{1w}\,u^{(e)} + H'_{1w}\,u^{(H)} + u^{(d)} \tag{23.50}$$

where a 2 subscript is not used on the right side for notational simplicity. The superscripts represent, in order, longitudinal curvature, transverse curvature, velocity slip, temperature jump, entropy gradient, stagnation enthalpy gradient, and displacement. The constant coefficients $\sigma,\ldots,H'_{1w}$ identify five of the terms. Longitudinal curvature cannot be identified in this manner because the curvature, *k*, is generally a function of *s*. Similarly, there is no constant coefficient that is uniquely associated with displacement. For the coefficient of $u^{(T)}$ Van Dyke (1969) uses $c_1 - a_1$, which, however, is not always convenient (Fannelop and Flügge-Lotz, 1965). Relations analogous to Equation (23.50) can be written for v_2, p_2, ρ_2, and t_2. Both p_2 and t_2, however, will be eliminated by utilizing Equations (23.49). Hence, only three relations of the form of Equation (23.50) are required. In summary, this decomposition replaces u_2, v_2, and ρ_2 with 21 new dependent variables.

The first three equations of (23.14) are now written as

$$A(u_2, v_2, \rho_2) = k\left[\bar{n}\frac{\partial(\rho_1 u_1)}{\partial s} - r_w^\sigma \rho_1 v_1\right] + \frac{\sigma\cos\theta}{r_w}(\bar{n}\rho_1 u_1 \sin\theta - r_w\rho_1 v_1) \tag{23.51a}$$

$$B(u_2, v_2, \rho_2, t_2) = k\left[\frac{\partial}{\partial\bar{n}}(\bar{n}\mu_1)\frac{\partial u_1}{\partial\bar{n}} - \rho_1 v_1\frac{\partial}{\partial\bar{n}}(\bar{n}\mu_1) - \mu_1' u_1\frac{\partial t_1}{\partial\bar{n}}\right] + \frac{\sigma\cos\theta}{r_w}\mu_1\frac{\partial u_1}{\partial\bar{n}} \tag{23.51b}$$

$$C(u_2, v_2, \rho_2, t_2) = k\left\{-\bar{n}\rho_1 v_1\frac{\partial t_1}{\partial\bar{n}} + \frac{1}{Pr}\frac{\partial}{\partial\bar{n}}\left(n\mu_1\frac{\partial t_1}{\partial\bar{n}}\right)\right.$$
$$\left.+(\gamma-1)M_\infty^2\left[\left(\bar{n}\frac{\partial u_1}{\partial\bar{n}} - 2u_1\right)\mu_1\frac{\partial u_1}{\partial\bar{n}} + \rho_1 u_1^2 v_1\right]\right\} + \frac{\sigma\cos\theta}{r_w}\frac{\mu_1}{Pr}\frac{\partial t_1}{\partial\bar{n}} \tag{23.51c}$$

where the unknowns are $u^{(\ell)}, v^{(\ell)}, \rho^{(\ell)}, \ldots, t^{(d)}$. In this formulation, only longitudinal and transverse curvature terms appear on the right sides. The other five effects enter through the wall or the infinity conditions. Thus, only three forms of these equations need to be considered. The first is

$$A = 0, \quad B = 0, \quad C = 0 \tag{23.52}$$

where the arguments of A, B, and C are $u^{(s)}, u^{(T)}, \ldots, t^{(d)}$, and these equations are used for all effects other than curvature. The second form is used for longitudinal curvature:

$$A(u^{(\ell)}, v^{(\ell)}, \rho^{(\ell)}) = k\left[\bar{n}\frac{\partial(\rho_1 u_1)}{\partial s} - r_w^\sigma\rho_1 v_1\right] \tag{23.53a}$$

$$B(u^{(\ell)}, v^{(\ell)}, \rho^{(\ell)}, t^{(\ell)}) = k\left[\frac{\partial}{\partial\bar{n}}\left(\bar{n}\mu_1\frac{\partial u_1}{\partial\bar{n}}\right) - \rho_1 v_1\frac{\partial}{\partial\bar{n}}(\bar{n}u_1) - \mu_1' u_1\frac{\partial t_1}{\partial\bar{n}}\right] \tag{23.53b}$$

$$C\left(u^{(\ell)}, v^{(\ell)}, \rho^{(\ell)}, t^{(\ell)}\right) = k\left\{-\bar{n}\rho_1 v_1\frac{\partial t_1}{\partial\bar{n}} + \frac{1}{Pr}\frac{\partial}{\partial\bar{n}}\left(\bar{n}\mu_1\frac{\partial t_1}{\partial\bar{n}}\right)\right.$$
$$\left.+(\gamma-1)M_\infty^2\left[\left(\bar{n}\frac{\partial u_1}{\partial\bar{n}} - 2u_1\right)\mu_1\frac{\partial u_1}{\partial\bar{n}} + \rho_1 u_1^2 v_1\right]\right\} \tag{23.53c}$$

while the last form holds for transverse curvature:

$$A(u^{(t)}, v^{(t)}, \rho^{(t)}) = \frac{\cos\theta}{r_w}(\bar{n}\rho_1 u_1\sin\theta - r_w\rho_1 v_1) \tag{23.54a}$$

$$B(u^{(t)}, v^{(t)}, \rho^{(t)}, t^{(t)}) = \frac{\cos\theta}{r_w}\mu_1\frac{\partial u_1}{\partial\bar{n}} \tag{23.54b}$$

$$C(u^{(t)}, v^{(t)}, \rho^{(t)}, t^{(t)}) = \frac{\cos\theta}{r_w}\frac{\mu_1}{Pr}\frac{\partial t_1}{\partial\bar{n}} \tag{23.54c}$$

Note the presence of r_w^σ in A and on the right side of Equation (23.53a). Hence, a transverse curvature parameter appears in the longitudinal curvature problem.

Before discussing the complete problem for each effect, we interject several remarks of a general nature. Since p_2 is given by Equation (23.49b), boundary conditions for it are unnecessary. Moreover, p_2 does not have a prescribed, or imposed, value at the wall; its $\bar{n} \to \infty$ value has already been used when writing Equation (23.49b). This equation further removes the need for boundary conditions, at the wall and at infinity, for ρ_2, if t_2 boundary conditions are imposed, as will be done in this section. When using Equation (23.49a) for $p^{(\ell)}$, e.g., we replace ρ_2 and t_2 with $\rho^{(\ell)}$ and $t^{(\ell)}$, respectively.

As previously indicated, the A, B, and C operators are fifth-order with respect to $\bar{n}$ derivatives, thus requiring five boundary conditions for u, v, and t for each of the seven effects. As in first-order theory, there is no infinity condition for v. A unique solution, in terms of $\bar{n}$, for each effect is thereby provided by wall conditions for u, v, and t (i.e., $u^{(\ell)}, v^{(\ell)}, \ldots, t^{(d)}$) and infinity conditions for u and t. These conditions are constructed from Equations (23.35b), (23.37b), (23.38b), and (23.44).

We start with longitudinal curvature, where Equations (23.53) apply. The wall conditions are

$$u^{(\ell)}(s,0) = v^{(\ell)}(s,0) = t^{(\ell)}(s,0) = 0 \tag{23.55a}$$

and the infinity conditions are

$$u^{(\ell)}(s,\bar{n}) \sim -\bar{n}kU_1(s,0), \quad \bar{n} \to \infty \tag{23.55b}$$

$$t^{(\ell)}(s,\bar{n}) \sim (\gamma-1)M_\infty^2 \bar{n}kU_1^2(s,0), \quad \bar{n} \to \infty \tag{23.55c}$$

For instance, the $u^{(\ell)}(s,\infty)$ result in Equation (23.55b) stems from the right side of Equation (23.44a). The other terms in Equation (23.44a) will appear in the u_2 infinity conditions for the entropy gradient, stagnation enthalpy gradient, and displacement, respectively.

For transverse curvature, Equations (23.54) are utilized in conjunction with

$$u^{(t)}(s,0) = v^{(t)}(s,0) = t^{(t)}(s,0) = 0$$

and

$$u^{(t)}(s,\infty) = t^{(t)}(s,\infty) = 0$$

It would appear that transverse curvature enters only through the inhomogeneous terms in Equations (23.54). This is not correct, however, since a transverse curvature parameter, r_w^σ, also appears in the A operator and in the inhomogeneous terms in Equation (23.53a), and it will shortly appear in the boundary condition for the entropy and stagnation enthalpy gradients. Although there is a distinct problem for the transverse curvature, this effect is nevertheless coupled with the others.

The remaining five effects all utilize Equations (23.52); only the boundary conditions differ. For velocity slip and temperature jump these are

$$u^{(s)}(s,0) = \gamma^{1/2}M_\infty\left(\frac{\mu_1 t_1^{1/2}}{p_1}\frac{\partial u_1}{\partial \bar{n}}\right)_w \tag{23.56a}$$

$$v^{(s)}(s,0) = t^{(s)}(s,0) = 0 \tag{23.56b}$$

$$u^{(s)}(s,\infty) = t^{(s)}(s,\infty) = 0 \tag{23.56c}$$

$$u^{(T)}(s,0) = t^{(T)}(s,0) = 0 \tag{23.57a}$$

$$t^{(T)}(s,0) = \gamma^{1/2} M_\infty \left(\frac{\mu_1 t_1^{1/2}}{p_1} \frac{\partial t_1}{\partial n} \right)_w \tag{23.57b}$$

$$u^{(T)}(s,\infty) = t^{(T)}(s,\infty) = 0 \tag{23.57c}$$

For the remaining three effects the wall conditions are

$$u(s,\infty) = v(s,0) = t(s,\infty) = 0$$

while the respective infinity conditions are

$$u^{(e)} \sim -\frac{\bar{n} r_w^\sigma}{(\gamma-1)M_\infty^2} P_1(s,0), \quad \bar{n} \to \infty$$

$$t^{(e)} \sim \bar{n} r_w^\sigma (U_1 P_1)_w, \quad \bar{n} \to \infty$$

$$u^{(H)} \sim \bar{n} r_w^\sigma \frac{X_\infty}{(\gamma-1)M_\infty^2} R_1(s,0), \quad \bar{n} \to \infty$$

$$t^{(H)}(s,\infty) = 0$$

$$u^{(d)}(s,\infty) = U_2(s,0) \tag{23.58a}$$

$$t^{(d)}(s,\infty) = T_2(s,0) \tag{23.58b}$$

The only place the second-order outer solution enters is in the displacement boundary conditions, Equations (23.58), and in Equation (23.49b). To obtain $U_2(s,0)$, $R_2(s,0)$, and $T_2(s,0)$, Equations (23.32) are to be solved subject to the wall condition given by Equation (23.46). We finally see precisely how the displacement speed $V_2(s,0)$ enters into the analysis.

Surface Properties

The shear stress is normalized with $\mu_\infty U_\infty/\ell$ and we write

$$\tau_w = \frac{1}{\varepsilon}\left(\mu \frac{\partial u}{\partial \bar{n}}\right)_w = \frac{1}{\varepsilon}(\mu_1 + \varepsilon \mu_1' t_2 + \cdots)_w \left(\frac{\partial u_1}{\partial \bar{n}} + \varepsilon \frac{\partial u_2}{\partial \bar{n}} + \cdots\right)_w$$

$$= \frac{1}{\varepsilon}\left(\mu_1 \frac{\partial u}{\partial \bar{n}}\right)_w + \left(\mu_1 \frac{\partial u_2}{\partial \bar{n}} + u_1' t_2 \frac{\partial u_1}{\partial \bar{n}}\right)_w + \cdots$$

Hence, set

$$\tau_w \sim \frac{1}{\varepsilon}\tau_{w1} + \tau_{w2} + \cdots$$

where

$$\tau_{w1} = \left(\mu_1 \frac{\partial u_1}{\partial \bar{n}}\right)_w$$

$$\tau_{w2} = \left(\mu_1 \frac{\partial u_2}{\partial \bar{n}} + \mu_1' \, t_2 \frac{\partial u_1}{\partial \bar{n}}\right)_w \tag{23.59a}$$

Since the only nonzero $t_2(s,0)$ quantity is given by Equation (23.57b), we can write

$$\tau_{w2} = \left(\mu_1 \frac{\partial u_2}{\partial \bar{n}}\right)_w + \left(\frac{\mu_1'}{\mu_1}\right)_w c_1 \tau_{w1} t^{(T)}(s, 0) \tag{23.59b}$$

where the c_1 constant stems from Equation (23.38b). All seven effects can contribute to the first term on the right side, whereas the temperature jump term can provide an additional contribution. A local skin-friction coefficient is introduced as

$$c_f = \frac{2\tau_w}{\rho_\infty U_\infty^2}$$

where τ_w is dimensional, to obtain

$$c_f \sim 2\varepsilon\tau_{w1} + 2\varepsilon^2\tau_{w2} + \cdots \tag{23.60}$$

The parameters τ_{w1} and τ_{w2} are provided by Equations (23.59).

The dimensional wall heat transfer is given by

$$q_w = -\left(\kappa \frac{\partial T}{\partial n} + \mu u \frac{\partial u}{\partial n}\right)_w$$

where the rightmost term is due to sliding friction (Van Dyke, 1969; Maslen, 1958), which occurs when there is velocity slip. We normalize q_w with $(\mu c_p T)_\infty/\ell$ and write

$$q_w \sim \frac{1}{\varepsilon} q_{w1} + q_{w2} + \cdots$$

With this normalization, we obtain

$$q_{w1} = -\frac{1}{Pr}\left(\mu_1 \frac{\partial t_1}{\partial \bar{n}}\right)_w \tag{23.61a}$$

for q_{w1}, while for q_{w2}, we have

$$q_{w2} = -\frac{1}{Pr}\left(\mu_1 \frac{\partial t_2}{\partial \bar{n}} + \mu_1' \, t_2 \frac{\partial t_1}{\partial \bar{n}}\right)_w + (\gamma - 1)M_\infty^2\left(\mu_1 u_1 \frac{\partial u_2}{\partial \bar{n}} + \mu_1 \frac{\partial u_1}{\partial \bar{n}} \mu_2 + \mu_1' \, t_2 u_1 \frac{\partial u_1}{\partial \bar{n}}\right)_w$$

Note that $(u_1)_w = 0$ and with Equations (23.56a) and (23.57b), we can write

$$q_{w2} = -\frac{1}{Pr}\left(\mu_1\frac{\partial t_2}{\partial n}\right)_w + \left(\frac{\mu_1'}{\mu_1}\right)_w c_1 q_{w1} t^{(T)}(s,0) - (\gamma-1)M_\infty^2 a_1 \tau_{w1} u^{(s)}(s,0) \qquad \textbf{(23.61b)}$$

where the a_1 constant stems from Equation (23.37b). Thus, temperature jump and velocity slip each now contribute two terms to q_{w2}, where two terms stem from $(\partial t_2/\partial n)_w$. The contribution of the once controversial sliding friction term yields the rightmost term in Equation (23.61b). In terms of a Stanton number, Equation (21.66), we have

$$St = \frac{q_w}{(h_{ow} - h_{oe})\rho_\infty U_\infty} = \frac{q_w}{X_\infty(1-g_w)Re} = \frac{1}{X_\infty(1-g_w)}(\varepsilon q_{w1} + \varepsilon^2 q_{w2} + \cdots) \qquad \textbf{(23.62)}$$

where $g_w \neq 1$. (The wall is adiabatic when $g_w = 1$; the above formulation of the boundary conditions does not include this case.) The quantities in the second term from the left are dimensional, and g_w, originally defined by Equation (21.27), is given by

$$g_w(s) = \frac{T_w(s)}{X_\infty}$$

where T_w is normalized with T_∞.

A wall pressure coefficient is defined as

$$c_p(s) = \frac{p_w(s) - p_\infty}{\frac{1}{2}\rho_\infty U_\infty^2}$$

or nondimensionally as

$$c_p(s) = \frac{p(s,0) - 1}{\gamma M_\infty^2/2} \qquad \textbf{(23.63)}$$

where p is given by its inner expansion, Equation (23.8c). Second-order boundary-layer thicknesses can also be defined (Werle, 1968; Werle and Davis, 1970), although we will not do so.

23.6 EXAMPLE: FIRST-ORDER SOLUTION

General Discussion

The theory is illustrated by considering steady flow over two surfaces that differ only in their values for r_w^σ. The first case is the ubiquitous, semi-infinite flat plate. The other surface is a circular cylinder with zero wall thickness in which the upstream velocity is parallel to the cylinder's axis. As with the flat plate, the cylinder has a sharp leading edge and the flow inside it is assumed inviscid. We will be interested only in the flow far downstream of the leading edge, along the upper surface of the flat plate and along the outer surface of the cylinder. The distance s, in each case, is measured in the flow direction from the leading edge. By making the cylinder's interior hollow and the adjacent flow inviscid, we minimize any disturbance due to bluntness.

The inviscid flow external to the boundary layer may be subsonic or supersonic; if it is the latter, any weak shock wave that may occur near the leading edge is ignored. Nonuniform freestream conditions are considered by invoking the substitution principle of Chapter 8, which provides an exact, nonuniform solution of the Euler equations. The first-order (inviscid) outer flow is thus parallel to the surface but, at the surface, possesses entropy, stagnation enthalpy, and vorticity gradients normal to the surface. For the normalization, freestream values at the leading edge of the surface are used. The wall temperature is taken as a constant; hence, an adiabatic wall is not considered.

The two geometries thus possess

$$r^{\sigma} = (r_w + n)^{\sigma}, \quad r_w = \text{constant}, \quad k = 0, \quad h_s = 1, \quad \theta = 0$$

and a single formulation holds for both configurations. (In the axisymmetric case, by choosing ℓ equal to r_w we can set $r_w^{\sigma} = 1$. For purposes of clarity, this will not be done.) With these two configurations, six of the second-order effects are involved; only longitudinal curvature is not present. For the first-order inner solution, the similarity theory of Chapter 21 is used. Unity values for the Prandtl number and Chapman–Rubesin parameter are thus assumed.

The treatment of the second-order inner and outer problems is not as complete as that given for the first-order problems. For the second-order outer problem, a stream function equation is derived for which special-case solutions are provided. For instance, a flat plate solution with h_o' small is obtained. Boundary conditions, including the displacement effect, are thus obtained for the second-order boundary-layer equations. Since these boundary-layer equations are rather complex — they require a numerical solution — only limited results are provided.

A number of papers (Maslen, 1963; Gersten and Gross, 1973, 1976) have previously examined second-order theory for either the flat plate or the circular cylinder. (These papers, along with Van Dyke, 1969, can be consulted for additional references.) Nevertheless, our approach is quite different from these earlier studies; e.g., we use the substitution principle to construct the first-order outer flow.

First-Order Outer Flow

The baseline flow is uniform, and with the current normalization, Equations (23.3), we have for Equation (8.16)

$$\lambda = h_o(n)$$

The nondimensional stagnation enthalpy function is arbitrary but satisfies $h_o(0) = 1$ and is positive. For analytical simplicity, we assume a linear relation

$$h_o = 1 + \left.\frac{dh_o}{dn}\right|_w n = 1 + h_o' n$$

where the gradient at the wall, h_o', is a positive or negative constant. If negative, n cannot be too large since $h_o(n) > 0$.

From Section 8.3, we obtain the first-order nondimensional solution

$$U_1(s, n) = (1 + h_o' n)^{1/2} \tag{23.64a}$$

$$V_1(s, n) = 0 \tag{23.64b}$$

$$P_1(s,n) = 1 \tag{23.64c}$$

$$R_1(s,n) = \frac{1}{1 + h'_o n} \tag{23.64d}$$

$$T_1(s,n) = 1 + h'_o n \tag{23.64e}$$

With the above and Equations (23.21b), we obtain for the first-order outer stream function

$$\frac{d\Psi_1}{dn} = \frac{(r_w + n)^\sigma}{(1 + h'_o n)^{1/2}} \tag{23.65a}$$

which integrates to

$$\Psi_1(s, n) = \frac{2r_w^\sigma}{h'_o}[(1 + h'_o n)^{1/2} - 1] + \frac{4\sigma}{3h_o'^2}\left[1 - \left(1 - \frac{1}{2}h'_o n\right)(1 + h'_o n)^{1/2}\right] \tag{23.65b}$$

Thus, H_1 and the parameter H'_{1w} are given by

$$H_1(s,n) = 1 + h'_o n \tag{23.66a}$$

$$H'_{1w} = \left.\frac{dH_1}{d\Psi_1}\right|_{\Psi_1=0} = \left.\frac{dH_1}{dn}\frac{dn}{d\Psi_1}\right|_{\Psi_1=0} = \frac{h'_o}{r_w^\sigma} \tag{23.66b}$$

where H_1 stems from Equation (23.25a). By means of Equation (8.17), we obtain

$$S_1(s, n) = \ln\lambda = \ln(1 + h'_o n) \tag{23.67a}$$

and

$$S'_{1w} = \left.\frac{dS_1}{d\Psi_1}\right|_{\Psi_1=0} = \left.\frac{dS_1}{dn}\frac{dn}{d\Psi_1}\right|_{\Psi_1=0} = \frac{h'_o}{r_w^\sigma} \tag{23.67b}$$

for S_1 and S'_{1w}. Finally, Equation (23.29) yields for the first-order vorticity

$$\Omega_1(s, n) = -\frac{h'_o}{2(1 + h'_o n)^{1/2}} \tag{23.68}$$

Equations (23.64) to (23.68) fully describe the first-order outer flow. If $h'_o = 0$, this flow is uniform, and only transverse curvature, velocity slip, temperature jump, and the displacement effects are present. When $h'_o \neq 0$, there are simultaneous gradients of H_1 and S_1 at the wall and Ω_1 is nonzero. Observe that this solution is entirely independent of s; thus, the substantial derivative defined by Equation (23.19a) simplifies to

$$\frac{D}{D_1 t} = (1 + h'_o n)^{1/2}\frac{\partial}{\partial s}$$

First-Order Inner Flow

Aside from unity values for *Pr* and *C*, we also have

$$\beta = 0$$

$$g_w = h_{o1w} = \frac{t_{1w}}{X_\infty} = \frac{T_w}{X_\infty}$$

and

$$p_1(s, \bar{n}) = 1$$

$$\mu_1(s, \bar{n}) = t_1(s, \bar{n})$$

$$\rho_{1e} = \mu_{1e} = u_{1e} = h_{o1e} = 1$$

Recall that quantities, such as ρ, μ, u, ξ, $\bar{x}$, and δ, in Chapter 21 are dimensional. For the boundary-layer coordinate transformation, we thus have

$$\xi = (\rho\mu U)_\infty \ell^{2\sigma+1} r_w^{2\sigma} s \tag{23.69a}$$

$$\Upsilon = \frac{1}{\ell}\left(\frac{Re}{2s}\right)^{1/2} \tag{23.69b}$$

$$\eta = \frac{1}{(2s)^{1/2}} \int_0^{\bar{n}} \rho_1(s, \bar{n}) d\bar{n} \tag{23.69c}$$

$$\frac{\partial}{\partial \bar{n}} = \frac{\rho_1}{(2s)^{1/2}} \frac{\partial}{\partial \eta} \tag{23.69d}$$

where *Re* is still defined by Equation (23.2) and is a constant. Most of the solution in Chapter 21 is provided by

$$u_1 = f'(\eta)$$

$$h_{o1} = g(\eta)$$

$$G = \frac{g - g_w}{1 - g_w} = f'(\eta)$$

$$t_1 = \frac{1}{\rho_1} = \frac{g_w + (1 - g_w) f' - S f'^2}{1 - S}$$

where the speed parameter *S* (defined by Equation 21.48) should not be confused with the entropy, and where the solution for v_1 is provided later. The function *f*, of course, is the solution of the Blasius equation, Equation (18.10). As expected, there is no dependence on σ in the first-order inner solution.

By means of the tables in Chapter 21, we have

$$C_v = C_t = 1.2168$$

$$\eta_{ev} = \eta_{et} = 3.4717$$

$$f''_w = G'_w = 0.4696$$

when $\beta = 0$. The first-order boundary-layer thicknesses, skin-friction coefficient, and Stanton number are denoted with a unity subscript. They are

$$\left(\frac{Re}{2s}\right)^{1/2}\delta_1 = \left(\frac{Re}{2s}\right)^{1/2}\delta_{1t} = \eta_{ev} + \frac{1}{1-S}[Sf''_w - (1-S-g_w)C_v] = \eta_{ev} + C_1$$

$$\left(\frac{Re}{2s}\right)^{1/2}\delta_1^* = \frac{1}{1-S}(Sf''_w + g_w C_v) = C^*$$

$$\left(\frac{Re}{2s}\right)^{1/2}\theta_1 = \left(\frac{Re}{2s}\right)^{1/2}\psi_1 = f''_w$$

$$c_{f1} = \frac{2}{Re^{1/2}}\tau_{w1} = \left(\frac{2}{Re\, s}\right)^{1/2} f''_w$$

$$St_1 = -\frac{q_{w1}}{(1-g_w)X_\infty Re^{1/2}} = \frac{G'_w}{2(Re\, s)^{1/2}}$$

where C_1 and C^* are newly defined constants, and C^* is associated with the first-order displacement thickness. This later parameter will frequently appear in the subsequent analysis; it is referred to as the displacement parameter.

In contrast to a flat plate, the axisymmetric case has a naturally occurring length, ℓr_w. In this circumstance, the ratio δ_1/r_w, given by

$$\frac{\delta_1}{r_w} = \left(\frac{2s}{r_w^2 Re}\right)^{1/2}(\eta_{ev} + C_1)$$

must be small compared to unity, say less than 0.05. (Remember that δ_1, r_w, and s are all normalized with ℓ.) Thus, at a sufficient distance downstream of its leading edge, boundary-layer theory, of any order, becomes invalid for the flow along a circular cylinder.

With the aid of Problem 23.6 and Equation (23.69d), we obtain, when $\beta = 0$,

$$v_1(s, \bar{n}) = \frac{\bar{n}f'}{2s} - \frac{f}{(2s)^{1/2}\rho_1}$$

and

$$\frac{\partial v_1}{\partial \bar{n}} = \frac{\bar{n}\rho_1 f''}{(2s)^{3/2}} + \frac{\rho'_1}{\rho_1}\frac{f}{2s}$$

where $\rho'_1 = d\rho_1/d\eta$. Equation (23.46) is now utilized, with the result

$$V_2(s, 0) = \frac{C^*}{(2s)^{1/2}} = \frac{Re^{1/2}\delta_1^*}{2s} \tag{23.70}$$

which establishes the connection between $V_2(s, 0)$ and the first-order displacement thickness, δ_1^*. For the problem at hand, we see that the displacement speed is positive and decreases as $s^{-1/2}$.

23.7 EXAMPLE: SECOND-ORDER OUTER SOLUTION

Derivation of the Governing Equation and Its Boundary Conditions

Since the first-order outer flow is rotational, there is no reason to expect its second-order counterpart to be irrotational. Hence, a second-order velocity potential (Gersten and Gross, 1973) is not considered. Instead, a stream function, defined in Problem 23.2, is utilized that satisfies Equation (23.32a). The equation to be obtained is a special case of the Ψ_2 equation discussed in the last paragraph of Section 23.3. With the aid of Problem 23.4, we write $U_2(s, n)$ and $V_2(s, n)$ as

$$U_2 = -\frac{1}{M_\infty^2 - 1}\frac{1 + h_o'n}{(r_w + n)^\sigma}\left[\frac{\partial \Psi_2}{\partial n} + \left(1 - \frac{1}{2}M_\infty^2\right)\frac{h_o'}{1 + h_o'n}\Psi_2\right] \quad \textbf{(23.71a)}$$

$$V_2 = -\frac{1 + h_o'n}{(r_w + n)^\sigma}\frac{\partial \Psi_2}{\partial s} \quad \textbf{(23.71b)}$$

These relations are introduced into Equations (23.32b,c) and P_2 is then eliminated by cross-differentiation. We thus obtain

$$(M_\infty^2 - 1)\frac{\partial^2 \Psi_2}{\partial s^2} - \frac{\partial^2 \Psi_2}{\partial n^2} = E\frac{\partial \Psi_2}{\partial n} + F\Psi_2 + G(n) \quad \textbf{(23.72)}$$

after one s integration, where G is the function of integration. The E and F coefficients are given by

$$E(n) = \frac{(r_w + n)^\sigma}{(1 + h_o'n)^{1/2}}\frac{d}{dn}\left[\frac{(1 + h_o'n)^{1/2}}{(r_w + n)^\sigma}\right] + \frac{h_o'}{2(1 + h_o'n)} = \frac{h_o'}{1 + h_o'n} - \frac{\sigma}{r_w + n} \quad \textbf{(23.73a)}$$

$$\begin{aligned} F(n) &= \frac{h_o'}{2}\frac{(r_w + n)^\sigma}{(1 + h_o'n)^{1/2}}\frac{d}{dn}\left[\frac{1}{(r_w + n)^\sigma(1 + h_o'n)^{1/2}}\right] \\ &= -\frac{h_o'}{4(1 + h_o'n)}\left(\frac{h_o'}{1 + h_o'n} + \frac{2\sigma}{r_w + n}\right) \end{aligned} \quad \textbf{(23.73b)}$$

Observe that Equation (23.72) is hyperbolic when $M_\infty^2 > 1$ and elliptic when $M_\infty^2 < 1$.

We eliminate V_2 from Equations (23.70) and (23.71b). The result is integrated with respect to s, to obtain

$$\Psi_2(s, 0) = -r_w^\sigma C^*(2s)^{1/2} \quad \textbf{(23.74)}$$

where the constant of integration is set equal to zero. This relation is the only boundary condition on Ψ_2, aside from the condition that U_2 and V_2 should vanish far from the wall. In view of this, we take the function of integration, G, in Equation (23.72) to be zero. (Problem 23.14 formally demonstrates that G is zero.)

Discussion

One purpose of Equations (23.72) to (23.74) is to provide $U_2(s, 0)$, which is required for the wall pressure and for boundary conditions for the second-order inner-layer equations. In view of Equations (23.71a) and (23.74), we have

$$U_2(s,0) = \frac{1}{(M_\infty^2 - 1)r_w^\sigma}\left[-\frac{\partial \Psi_2}{\partial n}\bigg|_w + \left(1 - \frac{1}{2}M_\infty^2\right) h_o' r_w^\sigma C^*(2s)^{1/2}\right]$$

and the quantity actually required from Equations (23.72) to (23.74) is $(\partial \Psi_2/\partial n)_w$.

Equations (23.72) to (23.74), with $G = 0$, are exact for the assumed flow model. Four second-order effects are still present: (a) transverse curvature via σ, (b) entropy and stagnation enthalpy gradients, both of which are represented by h_o', and (c) displacement, represented by C^*. Velocity slip and temperature jump are not present in the problem for Ψ_2.

Uniform, Supersonic Flow over a Flat Plate

For purposes of simplicity, let us initially consider a freestream that is supersonic, two-dimensional, and uniform. In this case we have $h_o' = \sigma = 0$, $M_\infty > 1$, and the only second-order effect is displacement. Equation (23.72) reduces to the wave equation

$$\lambda^2 \frac{\partial \Psi_2}{\partial s^2} - \frac{\partial^2 \Psi_2}{\partial n^2} = 0$$

where $\lambda^2 = M_\infty^2 - 1$, and Equation (23.74) remains as the sole boundary condition at the wall. The general solution of this equation is easily shown to be

$$\Psi_2(s,n) = f(s - \lambda n) + g(s - \lambda n)$$

where f and g are arbitrary functions. As usual, we assume there are no incoming waves; i.e., $g = 0$. With Equation (23.74), we readily obtain

$$\Psi_2(s,n) = -C^*[2(s - \lambda n)]^{1/2} \qquad \textbf{(23.75a)}$$

and by differentiation

$$U_2(s,n) = -\frac{C^*}{\lambda}\frac{1}{[2(s - \lambda n)]^{1/2}} \qquad \textbf{(23.75b)}$$

$$V_2(s,n) = \frac{C^*}{[2(s - \lambda n)]^{1/2}} \qquad \textbf{(23.75c)}$$

Hence, the desired wall speed is

$$U_2(s,0) = -\frac{C^*}{\lambda}\frac{1}{(2s)^{1/2}}$$

Since C^* is positive, $U_2(s,0)$ is negative and is scaled by the usual $(M_\infty^2 - 1)^{-1/2}$ factor. Moreover, it is easy to show that $|U_2(s,0)| \sim M_\infty$ when $M_\infty >> 1$.

Perturbation Solution

The above result for a flat plate can be extended to a rotational, supersonic flow by assuming h_o' is a small parameter compared to unity. For small h_o', E and F become

$$E \cong h_o', \qquad F \cong O(h_o'^2)$$

and a perturbation stream function, Ψ, is introduced as

$$\Psi_2 = \Psi^o + h_o'\Psi = -C^*[2(s-\lambda n)]^{1/2} + h_o'\Psi$$

where Ψ^o is given by the right side of Equation (23.75a). This relation is substituted into Equation (23.72), with the result

$$\lambda^2 \frac{\partial^2\Psi}{\partial s^2} - \frac{\partial^2\Psi}{\partial n^2} = \frac{C}{(s-\lambda n)^{1/2}} \qquad \textbf{(23.76a)}$$

The constant C is

$$C = \frac{1}{2^{1/2}}\lambda C^*$$

and the wall boundary condition on the perturbation stream function is

$$\Psi(s,0) = 0 \qquad \textbf{(23.76b)}$$

To assist in obtaining a solution of Equation (23.76a), we change variables to

$$\xi = s - \lambda n, \qquad \eta = s + \lambda n$$

And the second order derivatives are given by

$$\frac{\partial^2}{\partial s^2} = \frac{\partial^2}{\partial \xi^2} + 2\frac{\partial^2}{\partial \xi^2 \partial \eta} + \frac{\partial^2}{\partial \eta^2}$$

$$\frac{\partial^2}{\partial n^2} = \lambda^2\left(\frac{\partial^2}{\partial \xi^2} - 2\frac{\partial^2}{\partial \xi \partial \eta} + \frac{\partial^2}{\partial \eta^2}\right)$$

Equation (23.76a) thus becomes

$$\frac{\partial^2\Psi}{\partial \xi \partial \eta} = \frac{C}{4\lambda^2 \xi^{1/2}}$$

which can be integrated twice, to yield

$$\Psi = \frac{C\eta\xi^{1/2}}{2\lambda^2} + \int g_1(\xi)d\xi + g_2(\eta)$$

where g_1 and g_2 are functions of integration. Incoming waves are excluded; hence, we set $g_2(\eta) = 0$. As a consequence of Equation (23.76b), the g_1 integral equals

$$\int g_1(\xi)\,d\xi = -\frac{C\xi^{3/2}}{2\lambda^2}$$

Finally, after returning to the original variables and simplifying, the perturbation stream function is

$$\Psi = \frac{1}{2}C^* n[2(s-\lambda n)]^{1/2}$$

By combining this result with Equation (23.75a), we obtain

$$\Psi_2(s,n) = -C^*\left(1 - \frac{1}{2}h_o' n\right)[2(s-\lambda n)]^{1/2} \tag{23.77a}$$

$$U_2(s,n) = -\frac{C^*}{\lambda[2(s-\lambda n)]^{1/2}}\left(1 + h_o'\left[-\frac{1}{2}n + \lambda(s-\lambda n)\right]\right) \tag{23.77b}$$

and for the desired speed at the wall

$$U_2(s,0) = -\frac{C^*}{\lambda[(2s)]^{1/2}}(1 + \lambda h_o' s) \tag{23.77c}$$

Equations (23.77) are correct to $O(h_o')$.

For later use, we obtain $P_2(s,0)$, $R_2(s,0)$, and $T_2(s,0)$. These relations, along with equation (23.77c), are required for the second-order inner-layer boundary conditions. With the assistance of Equations (23.66b), (23.67b), and (23.77c), we obtain from Equations (23.34)

$$P_{2w} = P_2(s,0) = \frac{\gamma M_\infty^2 C^*}{\lambda(2s)^{1/2}} \tag{23.78a}$$

$$R_2(s,0) = \frac{M_\infty^2 C^*}{\lambda(2s)^{1/2}}\left(1 + \frac{2\lambda}{M_\infty^2}h_o' s\right) \tag{23.78b}$$

$$T_2(s,0) = \frac{(\gamma-1)M_\infty^2 C^*}{\lambda(2s)^{1/2}}\left[1 - \frac{2\lambda}{(\gamma-1)M_\infty^2}h_o' s\right] \tag{23.78c}$$

Thus, to first-order in h_o', P_{2w} is independent of h_o' and varies as $(2s)^{-1/2}$. The wall values for U_2, R_2, and T_2 consist of two terms, one of which is proportional to $(2s)^{-1/2}$, the other to $h_o'(2s)^{1/2}$. This s dependence will be of importance in the next section, when a second-order inner solution is obtained and the skin friction and heat transfer are discussed.

Uniform Axisymmetric Flow

The axisymmetric problem, with $h_o' = 0$, is given by

$$\lambda^2 \frac{\partial^2 \Psi_2}{\partial s^2} - \frac{\partial^2 \Psi_2}{\partial n^2} = -\frac{1}{r_w + n} \frac{\partial \Psi_2}{\partial n} \tag{23.79}$$

with Equation (23.74) as the boundary condition. This equation can be put in a standard form, used in the theory for slender bodies of revolution, with the substitution

$$\hat{n} = n + r_w$$

A potential function is normally used in slender-body theory, whereas we have used a stream function. This is the reason the signs in Equation (23.79) are different from those in slender-body theory. Presentations of this theory can be found, e.g., in Liepmann and Roshko (1957, Chapter 9) and in Ward (1955, Chapter 9). It can be used for both subsonic and supersonic flows.

23.8 EXAMPLE: SECOND-ORDER INNER EQUATIONS

Wall Pressure

The second-order boundary-layer equations are formulated for nonuniform, supersonic flow over a flat plate in which the magnitude of h_o' is small compared to unity. In this circumstance, the second-order outer solution, at the wall, is given by Equations (23.77c) and (23.78). The first-order boundary-layer solution is provided in Section 23.6; it is the $\beta = 0$ similar solution for a compressible flow.

From Section 23.6, we have

$$U_{1w} = P_{1w} = R_{1w} = T_{1w} = 1, \qquad H_{1w}' = S_{1w}' = h_o'$$

and Equation (23.49b) reduces to

$$p_2(s, \bar{n}) - (T_2 + R_2)_w = P_{2w}$$

With Equations (23.78), this becomes

$$p_2(s, \bar{n}) = \frac{\gamma M_\infty^2 C^*}{\lambda (2s)^{1/2}}$$

and the nondimensional pressure, to second order, is

$$p(s, \bar{n}) = 1 + \frac{\gamma M_\infty^2 C^*}{\lambda (2Re_s)^{1/2}}$$

where

$$Re_s = (Re)s \tag{23.80}$$

Hence, the second-order pressure is uniform across the boundary layer, and it slightly exceeds its first-order value, with the increment decreasing with s. A second-order pressure coefficient, Equation (23.63), can be written as

$$c_p = \frac{2C^*}{\lambda(2Re_s)^{1/2}} = \frac{\delta_1^*}{(M_\infty^2 - 1)^{1/2} s} \tag{23.81}$$

which shows its dependence on the first-order displacement thickness and the λ factor. Although h_o' is not zero, the wall pressure does not depend on h_o', at least when h_o' is small.

Second-Order Inner Equations

The term

$$\frac{\partial}{\partial s}(t_1\rho_2 + \rho_1 t_2)$$

appears in both Equations (23.48b,c). With the aid of Equations (23.49), we have

$$t_1\rho_2 + \rho_1 t_2 = P_{2w} \tag{23.82a}$$

where P_{2w} is given by Equation (23.78a), and

$$\frac{\partial}{\partial s}(t_1\rho_2 + \rho_1 t_2) = -\frac{P_{2w}}{2s} = -\frac{\gamma M_\infty^2 C^*}{\lambda(2s)^{3/2}} \tag{23.82b}$$

which will be used to simplify Equations (23.48). Equation (23.82a) is first used to eliminate t_2, with the result

$$t_2 = \frac{P_{2w}}{\rho_1} - \frac{\rho_2}{\rho_1^2} \tag{23.82c}$$

since $\rho_1 t_1 = 1$. Recall that this elimination was discussed beneath Equations (23.49a,b.)

Thus far, only easily procured results have been obtained. Further progress requires that we decide on a choice of coordinates for the second-order boundary-layer equations. We could use ξ or s and, for the transverse direction, η or $\bar{n}$. Since the boundary conditions, both at the wall and at infinity, are in terms of s, we will utilize this variable. Because the coefficients in Equations (23.48) stem from the first-order inner solution, the other coordinate to be used is η. It is worth noting that we are not necessarily trying to find a second-order similarity solution. At this time, our motivation for choosing s, η coordinates is to obtain from Equations (23.48) as simple a set of equations as is possible.

We shall need to replace $\bar{n}$ with η in these equations as well as in the $\bar{n} \to \infty$ boundary conditions. Problem 21.9, with $\beta = 0$, yields

$$\Upsilon n = \frac{1}{1-S}[g_w\eta + (1-g_w)f - S(f'' + ff' - f_w'')]$$

which here becomes (see Equation 23.69b)

$$\frac{\bar{n}}{(2s)^{1/2}} = N$$

where N is defined as

$$N(\eta; g_w, S) = \frac{1}{1-S}[g_w\eta + (1-g_w)f - S(f'' + ff' - f''_w)]$$

We also have

$$\frac{\partial}{\partial \bar{n}} = \frac{\rho_1}{(2s)^{1/2}} \frac{\partial}{\partial \eta}$$

and with $f \sim \eta - C_v$, $f' \sim 1$, $f'' \sim 0$

$$\frac{\bar{n}}{(2s)^{1/2}} \sim \eta, \qquad \bar{n}, \eta \to \infty \tag{23.83}$$

With the aid of Problem 23.6, we can write

$$v_1 = \frac{1}{(2s)^{1/2}\rho_1}(N\rho_1 f' - f)$$

and

$$\frac{\partial v_1}{\partial \bar{n}} = \frac{1}{2s}\left(N\rho_1 f'' + \frac{\rho_1'}{\rho_1} f\right)$$

The combination, $N\rho_1 f' - f$, that appears in v_1 frequently occurs in the analysis. Problem 23.8 shows that it equals

$$N\rho_1 f' - f = \frac{\rho_1}{1-S}[g_w(\eta f' - f) - S(f'' - f''_w) f'] \tag{23.84}$$

We also note that

$$\mu_1 = t_1 = \frac{1}{\rho_1} = \frac{g_w + (1-g_w)f' - Sf'^2}{1-S}$$

and

$$\mu_1' = \frac{d\mu_1}{dt_1} = 1$$

Hence, we have

$$\frac{\partial t_1}{\partial \bar{n}} = \frac{\rho_1}{(2s)^{1/2}} \frac{d}{d\eta}\left(\frac{1}{\rho_1}\right) = -\frac{1}{(2s)^{1/2}} \frac{\rho_1'}{\rho_1}$$

where

$$\frac{\rho_1'}{\rho_1} = -\frac{(1-g_w-2Sf')f''}{g_w+(1-g_w)f'-Sf'^2}$$

A number of other derivatives of consequence are

$$\frac{\partial u_1}{\partial \bar{n}} = \frac{\rho_1 f''}{(2s)^{1/2}}$$

$$\frac{\partial^2 u_1}{\partial \bar{n}^2} = \frac{\rho_1^2}{2s}\left(\frac{\rho_1'}{\rho_1} - f\right)f''$$

$$\frac{\partial u_2}{\partial \bar{n}} = \frac{\rho_1}{(2s)^{1/2}} \frac{\partial u_2}{\partial \eta}$$

$$\frac{\partial^2 u_2}{\partial \bar{n}^2} = \frac{\rho_1^2}{2s}\left(\frac{\partial^2 u_2}{\partial \eta^2} + \frac{\rho_1'}{\rho_1}\frac{\partial u_2}{\partial \eta}\right)$$

$$\frac{\partial t_2}{\partial \bar{n}} = \frac{1}{(2s)^{1/2}}\left(\frac{2\rho_1'\rho_2}{\rho_1^2} - \frac{1}{\rho_1}\frac{\partial \rho_2}{\partial \eta} - P_{2w}\frac{\rho_1'}{\rho_1}\right)$$

where Equation (18.10) is used to simplify the $\partial^2 u_1/\partial \bar{n}^2$ equation.

The inner region substantial derivatives, Equation (23.11a), become

$$\frac{d}{d_i t} = u_i \frac{\partial}{\partial s} + \frac{\rho_1 v_i}{(2s)^{1/2}} \frac{\partial}{\partial \eta}, \qquad i = 1, 2$$

We thereby obtain:

$$\frac{du_1}{d_1 t} = \frac{1}{2s}(N\rho_1 f' - f)f''$$

$$\frac{du_1}{d_2 t} = \frac{\rho_1 f''}{(2s)^{1/2}}$$

$$\frac{du_2}{d_1 t} = f'\frac{\partial u_2}{\partial s} + \frac{1}{2s}(N\rho_1 f' - f)\frac{\partial u_2}{\partial \eta}$$

$$\frac{dt_1}{d_1 t} = -\frac{1}{2s}\frac{\rho_1'}{\rho_1^2}(N\rho_1 f' - f)$$

$$\frac{dt_1}{d_2 t} = -\frac{1}{(2s)^{1/2}}\frac{\rho_1'}{\rho_1}v_2$$

$$\frac{dt_2}{d_1 t} = -\frac{P_{2w}}{2s\rho_1}\left[f' + \frac{\rho_1'}{\rho_1}(N\rho_1 f' - f)\right] + \frac{2}{(2s)}\frac{\rho_1'}{\rho_1^3}(N\rho_1 f' - f)\rho_2 - \frac{f'}{\rho_1^2}\frac{\partial \rho_2}{\partial s}$$
$$- \frac{1}{2s\rho_1^2}(N\rho_1 f' - f)\frac{\partial \rho_2}{\partial \eta}$$

With the assistance of the foregoing relations (see Problem 23.8), Equations (23.52) now take the form

$$\frac{\rho_1 \rho_1' v_2}{(2s)^{1/2}} + \frac{1}{2s}\left(N\rho_1 f'' + \frac{\rho_1'}{\rho_1} f\right)\rho_2 + \rho_1 \frac{\partial u_2}{\partial s} + f'\frac{\partial \rho_2}{\partial s} + \frac{\rho_1^2}{(2s)^{1/2}}\frac{\partial v_2}{\partial \eta} + \frac{1}{2s}(N\rho_1 f' - f)\frac{\partial \rho_2}{\partial \eta} = 0 \tag{23.85a}$$

$$\frac{\rho_1^2 f'' v_2}{(2s)^{1/2}} + \frac{f''}{2s}\left(N\rho_1 f' - 2f - \frac{\rho_1'}{\rho_1}\right)\rho_2 + \rho_1 f'\frac{\partial u_2}{\partial s} + \frac{\rho_1}{2s}(N\rho_1 f' - f)\frac{\partial u_2}{\partial \eta} + \frac{f''}{2s}\frac{\partial \rho_2}{\partial \eta} - \frac{\rho_1}{2s}\frac{\partial^2 u_2}{\partial \eta^2} = \frac{P_{2w}}{2s}\left(\frac{1}{\gamma M_\infty^2} - \rho_1 f f''\right) \tag{23.85b}$$

$$\begin{aligned}
&-\frac{\rho_1' v_2}{(2s)^{1/2}} + \frac{1}{2s\rho_1}\left[\frac{\rho_1'}{\rho_1}(N\rho_1 f' - f) - 3\frac{\rho_1''}{\rho_1} + 9\left(\frac{\rho_1'}{\rho_1}\right)^2 + (\gamma - 1)M_\infty^2 \rho_1 f''^2\right]\rho_2 - \frac{f'}{\rho_1}\frac{\partial \rho_2}{\partial s} \\
&- \frac{2(\gamma-1)M_\infty^2 \rho_1 f''}{(2s)}\frac{\partial u_2}{\partial \eta} - \frac{1}{2s\rho_1}\left(N\rho_1 f' - f + 5\frac{\rho_1'}{\rho_1}\right)\frac{\partial \rho_2}{\partial \eta} + \frac{1}{2s\rho_1}\frac{\partial^2 \rho_2}{\partial \eta^2} \\
&= \frac{P_{2w}}{2s}\left[\frac{1}{\gamma}f' + \frac{\rho_1'}{\rho_1}(N\rho_1 f' - f) - 2\frac{\rho_1''}{\rho_1} + 4\left(\frac{\rho_1'}{\rho_1}\right)^2 + (\gamma-1)M_\infty^2 \rho_1 f''^2\right]
\end{aligned} \tag{23.85c}$$

These are the flat plate equations for u_2, v_2, and ρ_2. Although Equations (23.52) are homogeneous, the replacement of t_2 with ρ_2 introduces inhomogeneous terms that are proportional to P_{2w}.

Boundary Conditions

Boundary conditions at the wall and at infinity are required for u_2, v_2, and ρ_2. Since Equations (23.85) are fifth order with respect to η in these variables, five η boundary conditions are required. In Section 23.5, these conditions were formulated for t_2 rather than ρ_2. Equation (23.82a) is used to transfer them to ρ_2. With

$$\rho_{1w} = \frac{1}{t_{1w}} = \frac{1-S}{g_w} = \frac{1}{X_\infty g_w}, \qquad \rho_1(\infty) = \frac{1}{t_1(\infty)} = 1$$

we obtain

$$\rho_2(s,0) = \frac{c_{-1}^{(s)}}{(2s)^{1/2}} - \frac{t_2(s,0)}{(X_\infty g_w)^2} \tag{23.86a}$$

$$\rho_2(s,\infty) = \frac{\gamma M_\infty^2 C^*}{\lambda (2s)^{1/2}} - t_2(s,\infty) \tag{23.86b}$$

where the frequently encountered constant is

$$c_{-1}^{(s)} = \frac{\gamma M_\infty^2 C^*}{\lambda X_\infty g_w} \tag{23.86c}$$

(This parameter is associated with several effects, not just velocity slip.)

The decomposition in Section 23.5 is utilized. Hence, the velocity slip wall condition for u_2 stems from Equation (23.56a) and

$$\left(\frac{\partial u_1}{\partial \bar{n}}\right)_w = \left[\frac{\rho_1 f''}{(2s)^{1/2}}\right]_w = \frac{f''_w}{g_w X_\infty (2s)^{1/2}}$$

The density conditions stem from Equations (23.56b,c) and (23.86). We thus obtain for velocity slip

$$u_2^{(s)}(s,0) = \gamma^{1/2} M_\infty (X_\infty g_w)^{1/2} \frac{f''_w}{(2s)^{1/2}}$$

$$v_2^{(s)}(s,0) = 0$$

$$\rho_2^{(s)}(s,0) = \frac{c_{-1}^{(s)}}{(2s)^{1/2}}$$

$$u_2^{(s)}(s,\infty) = 0$$

$$\rho_2^{(s)}(s,\infty) = \frac{\gamma M_\infty^2 C^*}{\lambda (2s)^{1/2}}$$

The temperature jump conditions are

$$u_2^{(T)}(s,0) = v_2^{(T)}(s,0) = 0 \tag{23.87a}$$

$$\rho_2^{(T)}(s,0) = \frac{1}{(2s)^{1/2}}\left[c_{-1}^{(s)} - \frac{\gamma^{1/2}}{g_w^{3/2} X_\infty^{1/2}} M_\infty (1-g_w) f''_w\right] \tag{23.87b}$$

$$u_2^{(T)}(s,\infty) = 0 \tag{23.87c}$$

$$\rho_2^{(T)}(s,\infty) = \frac{\gamma M_\infty^2 C^*}{\lambda (2s)^{1/2}} \tag{23.87d}$$

where the quantity within the square brackets in Equation (23.87b) is a constant that depends only on γ, M_∞, and g_w.

For each of the effects associated with the entropy gradient, stagnation enthalpy gradient, and displacement, we have at the wall

$$u_2^{(e)}(s,0) = u_2^{(H)}(s,0) = u_2^{(d)}(s,0) = 0$$

$$v_2^{(e)}(s,0) = v_2^{(H)}(s,0) = v_2^{(d)}(s,0) = 0$$

$$\rho_2^{(e)}(s,0) = \rho_2^{(H)}(s,0) = \rho_2^{(d)}(s,0) = \frac{c_{-1}^{(s)}}{(2s)^{1/2}}$$

For the infinity conditions, Equation (23.83) is utilized. Hence, the entropy gradient conditions are

$$u_e^{(e)} \sim -\frac{(2s)^{1/2}\eta}{(\gamma-1)M_\infty^2}, \qquad \eta \to \infty \tag{23.88a}$$

$$\rho_2^{(e)} \sim -(2s)^{1/2}\eta, \qquad \eta \to \infty \tag{23.88b}$$

For the stagnation enthalpy gradient, the infinity conditions are given by

$$u_2^{(H)} \sim \frac{X_\infty}{(\gamma-1)M_\infty^2}(2s)^{1/2}\eta, \qquad \eta \to \infty \tag{23.89a}$$

$$\rho_2^{(H)}(s,\infty) = \frac{\gamma M_\infty^2 C^*}{\lambda(2s)^{1/2}} \tag{23.89b}$$

Finally, for displacement, we have

$$u_2^{(d)}(s,\infty) = U_2(s,0) = -\frac{C^*}{\lambda(2s)^{1/2}} - \frac{1}{2}C^* h_o' (2s)^{1/2}$$

$$\rho_2^{(2)}(s,\infty) = \frac{\gamma M_\infty^2 C^*}{\lambda(2s)^{1/2}} - T_2(s,0) = \frac{M_\infty^2 C^*}{\lambda(2s)^{1/2}} + C^* h_o' (2s)^{1/2}$$

In contrast to other infinity conditions, $u_2^{(d)}$ and $\rho_2^{(d)}$ are proportional to both $(2s)^{-1/2}$ and $h_0'(2s)^{1/2}$ terms.

Discussion

Examination of Equations (23.85) and their boundary conditions reveals that $(2s)^{1/2}$ only appears with a positive or negative integer exponent. In view of Equation (23.78a), the nonzero right sides of Equations (23.85) contain the multiplicative factor $(2s)^{-3/2}$. On the left side of these equations, s appears in the following combinations:

$$\frac{\partial u_2}{\partial s}, \qquad \frac{1}{2s}\frac{\partial u_2}{\partial \eta}, \qquad \frac{1}{2s}\frac{\partial^2 u_2}{\partial \eta^2}$$

$$\frac{v_2}{(2s)^{1/2}}, \qquad \frac{1}{(2s)^{1/2}}\frac{\partial v_2}{\partial \eta}$$

$$\frac{1}{2s}\rho_2, \qquad \frac{\partial \rho_2}{\partial s}, \qquad \frac{1}{2s}\frac{\partial \rho_2}{\partial \eta}, \qquad \frac{1}{2s}\frac{\partial^2 \rho_2}{\partial \eta^2}$$

Hence, a solution of the form

$$u_2(s,\eta) = a_1(2s)^{1/2}\tilde{u}_1(\eta) + \frac{a_{-1}}{(2s)^{1/2}}\tilde{u}_{-1}(\eta) \tag{23.90a}$$

$$v_2(s,\eta) = \tilde{v}_1(\eta) + \frac{1}{2s}\tilde{v}_{-1}(\eta) \tag{23.90b}$$

$$\rho_2(s,\eta) = c_1(2s)^{1/2}\tilde{\rho}_1(\eta) + \frac{c_{-1}}{(2s)^{1/2}}\tilde{\rho}_{-1}(\eta) \tag{23.90c}$$

is appropriate. The a_i and c_i are constants that will be chosen to simplify some of the boundary conditions for the $\tilde{u}_i$ and $\tilde{\rho}_i$, $i = \pm 1$. (The a_1 and c_1 parameters should not be confused with their similarly denoted counterparts in Section 23.4. In addition, $\tilde{\rho}_i$, $\tilde{u}_i$,... should not be confused with ρ_i, u_i,..., which are first-order variables.)

The above relations reduce Equations (23.85) to a system of ODEs, with one system per second-order effect. The ODEs will then satisfy boundary conditions that stem from the ones previously derived. We thus obtain a similarity solution; initial conditions, at some upstream location, are therefore unnecessary. Generally, the second-order boundary-layer equations do not possess a similarity solution. In the present example, such a solution is obtained because β and g_w are independent of s and the first-order boundary-layer equations possess a similarity solution.

Appendix J

Equations (23.90) are substituted into Equations (23.85), with the result that each term in Equations (23.85) is proportional to either $(2s)^{-1/2}$ or to $(2s)^{-3/2}$. The unity-subscripted variables appear with a $(2s)^{-1/2}$ coefficient, while the 21 subscripted variables appear with a $(2s)^{-3/2}$ coefficient. For instance, the leftmost term in Equation (23.85a) becomes

$$\frac{\rho_1\rho_1'}{(2s)^{1/2}}\tilde{v}_1 + \frac{\rho_1\rho_1'}{(2s)^{3/2}}\tilde{v}_{-1}$$

Each PDE is thereby reduced to two ODEs; these are shown in Appendices J1 and J2. Hence, we have six linear ODEs in which the $\tilde{u}_1, \tilde{v}_1, \tilde{\rho}_1 (\tilde{u}_{-1}, \tilde{v}_{-1}, \tilde{\rho}_{-1})$ set is fifth-order. The two sets are not coupled and only the J2 set possesses inhomogeneous terms, as shown on its right side. The formulation is completed in J3–J7, which provide values for the constants in Equations (23.90) and the requisite boundary conditions.

For velocity slip and temperature jump, Appendices J3 and J4, the boundary conditions only involve $(2s)^{-1/2}$. As a consequence of this and the fact that the two sets of ODEs are uncoupled, we can set $\tilde{u}_1$, $\tilde{v}_1$, $\tilde{\rho}_1$ equal to zero for these two effects. The corresponding values for $a_1^{(s)}, c_1^{(s)}, a_1^{(T)}$, and $c_1^{(T)}$ are unnecessary. As indicated, the a_{-1}, and c_{-1} coefficients in Equations (23.90) are written with an s or T superscript. The particular values shown in J3 stem from Equations (23.90) and the appropriate boundary conditions, written as

$$u_2^{(s)}(s,0) = \frac{a_{-1}^{(s)}}{(2s)^{1/2}}\tilde{u}_{-1}^{(s)}(0) = \gamma^{1/2}M_\infty(g_w X_\infty)^{1/2}\frac{f''_w}{(2s)^{1/2}}$$

$$u_2^{(s)}(s,\infty) = \frac{a_{-1}^{(s)}}{(2s)^{1/2}}\tilde{u}_{-1}^{(s)}(\infty) = 0$$

$$\rho_2^{(s)}(s,0) = \frac{c_{-1}^{(s)}}{(2s)^{1/2}}\tilde{\rho}_{-1}^{(s)}(0) = \frac{c_{-1}^{(s)}}{(2s)^{1/2}}$$

$$\rho_2^{(s)}(s,\infty) = \frac{c_{-1}^{(s)}}{(2s)^{1/2}}\tilde{\rho}_{-1}^{(s)}(\infty) = \frac{\gamma M_\infty^2 C^*}{\lambda(2s)^{1/2}}$$

where the subscript 2 is now shown for purposes of clarity.

Appendices J5 to J7 deal with the entropy gradient, enthalpy gradient, and displacement, respectively. A boundary condition of the form $\tilde{u}_1(\eta) \sim \eta$ implies that $\eta \to \infty$. This asymptotic condition can be removed by replacing the $\tilde{u}_1(\eta)$ term in Equation (23.90a) with $\eta\tilde{u}_1(\eta)$, and the infinity condition now becomes $(\tilde{u}_1(\infty) = 1)$. Unfortunately, the corresponding wall condition has $\tilde{u}_1(0)$ equal to an unknown constant. This replacement is, therefore, of no advantage.

The solution of the J1 equations, which are homogeneous, is not expected to be unique. Additional solutions, termed eigensolutions, should occur. As discussed in Van Dyke (1975) with regard to the Blasius equation, this nonuniqueness is to be expected for similarity solutions, since initial conditions, which are required for a unique solution, are not imposed. In view of this, we ignore all eigensolutions; i.e., we adopt (Van Dyke, 1975) the "principle of minimum singularity." In other words, among the class of solutions that satisfy a given equation and its boundary conditions, only the least singular one is chosen. This is the solution normally obtained without concern about the presence of eigensolutions.

Discussion

Values for γ, M_∞, and g_w are required for all five effects. With these parameters and $\beta = 0$, all the constants in Appendices J3 to J7, except h_o', are known; i.e., we can determine f_w'', $c_{-1}^{(s)}$, C^*,…. Observe that the first-order displacement parameter, C^*, appears in each of these appendices, whereas h_o' appears only in the displacement problem, Appendix J7. This is somewhat surprising, since h_o' is a measure of the entropy and stagnation enthalpy gradients. In view of Equation (23.50), h_o' will appear with these two effects, e.g., when we evaluate the skin friction and heat transfer at the wall. The point of these remarks is that, despite the use of superposition, physical phenomena couple. (This was also noted in Section 23.5 with regard to the transverse curvature.) For instance, the gradient h_o' will affect the skin friction and heat transfer associated with displacement. Similarly, the displacement parameter C^* appears in all five second-order effects.

As noted in Section 23.2, the second-order inner equations are not unique. Equally important, the decomposition used with these equations is also not unique. For instance, Van Dyke (1962) does not eliminate either ρ_2 or t_2; hence, his analysis utilizes four differential operators instead of the three we use. As a consequence, numerical results for, say, $u^{(d)}$ in Equation (23.50) will differ between approaches. It is not always possible, therefore, to compare results among different authors for what is presumably the same second-order effect (Adams, 1968). Moreover, as observed in the preceding paragraph, the skin friction and heat transfer associated with displacement, for instance, is affected by h_o', while the two gradient effects depend on C^*.

Composite second-order quantities, such as u_2, p_2, the net skin friction, and the net heat transfer, should be independent of the specific mathematical (Adams, 1968) or numerical approach. Again, there is an exception to this rule, since the specific treatment of the nonunique, second-order inner equations will alter composite results. Nevertheless, it is composite quantities that should be compared among authors, with CFD results, and with experiments. Of course, we can still attribute a certain fraction of, say, the second-order skin-friction coefficient to velocity slip or displacement, with the understanding that this attribution is only conceptual and is limited to the particular method of decomposition.

As usual in boundary-layer theory, the second-order equations constitute a two-point boundary value problem, and an iterative numerical approach is required. In this procedure, two wall values are guessed, i.e., $\tilde{u}_1(\tilde{u}_{-1}')$ and $\tilde{\rho}_1'(\tilde{\rho}_{-1}')$ for the Appendix J, Section 1 (Section 2) equations, and these equations are then integrated starting at the wall. The iteration terminates when the two replaced infinity conditions are simultaneously satisfied. In contrast to the nonlinear first-order equations, convergence should be rapid because a linear combination of previous guesses can be used. (In this regard, we note that a numerical solution of the equations in Appendix J thus far, has not been attempted.) The principal results of the integration are numerical values for $\tilde{u}_{-1w}^{(s)\prime}, \tilde{u}_{-1w}^{(T)\prime}, \ldots, \tilde{u}_{-1w}^{(e)\prime}, \ldots, \tilde{u}_{-1w}^{(H)\prime}, \ldots$, which are needed for the skin-friction coefficient and Stanton number.

Another complicating factor is the need to simultaneously obtain a solution to Equations (18.10) and (18.16). For a flat plate, this complication can be avoided by using the approximate solution

(Bush, 1961) outlined in Problem 21.3. This solution has the form

$$f = \eta\,\mathrm{erf}(a\eta) + \frac{1}{\pi^{1/2}a}(e^{-a^2\eta^2} - 1) \quad \textbf{(23.91a)}$$

$$f' = \mathrm{erf}(a\eta) \quad \textbf{(23.91b)}$$

$$f'' = \frac{2a}{\pi^{1/2}}e^{-a^2\eta^2} \quad \textbf{(23.91c)}$$

where, by setting $a = 0.4162$, f'' equals its exact wall value of 0.4696. Equations (23.91) satisfy all the requisite boundary conditions and also provides an exponential behavior as η becomes large.

More generally, if the first-order boundary layer is nonsimilar, the simultaneous solution of the first- and second-order equations is particularly difficult. In this case, as discussed in Section 23.2, the numerical solution of the equations that combine first- and second-order terms is advantageous. A comparison of first- and second-order results can still be made by solving Equations (23.7) with and without the second-order terms.

Skin-Friction Coefficient

We have already evaluated the pressure coefficient c_p, Equation (23.81), to second order. The skin-friction coefficient is now similarly evaluated. In comparison to c_p, the formulas to be established for c_f (and *St*) are far more complicated. For the skin-friction coefficient, given by Equation (23.60), we first note that

$$c_{f1} = 2\varepsilon\tau_{w1} = \left(\frac{2}{Re_s}\right)^{1/2} f_w''$$

where Re_s is given by Equation (23.80) and

$$\tau_{w1} = \frac{f_w''}{(2s)^{1/2}}$$

The quantity $t^{(T)}(s, 0)$ in τ_{w2}, Equation (23.59b), is given by Equation (23.57b). It involves

$$\left(\frac{\mu_1 t_1^{1/2}}{p_1}\right)_w = (X_\infty g_w)^{3/2}$$

and

$$\left(\frac{\partial t_1}{\partial \bar{n}}\right)_w = -\frac{1}{X_\infty g_w(2s)^{1/2}}\left(\frac{\rho_1'}{\rho_1^2}\right)_w = \frac{(1 - g_w)f_w''}{g_w(2s)^{1/2}}$$

With the aid of Equation (23.90a) and Appendix J, after some effort (see Problem 23.9a), we obtain

$$\overset{*}{\tau}_{w2} = 2s\tau_{w2} = \Bigg\{ a_1\gamma^{1/2}M_\infty(X_\infty g_w)^{1/2} f''_w \, \tilde{u}_{-1}^{(s)'} + c_1\left[\tilde{u}_{-1}^{(T)'} + \left(\frac{\gamma}{g_w^3 X_\infty}\right)^{1/2} \times M_\infty(1-g_w)f''^2_w\right] + h'_o\left(\tilde{u}_{-1}^{(e)'} + \tilde{u}_{-1}^{(H)'}\right) - \frac{C^*}{\lambda}\tilde{u}_{-1}^{(d)'}$$

$$- \frac{h'_o}{(\gamma-1)M_\infty^2}\left[\tilde{u}_1^{(e)'} - X_\infty \tilde{u}_1^{(H)'} + \frac{\gamma-1}{2}M_\infty^2 C^* \tilde{u}_1^{(d)'}\right](2s)\Bigg\}_w \qquad \textbf{(23.92)}$$

where a_1 and c_1 are the velocity slip and temperature jump coefficients of Section 23.4, respectively, and $\tilde{u}_{-1w}^{(s)'}$, e.g., is the η derivative of the velocity slip component of $\tilde{u}_{-1}$, evaluated at the wall. Consequently, to second order the skin-friction coefficient is

$$c_f = \left(\frac{2}{Re_s}\right)^{1/2} f''_w + \frac{1}{Re_s}\overset{*}{\tau}_{w2} \qquad \textbf{(23.93)}$$

Stanton Number

For the Stanton number evaluation, Equation (23.62), remember that t_2 is written as

$$t_2 = a_1 t^{(s)} + c_1 t^{(T)} + h'_o(t^{(e)} + t^{(H)}) + t^{(d)}$$

Equation (23.82a) is used to replace t_2 with ρ_2. For the evaluation of q_{w2}, Equation (23.61b), we use Equation (23.90c), Appendix J, and set $Pr = 1$. We finally obtain [see Problem 23.9(b)], again to second order,

$$St = \frac{f''_w}{(2Re_s)^{1/2}} - \frac{1}{2Re_s}\overset{*}{q}_{w2} \qquad \textbf{(23.94)}$$

where

$$\overset{*}{q}_{w2} = \frac{2sq_{w2}}{X_\infty(1-g_w)} = \sum_{i=1}^{4} q_i \qquad \textbf{(23.95)}$$

and the q_i are

$$q_1 = \frac{\gamma M_\infty^2 C^* f''_w}{\lambda}(a_1 + c_1 + 2h'_o + 1)$$

$$q_2 = -\gamma^{1/2}M_\infty X_\infty^{1/2} f''_w\left[a_1\frac{(\gamma-1)M_\infty^2 g_w^{3/2}}{1-g_w} + 3c_1\frac{1-g_w}{g_w^{1/2}}f''_w\right]$$

$$q_3 = \frac{X_\infty g_w^2}{1-g_w}\Bigg\{c_{-1}^{(s)}\left[a_1\tilde{\rho}_{-1}^{(s)'} + c_1\tilde{\rho}_{-1}^{(T)'} + h'_o\left(\tilde{\rho}_{-1}^{(e)'} + \tilde{\rho}_{-1}^{(H)'}\right) + \frac{X_\infty g_w}{\gamma}\tilde{\rho}_{-1}^{(d)'}\right]$$

$$- c_1\left(\frac{\gamma}{g_w^3 X_\infty}\right)^{1/2} M_\infty(1-g_w)\, f''_w\, \tilde{\rho}_{-1}^{(T)'}\Bigg\}_w$$

$$q_4 = \frac{h'_o X_\infty g_w^2}{1-g_w}\left[-\tilde{\rho}_1^{(e)'} + c_{-1}^{(s)} X_\infty g_w \tilde{\rho}_1^{(H)'} + C^*\tilde{\rho}_1^{(d)'}\right]_w (2s)$$

Discussion

Both τ_{w2}^* and q_{w2}^* explicitly contain terms that are independent of s or linearly depend on $h_o'(2s)$. For instance, the $h_o'(2s)$ terms in q_{w2}^* are provided by q_4. Observe that τ_{w2}^* and q_{w2}^* are divided by Re_s in c_f and St, respectively. Hence, τ_{w2}^* and q_{w2}^* contribute terms to c_f and St that decay as $1/s$, whereas the $h_o'(2s)$ terms are constant. Since c_{f1} and St_1 vary as $(2s)^{1/2}$, the $h_o'(2s)$ terms will ultimately dominate c_f and St, for large s. There was some consternation when this phenomenon was first encountered, since second-order terms were not expected to dominate. Asymptotic expansions, however, need not be convergent. The troublesome $h_o'(2s)$ terms are associated with the $\tilde{u}_1^{(e)'}, \tilde{u}_1^{(H)'}, \tilde{u}_1^{(d)'}$ and $\tilde{\rho}_1^{(e)'}, \tilde{\rho}_1^{(H)'}, \tilde{\rho}_1^{(d)'}$ coefficients and stem from the cumulative effect of external entropy and stagnation enthalpy gradients. These gradients initially have little effect, but they continue to alter the boundary layer as it thickens and engulfs more of the inviscid flow containing these gradients. Hence, the downstream dominance of the $h_o'(2s)$ terms may not be due to nonconvergence; instead, it may be the correct physical picture. Of course, this discussion is for a wall without curvature. In the following subsection, more general flows are briefly discussed.

Evidently, the second-order results for c_f and St are too complicated to be easily interpreted. In particular, the values of $\tilde{u}_{-1w}^{(T)'}, \tilde{u}_{-1w}^{(e)'}, \ldots$ are presently unknown, as are the signs of the various terms. Of course, there is a considerable simplification if we focus on a single second-order effect. For instance, to focus on displacement, set all $\tilde{u}$ and $\tilde{\rho}$ terms equal to zero, except those associated with displacement, and set

$$a_1 = c_1 = h_o' = 0$$

We now obtain

$$\tau_{w2}^{(d)*} = -\frac{C^*}{\lambda}\tilde{u}_{-1w}^{(d)'} \tag{23.96a}$$

and

$$q_{w2}^{(d)*} = \frac{M_\infty^2 C^*}{\lambda}\left[f_w'' + \frac{X_\infty g_w^2}{1-g_w}\tilde{\rho}_{-1w}^{(d)'}\right] \tag{23.96b}$$

The derivatives of $\tilde{u}_{-1}^{(d)}$ and $\tilde{\rho}_{-1}^{(d)}$, evaluated at the wall, stem from solving the equations in Appendix J2, while the displacement constants and boundary conditions are provided in Appendix J7. With these equations, we obtain (see Problem 23.10)

$$\tilde{v}_{-1w}^{(d)'} = \frac{\gamma M_\infty^2 C^* X_\infty (1-g_w) f_w''}{\lambda} \tag{23.97a}$$

$$\tilde{u}_{-1w}^{(d)''} + M_\infty^2 X_\infty g_w f_w'' \tilde{\rho}_{-1w}^{(d)'} = X_\infty g_w \left[1 - \frac{\gamma M_\infty^2 - (1-g_w) f_w''^2}{X_\infty g_w^2}\right] \tag{23.97b}$$

$$2(\gamma-1)\tilde{u}_{-1w}^{(d)'} + 5X_\infty^2 g_w (1-g_w)\tilde{\rho}_{-1w}^{(d)'} + \frac{(X_\infty g_w)^2}{f_w''}\tilde{\rho}_{-1w}^{(d)''} =$$

$$3\gamma\left[X_\infty \frac{(1-g_w)^2}{g_w} f_w'' - (\gamma-1)M_\infty^2\right] \tag{23.97c}$$

for various derivatives at the wall. While these equations are useful for checking purposes, they do not provide insight with regard to the derivatives that appear in Equations (23.96).

Discussion of General Trends

We conclude with a brief discussion of a few trends. For instance, Hayes and Probstein (1959) indicate that if external vorticity causes U_1 to increase with n, then vorticity interaction increases both the skin friction and the heat transfer. In our analysis this would correspond to h_o' being positive; see Equation (23.64a). As a general conclusion, it is relatively self-evident, since vorticity, in this circumstance, is accelerating the flow in the boundary layer.

In view of the previous emphasis on flat-plate flow, the remaining discussion is limited to a blunt body in a hypersonic flow. Aside from a sphere (Van Dyke, 1962; Davis and Flügge-Lotz, 1964a) or cylinder (Fannelop and Flügge-Lotz, 1966), two other axisymmetric shapes (Adams, 1968; Davis and Flügge-Lotz, 1964a) have received some attention. These are the paraboloid

$$x = \frac{1}{2}r^2 \tag{23.98a}$$

and the hyperboloid

$$(\tan\theta_\infty)^4 x^4 = 1 + (\tan\theta_\infty)^2 r^2 \tag{23.98b}$$

(A paraboloid or hyperboloid is simply a parabola or hyperbola that is rotated about its symmetry axis.) In the above, all distances are normalized by the radius of curvature of the surface at the stagnation point, the freestream velocity is in the direction of positive x, and r is the radial coordinate. The hyperboloid surface is asymptotic to a cone of half-angle θ_∞ as x becomes infinite. For subsequent hyperboloid results, θ_∞ equals 22.5°. For the paraboloid, the shock steadily weakens until it becomes a Mach wave far downstream. On the other hand, the hyperboloid shock asymptotically weakens to that of a shock for a cone of half-angle θ_∞. As a consequence, we anticipate different magnitudes for the second-order effects for the two bodies as s increases.

Without exception, all such analysis assumes air as a perfect gas with $\gamma = 1.4$. Since the upstream flow is uniform, there is no stagnation enthalpy gradient. However, the curved, detached bow shock generates an entropy gradient or, alternatively, external vorticity. Problem 23.11, however, shows that a symmetric, two-dimensional flow has a zero value for the entropy gradient (or the vorticity) at the stagnation point and along the downstream wall. Hence, only an axisymmetric surface exhibits a second-order entropy gradient effect.

In the study by Adams (1968), which utilizes Equations (23.98), the freestream Mach number is 10, the Reynolds number, based on the radius of curvature of the nose, is 400, and the wall is either highly cooled ($g_w = 0.2$), or moderately cooled ($g_w = 0.6$). Both the local skin-friction coefficient and Stanton number are increased by second-order effects at the stagnation point and along the downstream wall. The increase is sometimes modest, in part because some effects are small and others are of opposite sign and tend to cancel each other. (This is not always the case, as discussed shortly.) For the skin friction, for instance, the entropy gradient and displacement effects are sometimes of comparable magnitude but of opposite sign; the other four effects are relatively small. On the other hand, displacement and longitudinal curvature are of minor importance for the heat transfer, whereas the other effects are roughly of a comparable magnitude.

On the hyperboloid, downstream of about 4 nose radii as measured from the stagnation point, the second-order skin-friction coefficient starts to substantially exceed its first-order counterpart (Van Dyke, 1969; Adams, 1968). To a lesser extent, the same behavior occurs for a paraboloid downstream of about 16 nose radii. [For a cold wall ($g_w = 0.2$), there is also a significant difference in the downstream first- and second-order Stanton numbers for a hyperboloid.] This phenomenon is due to the presence of terms in c_{f2} that are independent of s or grow with s.

Asymptotic expansions are not expected to be convergent. Moreover, it is pointless to discuss convergence or nonconvergence of expansions that are truncated after only two terms. In the preceding flat plate discussion of the skin friction and heat transfer, we noted that second-order effects may dominate at large s, since s may enter through a local Reynolds number as well as separately. This trend suggests the question: under what conditions are second-order results more accurate than first-order results with respect to a solution of the Navier–Stokes equations? At this time, CFD should be able to provide an answer. It may depend on M_∞, g_w, the wall configuration, etc. In particular, the analyticity of the body, i.e., a sharp-edged surface vs. a smooth one (flat plate vs. a paraboloid), may be an important difference. The answer may also depend on s. When s and the corresponding Reynolds number are relatively small, second-order terms only contribute a small perturbation to first-order results. In this circumstance, the second-order theory should be more accurate than the first-order theory.

REFERENCES

Adams, J.C., Jr., "Higher Order Boundary-Layer Effects on Analytic Bodies of Revolution," AEDC-TR-68-57 (April 1968).

Afzal, N., "Second-Order Effects in Self-Similar Laminar Compressible Boundary-Layer Flow," *Int. J. Eng. Sci.* 14, 415 (1976).

Back, L.H., "Transonic Laminar Boundary Layers with Surface Curvatures," *Int. J. Heat Mass Transfer* 16, 1745 (1973).

Bush, W.B., "A Method of Obtaining an Approximate Solution of the Laminar Boundary-Layer Equations," *J. Aerosp. Sci.* 28, 350 (1961).

Chapman, S. and Cowling, T.G., *The Mathematical Theory of Non-Uniform Gases,* Cambridge University Press, London, 1960.

Davis, R.T. and Flügge-Lotz, I., "The Laminar Compressible Boundary-Layer in the Stagnation-Point Region of an Axisymmetric Blunt Body Including the Second-Order Effect of Vorticity Interaction," *Int. J. Heat Mass Transfer* 7, 341 (1964).

Davis, R.T. and Flügge-Lotz, I., "Second-Order Boundary-Layer Effects in Hypersonic Flow Past Axisymmetric Blunt Bodies," *J. Fluid Mech.* 20, 593 (1964a).

Emanuel, G., "Bulk Viscosity of a Dilute Polyatomic Gas," *Phys. Fluids A* 2, 2252 (1990).

Emanuel, G., "Effect of Bulk Viscosity on a Hypersonic Boundary Layer," *Phys. Fluids A* 4, 491 (1992).

Fannelop, T.K. and Flügge-Lotz, I., "Two-Dimensional Hypersonic Stagnation Flow at a Low Reynolds Numbers," *Z. Flugwiss.* 13, 282 (1965).

Fannelop, T.K. and Flügge-Lotz, I., "Viscous Hypersonic Flow over Simple Blunt Bodies: Comparison of a Second-Order Theory with Experimental Results," *J. Méchanique* 5, 69 (1966).

Gersten, K. and Gross, J.F., "The Second-Order Boundary Layer Along a Circular Cylinder in Supersonic Flow," *Int. J. Heat Mass Transfer* 16, 2241 (1973).

Gersten, K. and Gross, J.F., "Higher Order Boundary Layer Theory," *Fluid Dyn. Trans.* 7 (II), 7 (1976).

Hayes, W.D. and Probstein, R.F., *Hypersonic Flow Theory,* Academic Press, New York, 1959, p. 339.

Kleinstreuer, C. and Eghlima, A., "Analysis and Simulation of New Approximation Equations for Boundary-Layer Flow on Curved Surfaces," *Math. Comp. Simul.* 27, 307 (1985).

Liepmann, H.W. and Roshko, A., *Elements of Gasdynamics,* John Wiley, New York, 1957.

Maslen, S.H., "On Heat Transfer in Slip Flow," *J. Aeronaut. Sci.* 25, 400 (1958).

Maslen, S.H., "Second-Order Effects in Laminar Boundary Layers," *AIAA J.* 1, 33 (1963).

Rott, N. and Leonard, M., "Vorticity Effect on the Stagnation Point Flow of a Viscous Incompressible Fluid," *J. Aerosp. Sci.* 26, 542 (1959).

Schlichting, H., *Boundary-Layer Theory,* 7th ed., McGraw-Hill, New York, 1979.

Street, R.E., "A Study of Boundary Conditions in Slip-Flow Aerodynamics," in *Rarefield Gas Dynamics,* edited by F.M. Devienne, Vol. 3, Pergamon Press, New York, 1960, p. 276.

Van Dyke, M., "Second-Order Compressible Boundary Layer Theory with Application to Blunt Bodies in Hypersonic Flow," in *Hypersonic Flow Research,* edited by F.R. Riddell, Academic Press, New York, 1962.

Van Dyke, M., "Higher Approximations in Boundary-Layer Theory. Part I. General Analysis," *J. Fluid Mech.* 14, 161 (1963).

Van Dyke, M., "Higher-Order Boundary-Layer Theory," *Annu. Rev. Fluid Mech.* 1, 265 (1969).

Van Dyke, M., *Perturbation Methods in Fluid Mechanics,* The Parabolic Press, Stanford, CA, 1975.

Van Tassell, W.F. and Taulbee, D.B., "Second-Order Longitudinal Curvature Effects in Compressible Laminar Boundary Layers," *AIAA J.* 9, 682 (1971).

Ward, G.N., *Linearized Theory of Steady High-Speed Flow,* Cambridge University Press, London, 1955.

Werle, M.J., "Solutions of Second-Order Boundary-Layer Equations for Laminar Incompressible Flow," Naval Ordinance Laboratory NOL-TR68-19 (January 1968).

Werle, M.J. and Davis, R.T., "Self-Similar Solutions to the Second-Order Incompressible Boundary-Layer Equations," *J. Fluid Mech.* 40, 343 (1970).

PROBLEMS

23.1 **(a)** Derive Equations (23.24) and (23.25).

(b) Prove that S_1 and H_1 are integrals of Equations (23.18).

(c) Derive Equation (23.29).

23.2 **(a)** Start with Equation (23.20) and show that the equations

$$\frac{\partial \Psi_2}{\partial n} = r^{\sigma}(U_1 R_2 + R_1 U_2), \qquad \frac{\partial \Psi_2}{\partial s} = -h_s r^{\sigma}(V_1 R_2 + R_1 V_2)$$

which satisfy Equation (23.32a), are consistent with Equations (23.17).

(b) Show that

$$\frac{D\Psi_2}{D_1 t} + \frac{D\Psi_1}{D_2 t} = 0$$

(c) Derive Equations (23.34).

23.3 Show that Equations (23.34) satisfy the energy equation, Equation (23.32d).

23.4 Derive equations for u_2 and V_2 in terms of the second-order sream function (see Problem 23.2 for the definition of Ψ_2) that have the functional form

$$r^{\sigma} R_1 U_2 = g(\Psi_2; \hat{U}_1, \hat{V}_1, \chi)$$

$$r^{\sigma} R_1 V_2 = f(\Psi_2; \hat{U}_1, \hat{V}_1, \chi)$$

where

$$\hat{U}_1 = \frac{M_\infty U_1}{T_1^{1/2}}, \qquad \hat{V}_1 = \frac{M_\infty V_1}{T_1^{1/2}}$$

$$\chi = \frac{r^{\sigma} R_1}{(\gamma - 1) M_\infty T_1^{1/2}} (\gamma T_1 S_1' - X_\infty H_1')$$

23.5 Write Equations (23.49) in terms of the boundary-layer thicknesses of Chapter 21 when $\bar{n} = 0$.

23.6 **(a)** Use the theory in Chapter 21, assuming $\beta = 0$, to show that

$$v_1(s, \bar{n}) = \frac{\bar{n} f}{2s} - \frac{f}{(2s)^{1/2} \rho_1}$$

(b) Start with Equation (23.46) and derive Equation (23.70).

23.7 **(a)** Derive Equations (23.71).
(b) Derive Equations (23.72) to (23.74).

23.8 **(a)** Derive Equations (23.84).
(b) Derive Equations (23.85).

23.9 **(a)** Derive Equations (23.92) and (23.93).
(b) Derive Equations (23.94) to (23.95).

23.10 Derive Equations (23.97).

23.11 Use the theory in Section 5.6 for the steady supersonic, inviscid, and isoenergetic flow about a blunt, symmetric, two-dimensional or axisymmetric body. For simplicity, consider the stagnation point region by assuming that the shock shape in this region is part of a circular cylinder or a sphere. Derive equations for $ds/d\psi$ and ω on the surface of the body. In so doing, show that both quantities are zero when the flow is two-dimensional and nonzero when it is axisymmetric.

23.12 **(a)** Start with Equations (23.18e) and (23.24a) and derive an equation involving R_1, Ψ_1 (and functions of Ψ_1), and the s and n first derivatives of Ψ_1.
(b) Start with Equations (23.18b,c) and derive a second PDE with R_1 and Ψ_1 as the dependent variables. These two PDEs are the general first-order outer equations.
(c) Use Appendix H to determine k, h_s, and r in terms of r_w and its derivatives.
(d) Use the results of part (c) to simplify the coefficient of Ψ_{1n}^2 in part (b).

23.13 Consider the cylinder-wedge and sphere-cone geometry of Problem 6.18. Remember that all variables and parameters in this chapter are nondimensional and, as in Problem 6.18, use R_b for the characteristic length. Denote the tangency point where the sphere-cone (cylinder-wedge) meet with a t subscript.
(a) Determine separate equations for the two configurations of the form $r_w = r_w(s)$ and for the location of the tangency point s_t.
(b) Utilize the results in part (c) of Problem 23.12 to determine equations for k, h_s, and r for both configurations.
(c) Write out the equation derived in part (b) of Problem 23.12 for the cylinder's geometry.

23.14 The purpose of this problem is the derivation of a single, second-order, general PDE for Ψ_2, which is defined in Problem 23.2. Utilize results from Problem 23.4 and Equations (23.34a,b) for P_2 and R_2. On the other hand, leave (V_{1s}, V_{1n}), and $DV_1/D_1t - kU_1^2/h_s$ as is. For this task, either Equation (23.32b) or (23.32c) could be used. Since it is somewhat simpler, use Equation (23.32c). Equations (23.32a,d,e) cannot be used, since they were utilized in Problem 23.4 and for Equation (23.34). Their use would only result in an identity.
(a) Develop equations for

$$U_1U_2 + V_1V_2, \quad P_2, \quad P_{2n}, \quad R_2, \quad \frac{DV_1}{D_2t}, \quad \frac{DV_2}{D_1t}$$

that depend only on order unity parameters, such as $R_1, T_1, \hat{U}_1, \hat{V}_1, \chi_a, \chi_b, S_1', \ldots,$ and Ψ_2 and its s and n partial derivatives, where

$$\chi_a = r^\sigma R_1(1 - \hat{U}_1^2 - \hat{V}_1^2), \qquad \chi_b = \frac{r^\sigma R_1}{(\gamma-1)M_\infty T_1^{1/2}}(\gamma T_1 S_1' - X_\infty H_1')$$

(b) Substitute the above into Equation (23.32c) and write your result in the form

$$A\Psi_{2ss} + B\Psi_{2sn} + C\Psi_{2nn} + D\Psi_{2s} + E\Psi_{2n} + F\Psi_2 = 0 \tag{1}$$

where A, B, and C only depend on $\hat{U}_1$, $\hat{V}_1$, and h_s. The D, E, and F coefficients are much more involved; e.g., D contains factors such as

$$\left(\frac{T_1^{1/2}\hat{U}_1\hat{V}_1}{h_s\chi_a}\right)_n, \quad (T_1^{1/2}\hat{V}_1)_s, \quad \left(\frac{1-\hat{U}_1^2}{h_s\chi_a}\right)_s, \quad \gamma T_1 S_1' - X_\infty H_1'$$

Simplify your A, B,…,F results as much as possible.

23.15 This is a continuation of Problem 23.14. As a consequence of this problem, we see that $G(n)$ in Equation (23.72) is identically zero.

(a) What is the physical condition for Equation (1) in Problem 23.14 to be hyperbolic?

(b) Use the results of Problem 23.14 to derive Equations (23.72) and (23.73).

(c) Assume the outer inviscid flow is homentropic and homenergetic. Derive the governing PDE for Ψ_2 without further assumptions.

Appendix A
Summary of Equations from Vector and Tensor Analysis

TABLE 1. VECTOR PRODUCTS

1. Scalar multiplication:

$$a\vec{A} = \vec{A}a$$

2. Dot product:

$$\vec{A} \cdot \vec{B} = \vec{B} \cdot \vec{A}$$

3. Cross product:

$$\vec{A} \times \vec{B} = -\vec{B} \times \vec{A}$$

4. Distributive law for the dot product:

$$\vec{A} \cdot (\vec{B} + \vec{C}) = \vec{A} \cdot \vec{B} + \vec{A} \cdot \vec{C}$$

5. Distributive law for the cross product:

$$\vec{A} \times (\vec{B} + \vec{C}) = \vec{A} \times \vec{B} + \vec{A} \times \vec{C}$$

6. Scalar triple product:

$$\vec{A} \cdot (\vec{B} \times \vec{C}) = (\vec{A} \times \vec{B}) \cdot \vec{C}$$

7. Vector triple product:

$$\vec{A} \times (\vec{B} \times \vec{C}) = (\vec{A} \cdot \vec{C})\vec{B} - (\vec{A} \cdot \vec{B})\vec{C}$$
$$(\vec{A} \times \vec{B}) \times \vec{C} = (\vec{A} \cdot \vec{C})\vec{B} - (\vec{B} \cdot \vec{C})\vec{A}$$

8. Other vector products:

$$(\vec{A}\times\vec{B})\times(\vec{C}\times\vec{D}) = (\vec{A}\cdot\vec{B}\times\vec{D})\vec{C}-(\vec{A}\cdot\vec{B}\times\vec{C})\vec{D}$$
$$= (\vec{A}\cdot\vec{C}\times\vec{D})\vec{B}-(\vec{B}\cdot\vec{C}\times\vec{D})\vec{A}$$
$$(\vec{A}\times\vec{B})\cdot(\vec{C}\times\vec{D}) = (\vec{A}\cdot\vec{C})(\vec{B}\cdot\vec{D})-(\vec{A}\cdot\vec{D})(\vec{B}\cdot\vec{C})$$
$$(\vec{A}\times\vec{B})\cdot(\vec{B}\times\vec{C})\times(\vec{C}\times\vec{A}) = (\vec{A}\cdot\vec{B}\times\vec{C})^2$$

TABLE 2. BASIC VECTOR RELATIONS

1. Basis in three-dimensional space:

$$\vec{e}_i \text{ where } \vec{e}_1\cdot\vec{e}_2\times\vec{e}_3\neq 0$$

2. Normalized basis:

$$\hat{e}_i = \frac{\vec{e}_i}{|\vec{e}_i|}$$

3. Orthogonal basis:

$$\vec{e}_i\cdot\vec{e}_j = 0, \qquad i\neq j$$

4. Orthonormal basis:

$$\hat{e}_i\cdot\hat{e}_j = \delta_{ij} = \begin{cases} 0, & i\neq j \\ 1, & i=j \end{cases}$$

5. Cartesian basis:

$$\hat{\imath}_i\cdot\hat{\imath}_j = \delta_{ij}, \qquad \frac{\partial\hat{\imath}_i}{\partial x^k} = 0$$

6. Dual basis and physical components $\hat{A}_i$ or $\hat{A}^j$:

$$\vec{e}_i = \frac{\vec{e}^{\,j}\times\vec{e}^{\,k}}{\mathrm{E}^{ijk}} = |g|^{1/2}\vec{e}^{\,j}\times\vec{e}^{\,k}, \qquad i,j,k \text{ are cyclic}$$

$$\vec{e}^{\,i} = \frac{\vec{e}_j\times\vec{e}_k}{\mathrm{E}_{ijk}} = \frac{\vec{e}_j\times\vec{e}_k}{|g|^{1/2}}, \qquad i,j,k \text{ are cyclic}$$

$$\vec{e}_i = g_{ij}\vec{e}^{\,j}, \qquad \vec{e}^{\,j} = g^{ij}\vec{e}_i$$

$$\vec{e}^{\,i}\cdot\vec{e}_j = \vec{e}_j\cdot\vec{e}^{\,i} = \delta^i_j$$

$$\vec{A} = A^i\vec{e}_i = g_{ii}^{1/2}A^i\hat{e}_i = \hat{A}^i\hat{e}_i$$

$$\vec{A} = A_j\vec{e}^{\,j} = (g^{jj})^{1/2}A_j\hat{e}^j = \hat{A}_j\hat{e}^j$$

7. Alternating symbols:

$$\varepsilon_{ijk} = \begin{cases} 0, & \text{repeated indices} \\ 1, & \text{cyclic order} \\ -1, & \text{not in cyclic order} \end{cases}$$

i	1	1	2	2	3	3
j	2	3	1	3	1	2
k	3	2	3	1	2	1
ε_{ijk}	1	−1	−1	1	1	−1

$$\varepsilon_{ijk}\varepsilon_{imn} = \delta_{jm}\delta_{kn} - \delta_{jn}\delta_{km}$$

$$\mathrm{E}_{ijk} = \vec{e}_i \times \vec{e}_j \cdot \vec{e}_k = \varepsilon_{ijk}|g|^{1/2}, \qquad \vec{e}_i \times \vec{e}_j = \mathrm{E}_{ijk}\vec{e}^{\,k}$$

$$\mathrm{E}^{ijk} = \vec{e}^{\,i} \times \vec{e}^{\,j} \cdot \vec{e}^{\,k} = \frac{\varepsilon_{ijk}}{|g|^{1/2}}, \qquad \vec{e}^{\,i} \times \vec{e}^{\,j} = \mathrm{E}^{ijk}\vec{e}_k$$

8. Physical components and an orthonormal basis:

$$\hat{e}_i = \frac{\vec{e}_i}{g_{ii}^{1/2}} = \frac{\vec{e}_i}{h_i} \qquad \text{(no sum)}$$

The dual basis, $\hat{e}^i$, is identical to the original orthonormal $\hat{e}_i$ basis.

$$\vec{A} = \hat{A}^i\hat{e}_i, \quad \vec{B} = \hat{B}^j\hat{e}_j$$

$$\vec{A} \cdot \vec{B} = \hat{A}^i\hat{B}^i$$

$$\hat{e}_i \times \hat{e}_j = \varepsilon_{ijk}\hat{e}_k$$

$$\vec{A} \times \vec{B} = \hat{A}^i\hat{B}^j\varepsilon_{ijk}\hat{e}_k = \begin{vmatrix} \hat{e}_1 & \hat{e}_2 & \hat{e}_3 \\ \hat{A}^1 & \hat{A}^2 & \hat{A}^3 \\ \hat{B}^1 & \hat{B}^2 & \hat{B}^3 \end{vmatrix}$$

$$\vec{A} \cdot \vec{B} \times \vec{C} = \hat{A}^i\hat{B}^j\hat{C}^k\varepsilon_{ijk} = \begin{vmatrix} \hat{A}^1 & \hat{A}^2 & \hat{A}^3 \\ \hat{B}^1 & \hat{B}^2 & \hat{B}^3 \\ \hat{C}^1 & \hat{C}^2 & \hat{C}^3 \end{vmatrix}$$

$$\boldsymbol{\varepsilon}_{ijk} = h_1h_2h_3\varepsilon_{ijk}, \qquad \boldsymbol{\varepsilon}^{ijk} = \frac{\varepsilon_{ijk}}{h_1h_2h_3}$$

9. Cogredient/contragredient relations:

$$\vec{A} = A^i\vec{e}_i = A_j\vec{e}^{\,j}$$

$$A^i = \text{contragredient components}$$

$$A_j = \text{cogredient components}$$

10. Transformation equations:

$$\vec{A} = A^i\vec{e}_i = A_j\vec{e}^j = \bar{A}^i\vec{\bar{e}}_i = \bar{A}_j\vec{\bar{e}}^j$$

$$a_s^j = \vec{e}^j \cdot \vec{\bar{e}}_s$$

$$b_i^s = \vec{e}_i \cdot \vec{\bar{e}}^s$$

$$c_{is} = \vec{e}_i \cdot \vec{\bar{e}}_s$$

$$d^{js} = \vec{e}^j \cdot \vec{\bar{e}}^s$$

$$\bar{A}_s = a_s^j A^i, \quad \vec{\bar{e}}_s = a_s^j\vec{e}_j, \quad \text{cogredient components}$$

$$\bar{A}^s = b_i^s A^i, \quad \vec{\bar{e}}^s = b_i^s\vec{e}^i, \quad \text{contragredient components}$$

$$\bar{A}_s = c_{is}A^i, \quad \vec{\bar{e}}_s = c_{is}\vec{e}^i, \quad \text{mixed component}$$

$$\bar{A}^s = d^{js}A_j, \quad \vec{\bar{e}}^s = d^{js}\vec{e}_j, \quad \text{mixed component}$$

TABLE 3. CURVILINEAR COORDINATES

1. General coordinates:

$$x^j = x^j(q^i)$$

$$d\vec{r} = \frac{\partial\vec{r}}{\partial q^i}dq^i = \vec{e}_i dq^i$$

$$\vec{e}_i = \frac{\partial x^k}{\partial q^i}\hat{i}_k$$

2. Fundamental metric tensor relations:

$$g_{ij} = g_{ji} = \vec{e}_i \cdot \vec{e}_j$$

$$g^{ij} = g^{ji} = \vec{e}^i \cdot \vec{e}^j$$

$$g^{ij}g_{ik} = \delta_k^j$$

$$(ds)^2 = g_{ij}dq^i dq^j$$

$$|g| = |g_\ell| = \begin{vmatrix} g_{11} & g_{12} & g_{13} \\ g_{21} & g_{22} & g_{23} \\ g_{31} & g_{32} & g_{33} \end{vmatrix}$$

$$|g^u| = \begin{vmatrix} g^{11} & g^{12} & g^{13} \\ g^{21} & g^{22} & g^{23} \\ g^{31} & g^{32} & g^{33} \end{vmatrix} = \frac{1}{|g_\ell|}$$

In the following two equations, i,j,k and m,n,p are cyclic.

$$g_{mi} = \frac{1}{|g^{u}|}(g^{nj}g^{pk} - g^{nk}g^{pj})$$

$$g^{mi} = \frac{1}{|g_{\ell}|}(g_{nj}g_{pk} - g_{nk}g_{pj})$$

For an orthogonal coordinate system:

$$g_{ij} = g^{ij} = 0, \qquad i \neq j$$

$$g_{ii}g^{ii} = 1 \qquad \text{(no sum)}$$

$$h_i = g_{ii}^{1/2} \qquad \text{(no sum)}$$

$$(ds)^2 = h_i^2(dq^i)(dq^i)$$

3. Covariant/contravariant relations:

$$\vec{A} = A^i\vec{e}_i = A_j\vec{e}^{\,j}$$

$$A^i = \text{contravariant components}$$

$$A_j = \text{covariant components}$$

$$\hat{e}_i = \frac{\vec{e}_i}{g_{ii}^{1/2}}, \qquad \hat{e}^j = \frac{\vec{e}^{\,j}}{(g^{ij})^{1/2}} \qquad \text{(no sum)}$$

$$\vec{A} = A^i\vec{e}_i = \hat{A}^i\hat{e}_i \qquad \hat{A}^i = g_{ii}^{1/2}A^i$$

$$\vec{A} = A_j\vec{e}^{\,j} = \hat{A}_j\hat{e}^j, \qquad \hat{A}_j = (g^{ii})^{1/2}A_j$$

$$\vec{A} = (\vec{A}\cdot\vec{e}_j)\vec{e}^{\,j} = (\vec{A}\cdot\vec{e}^{\,i})\vec{e}_i$$

$$A^i = \vec{A}\cdot\vec{e}^{\,i} = g^{ij}A_j, \qquad \vec{e}^{\,i} = g^{ij}\vec{e}_j$$

$$A_j = \vec{A}\cdot\vec{e}_j = g_{ij}A^i, \qquad \vec{e}_j = g_{ji}\vec{e}^{\,i}$$

$$\vec{e}_i\times\vec{e}_j = \varepsilon_{ijk}|g|^{1/2}\vec{e}^{\,k}$$

$$\vec{e}^{\,i}\times\vec{e}^{\,j} = \frac{\varepsilon_{ijk}}{|g|^{1/2}}\vec{e}_k$$

$$\vec{e}_i\times\vec{e}^{\,j} = -\vec{e}^{\,j}\times\vec{e}_i = g_{ik}\frac{\varepsilon_{kjm}}{|g|^{1/2}}\vec{e}_m$$

4. Transformation equations:

$$\vec{\bar{e}}_s = \frac{\partial q^j}{\partial \bar{q}^s}\vec{e}_j, \qquad \vec{e}_j = \frac{\partial \bar{q}^s}{\partial q^j}\vec{\bar{e}}_s \qquad \text{(covariant)}$$

$$\bar{A}_s = \frac{\partial q^j}{\partial \bar{q}^s}A_j, \qquad A_j = \frac{\partial \bar{q}^s}{\partial q^j}\bar{A}_s \qquad \text{(covariant)}$$

$$\vec{\bar{e}}^s = \frac{\partial \bar{q}^s}{\partial q^j}\vec{e}^j, \qquad \vec{e}^j = \frac{\partial q^j}{\partial \bar{q}^s}\vec{\bar{e}}^s \qquad \text{(contravariant)}$$

$$\bar{A}^s = \frac{\partial \bar{q}^s}{\partial q^j}A^j, \qquad A^j = \frac{\partial q^j}{\partial \bar{q}^s}\bar{A}^s \qquad \text{(contravariant)}$$

$$\frac{\partial q^j}{\partial \bar{q}^s}\frac{\partial \bar{q}^i}{\partial q^j} = \delta^i_s$$

$$\frac{\partial \bar{q}^j}{\partial q^s}\frac{\partial q^i}{\partial \bar{q}^j} = \delta^i_s$$

TABLE 4. DIFFERENTIAL AND INTEGRAL RELATIONS

1. Arc length, area, volume:

$$\begin{aligned}
(ds)^2 &= g_{ij}dq^i dq^j \\
&= h_i^2 dq^i dq^i && \text{(orthogonal)} \\
d\vec{a} &= \varepsilon_{ijk}|g|^{1/2}\vec{e}^k dq^i_a dq^j_b \\
&= h_2 h_3 \hat{e}_1(dq^2_a dq^3_b - dq^3_a dq^2_b) + \cdots && \text{(orthogonal)} \\
dv &= \varepsilon_{ijk}|g|^{1/2}dq^i_a dq^j_b dq^k_c \\
&= h_1 h_2 h_3 \varepsilon_{ijk} dq^i_a dq^j_b dq^k_c && \text{(orthogonal)}
\end{aligned}$$

2. Unitary basis derivatives:

$$\begin{aligned}
\Gamma^k_{ij} &= \text{Christoffel symbol} \\
&= \Gamma^k_{ji} = \frac{\partial \vec{e}_i}{\partial q^j}\cdot \vec{e}^k = \frac{g^{kr}}{2}\left[\frac{\partial g_{jr}}{\partial q^i} + \frac{\partial g_{ir}}{\partial q^j} - \frac{\partial g_{ij}}{\partial q^r}\right]
\end{aligned}$$

$$\frac{\partial \vec{e}_i}{\partial q^j} = \Gamma^k_{ij}\vec{e}_k, \qquad \frac{\partial \vec{e}^k}{\partial q^j} = -\Gamma^k_{ij}\vec{e}^i$$

$$d\vec{e}_i = \Gamma^k_{ij}dq^j\vec{e}_k, \qquad d\vec{e}^k = -\Gamma^k_{ij}dq^j\vec{e}^i$$

3. Orthogonal basis derivatives:

TABLE 4.1
Orthogonal Forms for the Christoffel Symbol

Case	i, j, k	Γ_{ij}^{k}	
1	all different	0	
2	$k = j, \quad i \neq j$	$\frac{1}{h_j}\frac{\partial h_j}{\partial q^i}$	(no sum)
3	$k = i, \quad i \neq j$	$\frac{1}{h_i}\frac{\partial h_i}{\partial q^j}$	(no sum)
4	$k \neq j, \quad i = j$	$-\frac{h_i}{h_k^2}\frac{\partial h_i}{\partial q^k}$	(no sum)
5	$i = j = k$	$\frac{1}{h_i}\frac{\partial h_i}{\partial q^i}$	(no sum)

$$\Gamma_{ij}^{k} = \frac{1}{h_k^2}\left[\delta_{jk} h_j \frac{\partial h_j}{\partial q^i} + \delta_{ik} h_i \frac{\partial h_i}{\partial q^j} - \delta_{ij} h_i \frac{\partial h_i}{\partial q^k}\right] \qquad \text{(no sum)}$$

$$\frac{\partial \hat{e}_i}{\partial q^j} = \sum_{k \neq i}\left[\frac{\delta_{jk}}{h_i}\frac{\partial h_j}{\partial q^i} - \frac{\delta_{ij}}{h_k}\frac{\partial h_i}{\partial q^k}\right]\hat{e}_k$$

$$d\hat{e}_i = \sum_{j \neq i}\left[\frac{1}{h_i}\frac{\partial h_j}{\partial q^i}dq^j - \frac{1}{h_j}\frac{\partial h_i}{\partial q^j}dq^i\right]\hat{e}_j$$

4. Del operations:

$$\nabla = \text{gradient or del operator} = \vec{e}^{\,i}\frac{\partial}{\partial q^i} = g^{ij}\vec{e}_j\frac{\partial}{\partial q^i}$$

$$= \frac{\hat{e}_i}{h_i}\frac{\partial}{\partial q^i} \qquad \text{(orthogonal)}$$

$$\vec{A} = A^j\vec{e}_j$$

$$\nabla\vec{A} = A^k_{,j}\,\vec{e}^{\,j}\vec{e}_k$$

$$A^k_{,j} = \text{covariant derivative of the contravariant component}$$

$$= \frac{\partial A^k}{\partial q^j} + A^i\Gamma_{ij}^{i}$$

$$A_{k,j} = \text{covariant derivative of the contravariant component}$$

$$= \frac{\partial A_k}{\partial q^j} - A_i\Gamma_{jk}^{i}$$

$$\frac{d\phi}{ds} = \text{directional derivative} = \frac{d\vec{r}}{ds} \cdot \nabla\phi$$

$$\vec{A} = A^i\vec{e}_i = \hat{A}^i\hat{e}_i$$

$$\nabla \cdot \vec{A} = \text{divergence of } \vec{A}$$

$$= \frac{\partial A^i}{\partial q^i} + A^i\Gamma^j_{ij} = \frac{\partial A^i}{\partial q^i} + \frac{A^i}{2|g|}\frac{\partial |g|}{\partial q^i} = \frac{1}{|g|^{1/2}}\frac{\partial}{\partial q^i}\left[|g|^{1/2}A^i\right]$$

$$= \frac{1}{h_1h_2h_3}\frac{\partial}{\partial q^i}\left[\frac{h_1h_2h_3}{h_i}\hat{A}^i\right] \qquad \text{(orthogonal)}$$

$$\nabla \times \vec{A} = \text{curl of } \vec{A} = A^k_{,j}\vec{e}^j \times \vec{e}_k = \left[\frac{\partial A^k}{\partial q^j} + A^i\Gamma^k_{ij}\right]\frac{\varepsilon_{jmn}}{|g|^{1/2}}g_{km}\vec{e}_n$$

$$= \frac{1}{h_1h_2h_3}\begin{vmatrix} h_1\hat{e}_1 & h_2\hat{e}_2 & h_3\hat{e}_3 \\ \dfrac{\partial}{\partial q^1} & \dfrac{\partial}{\partial q^2} & \dfrac{\partial}{\partial q^3} \\ h_1\hat{A}^1 & h_2\hat{A}^2 & h_3\hat{A}^3 \end{vmatrix} \qquad \text{(orthogonal)}$$

$$\nabla^2\phi = \text{Laplacian of } \phi = \nabla \cdot (\nabla\phi) = (\nabla \cdot \nabla)\phi$$

$$= \frac{1}{|g|^{1/2}}\frac{\partial}{\partial q^i}\left[|g|^{1/2}g^{ij}\frac{\partial\phi}{\partial q^j}\right]$$

$$= \frac{1}{h_1h_2h_3}\frac{\partial}{\partial q^i}\left[\frac{h_1h_2h_3}{h_i^2}\frac{\partial\phi}{\partial q^i}\right] \qquad \text{(orthogonal)}$$

5. Stokes' theorem:

$$\oint_c d\vec{r}\begin{bmatrix}\phi \\ \cdot\vec{A} \\ \times\vec{A}\end{bmatrix} = \int_s (\hat{n} \times \nabla)\begin{bmatrix}\phi \\ \cdot\vec{A} \\ \times\vec{A}\end{bmatrix} ds$$

6. Divergence theorem:

$$\oint_s \hat{n}\begin{bmatrix}\phi \\ \cdot\vec{A} \\ \times\vec{A}\end{bmatrix} ds = \int_v \nabla\begin{bmatrix}\phi \\ \cdot\vec{A} \\ \times\vec{A}\end{bmatrix} dv$$

TABLE 5. VECTOR IDENTITIES INVOLVING THE DEL OPERATOR

$$
\begin{aligned}
\nabla(\phi + \psi) &= \nabla\phi + \nabla\psi \\
\nabla\cdot(\vec{A} + \vec{B}) &= \nabla\cdot\vec{A} + \nabla\cdot\vec{B} \\
\nabla\times(\vec{A} + \vec{B}) &= \nabla\times\vec{A} + \nabla\times\vec{B} \\
\nabla(\phi\psi) &= \phi\nabla\psi + \psi\nabla\phi \\
\nabla\cdot(\phi\vec{A}) &= (\nabla\phi)\cdot(\vec{A} + \phi\nabla\cdot\vec{A}) \\
\nabla\times(\phi\vec{A}) &= (\nabla\phi)\times\vec{A} + \phi\nabla\times\vec{A} \\
\nabla(\vec{A}\cdot\vec{B}) &= (\vec{A}\cdot\nabla)\vec{B} + (\vec{B}\cdot\nabla)\vec{A} + \vec{A}\times(\nabla\times\vec{B}) + \vec{B}\times(\nabla\times\vec{A}) \\
\nabla\cdot(\vec{A}\times\vec{B}) &= \vec{B}\cdot(\nabla\times\vec{A}) - \vec{A}\cdot(\nabla\times\vec{B}) \\
\nabla\times(\vec{A}\times\vec{B}) &= \vec{A}(\nabla\cdot\vec{B}) - \vec{B}(\nabla\cdot\vec{A}) + (\vec{B}\cdot\nabla)\vec{A} - (\vec{A}\cdot\nabla)\vec{B} \\
(\nabla\times\vec{A})\times\vec{B} &= \vec{B}\cdot[\nabla\vec{A} - (\nabla\vec{A})^t] \\
\nabla\times(\nabla\times\vec{A}) &= \nabla(\nabla\cdot\vec{A}) - (\nabla\cdot\nabla)\vec{A} = \nabla(\nabla\cdot\vec{A}) - \nabla^2\vec{A} \\
\nabla\cdot\nabla\phi &= (\nabla\cdot\nabla)\phi = \nabla^2\phi \\
\nabla^2\vec{A} &= (\nabla\cdot\nabla)\vec{A} \\
\nabla\cdot(\nabla\vec{A})^t &= \nabla(\nabla\cdot\vec{A}) \\
\nabla^2(\phi\psi) &= \phi\nabla^2\psi + 2\nabla\phi\cdot\nabla\psi + \psi\nabla^2\phi \\
\nabla\times(\nabla\phi) &= 0 \\
\nabla\cdot(\nabla\times\vec{A}) &= 0 \\
\nabla\cdot(\nabla\phi\times\nabla\psi) &= 0 \\
\nabla\cdot\vec{r} &= 3 \\
\nabla r &= \frac{\vec{r}}{r} \\
\nabla\vec{r} &= \overleftrightarrow{I} \\
\nabla\cdot\overleftrightarrow{I} &= 0 \\
\vec{A}\cdot\nabla\vec{r} &= \vec{A}
\end{aligned}
$$

With $\hat{n}$ as a unit vector

$$
\begin{aligned}
\hat{n}\cdot(\nabla\times\vec{A}) &= (\hat{n}\times\nabla)\cdot\vec{A} \\
(\hat{n}\times\nabla)\times\vec{A} &= (\nabla\vec{A})^t\cdot\hat{n} - \hat{n}(\nabla\cdot\vec{A})
\end{aligned}
$$

TABLE 6. DYADIC SUMMARY

1. Basic properties:

$$\overleftrightarrow{\Phi} = \text{dyadic} = \vec{A}_i\vec{B}_i$$

$$\vec{C}\cdot\overleftrightarrow{\Phi} = (\vec{C}\cdot\vec{A}_i)\vec{B}_i$$

$$\overleftrightarrow{\Phi}\cdot\vec{C} = \vec{A}_i(\vec{B}_i\cdot\vec{C})$$

$$\overleftrightarrow{\Phi} = \text{nonion form} = \phi^{jk}\vec{e}_j\vec{e}_k$$

$$\overleftrightarrow{\Phi}^t = \text{transpose of } \overleftrightarrow{\Phi} = \vec{B}_i\vec{A}_i$$

$$= \phi^{jk}\vec{e}_k\vec{e}_j = \phi^{kj}\vec{e}_j\vec{e}_k$$

$$\overleftrightarrow{\Phi} \text{ is symmetric if } \overleftrightarrow{\Phi} = \overleftrightarrow{\Phi}^t$$

$$\overleftrightarrow{\Phi} \text{ is antisymmetric if } \overleftrightarrow{\Phi} = -\overleftrightarrow{\Phi}^t$$

$$\overleftrightarrow{I} = \text{unit dyadic}$$

$$= g^{ij}\vec{e}_i\vec{e}_j = g_{ij}\vec{e}^i\vec{e}^j = g_i^j\vec{e}^i\vec{e}_j = g_i^j\vec{e}_i\vec{e}^j$$

$$g_i^j = \delta_i^j, \qquad g_j^i = \delta_j^i$$

$$\vec{A}\cdot\overleftrightarrow{I} = \overleftrightarrow{I}\cdot\vec{A} = \vec{A}$$

2. Transformation laws of a second-order tensor:

$$\overleftrightarrow{\Phi} = \phi^{ij}\vec{e}_i\vec{e}_j = \phi_{ij}\vec{e}^i\vec{e}^j = \phi^i{}_j\vec{e}_i\vec{e}^j = \phi_i{}^j\vec{e}^i\vec{e}_j$$

$$= \bar{\phi}^{mn}\vec{\bar{e}}_m\vec{\bar{e}}_n = \bar{\phi}_{mn}\vec{\bar{e}}^m\vec{\bar{e}}^n = \bar{\phi}^m{}_n\vec{\bar{e}}_m\vec{\bar{e}}^n = \bar{\phi}_m{}^n\vec{\bar{e}}^m\vec{\bar{e}}_n$$

$$\bar{\phi}^{mn} = \phi^{ij}\frac{\partial\bar{q}^m}{\partial q^i}\frac{\partial\bar{q}^n}{\partial q^j} \qquad \text{(contravariant)}$$

$$\bar{\phi}_{mn} = \phi_{ij}\frac{\partial q^i}{\partial\bar{q}^m}\frac{\partial q^j}{\partial\bar{q}^n} \qquad \text{(covariant)}$$

$$\bar{\phi}^m{}_n = \phi^i{}_j\frac{\partial\bar{q}^m}{\partial q^i}\frac{\partial q^j}{\partial\bar{q}^n} \qquad \text{(mixed)}$$

$$\bar{\phi}_m{}^n = \phi_i{}^j\frac{\partial q^i}{\partial\bar{q}^m}\frac{\partial\bar{q}^n}{\partial q^j} \qquad \text{(mixed)}$$

3. Multiplicative operations:

$$\vec{A}_i = A_i^j \vec{e}_j, \qquad \vec{B}_j = B_j^k \vec{e}_k$$

$$\overleftrightarrow{\Phi} = \vec{A}_i \vec{B}_i = \phi^{jk} \vec{e}_j \vec{e}_k$$

$$\overleftrightarrow{\Phi}_s = \text{scalar contraction}$$

$$= \vec{A}_i \cdot \vec{B}_i = A_i^j B_i^k g_{jk} = \phi^{jk} g_{jk}$$

$$\overleftrightarrow{\Phi}_v = \text{vector of a second-order tensor}$$

$$= \vec{A}_i \times \vec{B}_i = A_i^j B_i^k \vec{e}_j \times \vec{e}_k$$

$$= \varepsilon_{ijk} |g|^{1/2} \phi^{ij} \vec{e}^{\,k}$$

$$\vec{A} \cdot \left(\overleftrightarrow{\Phi} - \overleftrightarrow{\Phi}^t \right) = \overleftrightarrow{\Phi}_v \times \vec{A}$$

$$\vec{C} \times \overleftrightarrow{\Phi} = (\vec{C} \times \vec{A}_i) \vec{B}_i$$

$$\overleftrightarrow{\Phi} \times \vec{C} = \vec{A}_i (\vec{B}_i \times \vec{C})$$

$$\overleftrightarrow{\Psi} = \vec{C}_j \vec{D}_j$$

$$\overleftrightarrow{\Phi} \cdot \overleftrightarrow{\Psi} = (\vec{B}_i \cdot \vec{C}_j) \vec{A}_i \vec{D}_j$$

$$\overleftrightarrow{\Phi} \cdot \overleftrightarrow{\Psi} = (\vec{A}_i \cdot \vec{C}_j)(\vec{B}_i \cdot \vec{D}_j) = \phi^{ij} \psi^{mm} g_{jm} g_{kn}$$

$$\left(\vec{u} \times \overleftrightarrow{\Phi} \right) \cdot \overleftrightarrow{\Psi} = \vec{u} \times \left(\overleftrightarrow{\Phi} \cdot \overleftrightarrow{\Psi} \right) = \vec{u} \times \overleftrightarrow{\Phi} \cdot \overleftrightarrow{\Psi}$$

$$\left(\overleftrightarrow{\Phi} \cdot \overleftrightarrow{\Psi} \right) \times \vec{u} = \overleftrightarrow{\Phi} \cdot \left(\overleftrightarrow{\Psi} \times \vec{u} \right) = \overleftrightarrow{\Phi} \times \overleftrightarrow{\Psi} \cdot \vec{u}$$

$$\left(\vec{v} \times \overleftrightarrow{\Phi} \right) \cdot \vec{u} = \vec{v} \times \left(\overleftrightarrow{\Phi} \cdot \vec{u} \right) = \vec{v} \times \overleftrightarrow{\Phi} \cdot \vec{u}$$

$$\vec{v} \cdot \left(\overleftrightarrow{\Phi} \times \vec{u} \right) = \left(\vec{v} \cdot \overleftrightarrow{\Phi} \right) \times \vec{u} = \vec{v} \cdot \overleftrightarrow{\Phi} \times \vec{u}$$

$$\vec{v} \times \left(\overleftrightarrow{\Phi} \times \vec{u} \right) = \left(\vec{v} \times \overleftrightarrow{\Phi} \right) \times \vec{u} = \vec{v} \times \overleftrightarrow{\Phi} \times \vec{u}$$

$$\vec{v} \cdot \vec{u} \times \overleftrightarrow{\Phi} = (\vec{v} \times \vec{u}) \cdot \overleftrightarrow{\Phi}$$

$$\overleftrightarrow{\Phi} \times \vec{v} \cdot \vec{u} = \overleftrightarrow{\Phi} \cdot (\vec{v} \times \vec{u})$$

$$\overleftrightarrow{\Phi} \cdot \left(\vec{v} \times \overleftrightarrow{\Psi} \right) = \left(\overleftrightarrow{\Phi} \times \vec{v} \right) \cdot \overleftrightarrow{\Psi}$$

4. Differential operations:

$$\text{grad}\vec{A} = \nabla\vec{A} = A^k_{,j}\vec{e}^j\vec{e}_k = A_{k,j}\vec{e}^j\vec{e}^k$$

$$= \frac{\partial A^k}{\partial x^j}\hat{\imath}_j\hat{\imath}_k \qquad \text{(Cartesian)}$$

$$A^k_{,j} = \frac{\partial A^k}{\partial q^j} + A^i\Gamma^k_{ij}$$

$$A_{k,j} = \frac{\partial A_k}{\partial q^j} - A_i\Gamma^i_{kj}$$

$$(\nabla\vec{A})_s = \text{scalar contraction of } \nabla\vec{A}$$

$$= \frac{\partial A^k}{\partial q^k} + A^i\Gamma^k_{ik} = \nabla\cdot\vec{A}$$

$$(\nabla\vec{A})_v = \text{vector of the tensor gradient}$$

$$= |g|^{1/2}g^{ji}A^k_{,j}\varepsilon_{ikm}\vec{e}^m = \nabla\times\vec{A}$$

$$\overleftrightarrow{\Phi} = \phi^{jk}\vec{e}_j\vec{e}_k = \phi_{jk}\vec{e}^j\vec{e}^k = \phi^k_j\vec{e}^j\vec{e}_k = \phi^j_k\vec{e}_j\vec{e}^k$$

$$\nabla*\overleftrightarrow{\Phi} = \phi^{jk}_{,i}(\vec{e}^i*\vec{e}_j)\vec{e}_k = \phi_{jk,i}(\vec{e}^i*\vec{e}^j)\vec{e}^k = \phi_{j,i}{}^k(\vec{e}^i*\vec{e}^j)\vec{e}_k = \phi^j_{k,i}(\vec{e}^i*\vec{e}_j)\vec{e}^k$$

$$\phi^{jk}_{,i} = \frac{\partial\phi^{jk}}{\partial q^i} + \phi^{mk}\Gamma^j_{mi} + \phi^{mj}\Gamma^k_{mi}$$

$$\phi_{jk,i} = \frac{\partial\phi_{jk}}{\partial q^i} - \phi_{mk}\Gamma^m_{ji} - \phi_{jm}\Gamma^m_{ki}$$

$$\phi_{j,i}{}^k = \frac{\partial\phi_j{}^k}{\partial q^i} - \phi_m{}^k\Gamma^m_{ji} + \phi_j{}^m\Gamma^k_{mi}$$

$$\phi^k{}_{j,i} = \frac{\partial\phi^j{}_k}{\partial q^i} + \phi^m{}_k\Gamma^j_{mi} - \phi^j{}_m\Gamma^m_{ki}$$

$$\nabla\overleftrightarrow{\Phi} = \phi^{jk}_{,i}\vec{e}^i\vec{e}_j\vec{e}_k = \phi_{jk,i}\vec{e}^i\vec{e}^j\vec{e}^k = \phi_{j,i}{}^k\vec{e}^i\vec{e}^j\vec{e}_k = \phi^j{}_{k,i}\vec{e}^i\vec{e}_j\vec{e}^k$$

$$\nabla\cdot\overleftrightarrow{\Phi} = \phi^{jk}_{,i}\vec{e}_k = g^{ij}\phi_{jk,i}\vec{e}^k = g^{ij}\phi_{j,i}{}^k\vec{e}_k = \phi^i{}_{k,i}\vec{e}^k$$

$$\nabla\times\overleftrightarrow{\Phi} = \varepsilon_{imn}\frac{g_{jm}}{|g|^{1/2}}\phi^{jk}_{,i}\vec{e}_n\vec{e}_k = \varepsilon_{ijm}\frac{g_{mn}}{|g|^{1/2}}\phi_{jk,i}\vec{e}^n\vec{e}^k$$

$$= \varepsilon_{ilm}\frac{g_{mn}}{|g|^{1/2}}\phi_{j,i}{}^k\vec{e}^n\vec{e}_k = \varepsilon_{imn}\frac{g_{jm}}{|g|^{1/2}}\phi^j{}_{k,i}\vec{e}_n\vec{e}^k$$

5. Stokes' and the divergence theorems:

$$\hat{n}\cdot\left(\nabla\times\overleftrightarrow{\Phi}\right) = (\hat{n}\times\nabla)\cdot\overleftrightarrow{\Phi}$$

$$\oint_c d\vec{r} \begin{bmatrix} \vec{A} \\ \cdot \overleftrightarrow{\Phi} \\ \times \overleftrightarrow{\Phi} \end{bmatrix} = \int_s (\hat{n} \times \nabla) \begin{bmatrix} \vec{A} \\ \cdot \overleftrightarrow{\Phi} \\ \times \overleftrightarrow{\Phi} \end{bmatrix} ds$$

$$\oint_s \hat{n} \begin{bmatrix} \vec{A} \\ \cdot \overleftrightarrow{\Phi} \\ \times \overleftrightarrow{\Phi} \end{bmatrix} ds = \int_v \nabla \begin{bmatrix} \vec{A} \\ \cdot \overleftrightarrow{\Phi} \\ \times \overleftrightarrow{\Phi} \end{bmatrix} dv$$

6. Dyadic invariants:

In Cartesian coordinates

$$\overleftrightarrow{\Phi} \cdot \vec{A} = \lambda \vec{A}$$

becomes

$$\phi_{ij} A_j - \lambda A_j = 0, \qquad i = 1, 2, 3$$

which has a nontrivial solution when

$$\begin{vmatrix} \phi_{11} - \lambda & \phi_{12} & \phi_{13} \\ \phi_{21} & \phi_{22} - \lambda & \phi_{23} \\ \phi_{31} & \phi_{32} & \phi_{33} - \lambda \end{vmatrix} = 0$$

or

$$\begin{aligned} \lambda^3 - I_1 \lambda^2 - I_2 \lambda - I_3 &= 0 \\ I_1 &= \phi_{ii} \\ I_2 &= \phi_{12}\phi_{21} + \phi_{13}\phi_{31} + \phi_{23}\phi_{32} - \phi_{11}\phi_{22} - \phi_{22}\phi_{33} - \phi_{33}\phi_{11} \\ I_3 &= |\phi_{ij}| \end{aligned}$$

The roots of the cubic, λ_i, are the eigenvalues of the equation whose solutions are the corresponding eigenvectors. The I_i are the invariants of $\overleftrightarrow{\Phi}$. The symmetric dyadic $\overleftrightarrow{\Phi}$

has a diagonal form in its principal axis system

$$\overleftrightarrow{\Phi} = \lambda_m \hat{e}_m^* \hat{e}_m^*$$

where the $\hat{e}_m^*$ are normalized solutions of

$$\overleftrightarrow{\Phi} \cdot \hat{e}_m^* = \phi_{ij} \hat{i}_i (\hat{i}_j \cdot \hat{e}_m) = \lambda_m \hat{e}_m^*$$

Eigenvectors corresponding to distinct eigenvalues are orthogonal when $\overleftrightarrow{\Phi}$ is symmetric. If all the eigenvalues are distinct, then the normalized eigenvectors constitute an orthonormal basis. If $\overleftrightarrow{\Phi}$ is symmetric with real components, then all its eigenvalues are real.

Appendix B
Jacobian Theory

B.1 PRELIMINARY REMARKS

A common occurrence throughout engineering and the physical sciences is to change variables in an ODE or PDE, where the change may involve the dependent variables, independent variables, or both. This is certainly the case in fluid dynamics and thermodynamics. Such changes are required for a wide variety of reasons, from finding an analytical solution to recasting the equations in an appropriate form for numerical computation. Any change of variables can be regarded as a transformation. To expedite this type of manipulation, Jacobian theory is extremely useful.

Most calculus textbooks discuss Jacobians in an elementary fashion. The reason for this interest occurs when changing variables inside an integral. For instance, suppose we have the integral

$$I = \int f(x)\,dx$$

which we hope to evaluate by changing variables. We use the transformation

$$\begin{aligned} x &= g(y) \\ dx &= g'dy \end{aligned}$$

to obtain

$$I = \int f[g(y)]g'dy$$

where g' is the Jacobian of the transformation. This process generalizes for multiple integrals of the form

$$I = \int \ldots \int f(x_1, \ldots, x_n)\,dx_1 \ldots dx_n$$

as follows. If we change variables by means of the transformation

$$x_1 = x_1(\xi_1, \ldots, \xi_n), \ldots, x_n = x_n(\xi_1, \ldots, \xi_n) \tag{B.1}$$

then I becomes

$$I = \int \ldots \int f(x_1(\xi_1, \ldots, \xi_n), \ldots)|J|\,d\xi_1 \ldots d\xi_n$$

where the Jacobian J is given by the determinant

$$J(x_1,\ldots,x_n) = \frac{\partial(x_1,\ldots,x_n)}{\partial(\xi_1,\ldots,\xi_n)} = \begin{vmatrix} \dfrac{\partial x_1}{\partial \xi_1} & \dfrac{\partial x_1}{\partial \xi_2} & \cdots & \dfrac{\partial x_1}{\partial \xi_n} \\ \dfrac{\partial x_2}{\partial \xi_1} & \cdots & & \vdots \\ \vdots & & & \\ \dfrac{\partial x_n}{\partial \xi_1} & \cdots & & \dfrac{\partial x_n}{\partial \xi_n} \end{vmatrix} \tag{B.2}$$

Elementary Considerations

Courant (1936) provides an introduction to Jacobian theory, which we follow in this subsection. Consider the change in variables

$$x = \phi(\xi,\eta), \qquad y = \psi(\xi,\eta) \tag{B.3}$$

where we go from ξ,η to x,y by a one-to-one mapping. By a mapping, we mean that there are two functions g and h

$$\xi = g(x,y), \qquad \eta = h(x,y) \tag{B.4}$$

in the neighborhood of some point x_o,y_o, such that

$$\begin{aligned} x_o &= \phi(\xi_o,\eta_o), & y_o &= \psi(\xi_o,\eta_o) \\ \xi_o &= g(x_o,y_o), & \eta_o &= h(x_o,y_o) \end{aligned} \tag{B.5}$$

We assume that the functions ϕ, ψ, g, and h are continuous and possess continuous derivatives. We write Equations (B.3) as

$$\begin{aligned} x &= \phi[g(x, y), h(x, y)] \\ y &= \psi[g(x, y), h(x, y)] \end{aligned} \tag{B.6}$$

and differentiate each of these by the chain rule with respect to x and y:

$$\begin{aligned} 1 &= \phi_g g_x + \phi_h h_x = \phi_\xi g_x + \phi_\eta h_x \\ 0 &= \psi_\xi g_x + \psi_\eta h_x \end{aligned} \tag{B.7}$$

$$\begin{aligned} 0 &= \phi_\xi g_y + \phi_\eta h_y \\ 1 &= \psi_\xi g_y + \psi_\eta h_y \end{aligned} \tag{B.8}$$

Equations (B.7) are solved for g_x and h_x, and Equations (B.8) are solved for g_y and h_y, with the result

$$g_x = \frac{\psi_\eta}{J}, \qquad g_y = -\frac{\phi_\eta}{J}, \qquad h_x = -\frac{\psi_\xi}{J}, \qquad h_y = \frac{\phi_\xi}{J}$$

or, in a more convenient notation,

$$\xi_x = \frac{y_\eta}{J}, \qquad \xi_y = -\frac{x_\eta}{J}, \qquad \eta_x = -\frac{y_\xi}{J}, \qquad \eta_y = \frac{x_\xi}{J} \tag{B.9}$$

The Jacobian J of the transformation is given by

$$J(x,y) = \frac{\partial(x,y)}{\partial(\xi,\eta)} = \begin{vmatrix} \frac{\partial x}{\partial \xi} & \frac{\partial x}{\partial \eta} \\ \frac{\partial y}{\partial \xi} & \frac{\partial y}{\partial \eta} \end{vmatrix} = \begin{vmatrix} \frac{\partial x}{\partial \xi} & \frac{\partial y}{\partial \xi} \\ \frac{\partial x}{\partial \eta} & \frac{\partial y}{\partial \eta} \end{vmatrix} \tag{B.10}$$

Observe that

$$J(x,y) = -J(y,x) \tag{B.11}$$

If Equations (B.3) are known, the existence of a unique inverse requires that $J \neq 0$. The actual construction of the inverse functions, Equations (B.4), is not necessarily trivial. This construction may require the integration of Equations (B.9), where the right-hand sides are known functions of ξ and η.

Suppose we have two consecutive one-to-one mappings; i.e.,

$$\xi,\eta \rightarrow x, y \rightarrow u,v$$

We can show that the Jacobian of the resulting transformation

$$\xi,\eta \rightarrow u,v$$

is given by

$$\frac{\partial(u,v)}{\partial(\xi,\eta)} = \frac{\partial(u,v)}{\partial(x,y)} \frac{\partial(x,y)}{\partial(\xi,\eta)} \tag{B.12}$$

This relation generalizes, in an obvious way, to any number of one-to-one consecutive mappings. One corollary of this equation is that, if

$$u = \xi, \qquad v = \eta$$

then we obtain

$$\frac{\partial(\xi,\eta)}{\partial(x,y)} \frac{\partial(x,y)}{\partial(\xi,\eta)} = 1 \tag{B.13}$$

since the Jacobian on the right side is

$$\frac{\partial(\xi,\eta)}{\partial(\xi,\eta)} = \begin{vmatrix} \frac{\partial \xi}{\partial \xi} = 1 & \frac{\partial \xi}{\partial \eta} = 0 \\ \frac{\partial \eta}{\partial \xi} = 0 & \frac{\partial \eta}{\partial \eta} = 1 \end{vmatrix} = 1 \tag{B.14}$$

If the above is extended to second derivatives, such as

$$\frac{\partial^2 \xi}{\partial x^2} = \xi_{xx} = g_{xx}, \qquad \frac{\partial^2 \eta}{\partial x^2} = \eta_{xx} = h_{xx} \tag{B.15}$$

we merely differentiate Equations (B.7) with respect to x by the chain rule, to obtain

$$\begin{aligned} x_\xi \xi_{xx} + x_\eta \eta_{xx} &= -(x_{\xi\xi}\xi_x^2 + 2x_{\xi\eta}\xi_x\eta_x + x_{\eta\eta}\eta_x^2) \\ y_\xi \xi_{xx} + y_\eta \eta_{xx} &= -(y_{\xi\xi}\xi_x^2 + 2y_{\xi\eta}\xi_x\eta_x + y_{\eta\eta}\eta_x^2) \end{aligned} \tag{B.16}$$

First derivatives, such as ξ_x, are replaced using Equations (B.9). Upon solving the preceding equations, we obtain

$$\xi_{xx} = -\frac{1}{J^3}\begin{vmatrix} x_{\xi\xi}y_\eta^2 - 2x_{\xi\eta}y_\xi y_\eta + x_{\eta\eta}y_\xi^2 & x_\eta \\ y_{\xi\xi}y_\eta^2 - 2y_{\xi\eta}y_\xi y_\eta + y_{\eta\eta}y_\xi^2 & y_\eta \end{vmatrix} \tag{B.17a}$$

$$\eta_{xx} = \frac{1}{J^3}\begin{vmatrix} x_{\xi\xi}y_\eta^2 - 2x_{\xi\eta}y_\xi y_\eta + x_{\eta\eta}y_\xi^2 & x_\xi \\ y_{\xi\xi}y_\eta^2 - 2y_{\xi\eta}y_\xi y_\eta + y_{\eta\eta}y_\xi^2 & y_\xi \end{vmatrix} \tag{B.17b}$$

where $J = J(x,y)$. These equations are analogous to Equations (B.9), and similar equations hold for the other four second derivatives.

For second-order partial derivatives that are continuous, the order of differentiation is immaterial. Furthermore, we must not confuse fixed parameters with those being differentiated. Suppose we have

$$F = V^2T^2N$$

Then,

$$\frac{\partial F}{\partial T}_{VN} = 2V^2TN$$

If we now compute the partial derivative with respect to V, we do not obtain zero. Instead, we have

$$\left[\frac{\partial}{\partial V}\left(\frac{\partial F}{\partial T}\right)_{VN}\right]_{TN} = 4VTN$$

Thus, V is fixed for the first partial derivative but not for the second. Had we inverted the order of differentiation, the result would be the same.

B.2 GENERAL THEORY

Jacobian theory was first systematically applied to thermodynamics by Shaw (1935) and further developed by Crawford (1949, 1950). The subsequent discussion is based on these references.

SECOND-ORDER JACOBIANS

The basic relations are given in Table B.1, where x and y (or x_i) are functions of ξ and η. [Tables B.1 and B.2 stem from Shaw (1935).] Relation (1) in the table is the Jacobian definition, and Equation (2) has already been discussed. Equation (3) is the basis of many relations given later, while Equation (4) is a general expression for a second-order partial derivative. Equation (5) provides the derivative of J. As an example, we derive Equation (6), which is one of the most useful of the Jacobian relations.

The proof is simpler to visualize if we change notation and write Equation (6) as

$$J(x,y)\,dz + J(y,z)\,dx + J(z,x)\,dy = 0 \tag{B.18}$$

where x, y, and z are functions of ξ and η. We write

$$x = x(\xi,\eta), \qquad y = y(\xi,\eta), \qquad z = z(\xi,\eta)$$

and, consequently,

$$\xi = \xi(x,y)$$
$$d\xi = \frac{\partial \xi}{\partial x}\,dx + \frac{\partial \xi}{\partial y}\,dy$$
$$\eta = \eta(x,y)$$
$$d\eta = \frac{\partial \eta}{\partial x}\,dx + \frac{\partial \eta}{\partial y}\,dy$$

TABLE B.1
Basic Jacobian Equations

$$J(x,y) = \frac{\partial(x,y)}{\partial(\xi,\eta)} = \begin{vmatrix} \dfrac{\partial x}{\partial \xi} & \dfrac{\partial x}{\partial \eta} \\ \dfrac{\partial y}{\partial \xi} & \dfrac{\partial y}{\partial \eta} \end{vmatrix} \tag{1}$$

$$\frac{\partial(x,y)}{\partial(\xi,\eta)} = \frac{\partial(x,y)}{\partial(u,v)}\,\frac{\partial(u,v)}{\partial(\xi,\eta)} \tag{2}$$

$$J[f_a(x_1,x_2), f_b(x_3,x_4)] = \frac{\partial f_a}{\partial x_1}\frac{\partial f_b}{\partial x_3}J(x_1,x_3) + \frac{\partial f_a}{\partial x_1}\frac{\partial f_b}{\partial x_4}J(x_1,x_4) + \frac{\partial f_a}{\partial x_2}\frac{\partial f_b}{\partial x_3}J(x_2,x_3) + \frac{\partial f_a}{\partial x_2}\frac{\partial f_b}{\partial x_1}J(x_2,x_4) \tag{3}$$

$$\left[\frac{\partial}{\partial x_4}\left(\frac{\partial x_1}{\partial x_2}\right)_{x_3}\right]_{x_5} = \frac{J\{[J(x_1,x_3)/J(x_2,x_3)],x_5\}}{J(x_4,x_5)} = \frac{J(x_2,x_3)J[J(x_1,x_3),x_5] - J(x_1,x_3)J[J(x_2,x_3),x_5]}{J(x_4,x_5)[J(x_2,x_3)]^2} \tag{4}$$

$$\frac{d}{dt}[J(x,y)] = J\left(\frac{dx}{dt},y\right) + J\left(x,\frac{dy}{dt}\right) \tag{5}$$

$$J(x_1,x_2)dx_3 + J(x_2,x_3)dx_1 + J(x_3,x_1)dx_2 = 0 \tag{6}$$

TABLE B.2
Derived Jacobian Equations

$J(x, y) = -J(y, x) = J(-y, x) = J(y, -x)$	(1)
$J(x, x) = J(k, x) = 0, \qquad k = \text{constant}$	(2)
$J(\xi, \eta) = \dfrac{\partial(\xi, \eta)}{\partial(\xi, \eta)} = 1$	(3)*
$\dfrac{\partial(x, y)\partial(\xi, \eta)}{\partial(\xi, \eta)\partial(x, y)} = 1$	(4)
$\dfrac{\partial z}{\partial x_y} = \dfrac{J(z, y)}{J(x, y)}$	(5)
$J(x_1 + x_2, y) = J(x_1, y) + J(x_2, y)$	(6)
$J(x^n, y^m) = nmx^{n-1}y^{m-1}J(x, y)$	(7)
$J(xy, z) = xJ(y, z) + yJ(x, z)$	(8)
$J[f_a(x), f_b(y)] = \dfrac{df_a}{dx}\dfrac{df_b}{dy}J(x, y)$	(9)
$J[f_a(x, y), f_b(x, y)] = \left(\dfrac{\partial f_a}{\partial x}\dfrac{\partial f_b}{\partial y} - \dfrac{\partial f_a}{\partial y}\dfrac{\partial f_b}{\partial x}\right)J(x, y)$	(10)
$J\left[\int f_a(x)dx, \int f_b(y)dy\right] = f_a(x)f_b(y)J(x, y)$	(11)
$J[J(x, y), x] = -\left[\dfrac{\partial J(x, y)}{\partial y}\right]_x J(x, y)$	(12)
$J[J(x, y), y] = \left[\dfrac{\partial J(x, y)}{\partial x}\right]_y J(x, y)$	(13)
$J[J(x, \xi), \eta] = J[J(x, \eta), \xi]$	(14)*
$J(x, y) = J(x, \xi)J(y, \eta) - J(x, \eta)J(y, \xi)$	(15)*
$J(x, y)J(z, w) + J(y, z)J(x, w) + J(z, x)J(y, w) = 0$	(16)
$J(x, y)J[J(x, y), z] + J(y, z)J[J(x, y), x] + J(z, x)J[J(x, y), y] = 0$	(17)
$J[J(x, y), z] + J[J(y, z), x] + J[J(z, x), y] = 0$	(18)

* These equations require ξ and η as the independent variables.

For z, we use

$$dz = \frac{\partial z}{\partial \xi}d\xi + \frac{\partial z}{\partial \eta}d\eta = \left(\frac{\partial \xi}{\partial x}\frac{\partial z}{\partial \xi} + \frac{\partial \eta}{\partial x}\frac{\partial z}{\partial \eta}\right)dx + \left(\frac{\partial \xi}{\partial y}\frac{\partial z}{\partial \xi} + \frac{\partial \eta}{\partial y}\frac{\partial z}{\partial \eta}\right)dy$$

where $d\xi$ and $d\eta$ are replaced with their preceding expressions. We now replace $\partial\xi/\partial x, \ldots, \partial\eta/\partial y$ by means of Equations (B.9), to obtain

$$dz = \left(\frac{y_\eta}{J}z_\xi - \frac{y_\xi}{J}z_\eta\right)dx + \left(-\frac{x_\eta}{J}z_\xi + \frac{x_\xi}{J}z_\eta\right)dy$$

or

$$\begin{aligned} J(x,y)dz &= -(y_\xi z_\eta - y_\eta z_\xi)dx - (z_\xi x_\eta - z_\eta x_\xi)dy \\ &= -J(y,z)dx - J(z,x)dy \end{aligned}$$

which proves Equation (B.18).

Several useful corollaries are easily derived from Equation (B.18). For example, if z is a function of x and y, we have

$$dz = \frac{\partial z}{\partial x_y}dx + \frac{\partial z}{\partial y_x}dy$$

However, Equation (B.18) also yields

$$dz = -\frac{J(y,z)}{J(x,y)}dx - \frac{J(z,x)}{J(x,y)}dy = \frac{J(z,y)}{J(x,y)}dx + \frac{J(x,z)}{J(x,y)}dy$$

By comparison, we obtain

$$\frac{\partial z}{\partial x_y} = \frac{J(z,y)}{J(x,y)} \tag{B.19}$$

This is Equation (5) in Table B.2, which lists a number of useful derived Jacobian relations.

From Equation (B.11),

$$\frac{J(x,y)}{J(y,x)} = \frac{J(x,y)}{J(z,y)}\frac{J(z,y)}{J(x,z)}\frac{J(x,z)}{J(y,x)} = \frac{J(x,y)}{J(z,y)}\frac{[-J(y,z)]}{J(x,z)}\frac{[-J(z,x)]}{J(y,x)} = -1$$

or

$$\frac{J(x,y)}{J(z,y)}\frac{J(y,z)}{J(x,z)}\frac{J(z,x)}{J(y,x)} = -1$$

With the aid of Equation (B.19), this becomes

$$\frac{\partial x}{\partial z_y}\frac{\partial y}{\partial x_z}\frac{\partial z}{\partial y_x} = -1 \tag{B.20}$$

We also can write

$$\frac{J(x,w)}{J(z,w)}\frac{J(y,w)}{J(x,w)}\frac{J(z,w)}{J(y,w)} = 1$$

With Equation (B.19), this yields

$$\frac{\partial x}{\partial z_w}\frac{\partial y}{\partial x_w}\frac{\partial z}{\partial y_w} = 1 \tag{B.21}$$

Equations (B.20) and (B.21) are readily extended, providing all variables are functions of ξ and η.

The most general expression for a second-order partial derivative can involve up to five different variables, as shown on the left side of Equation (4) in Table B.1. As a check on this equation, we derive Equation (B.17a). This derivation requires setting $x_1 = \xi$, $x_2 = x_4 = x$, and $x_3 = x_5 = y$. Consequently, Equation (4) in Table B.1 becomes

$$\frac{\partial^2 \xi}{\partial x_y^2} = \frac{J(x,y)J[J(\xi,y),y] - J(\xi,y)J[J(x,y),y]}{[J(x,y)]^3} \tag{B.22}$$

For the terms in the numerator, we use

$$J(x,y) = x_\xi y_\eta - x_\eta y_\xi$$

$$J(\xi,y) = \begin{vmatrix} \dfrac{\partial \xi}{\partial \eta} = 1 & \dfrac{\partial \xi}{\partial \eta} = 0 \\ \dfrac{\partial y}{\partial \xi} & \dfrac{\partial y}{\partial \eta} \end{vmatrix} = y_\eta$$

$$J[J(\xi,y),y] = y_\eta y_{\xi\eta} - y_\xi y_{\eta\eta}$$

$$\frac{\partial J(x,y)}{\partial \xi} = y_\eta x_{\xi\xi} + x_\xi y_{\xi\eta} - y_\xi x_{\xi\eta} - x_\eta y_{\xi\xi}$$

$$\frac{\partial J(x,y)}{\partial \eta} = y_\eta x_{\xi\eta} + x_\xi y_{\eta\eta} - y_\xi x_{\eta\eta} - x_\eta y_{\xi\eta}$$

$$J[J(x,y),y] = \frac{\partial J(x,y)}{\partial \xi} y_\eta - \frac{\partial J(x,y)}{\partial \eta} y_\xi$$

When the above are substituted into the numerator of Equation (B.22), we obtain Equation (B.17a). Note that Equation (13) in Table B.2 can be used in place of the equation for $J[J(x,y),y]$, but that this does not simplify the derivation.

Table B.2 contains a partial summary of derived Jacobian equations, in which the variables x, y, z, and w are functions of ξ and η. This list is not complete. For instance, if

$$dx = f(g)dg \tag{B.23a}$$

then

$$x = \int f(g)dg + k \tag{B.23b}$$

where k is an arbitrary constant. Suppose $g = g(\xi,\eta)$; let us determine $J(x,y)$, first using Equations (2) and (6) in Table B.2:

$$J(x,y) = J\left(\int f(g)\,dg + k,\, y\right) = J\left(\int f(g)\,dg,\, y\right) + J(k,y) = J\left(\int f(g)\,dg,\, y\right)$$

We now use Equation (9) in the table for the Jacobian on the right side with $f_b(y) = 1$, $f_a = f$, and $x = g$, to obtain

$$J(x,y) = J\left(\int f(g)\,dg, y\right) = f(g)J(g,y) \tag{B.23c}$$

Equation (B.23b) can be viewed as the quadrature solution to Equation (B.23a) or (B.23c).

This result can be generalized in the following manner. Consider the relation

$$dx = f_a(g_a)\,dg_a + f_b(g_b)\,dg_b + \cdots \tag{B.24a}$$

Then by integration, we have

$$x = \int f_a dg_a + \int f_b dg_b + \cdots + k$$

The result that is equivalent to Equation (B.23c) is

$$J(x,y) = f_a(g_a)J(g_a, y) + f_b(g_b)J(g_b, y) + \cdots \tag{B.24b}$$

Equations (14), (17), and (18) in Table B.2 are useful when dealing with second-order partial derivatives. Observe that Equations (3), (14), and (15) in the table require ξ and η as the independent variables.

As an illustration, we prove Equation (9) of Table B.2. Starting with the Jacobian on the left, we have

$$J[f_a(x), f_b(y)] = \begin{vmatrix} \dfrac{df_a}{dx}\dfrac{\partial x}{\partial \xi} & \dfrac{df_a}{dx}\dfrac{\partial x}{\partial \eta} \\ \dfrac{df_b}{dy}\dfrac{\partial y}{\partial \xi} & \dfrac{df_b}{dy}\dfrac{\partial y}{\partial \eta} \end{vmatrix} = \frac{df_a}{dx}\frac{df_b}{dy}\begin{vmatrix} \dfrac{\partial x}{\partial \xi} & \dfrac{\partial x}{\partial \eta} \\ \dfrac{\partial y}{\partial \xi} & \dfrac{\partial y}{\partial \eta} \end{vmatrix} = \frac{df_a}{dx}\frac{df_b}{dy}J(x,y)$$

which is the desired result.

*n*TH-ORDER JACOBIANS

The Jacobian $J(x,y)$ associated with Equation (B.3) is of second order. For the transformation (Crawford, 1949)

$$Z_j = Z_j(X_1, \ldots, X_n), \quad J = 1, \ldots, n \tag{B.25}$$

we have the *n*th-order Jacobian

$$J(Z_1, \ldots, Z_n) = \frac{\partial(Z_1, \ldots, Z_n)}{\partial(X_1, \ldots, X_n)} = \begin{vmatrix} \dfrac{\partial Z_1}{\partial X_1} & \dfrac{\partial Z_2}{\partial X_1} & \cdots \\ \vdots & \vdots & \dfrac{\partial Z_n}{\partial X_n} \end{vmatrix} \tag{B.26}$$

where

$$\frac{\partial Z_1}{\partial X_1} = \left(\frac{\partial Z_1}{\partial X_1}\right)_{X_2,\ldots,X_n}, \text{ etc.}$$

For the Jacobian in Equation (B.26), we have the following self-evident rules:

(1) The sign of J is changed whenever any pair of X or Z is interchanged, provided the order of the rest is preserved. This property stems directly from the sign change of a determinant when two columns or two rows are interchanged.
(2) Whenever a common variable occurs among the Z and X, a reduction in order can take place.

Thus, we have

$$\frac{\partial(X_1, Z_2,\ldots, Z_n)}{\partial(X_1,\ldots, X_n)} = \frac{\partial(Z_2,\ldots, Z_n)}{\partial(X_2,\ldots, X_n)_{X_1}} \tag{B.27a}$$

where the X_1 subscript is held constant throughout. The subscript, if not written, is understood. Similarly,

$$\frac{\partial(X_1,\ldots, X_m, Z_{m+1},\ldots, Z_n)}{\partial(X_1,\ldots, X_n)} = \frac{\partial(Z_{m+1},\ldots, Z_n)}{\partial(X_{m+1},\ldots, X_n)_{X_1,\ldots,X_m}} \tag{B.27b}$$

and if $m = n - 1$, we have

$$\frac{\partial(X_1,\ldots, X_{n-1}, Z_n)}{\partial(X_1,\ldots, X_n)} = \frac{\partial Z_n}{\partial X_{n_{X_1,\ldots,X_{n-1}}}} \tag{B.27c}$$

Thus, a first-order Jacobian is merely an ordinary partial derivative. The converse also holds in that any first-order partial derivative can be written as an nth-order Jacobian with $n - 1$ common Z and X arguments. This rule represents the generalization of Equation (B.19).

(3) A necessary and sufficient condition that Equation (B.25) has a unique inverse is that

$$J(Z_1,\ldots, Z_n) \neq 0.$$

(4) The general transformation property

$$\frac{\partial(Z_1,\ldots, Z_n)}{\partial(X_1,\ldots, X_n)} = \frac{\partial(Z_1,\ldots, Z_n)/\partial(x_1,\ldots, x_n)}{\partial(X_1,\ldots, X_n)/\partial(x_1,\ldots, x_n)} = \frac{J(Z_1,\ldots, Z_n)}{J(X_1,\ldots, X_n)} \tag{B.28}$$

holds, where in both Jacobians on the right, the independent variables are $x_1,\ldots, x_n$. The leftmost term is not a ratio but is a unified symbol, whereas the rightmost term is, in fact, a ratio. Because of the equality, the leftmost term can be treated in algebraic manipulations as if it were a ratio.

An immediate and useful consequence of Equations (B.27) and (B.28) is

$$\left(\frac{\partial Z_n}{\partial X_n}\right)_{X_1,\ldots,X_{n-1}} = \frac{1}{(\partial X_n/\partial Z_n)_{X_1,\ldots,X_{n-1}}} \tag{B.29}$$

A relation such as Equation (6) in Table B.1 generalizes to

$$\begin{aligned} &J(Z_1, Z_2, \ldots, Z_{n-1})dZ_n + J(Z_2, Z_3, \ldots, Z_n)dZ_1 + J(Z_3, Z_4, \ldots, Z_n, Z_1)dZ_2 \\ &+ \cdots + J(Z_n, Z_1, \ldots, Z_{n-2})dZ_{n-1} = 0 \end{aligned}$$

Second derivatives are developed in a manner consistent with Equation (4) in Table B.1. For example, suppose that $n = 3$ in Equation (B.25), and we have the derivative

$$\begin{aligned} \frac{\partial^2 Z_1}{\partial X_1 \partial X_2} &= \frac{\partial}{\partial X_1}\left[\frac{\partial(Z_1, X_1, X_3)}{\partial(X_2, X_1, X_3)}\right] = -\frac{\partial}{\partial X_1}\left[\frac{\partial(Z_1, X_1, X_3)}{\partial(X_1, X_2, X_3)}\right] = -\frac{\partial}{\partial X_1}[J(Z_1, X_1, X_3)] \\ &= -\frac{\partial[J(Z_1, X_1, X_3), X_2, X_3]}{\partial(X_1, X_2, X_3)} = -J[J(Z_1, X_1, X_3), X_2, X_3] \end{aligned}$$

As is evident from this equation and Equation (B.22), the Jacobian method appears to be awkward when applied to second-order derivatives. However, this appearance is deceptive, since it is simpler and more straightforward than conventional methods, especially for complicated second-order derivatives.

REFERENCES

Courant, R., *Differential and Integral Calculus,* Vol. II, Wiley, New York, 142 (1936).
Crawford, F.H., "Jacobian Methods in Thermodynamics," *Am. J. Phys.* 17, 1 (1949).
Crawford, F.H., "Thermodynamic Relations in *n*-Variable Systems in Jacobian Form: Pt. I, General Theory and Application to Unrestricted Systems," *Proc. Am. Acad. Arts Sci.* 78, 165 (1950).
Shaw, A.N., "The Derivation of Thermodynamic Relations for a Simple System," *R. Soc. Lond. Philos. Trans.* Ser A 234, 299 (1935).

Appendix C
Oblique Shock Wave Angle

Let β be the shock wave angle and θ be the velocity turn angle. Both are measured with respect to the velocity upstream of the shock, as pictured in the sketch:

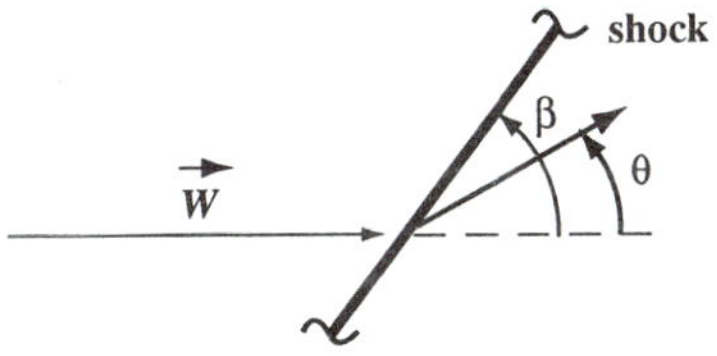

The two angles are related by the gas dynamic equation

$$\tan\theta = \cot\beta \frac{M^2\sin^2\beta - 1}{1 + \left(\frac{\gamma+1}{2} - \sin^2\beta\right)M^2} \tag{C.1}$$

where γ is the ratio of specific heats and M is the Mach number upstream of the shock. Equation (C.1) is an explicit relation for θ. Since θ represents the usually known wall turn angle, an explicit, computer-friendly equation for β is desirable. Indeed, the derivation of this relation is placed in a separate appendix because of its frequent usage in gas dynamics.

Thompson (1950) may have been the first to observe that Equation (C.1) could be written as a cubic in $\sin^2\beta$:

$$\sin^6\beta - \left(\frac{M^2+2}{M^2} + \gamma\sin^2\theta\right)\sin^4\beta + \left\{\frac{2M^2+1}{M^4} + \left[\left(\frac{\gamma+1}{2}\right)^2 + \frac{\gamma-1}{M^2}\right]\sin^2\theta\right\}$$
$$\times \sin^2\beta - \frac{\cos^2\theta}{M^4} = 0 \tag{C.2}$$

It is analytically convenient, however, to recast this relation as a cubic in $\tan\beta$, with the result

$$F(\beta) = \left(1 + \frac{\gamma-1}{2}M^2\right)\tan\theta\tan^3\beta - (M^2-1)\tan^2\beta + \left(1 + \frac{\gamma+1}{2}M^2\right)\tan\theta\tan\beta + 1 = 0 \tag{C.3}$$

Observe that the coefficients in this polynomial are appreciably simpler than those in Equation (C.2). We assume $M > 1$ and $\theta > 0$, and note that

$$F(-\pi/2) = -\infty, \qquad F(0) = 1, \qquad F(\pi/2) = \infty$$

From these values, we deduce that $F(\beta)$ has three real, unequal roots for an attached shock wave. The negative β root is not physical, while the two roots between $\beta = 0$ and $\pi/2$ correspond to the weak and strong shock solutions.

Equation (C.3) is next recast into the standard form

$$x^3 + ax + b = 0 \tag{C.4}$$

for solving a cubic equation, where

$$x = \tan\beta - \frac{M^2 - 1}{3\left(1 + \frac{\gamma-1}{2}M^2\right)\tan\theta} \tag{C.5}$$

$$a = \frac{3\left(1 + \frac{\gamma-1}{2}M^2\right)\left(1 + \frac{\gamma+1}{2}M^2\right)\tan^2\theta - (M^2-1)^2}{3\left(1 + \frac{\gamma-1}{2}M^2\right)^2\tan^2\theta}$$

$$b = \frac{-2(M^2-1)^3 + 18\left(1 + \frac{\gamma-1}{2}M^2\right)\left(1 + \frac{\gamma-1}{2}M^2 + \frac{\gamma+1}{4}M^4\right)\tan^2\theta}{27\left(1 + \frac{\gamma-1}{2}M^2\right)^3\tan^3\theta}$$

Because the roots are real and unequal, the trigonometric solution of a cubic equation is particularly convenient. This solution requires the quantity

$$\chi = \left(-\frac{27b^2}{4a^3}\right)^{1/2} = \frac{(M^2-1)^3 - 9\left(1 + \frac{\gamma-1}{2}M^2\right)\left(1 + \frac{\gamma-1}{2}M^2 + \frac{\gamma+1}{4}M^4\right)\tan^2\theta}{\left[(M^2-1)^2 - 3\left(1 + \frac{\gamma-1}{2}M^2\right)\left(1 + \frac{\gamma+1}{2}M^2\right)\tan^2\theta\right]^{3/2}} \tag{C.6}$$

The three solutions of Equation (C.4) are contained among the six expressions

$$\pm\left(-\frac{4a}{3}\right)^{1/2}\cos\frac{\phi}{3}, \quad \pm\left(-\frac{4a}{3}\right)^{1/2}\cos\left(\frac{\phi + 2\pi}{3}\right), \quad \pm\left(-\frac{4a}{3}\right)^{1/2}\cos\left(\frac{\phi + 4\pi}{3}\right)$$

where $\phi = \cos^{-1}\chi$.

In particular, the weak and strong solutions are given by

$$x_{weak} = \left(-\frac{4a}{3}\right)^{1/2}\cos\left(\frac{\phi + 4\pi}{3}\right), \qquad x_{strong} = \left(-\frac{4a}{3}\right)^{1/2}\cos\left(\frac{\phi}{3}\right)$$

where

$$\left(-\frac{4a}{3}\right)^{1/2} = \frac{2\left[(M^2-1)^2 - 3\left(1 + \frac{\gamma-1}{2}M^2\right)\left(1 + \frac{\gamma+1}{2}M^2\right)\tan^2\theta\right]^{1/2}}{3\left(1 + \frac{\gamma-1}{2}M^2\right)\tan\theta}$$

With the aid of Equation (C.5), a computationally convenient form for β is

$$\lambda = \left[(M^2-1)^2 - 3\left(1+\frac{\gamma-1}{2}M^2\right)\left(1+\frac{\gamma+1}{2}M^2\right)\tan^2\theta\right]^{1/2} \quad \textbf{(C.7a)}$$

$$\chi = \frac{(M^2-1)^3 - 9\left(1+\frac{\gamma-1}{2}M^2\right)\left(1+\frac{\gamma-1}{2}M^2+\frac{\gamma+1}{4}M^4\right)\tan^2\theta}{\lambda^3} \quad \textbf{(C.7b)}$$

$$\tan\beta = \frac{M^2-1+2\lambda\cos[(4\pi\delta+\cos^{-1}\chi)/3]}{3\left(1+\frac{\gamma-1}{2}M^2\right)\tan\theta} \quad \textbf{(C.7c)}$$

where the angle $(4\pi\delta + \cos^{-1}\chi)/3$ is in radians. The strong shock solution is provided by $\delta = 0$, while $\delta = 1$ yields the weak shock solution, and $|\chi| \le 1$ for an attached shock.

The author discovered the Equations (C.7) solution in March 1991. It was quickly submitted and accepted for publication by the AIAA journal. Before it was scheduled to appear, however, it was learned that the solution, in a different form, had already been published. In fact, it has repeatedly appeared in journals; e.g., see Mascitti (1969) and Wolf (1993).

REFERENCES

Mascitti, V.R., "A Closed-Form Solution to Oblique Shock-Wave Properties," *J. Aircraft* 6, 66 (1969).

Thompson, M.J., "A Note on the Calculation of Oblique Shock-Wave Characteristics," *J. Aeron. Sci.* 17, 774 (1950).

Wolf, T., "Comment on 'Approximate Formula of Weak Oblique Shock Wave Angle,'" *AIAA J.* 31, 1363 (1993).

Appendix D Conditions on the Downstream Side of a Steady Shock Wave in a Two-Dimensional or Axisymmetric Flow of a Perfect Gas

1. Jump Conditions

$$m \equiv M_1^2, \qquad w \equiv m \sin^2\beta, \qquad X \equiv 1 + \frac{\gamma - 1}{2} w,$$

$$Y \equiv \gamma w - \frac{\gamma - 1}{2}, \qquad (\rho w^2)_1 = \gamma p_1 m$$

$$u_2 = w_1 \cos\beta$$

$$v_2 = \frac{2}{\gamma + 1} w_1 \frac{X}{m \sin\beta}$$

$$p_2 = \frac{2}{\gamma(\gamma + 1)} (\rho w^2)_1 \frac{Y}{m}$$

$$\rho_2 = \frac{\gamma + 1}{2} \rho_1 \frac{w}{X}$$

$$M_2^2 = \left[1 + \left(\frac{\gamma + 1}{2} \frac{m \sin\beta \cos\beta}{X}\right)^2\right] \frac{X}{Y}$$

$$M_2^2 - 1 = \frac{\gamma + 1}{4} \frac{(\gamma + 1) m w + 2 + (\gamma - 3) w - 2\gamma w^2}{XY}$$

$$1 + \frac{\gamma - 1}{2} M_2^2 = \left(\frac{\gamma + 1}{2}\right)^2 \frac{[1 + (\gamma - 1) m/2] w}{XY}$$

$$p_{o,2} = \frac{2}{\gamma(\gamma + 1)} (\rho w^2)_1 \left(1 + \frac{\gamma - 1}{2} M_2^2\right)^{\gamma/(\gamma - 1)} \frac{Y}{m}$$

$$\tan\theta = \frac{1}{\tan\beta} \frac{w - 1}{(\gamma + 1) m/2 + 1 - w}$$

$$\sin(\beta - \theta) = \left[1 + \left(\frac{\gamma + 1}{2} \frac{m \sin\beta \cos\beta}{X}\right)^2\right]^{-1/2}$$

$$\cos(\beta - \theta) = \frac{\gamma + 1}{2} \left[1 + \left(\frac{\gamma + 1}{2} \frac{m \sin\beta \cos\beta}{X}\right)^2\right]^{-1/2} \frac{m \sin\beta \cos\beta}{X}$$

2. Tangential Derivatives

$$\left(\frac{\partial u}{\partial s}\right)_2 = -w_1\beta'\sin\beta$$

$$\left(\frac{\partial v}{\partial s}\right)_2 = -\frac{2}{\gamma+1}w_1\left(1-\frac{\gamma-1}{2}w\right)\frac{\beta'\cos\beta}{w}$$

$$\left(\frac{\partial p}{\partial s}\right)_2 = \frac{4}{\gamma+1}(\rho w^2)_1\beta'\sin\beta\cos\beta$$

$$\left(\frac{\partial \rho}{\partial s}\right)_2 = (\gamma+1)\rho_1\frac{\beta' m\sin\beta\cos\beta}{X^2}$$

$$\left(\frac{\partial M^2}{\partial s}\right)_2 = -\frac{(\gamma+1)^2}{2}\left(1+\frac{\gamma-1}{2}m\right)(1+\gamma w^2)\frac{\beta' m\sin\beta\cos\beta}{(XY)^2}$$

$$\left(\frac{\partial p_o}{\partial s}\right)_2 = -\frac{2}{\gamma+1}(\rho w^2)_1\left(1+\frac{\gamma-1}{2}M_2^2\right)^{\gamma/(\gamma-1)}\frac{\beta'(w-1)^2}{mX\tan\beta}$$

$$\left(\frac{d\theta}{ds}\right)_2 = \frac{\frac{\gamma+1}{2}m(1+w)+1-2w-\gamma w^2}{(\gamma+1)\left(1+\frac{\gamma+1}{4}m\right)w+1-2w-\gamma w^2}\beta'$$

$$\left(\frac{\partial \mu}{\partial s}\right)_2 = -\frac{1}{2M_2^2(M_2^2-1)^{1/2}}\left(\frac{\partial M^2}{\partial s}\right)_2$$

3. Normal Derivatives

$$\left(\frac{\partial u}{\partial n}\right)_2 = w_1\frac{g_1\beta'\cos\beta}{X}$$

$$\left(\frac{\partial v}{\partial n}\right)_2 = w_1\left[g_2-\frac{2}{\gamma+1}m(1+3w)\right]\frac{\beta'}{m(w-1)\sin\beta}-\left(\frac{2}{\gamma+1}\right)^2 w_1\frac{\sigma Y}{my\tan\beta}$$

$$\left(\frac{\partial p}{\partial n}\right)_2 = \frac{1}{\gamma+1}(\rho w^2)_1\frac{(mg_5+g_6)\beta'}{mX(w-1)}+\left(\frac{2}{\gamma+1}\right)^2(\rho w^2)_1\frac{\sigma Y\cos\beta}{my}$$

$$\left(\frac{\partial \rho}{\partial n}\right)_2 = \rho_1(mg_3+g_4)\frac{\beta' w}{X^3(w-1)}+\rho_1\frac{\sigma w\cos\beta}{yX}$$

$$\left(\frac{\partial M^2}{\partial n}\right)_2 = -(\gamma+1)\frac{\left(1+\frac{\gamma-1}{2}m\right)w}{XY^2}\left[\frac{(mg_7+g_8)\beta'}{4X^2(w-1)}+\frac{\sigma Y\cos\beta}{y}\right]$$

$$\left(\frac{\partial p_o}{\partial n}\right)_2 = (\rho w^2)_1\left(1+\frac{\gamma-1}{2}M_2^2\right)^{\gamma/(\gamma-1)}\left[\frac{(w-1)\cos\beta}{X}\right]^2\beta'$$

4. $g_i(\gamma, w)$

$$g_1 = \frac{1}{\gamma+1}[-(\gamma+5)+(3-\gamma)w]$$

$$g_2 = \frac{2}{(\gamma+1)^2}[(\gamma-1)-2(\gamma-1)w+(5\gamma+3)w^2]$$

$$g_3 = \frac{\gamma+1}{4}[2(\gamma+1)+(3-\gamma)w+3(\gamma-1)w^2]$$

$$g_4 = \frac{1}{2}[2-(\gamma^2+\gamma+6)w+(\gamma^2-4\gamma+1)w^2-(\gamma-1)(2\gamma+1)w^3]$$

$$g_5 = -(\gamma-1)+(\gamma+5)w+2(2\gamma-1)w^2$$

$$g_6 = \frac{2}{\gamma+1}[-(\gamma-1)+2(2\gamma-1)w+(\gamma^2-7\gamma-2)w^2-(3\gamma^2-1)w^3]$$

$$g_7 = \frac{\gamma+1}{2}[4-(\gamma-1)(\gamma+3)w+(\gamma^2+18\gamma-3)w^2-4\gamma(2-\gamma)w^3]$$

$$\begin{aligned} g_8 = {} & -2(\gamma-1)+2(\gamma-1)(3-\gamma)w+(9\gamma^2-14\gamma+1)w^2 \\ & +(\gamma^3-17\gamma^2-\gamma+1)w^3+\gamma(-3\gamma^2+4\gamma+3)w^4 \end{aligned}$$

Appendix E Method-of-Characteristics for a Single, First-Order PDE

There are many ways to introduce the method-of-characteristics (MOC). Here interest is limited to a single, first-order, linear or quasilinear partial differential equation (PDE). Our approach is thus specifically tailored for the task at hand.

For purpose of generality, we consider an inhomogeneous equation for the dependent variable f:

$$\sum_{i=0}^{n-1} a_i \frac{\partial f}{\partial x_i} + a_n = 0 \tag{E.1}$$

where n is a positive integer. This equation is assumed to be quasilinear, in which case $a_0, \ldots, a_n$ can depend on the x_j and f, but not on any derivative of f. We further simplify the equation by noting that if f is a solution, then

$$G(x_0, \ldots, x_n) = f(x_0, \ldots, x_{n-1}) + x_n \tag{E.2}$$

is a solution of the homogeneous equation

$$\sum_{i=0}^{n} a_i \frac{\partial G}{\partial x_i} = 0 \tag{E.3}$$

Thus, by adding a new independent variable, x_n, the inhomogeneous term in Equation (E.1) is incorporated into (E.3).

General Solution

Observe that G = constant is a solution of Equation (E.3). This constant may be taken as zero. We therefore seek a solution with the form

$$G(x_0, \ldots, x_n) = 0 \tag{E.4}$$

The remainder of the subsection provides this solution.

It is conceptually convenient to introduce an $(n + 1)$-dimensional Cartesian space that has an orthonormal basis $\hat{\imath}_i$. Thus, the gradient of G is

$$\nabla G = \sum_{i=0}^{n} \frac{\partial G}{\partial x_i} \hat{\imath}_i \tag{E.5}$$

and a vector $\vec{A}$ can be defined that is based on the a_i coefficients

$$\vec{A} = \sum_{i=0}^{n} a_i \hat{i}_i$$

Hence, Equation (E.3) becomes

$$\vec{A} \cdot \nabla G = 0 \tag{E.6}$$

Equation (E.4) represents a surface in an $(n + 1)$-dimensional space, and the gradient ∇G is everywhere normal to this surface. On the other hand, $\vec{A}$ is perpendicular to ∇G and therefore $\vec{A}$ is tangent to the surface. Thus, the solution of Equation (E.3) or (E.6) is a surface that is tangent to $\vec{A}$.

Consider a characteristic curve that lies on the surface given by Equation (E.4) and everywhere is tangent to $\vec{A}$. The surface can be viewed as consisting of an infinite number of these curves. Moreover, each of these curves constitutes a solution of Equation (E.3).

We need to construct a curve in the $(n + 1)$-dimensional space whose coordinates are $x_0,\ldots, x_n$. For example, in three dimensions a curve is determined by the intersection of two surfaces. More generally, the characteristic curve we seek is determined by the intersection of the n surfaces

$$\begin{aligned} u^{(0)}(x_0,\ldots,x_n) &= c_0 \\ u^{(1)}(x_0,\ldots,x_n) &= c_1 \\ &\vdots \\ u^{(n-1)}(x_0,\ldots,x_n) &= c_{n-1} \end{aligned} \tag{E.7}$$

where the c_j are constants and the first equation is sometimes written as $u = c$. We have a different curve for each choice of the c_j.

Since $\vec{A}$ is tangent to a characteristic curve, the differential change in the x_i along such a curve must stand in the same relationship to each other as the corresponding components of $\vec{A}$. Thus, on a characteristic curve, we have

$$\frac{dx_0}{a_0} = \frac{dx_1}{a_1} = \cdots = \frac{dx_n}{a_n} \tag{E.8}$$

As noted, G is constant along a characteristic curve. We therefore see from Equation (E.2) that dx_n can be replaced with $-df$. This change is usually convenient, since the a_i are functions of $x_0,\ldots, x_{n-1}$ and f. Equations (E.8) are n-coupled, first-order ODEs that relate the x_i along a characteristic curve. The unique solution of these equations is provided by Equation (E.7), where the c_j are constants of integration. We thus have reduced the problem of solving a first-order partial differential equation to solving n-coupled ODEs. As will become apparent, this reduction is advantageous whether Equation (E.3) is to be solved analytically or numerically.

We now see why Equation (E.3) is limited to being quasilinear. If one of the a_i depended on a derivative of f, then one of Equations (E.8) would not be an ODE, and the theory would collapse. Normally, the MOC applies only to hyperbolic equations. For Equation (E.3), this qualification is unnecessary. The only essential restriction is that it be quasilinear.

Note that $\vec{A} \cdot \nabla G$ is also the derivative of G along a characteristic curve. Equation (E.6) therefore means that G has a constant value along any particular characteristic curve. For this to be so, G can depend on the x_i only in combinations such that $dG = 0$ along any characteristic curve.

However, the $u^{(j)}$ depend on the x_i but have a constant value along any characteristic curve. Consequently, G is an arbitrary function of the $u^{(j)}$. The general solution of Equation (E.3) is thus

$$G(u^{(0)},u^{(0)},\ldots,u^{(n-1)}) = 0 \tag{E.9}$$

If one or more of the a_i depend on f, or if $a_n \neq 0$, then f explicitly appears in the $u^{(j)}$, and Equation (E.9) is a solution of Equation (E.1). On the other hand, if none of the a_i involves f and $a_n = 0$, then the general solution of Equation (E.1) can be written as

$$f = f(u^{(0)}, u^{(1)}, \ldots, u^{(n-2)})$$

Discussion

We verify that Equation (E.9) is a solution of Equation (E.3) by first evaluating $du^{(j)}$ with the aid of Equations (E.8)

$$du^{(j)} = \sum_{i=0}^{n} \frac{\partial u^{(j)}}{\partial x_i} dx_i = \frac{dx_0}{a_0} \sum_{i=0}^{n} a_i \frac{\partial u^{(j)}}{\partial x_i}$$

where we assume one a_i, say a_0, is nonzero. We next obtain

$$dG = \sum_{j=0}^{n-1} \frac{\partial G}{\partial u^{(j)}} du^{(j)} = \frac{dx_0}{a_0} \sum_{j=0}^{n-1} \frac{\partial G}{\partial u^{(j)}} \sum_{j=0}^{n} a_i \frac{\partial u^{(j)}}{\partial x_i}$$

$$= \frac{dx_0}{a_0} \sum_{j=0}^{n} a_i \sum_{i=0}^{n-1} \frac{\partial G}{\partial u^{(j)}} \frac{\partial u^{(j)}}{\partial x_i} = \frac{dx_0}{a_0} \sum_{i=0}^{n} a_i \frac{\partial G}{\partial x_i} = 0$$

in accordance with Equation (E.3).

The functional form of G is determined by an initial, or boundary, condition. Without loss of generality, this condition may be specified at $x_0 = 0$ as

$$G_0 = G[u^{(0)}(0,x_1,\ldots,x_n),\ldots,u^{(n-1)}(0,x_1,\ldots,x_n)]$$

where G_0 is the prescribed relation for G at $x_0 = 0$.

As we have mentioned, the unique solution of Equations (E.8) can be written as Equation (E.7). An analytical solution of Equations (E.8) may require inverting some of Equations (E.7). For example, suppose $n = 3$ and we have obtained a solution, $u = c$, to

$$\frac{dx_0}{a_0} = \frac{dx_1}{a_1}$$

Further, suppose a_2 depends on x_0, x_1, and x_2. If $u = c$ can be explicitly solved for x_0, we would then integrate

$$\frac{dx_1}{a_1} = \frac{dx_2}{a_2}$$

with x_0 eliminated. Similarly, if $u = c$ is more readily solved for x_1, we could obtain $u^{(1)}$ by integrating

$$\frac{dx_0}{a_0} = \frac{dx_2}{a_2}$$

instead. In either case, the elimination of x_0 (or x_1) from the dx_2 equation is consistent with obtaining a simultaneous solution of Equations (E.8).

Illustrative Example

As an example, we find the general solution to

$$xz\frac{\partial z}{\partial x} + yz\frac{\partial z}{\partial y} = xy$$

We first solve the characteristic equations

$$\frac{dx}{xz} = \frac{dy}{yz} = \frac{dz}{xy}$$

From the leftmost equation, we have

$$\frac{dx}{x} = \frac{dy}{y}$$

which integrates to

$$u = \frac{y}{x} = c$$

For a second equation, we use

$$\frac{dx}{z} = \frac{dz}{y}$$

or by elimination of y

$$\begin{aligned} cxdx &= zdz \\ cx^2 &= z^2 - c_1 \\ \left(\frac{y}{x}\right)x^2 &= z^2 - c_1 \\ u^{(1)} &= z^2 - xy = c_1 \end{aligned}$$

Hence, the general solution to the PDE is

$$G(z^2 - xy, y/x) = 0$$

which is readily verified by direct substitution. An alternate form for the solution can be written as

$$z^2 - xy = g(y/x)$$

or as

$$z = \pm[xy + g(y/x)]^{1/2}$$

where g is an arbitrary function of its argument.

Appendix F
Tangential Derivatives on the Downstream Side of a Shock in the ξ_1 and ξ_2 Directions

1. Equations for the Derivatives

$$\frac{\partial w_2^*}{\partial \xi_j} + w_1^* \frac{\cos\beta \sin(\beta-\theta)}{\cos^2(\beta-\theta)} \frac{\partial \theta}{\partial \xi_j} = A_1$$

$$\frac{\partial \rho_2}{\partial \xi_j} - \rho_1 \frac{\tan\beta}{\sin^2(\beta-\theta)} \frac{\partial \theta}{\partial \xi_j} = A_2$$

$$\frac{\partial p_2}{\partial \xi_j} - \frac{1}{2}(\rho w^{*2})_1 \frac{\sin 2\beta}{\cos^2(\beta-\theta)} \frac{\partial \theta}{\partial \xi_j} = A_3$$

$$h_{\rho,2}\frac{\partial \rho_2}{\partial \xi_j} + h_{p,2}\frac{\partial p_2}{\partial \xi_j} - w_1^{*2} \frac{\cos^2\beta \sin(\beta-\theta)}{\cos^3(\beta-\theta)} \frac{\partial \theta}{\partial \xi_j} = A_4$$

The inhomogeneous terms are:

$$A_1 = \frac{\cos\beta}{\cos(\beta-\theta)} \frac{\partial w_1^*}{\partial \xi_j} - w_1^* \frac{\sin\theta}{\cos^2(\beta-\theta)} \frac{\partial \beta}{\partial \xi_j}$$

$$A_2 = \frac{\tan\beta}{\tan(\beta-\theta)} \frac{\partial \rho_1}{\partial \xi_j} - \rho_1 \frac{\sin\theta\cos(2\beta-\theta)}{\cos^2\beta \sin^2(\beta-\theta)} \frac{\partial \beta}{\partial \xi_j}$$

$$A_3 = \frac{\partial p_1}{\partial \xi_j} + w_1^{*2} \frac{\sin\beta \sin\theta}{\cos(\beta-\theta)} \frac{\partial \rho_1}{\partial \xi_j} + 2(\rho w^*)_1 \frac{\sin\beta \sin\theta}{\cos(\beta-\theta)} \frac{\partial w_1^*}{\partial \xi_j}$$
$$+ \frac{1}{2}(\rho w^{*2})_1 \frac{\sin 2\theta}{\cos^2(\beta-\theta)} \frac{\partial \beta}{\partial \xi_j}$$

$$A_4 = h_{p,1}\frac{\partial p_1}{\partial \xi_j} + h_{\rho,1}\frac{\partial \rho_1}{\partial \xi_j} + w_1^* \frac{\sin\theta \sin(2\beta-\theta)}{\cos^2(\beta-\theta)} \frac{\partial w_1^*}{\partial \xi_j} - w_1^{*2} \frac{\cos\beta \sin\theta}{\cos^3(\beta-\theta)} \frac{\partial \beta}{\partial \xi_j}$$

2. Solution of the F1 Equations

$$\begin{aligned}
\Delta &= \text{determinant of the coefficients of the left side} \\
&= -w_1^{*2}\frac{\cos^2\beta\sin(\beta-\theta)}{\cos^3(\beta-\theta)} + \rho_1\frac{\tan\beta}{\sin^2(\beta-\theta)}h_{p,2} + \frac{1}{2}(\rho w^{*2})_1\frac{\sin 2\beta}{\cos^2(\beta-\theta)}h_{p,2}
\end{aligned}$$

$$\Delta\frac{\partial w_2^*}{\partial \xi_j} = \Delta A_1 + w_1^*\frac{\cos\beta\sin(\beta-\theta)}{\cos^2(\beta-\theta)}(h_{p,2}A_2 + h_{p,2}A_3 - A_4)$$

$$\begin{aligned}
\Delta\frac{\partial \rho_2}{\partial \xi_j} &= \frac{w_1^{*2}}{\cos^2(\beta-\theta)}\left[\frac{1}{2}\rho_1 h_{p,2}\sin 2\beta - \cos^2\beta\tan(\beta-\theta)\right]A_2 \\
&= \rho_1\frac{\tan\beta}{\sin^2(\beta-\theta)}(-h_{p,2}A_3 + A_4)
\end{aligned}$$

$$\begin{aligned}
\Delta\frac{\partial p_2}{\partial \xi_j} &= \left[\rho_1\frac{\tan\beta}{\sin^2(\beta-\theta)}h_{p,2} - w_1^{*2}\frac{\cos^2\beta\sin(\beta-\theta)}{\cos^3(\beta-\theta)}\right]A_3 \\
&\quad + \frac{1}{2}(\rho w^{*2})_1\frac{\sin 2\beta}{\cos^2(\beta-\theta)}(-h_{p,2}A_2 + A_4)
\end{aligned}$$

$$\Delta\frac{\partial \theta}{\partial \xi_j} = -(h_{p,2}A_2 + h_{p,2}A_3 - A_4)$$

Appendix G Conservation and Vector Equations in Orthogonal Curvilinear Coordinates ξ_i

$$\phi = \text{scalar function}$$

$$\vec{A} = \text{vector function} = A_i\hat{e}_i$$

$$x_j = x_j(\xi_1, \xi_2, \xi_3), \quad j = 1, 2, 3$$

$$h_i^2 = \sum_{j=1}^{3}\left(\frac{\partial x_j}{\partial \xi_i}\right)^2, \quad i = 1, 2, 3$$

$$(ds)^2 = (h_i d\xi_i)^2$$

$$\nabla\phi = \frac{1}{h_i}\frac{\partial\phi}{\partial\xi_i}\hat{e}_i$$

$$\nabla^2\phi = \frac{1}{h_ih_2h_3}\left[\frac{\partial}{\partial\xi_1}\left(\frac{h_2h_3}{h_1}\frac{\partial\phi}{\partial\xi_1}\right) + \frac{\partial}{\partial\xi_2}\left(\frac{h_1h_3}{h_2}\frac{\partial\phi}{\partial\xi_2}\right) + \frac{\partial}{\partial\xi_3}\left(\frac{h_1h_2}{h_3}\frac{\partial\phi}{\partial\xi_3}\right)\right]$$

$$\nabla\cdot\vec{A} = \frac{1}{h_1h_2h_3}\left[\frac{\partial}{\partial\xi_1}(h_2h_3A_1) + \frac{\partial}{\partial\xi_2}(h_3h_1A_2) + \frac{\partial}{\partial\xi_3}(h_1h_2A_3)\right]$$

$$\nabla\times\vec{A} = \frac{1}{h_2h_3}\left[\frac{\partial}{\partial\xi_2}(h_3A_3) - \frac{\partial}{\partial\xi_3}(h_2A_2)\right]\hat{e}_1 + \frac{1}{h_3h_1}\left[\frac{\partial}{\partial\xi_3}(h_1A_1) - \frac{\partial}{\partial\xi_1}(h_3A_3)\right]\hat{e}_2$$

$$+ \frac{1}{h_1h_2}\left[\frac{\partial}{\partial\xi_1}(h_2A_2) - \frac{\partial}{\partial\xi_2}(h_1A_1)\right]\hat{e}_3$$

$$\frac{\partial\hat{e}_j}{\partial\xi_j} = \sum_{k\neq i}\left(\frac{\delta_{ik}}{h_j}\frac{\partial h_k}{\partial\xi_j} - \frac{\delta_{ij}}{h_k}\frac{\partial h_j}{\partial\xi_k}\right)\hat{e}_k$$

$$\vec{w} = v_i\hat{e}_i$$

$$\frac{D\phi}{Dt} = \frac{\partial\phi}{\partial t} + \frac{v_i}{h_i}\frac{\partial\phi}{\partial\xi_i}$$

$$\frac{D\vec{w}}{Dt} = \frac{\partial v_i}{\partial t}\hat{e}_i + \frac{v_i}{h_i}\frac{\partial v_j}{\partial\xi_i}\hat{e}_j + \frac{v_iv_j}{h_i}\frac{\partial\hat{e}_j}{\partial\xi_i}$$

$$\varepsilon_{11} = \frac{1}{h_1}\left(\frac{\partial v_1}{\partial\xi_1} + \frac{v_2}{h_2}\frac{\partial h_1}{\partial\xi_2} + \frac{v_3}{h_3}\frac{\partial mph_1}{\partial\xi_3}\right)$$

$$\varepsilon_{22} = \frac{1}{h_2}\left(\frac{\partial v_2}{\partial\xi_2} + \frac{v_3}{h_3}\frac{\partial h_2}{\partial\xi_3} + \frac{v_1}{h_1}\frac{\partial h_2}{\partial\xi_1}\right)$$

$$\varepsilon_{33} = \frac{1}{h_3}\left(\frac{\partial v_3}{\partial \xi_3} + \frac{v_1}{h_1}\frac{\partial h_3}{\partial \xi_1} + \frac{v_2}{h_2}\frac{\partial h_3}{\partial \xi_2}\right)$$

$$\varepsilon_{12} = \varepsilon_{21} = \frac{1}{2}\left[\frac{h_2}{h_1}\frac{\partial}{\partial \xi_1}\left(\frac{v_2}{h_2}\right) + \frac{h_1}{h_2}\frac{\partial}{\partial \xi_2}\left(\frac{v_1}{h_1}\right)\right]$$

$$\varepsilon_{23} = \varepsilon_{32} = \frac{1}{2}\left[\frac{h_3}{h_2}\frac{\partial}{\partial \xi_2}\left(\frac{v_3}{h_3}\right) + \frac{h_2}{h_3}\frac{\partial}{\partial \xi_3}\left(\frac{v_2}{h_2}\right)\right]$$

$$\varepsilon_{31} = \varepsilon_{13} = \frac{1}{2}\left[\frac{h_1}{h_3}\frac{\partial}{\partial \xi_3}\left(\frac{v_1}{h_1}\right) + \frac{h_3}{h_1}\frac{\partial}{\partial \xi_1}\left(\frac{v_3}{h_3}\right)\right]$$

$$\nabla \cdot \vec{w} = \varepsilon_{ii}$$

$$\overleftrightarrow{\sigma} = \sigma_{ij}\hat{e}_i\hat{e}_j$$

$$\sigma_{ij} = 2\mu\varepsilon_{ij} + \delta_{ij}(-p + \lambda\nabla \cdot \vec{w})$$

$$\Phi = \mu\left[2\sum_{i=1}^{3}\varepsilon_{ii}^2 + 4(\varepsilon_{12}^2 + \varepsilon_{23}^2 + \varepsilon_{31}^2)\right] + \lambda(\nabla \cdot \vec{w})^2$$

Continuity:

$$\frac{1}{\rho}\frac{D\rho}{Dt} + \nabla \cdot \vec{w} = 0$$

Momentum:

$$\rho\frac{D\vec{w}}{Dt} = \nabla \cdot \overleftrightarrow{\sigma} = -\nabla p + \vec{F}^s + \nabla\lambda(\nabla \cdot \vec{w})$$

where

$$\vec{F}^s = F_i^s\hat{e}_i = 2\nabla \cdot (\mu\overleftrightarrow{e})$$

$$F_i^s = \frac{2}{h_1h_2h_3}\sum_{j=1}^{3}\frac{\partial}{\partial \xi_j}\left(\frac{h_1h_2h_3}{h_j}\mu\varepsilon_{ij}\right) + \frac{2\mu}{h_i}\sum_{j\neq 1}^{3}\frac{1}{h_j}\left(\varepsilon_{ij}\frac{\partial h_i}{\partial \xi_j} - \varepsilon_{jj}\frac{\partial h_j}{\partial \xi_i}\right), \qquad i = 1, 2, 3$$

Energy:

$$\rho\frac{Dh}{Dt} = \frac{Dp}{Dt} + \kappa\nabla^2 T + \sum_{i=1}^{3}\frac{1}{h_i^2}\frac{\partial \kappa}{\partial \xi_i}\frac{\partial T}{\partial \xi_i} + \Phi$$

Appendix H
Conservation Equations in Body-Oriented Coordinates

A. Assumptions

(i) Motionless wall

(ii) Fourier's equation

(iii) Newtonian fluid

(iv) Two-dimensional or axisymmetric flow (without swirl)

B. Geometrical Factors

Two-dimensional: $\sigma = 0, \quad \phi = z$

Axisymmetric: $\sigma = 1, \quad \phi =$ azimuthal angle

$$\xi_1 = s, \qquad h_1 = h_s = 1 + kn$$

$$\xi_2 = n, \qquad h_2 = h_n = 1$$

$$\xi_3 = \phi, \qquad h_3 = h_\phi = r^\sigma = (r_w + n\cos\theta)^\sigma$$

k, r_w, and θ are functions only of s

$$\sin\theta = \frac{dr_w}{ds} = r'_w, \qquad k = -\frac{d\theta}{ds} = -\theta', \qquad \cos\theta = -\frac{r''_w}{k}$$

$$r = r_w + n\cos\theta$$

$$\frac{\partial h_s}{\partial s} = \frac{dk}{ds}n, \qquad \frac{\partial h_s}{\partial n} = k$$

$$\frac{\partial h_\phi}{\partial s} = \sigma\frac{\partial r}{\partial s} = \sigma(1+kn)\sin\theta, \qquad \frac{\partial h_\phi}{\partial n} = \sigma\cos\theta$$

$$\frac{\partial \hat{e}_s}{\partial s} = -k\hat{e}_n, \qquad \frac{\partial \hat{e}_s}{\partial n} = 0, \qquad \frac{\partial \hat{e}_s}{\partial \phi} = \sigma\sin\theta\hat{e}_\phi$$

$$\frac{\partial \hat{e}_n}{\partial s} = k\hat{e}_s, \qquad \frac{\partial \hat{e}_n}{\partial n} = 0, \qquad \frac{\partial \hat{e}_n}{\partial \phi} = \sigma\cos\theta\hat{e}_\phi$$

$$\frac{\partial \hat{e}_\phi}{\partial s} = 0, \qquad \frac{\partial \hat{e}_\phi}{\partial n} = 0, \qquad \frac{\partial \hat{e}_\phi}{\partial \phi} = -\sigma(\sin\theta\hat{e}_s + \cos\theta\hat{e}_n)$$

C. Dynamic Factors

$$\vec{w} = u\hat{e}_s + v\hat{e}_n$$

$$\frac{D}{Dt} = \frac{\partial}{\partial t} + \frac{u}{h_s}\frac{\partial}{\partial s} + v\frac{\partial}{\partial n}$$

$$\frac{D\vec{w}}{Dt} = \left(\frac{Du}{Dt} + \frac{k}{h_s}uv\right)\hat{e}_s + \left(\frac{Dv}{Dt} - \frac{k}{h_s}u^2\right)\hat{e}_n$$

$$\nabla\cdot\vec{w} = \frac{1}{h_s r^\sigma}\left[\frac{\partial}{\partial s}(r^\sigma u) + \frac{\partial}{\partial n}(h_s r^\sigma v)\right]$$

$$= \frac{1}{h_s}\left(\frac{\partial u}{\partial s} + h_s\frac{\partial v}{\partial n} + kv\right) + \frac{\sigma}{r}(u\sin\theta + v\cos\theta)$$

$$\nabla p = \frac{1}{h_s}\frac{\partial p}{\partial s}\hat{e}_s + \frac{\partial p}{\partial n}\hat{e}_n$$

$$\varepsilon_{ss} = \frac{1}{h_s}\left(\frac{\partial u}{\partial s} + kv\right)$$

$$\varepsilon_{nn} = \frac{\partial v}{\partial n}$$

$$\varepsilon_{\phi\phi} = \frac{\sigma}{r}(u\sin\theta + v\cos\theta)$$

$$\varepsilon_{sn} = \varepsilon_{ns} = \frac{1}{2h_s}\left(h_s\frac{\partial u}{\partial n} - ku + \frac{\partial v}{\partial s}\right)$$

$$\varepsilon_{n\phi} = \varepsilon_{\phi n} = 0, \qquad \varepsilon_{\phi s} = \varepsilon_{s\phi} = 0$$

$$F_s^s = \frac{2}{h_s^2}\frac{\partial}{\partial s}\left[\mu\left(\frac{\partial u}{\partial s} + kv\right)\right] + \frac{1}{h_s}\frac{\partial}{\partial n}\left[\mu\left(h_s\frac{\partial u}{\partial n} - ku + \frac{\partial v}{\partial s}\right)\right]$$
$$+ \frac{\mu}{h_s}\left(\frac{k}{h_s} + \frac{\sigma}{r}\cos\theta\right)\left(h_s\frac{\partial u}{\partial n} - ku + \frac{\partial v}{\partial s}\right)$$
$$+ \frac{2\mu}{h_s}\left(\frac{\sigma}{r}\sin\theta - \frac{n}{h_s^2}k'\right)\left(\frac{\partial u}{\partial s} + kv\right)$$
$$- \frac{2\sigma\mu\cos\theta}{r^2}(u\sin\theta + v\cos\theta)$$

$$F_n^s = \frac{1}{h_s^2}\frac{\partial}{\partial s}\left[\mu\left(h_s\frac{\partial u}{\partial n} - ku + \frac{\partial v}{\partial s}\right)\right] + 2\frac{\partial}{\partial n}\left(\mu\frac{\partial v}{\partial n}\right)$$
$$+ \frac{\mu}{h_s}\left(\frac{\sigma}{r}\sin\theta - \frac{n}{h_s^2}k'\right)\left(h_s\frac{\partial u}{\partial n} - ku + \frac{\partial v}{\partial s}\right)$$
$$+ 2\mu\left[\left(\frac{k}{h_s} + \frac{\sigma}{r}\cos\theta\right)\frac{\partial v}{\partial n} - \frac{k}{h_s^2}\left(\frac{\partial u}{\partial s} + kv\right)\right]$$
$$- \frac{2\sigma\mu\cos\theta}{r^2}(u\sin\theta + v\cos\theta)$$

$$F_\phi^s = 0$$

$$\Phi = 2\mu\left[\frac{1}{h_s^2}\left(\frac{\partial u}{\partial s} + kv\right)^2 + \left(\frac{\partial v}{\partial n}\right)^2 + \frac{\sigma}{r^2}(u \sin\theta + v \cos\phi)^2\right]$$

$$+ \frac{\mu}{h_s^2}\left(\frac{\partial v}{\partial s} + h_s\frac{\partial u}{\partial n} - ku\right)^2 + \lambda(\nabla \cdot \vec{w})^2$$

$$\nabla^2 T = \frac{1}{h_s r^\sigma}\left[\frac{\partial}{\partial s}\left(\frac{r^\sigma}{h_s}\frac{\partial T}{\partial s}\right) + \frac{\partial}{\partial n}\left(h_s r^\sigma \frac{\partial T}{\partial n}\right)\right]$$

D. Conservation Equations

$$\frac{\partial}{\partial t}(h_s r^\sigma \rho) + \frac{\partial}{\partial s}(r^\sigma \rho u) + \frac{\partial}{\partial n}(h_s r^\sigma \rho v) = 0$$

$$\rho\left(h_s\frac{Du}{Dt} + kuv\right) = -\frac{\partial p}{\partial s} + \frac{\partial}{\partial s}(\lambda\nabla \cdot \vec{w}) + h_s F_s^s$$

$$\rho\left(\frac{Dv}{Dt} - \frac{ku^2}{h_s}\right) = -\frac{\partial p}{\partial n} + \frac{\partial}{\partial n}(\lambda\nabla \cdot \vec{w}) + F_n^s$$

$$\rho\frac{Dh}{Dt} = \frac{Dp}{Dt} + \kappa\nabla^2 T + \frac{1}{h_s^2}\frac{\partial\kappa}{\partial s}\frac{\partial T}{\partial s} + \frac{\partial\kappa}{\partial n}\frac{\partial T}{\partial n} + \Phi$$

$$\tau_w = \mu\left(\frac{\partial u}{\partial n}\right)_w, \qquad q_w = -\kappa\left(\frac{\partial T}{\partial n}\right)_w$$

E. Euler Equations

Set $\mu = \lambda = \kappa = 0$ in Section D.

F. Incompressible, Constant Property Conservation Equations

$$\nabla \cdot \vec{w} = \frac{\partial}{\partial s}(r^\sigma u) + \frac{\partial}{\partial n}(h_s r^\sigma v) = 0$$

$$\rho\left(h_s\frac{Du}{Dt} + kuv\right) = -\frac{\partial p}{\partial s} + h_s F_s^s$$

$$\rho\left(\frac{Dv}{Dt} - \frac{ku^2}{h_s}\right) = -\frac{\partial p}{\partial n} + F_n^s$$

$$\rho\frac{Dh}{Dt} = \frac{Dp}{Dt} + \kappa\nabla^2 T + \Phi$$

Appendix I
Summary of Compressible, Similar Boundary-Layer Equations

A. Assumptions
  (i) Steady, laminar, two-dimensional or axisymmetric flow without swirl
  (ii) Large Reynolds number flow over an impermeable wall
  (iii) A known solution of the Euler equation for the stream surface that wets the body; this stream surface has a constant entropy and a constant stagnation enthalpy
  (iv) First-order boundary-layer theory that provides a local similarity solution
  (v) Perfect gas, $\mu \sim T$, and $Pr = 1$

B. Special Symbols

$$(\ )_e = \text{Eular solution along the wall}$$
$$(\ )_o = \text{stagnation condition}$$
$$(\ )_t = \text{thermal boundary layer}$$
$$(\ )_v = \text{viscous boundary layer}$$
$$(\ )_w = \text{wall}$$
$$(\tilde{\ }) = \text{adiabatic wall}$$
$$(\ )' = \text{derivative with respect to } \eta$$

C. First-Order Boundary-Layer Equations in Terms of Primitive Variables

$$\frac{\partial}{\partial s}(r_w^\sigma \rho u) + \frac{\partial}{\partial n}(r_w^\sigma \rho v) = 0$$

$$\rho \frac{Du}{Dt} = -\frac{dp_e}{ds} + \frac{\partial}{\partial n}\left(\mu \frac{\partial u}{\partial n}\right)$$

$$\rho \frac{Dh}{Dt} = u \frac{dp_e}{ds} + \frac{\partial}{\partial n}\left(\kappa \frac{\partial T}{\partial n}\right) + \mu \left(\frac{\partial u}{\partial n}\right)^2$$

$$\frac{dp_e}{ds} = -\rho_e u_e \frac{du_e}{ds}$$

$$\frac{D}{Dt} = u \frac{\partial}{\partial s} + v \frac{\partial}{\partial n}$$

D. Transformation Equations

$$h_o = h + \frac{1}{2}u^2$$

$$\xi(s) = \int_0^s (\rho\mu u)_e r_w^{2\sigma} ds$$

$$\eta(s, n) = \frac{r_w^\sigma \rho_e u_e}{(2\xi)^{1/2}} \int_o^n \frac{\rho}{\rho_e} dn$$

For a similar solution, we introduce f, g, and G functions:

$$\frac{u}{u_e} = \frac{df}{d\eta} = f'(\eta)$$

$$\frac{h_o}{h_{oe}} = g(\eta)$$

$$G = \frac{g - g_w}{1 - g_w} = \frac{T_o - T_w}{T_{oe} - T_w}$$

$$\frac{\rho_e}{\rho} = \frac{T}{T_e} = \frac{h_{oe} g - (1/2)u_e^2 (f')^2}{h_{oe} - (1/2)u_e^2}$$

$$= \left(1 + \frac{\gamma - 1}{2} M_e^2\right) g - \frac{\gamma - 1}{2} M_e^2 (f')^2$$

$$= \frac{g_w + (1 - g_w)G - Sf'^2}{1 - S}$$

(See Section H for S.)

E. Similarity Equations

$$f''' + ff'' + \beta(g - f'^2) = 0$$

$$g'' + fg' = 0$$

$$f(0) = f_w = 0, \qquad f'(0) = f'_w = 0, \qquad f'(\infty) = 1$$

$$g(0) = g_w = \frac{T_w}{T_{oe}}, \qquad g(\infty) = 1$$

or

$$f''' + f'' + \beta[g_w + (1 - g_w)G - f'^2] = 0$$

$$G'' + fG' = 0$$

$$G(0) = G_w = 0, \qquad G(\infty) = 1$$

(See Section G for β.)

F. Solution for G or g

(i) g_w specified

$$G(\eta) = G'_w \int_o^{\eta} \exp\left(-\int_o^{\eta''} f\,d\eta'\right) d\eta''$$

$$G'_w = \left[\int_o^{\infty} \exp\left(-\int_o^{\eta''} f\,d\eta'\right) d\eta''\right]^{-1}$$

(ii) adiabatic wall

$$g(\eta) = 1$$

G is undefined

$$\frac{\rho_e}{\rho} = \frac{T}{T_e} = \frac{1 - Sf'^2}{1 - S}$$

(See Section H for S.)

G. β Parameter

$$\beta = \frac{2\xi}{u_e}\frac{du_e}{d\xi}\frac{T_{oe}}{T_e} = \frac{2\xi}{u_e}\frac{du_e}{d\xi}\left(1 + \frac{\gamma-1}{2}M_e^2\right)$$

$$= \frac{2}{M_e}\frac{dM_e}{ds}\frac{\int_o^s r_w^{2\sigma} F\,ds}{r_w^{2\sigma} F(M_e)}$$

$$F = \frac{M_e}{\left(1 - \frac{\gamma-1}{2}M_e^2\right)^{(3\gamma-1)/[2(\gamma-1)]}}$$

$$\frac{1}{\beta}\frac{d\beta}{ds} = \frac{1}{\frac{dM_e}{ds}}\frac{d^2M_e}{ds^2} + 2\left(\frac{1}{\beta} - \frac{1 - \frac{\gamma+1}{4}M_e^2}{1 + \frac{\gamma-1}{2}M_e^2}\right)\frac{1}{M_e}\frac{dM_e}{ds} - \frac{2\sigma}{r_w}\frac{dr_w}{ds}$$

H. Boundary-Layer Parameters

$$S = \frac{\gamma-1}{2}\frac{M_e^2}{1 + \frac{\gamma-1}{2}M_e^2}$$

$$M_e^2 = \frac{2}{\gamma-1}\frac{S}{1-S}$$

$$\bar{x} = \frac{\xi}{(\rho\mu u)_e r_w^{2\sigma}} = \frac{\int_o^s (\rho\mu u)_e r_w^{2\sigma} ds}{(\rho\mu u)_e r_w^{2\sigma}}$$

$$Re_{\bar{x}} = \frac{\rho_e u_e \bar{x}}{\mu_e}$$

$$\Upsilon = \left(\frac{Re_{\bar{x}}}{2}\right)^{1/2}\frac{1}{x} = \frac{(\rho u)_e r_w^{\sigma}}{(2\xi)^{1/2}}$$

$$\Upsilon n = \int_o^n \frac{\rho_e}{\rho} d\eta$$

$$q_w = -\kappa_w\left(\frac{\partial T}{\partial n}\right)_w = \frac{h_{ow} - h_{oe}}{(2Re_{\bar{x}})^{1/2}}\rho_e u_e G'_w$$

$$St = \frac{q_w}{(h_{ow} - h_{oe})\rho_e u_e} = \frac{G'_w}{(2Re_{\bar{x}})^{1/2}}$$

$$\tau_w = \mu_w\left(\frac{\partial u}{\partial n}\right)_w = \frac{\rho_e u_e^2}{(2Re_{\bar{x}})^{1/2}} f''_w$$

$$c_f = \frac{2\tau_w}{\rho_e u_e^2} = \left(\frac{2}{Re_{\bar{x}}}\right)^{1/2} f''_w$$

$$\frac{2St}{c_f} = \frac{G'_w}{f''_w} \qquad \text{(Reynolds' analogy)}$$

$$= Pr^{-2/3}\frac{G'_w}{f''_w} \qquad \text{(Colburn's analogy)}$$

$$C_v(\beta, g_w) = \eta - f, \qquad \eta \to \infty$$

$$C_t(\beta, g_w) = \eta - \int_o^{\eta} G d\eta, \qquad \eta \to \infty$$

I. Boundary-Layer Thicknesses for a Diabatic Wall

δ = velocity boundary-layer thickness

$\delta = n$ when $f' = 0.99$ and $\eta = \eta_{ev}(\beta, g_w)$

This definition is for a boundary layer without velocity overshoot.

$$\Upsilon\delta = \int_o^{\eta_{ev}} \frac{\rho_e}{\rho} d\eta$$

$$= \eta_{ev} + \frac{1}{(1-S)(1+\beta)}\{S(f''_w + C_v) - (1 - g_w)[1 + \beta(1-S)]C_t\}$$

δ_t = thermal boundary-layer thickness

$\delta_t = n$ when $G = 0.99$ and $\eta = \eta_{et}(\beta, g_w)$

$$\Upsilon\delta_t = \int_o^{\eta_{et}} \frac{\rho_e}{\rho} d\eta$$

$$= \eta_{et} + \frac{1}{(1-S)(1+\beta)}\{S(f''_w + C_v) - (1 - g_w)[1 + \beta(1-S)]C_t\}$$

δ^* = displacement thickness

$$= \int_o^{\infty}\left(1 - \frac{\rho u}{\rho_e u_e}\right) dn$$

$$\Upsilon\delta^* = \int_o^\infty \left[\frac{g_w + (1-g_w)G - Sf'^2}{1-S} - f'\right] dn$$

$$= \frac{1}{(1-S)(1+\beta)}\{Sf''_w + [1+(1-S)\beta][C_v - (1-g_w)C_t]\}$$

θ = momentum defect thickness

$$= \int_o^\infty \frac{\rho u}{\rho_e u_e}\left(1 - \frac{u}{u_e}\right) dn$$

$$\Upsilon\theta = \int_o^\infty f'(1-f')d\eta$$

$$= \frac{1}{1+\beta}\{f''_w - \beta[C_v - (1-g_w)C_t]\}$$

ϕ = stagnation enthalpy defect thickness

$$= \int_o^\infty \frac{\rho u}{\rho_e u_e}\left(1 - \frac{h_o - h_{ow}}{h_{oe} - h_{ow}}\right) dn$$

$$\Upsilon\phi = \int_o^\infty f'(1-G)d\eta$$

$$= G'_w$$

J. Boundary-Layer Thicknesses for an Adiabatic Wall

When the wall is adiabatic, the definitions and formulas for $\eta_{ev}, \delta, \delta^*, \theta$, and c_f are unchanged. (See Section F for ρ and T.) Since G is undefined, η_{et}, δ_t, ϕ, C_t, and St are also undefined. Instead, we use $\tilde{\eta}_{et}$, $\tilde{\delta}_t$, and $St = 0$, where

$$\tilde{\delta}_t = n \quad \text{when } (T_{oe} - \tilde{T}_{et})/(T_{oe} - T_e) = 0.9801$$

$$\Upsilon\tilde{\delta}_t = \tilde{\eta}_{et} + \frac{S}{(1-S)(1+\beta)}(f''_w + C_v)$$

$$\tilde{\eta}_{et} = \tilde{\eta}_{ev}(\beta,1)$$

For $\tilde{\phi}$, we have

$\tilde{\phi}$ = adiabatic wall temperature defect thickness

$$= \int_o^\infty \frac{\rho u}{\rho_e u_e}\left(1 - \frac{T_{oe} - T}{T_{oe} - T_e}\right) dn$$

$$\Upsilon\tilde{\phi} = \int_o^\infty f'(1-f'^2)d\eta$$

This integral has not been numerically evaluated.

Appendix J
Second -Order Boundary-Layer Equations for Supersonic, Rotational Flow over a Flat Plate

1. Equations for $\tilde{u}_1$, $\tilde{v}_1$, and $\tilde{\rho}_1$

$$a_1\rho_1\tilde{u}_1 + \rho_1\rho_1'\,\tilde{v}_1 + c_1\left(N\rho_1 f'' + f' + \frac{\rho_1'}{\rho_1}\right)\tilde{\rho}_1 + \rho_1^2\tilde{v}_1' + c_1(N\rho_1 f' - f)\tilde{\rho}_1' = 0$$

$$a_1\rho_1 f'\tilde{u}_1 + \rho_1^2 f''\tilde{v}_1 + c_1\left(N\rho_1 f' - 2f - \frac{\rho_1'}{\rho_1}\right)f''\tilde{\rho}_1$$
$$+ a_1\rho_1\,(N\rho_1 f' - f)\tilde{u}_1' - a_1\rho_1\tilde{u}_1'' + c_1 f''\tilde{\rho}_1' = 0$$

$$-\rho_1'\,\tilde{v}_1 + \frac{c_1}{\rho_1}\left[\frac{\rho_1'}{\rho_1}(N\rho_1 f' - f) - f' - 3\frac{\rho_1''}{\rho_1} + 9\left(\frac{\rho_1'}{\rho_1}\right)^2 + (\gamma-1)M_\infty^2\rho_1 f''^2\right]\tilde{\rho}_1$$
$$-2(\gamma-1)M_\infty^2 a_1\rho_1 f''^2\tilde{u}_1' - \frac{c_1}{\rho_1}\left(N\rho_1 f' - f + 5\frac{\rho_1'}{\rho_1}\right)\tilde{\rho}_1' + \frac{c_1}{\rho_1}\tilde{\rho}_1'' = 0$$

2. Equations for $\tilde{u}_{-1}$, $\tilde{v}_{-1}$, and $\tilde{\rho}_{-1}$

$$-a_{-1}\rho_{-1}\tilde{u}_{-1} + \rho_1\rho_1'\,\tilde{v}_{-1} + c_{-1}\left(N\rho_1 f'' - f' + \frac{\rho_1'}{\rho_1}\right)\tilde{\rho}_{-1}$$
$$+ \rho_1^2\tilde{v}_{-1}' + c_{-1}(N\rho_1 f' - f)\tilde{\rho}'_{-1} = 0$$

$$-a_1\rho_1 f'\tilde{u}_{-1} + \rho_1^2 f''\tilde{v}_{-1} + c_{-1}\left(N\rho_1 f' - 2f - \frac{\rho_1'}{\rho_1}\right)f''\tilde{\rho}_{-1}$$
$$+ a_{-1}\rho_1(N\rho_1 f' - f)\tilde{u}_{-1}' - a_{-1}\rho_1\tilde{u}_{-1}'' + c_{-1}f''\tilde{\rho}_{-1}'$$
$$= \frac{C^*}{\lambda}(1 - \gamma M_\infty^2\rho_1 ff'')$$

$$-\rho_1'\tilde{v}_{-1} + \frac{c_{-1}}{\rho_1}\left[\frac{\rho_1'}{\rho_1}(N\rho_1 f' - f) + f' - 3\frac{\rho_1''}{\rho_1} + 9\left(\frac{\rho_1'}{\rho_1}\right)^2 + (\gamma-1)M_\infty^2\rho_1 f''^2\right]\tilde{\rho}_{-1}$$
$$-2(\gamma-1)M_\infty^2 a_{-1}\rho_1 f''\tilde{u}_{-1}' - \frac{c_{-1}}{\rho'}\left(N\rho_1 f' - f + 5\frac{\rho_1'}{\rho_1}\right)\tilde{\rho}_{-1}' + \frac{c_{-1}}{\rho_1}\tilde{\rho}_{-1}''$$
$$= \frac{\gamma M_\infty^2 C^*}{\lambda}\left[\frac{1}{\gamma}f' + \frac{\rho_1'}{\rho_1}(N\rho_1 f' - f) - 2\frac{\rho_1''}{\rho_1} + 4\left(\frac{\rho_1'}{\rho_1}\right)^2 + (\gamma-1)M_\infty^2\rho_1 f''^2\right]$$

3. Velocity Slip

$$\tilde{u}_1^{(s)}(\eta) = \tilde{v}_1^{(s)}(\eta) = \tilde{\rho}_1^{(s)}(\eta) = 0$$

$$a_{-1}^{(s)} = \gamma^{1/2} M_\infty (X_\infty g_w)^{1/2} f_w'', \qquad c_{-1}^{(s)} = \frac{\gamma M_\infty^2 C^*}{\lambda X_\infty g_w}$$

$$\tilde{u}_{-1}^{(s)}(0) = 1, \qquad \tilde{u}_{-1}^{(s)}(\infty) = 0$$

$$\tilde{v}_{-1}^{(s)}(0) = 0$$

$$\tilde{\rho}_{-1}^{(s)}(0) = 1, \qquad \tilde{\rho}_{-1}^{(s)}(\infty) = X_\infty g_w$$

4. Temperature Jump

$$\tilde{u}_1^{(T)}(\eta) = \tilde{v}_1^{(T)}(\eta) = \tilde{\rho}_1^{(T)}(\eta) = 0$$

$$a_{-1}^{(T)} = 1, \qquad c_{-1}^{(T)} = c_{-1}^{(s)} - \frac{\gamma^{1/2}}{g_w^{3/2} X_\infty^{1/2}} M_\infty (1 - g_w) f_w''$$

$$\tilde{u}_{-1}^{(T)}(0) = 0, \qquad \tilde{u}_{-1}^{(T)}(\infty) = 0$$

$$\tilde{v}_{-1}^{(T)}(0) = 0$$

$$\tilde{\rho}_{-1}^{(T)}(0) = 1, \qquad \tilde{\rho}_{-1}^{(T)}(\infty) = \left[\frac{1}{X_\infty g_w} - \frac{1}{(\gamma g_w^3 X_\infty)^{1/2}} \frac{\lambda(1 - g_w) f_w''}{M_\infty C^*}\right]^{-1}$$

5. Entropy Gradient

$$a_1^{(e)} = -\frac{1}{(\gamma - 1) M_\infty^2}, \qquad a_{-1}^{(e)} = 1, \qquad c_1^{(e)} = -1, \qquad c_{-1}^{(e)} = c_{-1}^{(s)}$$

$$\tilde{u}_1^{(e)}(0) = 0, \qquad \tilde{u}_1^{(e)}(\eta) \sim \eta, \qquad \tilde{u}_{-1}^{(e)}(0) = 0, \qquad \tilde{u}_{-1}^{(e)}(\infty) = 0$$

$$\tilde{v}_1^{(e)}(0) = 0, \qquad \tilde{v}_{-1}^{(e)}(0) = 0$$

$$\tilde{\rho}_1^{(e)}(0) = 0, \qquad \tilde{\rho}_1^{(e)}(\infty) = 1, \qquad \tilde{\rho}_{-1}^{(e)}(0) = 1, \qquad \tilde{\rho}_{-1}^{(e)}(\infty) = 0$$

6. Enthalpy Gradient

$$a_1^{(H)} = \frac{X_\infty}{(\gamma - 1) M_\infty^2}, \qquad a_{-1}^{(H)} = 1, \qquad c_1^{(H)} = \frac{\gamma M_\infty^2 C^*}{\lambda}, \qquad c_{-1}^{(H)} = c_{-1}^{(s)}$$

$$\tilde{u}_1^{(H)}(0) = 0, \qquad \tilde{u}_1^{(H)}(\eta) \sim \eta, \qquad \tilde{u}_{-1}^{(H)}(0) = 0, \qquad \tilde{u}_{-1}^{(H)}(\infty) = 0$$

$$\tilde{v}_1^{(H)}(0) = 0, \qquad \tilde{v}_{-1}^{(H)}(0) = 0$$

$$\tilde{\rho}_1^{(H)}(0) = 0, \qquad \tilde{\rho}_1^{(H)}(\infty) = 0, \qquad \tilde{\rho}_{-1}^{(H)}(0) = 1, \qquad \tilde{\rho}_{-1}^{(H)}(\infty) = X_\infty g_w$$

7. Displacment

$$a_1^{(d)} = -\frac{1}{2}C^*h_o', \qquad a_{-1}^{(d)} = -\frac{C^*}{\lambda}, \qquad c_1^{(d)} = C^*h_o', \qquad c_{-1}^{(d)} = \frac{M_\infty^2 C^*}{\lambda}$$

$$\tilde{u}_1^{(d)}(0) = 0, \qquad \tilde{u}_1^{(d)}(\infty) = 1, \qquad \tilde{u}_{-1}^{(d)}(0) = 0, \qquad \tilde{u}_{-1}^{(d)}(\infty) = 1$$

$$\tilde{v}_1^{(d)}(0) = 0, \qquad \tilde{v}_{-1}^{(d)}(0) = 0$$

$$\tilde{\rho}_1^{(d)}(0) = 0, \qquad \tilde{\rho}_1^{(d)}(\infty) = 1, \qquad \tilde{\rho}_{-1}^{(d)}(0) = \frac{\gamma}{X_\infty g_w}, \qquad \tilde{\rho}_{-1}^{(d)}(\infty) = 1$$

Index

D

E

F

G

N

O

P

T

U

V

W